AF740078

Turbulence and Self-Organization

Astrophysics And Space Science Library

EDITORIAL BOARD

Chairman

W.B. BURTON, *National Radio Astronomy Observatory, Charlottesville, Virginia, U.S.A. (bburton@nrao.edu); University of Leiden, The Netherlands (burton@strw.leidenuniv.nl)*

F. BERTOLA, *University of Padua, Italy*
C.J. CESARSKY, *Commission for Atomic Energy, Saclay, France*
P. EHRENFREUND, *Leiden University, The Netherlands*
O. ENGVOLD, *University of Oslo, Norway*
A. HECK, *Strasbourg Astronomical Observatory, France*
E.P.J. VAN DEN HEUVEL, *University of Amsterdam, The Netherlands*
V.M. KASPI, *McGill University, Montreal, Canada*
J.M.E. KUIJPERS, *University of Nijmegen, The Netherlands*
H. VAN DER LAAN, *University of Utrecht, The Netherlands*
P.G. MURDIN, *Institute of Astronomy, Cambridge, UK*
B.V. SOMOV, *Astronomical Institute, Moscow State University, Russia*
R.A. SUNYAEV, *Space Research Institute, Moscow, Russia*

For further volumes:
http://www.springer.com/series/5664

Mikhail Ya Marov • Aleksander V. Kolesnichenko

Turbulence and Self-Organization

Modeling Astrophysical Objects

Springer

Mikhail Ya Marov
Department of Planetary Sciences
and Cosmochemistry
V.I. Vernadsky Institute of Geochemistry
and Analytical Chemistry
Russian Academy of Sciences
Moscow, Russia

Aleksander V. Kolesnichenko
Department of Planetary Science and Aeronomy
Keldysh Institute of Applied Mathematics
Russian Academy of Sciences
Moscow, Russia

ISSN 0067-0057
ISBN 978-1-4614-5154-9 ISBN 978-1-4614-5155-6 (eBook)
DOI 10.1007/978-1-4614-5155-6
Springer New York Heidelberg Dordrecht London

Library of Congress Control Number: 2012948282

© Springer Science+Business Media New York 2013
This work is subject to copyright. All rights are reserved by the Publisher, whether the whole or part of the material is concerned, specifically the rights of translation, reprinting, reuse of illustrations, recitation, broadcasting, reproduction on microfilms or in any other physical way, and transmission or information storage and retrieval, electronic adaptation, computer software, or by similar or dissimilar methodology now known or hereafter developed. Exempted from this legal reservation are brief excerpts in connection with reviews or scholarly analysis or material supplied specifically for the purpose of being entered and executed on a computer system, for exclusive use by the purchaser of the work. Duplication of this publication or parts thereof is permitted only under the provisions of the Copyright Law of the Publisher's location, in its current version, and permission for use must always be obtained from Springer. Permissions for use may be obtained through RightsLink at the Copyright Clearance Center. Violations are liable to prosecution under the respective Copyright Law.
The use of general descriptive names, registered names, trademarks, service marks, etc. in this publication does not imply, even in the absence of a specific statement, that such names are exempt from the relevant protective laws and regulations and therefore free for general use.
While the advice and information in this book are believed to be true and accurate at the date of publication, neither the authors nor the editors nor the publisher can accept any legal responsibility for any errors or omissions that may be made. The publisher makes no warranty, express or implied, with respect to the material contained herein.

Printed on acid-free paper

Springer is part of Springer Science+Business Media (www.springer.com)

Dedicated to Academician[1] Leonid Ivanovich Sedov

[1] The title *Academician* denotes a Full Member of an art, literary, or scientific academy. In this case, it refers to the Academy of Sciences of the USSR.

God the Almighty on the throne with a threateningly raised hand:

—Noli turba circulos meos![2]

Underneath there are myriads of stars, worlds moving over spheres. And Prometheus stretching his muscles to break the circle of human being. And someone who fell from the circle and is plunging into chaos.
And the threatening, warning Finger of the Almighty:

—Noli turba circulos meos!

From the book by Leonid Andreev S.O.S.

In mechanics, the meaning of the modeling of real bodies and phenomena using objects and processes invented in science is clear to everybody and we all understand well the meaning of our actions in these cases. It is inappropriate to say that "science discovered" an ideal incompressible fluid, or a perfectly rigid body, or a perfectly elastic body. We understand that these objects are absolutely necessary and useful scientific inventions.

From the book by L.I. Sedov
"Reflections on Science and Scientists".

Turbulent motion appears as a very complex motion in open systems emerging from a less ordered motion—"physical chaos"... The transition from a laminar state to a turbulent one may be considered to be an example of self-organization in a nonlinear open system.

From the book by Yu.L. Klimontovich
"An Introduction to the Physics of Open Systems".

[2] "Don't disturb my circles"—Archimedes' dictum.

Preface

There are notable words in the book of the renowned French scientist David Ruelle "Hasard et Chaos" (chance and chaos), "Mechanics was born from the desire to explain the world". The validity of this utterance, which refers to the history of science, acquires a new deep meaning at the present stage of the development of natural sciences, where not only the symbiosis and close intertwining of various disciplines occur but also the fundamentals of the cognition of natural laws are revealed, whose carrier is traditionally served by mechanics, along with its individual branches and numerous applications. The classical conservation laws and the differential equations of motion, including the motion of celestial bodies, the equations of continuum mechanics, dynamics of rarefied multicomponent gases, and physical kinetics formulated on their basis constitute the foundation for developing various physical, geophysical, and astrophysical models that are designed not only to explain the surrounding world, but also to understand the origins of its birth and evolution. The avalanche-like accumulation of knowledge about the Universe (or in modern notion, Multiverse) based on enormous observational data sets and the continuously improving methods of mathematical modeling, which use high-performance computing systems, open new vast horizons for the cognition of nature.

The rapidly expanding views about the surrounding regions of space near the Earth and far beyond, which is attributable primarily to enormous progress in astronomy and space research, led to a deeper penetration into the physical essence of the processes and phenomena in various natural and cosmic media in divers states of their constituent matter. This inspired increasingly complicated mathematical models for these media, which were in turn made possible by immense breakthroughs achieved in creating powerful computing systems – their architecture, performance, and software. Truly boundless opportunities open up on this path, because linked to this capacity of the computing systems is the prospect of formulating and solving complex multidimensional nonstationary geo- and astrophysical problems and analyzing evolutionary processes based on large-scale numerical experiments.

Turbulence is at the same time a very common and a complex natural phenomena associated with the emergence and development of an enormous number of vortices on all possible scales (organized vortex structures) under certain regimes of fluid motion in an essentially nonlinear hydrodynamic system. When the stability of a laminar flow defined by the critical Reynolds number is lost, an unsteady fluctuating flow emerges in the hydrodynamic system. A continuous distribution of the velocity fluctuations and other thermohydrodynamic parameters is produced in this flow through the stretching of vortices in a range of smallest wavelengths, determined by dissipative (viscous) forces, to the longest ones, determined by the flow boundaries. The conditions for emerging vorticity and structuring of the developed turbulence are influenced by the physical properties of the medium, such as the molecular transport coefficients to which the energy dissipation processes in a turbulent flow are related, as well as by the conditions at the boundary, where thin boundary vortex layers are observed, whose instability manifests itself in the generation of vortex tubes. Turbulization leads to a rapid mixing of particles in a continuous medium and to an increase in the efficiency of mass, momentum, and heat transfer; in multiphase multicomponent media, it also contributes to the acceleration of phase transitions and chemical reactions. As the knowledge about the various natural objects in which turbulence plays a significant and, in many cases, crucial role is accumulated, modeling this phenomenon and related hydrodynamic effects acquires a key significance. At the same time, a direct numerical simulation of turbulent flows involves great mathematical difficulties, while constructing a general theory of structured turbulence in a compressible fluid is unlikely to be possible, because the mechanisms according to which different-scale vortex structures interact are extremely complex.

This monograph is based on our studies of problems that arise in phenomenological modeling of developed turbulence in multicomponent mixtures of reacting gases and heterogeneous gas-dust media. The first steps of these studies are studying the natural environment and evolution mechanisms of the Earth and planets in the Solar System, protoplanetary gas-dust accretion disks, and turbulent heat and mass transport in the upper atmospheres of planets (the tenuous gaseous envelopes of celestial bodies lying in the boundary regions between the dense atmospheric layers and circumplanetary space). We are concerned here with investigating turbulent flows in gas and gas-dust media with complicated physical-chemical characteristics whose mathematical modeling requires taking into account the flow compressibility, the variability of thermophysical properties, heat and mass transfer, chemical reactions, and radiation, and the influence of gravitational and electromagnetic forces. The additional mechanical effects that emerge under these circumstances generally do not allow one to employ the results obtained in terms of the traditional description of turbulent homogeneous incompressible fluid flows that are used, for example, in meteorology. Therefore, to study natural media of this type, we need to develop new approaches to model turbulence that adequately describe the hydrodynamic motions of compressible mixtures, transport processes, and chemical kinetics in a fluctuating multiphase multicomponent continuum. Since the hydrodynamic and physical-chemical pattern of turbulent motion is complex, the theoretical approaches to solving this problem must be "semiempirical" in character.

We focus on the thermodynamic construction of continuum models for turbulized natural media in outer space, which are the basis for solving the formation and evolution problems of various astrophysical and geophysical objects. These primarily include turbulent motion models of multicomponent chemically active gases with diffusion, heat transfer, viscosity, and radiation processes; turbulent motion models of gas suspensions and various types of heterogeneous media with phase transitions; and models of structured turbulent flows as well as turbulent flows interacting with an electromagnetic field.

The book is devoted to the topical problem of self-organization in developed turbulent flows, which serves as a reflection of the most general concept of the relationship between order and chaos in natural processes. Submitting our work for the judgment of specialists in various areas of knowledge, we realize that the approach we develop is in some aspects unusual for mechanicians guided by "classical" yardsticks. For this reason, the history of a new understanding of turbulence is of interest. The question about the relative degree of order of laminar and turbulent motions was probably first discussed publicly during the "Synergetics 83" International Conference in the town of Pushchino near Moscow. In his report, Yu. L. Klimontovich formulated a general criterion, the so-called "S-theorem" (to which we will return later), which characterizes the relative degree of order of the states for open dissipative systems. This criterion was used by the speaker for a quantitative proof of the assertion that a turbulent motion is more ordered than a laminar one. There were very few supporters of this viewpoint in the hall. I. Prigogine, H. Haken, and B. Ebeling were among them. Curiously, the renowned hydromechanician G.I. Barenblatt was openly indignant during this report, saying: "All mechanicians know excellently well that a turbulent motion is more chaotic".

Nevertheless, contrary to the traditional viewpoint of mechanicians on turbulence, a large number of various vortex coherent structures (CSs) have been discovered and their topological characteristics have been firmly established in the last decade owing to the progressive development of the methods for a visual observation of turbulized fluid flows. As examples, we can name Taylor vortices, turbulent spots, vortex rings, vortex balls, hairpin-like vortices, burstings, vortex spirals, streaks, Brawn-Thomas structures, mushroom-shaped vortices, etc. The frequency of occurrence of a particular structure depends on the type of flow (a boundary layer, a mixing layer, a jet, etc.), geometry, and the regime of turbulized fluid motion. It is important to ascertain how these coherent structures can emerge. Nonlinear nonequilibrium thermodynamics, whose principles are widely used in the monograph, partly answers this question.

Basically, the distinctive feature of our studies consists of the proposed stochastic-thermodynamic approach to constructing semiempirical turbulence models in reacting multicomponent gases and gas-dust media and structured turbulence in a homogeneous fluid. These studies are oriented mainly toward solving a number of complex present-day problems in astrophysics and geophysics based on the methods of continuum mechanics. The developed approaches have a direct bearing on modeling the mechanisms that shape the properties of astrophysical and geophysical objects at various stages of their evolution; on investigating the problems of stellar and planetary cosmogony, including the formation of protoplanetary gas-dust accretion disks and the subsequent accumulation of planetary systems; on the early evolutionary stages of planets; on the formation and evolution of planetary and cometary atmospheres; and on the problems of ecology associated with the diffusion of pollutants and natural environmental protection that attract increasingly greater attention.

The book has nine chapters. Some questions are concretized in the appendix. We have attempted to make the chapters of the book independent of one another, where possible, although they are naturally unified by a common conceptual orientation.

In the *first chapter*, which is the introductory one, we briefly consider the properties of turbulent flows, the elements of stochastic nonlinear dynamics, and the relationship between order and chaos, including the synergetic aspects of the formation of ordered structures. The basic views of turbulence as a dynamical system are discussed. Since the emphasis in the monograph is on turbulent flows in natural inhomogeneous media with a variable density, the problem of modeling these media acquires entirely new facets. The classical views of turbulent motions in an incompressible fluid are intertwined with other areas of mechanics that combine a mixture of hydromechanics, thermodynamics, the theory of radiative transfer, and the kinetics of chemical reactions. In addition to velocity, density, and temperature fluctuations, the concentration of individual chemical mixture components acquires a considerable significance. As a result, we face one of the most complex problems in the mechanics of turbulized media, which requires semiempirical modeling of interrelated hydrodynamic, physical-chemical, and radiative processes and phenomena in a turbulent flow.

The possibility of emerging order in a cosmic medium and during the evolution of cosmic objects is discussed from the standpoint of stochastic dynamics and the theory of self-organization in open nonlinear dissipative systems. The questions of dynamical astronomy, the dynamics of the Solar System, the nature of galaxies, stars, planets, and small bodies in the Solar System, the questions of stellar-planetary evolution, including the formation of protoplanetary accretion disks, and some problems of the structure and evolution of the Universe are considered as examples. This discussion goes far beyond the narrower problems of turbulence that the monograph is devoted to. This serves, on the one hand, the purpose of giving examples of order based on present views of the cosmos and the nature of its inhabitants and, on the other hand, the purpose of reflecting the generality of the concept of macromolecular structure formation in cosmic and other natural media.

The *second chapter* is devoted to formulating general mass, momentum, and energy balance laws in a multicomponent chemically active gas mixture. It can be also regarded as auxiliary and is used as the basis for a more detailed consideration of the turbulence problems suggested in the subsequent chapters of the monograph. As is well known, the most complete and rigorous mathematical description of a multicomponent medium in a regular (laminar) flow can be given in terms of the kinetic theory of multicomponent mixtures of polyatomic ionized gases.

The system of generalized integro-differential Boltzmann equations for the distribution functions of particles of each type in the mixture (with the right-hand parts containing the collision and reaction integrals) supplemented by the radiative transfer equation and the Maxwell equations for electromagnetic fields serves as the basic one. This approach was developed in our monograph (Marov and Kolesnichenko 1987), where a generalized Chapman-Enskog method was used to derive the system of differential gas-kinetic equations for a reacting mixture. In addition, from the viewpoint of macroscopic properties, such a multicomponent gas mixture (e.g., the upper atmosphere of a planet) can be considered as a continuous medium and the methods of continuum mechanics for mixtures can be used to describe it adequately. Based on the principles of nonequilibrium thermodynamics, these methods allow one to obtain the system of hydrodynamic equations with all necessary closing relations. This phenomenological approach also allows semiempirical models of turbulent flows in reacting gas media to be developed by using extended irreversible thermodynamics.

The formalism of classical nonequilibrium thermodynamics is used to study the mass, momentum, and energy transfer processes. As is well known, a wide class of nonequilibrium transport processes in gases can be described by means of this formalism fully in accord with the experimental data. In particular, the technique for the thermodynamic derivation of generalized Stefan-Maxwell relations for multicomponent diffusion proposed here allows one to obtain a number of algebraic relations for the transport coefficients that relate, for example, the thermal diffusion ratios to the thermal diffusion and multicomponent diffusion coefficients, the true and partial thermal conductivities, and the multicomponent and binary diffusion coefficients. All these hydrodynamically derived relations agree completely with the results of the gas-kinetic theory for multicomponent mixtures of monatomic gases obtained in the second approximation of the Chapman-Enskog method. However, in contrast to this, the thermodynamic approach is not related to the postulation of a specific microscopic model for the interaction of molecules in the natural medium that is investigated, which is indicative of its universality.

In the *third chapter*, we derive a closed system of averaged hydrodynamic equations for a turbulized multicomponent chemically active gas mixture designed to describe a wide class of turbulent motions and physical-chemical processes in natural media. We analyze the physical meaning of the individual terms in these equations, including the energy transition rates between various energy balance components. Here, we systematically use the weighted-mean Favre averaging, which allows the form and analysis of the averaged equations of motion for chemically active gases with variable thermophysical properties to be simplified considerably, along with the traditional probability-theoretic averaging of fluctuating thermohydrodynamic parameters. Assessing the status of the first-order closure problem on the whole, it should be recognized that no general phenomenological theory of turbulent heat conduction and turbulent diffusion for multicomponent reacting mixtures has existed until now. Therefore, we consider in this chapter a thermodynamic approach to the closure of the averaged hydrodynamic equations for a mixture at the level of first-order turbulence models based on the methods of extended irreversible thermodynamics. Special attention is paid to deriving closing gradient relations for the Reynolds turbulent stress tensor and the turbulent heat and diffusion fluxes in a multicomponent mixture by thermodynamic methods. The Onsager formalism allows one to obtain the most general structure of these relations, including those in the form of generalized Stefan-Maxwell relations for multicomponent turbulent diffusion. At the closure level under consideration, these relations describe most comprehensively the turbulent heat and mass transport in a multicomponent medium. The turbulent exchange coefficients are determined with classical models dating back to Prandtl, Taylor, and Karman as well as with the more recent second-order closure models that are based on the differential balance equations for the turbulent energy and integral turbulence scale. For the convenience of the reader, all calculations are performed comprehensively and can be traced in all details.

In the *fourth chapter*, we consider the problem of constructing semiempirical second-approximation turbulence models for a multicomponent chemically active gas mixture with a variable density and variable thermophysical properties. We derive closing differential transfer equations for the various one-point (one-time) second-correlation moments of the fluctuating thermohydrodynamic parameters that appear in the averaged hydrodynamic equations of mean motion for a reacting mixture. The closure problem for a chemically active medium is generally highly complicated because one needs to average the nonlinear "source terms" of substance production in chemical reactions with an exponential behavior. Therefore, we propose an original procedure for averaging the rates of chemical reactions of any order and outline a scheme for semiempirical modeling of these additional correlations. Approximating expressions that contain universal empirical coefficients that need not be chosen again for each new flow are used in modeling the third-order correlations in the transfer equations. We emphasize that although these additional equations are semiempirical, the invariant models of fully developed turbulence in chemically active gases based on them are fairly flexible. In particular, they allow taking into account the influence of the mechanisms of convection, diffusion, formation, redistribution, and dissipation of stochastic turbulent characteristics for the field of fluctuating thermohydrodynamic parameters on the spatiotemporal distribution of averaged thermohydrodynamic parameters for the medium. Basically, the approach we developed is widely used in numerical simulations of real reacting turbulized fluid flows with a significant influence of the flow prehistory on the turbulence characteristics at a point. On the other hand, it is used to derive more accurate algebraic relations for the turbulent transport coefficients in multicomponent shear mixture flows (and as applied to the specificity of modeling natural media), which is embodied in this chapter of the book.

The *fifth chapter* describes a phenomenological model for the developed turbulence in a compressible homogeneous medium by taking into account nonlinear cooperative processes. The primary concept is to represent a turbulized fluid motion as a thermodynamic system consisting of two continua, the subsystem of averaged motion and the subsystem of turbulent chaos. This in turn is considered as a conglomerate of vortex structures of various spatiotemporal scales. We develop the ideas of a stationary nonequilibrium state of the dissipatively active subsystem of turbulent chaos that emerges due to the influx of negentropy from the external medium (the subsystem of averaged motion) and the appearance of relatively stable coherent vortex structures in the system when the flow control parameters are varied. This allows some of the turbulent field rearrangement processes to be considered as self-organization processes in an open system. The methods of the stochastic theory of irreversible processes and extended irreversible thermodynamics are used to derive the defining relations for the turbulent fluxes and forces that close the system of averaged hydrodynamic equations and describe the transport and self-organization processes in the stationary nonequilibrium case with completeness sufficient for practice.

Our original approach to the stochastic-thermohydrodynamic modeling of the subsystem of turbulent chaos is based on introducing a set of random variables into the model—fluctuating internal coordinates (such as the turbulent energy dissipation rates, the intrinsic vorticities of the velocity field fluctuations that refer to mesoscale vortex structures, etc.) that characterize the structure and temporal evolution of the fluctuating field of hydrodynamic flow parameters. This makes it possible to model the Richardson-Kolmogorov cascade process and to derive the kinetic Fokker-Planck-Kolmogorov (FPK) equations, which describe the evolution of the probability density function for small-scale turbulence characteristics with thermodynamic methods. These equations are a basis for analyzing the Markovian diffusion processes of the transition from one stationary-nonequilibrium state to another in the space of internal coordinates through a successive loss of stability (and increase in supercriticality) by the subsystem of turbulent chaos that is far from complete chaos of thermodynamic equilibrium. These transitions can be described as nonequilibrium "second-order phase transitions" in a vortex continuum, causing the internal coordinates at bifurcation points to change abruptly.

We also consider an alternative approach to investigating the mechanisms of such a transition. This is based on stochastic Langevin equations, which are closely related to the derived kinetic FPK equations. We analyze a cardinal problem of the approach being developed—the possibility that asymptotically stable stationary-nonequilibrium states may occur in the subsystem of turbulent chaos. We propose a nonequilibrium thermodynamic potential for the stochastic internal coordinates of turbulent chaos that generalizes the well-known Boltzmann-Planck relation for equilibrium states to stationary nonequilibrium states of the ensemble representing chaos and show that this potential is the Lyapunov function for stationary nonequilibrium states of the ensemble that corresponds to the subsystem of turbulent chaos.

The third section of Chap. 5 is devoted to the thermodynamic derivation of generalized fractional FPK equations that describe based on fractional dynamics how the internal coordinates of the subsystem of turbulent chaos evolve. Introducing fractional time derivatives into the kinetic FPK equation allows one to include the effects of intermittency in time in a unified mathematical formalism. These effects are associated with turbulent bursts against the background of less intense low-frequency background turbulence oscillations.

In the *sixth chapter*, we prove the so-called H-theorem for the Kullback entropy. This postulates that any initial probability distribution for the internal coordinates of the subsystem of turbulent chaos under known assumptions asymptotically approaches a certain stationary state after a sufficiently long time. Here, we demonstrate that self-organization (i.e., the emergence of ordered dissipative structures with a lower symmetry than that of the initial state) is in principle possible in the thermodynamically open subsystem of turbulent chaos in due corse of the temporal evolution of the quasi-equilibrium vortex system. This can happen when coherent structures can be generated that are associated with the effect of multiplicative-noise-induced nonequilibrium phase transitions in the subsystem of chaos. We show that if the multiplicative noise of chaos is intense enough, the extrema of the probability density that describe the stationary behavior of a stochastic vortex system differ significantly in both number and position from the stationary states that correspond to a deterministic system. Moreover, multiplicative noise can generate new stationary states, thereby changing the properties (in particular, the bifurcation diagrams) of the local stability of chaos themselves: the transition points can be displaced under the influence of intense noise in a turbulent fluid.

Based on the general concept of generating coherent vortex structures in the thermodynamically open subsystem of turbulent chaos (due to nonequilibrium phase transitions induced by the multiplicative noise of chaos), we also consider in this Chap. one of the specific mechanisms for the formation and evolution of mesoscale vortex structures. These are associated with the phase-frequency synchronization of the self-oscillations of those internal coordinates that refer to the coherent component of chaos. In addition, we study some of the scenarios for the dynamical influence that the incoherent component (fine-grained fluctuating field) of turbulent chaos has on the formation and evolution of vortex structures. These transitions are shown to interrelate with the self-organization of clusters with a lower symmetry than that of the initial state. We conclude that while the growth in size of solid particles during collisions in classical turbulence is hampered, they can coalesce and enlarge within such dissipative ordered structures. In other words, the emergence of vortex clusters facilitates the solution of a key problem of the evolution of accretion disks—the problem of solid particle enlargement through collisions at relatively low velocities. This encounters obvious difficulties in attempting to reproduce similar processes in laboratory experiments.

In the *seventh chapter*, our attention is focused on one of the fundamental problems in astrophysics—the formation of protoplanetary accretion disks around late-type stars, a special case of which is the origin of the Solar System. Particular attention is given to developing a semiempirical approach to modeling heterogeneous turbulence in the accretion disk that surrounded the proto-Sun at the early stage of its existence with the goal of reducing the number of assumptions in the models used. We formulate a complete system of equations of two-phase multicomponent mechanics by taking into account the relative motion of the phases, coagulation processes, phase transitions, chemical reactions, and radiation. Basically, it is designed for schematized formulations and numerical solution of specific model problems on mutually consistent modeling of the structure, dynamics, thermal regime, and chemical composition of the circumsolar disk at various stages of its evolution. The processes in the disk medium in the presence of the developed turbulent motions of a coagulating gas suspension are addressed, which eventually contribute to the formation of a dust subdisk near the equatorial plane of the proto-Sun within the model under consideration. We also consider the emerging hydrodynamic and then gravitational instability in it followed by dust-cluster formation.

For an adequate phenomenological description of turbulent flows in a gas-dust disk, we perform the probability-theoretic Favre averaging of the stochastic equations of heterogeneous mechanics and derive the defining gradient relations for the turbulent interphase diffusion and heat fluxes. We also derive the "relative" and Reynolds stress tensors, which are needed to close the hydrodynamic equations of mean motion. We investigate the influence of the inertial effects of dust particles on the characteristics of turbulence in the disk, particularly on the additional generation of turbulent energy by large particles. We propose a semiempirical modeling method for the turbulent viscosity coefficient in a two-phase disk by taking into account the inverse effects of dispersed phase and heat transport on the development of turbulence with the goal of modeling the vertically inhomogeneous thermohydrodynamic structure of the dust subdisk and the ambient gas.

For a steady motion when solid particles settle down to the central plane of the disk under gravity, we investigate a parametric method of moments for solving the Smoluchowski integro-differential coagulation equation for the particle size distribution function. This states that a priori the sought-for distribution function belongs to a certain parametric class of distributions. We also analyze to which degree fine dust particle saturation can limit the subdisk atmosphere. This is responsible for the intensification of various coagulation mechanisms in a turbulized medium. The results of this chapter open new possibilities for constructing improved (and more realistic) models of stellar-planetary cosmogony, thereby providing a new approach to solving the fundamental problem of the origin and evolution of the Solar System and planetary systems around other stars.

In the *eighth chapter*, the problem of hydrodynamic helicity and, in particular, the influence of helicity on the evolution of disk turbulence is investigated. We show that the relatively long decay of turbulence in the disk is associated with the absence of reflection symmetry (relative to the equatorial plane) in the anisotropic field of turbulent velocities. We formulate the concept of emerging mesoscale coherent vortex structures with a large energy content in the thermodynamically open subsystem of turbulent chaos related to the realization of an inverse cascade of kinetic energy in mirror-asymmetric disk turbulence. We furthermore show that the inverse cascade generates a hierarchical system of clumps with a fractal density distribution that eventually initiate the triggering cluster-formation mechanisms through energy release. In turn, the formation of vortex clusters leads to an intensification of the mechanical and physical-chemical interactions between matter particles. As a result, spontaneous emergence and growth of dust clusters, stimulation of the condensation processes and phase transitions, the processes of heat and mass exchange between various regions of a heterogeneous disk, and significant modification of the spectrum of oscillations (density waves) are possible.

We discuss the influence of helicity on the energy cascade in a rotating disk and negative viscosity (generated by the cascade of helicity in disk turbulence when inverse energy transfer from small vortices to larger ones takes place), which we describe with a phenomenological approach. We also provide the relationships between the shear and rotational viscosities in a turbulent disk.

Finally, in the concluding *ninth chapter*, we address the problem of reconstructing an evolving protoplanetary gas-dust disk including electromagnetic effects, which also applies mainly to the basic concepts of stellar-planetary cosmogony. A closed system of magnetohydrodynamic (MHD) equations of mean motion designed to model shear and convective turbulent flows in electrically conducting media in the presence of a magnetic field is derived in the approximation of single-fluid magnetohydrodynamics. These equations can be used to numerically solve the problems of mutually consistent modeling of powerful turbulent cosmic plasma flows in accretion disks and related coronas. There the magnetic field significantly affects the dynamics of occurring astrophysical processes. We also systematically used weighted Favre averaging in addition to the traditional probability-theoretic averaging of the MHD equations to develop a model of a conducting turbulized medium. Favre averaging considerably simplifies the form of the averaged equations of motion for a compressible electrically conducting fluid and the analysis of the mechanisms for the amplification of macroscopic fields by turbulent flows. For a clear physical interpretation of the individual components of the plasma-field energy balance, we derive various energy equations that allow us to trace the possible transitions of energy from one form to another and, in particular, to understand the transfer mechanisms of the gravitational and kinetic energies of mean motion to magnetic energy.

In this context we focus on deriving the closing relations for the total (including the magnetic field) kinetic turbulent stress tensor in an electrically conducting fluid and the turbulent electromotive force (or the so-called magnetic Reynolds tensor) within the framework of extended irreversible thermodynamics. This also allows us to analyze the constraints imposed by the entropy growth condition on the turbulent transport coefficients. We propose a technique for modeling the turbulent transport coefficients, in particular, the kinematic turbulent viscosity, which makes it possible to take into account the influence of a magnetic field and the inverse effect of heat transfer on developing turbulence in a differentially rotating electrically conducting accretion disk.

It obviously follows from this brief discussion of the content that the monograph is basically oriented toward solving problems that are traditionally related to astrophysics and geophysics by the methods of mechanics. Leaving aside the debatable question regarding the very artificial breakdown of science into divisions and branches, we will only note that the merits and advantages of our approach are justified if they ensure the most comprehensive modeling of natural phenomena. This equally applies to cosmic objects that are inaccessible to direct study or to an attempt to reconstruct the events responsible for their set-up. The appearance and evolution of turbulized gas-dust accretion disks that result in the formation of planetary systems is just an example of the modeling approach developed in the monograph. At the same time, the monograph content in general may be considered as a theoretical basis for numerical simulations of a wide class of phenomena in which the mechanics of inhomogeneous (multicomponent, multiphase) and structured turbulent media widespread in nature plays a crucial role. An important constituent of these studies is the fundamentally new stochastic-thermodynamic approach to modeling developed turbulent flows and structured turbulence in astro- and geophysical systems considered from the standpoint of stochastic dynamics of open dissipative systems, which we have developed for many years. This is why we describe turbulent flows in natural phenomena as a typical example of nonequilibrium nonlinear dynamics, and this is the pivot around which the exposition in the monograph is built.

The book is the joint work of two authors. We gratefully point out that the discussions of several key problems with L.I. Sedov had a profound influence on the basic concepts underlying the content as well as on our own works on the mechanics of cosmic and natural media. In this monograph, we have tried to reflect many of L.I. Sedov's ideas on model construction of continuous media with complicated physical-chemical and thermal properties, in particular, on modeling multiphase multicomponent hydromechanics and multicomponent turbulent media.

The materials of Chaps. 2–9 and partially of Chap. 1 are based entirely on our original studies. The results of these studies are used in the numerical models that we are developing in collaboration with our colleagues and students. Some of the results have been published previously in the papers presented in the list of references and in our monographs "An Introduction to Planetary Aeronomy", Nauka, Moscow, 1987; "Turbulence of Multicomponent Media", Nauka-Interperiodika, Moscow, 1998; "Mechanics of Turbulence of Multicomponent Gases", ASSL series, vol. 269, Kluwer Academic Publishers, Dordrecht/Boston/London, 2001; and "Astrophysical Discs" (Ed. by A.M. Fridman, M.Ya. Marov, and I.G. Kovalenko), ASSL series, vol. 337, Springer, 2006.

We are well aware that we have not managed to cover equally completely all questions in the extensive subject matter that the monograph is concerned with. This is primarily because, despite certain progress achieved in recent decades in studying such a complex phenomenon as turbulence, especially the turbulence of inhomogeneous media and structured turbulence, much still remains unclear and the emerging mathematical difficulties often seem insurmountable. Another not exhaustively studied area are the physical-chemical characteristics of a cosmic medium, to which the observed peculiarity of accretion disks is directly related, especially their structure-forming mechanisms and evolution, including the evolution of planetary systems and the formation of peculiar phenomena on the planets and their satellites. In particular, developing the theory of structured turbulence will play a key role in understanding the natural self-organization mechanisms. This is an appeal to find answers to these challenging questions, or at least to place more rigorous constraints on the developed models.

We are also aware that it is a desideratum to develop new original approaches that would allow one to efficiently model turbulent dynamics when describing complex natural phenomena, in particular, cosmic media. This is our goal in our studies.

A. Buckingham from the Livermore Laboratory apparently completely agrees with this concept. In his review of our book "Mechanics of Turbulence of Multicomponent Gases" mentioned above, he noted that at this juncture, it is more useful to develop "...a model that would allow one to calculate how turbulence affects the accompanying physical processes than to concentrate on a deeper understanding of the essence of turbulence – the goal that, while being academically attractive, is fraught with potential disappointments " (Appl. Mech. Rev, vol. 56, no.1. 2003).

We hope that our new book will be received with interest by a wide circle of specialists in the fields of astrophysics, geophysics, mechanics, plasma physics, and space research. Astrophysicists, geophysicists, and planetologists will find in it a fairly deep justification of the theoretical approaches and methods for mathematical modeling of turbulence used in describing various natural and cosmic media, in particular, disk structures widespread in the cosmos, from the standpoint of mechanics. In turn, mechanicians will find a new rapidly progressing area of knowledge and the possibility of applying this fundamental science to the fascinating prospects of penetrating into the deepest mysteries of nature.

Numerous discussions of the questions that are touched upon here with our colleagues from various organizations contributed to writing the monograph and to improving its content. We would like to express our gratitude primarily to A.M. Fridman, E.M. Galimov, A.B. Makalkin, V.I. Maron, and I.S. Veselovskii. We are also grateful to K.K. Manuilov and O.A. Devina for their technical assistance in preparing the electronic version of the monograph and V. Astakhov for translating it into English. The publication of the book in Russian became possible owing to financial support by the Russian Foundation for Basic Research (project no. 08-01-07033), for which we express our gratitude.

We are grateful to Springer for offering to publish the book in English. It is worth to note that no specific advancement of the problems concerned was made since it was published in Russian, except for some new factual data about the inner planets of the Solar System and exoplanets, which do not impact, however, on our basic knowledge about planetary systems described in Chap. 1. Most importantly, we underline that the theoretical ground and principal concepts we deal with correspond to the contemporary views on the problems of self-organization in turbulent media and their application to accretion disk evolution.

Any critical remarks on the content of the book are welcome.

Mikhail Ya Marov
Moscow, Russia Aleksander V. Kolesnichenko

Contents

Chapter 1
Turbulent Chaos and Self-Organization in Cosmic Natural Media

Natural objects evolve from initial chaotic motions to order through a fascinating internal self-organization, which is embedded in their structure. The dynamics of this process is the focus of this work. They can be subjected to temporal and spatial variations or retain their stability for a long time. Ordered structures surround us ubiquitously on Earth; numerous examples of self-organization are observed in space. Turbulent flows characterized by a great variety of dynamical processes are widespread in the surrounding world. We mainly focus on the problems of macroscopic modeling these natural flows.

Turbulence is a widespread and extremely complex physical phenomenon associated with the fluctuating motion of a fluid that is present in various engineering systems and natural media. The atmospheres of planets in the Solar System, including the outer gaseous envelopes of these celestial bodies that lie in the atmosphere-space boundary regions, are typical examples of turbulized cosmic natural media. Developed turbulence also plays an important role in forming the structure and properties of astrophysical objects – galaxies and stars at various evolutionary stages as well as protoplanetary clouds and accretion disks that serve as a basis for cosmogonic models. Numerous present-day experimental studies of the peculiar features of turbulent motions of liquid and gaseous media and the key mechanisms determining their nature create the necessary prerequisites for developing theoretical approaches aimed at constructing appropriate mathematical models of this phenomenon.

Although the theory of hydrodynamic turbulence has been developed for more than a century now, it is still far from complete. New advanced statistical and phenomenological approaches continue to appear. There are increasingly more various mathematical models developed for a better understanding of the formation and evolution of turbulent motions of a homogeneous fluid (the classical theory of turbulence) and fluids with complicated physical-chemical and thermal properties. Numerous semiempirical (engineering) turbulence models designed to solve practical problems based on large-scale numerical simulations are developed based on the concepts of turbulent exchange coefficients for various transferable substances. Our objective here is to familiarize the reader with some important ideas that inspire

M.Y. Marov and A.V. Kolesnichenko, *Turbulence and Self-Organization: Modeling Astrophysical Objects*, Astrophysics and Space Science Library 389, DOI 10.1007/978-1-4614-5155-6_1, © Springer Science+Business Media New York 2013

this progress, to demonstrate the potentials of present-day phenomenological approaches to modeling hydrodynamic turbulence, in particular, structured turbulence, and to point out some problems that have not yet been resolved.

In this introductory chapter, we discuss the basic principles of the classical theory of homogeneous fluid turbulence and the structural properties of turbulent motions that allow the transition from a laminar (less ordered) flow to a turbulent (more ordered) one to be interpreted as self-organization in an open dissipative system. We analyze possible approaches to the macroscopic modeling of turbulent motions, the questions of stochastic nonlinear dynamics, and the turbulence-chaos relationship, including the synergetic aspects of the formation and evolution of ordered vortex structures in developed turbulent flows. In addition, we provide examples of turbulized cosmic natural media for which the phenomenological models considered in the monograph were constructed.

Concurrently, based on the elements of stochastic dynamics, we outline the general idea of the order-chaos relationships in dynamical systems, with various natural and cosmic media serving as typical examples. For a better understanding of these media, it seems necessary to familiarize the reader, at least briefly, with the cosmic environment of our own planet and with the unique natural complexes formed on other celestial bodies, many of which may be considered as the reflection of self-organization processes on the long path of evolution. Chapters 7-9 provide more details on the original approaches to the construction of protoplanetary accretion disk models. These models are among the most important elements in the evolution of stars and gas-dust clouds, out of which the planetary systems are formed. Because of the specificity of transport processes in heterogeneous turbulent media, the influence of magnetohydrodynamic effects, and the cosmochemical constraints, comprehensive modeling of these complex cosmic objects will help in answering the most fundamental questions of the cosmogony and evolution of cosmic objects in the Universe.

1.1 Turbulent Fluid Motion. General Principles

All flows of compressible liquids and gases are divided into two sharply different types: a calm and smooth flow called laminar and its opposite – the so-called turbulent flow, in which the hydrodynamic and thermodynamic fluid characteristics (velocity, temperature, pressure, mass density, chemical component concentrations, refractive index, etc.) undergo chaotic fluctuations and, hence, vary highly irregularly in space and time. Numerous examples of recording the fluctuations (with different periods and amplitudes) of thermohydrodynamic parameters that abound in special literature on turbulence illustrate a complex internal structure of real turbulized flows, which in this respect differ sharply from calm laminar flows. Owing to the formation of numerous vortices with all possible sizes, turbulent flows have an enhanced ability to transport the momentum, energy, and mass of elementary fluid volumes. This leads to an increased force action on solid bodies immersed in a flow, to intense heat and mass exchange between individual flow layers, to an acceleration of chemical reactions, etc. A turbulent (outwardly disordered) regime of fluid motion arises as a laminar flow loses its stability when the dimensionless Reynolds number $\mathbf{Re} = UL/\nu$ (where U and L are the characteristic velocity and linear length scale of the flow, respectively, ν is the molecular kinematic viscosity) exceeds some critical value $\mathbf{Re}_{cr}$. The dimensionless number **Re**, which reflects the relationship between the inertial and viscous forces in a flow, is the most general characteristic of a turbulized fluid. Since turbulent motions are always dissipative, they cannot be maintained per se but must draw energy from the environment. The emergence of a turbulent regime of motion of a laminar fluid is usually preceded by the excitation of oscillations of one or more independent frequencies f and their harmonics (and occasionally subharmonics). Turbulence arises either from the growth of small perturbations in a laminar flow or from the convective instability of motion. The turbulence energy is extracted from the kinetic energy of large-scale shear flows in the former case and from the potential energy of a nonuniformly heated fluid in a gravitational field in the latter case.

Turbulization of motion is a characteristic feature of many natural phenomena in which hydrodynamic processes accompanied by mass, momentum, and energy transport take place. All possible effects of its manifestation are observed on spatial scales from centimeters to megaparsecs. These include for example the various hydrodynamic processes in the terrestrial atmosphere and hydrosphere, in the atmospheres and interiors of stars and planets, in interstellar gas-dust clouds (planetary nebulae and protoplanetary disks), in the galactic and intergalactic medium, and in space plasmas (magnetohydrodynamic or plasma turbulence). The meteorological processes involving the atmosphere-ocean interaction, evaporation from water surfaces, vertical and horizontal heat transport, and intense mixing of admixtures (including pollutants) are predominantly turbulent. Turbulence also emerges in many engineering devices during the motion of liquids, gases, or plasmas, in particular, in the boundary layers and turbulent wakes of solid bodies immersed in a flow, in jet streams and mixing layers, in channel and pipe flows, during the vortex excitation of oscillations of mechanical and acoustic oscillatory systems (the so-called aeolian tones), in plasma beams, etc.

1.1.1 Physical Nature of Turbulence and Scenarios for Its Generation

First of all, let us make several general remarks about the physical nature of turbulence emerging under certain conditions in a nonlinear dissipative liquid or gaseous medium with a very large number of degrees of freedom that can exchange energy with the environment. When turbulence is present in a fluid, an enormous number of degrees of freedom turn out to be always excited. As a result, the change of any hydrodynamic quantity in time and space is described by functions containing a huge number of Fourier components, i.e., having a very complex pattern. Therefore each individual fluid particle (i.e., a small fluid volume element containing a very large number of molecules) of such a system moves in an outwardly complex, intricate way, so that its coordinates and direction of motion change with time according to the laws of stochastic mechanics. In this case, the velocity correlations at any point of the flow are limited by short time-intervals that depend on the initial conditions, outside of which no causal relationship can be established between the velocity fields at various instants of time, including the correlation with the preceding motion. All of this supports the view that the fluctuations of the velocity are stochastic in nature, as are other physical parameters in a turbulent flow that result from the loss of stability of the laminar motion of a hydrodynamic system when external controlling parameters (e.g., the number **Re**) change. From this viewpoint, turbulent motion is more chaotic than laminar one, i.e., turbulence is identified with chaos.

In a more general sense, the turbulization of liquid or gas motion can be represented as the result of a change in the topology of phase trajectories leading to a rearrangement of attractors and to a qualitative change (bifurcation) of the state of motion in the phase space. Complete mixing of phase trajectories with different asymptotic behaviors and structures (topologies) of the domains of attractions (attractors) surrounding them reflects the stochastic nature of turbulence. Rapid decay of the correlation functions on long time scales and continuity of the frequency spectra are signatures of mixing. Such a behavior of the trajectories in the phase space implies that the system is ergodic, i.e., for almost all realizations of a random field, its time averages are equal to the corresponding statistical averages

$$\langle \mathrm{A}(\mathbf{r}) \rangle = \lim_{T \to \infty} \frac{1}{T} \int_0^T \mathrm{A}(\mathbf{r}, t) dt, \tag{1.1.2}$$

where everywhere below the angle brackets will denote an average over an ensemble of realizations. The ergodic property is probably one of the characteristic features of a stationary homogeneous small-scale turbulent field (see, e.g., Kampe de Feriet, 1959).

The complete statistical description of a turbulent motion is reduced to determining a probability measure on its phase space that consists of all possible random realizations of its hydrodynamic fields. Therefore, the statistical description based on the study of peculiar statistical regularities inherent in large ensembles of similar hydrodynamic systems seems most adequate in the theory of turbulence. Only statistical hydromechanics, which studies the statistical properties of ensembles of liquid or gas flows under macroscopically identical external conditions (see, e.g., Monin, Yaglom, 1992), can serve as a basis for describing a turbilized medium. Thus, the problem of phenomenological modeling-developed turbulence in natural media, which this book is devoted to, consists of selecting some average characteristics that adequately describe the properties of a system with an enormous number of degrees of freedom. Therefore this approach requires a particular way of limiting the number of degrees of freedom.

Thus, turbulence is one of the manifestations of the diversity of motions in open mechanical systems with a very large number of degrees of freedom and a high degree of nonlinearity. Chaotically distributed and chaotically oscillating structures of various scales emerge in these systems when some controlling parameter increases. Therefore, turbulence has long been associated with the embodiment of pure chaos. At the same time, as is now clear, there exist elements of order in a developed turbulent flow, when ordered spatiotemporal structures, the so-called coherent vortex structures, can be born against the background of a small-scale fluctuating fluid motion (see, e.g., Taunsend, 1956; Cantwell, 1981; Belotserkovskii, 1997; Klimontovich, 2002). For this reason, whereas the generation of hydrodynamic turbulence characterizes the transition from order to chaos, the inverse process also takes place in a developed turbulent flow (when $\mathbf{Re} \gg \mathbf{Re}_{cr}$) – the birth of order from chaos, which will be discussed in more detail below.

We mainly consider the problems of modeling already developed turbulence that emerges at fairly large Reynolds numbers **Re**. However, dealing with some problems, specifically those related to modeling structured turbulence, we will often have to turn to particular turbulence generation scenarios. We will note at once that no universal mechanism of the transition to turbulent chaos in hydrodynamic flows of various types has been found as yet. We recall here very briefly the four known mechanisms of the transition from a laminar flow to a turbulent one when the Reynolds number reaches its critical value $\mathbf{Re}_{cr}$. This question is presented in greater detail in Sections 30-32 of the third revised edition of "Fluid Mechanics" by Landau and Lifshitz (1988), to which we refer the interested reader. With the exception of Landau's earliest hypothesis, all other mechanisms are related to finite-dimensional turbulence models; the experimental data available to date do not allow the ultimate choice to be made between them, because the features of different mechanisms are generally present in the experiments.

Landau—Hopf Scenario for the Transition to Turbulence

The picture of turbulence generation proposed by Landau (see Landau and Lifshitz, 1944) and then independently by Hopf in 1948 (see Marsden, McCracken, 1976) is based on the concept of a hierarchy of quasi-periodic motions. At $\mathbf{Re}<\mathbf{Re}_{cr}$, the perturbations decay with time. On the contrary, when **Re** increases, the perturbations grow after it has reached its critical value $\mathbf{Re}_{cr}$; the stationary laminar fluid motion (with the limit point serving as its image in the phase space) becomes periodic (a limit cycle, an attractor with a certain domain of attraction in the state space, is born) with an oscillation frequency f_1. As the Reynolds number increases even more and reaches the second Andronov—Hopf instability threshold, this periodic regime, in turn, loses its stability; the second frequency f_2 incommensurate with f_1 (when f_1/f_2 is an irrational number) appears. The loss of stability through the periodic motion is accompanied by a certain qualitative rearrangement in the state phase (Andronov—Hopf bifurcation) of the trajectory behavior near the now unstable limit cycle. The additional destabilization of an increasingly large number of oscillatory modes gives rise to a complex and intricate quasi-periodic regime of fluid motion with a set of generally incommensurate frequencies $f_1, f_2, \ldots f_r, \ldots$. This regime at large r is perceived as a turbulent one. The corresponding attractor of the turbulent motion is an invariant torus T^r in a state space of large dimension r. The Fourier power spectrum remains discrete here. Although the system's motion acquires a complex and intricate pattern, the continuous spectrum and flow chaotization inherent in real turbulence arise only at an infinite number of Andronov—Hopf bifurcations. However, according to current views, this scenario does not correspond entirely well to the actual state of things: even a system with three degrees of freedom gives a continuous Fourier spectrum, which is a signature of its chaotic motion. As experiments (in particular, those related to Benard's convection) have shown, the chaotic regime of viscous fluid motion sets in already after the appearance of two fundamental frequencies.

Ruelle—Takens Scenario for the Chaotization of Fluid Motion: Strange Attractor

The Ruelle—Takens (1971) scenario for the generation of hydrodynamic turbulence, which is considerably shorter than Landau's, also relies on the concept of an Andronov—Hopf bifurcation. As a result of the latter, a limit cycle is born from a fixed point and a new oscillation frequency incommensurate with the preceding one and previously absent from the system appears in it. According to this mechanism, two consecutive Andronov—Hopf bifurcations initially occur, as in Landau's model, but subsequently, due to the system's nonlinearity, the three-frequency motion becomes unstable, is destroyed, and passes into a chaotic motion on a strange attractor[1] that corresponds to a complete filling of the Fourier spectrum.

A strange attractor is an attractor topologically different from a torus that attracts the entire set of unstable trajectories belonging to it in the state space of a dissipative system when $t \to \infty$. This can be not only an unstable cycle, but also open trajectories infinitely wandering inside a bounded region of multidimensional state space without emerging from it. This peculiar property of a strange attractor is supplemented moreover by a sensitive dependence of chaotic motion on initial conditions. This implies that close trajectories diverge exponentially. In other words, the system must forget the initial conditions because now there are small perturbations. It is this complex, irregular behavior of trajectories in the phase space that is associated with chaos, i.e., turbulent fluid motion.

[1] The two-period motion corresponds to a trajectory on the torus (two-dimensional tube) on which the appearance of chaos is forbidden by the well-known Poincare—Bendixson theorem.

Let us dwell on several more properties of a strange attractor related to its geometric structure. A characteristic feature of this set of points is scale invariance: as the size of some subdomain of a strange attractor increases, a geometric object similar in structure to the whole attractor is produced. These objects are called scaling structures in physics, while the geometric model of a self-similar structure in mathematics is a fractal. For practical purposes, it is important to find a topological characteristic describing the degree of complexity of this set of points, which would allow one to characterize the degree of chaotization of a motion, to distinguish a motion with mixing from a simply ergodic motion, etc. The Hausdorff—Besicovitch dimension of a fractal is such a quantitative characteristic (see, e.g., Schröder, 1990). Since we are dealing with dissipative systems in which the volume in phase space shrinks (in contrast to ergodic systems), the volume of a strange attractor in its state space must be equal to zero. But its dimension must then be smaller than that of n-dimensional space. This nonzero dimension (the Hausdorff—Besicovitch dimension) is usually defined as follows. If we break up the volume of the n-dimensional space of an attractor into small (hyper) cubes with an edge length η and volume η^n and denote the minimum number of cubes whose set completely covers it by $N(\eta)$, then the Hausdorff—Besicovitch dimension D is defined by the formula

$$D = \lim_{\eta \to 0} \frac{\ln N(\eta)}{\ln(1/\eta)}. \tag{1.1.2}$$

The existence of this limit means that the attractor volume in D-dimensional space is finite: at small η, we have $N(\eta) \approx V\eta^{-D}$ (where V is a constant), whence we see that $N(\eta)$ may be considered as the minimum number of D-dimensional cubes covering the volume V in D-dimensional space. The Hausdorff—Besicovitch dimension is therefore a generalization of the ordinary geometric dimension that allows fractal objects to be characterized. Obviously, the dimension defined by Eq. (1.1.2) cannot exceed the total dimension n of the state space, but it can be smaller than the latter and, in contrast to the conventional dimension, can be fractional; it is precisely such for the strange attractors of a turbulent field.

The fractal description of turbulence is applicable, in particular, to the field of the dissipation rates of turbulent energy (a key characteristic of locally isotropic turbulence)

$$\varepsilon = \varepsilon(\mathbf{r}, t) = \frac{\nu}{2} \sum_{ij} \left\langle \left(\partial u_i'/\partial x_j + \partial u_j'/\partial x_i \right)^2 \right\rangle$$

concentrated, as a rule, in the volumes of the dissipation (Kolmogorov) length scale $\eta = (\nu^3/\varepsilon)^{1/4}$. Let D be the fractal dimension of the dissipation field. After estimating the minimum scale η and the determination of the number $N(\eta)$ of (hyper) cubes with size η, the fractal dimension D calculated from Eq. (1.1.2) takes the form $\mathrm{D} = \ln N / \ln(L/\eta)$, whence $N = (L/\eta)^D$. Since the volume of the dissipation regions is $(L/\eta)^3$, the total volume of active dissipation is proportional to $(L/\eta)^{D-3}$. Using this result in establishing the relation

$$\mathrm{K} \propto (\mathbf{Re})^{3(3-D)/2}, \; K \equiv \langle(\partial u/\partial x)^4\rangle / \left\{\langle(\partial u/\partial x)^2\rangle\right\}^2$$

between the experimentally determined flattening coefficient K and the Reynolds number **Re**, Mandelbrot (1975) obtained an estimate for the fractal dimension of the dissipation field that did not exceed 2.66.

In conclusion, there is probably no complete classification of strange attractors as yet. But two types of strange attractors are observed in real physical systems: Lorenz attractors and quasi-attractors.

Transition to Turbulent Chaos Through a Sequence of Period-Doubling Bifurcations

The third model of the transition to turbulence was proposed by Feigenbaum (1978) and is associated with an infinite sequence of period-doubling bifurcations. The transition begins with Andronov—Hopf bifurcations from a fixed point to a limit cycle with frequency f_1. As the control parameter increases further, consecutive period-doubling bifurcations occur that lead to a periodic motion with frequencies $f_1/2, f_1/4, f_1/8$, etc. This sequence converges at a critical Reynolds number at which a strange quasi-attractor with a complex alternation of regions of chaos and order emerges.

Transition to Turbulence Through Intermittency

The fourth turbulence generating mechanism on which the Pomeau—Manneville *(1980)* model is based is associated with the transition to a chaotic motion through intermittency (alternation). According to this model, the trajectory at $\mathbf{Re} < \mathbf{Re}_{cr}$ reaches a stable cycle, i.e., a laminar periodic motion is established in the physical space. As the Reynolds number increases, $\mathbf{Re} > \mathbf{Re}_{cr}$, but near the critical point the periodic trajectory is destroyed and directly turns into a chaotic trajectory with intermittency as the result of a reverse tangential bifurcation. In the geometric space, this implies the appearance of a flow in which turbulent states alternate with laminar ones. The duration of the turbulent intervals cannot yet be determined theoretically. The duration of the laminar flow periods τ_{lam} has the form of a power law $\tau_{lam} \propto (\mathbf{Re} - \mathbf{Re}_{cr})^{-1/2}$ and determines the pattern of decrease in interval τ_{lam} with growing supercriticality. Note also that the question about both the pattern of intermittency formation and its role in the evolution of turbulence remains open in this scenario.

To conclude this subsection, note once again that all considered models describe only the emergence of turbulent chaos and exclude the properties of developed turbulence. At the same time, as has already been pointed out above, according to current views (see, e.g., Klimontovich, 2002), during the transition to extremely developed turbulence in an open hydrodynamic system, new macroscopic connections (attributable to the collective interaction of its constituent subsystems) can be established between separate regions, which enhances its internal order compared to the microscale fluctuations. In this case, the set of spatiotemporal scales on which this process develops corresponds to a coherent behavior of an enormous number of fluid particles. In particular, the appearance of mesoscale-ordered coherent (dissipative) structures with a certain degree of organization and the formation of regions of enhanced vorticity concentration in the form of vortex tubes, vortex layers, etc. against the background of a small-scale turbulent motion is associated with this. Hence, one can reach the at first glance paradoxical conclusion that a developed turbulent motion, despite its high complexity, corresponds to a state of higher order than a more symmetric laminar motion. This phenomenon, "showing how difficult it is to distinguish order from chaos in the case of complex motions" (Klimontovich, 1990), constitutes part of the general problem of self-organization (synergetics). Interest in this problem, which is found in various natural phenomena and areas of technology, has increased in recent years. According to the existing views (see, e.g., Nikolis, Prigogine, 1977; Haken, 1983), these self-organizing dissipative structures result from the stabilization of spatially inhomogeneous instabilities in an open nonlinear dynamical system due to the emergence of coherence (the synchronization of oscillations [see Ch. 6]) and, thus, the attainment of "order through fluctuations".

1.1.2 Developed Turbulence. Kolmogorov's Theory

Fluctuations with scales from the largest to very small ones are present in a developed turbulent flow. These flows are characterized by filled Fourier spectra, not only the temporal, but also the spatial ones. Large-scale fluctuations with a scale on the order of a characteristic length L, which defines the size of the region G where the turbulent motion takes place, play a major role in a turbulent flow. This quantity is called an external turbulence scale. The large-scale motions with the highest amplitudes have a velocity U comparable in order of magnitude to the changes in mean velocity over the distance L. In contrast, the small-scale fluctuations with much lower amplitudes are like a fine detailed structure superimposed on the main large-scale turbulent motions. Only a comparatively small part of the entire fluid kinetic energy is contained in the small-scale fluctuations. Developed small-scale turbulence is locally homogeneous and isotropic.

According to Kolmogorov's theory (see, e.g., Landau and Lifshitz, 1988), the small-scale structure of developed turbulence (in this section, for simplicity, we are dealing with an isothermal flow where $\rho = const, \quad \nu = const$) is determined by cascade energy transfer over the spectrum of vortices (turbulent fluctuations) with various spatiotemporal scales. The qualitative scheme of a Richardson-Kolmogorov cascade is as follows. Small vertices gain energy through successive fragmentation of large vortices as the Reynolds number **Re** grows. The largest energy-carrying vortices are formed through the loss of stability of an initial laminar flow, and their sizes $\lambda_1 \equiv L_1$ (Prandtl mixing length) are much smaller than the characteristic length scale L of the flow region G itself. Since the Reynolds number of large-scale fluctuations $\boldsymbol{Re}_{\lambda 1} \sim \lambda_1 u_1'/\nu$ corresponding to their scale λ_1 and relative velocity u'_1 is too large, these first-order perturbations are also unstable and are destroyed to produce second-order perturbations (λ_2, u'_2), which, in turn, give rise to even smaller vortices for the same reason, etc. The fragmentation of vortices stops when the forces of molecular viscosity in the fluid begin to play a significant role, which occurs for vortices with Reynolds numbers $\boldsymbol{Re}_\eta \sim 1$. In this case, the specific kinetic energy $\sim U^3/L$ of the carrier flow is continuously redistributed from large-scale vortices to smaller ones down to the smallest vortices with a characteristic size on the order of the internal turbulence scale $\eta = (\nu^3/\varepsilon)^{1/4}$, which characterizes the influence of viscous effects on the structure of small-scale turbulence. Within the limit of very large numbers **Re**, a quasi-stationary regime of inertial kinetic energy transfer from large vortices to smaller ones is established in which the energy is eventually converted through molecular viscosity ν into heat in hydrodynamically stable small vortices with a scale $\lambda_k \leq \eta$ (the viscous range). Kolmogorov assumed that there is a so-called inertial range ($\eta < \lambda_k < L_1$) of scales of vortices in which no noticeable energy production and dissipation takes place or, more precisely, the dissipation in them is negligible compared to the energy that they gain from larger vortices and transfer to smaller ones. If **Re** is high, a statistical regime of small-scale fluctuations lying in the spatiotemporal region G,which result from the randomness of the vortex fragmentation and of the chaoticity of the energy transport over the cascade, is locally isotropic, homogeneous, and quasi-stationary, although the averaged flow is anisotropic, inhomogeneous, and nonstationary. Of course, the quasi-stationary regime of turbulent flow in which an energy flux into the region of small scales is realized suggests certain boundary conditions at the boundary ∂G of the flow region G under consideration, which create a pump and a sink. The formula

$$\eta = (\nu^3/\varepsilon)^{1/4} \tag{1.1.3}$$

for the dissipation length scale η is a corollary of Kolmogorov's (1941) first similarity hypothesis. According to the latter, the statistical regime of small-scale, locally isotropic turbulence is uniquely defined by two dimensional parameters – the mean energy dissipation rate $\varepsilon(\mathbf{r}, t)$ and molecular viscosity ν. The quantity ε is the turbulent energy dissipation (per unit fluid mass per unit time) averaged over the ensemble of possible realizations of the fluid flow and simultaneously characterizes the transfer rate of the kinetic energy of a fluctuating motion over the hierarchy of vortices in the cascade process:

$$U^3/L \approx u_1^3/\lambda_1 \approx \ldots \approx u_\eta^3/\eta \cong \nu\, u_\eta^2/\eta^2 \equiv \varepsilon = const \qquad (1.1.4)$$

- a corollary of Kolmogorov's (1941) second similarity hypothesis, according to which the statistical regime of turbulence in the inertial range ($\eta<\lambda_k<L_1$) is defined by only one parameter ε. Here, $u_k^3 / \lambda_k \sim u_k^2 / t_k$ is the specific kinetic energy gained per unit time by vortices of the k-th order from vortices of the $(k-1)$-th order and transferred by them to vortices of the $(k+1)$-th order; $u_\eta \cong (\nu\,\varepsilon)^{1/4}$ and $t_\eta \cong \eta/u_\eta = \sqrt{\nu/\varepsilon}$ are the fluctuation velocity and time in vortices of the Kolmogorov dissipation scale η, respectively. It is also assumed in Kolmogorov's theory that the vortices continuously fill the entire region G on each scale (at each cascade step).

We now turn to an important relation concerning the viscosity-conditioned lower threshold – the Kolmogorov dissipation scale. It follows from (1.1.3) and (1.1.4) that

$$\mathrm{L}/\eta \propto \left(\nu^3/L^3U^3\right)^{-1/4} \propto \mathbf{Re}^{3/4}. \qquad (1.1.5)$$

In Kolmogorov's theory, the inertial range therefore encompasses the range of scales that grows with increasing Reynolds number as $\mathbf{Re}^{3/4}$. If a developed turbulent flow is numerically simulated on a uniform grid, then the minimum number of grid points in a cube with a side equal to the main scale length L must be $\mathrm{N} \propto \mathbf{Re}^{9/4}$. Generally, this provides bleak prospects for numerical simulations. However, because thin vortex filaments (or other coherent structures) thread much of the turbulent flow region in real turbulence, the number of degrees of freedom of developed turbulence can be much less than $\mathbf{Re}^{9/4}$.

The quantitative description of small-scale, locally isotropic turbulence in a region G (we assume $\Lambda \sim L_1$ throughout) is based on the use of structure functions

$$D_{ij}(\mathbf{r}) = \overline{(u'_{\mathrm{i}}(\mathbf{x}) - u'_{\mathrm{i}}(\mathbf{x}+\mathbf{r}))(u'_{\mathrm{j}}(\mathbf{x}) - u'_{\mathrm{j}}(\mathbf{x}+\mathbf{r}))}$$

and their spectra

$$E_{ij}(\mathbf{k}) = \int D_{ij}(\mathbf{r}) \exp(i\mathbf{k}\cdot\mathbf{r})d\mathbf{r},$$

where $i = \sqrt{-1}$, $\mathbf{k}$ is the wave number (Kolmogorov, 1941; Obukhov, 1941). From the first similarity hypothesis and the assumption that the parameters of large-scale turbulence change only slightly at distances on the order of $r = |\mathbf{r}|$ if $r<L_1 \ll L$, it follows that

$$D_{ij}(r) = (\varepsilon r)^{2/3}[f(r/\eta)r_i r_j r^{-2} + g(r/\eta)\delta_{ij}], \tag{1.1.6}$$

where f and g are arbitrary functions of the dimensionless argument (r/η). According to the second similarity hypothesis, the structure function in the inertial range $L_1 >> r >> \eta \sim L\mathbf{Re}^{-3/4}$ does not depend on viscosity ν, i.e., $f = const$ and $g = const$ at $r/\eta \gg 1$. From this follows one of the most important laws of small-scale turbulent motions (the two-thirds law): in any turbulent flow with a sufficiently large Reynolds number **Re**, the mean square of the velocity difference at two points separated by a distance r at moderately high values of r (comparable to the length scale Λ of the corresponding averaged flow) should be proportional to $r^{2/3}$, $D_{11}(r) = C(\varepsilon r)^{2/3}$, where C is a universal constant. The assertion formulated in terms of spectra equivalent to the two-thirds law for the structure function of the velocity field takes the form $E(k) = C^* \varepsilon^{2/3} k^{-5/3}$ at $1/\eta \gg k \gg 1/L$ (Obukhov, 1941). This law has now been sufficiently well confirmed experimentally for various turbulent flows (see Monin and Yaglom, 1996).

For the n th-order structure functions, Kolmogorov's similarity theory provides the relation $\overline{V^n} \sim (\varepsilon r)^{n/3}$, where $\mathrm{V} \equiv u'(\mathbf{r}) - u'(\mathbf{r} + \mathbf{r}^*)$, which, in general, is not confirmed experimentally, especially for $n \gg 1$. As is well known, this is because the similarity hypotheses for locally isotropic turbulence in their original form suggested that the constancy of the energy influx to small-scale perturbations lies in the inertial range, i.e., the constancy of Kolmogorov's parameter ε. However, for real turbulent flows, the energy dissipation $\varepsilon(\mathbf{r}, t)$ not only changes when passing from one point $\mathbf{r}$ (in region G) of the flow to another, but is also a random variable of the coordinates $\mathbf{r}$ and time t, fluctuating with the instantaneous velocity field $\mathbf{u}'(\mathbf{r}, t)$. The probability distributions of the energy dissipation $\varepsilon(\mathbf{r}, t)$ (which generally depend on hydrodynamic characteristics of the averaged fluid motion, in particular, on the Reynolds number **Re**) have an effect on the unconditional probability distributions for the small-scale characteristics of developed turbulence. Therefore they generally cannot be quite universal. Landau's well-known remark (see, e.g., Landau and Lifshitz, 1988) regarding Kolmogorov's original similarity hypotheses concerned precisely this effect. In connection with these difficulties, Kolmogorov's views of a random cascade were refined by Obukhov (1962), who suggested abandoning the condition $\varepsilon(\mathbf{r}, t) = const$ in region G (with the center at point $\mathbf{r}$ and a length scale $\Lambda \ll L$) and proceeding from the fact that the statistical characteristics of small-scale motions (e.g., the structure functions) are not defined by the probability-theoretic mean $\varepsilon(\mathbf{r}, t)$ of the random variable ε but depend on the values of the dissipation $\varepsilon_r(\mathbf{r}, t)$ averaged over some volume V_r with a characteristic size r that is small compared to the typical inhomogeneity scale of the averaged flow, $r \ll \Lambda$. If a sphere of radius r is chosen as the averaging region lying within G (the results obtained depend weakly on the shape of the averaging region), then the quantity ε_r is defined by the formula

$$\varepsilon_r(\mathbf{r},t) = \tfrac{6}{\pi r^3} \int\limits_{|\mathbf{r}^*| \le r} \varepsilon(\mathbf{r}+\mathbf{r}^*,t)d\mathbf{r}^*.$$

The original (unrefined) cascade theory is not absolutely accurate, either, because it does not explicitly take into account all small-scale coherent dissipative structures that the turbulent field can possess, except for the flow macrostructuring on large scales $\sim L$, admissible within its framework. Meanwhile, as has become clear especially in recent years, any adequate theory of turbulence should take into account the presence and dynamics of these dissipative structures as well as the nonuniformity of their spatiotemporal distribution in a chaotic flow (Crow, Champagne, 1971; Brown and Roshko, 1974).

On the other hand, the phenomenon of hydrodynamic internal intermittency, whereby the regions occupied by the so-called turbulent spots (where intense fluctuations of the velocity gradients are observed, $\varepsilon_r > 0$) are closely intertwined with the regions with a weakly turbulized or completely vortex-free fluid (where there are virtually no such fluctuations, $\varepsilon_r \cong 0$) is, in turn, associated with the chaotic nature of the kinetic energy transfer over the cascade because of the instability of dissipative structures. The existence of intermittency corresponds to the flow regime (at a small number **Re**) when the power of the constantly acting fluid turbulization mechanism that transfers energy to the flow on large scales is not yet sufficient to produce fully developed turbulence in the entire region G occupied by the fluid. Thus, the values of $\varepsilon_r(\mathbf{r},t)$ (or the related quantities quadratic in velocity gradients) can serve as an indicator of intermittency.

1.1.3 On the Spectrum of Developed Turbulence

The apparatus of mathematical statistics for quasi-stationary random processes, along with correlation and spectral analyses, is traditionally used to theoretically study the turbulent fluctuations in the velocity of a fluid motion (or other thermohydrodynamic parameters fluctuating in the flow). The most commonly used characteristics of the field of random variables $\mathrm{A}(\mathbf{r},t)$ include their means (below a bar over an averaged quantity is used to denote the means) and variances, various correlation and structure functions, spectral density functions, etc. Since the systems of motion of various spatiotemporal scales, as a rule, coexist in a turbulized medium, the thermohydrodynamic parameters characterizing it should, in general, be averaged (it is assumed that an individual realization of turbulent flows can be described by the hydrodynamic equations for instantaneous motion). The averaging can be performed either over a temporal (and/or spatial) interval Δt (e.g., over the fluctuation values measured at a fixed point during some period Δt) or over an ensemble of physically admissible realizations of the hydrodynamic system. To obtain representative statistical estimates for the defining parameters of a fluctuating flow, the averaging interval should be chosen properly. The larger the fluctuation scales, the larger should be this interval.

Accordingly, to choose the averaging scale, one should know the order of magnitude (spectrum) of the fluctuations for all thermohydrodynamic parameters of a turbulent medium (such as the wind velocity, pressure, temperature, etc.). The energy spectrum of the fluctuation field of any physical quantity $A(\mathbf{r}, t)$ consists of a series of curves describing the dependence of the square of the fluctuation amplitude $\langle \mathbf{A}'^2 \rangle$ of the random variable A on the angular frequency ω ($\omega = 2\pi f$, f is the oscillation frequency in s^{-1} or Hz) and/or linear sizes κ (κ is the wave number).

We particulary highlight the micrometeorological oscillations with periods from fractions of a second to minutes, which emerge directly in the surface layer of air and represent small-scale isotropic turbulence, which serves as the most important mechanism for viscous turbulent energy dissipation. These oscillations are part of the broad spectrum of oscillations (in time) of the above random variables that are characteristic of the terrestrial atmosphere with periods from fractions of a second to thousands of years (see Monin, 1969). The maximum of its energy spectrum ω $E(\omega)$ (where $E(\omega)$ is the spectral kinetic energy density of the flow) occurs at a period $\tau_m = 1/\omega \approx 1$ min. For a typical air velocity in synoptic processes, $V = 10$ m/s, this corresponds to the scale of horizontal turbulent inhomogeneities $L = V\tau_m \approx 600$ m. At $\omega > 1/\tau_m$, the wind velocity spectra obey the Kolmogorov—Obukhov "five-thirds law" (Obukhov, 1941) $E(\omega) \sim (\varepsilon_e^{\frac{2}{3}}/V)(\omega/V)^{\frac{-5}{3}}$, and the turbulence spectrum is abruptly cut off in the range of maximum turbulent fluctuation frequencies $\omega \sim V\varepsilon_e^{\frac{1}{4}}\nu^{-\frac{3}{4}}$ (Monin, 1969; 1988).

For an isotropic turbulent flow, it is convenient to analyze the cascade process of kinetic energy transport from large-scale vortex motions to small-scale ones in the space of wave numbers κ using the spatial spectral energy density $E(\kappa)$. The latter is defined as

$$E(\kappa) = \int\limits_{|\kappa'|=k} \Phi_{ii}(\kappa')\, d\kappa', \tag{1.1.6}$$

where the summation is performed over repetitive indices. Here, $\Phi_{ij}(\kappa)$ is the Fourier transform of the correlation function $R_{ij}(\mathbf{r}) = \overline{u'_i(\mathbf{x}+\mathbf{r})u'_j(\mathbf{x})}$ of the random velocity field; κ is the absolute value of the wave number, $\kappa = |\kappa|$. Note that $R_{ii}(0) = \overline{u'_i u'_i}$ coincides with the kinetic energy of the fluctuating motion of a unit fluid mass. Hence, using the definition of $E(\kappa)$, it follows that the turbulent energy is

$$b \equiv \frac{1}{2}\overline{u'_i u'_i} = \frac{1}{2}\int\limits_0^\infty E(\kappa)\, d\kappa. \tag{1.1.7}$$

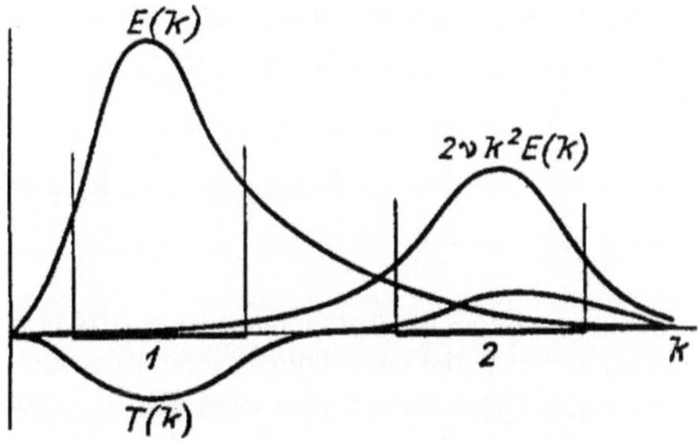

Fig. 1.1.1 Energy, $E(k)$, and energy dissipation, $2\nu k^2 E(k)$, spectra and energy redistribution function $T(k)$. Regions 1 and 2 on the horizontal scale denote the energy and dissipation intervals, respectively. According to Monin and Yaglom (1996)

Figure 1.1.1 shows the pattern of the energy distribution for a turbulent motion in various scales, i.e., the spatial spectrum of the quantity b. The deep physical meaning of the fluctuating motion of the velocity field follows from the "spectral form" of the Karman—Howart equation (Monin, Yaglom, 1996), which describes the time variation of the spectral turbulence energy distribution $E(\kappa, t)$:

$$\partial E(\kappa, t)/\partial t = T(\kappa, t) - 2\nu\kappa^2 E(\kappa, t). \tag{1.1.8}$$

Here, the first term $T(\kappa, t)$ describes the energy redistribution over the turbulence spectrum caused by the interactions of its spectral components with wave numbers κ and all other spectral components produced by the nonlinear "inertial terms" of the basic hydrodynamic equations. We emphasize that the energy is redistributed between the spectral components without any change in the total energy of the turbulent motion $\left(\int_0^\infty T d\kappa = 0\right)$. The second term $2\nu\kappa^2 E(\kappa)$ describes the energy dissipation due to viscosity and characterizes the decrease in the kinetic energy of the fluctuations of perturbations with a wave number κ proportional to the intensity of these perturbations multiplied by $2\nu\kappa^2$. This implies that, due to viscosity, the energy of long-wavelength perturbations (with small κ) decreases much more slowly than the energy of short-wavelength perturbations, thereby reflecting the proportionality of the fiction force to the velocity gradient. The energy, $E(\kappa)$, and energy dissipation, $2\nu\kappa^2 E(\kappa)$, spectra and the function $T(\kappa, t)$ defining the energy redistribution over the spectrum are schematically shown in Fig. 1.1.1. As can be seen, negative values of the function $T(\kappa)$ in the range of small κ, to which the maximum of the energy of large-scale motions in the curve $E(\kappa)$ corresponds, give way to positive values in the range of large κ, to which the maximum of the energy dissipation in the curve $2\nu\kappa^2 E(\kappa)$ corresponds. This confirms the scenario of cascade energy transport from large-scale components of motion to small-scale ones. The latter are characterized by steep local velocity gradients and, hence, the role of viscosity is important. Following Monin and Yaglom (1996), the first of these ranges, where up to 80-90% of the total turbulence energy (1.1.1) is concentrated, is called the energy range, while the second range, where the total energy dissipation takes place,

$$2\nu \int_0^\infty \kappa^2 E(\kappa)\, d\kappa \cong 2\nu \int_{\kappa_0}^\infty \kappa^2 E(\kappa)\, d\kappa,$$

is called the viscous dissipation range (here, κ_0 is an intermediate value of the wave number in the inertial range lying beyond the energy range and before the dissipation range). These authors also considered the problem of the "rate of energy redistribution over the spectrum" $T(\kappa)$ in more detail, using a spectral representation of the velocity field itself based on the approach developed by Batchelor (1953). Thus, we make sure once again that the initial energy of large vortices determining the dynamical and kinematical properties of a flow is expended on their fragmentation, which is expressed via turbulent viscosity, and, in the long run, determines the properties of viscous dissipation when the total energy balance condition is met.

1.1.4 Turbulent Diffusion

Modeling the propagation of minor admixtures (including chemically active gas components in planetary atmospheres and dispersed particles of various types and sizes that are also partly involved in chemical transformations and phase transitions) is very important for the of problems related to the theoretical consideration of turbulent heat and mass transport in a natural medium. The pattern of turbulent diffusion depends on the structure of the fluctuating velocity field and the turbulence energy distribution between fluctuations of various spatial scales, and it plays a crucial role in the transport of these admixtures and their mixing. The mean admixture concentrations $\langle Z_\alpha \rangle$ and their fluctuation deviations Z'_α, along with the mean velocity of the gas mixture $\langle \mathbf{u} \rangle$ and its fluctuations $\mathbf{u}'$, can be singled out when describing the diffusion processes in a turbulized medium. This allows one to pass from the diffusion equations for instantaneous concentrations to the diffusion equations for the scale of mean motion using conventional averaging techniques [see Eq. (2.1.7) and (3.1.23)]. Gradient-type diffusion produced by comparatively small vortices and non-gradient-type diffusion produced by large vortices are known to be distinguished in the theory of turbulent mixture diffusion, depending on the scales of the phenomenon (Monin and Yaglom, 1992). A phenomenological approach can be applied to gradient-type turbulent diffusion [see Ch. 3]. Since turbulence is a property of the regime of fluid flow and not of the fluid itself, the momentum (energy, mass) exchange mechanism resembles the molecular momentum exchange only remotely. Nevertheless, there exists a certain analogy between the diffusion of a mixture in the field of small-scale turbulence and molecular diffusion (which is also valid for contamination of the medium by a fine-grained admixture when the turbulence scale is small compared to the "contamination cloud" scales). This analogy assumes a proportionality between the turbulent flux of some diffusing substance and the gradient of its averaged concentration. Indeed, just as the chaotic molecular motion is characterized by some mean velocity of molecules v_m and mean free path l_m (so that the molecular diffusivity $\mathrm{D} \propto v_\mathrm{m} l_m$), the chaotic turbulent mixing can be described via the turbulent diffusivity $D^{\mathrm{turb}} \propto v_{turb} l_{turb}$, where v_{turb} is the characteristic magnitude of the turbulent velocity fluctuations and l_{turb} is the local turbulence scale (the so-called mixing length). However, in contrast to v_m and l_m, the parameters v_{turb} and l_{turb} define not just the properties of the medium, but the regime of fluid flow. Accordingly, D and D^{turb} are understood as the proportionality coefficients between the molecular diffusion flux $\mathbf{J}_\alpha$ of some substance α and the gradient of its concentration $\partial Z_\alpha / \partial \mathbf{r}$ ($\mathbf{J}_\alpha = -D\, \partial Z_\alpha / \partial \mathbf{r}$) in the former case and between the turbulent flux of this substance in the fluctuating velocity field $\mathbf{J}_\alpha^{turb} = \overline{\rho Z''_\alpha \mathbf{u}''}$ and the gradient of its averaged concentration $\partial \langle Z_\alpha \rangle / \partial \mathbf{r}$ in the latter case, i.e.,

$$\mathbf{J}_\alpha^T = -\bar{\rho} D^{turb} \cdot \partial \langle Z_\alpha \rangle / \partial \mathbf{r} \qquad (1.1.9)$$

(Monin, 1962). Since turbulent diffusion, in contrast to molecular one, is usually anisotropic in pattern, tensors should generally be described by D^{turb} and $\mathbf{l}_{turb}$.

If the particles of a diffusing substance retain their individual properties when moving in a fluctuating flow (a passive admixture), then it is most natural to use the Lagrangian concept [see Eq. (3.3.1)] as a basis for the turbulent mixing model. This makes it possible to consider this transport as an "ordinary motion" of fluid particles during which diffusion takes place without any detailed concretization of the mixing mechanism. In this case, introducing the Lagrangian turbulent fluctuation $Z''_{\alpha L}$ $(\equiv Z''_\alpha + \mathbf{l}_{turb} \cdot (\partial/\partial \mathrm{r}) \langle Z_\alpha \rangle \cong 0)$ of a conservative attribute Z_α allows one to define some length $l_{turb} = |\mathbf{l}_{turb}|$ (roughly proportional to the mean linear scale of the velocity fluctuations) in which the correlation between the initial and final velocities of a given Lagrangian particle disappears. One can now model the turbulent diffusivity

$$\mathbf{D}^{turb} = \overline{\mathbf{l}_{turb} \mathbf{u}''} \qquad (1.1.10)$$

in the entire time interval of the fluctuating motion. The possibility of some direct influence of molecular diffusion on turbulent diffusion transport must not be ruled out, provided that the amount of diffusing substance in a large-scale vortex changes noticeably in the time during which a certain correlation is retained between the vortex velocities. Moreover, these diffusion processes differ not only in nature, but apparently also have superposition properties.

When modeling the turbulent diffusivity D^{turb} for all possible atmospheric components (including pollutants), the factors that affect their mixing most strongly (dispersal) should be taken into account. In particular, an *a priori* estimate of the mixing length (turbulence scale) or its approximation is needed primarily, with allowance made for the thermal (and/or concentration) stratification of the atmosphere that ultimately determines the pattern of diffusion processes. As has already been mentioned, in the general case of an anisotropic fluctuating velocity (concentration) field, the diffusion transport of minor components is characterized by the turbulent diffusion tensor $\mathbf{D}^{\mathrm{turb}}$ (or the viscosity tensor ν^{turb}). For flows with a well-defined direction of inhomogeneity (e.g., in an atmosphere stratified in a gravity field), the turbulent velocity fluctuations are superimposed on its mean (wind) component in the horizontal direction and are enhanced if there is a wind shear. Various simplified approximations for the vertical turbulent viscosity ν^{turb} (such as Prandtl's mixing length model [see Eq. (3.3.15)] or Heisenberg's hypothesis about the role of small vortices in estimating ν^{turb} when the gradient approximation is admissible) are successfully used for these flows. However, these approximations are valid only if there is a local equilibrium of the turbulent field that provides a balance between the generation and dissipation of turbulent energy at each spatial point of the medium. For all that, these approaches are, in general, inapplicable to multicomponent chemically active media, because chemical kinetics breaking the Lagrangian invariance of the transferable substance can play a significant role in this case. Here, we again face the necessity of developing new approaches to modeling the turbulent transport coefficients for a mixture that would allow the specificity of these media to be taken into account.

1.1.5 Geophysical Turbulence

Atmospheric stratification (the distribution of the mass density ρ and other thermohydrodynamic parameters in the direction of gravity) and the Earth's rotation (with an angular velocity $\Omega = 7.29 \cdot 10^{-5} s^{-1}$) exert a peculiar influence on the pattern of geophysical turbulence. In addition, the multicomponent structure of the real atmosphere often leads to mixture baroclinicity owing to the dependence of ρ not only on the pressure p (as in barotropic media), but also on the temperature T and/or the concentration Z_α of its individual components.

Baroclinicity is dynamically important, because it gives rise to a source term in the well-known Friedman equation for vorticity [see Eq. (1.3.1)]. The source of turbulent convection is the accelerating action of the buoyancy force, and it develops in the atmosphere if its stratification is unstable. The Earth's rotation forms turbulent boundary (Ekman) layers near the land surface in the atmosphere and near the floor surface in the ocean. Rossby-Blinova waves emerge due to a global change in the Coriolis parameter $f = 2\Omega \cos\theta$ (where θ is the complement of the latitude ϕ to $\pi/2$; $f = 10^{-4} s^{-1}$ everywhere except in the regions close to the equator) in the meridional direction. They generate cyclones and anticyclones in the atmosphere and synoptic eddies in the ocean, exemplifying two-dimensional macroturbulence. In turn, macroturbulence resulting from large-scale inhomogeneities of the heat influx to the atmosphere from the underlying surface generates intense microturbulence due to the hydrodynamic instability of the vertical wind velocity gradients. The development of comparatively weak microturbulence (with small numbers **Re**) in the ocean is associated with the presence of thin quasi-homogeneous layers separated by temperature and salinity discontinuity surfaces (Monin, 1988). In studying geophysical turbulence and in some other cases, the intensity of turbulent transport can be estimated from the turbulent viscosity ν^{turb} (or diffusivity D^{turb}).

The influx of solar heat to the atmosphere and surface of the Earth (and other terrestrial planets) is the main energy source of atmospheric physical-chemical processes. A relatively small fraction of this energy is converted into the kinetic energy of vortex motions of all scales. The characteristic size of large energy-carrying vortices in which the kinetic energy of the system's turbulent motion emerges because of a change (shear) of the mean hydrodynamic velocity V, determines the external (integral) turbulence scale L. Vortices with sizes larger than L are, as a rule, anisotropic, while vortices with sizes smaller than the external turbulence scale are approximately isotropic. A cascade process of energy transport from large-scale to small-scale vortex motions takes place in the range of scales from millimeters to thousands of kilometers, $l<<r<<L$ (where l is the so-called internal turbulence scale proportional to the Kolmogorov micro-scale η), spanning almost the entire spectrum of dynamical processes in the atmosphere. The quasi-stationary regime of this cascade mechanism in the Earth's turbulized atmosphere is characterized by some approximately constant (independent of the vortex scale) quantity $\varepsilon_e \approx 3\ \text{cm}^2/\text{s}^3$. On the one hand, this is the transfer rate of the kinetic energy of vortex rotation from large atmospheric vortices to smaller ones; on the other hand, it characterizes the specific dissipation rate of turbulent (kinetic) energy into heat through molecular viscosity that occurs in vortices of the Kolmogorov scale η.

In the inertial range of scales in which the above cascade process takes place in the atmosphere, the turbulent viscosity ν^{turb} corresponding to the empirical Richardson-Obukhov "four-thirds law" (this law also follows from dimensional and similarity considerations) has the form $\nu^{turb}(r) \propto \varepsilon_e^{1/3} r^{4/3}$ (Richardson, 1926; Obukhov, 1941). The effective vertical diffusivity $\nu(L)$ is plotted against the local turbulence scale $L(z)$ in Fig. 1.1.2, where the region of laminar motion at small L with molecular viscosity $\nu = 0{,}16\ \text{cm}^2/\text{s}$ and the region of free turbulence (the inertial range) at large L are highlighted.

We recall that at these values of L there is no production or dissipation of kinetic energy, but it is only transported to increasingly small scales r (down to the scale $r = \eta$), corresponding to the growth of the wave number κ in the spectral energy density curve $E(\kappa)$ of the velocity field (see Fig. 1.1.1). The maximum turbulent viscosity ν^{turb} in the atmosphere corresponds to a typical length scale for synoptic processes $L_0 = a/f$ (Obukhov, 1949), where a $= \sqrt{gH}$ is the isothermal speed of sound. At temperate latitudes of the Earth, $L_0 \sim 3000$ km and $\nu^{\text{turb}} \approx 10^{11}\text{cm}^2/\text{s}$. Obviously, in this case, the degeneration time of the atmospheric kinetic energy caused by turbulent viscosity $\tau(L) {}^{\sim} L^2/\nu^{\text{turb}}(L) {}^{\sim} \varepsilon_e^{-\frac{1}{3}} L^{\frac{2}{3}}$ corresponds to a typical

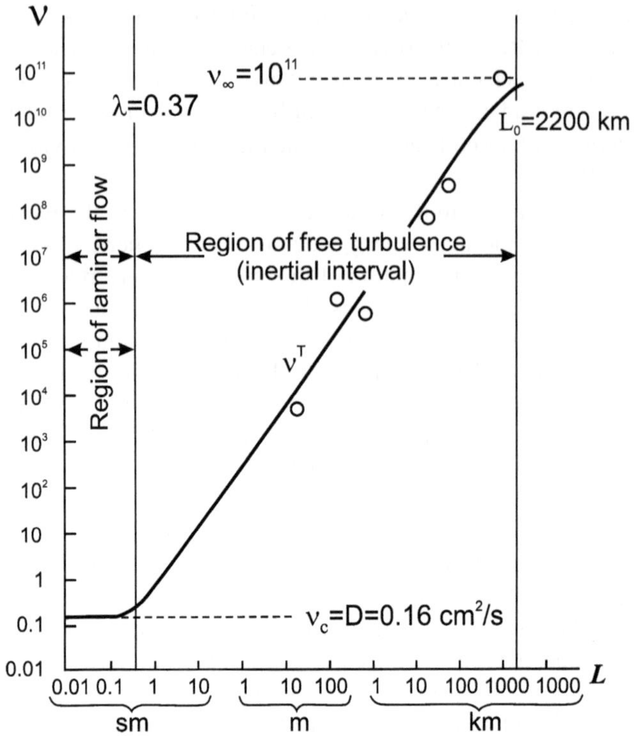

Fig. 1.1.2 Vertical diffusivity versus turbulence scale *L*. The regions of laminar motion and free turbulence (inertial range) are highlighted. The empirical points corresponding to the Richardson-Obukhov law are shown. According to Monin (1969)

conversion time of the potential energy of synoptic processes $\tau = (E^{-1}\partial E/\partial t)^{-1} \cong 5 \cdot 10^5$ s, i.e., it is approximately equal to one week (here, $E \approx 10^{21}$ J is the total kinetic energy of the atmosphere; $\partial E/\partial t \approx 1 - 2 \cdot 10^{12}$ kW is the conversion rate of potential energy into kinetic one) (Monin, 1969).

Clearly, turbulence with turbulent viscosity ν^{turb} may be considered to be an efficient kinetic energy dissipation mechanism for synoptic-scale motions or large-scale vortices with a characteristic size *L*. In this case, the minimum scale of synoptic motions capable of overcoming viscosity is estimated as $L_{\min} \propto \varepsilon_e^{-1/4} (\nu^{turb})^{3/4}$ (Monin, 1969). This approach is consistent with the views that there are universal relations between the characteristics of large-scale and small-scale fluctuations for turbulent flows determined for $\mathbf{Re} \to \infty$ by small-scale fluctuations and not by large-scale fluctuations of the velocity, which define the mean values of all thermohydrodynamic quantities and are essentially independent of $\mathbf{Re}$. The hypothesis by Batchelor (1953) about the statistical independence of large and small-scale motions is based onthis fact.

1.1.6 Turbulence Modeling Methods

In addition to the statistical approach to turbulence modeling, direct numerical turbulence simulation is increasingly widely used. This is approached with different methods that are based on the solution of a nonstationary system of three-dimensional Navier-Stokes equations, although only averaged characteristics of motion can actually be obtained, because this phenomenon is stochastic. Nevertheless, this occasionally allows one to trace not only the time evolution of the formation of various spatial structures, but also to study the general dynamics and nature of the turbulence development. For example, the results of numerical simulations of the "overthrow" phenomenon in a hydrodynamic system (constructed as a multitier model of the engagement of the simplest elements - triplets) illustrate the cascade process of energy transport in a developed turbulent flow corresponding to the Kolmogorov-Obukhov law (Gledzer et al. 1981) and corroborate the assumptions of the general properties in the behavior of dynamical systems. Remarkably, investigating the stochastization of dynamical systems and scenarios for the transition to chaos in numerical simulations of turbulence serves as an analog of the solution of ill-posed problems using an averaging and "parametric extension" operator (Tikhonov, Arsenin, 1986). In this approach, the ordered structure of a turbulent flow, which is defined as the attractor of an asymptotically stable solution for averaged quantities, is its regularized description (Belotserkovskii, 1997). However, the methods of direct numerical turbulence simulation are often difficult to use for solving practically important problems (especially those related to the calculations of turbulent heat and mass transport in multicomponent, chemically active mixtures). Moreover, the complete solution of the hydrodynamic equations for fluctuating parameters (owing to the very large number of turbulent degrees of freedom) would contain too much information to use it fully.

Therefore, the problem of an apprpoximate (semiempirical) description of turbulence comes to the fore. When deriving the averaged equations of fluid motion, which form the basis for the present-day hydrodynamic theory of turbulence, most researchers rely on hydrodynamic equations in the form of Navier-Stokes equations. The latter are assumed to describe a fluid flow in a turbulized regime as well, even at extremely large dimensionless parameters. However, the very possibility of choosing these equations as the starting ones to pass to the averaged Reynolds equations is not indisputable, if only because they can only be derived under the fairly strong assumption about a linear relation between the viscous stress tensor Π_{ij} and the first derivatives of the velocity field $\partial u_i/\partial x_j$. Although this assumption holds in laminar and weakly supercritical flows, a more complex dependence of the viscous stress tensor on the strain rate tensor must not be ruled out in strongly nonlinear turbulent fluid motions. Therefore, we will adduce another argument for these doubts by quoting from the book by O.M. Belotserkovskii (1994): "Thus, the hypothesis that turbulence is completely described by the Navier-Stokes equations is mathematically unjustified, because there is no general theorem guaranteeing the global existence of solutions to the Navier-Stokes equations as a problem with initial conditions. It is quite probable (though there are no such examples as yet) that these solutions can become singular, so that the equations cease to be valid and new physical principles outside the scope of classical hydrodynamics are apparently needed to construct a complete theory".

To overcome the difficulty of proving the existence of a solution to the Navier-Stokes equations, O. Ladyzhenskaya regularized these equations by introducing a nonlinear viscosity. This implied the following redefinition of the viscous stress tensor:

$$\Pi_{\rm ij} = -\nu(\partial u_i/\partial x_j) \to -\nu\left[1 + \varepsilon(\partial u_i/\partial x_j)^2\right](\partial u_i/\partial x_j),$$

where ε is an arbitrary small coefficient. The nonlinear contribution to viscosity smoothes out the steep velocity gradients, thereby significantly affecting the pattern of hydrodynamic motion on small scales. However, since the number of collective (macroscopic) degrees of freedom is very large, even the regularized Navier-Stokes equations can hardly be solved completely.

We use mainly the classical approach to modeling shear turbulence, which is based on Reynolds' idea of averaging the hydrodynamic equations over an ensemble of identical flows (possible realizations) or through another equivalent procedure. Because the original hydrodynamic equations are nonlinear, the equations for the scale of mean motion obtained in this way contain indefinite correlation terms (such as the turbulent diffusion and heat vectors, or the Reynolds turbulent stress tensor) and, hence, are not closed. The Reynolds-averaged hydrodynamic equations for a mixture are usually closed using particular semiempirical turbulence models, for example, by invoking a sequence of differential equations for the moments of fluctuations (to which the third chapter of this monograph is devoted). Yet we caution already at this point that this approach has a fundamental shortcoming: the Reynolds averaging is performed over all spatial turbulence scales, i.e., the modeling based on semiempirical closure hypotheses is necessarily carried out simultaneously over the entire spectrum of multiscale vortex structures.

Considering that in contrast to the essentially universal spectrum of small-scale fluctuations, the largse-scale vortex structures are significantly different for different flows (see Fig.1.1.3), it becomes obvious that the construction of universal semiempirical turbulence models suitable for describing turbulent flows of various types has no prospects. For this reason, the task consists mainly of establishing the boundaries of applicability for a particular semiempirical turbulence model. Nevertheless, there is reason to hope that invoking multiparameter approximations based on the evolutionary transfer equations for the higher moments of thermohydrodynamic parameters fluctuating in a flow will allow some progress to be made in constructing universal turbulence models with complicated properties that can describe a fairly large number of various turbulized natural media.

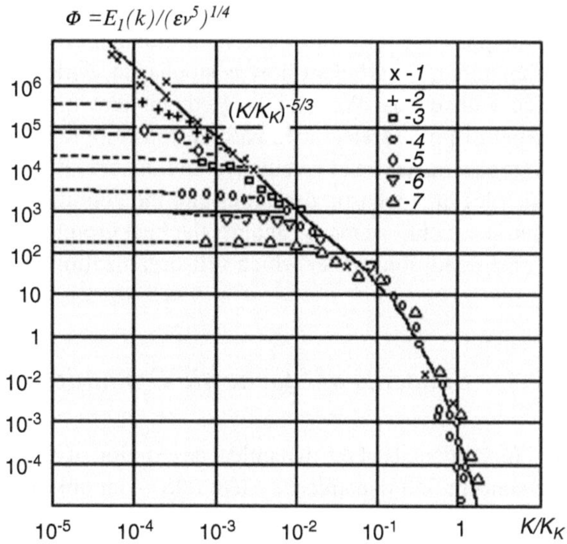

Fig.1.1.3 Spectrum of the longitudinal turbulent pulsation component for various flows: 1 - tide-ebb channel; 2 - circular stream; 3 - tube flow; 4 - permanent shear flow; 5 - wake behind a cylinder; 6 - turbulence behind a grid; 7 - boundary layer. According to Chapman (1980)

At the same time, the highly topical problem of modeling structured turbulence poses the question about the search for other basic equations for which there are physical grounds for solving the corresponding closure problem. Describing this turbulence can be made possible, in particular, through the construction of kinetic Fokker-Planck-Kolmogorov equations similar to those used in the theory of second-order phase transitions [see Ch.5]. We are confident that this way of constructing the developed turbulence theory will also be efficient.

1.2 Chaos and Self-Organization in Dynamical Systems

Stochasticity is one of the most important properties of nonlinear dynamical systems that can execute both regular and irregular chaotic motions, and the stochastic evolution of the system is the result of this integral intrinsic property of motion in the form of instability. The problem of modeling chaos in the surrounding nature and the Universe has increasingly become topical in the past decades. These systems are characterized by extreme complexity and irregularity, which are eventually attributable to a set of various nonlinear interactions that lead to chaoticity, i.e., to the possibility for a separate trajectory to be random even in without any action of explicit random forces on the system. The methods of nonlinear dynamics may be said to be universal, which suggests that the processes that take place in a great variety of systems are common phenomena. This, in turn, predetermines a commonality of the approaches that will solve a wide range of model problems that are of interdisciplinary interest.

While talking about the physical nature of turbulence in the preceding section, we emphasized that the pattern of fluctuations in velocity and other physical parameters in a turbulent flow is stochastic, which is why we may identify turbulence with chaos. We also pointed out that in a more general sense, complete mixing of phase trajectories with various topologies of attractors is a reflection of the stochastic nature of turbulence. In this section, we briefly consider some general principles of stochastic dynamics and the conditions under which relatively stable ordered structures emerge against the background of a completely chaotic motion, i.e., the conditions under which self-organization takes place.

1.2.1 Elements of Stochastic Dynamics

The fundamentals of the dynamical description of evolutionary processes are based on the concept of a dynamical system (DS) – an object of any nature whose state $\{X_i\}$ changes with time under the action of a deterministic evolution operator that specifies a correspondence between the initial state of the system and the unique state at each succeeding instant of time. The set of N dynamical variables X_i ($\mathrm{i} = 1, 2, \ldots, \mathrm{N}$) or the N-dimensional vector $\mathbf{X}$ is associated with a point in state space $\mathbf{R}^{\mathrm{N}}$ (the phase space of DS).

Let the dynamical system under study be described by the differential equations

$$\frac{\partial}{\partial \mathrm{t}} \mathbf{X}(\mathbf{r}, t) = \mathbf{f}_\alpha(\mathbf{X}(\mathbf{r}, t)) \tag{1.2.1}$$

where $\mathbf{X}(\mathbf{r}, t)$ and $\mathbf{f}_\alpha(\mathbf{X}(\mathbf{r}, t))$ are the vectors whose components are the state parameters X_i and the functional dependences that express the local evolution of the parameters X_i in time t and space $\mathbf{r}$, respectively. As a rule, the functionals $\mathbf{f}_\alpha$ contain partial derivatives with respect to the spatial variables $\mathbf{r}$ and are nonlinear, for example, because of transport phenomena (such as convection in hydrodynamics). They can also depend on some set of control parameters α that reproduce the force exerted on the system by the outside world. The motion of the phase point corresponds to the evolution of the DS state with time. For the problem (2.1.1) to be a well-posed one, the control parameters and boundary conditions maintained at the DS boundary Σ must be known. The boundary conditions are usually either the Dirichlet conditions, which specify the state parameters $\{X_i^\Sigma\}$ on the surface, or the von Neumann conditions, which specify the fluxes on the surface $\{\mathbf{n} \cdot \nabla X_i^\Sigma\}$ ($\mathbf{n}$ is the normal vector to the surface). Generally, the specific trajectory $\mathbf{X}(\mathbf{r}, t)$ in phase space that is described by the phase point $\mathbf{X}$ when $t \to \pm\infty$ corresponds to the regime of DS evolution from some given initial state $\mathbf{X}_0 = \mathbf{X}(\mathrm{t}_0)$.

Two classes of DS are distinguished – conservative systems, for example, Hamiltonian systems with dynamics conserving the initial volume of phase space, and dissipative systems for which the cloud of points in phase space $\mathbf{R}^N$ shrinks with time and, in the end, gathers on one or more attractors (the attracting set of points in phase space that all neighboring trajectories approach after the damping of transient processes). An equilibrium, a periodic motion, or a limit cycle, a quasi-periodic motion, are simple examples of classical attractors. Classical geometric objects in phase space correspond to classical attractors: a point, a closed curve, and a limit N-dimensional torus in phase space correspond to an equilibrium state, a periodic motion, or a limit cycle, and a quasi-periodic motion, respectively (see Poincaré, 1971, 1972).

The discovery of chaotic dynamics was an achievement of the DS theory. The emergence of chaos is directly related to the instability of dynamical systems. Poincaré, who initiated the studies on nonlinear dynamics, already knew in 1882 (see Poincaré, 1971, 1972) that a number of mechanical problems involving several bodies deal with instability, for example, the three-body problem in celestial mechanics. In unstable (stochastic) regions of phase space, the divergence between two initially infinitely close trajectories increases exponentially with time. Modern science defines systems of this kind as chaotic. At first glance, the emergence of chaos seems incompatible with the definition of DS, which implies the possibility of an unambiguous prediction of the final state from the initial state. Nevertheless, the chaotic regime of DS motion, in which an arbitrarily small uncertainty in the initial conditions grows rapidly with time, so that the predictability of the DS behavior becomes unachievable on fairly long time scales, is indeed admissible. Regimes of this kind are characterized by an irregular, chaotic change in dynamical variables with time. In the phase space of dissipative systems, strange or chaotic attractors, which are the images of deterministic chaos, correspond to these processes. New (with respect to classical geometry) geometric objects called fractal sets are associated with these attractors. The properties of strange attractors have not been completely investigated yet. An important property is the dimension of a strange attractor, which is usually fractional (fractal), but not integer. In the special case of three-dimensional phase space, the fractal set of a strange attractor appears as a set of an infinite number of layers or parallel planes, with the separation between some of them becoming infinitely small. The existence of a fractal structure in these attracting sets when the phase trajectories asymptotically approach them can be attributed to one of the paradigms of nonlinear dynamics.

The existence of chaotic dynamics is closely related to the instability inherent in the phase trajectories of DS. Thus, one of the most important tasks in the stochastic DS theory is to analyze the instability of a specific regime of motion. There are several different definitions of stability, namely, the Lyapunov stability, the asymptotic stability, the orbital stability, and the Poisson stability (see, e.g., Anishchenko, 1990). The phase trajectory $\mathbf{x}^*(\mathrm{t})$ that we investigate is Lyapunov-stable against a perturbation of the initial data if a $\delta(\varepsilon)>0$ can be found from a given $\varepsilon>0$ in a way that the inequality $\|x(t)-x^*(t)\|<\varepsilon$ will hold at $t\geq t_0$ for all trajectories $\mathbf{x}(\mathrm{t})$, such that $\|x(t_0)-x^*(t_0)\|<\delta$. The sign $\|\ldots\|$ denotes any norm in $\mathbf{R}^N$ (e.g., $\sqrt{(\mathbf{x},\mathbf{x})}$,

$\max_i |x_i|$ or $\sum_i |x_i|$). Accordingly, a small initial perturbation of Lyapunov-stable phase trajectories does not grow with time. The stability is called asymptotic if, in addition, $\|x(t) - x^*(t)\| \to 0$ when $t \to \infty$.

The simplifying assumption that the medium is constant in time is often made when the stability is analyzed; then, $\mathbf{f}_\alpha = \mathbf{f}_\alpha(\mathbf{X}(\mathbf{r}))$ (an autonomous system). Clearly, this assumption requires the constancy of all control parameters α and boundary conditions in time. In this case, the emergence of an unstable regime in the system can for instance be investigated as follows. Because the medium is constant, we can assume that there exists at least one time-independent solution $\bar{\mathbf{X}}$ ($\mathbf{r}$) of Eq. (1.2.1), i.e., a trivial solution for which $\mathbf{f}_\alpha(\bar{\mathbf{X}}) = 0$, which we will take below as the reference state. The simplest way of investigating the sensitivity to initial conditions is, first of all, to check the stability of the reference state $\bar{\mathbf{X}}$ against small perturbations. For this purpose, it is necessary to assume that

$$\mathbf{x}(\mathbf{r},\mathrm{t}) = \bar{\mathbf{x}} + \mathbf{x}(\mathbf{r},\mathrm{t}), \text{where} |x_i/\bar{X}_i| \ll 1, \tag{1.2.2}$$

and to substitute it into (1.2.1). The time evolution of the perturbation $\mathbf{x}(\mathbf{r},\mathrm{t})$ is then specified by the solution of the system of equations

$$\partial \mathrm{x_i}/\partial t = \sum_j A_{ij}\mathrm{x_j} \tag{1.2.3}$$

which is obtained when the vector equation (1.2.1) is linearized. Here, A_{ij} is the matrix $A_{\mathrm{ij}} = \partial f_i/\partial x_j\big|_{x(t)=\bar{x}}$ composed of the partial derivatives of the vector function $f_a(X(r))$, which is characterized by the eigenvectors $\mathbf{e}^k$ and eigenvalues ω_k: $\mathbf{Ae}^k = \omega_k\mathbf{e}^k$ $(k = 1, 2, \ldots, N)$. Consequently, the system of equations (1.2.3) admits solutions of the form

$$x_i(\mathbf{r},\mathrm{t}) = x_i^k(\mathbf{r}, t_0)\exp[\omega_k(t - t_0)], \tag{1.2.4}$$

where the eigenvalues ω_k of the matrix A_{ij} are the roots of the characteristic equation $\mathrm{Det}[\mathbf{A} - \omega\mathbf{U}]$; here, $\mathbf{U}$ is a unit matrix. The increase or decrease in magnitude of the perturbation $\|x_i(\mathbf{r},\mathrm{t})\|$ is determined by the sign of the real part of ω_k. Clearly, if the reference state $\bar{\mathbf{X}}$ is stable, then all $\mathrm{Re}\{\omega_k\}$ must be negative. Therefore one should investigate the behavior of $\mathrm{Re}\{\omega_k\}$ as a function of the control parameter α and the boundary conditions imposed on the system to analyze the properties of phase trajectories in a local neighborhood of the equilibrium state.

For an unsteady reference state $\bar{\mathbf{X}}(\mathbf{r},\mathrm{t})$, the matrix $\mathbf{A}(t)$ obtained by linearizing Eq. (1.2.1) will depend on time. For this reason, its eigenvalues ω_k and eigenvectors $\mathbf{e}^k$ will also change with time. In this case, the exponent $\omega_k(t)$ takes on various values as t changes. Consequently, a situation is possible where some small perturbation $\mathbf{x}(\mathrm{t}) = \sum_{k=1}^{N} \mathbf{x}^k(\mathrm{t})$ increases exponentially at some points of the trajectory $\bar{\mathbf{X}}(\mathbf{r},\mathrm{t})$ and decreases at other points (here, $\{\mathbf{x}^k(\mathrm{t})\}$ is the fundamental system of linearly independent solutions of Eq. (1.2.3)). To analyze this case, consider, for example, the evolution of the component of a small perturbation $\mathbf{x}^k(\mathrm{t})$ directed along the k-th eigenvector $\mathbf{e}^k$ of the matrix $\mathbf{A}$. The so-called characteristic Lyapunov exponents calculated from the known equations of motion are known to serve as a quantitative measure of the stochasticity of Hamiltonian and non-Hamiltonian dissipative systems in the DS theory. In this case, the stability of the trajectory along the eigenvector $\mathbf{e}^k(t)$ is defined by the characteristic Lyapunov exponent λ_k (see, e.g., Bylov et al., 1966):

$$\lambda_k = \overline{\lim_{k\to\infty}} \frac{1}{t-t_0} \ln\left\| \frac{\mathbf{x}^k(\mathrm{t})}{\mathbf{x}^k(\mathrm{t}_0)} \right\|, \quad (k = 1, 2, ..., \mathrm{N}), \tag{1.2.5}$$

where the overbar denotes the upper limit. For the N-dimensional phase space under consideration, the stability of the trajectory is defined by a set of N Lyapunov exponents that if they are arranged in decreasing order $\lambda_1 \geq \lambda_2 \geq \ldots \geq \lambda_N$, form the so-called characteristic Lyapunov exponent (CLE) spectrum for the phase trajectory $\bar{\mathbf{X}}(\mathbf{r},\mathrm{t})$. Note that the CLE spectrum does not depend on the choice of the fundamental system $\{\mathbf{x}^k(\mathrm{t})\}$ and is a characteristic of the linear system (1.2.3) as a whole. Lyapunov exponents are used to check the sensitivity of DS to variations in initial conditions. The Lyapunov exponent showed that the exponent λ_k could be treated as the real part of the eigenvalue $\omega_k(t)$ of the linearization matrix $\mathbf{A}(t)$ averaged along the trajectory under study:

$$\lambda_k = \overline{\lim_{k\to\infty}} \frac{1}{t-t_0} \int_{\mathrm{t}_0}^{t} \mathrm{Re}\omega_k(t')dt' \tag{1.2.6}$$

This expression shows what happens on average to the corresponding component of the initial perturbation along the trajectory. If the trajectories $\bar{\mathbf{X}}(\mathbf{r},\mathrm{t})$ are Lyapunov-stable, then an arbitrary initial perturbation $\mathbf{x}(\mathrm{t}_0)$ along the trajectory does not grow, on average. For this, it is necessary and sufficient that the CLE spectrum contains no positive exponents; a chaotic motion of the system means that at least one of them is positive. Accordingly, the criterion of chaos in terms of the Lyapunov exponent is

$$\begin{cases} \lambda > 0 - \text{chaotic motion in DS}, \\ \lambda \leq 0 - \text{regular motion in DS}. \end{cases} \tag{1.2.7}$$

If the particular solution $\bar{\mathbf{X}}(\mathbf{r},\mathrm{t})$ of the autonomous system (1.2.1) is an equilibrium state, i.e., $\mathbf{f}_\alpha(\bar{\mathbf{X}}) = 0$, then the Lyapunov exponents are equal to the real parts of the eigenvalues $\lambda_k = \mathrm{Re}\omega_k$. Given the Lyapunov exponents, it is easy to determine which type of limit sets the equilibrium state that we investigate belongs to. The equilibrium position is an attractor if it is asymptotically stable in all directions, i.e., the CLE spectrum consists only of negative exponents (a stable node or focus).

If the equilibrium state is unstable in all directions, then it is a repeller (an unstable node or focus). If the CLE spectrum includes both positive and negative exponents, then the equilibrium state is a saddle-type one (a simple saddle or saddle focus). Lyapunov exponents also allow one to investigate the behavior of phase trajectories in the neighborhood of periodic, quasi-periodic, or chaotic solutions of system (1.2.1) (see, e.g., Anishchenko et al., 2005). The nonperiodic solutions of a system (1.2.1) can correspond to chaotic attractors of a complex geometric structure that have at least one positive Lyapunov exponent and, as a result, a fractal Hausdorff dimension.

The so-called Kolmogorov-Sinai entropy (*K*-entropy) for dynamical systems (Kolmogorov, 1959; Sinai 1970) is closely related to the Lyapunov exponents that define a measure of divergence between trajectories. A rigorous definition of this quantity is beyond the scope of our book, but, in the most general case, it is the sum of positive Lyapunov exponents characterizing a quantitative measure of chaotization. In this case, a positive *K*-entropy points to the origination of chaoticity or, in other words, to exponential divergences of neighboring trajectories whose local instability, in physical essence, characterizes the statistical properties of DS (to be more precise, its mathematical model). Note that A.N. Kolmogorov and Ya.G. Sinai initiated a consistent theory of chaotic dynamical systems in the works cited above. There they give an explanation of chaoticity and its origination in deterministic systems.

There close relation between the Lyapunov exponents λ_j and the fractal Hausdorff dimension of an attractor can be estimated, for example, from the Kaplan-Yorke (1978) formula

$$\mathrm{D_L} = \mathrm{k} - \frac{\sum_{j=1}^{k} \lambda_j}{|\lambda_{k+1}|}, \tag{1.2.8}$$

where k is the greatest integer for which $\lambda_1 + \lambda_2 + \ldots + \lambda_k \geq 0$. The dimension D_L calculated from Eq. (1.2.8) is called the Lyapunov dimension and serves as a lower limit for the metric dimension of attractors. The Lyapunov dimension is of great practical importance, because only the CLE spectrum is required for its calculation. In particular, $\mathrm{D_L} = 0$ for an equilibrium state, $\mathrm{D_L} = 1$ for a limit cycle, and $\mathrm{D_L} = N$ for an N-dimensional torus, i.e., the fractal dimension is equal to the metric dimension of these regular attractors in all these cases.

Given the Lyapunov exponents, we can also estimate the fractal dimension of hyperbolic strange attractors whose existence for the rough and hyperbolic dynamical system (1.2.1), whose phase trajectories are all saddle trajectories and retain their properties under small perturbations, was proven by Ruelle and Takens (1971). In this case, the CLE spectrum contains at least one positive exponent. A hyperbolic strange attractor consists of a continuum of unstable "sheets'' or curves that are dense everywhere in the attractor, along which close phase trajectories diverge exponentially (the minimum dimension of the phase space into which a strange attractor can be embedded is 3). Thus, on the phase plane where the set of all possible trajectories forms stable and unstable manifolds, the Lyapunov exponents characterize the average divergence rate of close trajectories.

We recall that the regions of temporal (and/or spatial) chaos can emerge almost in any nonlinear dynamical system even at a small number of degrees of freedom. Therefore, the original views of the random processes that characterize statistical ensembles and large dynamical systems turn out to be applicable even at a number of degrees of freedom $N \geq 3/2$, for example, when a periodic force acts on a system with one degree of freedom (see, e.g., Zaslavsky, 2004). In particular, for a perturbed motion of DS described by the Hamiltonian $H = H_0(I) + \varepsilon V(I, \theta, t)$ (where the function V is periodic in time with a period $T = 2\pi/\nu$) in the phase space of canonical action-angle variables I, θ, random trajectories can, nevertheless, appear in it even at low values of the dimensionless perturbation parameter $\varepsilon \ll 1$ and in the absence of random forces. This means that the dynamics of the system will be stochastic in some region of phase space and control parameters. This is the result of the complexity of the motion that emerges during the interaction of the possible, in this case, nonlinear resonances and the chain of islands (the regions of initial conditions in which stable trajectories originate). The resonance terms in this Hamiltonian system generate the separatrices of a nonlinear resonance (which, because they are not the real trajectories of the image point, separate the regions of the phase plane with phase trajectories of various types) and the related systems of elliptic and hyperbolic singular points. The stability inside each cell of the separatrix is successively reduced and the cell is "dressed" with a stochastic layer of some width with a complex temporal behavior near the separatrix, which leads to a hierarchy of structures in phase space. The appearance of a stochastic layer near the separatrix has a strong influence on the dynamics of nonlinear systems. If the separatrices of the resonances overlap, then the stochastic layers (or the so-called hemoclinic structures that emerge when the stable and unstable manifolds of one saddle point intersect) of various resonances can combine to form wider regions of stochasticity. The formation of stochastic layers followed by their destruction (accompanied by the destruction of the integrals of motion) strongly affects the dynamics of nonlinear systems, giving rise to separatrix chaos (see Zaslavsky, 2004; Loskutov, 2007). Obviously, the appearance of chaoticity in the dynamics of more complex physical systems with many degrees of freedom is even more probable. Hydrodynamic turbulence can serve as a characteristic example of these systems.

In contrast to the resonant (and some of the nonresonant) tori to which chaotic dynamics corresponds, most of the nonresonant tori to which quasi-periodic motions correspond are conserved in completely integrable Hamiltonian systems that are subjected to small perturbations. This important assertion forms the basis for the famous Kolmogorov-Arnold-Moser (KAM) theorem that was the starting point in understanding the origin of chaos in nonlinear Hamiltonian systems (Kolmogorov, 1954; Arnold, 1963; Moser, 1962). In particular, it follows from this theorem that for systems with two degrees of freedom, the invariant *2D* tori divide the *3D* energy space $H = E$ into nonoverlapping regions and the system retains its stability under small perturbations. At the same time, the invariant tori at $N>2$ no longer divide the $(2N - 1)$-dimensional hypersurface into nonoverlapping parts, causing a profound change in the topology of phase trajectories and giving rise to a chaotic motion.

Approximately the same features of the birth of chaos are also characteristic of non-Hamiltonian systems. However, the dissipation that takes place in this case causes the system's phase volume to decrease; an important condition for these systems is finiteness of their dynamics in phase space, while their oscillatory dynamics can be absent altogether. A distinctive property of dissipative systems is the possible appearance of the previously mentioned various invariant sets of points of a fractal nature – attractors (repellers) attracting (repelling) the phase trajectories and even more complex attracting sets in a finite region of phase space on which stochastic dynamics is realized – stochastic attractors (with the mixing property) and quasi-attractors. Under conditions of strong instability, the entire phase space of the system, irrespective of the perturbation amplitude, is covered with a stochastic web over which the phase points walk (Arnold diffusion). Moving along the threads of this web, the phase point on a hypersurface of a given energy can come arbitrarily close to any point of this hypersurface. These structures constitute a fundamental property of both Hamiltonian and non-Hamiltonian (dissipative) systems and are attributable to the local instability and topology of the system (for more details, see, e.g., Loskutov, 2007).

In turn, the mixing property in some finite region of phase space that all nonlinear (both conservative and nonconservative) dynamical systems have that execute a finite motion (Krylov, 1950) is related to the criterion of local instability for dynamical systems that are not influenced by random fields. Owing to the presence of this instability, the initial volume element is strongly deformed while moving over the entire energetically accessible hypersurface. Mixing can be imagined as a peculiar kind of drift and intertwining of the phase trajectories of a nonlinear DS when small perturbations of the initial conditions emerge. Three very important consequences observed in real systems follow from mixing: unpredictability, uncoupling of temporal correlations, and irreversibility inherent in closed systems.

A mixing flow emerges when initially close points in phase space move along exponentially diverging trajectories. In other words, if the initial position of the phase point is known with a finite accuracy, then it cannot be said precisely at which position it will be after a fairly long time interval. Thus, unpredictability arises in systems with mixing. The sensitivity of the system to small perturbations of the initial conditions plays a crucial role here. Note that the mathematical theory of chaos strongly depends on the initial conditions. Lorentz (1963) drew attention to this almost 40 years ago when he observed that it is impossible to make reliable long-term weather forecasts because there are always infinitesimally small errors in the current weather parameters. He called this sensitivity to initial conditions the butterfly effect.

In addition, mixing is characterized by the uncoupling of temporal correlations when $t \to \infty$. In other words, a system with mixing "forgets'' its initial conditions in the course of time; the correlations can decay exponentially with some characteristic correlation uncoupling time τ_c. The uncoupling of temporal correlations is a very important consequence of chaoticity. This is that after a time interval equal to the correlation uncoupling time τ_c, the values of the phase variables turn out to be statistically independent. This mixing property is commonly used as a criterion of chaoticity.

Finally, the ergodicity of a system, i.e., the equality of the time and phase means for some integrable function in the region of DS phase space, automatically follows from the mixing condition. In particular, a characteristic feature of the ergodic motion of a Hamiltonian system is the shape invariability of a small region that in the course of time walks over the torus, crossing each of its segments an infinite number of times. Accordingly, ergodicity by no means implies chaoticity, i.e., not all ergodicity automatically causes mixing in a system that would in turn result in chaotic behavior. At the same time, ergodicity explains why the temporal evolution of systems described by reversible dynamical equations of motion are irreversible (e.g., the Hamilton equations). Indeed, if the system is ergodic, then it will be in all energetically possible states in the course of time and its phase trajectory will visit all accessible regions of phase space. In this case, the residence time in a specific region is proportional to the volume of this region. When $t \to \infty$, any of these states will be observable. If the number of degrees of freedom of the system is large enough, then the time during which the system will pass to some selected (taken as initial) state may turn out to be very long. From this it obviously follows that closed microscopic systems with a large number of degrees of freedom are irreversible. Therefore, the macroscopic irreversibility of a dynamical system, which is a prerequisite for the processes of self-organization to take place in the system, can be explained by the instability of the microscopic motion, i.e., chaos of the microworld. However, we emphasize that the ultimate solution of the macroscopic irreversibility problem is yet to be found, and it must necessarily include the cosmological aspect of the problem.

In summary, whereas distinct differences exist between regular (periodic or quasi-periodic) and stochastic motions in the literature on chaotic dynamics, which follow from the analysis of the fundamental differential equations of classical mechanics, the conditions of the corresponding theorems for real mechanical systems are not always met. Indeed, the DS theory, as a rule, deals with the asymptotic behavior of the system as the computational time tends to infinity. This means that, for example, finding the characteristic Lyapunov exponent requires that the computational time tends to infinity. Meanwhile, the characteristic observation times of some systems such as astronomical ones are generally much shorter than their lifetimes. Since one cannot reach the limit $t \to \infty$ in many practically important astrophysical problems, the question arises whether the estimate of the characteristic Lyapunov exponent obtained in this case, for example, to study the dynamical properties of the gaseous disk in a spiral galaxy, is reliable. This estimate is needed, in particular, for understanding the stochastic nature of the dynamics of real galaxies. The finiteness of the observation time interval requires developing appropriate computational algorithms for numerical calculations of the Lyapunov exponent for various families of trajectories near the separatrices of giant vortices and the saddle points of the galactic velocity field (Fridman et al.. 2003).

1.2.2 Relationships between Order and Turbulent Chaos

Let us now consider the self-organization of a chaotic system that spontaneously structures itself, i.e., "per se", without any external control. Here, the word "structure" is understood as the type of ordering of the elements of a chaotic system. The instability of a dynamical system triggers a process of self-organization that produces new coherent structures that show precisely the irreversibility in the system that then creates the prerequisites for the emergence of structure formation in it. Note that the highly organized structures that appear as a result of self-organization emerge as a rule through an abrupt reduction in the system's symmetry, which, in this case, may be treated as a certain second-order "phase transition" (see Kadomtsev and Ryazanov, 1983). The result of self-organization is the emergence of a new, relatively stable evolutionary state that may turn out to be the beginning of the next cycle. During the transitions between the self-organization cycles that have the pattern of bifurcations, the dynamical system near the phase transitions exhibits some peculiarities, including powerful fluctuations, long relaxation times, extensive spatial correlations, structures on many scales, and peculiar noise spectra (see Ebeling and Feistel, 2005).

In the class of open hydrodynamic systems, the capability for self-organization is realized only under specific conditions. These enable some supercritical domain of state parameters that bring the system out of local equilibrium and remove entropy. New solution branches of the dynamical system (1.2.1) with a lower symmetry can simultaneously emerge within this domain, which is determined by the efficiency of the internal feedbacks if the control parameters α, which reproduces the action of the outside world on the system, reach a certain threshold value. It is from these positions that we here approach the transition (which is related in the long run to the dynamical instability of the fluid motion) from a laminar flow to a turbulent one, which is usually deemed to be a chaotic process, but which may also be considered as self-organization (see Progogine, Stengers, 1984; Klimontovich, 2002). In other words, according to this concept, the complex dynamics of turbulent flows should not be perceived *a priori* as disorder. On the contrary, hydrodynamic turbulence ought to be considered as a flow with a fairly high degree of organization in which that part of the energy of a chaotic system that is attributable to mesoscale turbulent fluctuations transforms into a highly organized motion of coherent dissipative structures. In particular, the cascade fragmentation of vortices in developed turbulence can be interpreted as an unbounded sequence of self-organization processes. Based on this approach, it is of primary interest to construct a macroscopic model of structured turbulence that includes the nonlinear cooperative processes in the flow. To solve this problem, it is appropriate to use the methods of stochastic thermodynamics of irreversible processes, which in recent years have proven to be so efficient in studying various nonequilibrium dissipative structures far from thermodynamic equilibrium (see, e.g., Kaizer, 1987). We conceptually consider all these questions in the fifth and sixth chapters of the monograph.

We now consider the following key question: Is a stationary developed turbulent fluid motion chaos, or order? In this section, we will attempt to present a number of arguments that, in our view, allow the transition from a laminar motion to a developed turbulent one to be treated as self-organization, i.e., as a nonequilibrium (second-order) "phase transition" from a more symmetric (laminar) motion to a less symmetric (turbulent) one.

First of all, we will list the characteristic features of a developed turbulent motion, the collection of which serves as the physical definition of the concept of turbulence (Klimontovich, 2002):

- A turbulent fluid motion is the consequence of the instability of a laminar flow or, in other words, a consequence of the dynamical instability of a motion described by hydrodynamic equations. This is a nonequilibrium process in a continuous dissipative medium that is characterized by a very large number of excited collective (macroscopic) degrees of freedom (an example are the Fourier expansion modes for the velocity field).
- A specific realization of a turbulent motion changes significantly when the initial and boundary conditions are slightly modified. Because of the great complexity of the motion related to the excitation of a large number of macroscopic degrees of freedom, a turbulent flow is virtually unpredictable at the dynamical level of description.
- The transition from a laminar flow to a turbulent one is accompanied by a reduction in symmetry and may be considered as a nonequilibrium second-order phase transition (when the critical Reynolds number corresponds to the critical temperature). In this sense, the dynamical instability of a laminar flow plays a constructive, positive role.
- A developed turbulent flow is an example of self-organization in an open system that reduces its entropy. In other words, a turbulized fluid is characterized by self-organization as the process of formation and complication of mesoscale coherent vortex structures far from equilibrium in an initially symmetric small-scale fluctuating medium. Mesoscale hydrodynamic fluctuations (external to the distinguished degrees of freedom) then play a crucial role.

A significant increase in turbulent viscosity compared to the corresponding molecular viscosity in the laminar flow regime may also be considered as circumstantial evidence for an increase in the degree of order during the transition to a turbulent motion. Indeed, in a laminar flow, momentum is transferred from layer to layer by independent changes in the momentum of individual gas molecules. In contrast, in a turbulent flow, the momentum transfer from layer to layer is a collective process and, hence, is related to a more organized motion.

Moreover, according to the currently available experimental data (a comprehensive overview of the corresponding publications is given, e.g., in the monograph by Khlopkov et al., (2002)), the presence of localized coherent vortex structures that are randomly distributed in space and time is a characteristic feature of many, if not all, developed turbulent flows. As has now become clear, developed turbulence is associated with the formation and evolution of highly organized dissipative structures – vortex coherent structures (CSs) of all possible spatial, temporal, or spatiotemporal scales – in certain regimes of fluid flow in an essentially nonequilibrium open system. The part of the system's energy that was in the thermal motion of molecules in a laminar flow transforms into a macroscopic organized CS motion. In this case, a coherent vortex structure can be defined, in particular, as a connected, liquid mass with phase-correlated (i.e., coherent) vorticity in the entire region of coordinate space occupied by the structure [see Sect. 6.2].

Thermodynamic Proof of the Possibility of Structuring of Small-Scale Turbulent Chaos

According to thermodynamic concepts, the existence of a certain relationship between the production of entropy inside the system and the exchange of entropy with the ambient medium is a prerequisite for the formation of ordered coherent structures in open dynamical dissipative systems. In this case, the formation of coherent structures during irreversible processes is related to certain conditions – here, just as for the equilibrium, a qualitative jump (similar to the phase transition) is observed when the critical parameters reach their threshold values. The most important physical condition for the emergence of an ordered state in dissipative systems that manifests itself in a coherent behavior of molecules is that this process takes place only far from thermodynamic equilibrium, with the motion of the system's elements (molecules, waves, etc.) being synchronized by the order parameters. Nonlinearity is an important and common feature of the processes far from equilibrium. The fundamental priniciple according to which the ordered structures are organized is that the system's self-organization reduces its entropy. This is probably expressed by the cooperative behavior of the molecules.

In particular, the so-called convective Bénard cells are a typical example of dissipative structures in hydrodynamics. Compared to a uniform equilibrium velocity distribution, the convective cells are better organized structures that result from a cooperative motion of fluid molecules. In the stationary case, a nonuniformly heated fluid gives up entropy (or derives negentropy from the ambient medium) just as much as this is produced inside the system through the losses by internal friction (viscosity) and heat exchange between its various parts. This is how the structure formation mechanism at a large deviation from equilibrium is identified for our specific case.

As has already been mentioned, the transitions from one dissipative structure to another are similar in properties to equilibrium phase transitions if the concept of a phase transition is extended to this class of nonequilibrium phenomena. An initially turbulent motion in a closed system is known to decay monotonically with time (to degenerate), i.e., it degrades to a laminar one. This once again shows that self-organization can take place only in a closed (half-closed) hydrodynamic system that has a control parameter whose role is generally played by the Reynolds number. Closed systems that exchange only heat (radiation) with the environment are also capable of "exporting" and "importing" entropy, with order emerging in the system if more entropy is removed than is produced inside it (i.e., the entropy within the system decreases). Stable mesoscale coherent structures against a background of small-scale turbulent chaos can result, in particular, from a continuous redistribution of the energy entering the system from the outside (i.e., into the subsystem of small-scale turbulent chaos from an averaged fluid motion [see Ch. 3]) through dissipation, from the processes of transport by diffusion or convection, and from a nonlinear interaction between various small-scale vortex structures. We emphasize once again that dissipation plays a constructive role in the turbulence structuring processes. Stable spatiotemporal structures whose sequences constitute the self-organization process cannot be formed without its involvement.

Consequently, it is necessary to develop quantitative criteria for the relative degree of order of the various nonequilibrium states of closed systems and the direction of their evolution. In his works on nonequilibrium thermodynamics, I. Prigogine showed how significant and diverse the role of the concepts of entropy and entropy production is in describing the pattern of evolution of a dissipative system. The "principle of minimum entropy production in stationary states" formulated by him (see, e.g., Prigogine, Defay, 1964) is a very important tool for investigating thermodynamic processes in a hydrodynamic system. According to this principle proven only for linear irreversible processes, the state of any closed system with time-independent boundary conditions always changes in the direction of decreasing entropy production until the stationary state in which the entropy production is minimal will eventually be reached. However, the validity of this principle is limited only to the region of linear processes, i.e., to relatively small deviations from equilibrium. The authors of the monograph cited above showed that the stationary states of a linear system are always stable against small deviations, i.e., unstable stationary states are fundamentally impossible in the linear region. Hence it follows that an oscillating fluid motion that leads to the formation of coherent vortex structures from turbulent chaos under certain conditions (as has been noted above) is also impossible in the vicinity of a linear stationary state.

Since the formation of, for example, temporal structures similar to periodic oscillations or the transitions from unstable stationary states to stable ones (which is characteristic of the emergence of turbulence) is not possible in the linear region, it is necessary to turn to the nonlinear region, because a completely different situation is admissible for them. In the nonlinear region, the theorem regarding minimum entropy production in a stationary state is modified – it transforms into the well-known Glansdorff-Prigogine principle of evolution (see, e.g., Glansdorff and Prigogine, 1973). When some system deviates strongly from equilibrium, the state variables characterizing it generally no longer satisfy linear equations but more accurate nonlinear ones. Far from equilibrium, qualitatively new features manifest themselves in the behavior of closed systems. These are for instance a self-oscillatory behavior such as the well-known Belousov-Zhabotinsky reaction (where after the introduction of high-quality chemicals into a solution the cerium and bromine ions and malonic acid molecules enter into a reaction to produce spatial and temporal structures), the formation of hierarchically subordinate structures, etc. We emphasize once again that coherent structures can emerge only far from equilibrium under special external and internal conditions: the system is closed and possesses nonlinear internal dynamics, while its external parameters have supercritical values. We recall that it was Prigogine who called these spatial, temporal, or spatiotemporal structures "dissipative". They can emerge far from equilibrium in the nonlinear region, where the system parameters exceed their critical values. The nonlinear thermodynamics of irreversible processes gives a clue to a deeper understanding of the processes in a nonlinear system, which is just what turbulence is.

In the set of problems to which this monograph is devoted, we are interested in quasi-stationary nonequilibrium states of a homogeneous turbulent field whose evolution can produce coherent structures. However, whereas the entropy maximum condition can be used for stationary equilibrium states, there is no such universal extremal principle for a quasi-stationary nonequilibrium system. In nonlinear nonequilibrium thermodynamics, there are no state functions that would have an extremum in a stationary state; in particular, the entropy production by no means should be minimal. It is also important to keep in mind that the stability of a stationary state in the nonlinear region is not proven but exists by definition. This raises the following question: Can Prigogine's principle be generalized to the nonlinear thermodynamics of irreversible processes? The theoretical studies carried out by Stratonovich (1985) showed that there is generally no such possibility.

The Reverse Cascade and the Fractality of Coherent Structures

At the same time, based on the specific physical process of the transition from a laminar flow to a turbulent one, Klimontovich (2002) formulated the "principle of minimum entropy production in self-organization processes" that allowed order to be distinguished from chaos. Basically, this principle consists in the following: the self-organization process is represented as a "phase transition" (or a sequence of phase transitions) as a result of which the transition to a more ordered state occurs corresponding to a lower symmetry. The principle asserts that the entropy production in the new, less symmetric state that results from the next phase transition is less than the entropy production in the preceding state, which can be mentally extended into the unstable region. The author of this principle demonstrated its efficiency particularly for the transition from a steady laminar Poiseuille flow in a tube to a steady turbulent flow. His calculation based on comparing the values of the Boltzmann-Gibbs-Shannon entropy (see Klimontovich, 1990) renormalized to a given mean effective energy of an open system showed that this transition proceeds with a reduction in entropy and, hence, is the phenomenon of self-organization.

Note that this example is only one of the many other arguments for Prigogine's viewpoint on turbulence accepted here. The relative degree of order of a turbulent motion can be judged from various criteria: from the Lyapunov exponents, the Kolmogorov-Sinai entropy, the fractal dimension of the effective phase space, etc. In Ch. 5 of this book we also consider the thermodynamic proof of the possibility of small-scale turbulent chaos structuring.

We already saw that some of the turbulent motions are characterized by the energy transfer from small-scale motions to large-scale ones (the reverse cascade) (see, e.g., Bershadskii, 1988, 1989). The physical mechanism of the reverse cascade can be a combination of small-scale vortices into larger-scale coherent structures, fractal vortex clusters. Flows of this kind include two-dimensional turbulence, turbulence in a rotating fluid, magnetohydrodynamic turbulence, etc. The general circulation in the global system of atmospheric motions in particular can consume the energy of cyclones through the reverse cascade. The reality of the fragmentation process and the inverse process of vortex combination is also confirmed by direct observations (see, e.g., Offen and Kline, 1975). In Ch. 8 we will formulate the general concept of energy feeding of mesoscale coherent vortex structures in a thermodynamically open subsystem of turbulent chaos that is related to the realization of the reverse cascade of kinetic energy in mirror-asymmetric disk turbulence.

A fractal vortex cluster realizes an intermittent turbulent motion. In this case, the physical and chemical properties of these clusters differ significantly from those of both an equivalent homogeneous particle and a system of independent small particles. A fractal vortex cluster has a porous structure whose fractal dimension can be considerably lower than that of the coordinate space in which this cluster is formed. A typical fractal cluster has sizes of up to several tens of micrometers and includes up to 10^4 particles. The cluster stability decreases with increasing size. Fractal aggregates are characterized by a decreasing density, a developed structure of its pores, and a high specific surface area. As the cluster grows, the size of its voids increases and its mean density decreases. Note that in a gravitational field fractal clusters settle down considerably more slowly than compact particles of the same mass.

1.2.3 The Emergence of Order in Turbulent Flows. A Stochastic-Thermodynamic Model

Since the energy of turbulent motions is continuously dissipated through viscosity, the local thermodynamic equilibrium state of turbulent chaos generally cannot be attained. The thermodynamic self-organization threshold of a system is reached when the thermodynamic branch undergoes the first bifurcation. At the bifurcation point (at the critical point induced by the natural noise of the subsystem of turbulent chaos associated with its "thermal" structure), the system's dynamics is determined by the nonlinearities existing in it. In other words, beyond the thermodynamic self-organization threshold we enter the region of synergetics where the huge number of degrees of freedom of a turbulized fluid is reduced sharply and, according to the terminology of Haken (1988), trillions of molecules turn out to be "subordinate" to relatively few modes.

In this case, it is important to keep in mind that turbulent fluctuations (the natural noise of chaos) can not only lead to fluctuations in the system's thermodynamic parameters, but also cause a qualitative rearrangement of its regime, namely they can give rise to new stationary states, undamped periodic oscillations, etc. Thus, a new class of effects is important in the problem under consideration – nonequilibrium "phase transitions" induced by the natural noise of chaos.

The theory of noise-induced phase transitions is based on the mathematical theory of random processes (see Horsthemke and Lefever, 1984). The Fokker-Planck-Kolmogorov (FPK) equations for Markovian diffusion processes and stochastic differential equations are the mathematical apparatus for describing nonlinear phenomena in a random medium (in our case, the subsystem of turbulent chaos). One can trace the evolution of the fluctuating characteristics of turbulent chaos with the stochastic-thermodynamic approach to constructing a phenomenological macroscopic model of developed turbulence in a compressible fluid that includes the formation of stable spatiotemporal dissipative structures in it. At the same time, one can deduce the kinetic FPK equations in the space of configurations with the methods of stochastic nonequilibrium thermodynamics by including the set of random variables in the model of structured turbulence as the internal parameters of the subsystem of turbulent chaos (which refer, e.g., to the cascade fragmentation of large-scale vortex structures or temperature inhomogeneities in the system). This space of configurations can be, for example, in the space of the fluctuating energy of turbulent dissipation. These equations are designed to determine the time evolution of the conditional probability density of the state parameters of structured chaos and to analyze the Markovian stochastic processes of the transition from one stationary nonequilibrium state to another through a successive loss of stability as the control parameters (e.g., the Reynolds number) change. An alternative method of describing the mechanisms of transitions of this kind, which is closely related to the thermodynamically derived kinetic FPK equations, is based on Langevin-type stochastic equations.

All these questions are considered in detail in Ch. 5, where a thorough justification is given for several physical and mathematical postulates and assumptions used in the stochastic-thermodynamic model of structured turbulence. In particular, we analyze a cardinal problem of the approach we develop here – the possibility of the existence of asymptotically stable states in the subsystem of turbulent chaos, and we prove the corresponding H-theorem for the FPK equations, which leads us to conclude that any solution of the FPK equations for $t \to \infty$ tends to the stationary regime. We obtain a nonequilibrium probability potential that generalizes the well-known Boltzmann-Planck relation for thermodynamic equilibrium states to stationary nonequilibrium states of the representing ensemble and show that it is the Lyapunov function for these states.

Finally, we consider in the sixth chapter one of the possible formation mechanisms of mesoscale coherent vortex structures related to the frequency-phase synchronization of self-oscillatory modes as the formation mechanism of mesoscale vortex structures in high-dimension systems. This mechanism suggests an investigation of the synchronization of the elements of an ensemble that consists of very many interacting stochastic oscillators and that models the subsystem of turbulent chaos. This last is important, because synchronization in an ensemble of self-excited oscillators is known to restrain the growth of the attractor dimension, which gives rise to stable spatiotemporal structures. In addition, we consider the effects of frequency cluster formation (fluid particles composed of phase-synchronized vortices) in inhomogeneous chains of quasi-harmonic oscillators and investigate the influence of noise on the regime of cluster synchronization.

1.3 Cosmic Objects and Self-Organization

The subject matter of this monograph is an elaboration of the synergetic approach in developing the models of turbulence and self-organization processes with regard to various astrophysical and geophysical systems. At the same time, we provide several characteristic examples of these systems, the description and dynamical properties of which far exceed the scope of the narrower problem of self-organization in developed turbulence. The problem of order and chaos in the surrounding nature, especially in cosmic media, is extremely varied. The goal of this section is to show the specificity and complexity of these objects with several well-chosen examples, while at the same time we outline that there is a certain generality in the conceptual approaches of constructing these models, which is important for the completeness of the synergetic concepts that are being developed. From the viewpoint of self-organization in a dynamical system, the very understanding of natural complexes and their governing evolutionary processes is to a certain extent subjective. It reflects the position of the authors, who are well aware that by no means all researchers of natural processes and natural and cosmic media adhere to this approach. Likewise, we by no means deny the possibility that a more complex functional level of order dominates in nature, especially in the part that concerns the evolution processes that accompany the origin of life and natural selection (see, e.g., Galimov, 2001).

1.3.1 Dynamical Astronomy. General Principles.

The problem of order and chaos has an important application in dynamical astronomy (Contoupolos, 2002). Many of the key concepts of dynamical systems are embodied in the present views of galactic, stellar, and planetary structures and their evolution. They are backed by well-known examples from celestial and statistical mechanics, including the N-body problem, using both analytical and particularly numerical methods. Therefore, dynamical astronomy can be said to be the area where the fundamental concepts of order and chaos clearly manifest themselves, along with other numerous areas of modern science, such as hydrodynamics, geophysics, biology, plasma physics, cosmology, etc.

In addition to the completely integrable dynamical systems corresponding to an integrable Hamiltonian H_0, the study of chaotic and ordered structures in dynamical astronomy also incorporates the systems $H = H_0 + \varepsilon H_1$, which differ from the integrable ones by small perturbations ($\varepsilon \ll 1$). This also includes the important problem of convergence of the canonical expansions of the Hamiltonian $H(I,\theta,t)$ for the regular and resonant regimes of motion expressed in action-angle (I,θ) variables, which are widely used in celestial mechanics. For periodically perturbed Hamiltonian systems $H_1(I,\theta,t+\tau) = H_1(I,\theta,t)$ the KAM theorem on the existence of invariant tori of dimension $N+1$ mentioned above is applicable for stable orbits in $(2N+1)$-dimensional phase space, while the nondegeneracy condition

$$det\left|\partial^2 H_0/\partial I_i \partial I_j\right| \neq 0$$

ensures that most of them are retained at a small perturbation $\varepsilon \ll 1$. At $N = 1$ the fulfillment of this condition also guarantees the "stability" of the action variables that at fairly small ε remain close to their initial values over the entire time t (Nekhoroshev's generalized theorem corresponds to the more general case of orbital stability in time). As we already mentioned above, the invariant KAM tori for $N>2$ do not divide the energy hypersurface into nonoverlapping parts, which leads to the replacement of ordered orbits by chaotic ones.

In the phase space of conservative systems for the families of plane orbits (important for astronomical applications) that are periodically dependent on time, changes in the pattern of motion (bifurcations) are observed on a long time scale, while for perturbed real Hamiltonian systems resonant perturbations can produce chains of islands in the phase space and can increase the number of elliptic and hyperbolic points, with the overlap between the resonant chains of islands leading to chaotic dynamics. To investigate a Hamiltonian gradually changing with time, it is efficient to invoke the theory of so-called adiabatic invariants (functions of phase variables and parameters whose values change little as the parameters change significantly).

In dynamical astronomy, stellar dynamics is attributed to conservative systems, while the orbits of artificial and some natural bodies should be considered in terms of dissipative systems, which are characterized by chaotic domains with possible stability islands. Irregularity (chaotic behavior) of orbits is most commonly observed if there are perturbations produced by the heteroclinic intersection of orbits that were initially in different regions of phase space. This case has much in common with linear ergodic systems, and it can be implicitly likened to polygonal billiards. The transition to chaos can also arise if the system's Hamiltonian is limited by some critical energy and the orbits can go to infinity. These dynamical systems have a potential maximum to which unstable Lyapunov orbits correspond (Arnold, 1963a, 1963b).

Analyzing periodic and/or chaotic orbits in systems with three or more degrees of freedom is very important in dynamical astronomy. It is often more difficult there to separate stable systems from unstable ones (for this purpose, the well-known Krein-Moser theorem is used in systems with $N \geq 3$) and to distinguish the manifestation of real chaos from disorder and noise. The characteristic Lyapunov exponents grow with number N, although this growth slows down when N reaches $\sim$10. Periodic orbits on the phase plane are distinguished by great complexity and instability, including the collisions and bifurcations as well as the manifestation of ordered and chaotic domains. The behavior of orbits can occasionally be traced by peculiar shapes of the power spectra (when the trajectories are represented by means of the Fourier transform) in various systems. The behavior of these systems can also be studied based on Arnold's diffusion, as mentioned in Section 1.2.1. This allows the dimension of a system to be reduced. Since numerical simulations are currently widely used to analyze the properties of orbits, a qualitative verification of the generality of the results obtained for particular astronomical applications is of paramount importance. Note also that numerical methods are used especially efficiently to analyze systems of particles with N degrees of freedom under the action of linear or nonlinear forces. It is important to find the relationship between Arnold's diffusion and various channels of separation of the system's total energy, taking into account the form of the particle interaction potential and the contribution of resonant and nonresonant modes.

Some general concepts of dynamical astronomy are backed by observations of stellar-galactic and planetary systems. A combination of order and chaos should be considered to be the most important mechanism for the formation of the structure and dynamics of galaxies (Diagu and Holmes, 1997; Binney and Merrifield, 1998). Only a limited number of galaxies are described by a spherical potential, but many of them have an axial or triaxial symmetry. In the latter case, the potential is strongly deformed, and strongly disordered chaotic motions take place. The main family of orbits in *2D*-galaxies contains epicyclic, resonant, and nonresonant orbits inside and outside the corotation regions as well as short- and long-period orbits, along with nonperiodic and vanishing (going to infinity) orbits. Planar rings, shock waves, and vortices are also distinguished in the symmetry plane of *2D*-galaxies. At the same time, unstable chaos-generating orbits that usually do not combine with nonperiodic ones and orbits going beyond the corotation region are a characteristic feature of the structures of galaxies with a noticeable z-component. Unstable orbits can emerge from planar ones as a result of multiple bifurcations along the z axis, and they are directly related to Lindblad resonances, which are limited by the z-oscillation frequency. *3D*-galaxies also contain polar rings, just as *2D*-ones, while the families of *3D*-orbits themselves are most characteristic of warped, twisted, and peculiar galaxies that resemble a peanut or a box in shape.

Order and chaos in the structure of galaxies also emerge during the interaction of the resonances produced by a high-amplitude perturbing force, which can lead to chaotic motion. At the same time, the arms of spiral galaxies are characterized by the propagation of nonlinear density waves that affect the periodic orbits, as described in the theory of spiral density waves developed by B. Lindblad (Lindblad, 1941). These interactions can apparently take place at the earliest formation stage of stellar systems, which gave rise to certain order in the structure of matter in the Universe in the course of evolution. Attempts to reproduce this process are made within the framework of self-consistent evolutionary models based on the Liouville and Poisson equations, which describe the initial spatial density distribution function, but it can also be reproduced by direct numerical simulations for the N-body problem using various models of collision-relaxation processes and Lynden-Bell statistics.

In conclusion, we note an important difference between stellar dynamics and statistical physics and the inapplicability of the concept of thermodynamic equilibrium to any self-gravitating (stellar) system. In particular, there is no analogy between the description of a given state in accordance with the Gibbs distribution and the probability of finding a certain number of stars in a given orbit. This implies that whereas there is a tendency for equilibrium to be established in a thermodynamic system, no such process takes place in a self-gravitating system with a dense center (Binney, Merrifield, 1998; EAA, 2001). Thus, in contrast to the star itself with its thermal equilibrium and a certain temperature at each radius, the stars in a stellar system move with a constant energy in a wide range of distances from the gravitating center, and the stellar velocity distribution at a given radius is not a temperature characteristic. Irrespective of the specific configuration of a stellar system, the entropy in it grows through the transfer of energy from the dense central region to the much more rarefied outer region.

1.3.2 The Solar System: Dynamical Properties

One of the most important fields of dynamical astronomy is the dynamics of planetary systems, and primarily the planets of the Solar System. Studying it over several millennia, from C. Ptolemy's Almagest to the revolutionary change of the world view by N. Copernicus, laid the foundations of celestial mechanics that was originated by I. Newton, I. Kepler, and J.-P. Laplace. Only at the very end of the past century did one discover the first planetary systems around other stars. This immeasurably extended the views on the properties and diversity of the immediate neighborhoods of the stellar population. Slightly earlier, detailed images made the truly fantastic world of our own star accessible – the Sun and its nearest regions, in which individual ordered structures resulting from the plasma-magnetic field interaction can be distinguished against the background of chaotic dynamics (Fig. 1.3.1).

The major planets and their satellites, minor planets (asteroids), comets, meteoroids, and meteoric dust belong to the population of celestial bodies in the Solar System (Fig. 1.3.2). The plane-restricted three-body problem serves to describe the characteristic features of their orbital motions. Many works are devoted to the study of periodic solutions of this problem, including the degenerate solutions where the mass of one of the two bodies rotating in circular orbits around their common center of mass approaches zero (see, e.g., Marchal, 1990; Murray and Dermott, 1999; Bruno, 1990). A classic example is the dynamics of the Trojan asteroids at the Lagrangian points L_4 and L_5 in Jupiter's orbit, which have a potential maximum but retain their stability through a Coriolis component in a rotating frame of reference. The approaches to studying the general three-body problem, including regular and chaotic motions, periodic orbits, collisions, etc., are based on Hopf's theorem, which is a special case of Poincaré's recurrence theorem (see Barrow-Green, 1996).

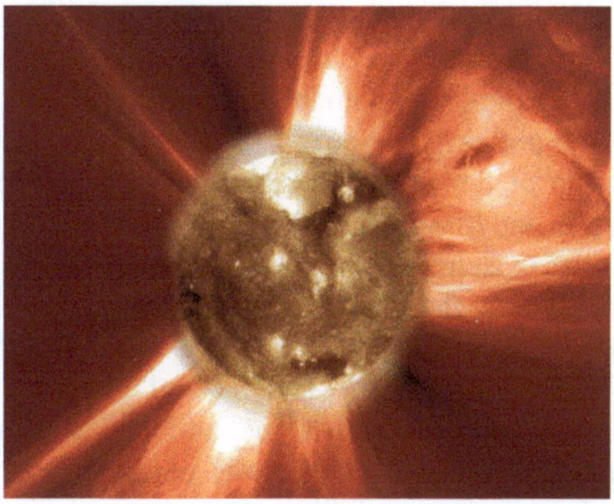

Fig. 1.3.1 Image of the Sun during powerful flare (Coronal mass ejections - CME). Numerous eruptive filaments connected with active regions on the solar photosphere and huge plasma bubbles extending to distances of more than two million km are observed. Self-organization processes occur in them under the influence of a magnetic field. Such frequently occurring events on the Sun give rise to magnetic storms on Earth. The image was obtained in 2002 by SOHO spacecraft (ESA courtesy)

Fig. 1.3.2 Population of the Solar system: the major planets and representatives of the minor bodies – comets and asteroids. Courtesy of NASA

The thoroughly developed and widely used exact and approximate analytical methods of celestial mechanics allow one to solve the problems of determining the orbits and their evolution as well as to establish the stability and instability regions of the solutions for various classes of motions. However, the methods of direct numerical integration gained the widest acceptance in the past decades.

These methods, along with the numerical-analytical ones, turned out to be most efficient for investigating order and chaos in the Solar System. They led to the conclusion that the orbits of planets with low eccentricities and inclinations are only slightly chaotic and do not have any noticeable secular component on time scales comparable to the age of the Solar System (Laskar, 1989; 1990; Sussman and Wisdom, 1992; Duncan and Lissauer, 1999; Murray, 1999). Meanwhile, resonances and commensurabilities exist in the motions of planets. The best known of them are the spin-orbit resonance in the motion of Mercury, which leads to a unique relationship between the Mercurian year and the solar day, and the characteristic commensurability in the motions of Neptune and Pluto, owing to which the collision between these bodies is ruled out, although accurate calculations of Pluto's orbit revealed certain chaoticity in its motion (Sussman and Wisdom, 1988). However, whereas the possibility of close encounters between planets capable of affecting their motion through the mutual attractions is ruled out, the comets and asteroids undergo strong perturbations, especially those that are subjected to the tidal effects from planets. The orbits of comets are subjected to the strongest chaotization. This makes it difficult to predict their motions and to accurately determine their ephemerides. The calculations of characteristic Lyapunov numbers for certain types of comets and asteroids with chaotic orbits confirm their extremely short (on cosmic scales) lifetime (see Wisdom, 1987; Fernandez, 1999).

Very many small bodies are captured into resonant orbits with the nearest planets. The Kirkwood gaps in the Main Asteroid Belt between the orbits of Mars and Jupiter in the region 2.7-3.2 AU, which are attributable to the resonances of the orbital periods of asteroids with Jupiter's period (4:1; 3:1; 5:2; 2:1; 3:2), serve as a remarkable example. The trans-Neptunian bodies in the Kuiper belt serve as another example. Here were found the resonances of the mean motion (commensurabilities of periods) with Neptune (4:3; 3:2; 2:1) and the secular resonances (precession of orbits) due to the commensurabilities of the longitude of ascending node and the argument of perihelion. Interestingly, the orbits of these bodies are stable (outside the resonances) on a time scale $t \sim 10^8$ yr, but the "accumulation of instability" and the sharp growth in eccentricity owing to the gravitational influence of Neptune led to their dissipation from the Kuiper belt (Ward and Hahn, 1998; Morbidelli, 1998; Levison, Weismann, 1999).

Because of its secular perturbations, the region between the "inner" and "classical" belts ($\sim$ 40 - 43 AU) is distinguished by the greatest instability. The bodies from this zone migrate into the Solar System and are initially captured predominantly into Jupiter-crossing orbits (JCOs), while a certain fraction of them subsequently migrate toward the Sun, replenishing the Main Asteroid Belt and three groups of asteroids (Amur, Apollo, Aton) crossing the orbits of terrestrial planets that are generically called Near Earth Objects (NEOs). The orbits of these objects, especially the Earth-approaching Apollo asteroids and even the Aton asteroids going inside the Earth's orbit, are subjected to the greatest chaotization and pose the main hazard of colliding with the Earth (McFadden, 1999). Comets also contribute to the asteroid hazard and pose a threat to our civilization.

In general, we emphasize that the collisions of small bodies with planets served as one of the most important factors of planetary evolution. The prevalence and importance of these processes are clearly demonstrated by the heavy cratering of planet surfaces and their satellites, especially those on which the traces of impact bombardment were not wiped out by the subsequent erosion processes through the influence of an atmosphere and hydrosphere. The Moon, Mercury, satellites of the giant planets, and even the asteroids themselves, whose surfaces are literally pockmarked with craters, are most characteristic here. In particular, wide basins (maria) of more than 1000 km in diameter on the Moon were produced by the impacts of asteroids several hundred kilometers in size. At the same time, the lunar crater size spectrum extends from centimeters to thousands of kilometers. The lower threshold of crater sizes on Earth, Mars, and particularly Venus with atmospheres corresponding to the size cutoff threshold for bodies capable of penetrating to the surface is considerably higher. Bodies with sizes ≤ 10 m are destroyed in the Earth's atmosphere and only their fragments reach the surface in the form of meteorites (Ivanov, 2003). Meanwhile, there is ample evidence for the collisions of even large asteroids and comets with the Earth over at least the last several hundred million years of its history, which led to global cataclysms. One example is the *Chicxulub* event, which presumably had catastrophic global consequences for the terrestrial biosphere: An asteroid of about 10 km in size collided with the Earth, leaving a crater about 170 km in diameter on the Yucatán Peninsula in Mexico at the turn of the Cretaceous and Neogene ~ 65 million years ago. Another example is the local-scale catastrophe near the Tunguska River in Siberia in 1908 caused by the collision of an assumed comet fragment with the Earth. The fragment's size is estimated to be ~ 60 m. The fall of 22 fragments of about 1 km diameter from comet P/Shoemaker-Levy-9, which was torn apart by tidal forces during its close encounter with Jupiter, onto the planet in 1994 is convincing evidence that these events occur more or less regularly in the Solar System.

The populations of comets and asteroids change continuously under the gravitational attraction of planets. This is responsible for the predominant streams of meteorites originating in the resonant zones of the Main Belt and the Kuiper belt, and it makes a laboratory study of extraterrestrial matter possible. Close encounters with planets have the greatest influence on the dynamics of comets, which can lead to their transition to hyperbolic orbits and their "evaporation" into the interstellar medium. On the other hand, comets from the Oort Cloud can be thrown into the Solar System under the action of tidal perturbations from the nearest stars, thereby providing its connection with the Galaxy and, in particular, as it were, the "probing" of the material of the protosolar nebula. Thus, the interactions of comets with planets open up possibilities for a better understanding of the role of these primitive bodies in planetary cosmogony and galactic evolution.

At the crossroads of astronomy and geophysics, there is a growing understanding of impact phenomena during the interaction of small bodies with planets throughout the history of the Solar System. The importance of these phenomena in the evolution of planets manifests itself not only in the obvious fact that small bodies leave "scars" in the form of craters on the planetary surfaces, but also in the growing evidence that the stochastic processes produced by them had a significant influence on the formation of natural conditions that are directly related to the existence of an atmosphere and hydrosphere on the planet, among others. For example, chaotization of the orbits of trans-Neptunian objects and/or Main Belt asteroids, which are considered to be relics of the bodies (planetesimals) from which the planets were formed, could play an important role in supplying volatiles at the initial stage of their evolution, especially in the period of heavy bombardment of the planets about 4 billion years ago (Marov, Ipatov, 2001; 2005; 2006), as will be discussed in more detail below. In recent years, the origin of life has been associated with the transport of matter inside and outside the Solar System through migration-collision processes (see McKay and Davis, 1999; Marov, 2005). These events may be considered as a product and a consequence of the mechanism of local instability in a nonlinear chaotic system with a large number of degrees of freedom, which lead to a sequence of state bifurcations and to the possibility of self-organization to which a different set and interaction of natural conditions on the planet correspond.

Self-organization may possibly play a key role in the processes of prebiological evolution.

In this approach, the growing order of an initial (chaotic) system involves a sequence of bifurcations, from the appearance of primitive polymer structures through the development of a universal catalytic function of peptides to the emergence of nucleotide sequences (codons) involved in protein synthesis and a genetic code. The implementation of the model requires the presence of a reducing environment, and a key role may play the adenosine triphosphate (ATP) molecule (Galimov, 2001; 2005). In a reducing environment, the atmosphere and hydrosphere exist separately and phosphates are available and mobile, which is consistent with the existing views of the natural conditions on Earth in the period with which the emergence of the first life forms is associated (about 3.8 billion years ago). Another approach is related to an ancient RNA world as a precursor of life emergence, which, in particular, invokes comets as an important carrier of primordial life form structures to satisfy the required time frame of the world's original development/evolution (Spirin, 2005; 2007; 2010). Choosing between the alternative models that either postulate the origin of life and biosphere directly on Earth or hold that this involved an external source is one of the most topical and intriguing problems in modern natural science.

1.3.3 The Solar System: Nature of the Planets and Satellites

From the standpoint of self-organization processes in natural dynamical complexes, the entire set of celestial bodies populating the Solar System is of primary interest – the planets, their satellites, comets, asteroids, and meteoroids. Each of them is a unique object in its own way and reflects certain characteristic features of cosmic media.

While there is a huge variety of these objects, we can nevertheless reveal some common patterns that apparently point to a unified origin of the Solar System and, concurrently, study the mechanisms that determined the paths of evolution that were different from the Earth's evolution scenario. The relationship between order and chaos undoubtedly played an important role in the formation of various natural objects. A considerably closer symbiosis of astrophysical, planetary, and geophysical studies at the turn of the last century aimed at answering the key question of cosmology and cosmogony, that is, the origin and evolution of the Universe and planetary systems. The research on the origin of life (astrobiology) contributes to understanding the processes that formed an overall evolution of matter.

In the complex of problems related to planetary research that are united by the very comprehensive concept of planetology, three main directions can be distinguished: the origin and evolution of the Solar System; the nature of the Moon, planets, and their satellites; and the nature of small bodies and their relationship to the evolution of planets and the origin of life. Obviously, the key problem among them is investigating and modeling the main physical-chemical processes that ensured the birth of the Solar System and its subsequent evolution. This is studied by reconstructing these processes based on the available observational data and by developing models. In contrast to stars, the views of whose properties are based on the possibility of comparing many objects at various stages of their evolution, until recently we had only one example of a planetary system – our Solar System. Therefore it was impossible to analyze it with a statistical approch, as for the planetary population.

The discovery of circumstellar disks (Hildebrand, 1983; Artymowicz, 1997; O'Dell, 1998; Schneider et al., 1999; Padgett et al., 1999; Lagrange et al., 2000; Watson et al., 2007) and planets around other stars (Mayor and Queloz, 1995; Marcy et al., 2000) through high-resolution optical, infrared, and submillimeter observations considerably extended our views of the properties of planetary systems and the great variety of their configurations. Unique opportunities for understanding their formation appeared. However, using essentially the only available methods of Doppler spectrometry and transits (the passage of a planet over the stellar disk) restricts the mass limit for the planets that can be discovered mostly to the Jupiter-Saturn masses at a distance closer to $\sim$ 10 AU from the parent star. As a result, we are not yet able to detect Earth-type planets (the best threshold yet achieved is at $\sim 5M_E$) and to reproduce the configuration of the entire planetary system (Black, 1999; Nagasawa et al., 2007). Meanwhile, this is of primary interest, especially considering that the super-Jupiters discovered in orbits in the immediate neighborhood of a star make exotic configurations that are fundamentally different from what we observe in the Solar System. This acutely poses the question about the stability of these planetary systems.

Recently, a model was proposed according to which the formed Earth-type planets migrate into the system because of their interaction with the residual disk gas (dynamical friction), which severely restricts their lifetime (Papaloizou et al., 2007). An alternative model initially assumes an important role of the massive planetesimals left in the neighborhood of the formed planet, which exert a significant gravitational influence on the evolution of its initial orbit, the migration of the planet itself, and the swarm of planetesimals to considerable distances toward and away from the star, following the condition for the conservation of orbital energy and angular momentum of the planet and the disk. In other words, migration controls the structure of a forming planetary system with a dynamical instability and is eventually responsible for the configuration acquired by it (Levison et al., 2007). It is quite possible that the migration mechanism affected the evolution of planetary orbits in the Solar System, at least in its outer regions, which is responsible for the displacement of Uranus and Neptune farther from the Sun from the zone of their initial formation closer to Jupiter-Saturn and the formation of the Kuiper belt (Eneev, 1981; Malhotra, 1995). The fact that times exceeding the age of the Solar System would have been required for the accretion of the farthest planets in their present-day orbits, along with a number of geochemical considerations, serves as a justification for this assumption.

Turning to our immediate cosmic environment, we naturally wish not just to understand how the Solar System came into being but what is responsible for its stable configuration. We also aim to understand what distinguishes the Earth with its unique natural conditions from other terrestrial planets, primarily its nearest neighbors Venus and Mars, and what the limits are of the existing regulation mechanisms for the feedback on Earth, which prevent unfavorable scenarios for its subsequent evolution. In other words, whether, for example, the accumulation of negative anthropogenic impacts on the surrounding natural medium will lead to a radical change of conditions associated with the bifurcation of the state of an open nonlinear dissipative system. We need to integrate Earth and planetary sciences aimed at a better understanding of the present, past, and future of the Earth based on comparative planetology to give an answer to this cardinal question. Concurrently, this must help solve the cardinal problems of planetary cosmogony and, in particular, to impose much more stringent constraints on the range of parameters used in developing models for the origin of the Earth (see, e.g., Wetherill, 1990).

In this section, we will mainly address the nature of the planets and satellites. This seems necessary because their natural properties and some peculiarities reflect the general principles of self-organization, and we wish to draw attention to precisely this fact. This will allow us also to better understand the emergence of order in the gaseous envelopes of these bodies, whose atmospheres we will investigate in a special section of this chapter. They very clearly show the synergetic processes in turbulent natural media, which are our main focus in the subsequent chapters.

The Moon

As the body nearest to the Earth, the Moon (Fig. 1.3.3) is of primary interest for planetology and Earth sciences - geophysics, geology, and geochemistry. Unfortunately, many problems remain unsolved despite a huge breakthrough in our knowledge of the Moon provided by the unprecedented, in scale and cost, American and Soviet space programs in the 1960s-1970s. After quite a long break, interest in

Fig. 1.3.3 The near (left) and far (right) sides of the Moon

studying the Moon resumed in the 1990s with the flights of American spacecraft Clementine and Lunar Prospector, followed by the Japanese Kaguya, Chinese Chang-E, and Indian Chandrayaan-1 in the early 2000s. These flights were focused mostly on studying the lunar relief morphology and gravity field, but Chandrayaan-1 brought also evidence about H_2O/OH distribution on the Moon's surface (Pieters et al., 2009). The new breakthrough was most recently accomplished by the launch of the American Lunar Reconnaissance Orbiter (LRO) and LCROSS – the final stage of the rocket that impacted the lunar surface to release volatiles. These experiments brought evidence that the surface of the Moon in polar regions has a significant abundance of hydrogen-bearing compounds associated with water ice (up to a few percent) and dramatically changed the existing concept of the Moon as an extremely dry body, as had been assumed until most recently (Mitrofanov et al., 2010; Paige et al., 2010). Especially convincing results were obtained by the observations of a plume that accompanied the LCROSS impact (Colaprete et al., 2010; Shultz et al., 2010; Gladstone et al., 2010; Hayne et al., 2010).

Studying the Moon is of crucial importance primarily for understanding its origin and a better understanding of the Earth's geological history, because the most ancient rocks are preserved only on the Moon's surface, whereas on Earth they were eroded by the hydrosphere-atmosphere-biosphere set-up. It is necessary first of all to select one of two main hypotheses of the Moon's formation: it was formed either through a mega impact - the collision of the early Earth with a Mars-size body, or it was formed together with the Earth from partially differentiated proto-planetary disk material (Hartmann, Davies, 1975; Cameron, Ward, 1976; Melosh, Sonett, 1986; Taylor, 1999; Galimov et al., 2005). The latter scenario agrees with some cosmochemical constraints, in particular certain radio isotopes ratios (Galimov, 2010). However, the very fact of the Moon's origin can be associated either with the emergence of certain order in the original structure of the protoplanetary material in the Earth's orbit, which governed the subsequent contraction of two separate gas-dust clumps, or with the initial chaoticity of the orbits of large planetesimals (protoplanets) near the formed Earth and an occurrence of catastrophic collision of one of these bodies with it, which led to a huge ejection of mantle material and to the formation of a new planet-satellite configuration in the process of self-organization.

To choose between these two hypotheses, a detailed study of the Moon's geological history is also necessary, in particular, an analysis of the abundance of refractory and siderophilic elements in lunar rocks, a comparison of the heat flux on the Moon and Earth, and seismic sounding to solve the problem of the Moon's interior, first of all whether it has retained a liquid core, its size, and the degree of differentiation of the bulk of lunar material. These questions should be considered in the context of the formation of all planets and their satellites, including the distance from the Sun, the early evolution with allowance made for the contents of high- and low-temperature condensates and gases, and the reserves of radiogenic elements, which underlies the self-organization processes that predetermined order in the structure of the interiors (Fig. 1.3.4). The Moon gives quite a definite approach to solving these problems (see Vinogradov, 1975; Galimov; 2004; Wilhelms, 1993; Canup, Asphaug, 2001).

The absolute age and dating of events determined from the U/Pb, Rb/Sr, and K/Ar isotopic ratios in lunar rock samples led to the conclusion that the Moon was formed 50-70 million years after the Solar System's formation (Kleine et al., 2008a). This last is dated by the age of refractory CAIs inclusions in the meteorites (CV3 chondrites) (Allende and Efremovka) and varied from 4567.1 $\pm$ 0.1 Ma (Amelin et al., 2002*; 2006)* to 4568.5 $\pm$ 0.5 Ma (Bouvier et al., 2007; 2008). The value currently referred to as the most probable one is 4567.22 $\pm$ 0.21 Ma (Amelin et al., 2010; Bouvier et al., 2009; 2010).

The Moon appears to have experienced interior differentiation with core separation within the subsequent $\sim$ 200 Myr. The phase of intense impact bombardment by planetesimals left from the accretion stage in the region of the inner planets' formation left huge basins and multitiered structures on its surface. Differentiation and impact processes left behind significant height changes on the lunar surface, reaching 18 km in polar regions on the far side. The impact energy and mantle heating by radiogenic isotopes led to the outflow of lava from the mantle onto the surface from a depth of more than 100 km and to the filling-up of lunar maria on the visible side where the crust was considerably thinner. Some peculiarities of the lunar morphology are associated with lava cooling, which gave rise to faults, highlands, and valleys.

The endogenic processes ceased 3.18 Gyr ago and the Moon was subsequently subjected only to impact bombardment by bodies mostly migrating from the outer Solar System, which we discuss in more detail in Section 1.3.5. Meteorite processing of the surface material, which is responsible for the origin of regolith, has a very long time scale. As a result, the Moon stores the chronicles of events, including its interaction with the Sun and the interplanetary medium, over several billion years on its surface and in the layer of regolith. Interestingly, the tracks of astronauts and lunar rover wheels will be retained without noticeable erosion at

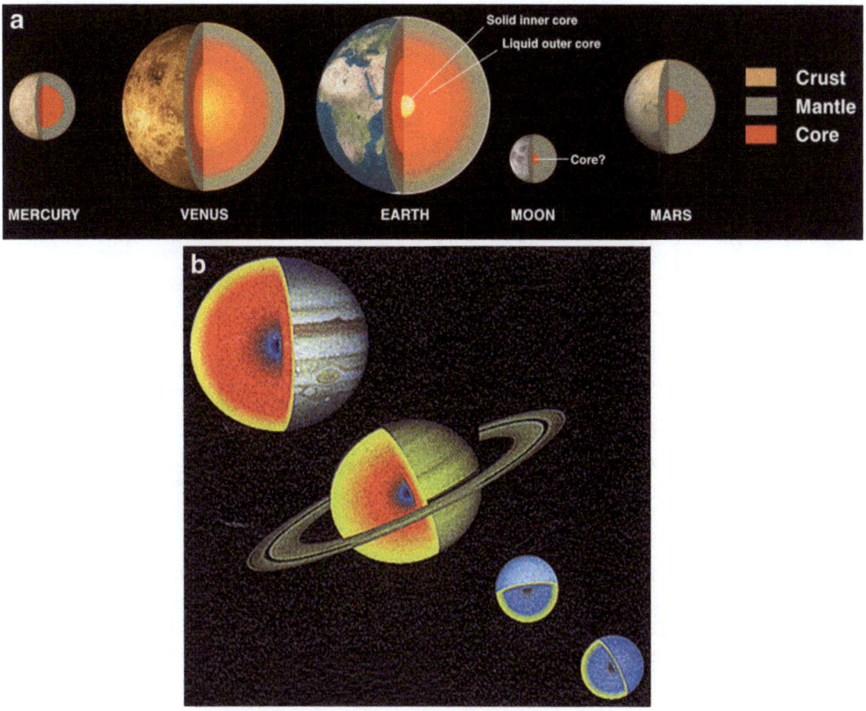

Fig. 1.3.4 Internal structure of the terrestrial planets and the Moon (a) and the giant planets (**b**). The order in the arrangement of main regions (core, mantle, crust) is a consequence of the differentiation of their constituent matter into shells, their extension depending on the size (mass) of the planet, the abundances of major components, and the condensation temperature in the formation zone. In the case of terrestrial planets and the Moon mass of the body predetermines the core state and the crust (lithosphere) thickness whereas the constituent gases and ices ratio mostly defines the giant planets interiors. Courtesy of NASA and Wikipedia

least for several million years, as we can determine by the average level of meteorite bombardment.

Let us now consider in more detail the most relevant problems that pertain to the key questions of the Moon's genesis and internal structure.

Basically, it was concluded from isotopic analysis, first of all from shifts in the oxygen isotopes ratio ^{16}O-^{17}O-^{18}O (Fig. 1.3.5), that the Earth and the Moon were formed from a single reservoir in which the isotopic ratio differed from that in the parent bodies of meteorites. This supports the hypothesis that the Moon originated jointly with the Earth from a partially differentiated protoplanetary gas-dust cloud. The lunar minerals are generally similar to the terrestrial ones, although they have much lower volatile abundances. The Moon is depleted in alkaline elements, in particular potassium, which is a constituent of the terrestrial granites, but, on the other hand, there are many more refractory elements (Al, Ca, Ti) in the lunar crust. These elements are usually not separated within this group, but the entire group can be separated from other classes of elements at the pre-accretion stage of the planet.

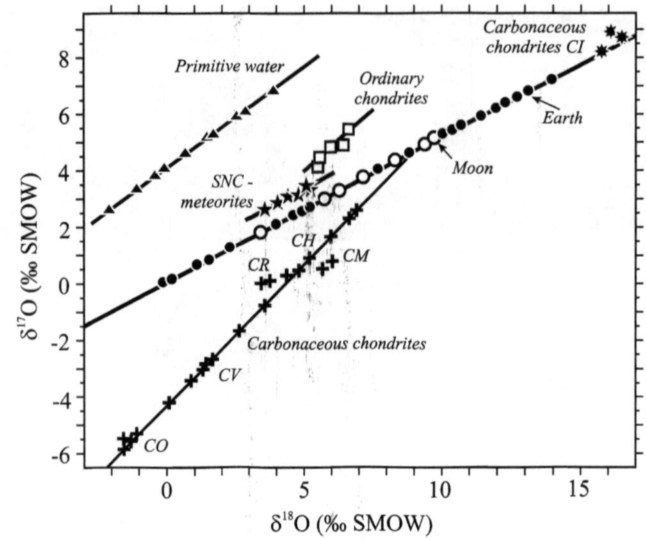

Fig. 1.3.5 Diagram of oxygen isotopes $\delta^{17}O$ and $\delta^{18}O$ ratio. This diagram characterizes shifts in the oxygen isotopes ratios $^{17}O/^{16}O$ and $^{18}O/^{16}O$ relative to adopted standard SMOW (*Standard Model Ocean Water*). Earth and the Moon matters fit the same line in contrast to different types of chondrites and SNC meteorites

The fundamental question is whether and how partial evaporation of the refractory and more volatile elements occurred in a high-temperature process at the Moon's formation stage.

The geochemical data cannot be directly used to estimate the degree of segregation of the metallic fraction and, accordingly, the core size, because this estimate depends on the unknown initial abundances of siderophilic elements (iron satellites) and the degree of differentiation of the lunar material. Therefore, as for refractory elements, the solution of the deficit problem of siderophiles in the lunar material depends fundamentally on which model of the deep structure is used, primarily on the presence and size of the core limited by the Moon's dimensionless moment of inertia, according to which it does not exceed $R \sim 300$ km. The problem is even more complicated because one needs to take into account to which degree the elements in the silicate-metal system are extracted if there is solid-phase silicate in the solution, and moreover, the degree of mantle melting is also uncertain.

The magnesium number Mg/(Mg+Fe) serves as an important indicator of the initial composition of a silicate planet and the degree of its differentiation. According to geochemical and petrological data, the magnesium number is limited to 0.89 for the terrestrial mantle. An estimate of 0.80 was obtained for the Moon, but the error reaches 30%, so that the upper threshold can be reconciled with the terrestrial value. Consequently, this estimate is not adequate for characterizing the composition of the material of the entire Moon. This difference, if confirmed, would only imply either that the Moon was formed from the Earth or that the mechanism of its subsequent evolution after "independent" accretion included complex chemical transformation processes. Therefore, several assumptions need to be made to interpret the measurements of this ratio in the surface rocks, and it is so far hardly possible to unambiguously determine the degree of differentiation of the lunar material.

As we said above, the high-temperature process through which the Moon went during its evolution could be responsible for its enrichment with refractory elements and the loss of iron. However, it is by no means necessary to resort to the model for the contraction of an initial gas-dust clump to justify the required heating, because even the accretion model of the terrestrial mantle material provides sufficient energy release through the collisions of numerous parent bodies. The enrichment of the lunar crust with plagioclase in feldspars is naturally associated with intense heating, suggesting initial differentiation in depth, at least in the first several hundred kilometers. However, the numerical estimates for the thickness of the melting layer above the non-differentiated interior material are very contradictory: several geochemical arguments provide evidence for the model of a completely differentiated Moon based on the volume content of the parent material needed to produce the existing concentration of elements in the crust, while the model of non-differentiated interiors appears to be supported by the interpretation of volcanic gas contents in mare regions. In addition, the existence of a primordial magma ocean can be explained by the mechanism of magma intrusion into the primordial crust. Clearly, the question about the degree of initial melting of the Moon and the heating sources remains problematic and in some measure resembles the situation with the parent bodies of partially differentiated meteorites.

Some of the refractory elements produced by partial melting of the silicate fraction are known to be concentrated more easily in the less dense liquid phase than in the solid one. This leads to their enhanced abundances in the surface rocks as the material drifts from the interiors. The abundance of uranium $U \sim 35$ ppb for the Moon, compared to U=20 ppb for the Earth, was obtained using Apollo data based on the heat flux measurements at two points that vary between 16-20 and 10-15 erg cm^{-2} s^{-1}. This uncertainty prevents one from choosing between the alternative hypotheses of the Moon's origin.

This choice, as well as the Moon's interior structure, also depends on how the pattern of seismic wave propagation velocities in the mantle is interpreted. Therefore, seismic sounding is a key in any future lunar research program. Unfortunately, the models for the propagation velocity of seismic waves even in the lunar crust have a large error because the exact crust composition is unknown: the range reaches from gabbro and norites to anorthosites. A strongly absorbing zone in the mantle is known to make it impossible to investigate the core. The focal zone of moonquakes lies at a depth of between 800 to 1100 km, which fundamentally limits the maximum depth to which the velocities of seismic waves can be determined at present.

One also needs to consider that the geochemical measurements within a relatively low depth on the lunar surface are restricted by the upper layer of regolith, whose composition can differ from the underlying layer of igneous rocks, primarily because of the unknown abundance ratio of the Moon's material and the meteorites bombarding its surface. This consideration is particularly important when discussing the hypothesis of the Moon's detachment from the Earth, given the great uncertainty in the estimates of the relative contribution from the target material (terrestrial mantle) or the impactor (cosmic body) to the Moon's formation. Indeed, one should bear in mind that the latter could constitute up to $\sim$ 80% of the ejected matter to form the Moon.

Of great interest is the region on the far side of the Moon near the South Pole, a multitiered structure extending 2500 km from the Aitken crater 135 km in diameter and 13 km in depth, which was formed when a large asteroid fell onto the Moon 4.1 Myr ago. There is convincing evidence that water ice deposits might be inside this basin in its permanently shadowed region, brought there by the later collision of a comet with the Moon, which are estimated to be $\sim$ 3x10^8 tonn by mass. Basically, water ice deposits by comets are among the plausible sources of water in the lunar soil (as are inter-particles frost and crystalline water), together with solar wind protons and plausible exiling from the interior (Lucey, 2009). The Moon is also thought to preserve in its surface regolith different volatiles and organics delivered through impact processes. Their funding is the goal of future Moon explorations and is closely related to the intriguing astrobiology problems as well.

Various possibilities are linked to future lunar exploration and utilization (see, e.g., Spudis, 1996). These are a possible development and industrial use of minerals containing Fe, Al, Si, Ti, which serve as a source of building materials, but also H_2 and O_2 contained in rocks as a raw material for air and water production. A key significance could have likewise the estimates of local water reserves, as well as the claimed prospects for the practical use of ^{3}He as an efficient energy source (Galimov, 2006), whose abundance in ilmenite is estimated to be $\sim$ 10 ppb.

Venus

Venus stands out among the planets of the Solar System primarily because of its massive gaseous envelope and thermal regime. The most topical question is why the evolution of Venus, which has many similar features with the Earth and is only $\sim$40 million kilometers away from us, followed a different scenario than the Earth. This ultimately determined the formation of a hot carbon dioxide atmosphere on this planet with a mean temperature near the surface of 475°C and a pressure of 92 atm, whose density is less than that of water by only about an order of magnitude. These unique properties fundamentally distinguish Venus from the other two terrestrial planets with atmospheres – the Earth and Mars. As the closest analog of the Earth, it is natural to consider Venus as one of its limiting models (Fig. 1.3.6.)

The unusual and in many respects exotic climate of Venus is predetermined by many factors, primarily by the runaway greenhouse effect that probably resulted from the loss of water and the decomposition of carbonates in surface rocks. At the same time, although the insulation on Venus is approximately twice that on Earth owing to its proximity to the Sun, its total spherical albedo A is also approximately

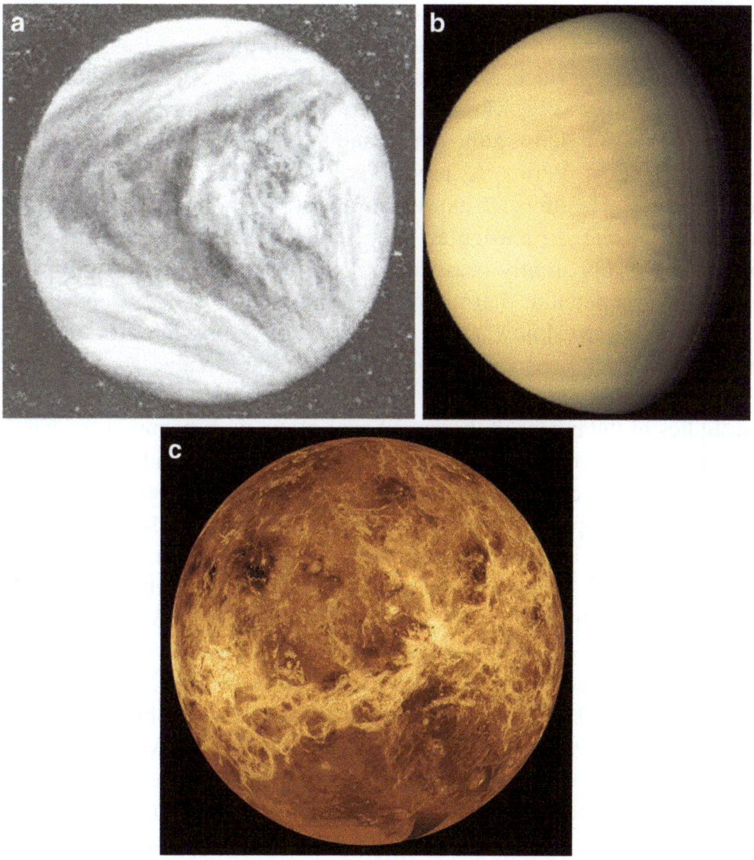

Fig. 1.3.6 Images of Venus in the ultraviolet (**a**) and visible (**b**) spectral ranges as well as in the centimeter radio wavelength (**c**) obtained by aperture synthesis radar. Characteristic structures are observed in the ultraviolet due to the presence of an absorber (probably sulfur particles) at the upper cloud boundary. These structures move over the Venusian disk with a speed of ~ 100 m/s that exceeds the wind speed near the surface by two orders of magnitude due to the superrotation of the atmosphere. The visible range contains little information; the isolated bands are associated with the inhomogeneous structure of sulfuric-acid clouds. The surface of Venus can be seen only in the radio wavelength for which the atmosphere and clouds are transparent. Radio mapping has revealed many relief features and peculiarities of the Venusian surface. (**a**) An image from the Earth on which ordering structure of ultraviolet clouds is clearly seen; (**b**) an image from the Galileo spacecraft during the Venus flyby; (**c**) mosaic of images of the Venus surface returned by the Magellan spacecraft; more or less ordered structures can be distinguished in the chaotic pattern of the relief . Courtesy of NASA

twice that of the Earth. As a result, both planets derive comparable radiant energies from the Sun at the present epoch, and the cause of the Venusian surface heating should obviously be sought in the formation of specific conditions at the earliest stages of its evolution as another example of self-organization in nature. We can hardly reproduce this process, but we can study in sufficient detail the physical mechanism of the formed stable thermal equilibrium state on the planet at the present epoch based on a tremendous breakthrough in our knowledge of the atmosphere, clouds, surface, and internal structure of the nearest planet that is so unlike Earth. Radio waves, which, in contrast to optical emission, are capable of penetrating through a dense atmosphere and a cloud layer, allowed one to investigate the surface of Venus, to reveal numerous features of its relief, and to reconstruct certain features of its evolution that are both similar and dissimilar to the terrestrial one. In particular, it was found that the widespread volcanic activity on Venus apparently ceased relatively recently ($\sim$ 10-100 Myr ago) and the formation of a peculiar thermal regime is in many respects associated with it (Fig. 1.3.7). What we know about Venus today is presented in detail in the monograph by Marov and Grinspoon (1998) and the reviews by Hunten (1999) and Head and Basilevsky (1999).

The studies of the nature of the greenhouse effect were begun in the late 1960s, when the first models of radiative-convective heat exchange in the lower Venusian atmosphere based on direct measurements of its vertical temperature and pressure profiles were developed (Avduevsky et al., 1968; 1971; Marov, 1971; 1972). Additional progress in this study was achieved primarily through direct measurements of the integrated and monochromatic solar radiation fluxes in the atmosphere and on the surface of Venus (Marov et al., 1973; Avduevsky et al., 1976), measurements of the altitude profile and microphysical properties of Venusian clouds, which are composed of sulfuric acid (Marov et al., 1976; Marov et al., 1980; Esposito et al., 1983), and a refinement of the altitude profiles for atmospheric parameters with allowance made for their diurnal variations (Marov, 1978; Marov et al., 1989a; Seiff et al., 1985). In addition, comprehensive theoretical and laboratory studies of the opacity characteristics for the main absorbing gases in the lower Venusian atmosphere (carbon dioxide, water vapor, and sulfur dioxide) were carried out in a wide range of pressures and temperatures, the technique for numerical simulations of radiative heat transfer was developed (Marov, Shari, 1973; Marov et al., 1984, 1985, 1989b), and the first measurements of the effective heat fluxes were made in several regions of the planet (Revercomb et al., 1985). This allowed one to pass from comparatively simple estimations to a more comprehensive analysis of the thermal regime and the construction of a radiation model by taking into account the available data on the content of minor admixtures in the Venusian atmosphere (Marov et al., 1985, 1989b, c; Tomasko et al., 1983; Marov and Grinspoon, 1998). The exotic extended ($\sim$ 20 km thickness) Venus cloud deck composed of sulfuric acid droplets of micron size (Esposito et al., 1983; Volkov et al., 1989) also contributes to the peculiar thermal regime formation. Below, we consider the pattern of radiative heat exchange in the Venusian atmosphere by also touching upon (where necessary) the main properties of planetary circulation that contribute to the energy balance on the planet. Other peculiarities are considered in the section devoted to the dynamics of planetary atmospheres.

In contrast to stellar atmospheres, the influx of solar radiation absorbed by the planetary atmosphere, which is a permanently acting factor, leads to heat flux variability with altitude. The interaction of radiation with the gaseous medium

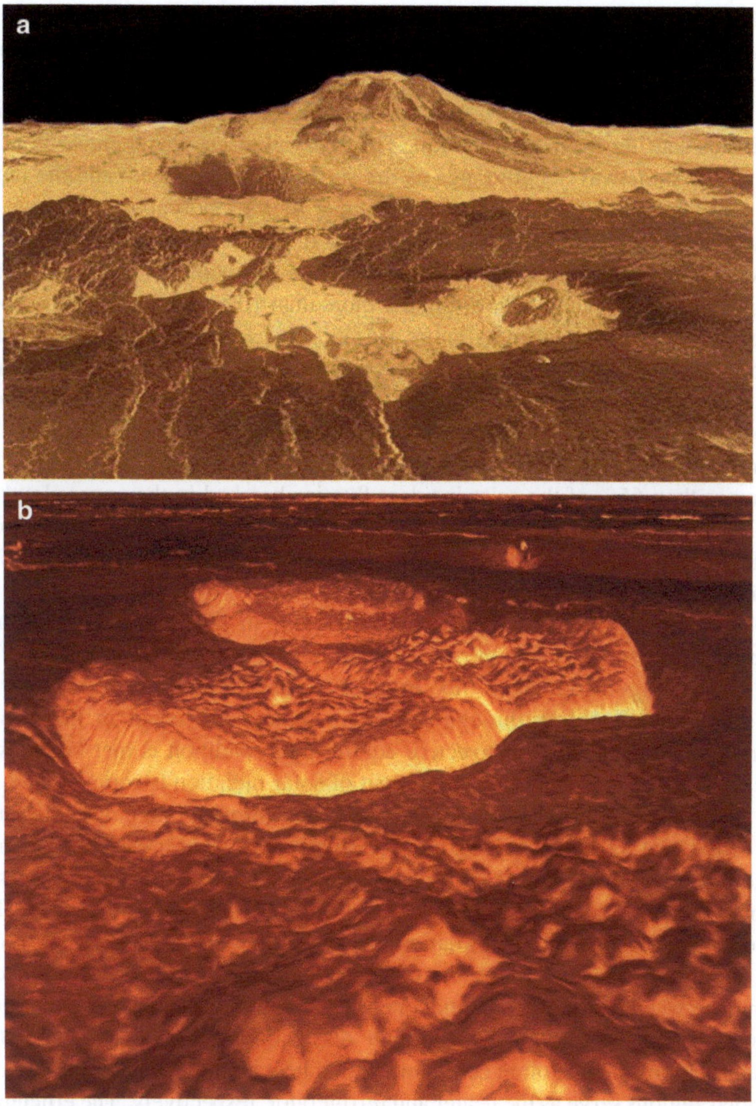

Fig. 1.3.7 **(a)** Image of the volcanic Maat Mons (Alta Regio – Aphrodite Terra) on Venus in perspective projection. The height of the mountain is ~ 5 km; the traces of lava flows are clearly seen. The vertical scale was enlarged by a factor of 10 for clarity. **(b)** An image of the surface outpouring of volcanic lava ("pancakes") in perspective projection from the radar mapping of Venus from the Magellan spacecraft. Courtesy of NASA

characterizes the energy release. Thus, to understand the main features of the thermal regime in the Venusian atmosphere, it is necessary to answer the questions of how the outgoing radiation (net heat flux) is formed and which interrelationship exists between the radiative and dynamical heat exchanges.

Measurements of the monochromatic and integrated solar radiation fluxes as a function of the solar zenith angle serve as a key limiting factor for thermal balance estimations. Since no direct solar rays reach the surface of Venus because of the high optical thickness of its atmosphere and clouds, the scattered radiation dominates starting from about 60 km. This then increases the attenuation of solar light in the short-wavelength range as the surface is approached, and the red rays that paint the sky of Venus and the features of its landscape in red (or, more precisely, orange) dominate near the surface. If the measured solar radiation flux is averaged over the entire planetary surface, then the incident flux per unit surface area is 16.8 ± 2.3 W/m^2. This is slightly more than 10% of the energy absorbed by Venus when averaged over its entire surface (157 ± 6 W/m^2). Since the surface albedo estimated by analyzing the photometric measurements and panoramas of Venus does not exceed 10% (Marov, 1978), this implies that the surface absorbs almost 90% of the solar radiation flux reaching it. In turn, it can be concluded from several additional measurements and theoretical estimations (Moroz et al., 1985) that more than 65% of the solar incident flux is absorbed in the upper cloud layer and the upper cloud haze (in the range of altitudes 60-90 km), about 8% is absorbed in the middle and lower cloud layers (49-60 km), and approximately 27% is absorbed by the lower atmosphere and the surface.

Obviously, if there is a radiative equilibrium in the atmosphere (as in the terrestrial stratosphere), then the solar radiation energy is compensated for at each level by the outgoing radiation flux. In other words, the rates of solar energy input and atmospheric cooling through infrared radiation are equal in this case. If, alternatively, the radiative-convective equilibrium condition is met (as is usually the case in the troposphere), then the input of solar radiation is compensated for at each level in the atmosphere by the net heat flux through outgoing radiation and convective heat transport. Indeed, these alternatives serve as a local approximation to the more complete picture of heat transport including dynamical processes of various spatial scales (planetary circulation, convection, turbulence). These motions result from the transformation of part of the radiant energy flux from the Sun. In the most general sense, the planetary dynamics can be said to reflect the balance between the rate of potential energy generation through solar radiation and the rate of kinetic energy loss through dissipation. From this viewpoint, the planetary atmosphere is often compared with a heat engine in which the equatorial and polar regions serve as the heater and refrigerator, respectively. The efficiency of such an engine is very low, no more than a few percent (see, e.g., Kuzmin, Marov, 1974).

As we see, the inequality between the input and output energies in some regions of the planet, with the condition for global-scale heat balance characterized by the effective temperature near the upper boundary of clouds, $T_e = 228$ K for Venus, being rigorously met, serves as the source of motions with various spatial scales. In other words, the emergence of thermal inhomogeneities (horizontal temperature gradients due to differential heating) is compensated for by the development of large-scale motions with a broad spectrum of spatial sizes, including the cascade fragmentation of turbulent vortices to sizes at which the dissipation of mechanical energy into heat takes place (Kolesnichenko, Marov, 1999; Marov and Kolesnichenko, 2001). The dynamical transport processes on Venus are very efficient, which is responsible for the well-known phenomenon of superrotation (4-day circulation) and the formation of huge vortex structures at the poles (Limaye et al., 1982; Marov and Grinspoon, 1998; Titov et al., 2004). Unfortunately, the quantitative estimates of the heat exchange components in the total energy budget of the atmosphere are quite limited, especially in the part related to the role of planetary dynamics.

We need to know the optical properties of the main atmospheric components, including the contents of minor admixtures, and their altitude distribution for radiative transfer calculations. On the other hand, it allows us to understand how a stable temperature regime on Venus is established and supported, implying again the unique mechanism of self-organization in nature.

Infrared radiation is absorbed by almost all gases of the Venusian troposphere. As has already been noted, the main component is carbon dioxide, while nitrogen, whose relative abundance does not exceed 3%, can contribute only at very high pressures through the emergence of an induced dipole moment or through the formation of simultaneous transitions with CO_2-type triatomic molecules. Minor admixtures, primarily water vapor and sulfur-bearing compounds, contribute considerably more, although their relative content (mixture ratio) accounts for hundredths or even thousandths of a percent. Given the concentration and characteristics of the infrared spectrum, the following sequence of gases is efficient in degree of contribution to the total opacity of the Venusian atmosphere: CO_2, H_2O, SO_2, COS, CO, H_2S; in other words, apart from CO_2, H_2O and SO_2 play the most important role (Marov et al., 1989c). However, the influence of these gases on the transfer of thermal radiation is determined by whether their absorption bands fall into the regions of the infrared spectrum where the opacity of the carbon dioxide atmosphere is comparatively low, thereby blocking the CO_2 "transparency windows". The mixture ratios in a homogeneous medium are kept constant only for the components that are not subjected to any chemical or phase transformations in the range of temperatures under consideration. These include water vapor and sulfur dioxide, whose contents change with altitude, especially in the cloud layer.

The optical properties of CO_2 are determined mainly by a number of strong - vibrational-rotational bands and several weaker bands that are produced by asymmetric isotopes. Because of a significant CO_2 content in the path of the ray ($\sim 10^9$ atm · cm) and a high temperature, the Venusian atmosphere is opaque in the region of intense bands and radiation is transferred between them in the transparency windows mentioned above. The absorption in them is attributable to the far wings of strong bands and to the weak bands falling into these windows (Marov et al., 1989b). The induced CO_2 absorption spectrum whose bands fall within the segments without absorption under normal conditions (except for the weak bands of asymmetric CO_2 molecular isotopes) also exerts a strong influence.

The optical properties of water vapor are also determined by several bands, among which the 6.3 and 2.7 μm bands have the greatest integrated intensity. Since, in contrast to the CO_2 molecule, the nonlinear H_2O molecule has a constant dipole moment, the vibrational band located in the spectral region near 50 μm is also active in absorption. Although the detailed calculations of the intensities of rotational water vapor lines and their full widths at half maximum are available, the transmission models in relatively narrow spectral ranges are most suitable for the conditions of the Venusian atmosphere.

Sulfur dioxide contributes less to the optical properties of the Venusian atmosphere than water vapor; nevertheless, it should be taken into account. Just like the H_2O molecule, the SO_2 molecule is an asymmetric top and several of its characteristic absorption bands lie in the spectral range from 3.5 to 19 μm. Unfortunately, the available data on the characteristics of rotational lines are barely suitable for the Venusian conditions and, therefore, one has to introduce several assumptions and use an approximate method to calculate the SO_2 transmission spectrum, just as for H_2O (Marov et al., 1989c; Marov and Grinspoon, 1998).

The pattern of heat exchange in the planetary atmosphere is also strongly affected by cloudiness, as on Earth. Meanwhile, as calculations showed, the contribution of cloudiness to the effective heat flux near the surface and in the lower troposphere of Venus may be neglected in regions where the molecular Venusian atmosphere is highly opaque. At the same time, it should be taken into account at high altitudes (in the cloud layer and near its lower boundary). The calculation of radiation transfer in clouds is determined by the properties of aerosols, which is an independent, fairly complex problem. When using the currently available, fairly reliable data on the microphysical properties of three modes of cloud particles in the range of sizes approximately from 1 to 8 μm (Marov et al., 1976; Marov et al., 1980; Esposito et al., 1983; Grinspoon et al., 1993), these calculations allow the profiles of the heat fluxes inside the cloud layer to be refined. Parameterization of the optical characteristics of clouds considered as a surface with a certain degree of reflection or thermal radiation serves as an alternative approach. This parameter can be arbitrarily attributed to the lower cloud boundary that is located at an altitude of 48-49 km.

As we see, modeling the transfer of infrared radiation in the Venusian atmosphere for determining its thermal regime involves great difficulties mainly because of the insufficiently complete data on the medium's opacity in a wide range of temperatures and pressures. Nevertheless, we were able to ensure an acceptable accuracy in determining the absoprtion spectrum for the main atmospheric components through the methods we developed (Marov and Shari, 1973; Marov et al., 1984, 1985, 1989b) and comparison of our series of calculations with the measurements and approaches of other authors (Pollack et al., 1980; Crisp and Titov, 1997).

We calculated the transfer of thermal radiation in the mid-latitude Venus atmosphere in more than 500 wavelength ranges with a high spectral resolution. Some results of these calculations based on the above cited works by Marov and collaborators are presented below. In recent years, the algorithms for calculating the form factor of a spectral line in CO_2 bands were developed owing to a considerable increase in computing power. The transmission functions that were successfully used in the calculations of the Venusian thermal regime were determined on this basis by taking into account the far wings of the bands (Afanasenko and Rodin, 2005). However, they did not change the main results we derived from our calculations because these are based on the only direct measurements made by the Venera landers available to date. Similarly, our calculations were not affected by the new set of measurements from the Venus Express orbiter, which mostly concerned the upper cloud atmosphere.

Almost the entire equilibrium atmospheric radiation at a surface temperature $T_s = 735$ K is concentrated in the infrared spectral range from 1.5 to 1000 μm, with the intensity peak being at 4 μm and shifting to 8.2 μm near the lower cloud boundary at a temperature of 365 K. In the spectral range under consideration, the equilibrium radiation net flux accounts for more than 99.7% of the total radiation. As calculations showed, in a purely carbon dioxide atmosphere, the outgoing flux of thermal energy from the surface and the lower atmosphere turns out to be very high, exceeding considerably the incident flux of energy to a rotating planet from the Sun at maximum (157±6 W/m^2 at the solar constant $E =$ 2621 W/m^2 and $A =$ 0.76±0.01), which contradicts the physical meaning. Hence it follows that the absorption of thermal radiation in the subcloud Venusian atmosphere is also determined by other components than CO_2, primarily H_2O vapors.

The efficiency with which the transparency windows of a purely carbon dioxide atmosphere are blocked by water vapor at its relative contents of 10^{-5}, 10^{-4}, and 10^{-3} and by sulfur dioxide at its maximum relative content of $2 \cdot 10^{-4}$ corresponding to the measurements within the range (0,4-2) .N>2. 10^{-4} was investigated at three levels of altitudes corresponding to temperatures of 300, 500, and 700 K (Bezard et al., 1993; Pollack et al., 1993; Jenkins et al., 2002). As expected, the presence of H_2O vapors affects the degree of atmospheric opacity for the outgoing thermal radiation particularly strongly. The contribution of a minor SO_2 admixture to the total absorption is much less than that of H_2O vapors. The role of H_2O is very significant for increasing the absorption in the ranges from 1.7 to 1.9 μm, from 2.3 to 2.7 μm, from 2.9 to 3.8 μm, from 4.9 to 9.1 μm, and from 20 to 200 μm. In this case, the influence of H_2O decreases as the surface is approached (see Fig. 1.3.8). Concurrently, the influence of the latitude dependence of the atmosphere's physical parameters is reduced at altitudes below 33 km.

Calculations for a wet carbon dioxide Venusian atmosphere revealed five main spectral ranges in which the heat flux is removed even at the maximum possible relative water vapor content of 0.1%: 1.6-1.92, 2.08-2.67, 2.86-4.00, 4.65-9.09, and 18.2-1000 μm. Water vapors screen the energy efflux in the 4th and 5th windows even at a relative content of 10^{-5} considerably, while such an H_2O concentration still has virtually no effect on the radiation transfer in the 2nd and 3rd windows. The 4th window makes a decisive contribution to the flux at altitudes above 35 km at all H_2O contents considered up to 5×10^{-3}. This contribution decreases toward the planetary surface and vanishes close to it at an H_2O content of $\sim 10^{-3}$. The maximum contribution to the total flux in the 5th window does not exceed 5% in a wet atmosphere and is virtually zero below 30 km. The fluxes in the 2nd and 3rd windows are close below 20 km and differ maximally at ~35 km. The fluxes in the 1st window decrease similarly with increasing water vapor content in the atmosphere. The constraints on the limiting content of water vapor in the lower Venusian atmosphere and its altitude distribution (Marov et al., 1985; Marov et al., 1989b)

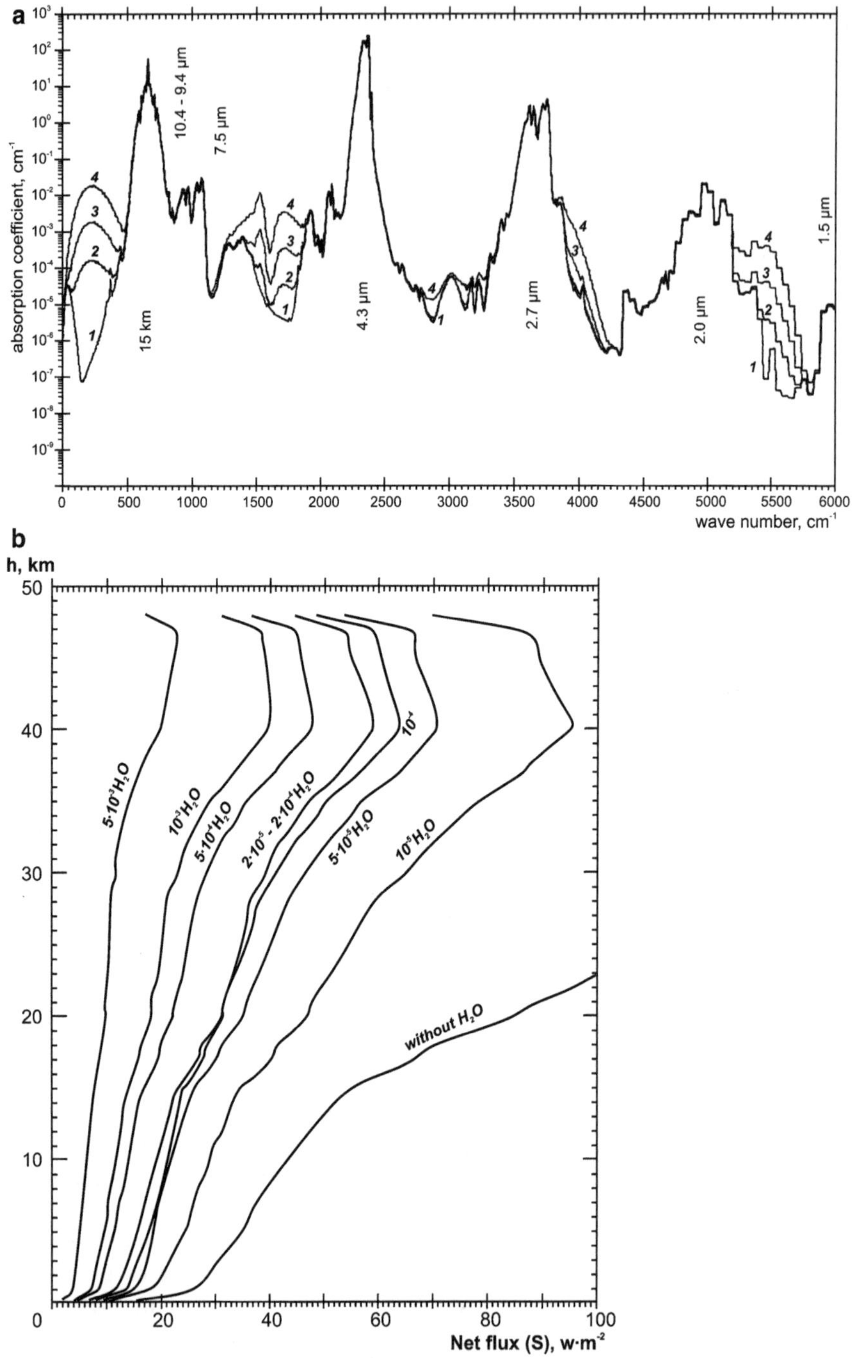

Fig. 1.3.8 Modeling the thermal regime of the Venusian atmosphere. (**a**) The opacity of the Venusian carbon dioxide atmosphere in the thermal infrared spectral range under temperature 500 K. The main absorption bands and transparency windows partially overlapped by the addition of water vapors (1-4) are distinguished. (**b**) The fluxes of thermal radiation in the lower atmosphere for a purely carbon dioxide atmosphere and at various relative contents of water vapor. (c) The contribution from individual CO_2 bands to the transfer of thermal radiation (different bands 1(1a) – 5(5a) within wave numbers 10-6000 cm^{-1}). (**d**) Comparison of the calculated net radiation fluxes with those measured by the Pioneer-Venus probes. Curves 1-5 correspond to the water vapor admixtures 10^{-5}, 5 10^{-5}, 10^{-4}, 5 10^{-4}, and 10^{-3}

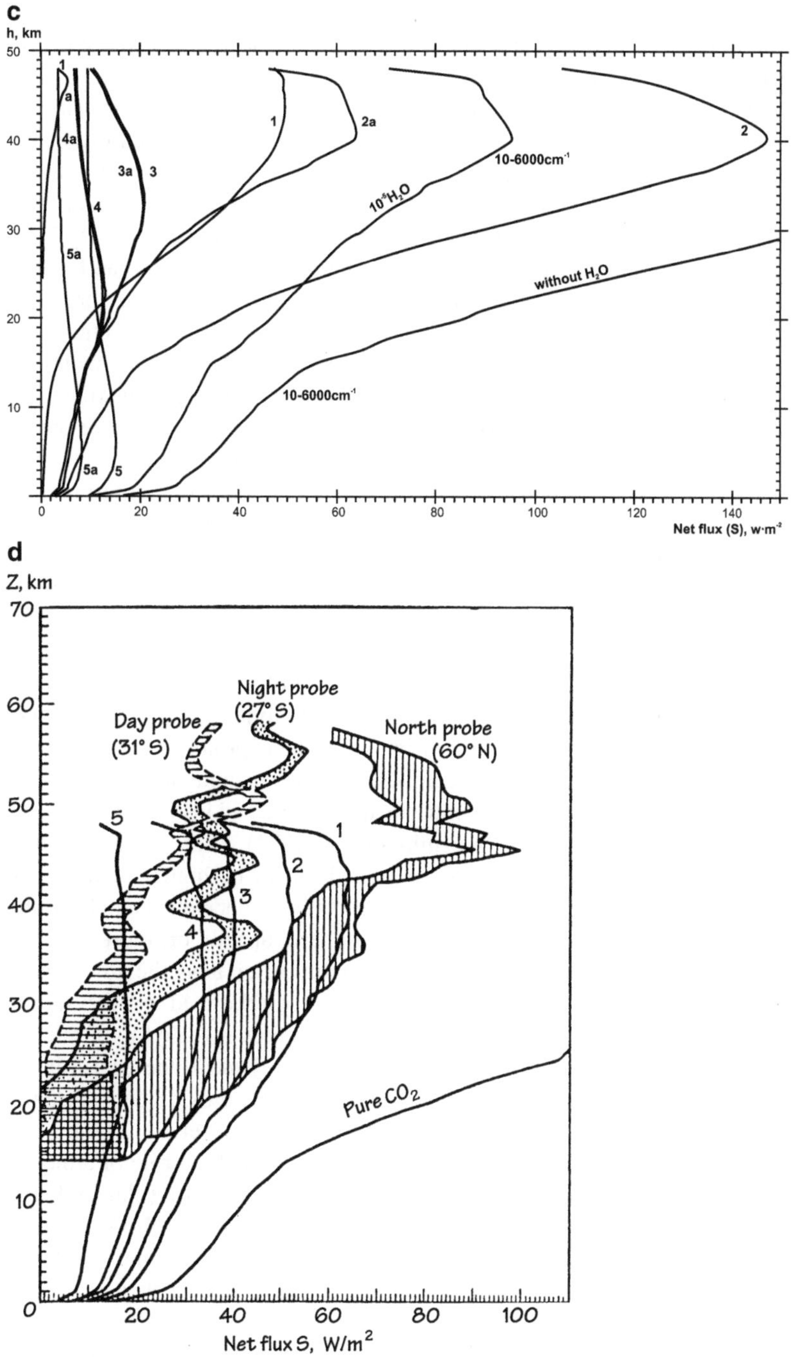

Fig. 1.3.8 (continued)

were obtained by comparing the calculated total thermal radiation profiles at various H_2O:CO_2 mixture ratios with allowance made for the ambiguity in the measured values (Oyama, 1980; Suomi et al., 1980; Moroz et al., 1983).

The contribution of sulfur dioxide to the heat flux screening depends strongly on altitude. In lower layers, up to an altitude of 20 km, the influence of SO_2 on the flux is weak. At the same time, the flux decreases by several times at altitudes above 40 km, which leads to its more monotonic dependence on altitude. This is because the main contribution to the atmospheric opacity is made by the SO_2 7.35 and 8.69 μm bands located in the region of the induced CO_2 7.5 μm band, while the absorption in it decreases with altitude proportionally to the pressure squared. Accordingly, the opacity of the 0-20 km atmospheric layer is attributable mainly to CO_2 and H_2O, while SO_2 makes a significant contribution above this level.

As we see, the modeling results have a fairly high accuracy. Nevertheless, they allow mainly only qualitative conclusions about the pattern of heat exchange on Venus. The reasons for this include the remaining uncertainties in the opacity estimates for a carbon dioxide atmosphere with admixtures of minor components in a wide range of partial pressures and temperatures and the optical properties of cloud aerosols. The nature of the imbalance between the calculated heat fluxes and the measured altitude profile of the incoming solar radiation in the middle troposphere can be interpreted in different ways even within $\sim$ 10-20% error limits. The radiometric heat flux measurements on the large and three small Pioneer-Venus probes did not clarify the situation either because of large errors of the onboard radiometers (Revercomb et al., 1985). At the same time, these measurements did not reveal any noticeable differences in the altitude profiles of the heat fluxes measured on daytime and nighttime probes. This seems quite obvious, considering the enormous heat content of the Venusian atmosphere and insignificant heat losses during night (58.5 Earth's days) (Kuzmin, Marov, 1974). At the same time, the heat flux measured on a high-latitude probe turned out to be higher than that on an equatorial one. This suggests the existence of a latitude dependence in heat transfer that is attributable to the differences in atmospheric opacity or, more specifically, that the high-latitude atmosphere is considerably drier than the low- and mid-latitude one. The validity of this conclusion is corroborated by the significant latitude dependence of the vertical temperature profiles revealed by a microwave sounding of the Venusian atmosphere (Kliore, Patel, 1980). The heat flux profile near the lower cloud boundary may be affected by sulfur dioxide, whose content in the polar regions is half that in the equatorial ones (Jenkins et al., 2002).

To judge the specificity of radiative heat exchange and the role of atmospheric dynamics, it is important to compare the altitude profiles of solar and thermal radiation, each of which is the downflow-upflow difference (net flux). The obvious difference in heat influx from the Sun to the equator and poles on Venus is apparently not compensated for by the outgoing radiative heat flux caused by the change in the contents of H_2O and SO_2 with latitude. In other words, the heat losses at low latitudes are smaller at a higher influx of solar radiation, while the reverse is true at high latitudes. If we take into account the fact that the diurnal variations in the thermal regime of the lower atmosphere are small, then an explanation of this phenomenon can be found in the mechanism of planetary circulation through which a redistribution of heat and, accordingly, additional heating or cooling of the lower atmosphere occurs.

In addition, by comparing the measurements of solar incident radiation and outgoing thermal radiation with model calculations, it is easy to verify that the latitude gradient of infrared cooling in the subcloud atmosphere considerably exceeds the latitude gradient of solar heating at the same altitudes. This can serve as an important energy factor of the planetary circulation mechanism, whose distinctive feature is the global superrotation of the Venusian atmosphere. Indeed, the radiative heat transfer reflects the pattern of heat exchange between the equator and poles with the formation of Hadley cells and upward motions near the equator (Rossow et al., 1980; Limaye et al., 1982; Schubert et al., 1983). In this case, the significant heat efflux at high latitudes can be caused by downward motions of atmospheric gas accompanied by backward flows in the lower atmosphere in the meridional direction. The emergence of superrotation and the formation of vortices near the poles, in which the angular momentum for atmospheric circulation is generally conserved, must be associated with the same mechanism.

Let us now return to the question with which we began our discussion of the formation of the thermal regime: Why did Venus evolve in a different way than Earth?

According to the existing views (see, e.g., Marov, 1986; Marov and Grinspoon, 1998), Earth is located in a comparatively narrow zone of circumsolar space where the development of favorable (for the existence of life) climatic conditions is possible. Its inner boundary lies closer to the Sun only by 10-15 million km, while its outer boundary reaches approximately the orbit of Mars. The orbit of Venus turns out to be outside this zone, at a distance that is almost triple the critical value. In other words, if the Earth were moved to the place of Venus, it would probably evolve according to the Venusian scenario.

Indeed, if the Earth's initial albedo were assumed to be entirely determined by the surface, and if it corresponded to the lunar one (A=0.07), its equilibrium temperature at the current solar luminosity would not exceed 275 K. At this temperature and at a comparatively low pressure ($\sim$ 0.005 atm), the Earth could retain the water that was concentrated in the primordial water reservoirs and atmosphere. In turn, carbon dioxide would accumulate in the terrestrial hydrosphere as would carbonates of sedimentary rocks through its binding with metal oxides incorporated into the ocean crust and upper mantle (with the formation of aqueous silicates) and biogenically, through the deposits of the lime skeletons of sea organisms.

Based on the average annual balance between the solar radiation absorbed by the terrestrial surface and the thermal (bolometric) radiation emitted by the Earth at the reduced luminosity of the young Sun, the effective temperature turns out to be 255 K, which is below the freezing point of even salt water. Consequently, we can assume that the greenhouse effect noticeably contributed already at the earliest evolutionary stages of the terrestrial atmosphere. At the same time, the mean temperature near the terrestrial surface at the present epoch due to the greenhouse effect is 288 K (Mitchel, 1989). A similar situation may have existed on early Mars if the hypothesis of its initially favorable climate is adopted; otherwise, Mars would be unable to retain liquid water on its surface, because its equilibrium temperature does not exceed 220 K.

For Venus, at the same postulated initial albedo, the equilibrium temperature turns out to be at least 325 K, which is above the water boiling temperature down to a pressure of 0.2 atm (Marov, 1986). Thus, to retain water on its surface, Venus would have needed to possess an initial atmosphere denser than that of the Earth by almost two orders of magnitude. At equal rates of mantle material degassing and atmospheric dissipation into the surrounding space, this is unlikely. More likely, we can assume that carbon dioxide gradually accumulated in the atmosphere together with water vapors. This in turn contributed to an additional rise in surface temperature through the greenhouse effect and to the transport of increasingly large amounts of CO_2 and H_2O into the atmosphere, up to some equilibrium state that is characterized by certain relationships between the mineral phases and volatiles on the surface, of which the carbonate-silicate interaction in the upper layer of the planetary crust is the most important.

In other words, it could happen what is characteristic of a system with a positive feedback when the system is unstable and the initial perturbation is not suppressed but, is instead enhanced fairly rapidly. Therefore, this development of events on Venus, which eventually led to such a high surface temperature, was not called a runaway greenhouse effect by chance. At this temperature, carbon dioxide turned out to be not bound in carbonates of sedimentary rocks, as on Earth (and, probably, on Mars), but was entirely released into the atmosphere, which produced this high pressure (Vinogradov and Volkov, 1971). According to the existing estimates, the amount of carbon dioxide locked in the Earth's sedimentary cover is comparable to the content of CO_2 in the Venusian atmosphere. It cannot be ruled out that the carbonates decompose and the atmospheric pressure rises sharply in consequence if an unfavorable scenario for the Earth's heating actually comes true.

Whereas the abundance of carbon dioxide on Venus can be explained in terms of a fairly simple equilibrium model (though the real geochemical processes involving sulfur-bearing and other compounds are more complex), the situation with water is much more difficult to understand. Under the assumption of a "geochemical similarity" between the evolution processes involving endogenic and exogenic processes, the volumes of volatiles and, accordingly, the reserves of water on Venus must have corresponded to the volume of the terrestrial hydrosphere, which is approximately 1370 million km^3 or more than $1.37x10^{24}$ g. Meanwhile, water is not preserved on the Venusian surface at a temperature above the critical one (647 K), which also remains valid for aqueous solutions (brines) with a slightly higher critical temperature (675-700 K). As regards the atmosphere, the amount of water in it at an average relative water vapor content of $5x10^{-4}$ does not exceed $3.5x10^{20}$ g. This is considerably higher than the total water content in the terrestrial atmosphere but is less than the reserves of water in the hydrosphere by almost four orders of magnitude.

Clearly, whether water was present on Venus and in what amounts, and if it was present, when and how it was lost, remains a key question in the problem of the evolution of Venus and its atmosphere. The argumentation seems quite justified that the Earth and Venus received approximately the same reserves of volatiles, including water, through degassing from the mantle that accompanied the interior differentiation process as well as from heterogeneous accretion through the migration of planetesimals from the formation zone of the giant planets and their fall onto the surfaces of inner planets in the form of comets and asteroids (Marov, Ipatov, 2001). Accordingly, one might assume Venus to have possessed an ocean at the initial phase of its evolution that was comparable in volume to the terrestrial one, and one might furthermore assume that it was lost subsequently. The deuterium-to-hydrogen ratio measured in the Venusian atmosphere, which turned out to be higher than that in the Earth's atmosphere by two orders of magnitude (Donahue et al., 1982), serves as a confirmation of this concept. This high degree of enrichment can be explained by an efficient thermal escape of the lighter hydrogen isotope during the evaporation of the primordial ocean and the dissociation of water molecules by solar ultraviolet radiation. However, the ocean loss mechanism itself is more difficult to explain because it requires the evacuation of enormous masses of hydrogen from the atmosphere and the binding of oxygen that is liberated by surface rocks.

Yet a different assumption is that Venus was initially formed as a "dry" planet. However, the introduction of volatiles through the material of comets and carbonaceous chondrites is assumed to play an important role in this case as well. In this scenario, the water content was essentially constant throughout the geological history of Venus, remaining approximately at the current level, while the efficiencies of heterogeneous accretion and hydrogen dissipation from the atmosphere were considerably lower. The observed degree of enrichment of the Venusian atmosphere with deuterium during the separation of isotopes in the process of their dissipation from the atmosphere can also be reproduced within this scenario (Donahue et al., 1997; Marov and Grinspoon, 1998).

Evidently, it is difficult at present to answer the question of whether Venus had a primordial ocean and following from this, to choose between the scenarios for its evolution because there are only limited possibilities for backing the models by experimental data. Both scenarios basically lead to the development of a runaway greenhouse effect and the present-day climatic conditions. At the same time, the possibility that Venus possessed a more temperate climate at the earliest stages of its evolution must not be excluded because there may have been a negative feedback that stabilized some equilibrium state instead of the positive feedback that enhanced the greenhouse effect. The negative feedback may be attributable to the atmosphere-lithosphere interaction pattern controlled by the carbonate-silicate cycle (Kasting et al., 1988) and may have existed on Venus in a certain period, before the luminosity of the Sun rose by 20% (when it passed to the main sequence on the Hertzsprung-Russell diagram) and before some atmospheric humidity threshold was reached. The mechanism itself was called a "wet" greenhouse effect. As calculations showed, the transition from the wet greenhouse effect to the runaway would have been possible but may not have happened if the increase in the planet's albedo to its current value occurred earlier than the rise in solar luminosity and thus compensated for the increase in the influx of solar energy.

Venus might have possessed a wet carbon dioxide atmosphere with a pressure near the surface of several atmospheres and a temperature of less than one hundred degrees Celsius. Such a planet could be suitable for the emergence of primitive life forms on it.

This consideration again leads us to the idea that some spectrum of possibilities containing both deterministic and stochastic components existed in the evolution of Venus, which eventually led to the runaway greenhouse scenario. Several random processes that led to the system's bifurcation from the initial state to the existing one can be distinguished: The sequence starting from the planet's accretion at a certain distance from the Sun, the orientation of the rotation axis in space, the decomposition of carbonates and the loss of water, the retention of minor components in the atmosphere that are maintained by volcanic activity and provide its high opacity, the circulation of sulfur-bearing compounds and halogens between the surface, atmosphere, and clouds, the pattern of geological evolution and lithosphere-atmosphere interaction, and the formation of peculiar features of the planetary circulation. It is believed that in the absence or disruption of any of the listed components, this nonlinear dissipative system would pass to a different stable state with natural conditions different from the current ones.

Unfortunately, the current state of Venus seems more stable than the climatic state of the Earth. On the one hand, the concerns for the consequences of an unfavorable impact of mankind on the natural environment and, on the other hand, the doubtful but nevertheless widely discussed (see, e.g., Dyson, Freeman, 1989; Pollack, Sagan, 1991; Gillett, 1991) possibility of attempting to change the climate (terraforming) of Venus in the distant future are justifiably associated with this fact. In this connection, we recall the statement of the Nobel laureate A. Bergson as quoted by I. Prigogine in the book "The End of Certainty: Time's Flow and the Laws of Nature" (1997), that the human existence consists of the "continuous creation of unpredictable innovation". In other words, this is an example of a nonequilibrium situation in which dissipative structures and a new level of self-organization can emerge.

Mars

From the viewpoint of evolution, Mars is the antipode of Venus and is another opposite of the Earth, although it is more similar in its natural properties to our planet and may have been a close analog of the Earth at early evolutionary stages. The most topical problems that now face the researchers of Mars include studying the short-term and long-period variations in climate, the geological and chemical composition of the rocks forming the surface layer of the planetary crust, the seasonal cycles of the water and carbon dioxide contents in the atmosphere, searching for the distribution of various forms of water both at depth and in the subsurface layer – liquid, bound, and water ice. In general, the construction of various climatic, geological, and evolutionary models for this planet and the presence of possible traces of primitive extant or extinct life (bacterial fossils) are associated with the problem of water in the geological past and present of Mars.

Naturally, considering numerous aspects of the nature of Mars in the studies of which enormous progress has been achieved in the last decades by space missions is not among our objectives. This was primarily contributed by the successful flights of the Mars Global Surveyor, Mars Reconnaissance Orbiter, Mars Odyssey, and Mars Express, and the long work of the Spirit and Opportunity Martian rovers on the surface. An overview of their results and developed model approaches would require a considerable expansion of the list of literature. Therefore, we will restrict ourselves in the references to the main reviews and monographs (see, e.g., Kieffer et al., 1992; Carr, 1981; 1990; 1999a, b; 2000; Jakosky and Jones, 1997; Fanale, 1999; Hartmann, 2003; Hartmann and Neukum, 2001) where the necessary source material can be found. Here our primary interest is focused on the evolution of Mars as an open nonlinear dissipative system with the account for influence of separate factors or their collection on the transition to a stable state of the natural environment.

A main concern is answering the question of which mechanisms might have had a critical impact on the natural conditions that presumably existed on Mars at an early epoch, which is backed by the morphological features preserved on the surface. They suggest the presence of liquid water flows on the surface in the case of a considerably denser atmosphere and, consequently, a much more favorable climate. If these conditions actually existed, then the nature of the present-day dry (the content of H_2O in the atmosphere does not exceed 10 μm of precipitated water) and cold (the mean temperature is 210 K) Mars cannot be explained by the fact that it is 0.5 AU farther away from the Sun than the Earth. It should more likely be associated with its size, which is approximately half that of the Earth. Accordingly, the mass of Mars is almost an order of magnitude lower than that of the Earth, which should have led to early depletion of radiogenic isotopes. These had a strong influence on Mars' thermal history, the geology and the eventual collapse of the atmosphere. In contrast to Venus, the preserved thin carbon dioxide atmosphere of Mars (95% CO_2, 2.7% N_2 and 1.6% Ar by volume) with an average surface pressure of 6.1 mbar (a factor of 160 lower than the Earth's one) produces no noticeable greenhouse effect. Therefore, the seasonal-diurnal temperature variations exceed 150 degrees, from +20°C in some regions near the equator to -130°C on the winter polar cap, where dry CO_2 ice condenses. However, an assumed ancient atmosphere comparable in density to the terrestrial one could well provide a sufficiently high surface temperature capable of retaining liquid water and even its circulation between the surface and atmosphere (Squyres and Kasting, 1994; Carr, 1999,a,b). Interestingly, the present-day tenuous atmosphere raises the surface temperature through the greenhouse effect only by 4 K and hence strong climatic variations occur (Rodin and Wilson, 2006).

The present-day Mars is a cold desert world with a plethora of craters, systems of mountain ridges, plateaus, uplands and valleys that retained the traces of a paleomagnetic field, ancient wide-spread volcanism, destroyed igneous rocks, and the influence of atmospheric dynamics on the surface landscapes – weathering

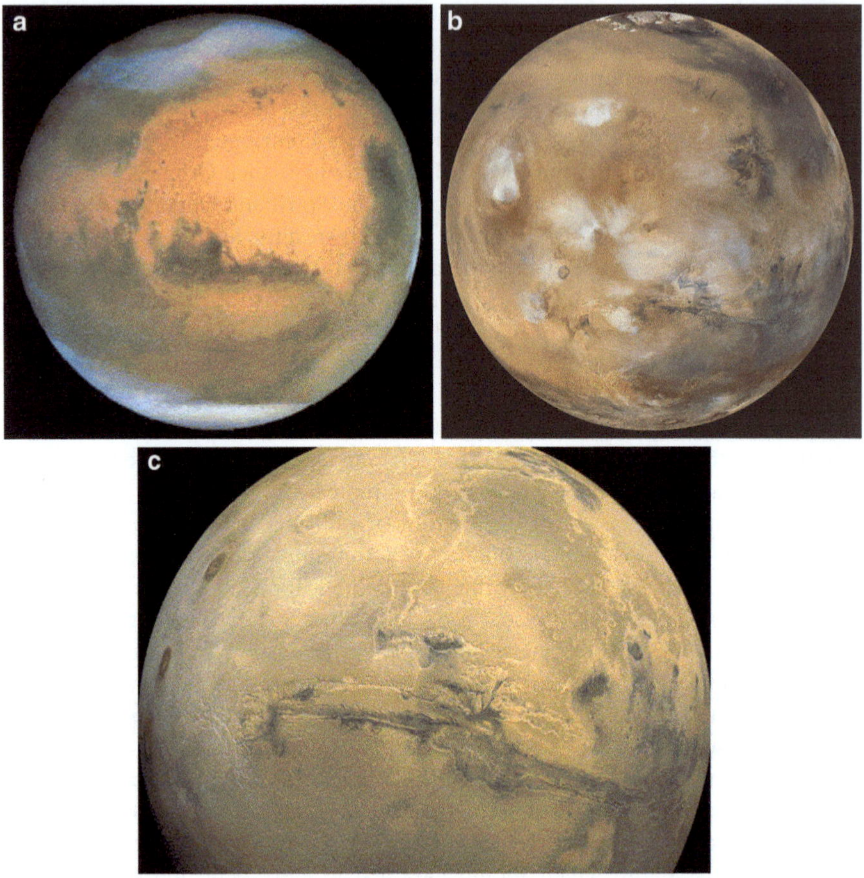

Fig. 1.3.9 Images of Mars from spacecraft. **(a)** Image of the Martian surface on which differences in the albedo of the Northern and Southern Hemispheres, heavily cratered regions near the equator, and clouds above the polar regions are clearly seen. **(b)** Clouds above the huge shield volcanoes in the Tharsis region, relief of the Northern polar region, and Valles Marineris rift zone are distinguished in this image. **(c)** Mosaic of images in which the main features of the Martian relief, primarily the highest in the Solar system shield volcanoes with a height up to 26 km above the mean surface level in the northwest and the huge tectonic fracture Valles Marineris extending for more than 3000 km nearly along the equator having width of more than 100 km, and depth up to 8 km, are clearly seen. Courtesy of NASA

(Figs. 1.3.9 and 1.3.10). There is no doubt that the surface and atmosphere of Mars changed over its geological history as a result of intense impact bombardment and volcanic, tectonic, and erosion processes. Tectonics probably affected the evolution of the ancient Mars (Anderson et al., 2001). The appearance of a hydrological cycle and secondary atmosphere 3.9-3.8 Gyr ago is associated with tectonic and volcanic processes. Under the action of hydrologic and glaciologic processes, along with atmospheric weathering, the craters severely eroded and the Martian landscapes were subjected to modification. In particular, the degree of crater erosion suggests

Fig. 1.3.10 Panorama of the Martian surface at the Pathfinder landing site. Courtesy of NASA

that it occurred under conditions of a dense atmosphere – the present-day atmosphere could not have such a destructive impact. The craters were destroyed most strongly by weathering processes involving wind and water. This took place more than 3.5 Gyr ago, before the catastrophic collapse of the atmosphere, so that the oldest craters were virtually wiped off the face of Mars (Brain and Jakosky, 1998; Golombek and Bridges, 2000).

The huge layers of dust-sand material hid many of the original structures, including large deposits of subsurface ice formed after the climate change (Mangold and Allemand, 2001). The presence of these deposits at a depth of about 1 m predominantly at high latitudes was confirmed by neutron spectrometry measurements during a monitoring from the Mars Odyssey satellite orbit (Mitrofanov et al., 2004; Litvak et al., 2006; Tokano, 2005; Kuzmin et al., 2007). The ice content in rocks reaches 50% by mass. The individual relief features, especially those on slopes, were probably formed through water flows, among others, and there are features detected on several surface structure that one would like to associate with periodic seepage of subsurface water in the recent past, see Fig. 1.3.11 (Malin and Edgett, 2000). With a high probability, one may expect even larger reserves of water in the form of ice lenses and interlayers to have been preserved at a depth of several hundred meters. It cannot be excluded that there may be liquid water near the lower surface of a thawing lens due to the accumulation of heat through an internal heat flux and because of very low thermal conductivity of ice (Marov, 1986; Clifford and Parker, 2001).

Many of the signatures that are an integral feature of Martian geology suggest that water furrowed the surface of Mars for several hundred million years of its history (probably during a period from about 3.8 to 3.5 Gyr ago). First of all, these include the systems of valleys and gullies that resemble the beds of dried-up rivers with numerous tributaries extending for hundreds of kilometers, some of which can

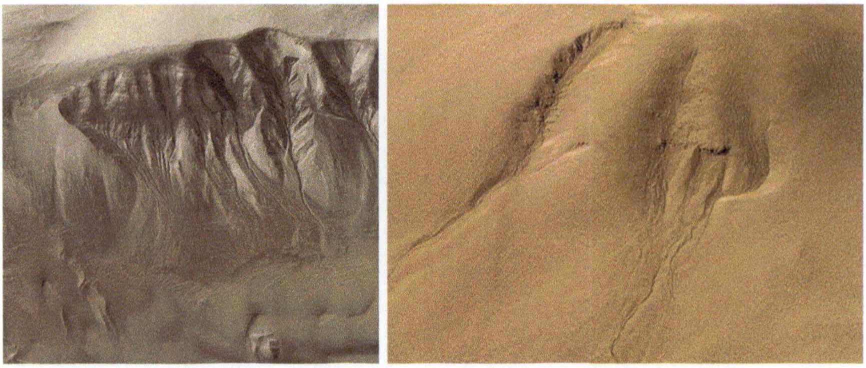

Fig. 1.3.11 Examples of water erosion on Mars. The fissures that were most probably formed by water flows are seen on the left; the gullies that could be formed due to the emergence of water on the surface (seepage) are on the right. Courtesy of NASA

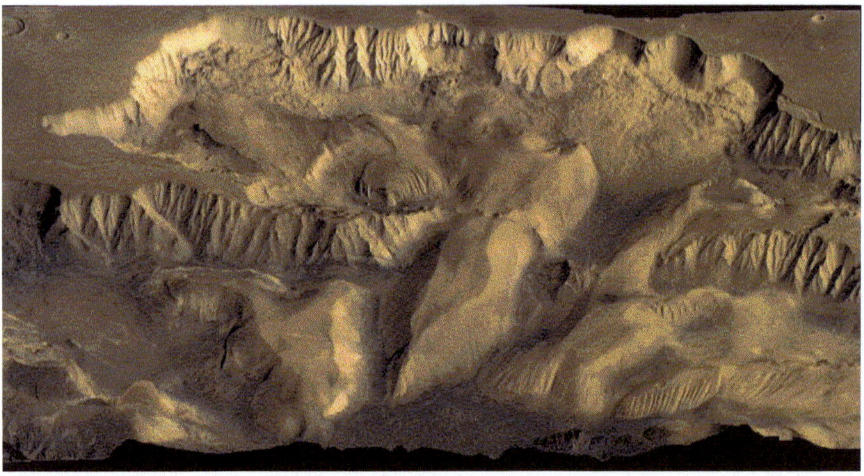

Fig. 1.3.12 The Candor region in the central part of Valles Marineris on Mars. Isolated more ordered structures whose formation can obviously be associated with giant water flows are distinguished against the background of chaotic morphology attributable to the early tectonic-volcanic processes that produced large canyons, mensae, and plains. Courtesy of the US Geological Survey and NASA

be associated with vigorous water flows produced by the melting of subsurface ice during the denudation of ice lenses or can be likened to the motions of Antarctic glaciers (Gulick, 2001; Lucchitta, 2001). The region to the north of Elysium, where water might have risen to the surface through faults in the Martian crust, serves as an example of such structures. In turn, the eroded traces of flows on slopes that might have been produced by the fall of heavy rains or even torrents and where order was formed against a general chaotic background of the relief are detected in regions of ancient tectonic activity (Fig. 1.3.12).

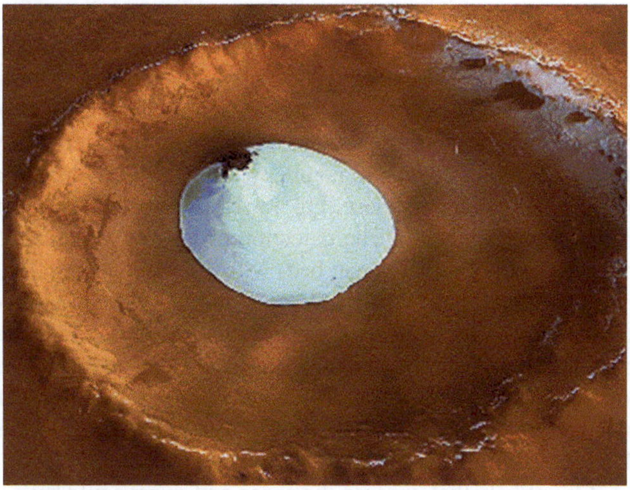

Fig. 1.3.13 Icy lake at the bottom of a Martian crater. An image from the Mars Express spacecraft (ESA courtesy)

Meanwhile, younger gullies with an estimated age of not billions, but millions of years were also detected on the inner slopes of impact craters or on the walls of deep troughs, mainly on their southern side. These relief features may be associated with periodic changes in the inclination of the Martian rotation axis within the range from 15° to 35° on a time scale of $\sim$10 Myr attributable mainly to the influence of Jupiter's strong gravitational field. On even longer time scales, the perturbations might have become chaotic in nature, so that the inclination of the rotation axis changed from 0 to 60°. This creates prerequisites for enormous climate changes caused by a change in insolation at the poles and in the intensity of the transfer of volatiles into the atmosphere from the polar caps that are composed of water ice, along with the deposits of carbon dioxide, regolith, and permafrost in the winter hemisphere (Ward et al., 1974; Ward and Rudy, 1991; Phillips et al., 2001). The relief of high-latitude regions is consistent with this possibility. In particular, it was found that the surface between the poles and the equator is overlaid with sedimentary rocks with a thickness of 4-6 km in the north and 1-2 km in the south, while the surface itself is crisscrossed by cliffs and fractures that, as it were, are wound around the poles. The sedimentary cover itself has a layered structure, backing the assumption about periodic climate changes. Finally, a number of morphological features can be associated with the appearance of liquid water on the surface considerably later than 3.5 Gyr, which requires the presence of a greenhouse effect in a much denser atmosphere than the present-day one. The detection of a large ice lake inside one of the craters on an image from the Mars Express spacecraft serves as a convincing example (Fig. 1.3.13).

The relief structures in Martian northern regions revealed by observations from the Mars Global Surveyor satellite resemble the coastlines of an ancient ocean bounding the regions of constant surface height. This can be explained by uniform precipitation accumulation in large volumes of water on Mars' northern plains. According to the estimates of geologists, the total mean depth of the ocean could reach 0.5 km (Carr, 1999a,b; Head et al., 1999; Ivanov and Head, 2001; Malin and Edgett, 2001) and this value apparently restricts the maximal reserves of water that could be preserved on Mars, minus the losses (~30%) through aeronomy processes, as was shown by calculations of the nonthermal escape of hydrogen and oxygen atoms from the planetary atmosphere, Fig 1.3.14 (Krestyannikova, Shematovich, 2006; Shematovich et al., 2007). The detection of layered structures (resembling sedimentary rocks in the Earth's ocean bed) with a high concentration of chlorine and bromine salts at Meridiani Planum by the Opportunity Martian rover serves as confirmation for the idea of an ancient ocean on Mars (Fig. 1.3.15). These deposits can even be associated with the coastline of an ancient ocean where the cyclic processes of evaporation and/or freezing took place on shallow water, leaving behind layered sedimentary rocks rich in chlorine and bromine salts (Squyres et al., 2004; Rieder et al., 2004).

However, it is unlikely that water in the ocean and rivers could have appeared on the surface in periods of comparatively short-term climatic warmings. This is evidenced by the age of the craters near these relief shapes, which most likely date back to the first billions of years of the planet's evolution (Hartmann and Berman, 2000). At the same time, it is hard to reconcile the above data on the layered structure of sedimentary rocks at mid-latitudes with these views, so that certain contradictions remain. Moreover, they by no means rule out an episodic hydrologic activity on present-day Mars, along with preserved residual volcanism. A mineralogical mapping of the Martian surface from the Mars Express satellite contributed to resolving these contradictions (Bibring et al., 2005).

If we adopt the hypothesis about a favorable climate of the ancient Mars, then the following fundamental question immediately arises: were its present-day natural conditions formed as a result of long and complex evolution or did these changes occur essentially suddenly on a geological time scale? The composition of SNC-meteorites, particularly the D/H ratio, whose origin is associated with Mars, imposes certain constraints on the standard model of thermal evolution (McSween, 1994; Watson et al., 1994; Mathew and Marti, 2001; see also Carr, 1996). The model suggests the separation of an iron-sulfide core shortly after the end of accretion, the differentiation of constituent material into envelopes (though less complete than that for the Earth), convective heat transport in the mantle, which provided early volcanism, and the generation of a magnetic field through a dynamo in the core, as long as it remained liquid (Zuber et al., 2000; Zharkov and Gudkova, 2005). Meanwhile, it cannot be answered yet whether Mars was formed just like other terrestrial planets from the most ancient primordial matter that is assumed to constitute chondrite meteorites, nor do we know whether the ratio of the iron and silicon contents in it corresponds to the terrestrial one, which is 1.7 (McSween, 1999). Should it turn out that the composition of Mars differs from the chondrite model, for which assumption there are certain grounds, then this will be fundamentally important for the cosmogony of the Solar System. The magnetic anomalies (paleomagnetism) recorded in ancient surface rocks of the Southern

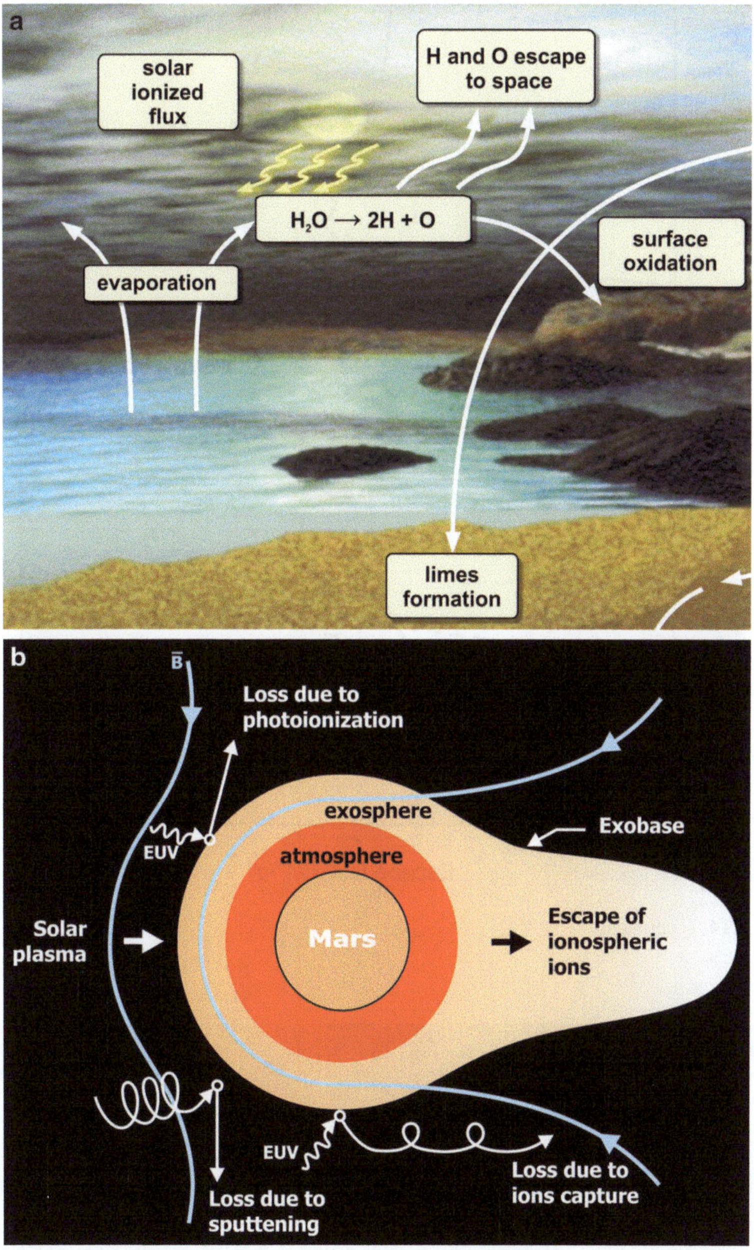

Fig. 1.3.14 **(a)** Model for the loss of water from the Martian atmosphere caused by the aeronomy processes (superthermal particles formation and escape). This mechanism could be responsible for approximately a third of the Mars ancient ocean over the geological time. **(b)** Detailed diagram of the mechanisms of hydrogen and oxygen dissipation from the atmosphere, with the formation of superthermal oxygen atoms with an energy sufficient for their escape from Mars. **(c)** The calculated fluxes and density of superthermal oxygen atoms in the upper Martian atmosphere as a consequence of the photolysis of water molecules and deactivation of the excited O atoms

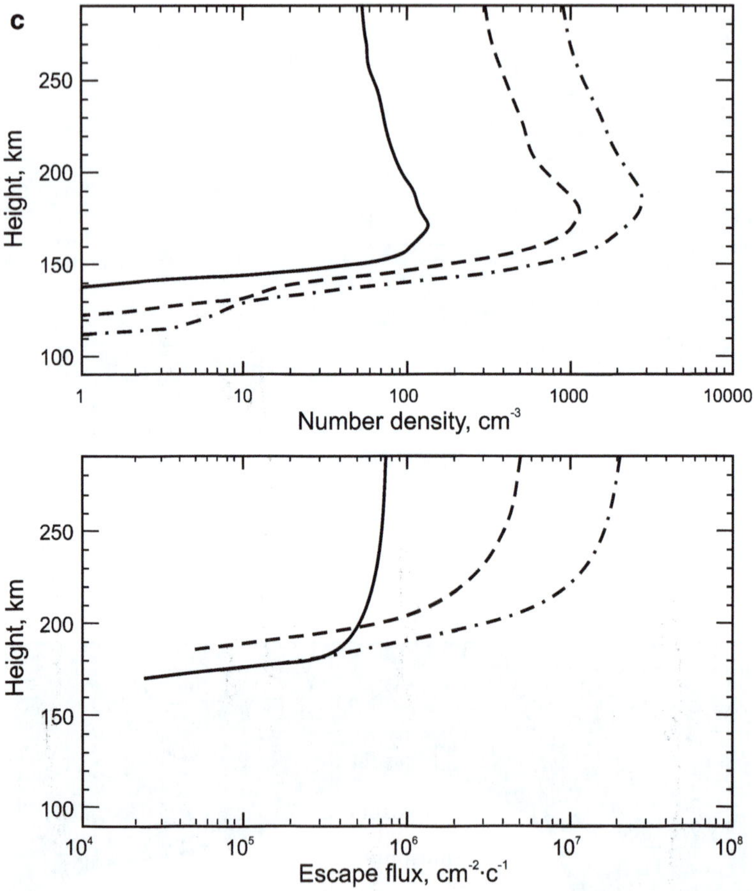

Fig. 1.3.14 (continued)

Fig. 1.3.15 These rocks at the Martian Meridiani Planum resemble those on the Earth's ocean floor. High concentrations of chlorine and bromine salts detected in them also relate their emergence with the Mars ancient oceans. The region in the images may be associated with a coastline where cyclic evaporation and/or freezing occurred on shallow water. Images from the Opportunity Mars rover. Courtesy of NASA

Hemisphere are relicts of the ancient magnetic field on Mars. These regions are enriched with gematite, which, in turn, confirms the presence of subsurface ice (Acuna et al., 1999; Christensen et al., 2000; Mellon et al., 2004). The dissipation of heat during accretion and differentiation, along with the heat released during the decay of long-lived radioactive isotopes in the interiors, obviously served as energy sources. Volcanism ceased because Mars began to cool down due to a limited reserve of radiogenic isotopes, which probably contributed to the atmosphere's reduction.

This model seems to us to be reasonably justified. At the same time, it was also hypothesized that volcanism on Mars resulted from its collision with a large asteroid that produced the Hellas basin, which measures about 2000 km across and is situated in the region antipodal to the Tharsis upland, on which the largest shield volcanoes in the Solar System are located, which are up to 26 km high. In other words, the release of magmatic rocks through a young Martian crust were triggered by just such a catastrophic event. In turn, this led to a radical change of the relief in the Northern Hemisphere and may have contributed to the formation of an atmosphere. The loss of the atmosphere and the transition of Mars from an evolutionary path similar to that of the Earth in the first billion years to a completely different scenario of development are associated with another more recent global catastrophe that was also produced by the collision with a large asteroid. It should be recognized that both these hypotheses are largely speculative, although, of course, the possibility of catastrophic events should not be ruled out.

At the same time, even without invoking such events, that the evolution of Mars was different from that of Venus and Earth was probably predetermined by its formation in a ring clump of a turbulent gas-dust cloud closer to Jupiter, where the solar constant is lower. It was also determined by its orbit having fairly high orbital eccentricity, which may be subjected to short- and long-period variations caused by the influence of Jupiter. The absence of a large satellite such as the Moon around Mars, which would be capable of stabilizing the position of the rotation axis in space (the two small asteroid-like satellites of Mars, Phobos and Deimos, cannot perform this role) should also be added to this. Finally, the higher efficiency of impact bombardment near Jupiter and the comparatively weak gravitational field of Mars, which limited the possibility of retaining a dense atmosphere, may have contributed as well. In other words, all these factors create prerequisites for a lower degree of stability of the system and its greater vulnerability to external impacts. The influence of these impacts, which undoubtedly played an important role in the history of Mars, facilitated the sequence of bifurcations of the system that is associated with forming some natural complex from a set of provisional options at a certain evolutionary stage and its follow-up transition to a new state. Mars appears to represent a perfect example of a sequence of these self-organization processes.

Giant Planets and their Satellites

In contrast to the terrestrial planets, which were formed near the Sun from the heavy high-temperature fraction of the original protoplanetary cloud material, the giant planets (Jupiter, Saturn, Uranus, and Neptune) accumulated considerably lighter elements that corresponded to the cosmic abundance of elements. Accordingly, Jupiter and Saturn are composed mainly of the most abundant hydrogen and helium; i.e., their composition is nearly solar. Only the very massive planets were capable of accreting these light gases from the protoplanetary cloud and retaining them in their composition – according to the existing estimates (Wetherill, Steward, 1989; Lissauer, 1993), a threshold mass $M > 10\ M_E$ (M_E –mass of Earth) corresponds to this scenario. Both the composition of the interiors and the composition of the protoplanetary disk material can be judged from the measurable chemical composition of the atmosphere. At the same time, the atmospheres of the giant planets are enriched with heavier elements, particularly carbon, whose abundance exceeds the solar one by a factor of 3 on Jupiter and even by a factor of 30 on Uranus and Neptune. This may be explained because they were bombarded by planetesimals for a long time after the completion of the main accretion phase (Owen et al., 1999; Atreya et al., 2003).

Hydrogen-bearing compounds (water, ammonia, methane "ices"), which entered mainly into the composition of Uranus and Neptune in various ratios, condensed most efficiently at very low temperatures in the Solar System regions between 10 and 30 AU. Even heavier elements and compounds concentrated only in the cores of the giant planets onto which lighter elements and compounds accreted (see Fig. 1.3.4b). In Jupiter and Saturn, they account for a relatively small fraction of the planetary mass, only from 3-15%, reaching 60-80% in Uranus and Neptune. This is apparently explained by a lower content of primordial gases in this disk region at a later formation stage of these planets. For example, the extrasolar planet HD 149026b, which is comparable in size to Saturn (see Lissauer, 2007), is assumed to have a similar ratio of the heavy and light fractions of elements. On the whole, however, the giant planets are gas-liquid and icy bodies with a low mean density, despite an enormous pressure in their interiors, and have no solid surface, while the extension of their interiors, i.e., the outer layer of the upper mantle, may be called an atmosphere. The properties and dynamics of planetary atmospheres are discussed in more detail in Sect. 1.3.4.

Jupiter, with a mass of 0.1% M_S (M_S – mass of the Sun) was only about three orders of magnitude short of the mass needed for the thermonuclear reactions in the hydrogen-deuterium ($M = 0.11\ M_S$) and hydrogen-helium ($M = 0.8\ M_S$) cycles to begin. Therefore, it is believed that Jupiter occupies a position near the lower boundary of stellar evolution, which is close to that of brown dwarfs [see Sect. 1.3.6].

All giant planets have satellites and rings. Owing to enormous progress achieved in the observing methods and technologies of ground-based astronomy, the number of satellites discovered in recent decades has much increased. At the time of writing,

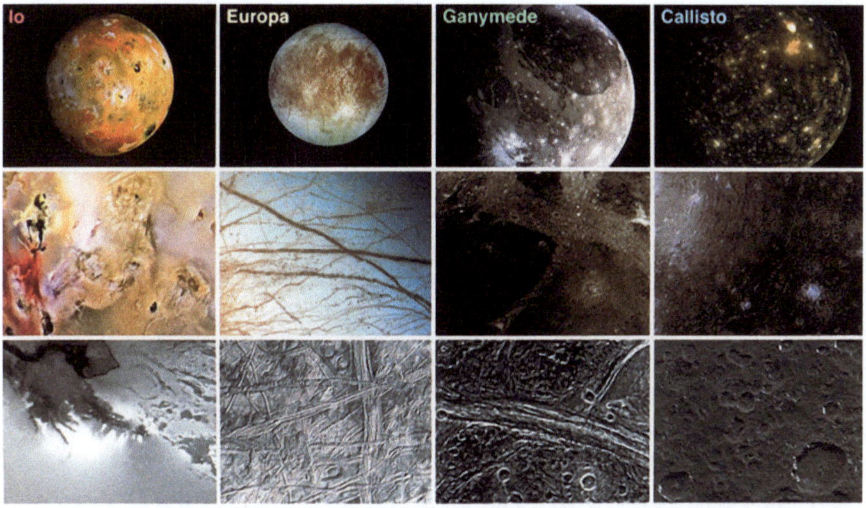

Fig. 1.3.16 Galilean satellites Io, Europa, Ganymede, and Callisto. Most prominent unique features of the satellites are shown in the images of the surface areas under each of them. Images from the Voyager and Galileo spacecraft. Courtesy of NASA

162 satellites were known: 62 for Jupiter, 60 for Saturn, 27 for Uranus, and 13 for Neptune (Vashkov'yak and Teslenko, 2008). The sizes of most satellites do not exceed tens and hundreds of kilometers, but some of them are comparable to the Moon and even Mercury. Nevertheless, the ratio of the total mass of the satellites to the planetary mass is no more than 0.01%. In contrast to the planet itself, its satellites are solid bodies with a surface, although their relatively low mean masses suggest a significant fraction of ices, primarily water ice, in their composition. Obviously, most large satellites were formed from the material of the same disk as the planet and are much closer to it, increasing their radial distances subsequently through tidal interactions with Jupiter. The satellites considerably farther from the planet are mostly captured asteroids and cometary nuclei. The system of satellites itself around each of the planets lying near the plane of its equator resembles the Solar System in miniature.

First of all, the Galilean satellites of Jupiter attract attention – Io, Europa, Ganymede, and Callisto (Figs. 1.3.16 and 1.3.17). The flights of the Voyager and Galileo spacecraft contributed decisively to the understanding of these satellites' nature (Morrison, Matthews, 1982; Burns and Matthews, 1986; Buratti, 1999). The Galilean satellites were probably formed simultaneously with, and close to, the planet itself and subsequently increased their radial distances due to tidal effects from Jupiter (see Spencer, 2001). The theory of motion for these satellites has a characteristic feature found long ago by Laplace – there is a triple (1:2:4) resonance in the system of these satellites, i.e., the revolution periods of Io, Europa, and Ganymede around Jupiter are in this multiple ratio. In addition, there exists a relationship between the mean motions that causes libration of the satellites, while

Fig. 1.3.17 Galilean satellites against the background of Jupiter's Great Red Spot (montage). Courtesy of NASA

they themselves experience perturbations from a strong gravitational interaction with each other (Lieske, 1977). As a result, certain unique features are inherent in each of the Galilean satellites, but Io and Europa stand out. They are closest to Jupiter with revolution periods of 1.77 and 3.55 days, respectively. The following discussion provides convincing evidence that the formation of unique features in both these and many other satellites of the giants is a consequence of the self-organization processes that initially gave rise to resonances and, in the course of subsequent evolution, to peculiar natural complexes.

Global large-scale volcanic activity continuing at the present epoch was detected on Io, whose radius (1821 km) is comparable to that of the Moon. As a rule, several active volcanoes, the largest of which were named Prometheus and Pele, are simultaneously observed on its surface, which is virtually devoid of any traces of impact craters because of its continuous refreshment at a rate up to $\sim$1 cm/yr. The surface itself in the region of widespread valleys is covered with deposits of sulfur and its allotropes. This is produced by phase transitions and imparts the characteristic yellow-orange hue to Io (Fig. 1.3.18). Apart from the lava itself, the condensation of volatiles and pyroclastics probably contributed to the formation of these properties. In addition, there are mountains (up to 15 km in height) and numerous calderas from 10 to 200 km in width and up to 1 km in depth, while the traces of lava flows extend to hundreds of kilometers (Johnson and Soderblom, 1983; Nash et al., 1986). There are numerous thermal anomalies against the background of the surrounding cold surface among which the Loki Patera region is most remarkable.

The eruption of lava and gases into a vacuum with a speed of about 1 km/s, forming plumes more than 200-400 km in height under low-gravity conditions is a grandiose process per se (Fig. 1.3.19). The lava is composed mainly of silicates together with sulfur compounds, so that the volcanism on Io differs little in nature

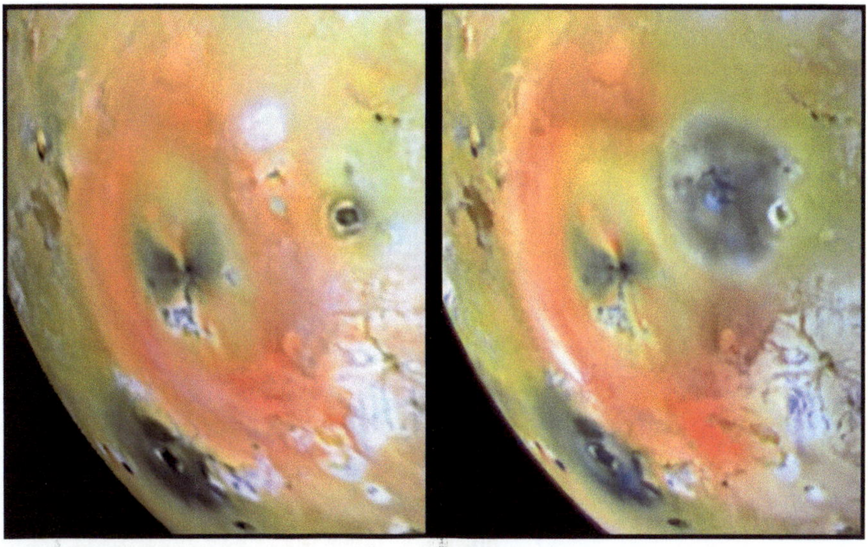

Fig. 1.3.18 Portion of Io's surface with an active volcano in the images obtained by Voyager in 1981 (left) and Galileo in 1996 (right). The surface underwent small changes; an absence of impact craters is indicative of its young age attributable to active volcanism. Courtesy of NASA

from the silicate explosive volcanism on terrestrial planets, with the only difference that it is generated not by H_2O and CO_2 gases, but by SO_2 (Keszthelyi and McEwen, 1997; McEwen et al., 1998a, 1998b; Wilson, 1999). The recorded high lava temperature (more than 1500 K) is consistent with silicate volcanism, as is the model for the interiors of Io that assumes Io to be composed of an iron or iron-sulfide core ($\sim$ 20 % by mass), a partially melted convective silicate mantle (possibly enriched with magnesium), and a lithosphere $\sim$ 30 km in thickness. However, in contrast to Earth, Venus, and Mars, where the volcanism is attributable to radiogenic heat, the radioactive isotopes on Io, just as on the Moon, have long been exhausted because of its small size. The source of eruptions is the dissipation of tidal energy due to the above-mentioned gravitational interaction of Io with other Galilean satellites during the orbital motion in Jupiter's gravitational field, causing periodic strong deformations of Io's shape (Peale et al., 1979). According to estimates, this mechanism is more efficient than other possible energy sources by approximately two orders of magnitude. The fairly high mean density of Io (3.53 g/cm^3) suggests that it consists almost entirely of rocks, while its ice-water envelope (if it ever existed) was apparently lost through the heating of its interiors already at an early evolutionary stage. Interestingly, the energy emitted by Io into the surrounding space (2.5 - 5) exceeds the internal heat fluxes on Earth (0.08 W/m^2) and Moon (0.02 W/m^2) by approximately two orders of magnitude. Since, in addition, this energy at the present epoch is almost an order of magnitude higher than the equilibrium value (0.8 W/m^2), this may suggest that the tidal energy dissipation rate underwent noticeable changes (Matson and Blaney, 1999).

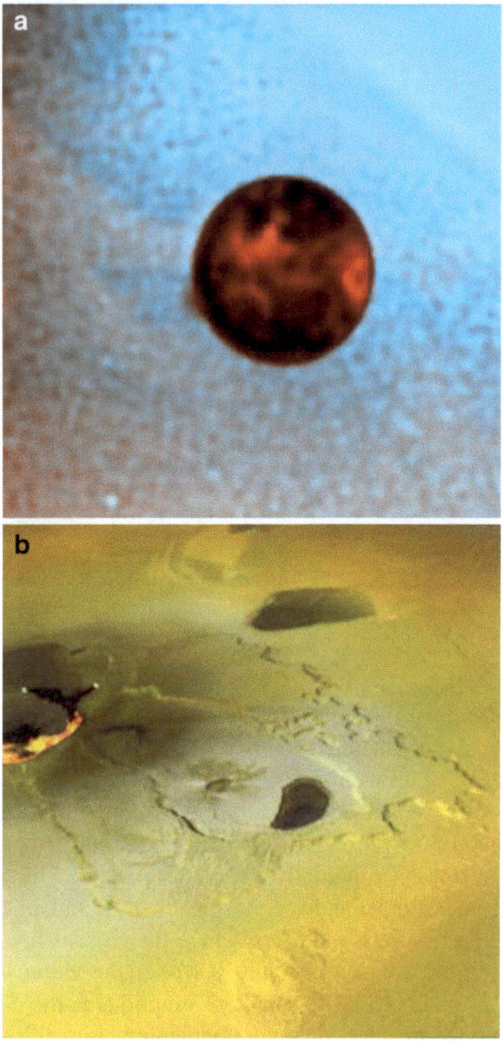

Fig. 1.3.19 (**a**) Volcanic eruption (plumes) on Io's limb. An image from the Voyager spacecraft. (**b**) Lava outflow on the Io surface. An image from the Galileo spacecraft. Courtesy of NASA

Owing to its volcanic activity, Io possesses a sulfur dioxide atmosphere, though very tenuous, and numerous neutral clouds. Because of strong sunlight scattering, the sodium emission is most characteristic of these clouds, although sulfur and oxygen dominate in them. The same components in an ionized state form a ionosphere and a plasma torus along the orbit of Io, which actively interacts with Jupiter's massive magnetosphere (Schneider and Trauger, 1995; Krimigis et al., 2002). This interaction leads to the well-known modulation of Jupiter's decametric emission recorded back in the middle of the past century. Io located at a radial distance of 5.91 R_j is literally immersed into its magnetosphere. As a result, neutral particles are knocked out of the surface with a high velocity by energetic magnetospheric ions *(sputtering)*, which is the main mechanism for the loss of material from the satellite and the replenishment of atmosphere and torus. The interaction of the plasma surrounding Io with Jupiter's magnetosphere generates currents that can be as strong as a million amperes projected from the torus into the planetary ionosphere.

Europa (radius 1561 km, density 3 g/cm^3) experiences a similar, but weaker tidal effect. It is slightly farther away from Jupiter and is smallest in size among the Galilean satellites, but may be most intriguing satellite in the Solar System. However, whereas intense heating and related volcanism led to the loss of Io's icy envelope, as on other Galilean satellites, on Europa it presumably turned into a water ocean $\sim$ 50 km in depth beneath an ice shell $\sim$ 10 - 20 km in thickness (Fig. 1.3.20). An estimate of the heat flux through the dissipation of tidal energy of about 5 K/km makes the appearance of liquid water at such a depth possible. This possiblilty is also supported by the tidally driven asynchronous (with respect to the interiors) rotation of Europa's outer envelope, the presence of numerous cracks in ice, the formation of regions with a chaotic pile-up of blocks that can be likened to icebergs, and the unusual shape of impact craters (Spencer, 2001; Carr et al., 1998). Moreover, the ocean can warm up through periodic energy release in the satellite's interiors and may accordingly even be a suitable biogenic medium. Interestingly, if these estimates are correct, the volume of water in Europa's ocean considerably exceeds the Earth's reserves of water.

Another confirmation of an ocean on Europa are the unique smoothness of Europa's figure and the fact that its icy surface is literally crisscrossed by countless low ridges and faults up to 3000 km in length, up to 70 km in width, and several hundred meters in depth, as can be seen from the images obtained during several encounters of the Galileo spacecraft with Europa. These geological features of the relief devoid of any order were apparently formed comparatively recently, as suggested by the absence of ancient impact craters on the surface. The origin of faults can be attributed to convective and other dynamical processes in the ocean that trigger ice motions and, possibly, a partial outflow of water outward where the forming cracks are filled with fresh brash ice. Obviously, the fairly fast erosion (smoothing) of the traces of Europa's meteorite bombardment is associated with the same processes. Another important argument is the comparatively weak magnetic field that most likely owes its origin to an electrically conducting salty ocean where electric currents are induced that noticeably perturb Jupiter's magnetosphere.

Magnetic fields were also detected on Ganymede, the largest satellite in the Solar System, larger even than Mercury, and on the slightly smaller Callisto (Kivelson et al., 1996, 1999). On Ganymede one can distinguish light and dark regions (Fig. 1.3.21), which probably owe their origin to the denudation of water ice or the deposits of rocks under meteorite bombardment of the surface, including large asteroids and comets (Fig. 1.3.21). In addition, there are systems of troughs and fractures probably associated with tectonic activity that, apart from the influx of tidal energy during the resonant interaction with Europa and Io, may also be attributable to the preserved source of radiogenic heat in a silicate mantle controlled by convective transport. These energy sources led to the differentiation of Ganymede's interiors, which is obviously responsible for its magnetic field. This can be generated

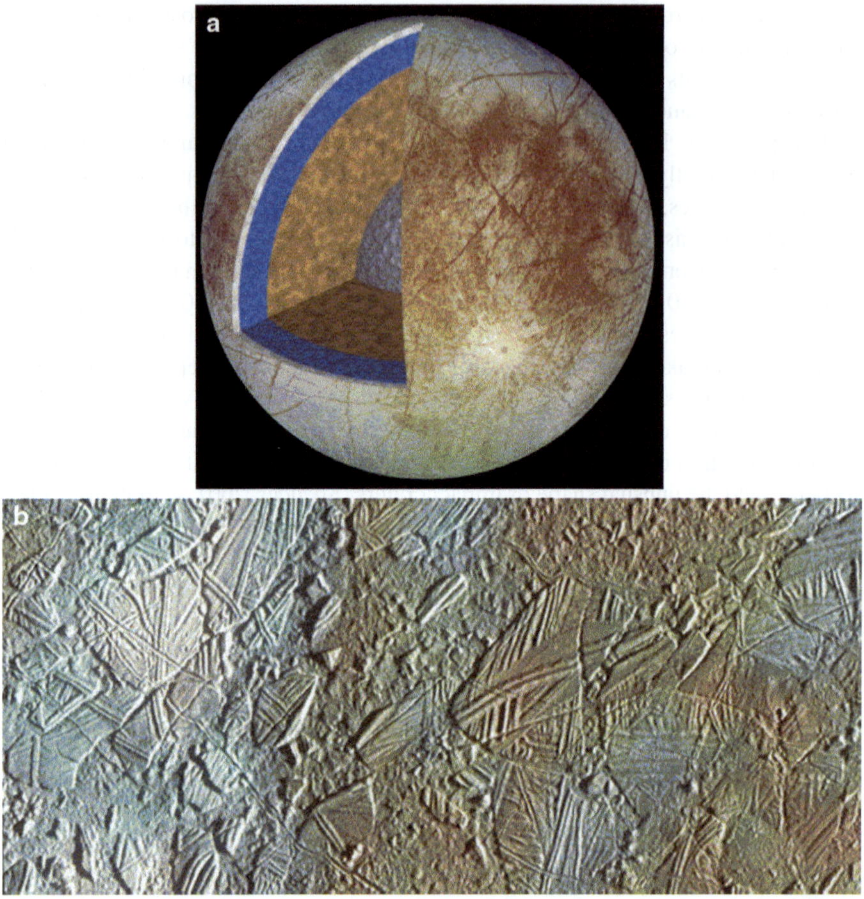

Fig. 1.3.20 (**a**) Jupiter's Galilean satellite Europa. The surface is crisscrossed by ridges, troughs, and faults whose relief does not exceed several hundred meters in height. The absence of craters is indicative of a young surface. The section on the left corresponds to the present-day model of Europa's internal structure: there is a water ocean ~50 -100 km in depth under a relatively thin ice crust ~10 -15 km in thickness and a silicate mantle and a core composed of rocks lie below it. An image of the surface from the Galileo spacecraft taken on December 17, 1996. (**b**) A 70 km x 30 km area of Europa's surface (the Conamara region). The colors are enhanced to emphasize the relief features; the Sun is on the right. The white and blue regions correspond to a fresh surface partially covered with dust, while the brown ones probably owe their origin to mineral deposits. The areas ~ 10 km in size bear the traces of displacements of the upper ice crust layer, which can be associated with the presence of water or soft ice at a comparatively small depth. An image from the Galileo spacecraft. Courtesy of NASA

both by a dynamo in a partially melted iron or iron-sulfide core whose radius is estimated to be $\sim$ 1000 km or by induction in an assumed water-ice outer envelope $\sim$ 800 km in extent (Anderson et al., 1996; Kivelson et al., 1996; 1999; Schubert et al., 1996). A deep water ocean beneath an ice shell is also not excluded.

Fig. 1.3.21 Jupiter's satellite Ganymede. Bright white regions attributable to relatively fresh ice crust denudations under impact bombardment are distinguished on the dark heavily cratered icy surface covered with mineral dust with numerous ridges, troughs, and large depressions. Ganymede is assumed to contain water ocean beneath its icy-rocky surface. An image from the Galileo spacecraft. Courtesy of NASA

Callisto is the least modified of all Galilean satellites (Anderson et al., 1998). The degree of its surface cratering is close to saturation, suggesting a high efficiency of impact bombardment in the vicinity of Jupiter. Large old craters degraded much less than those on Ganymede were preserved on Callisto. The Valhalla region is particularly remarkable. This is a system of concentric rims and crests several hundred meters in height and several thousand kilometers in extent, which were probably produced by the fall of a large asteroid that formed a vast basin similar to those on the Moon. However, in contrast to the Moon, the icy surface of this satellite retained the sequence of the wave propagation from the explosion's epicenter in its ice crust. It is covered with layers of darker material (probably, of an exogenic origin) that contains carbon-bearing compounds such as CH, CO_2, CN, SO_2, and possibly SH radicals.

Because in contrast to the other three Galilean satellites, Callisto is not subjected to tidal heating due to the Laplace orbital dynamical resonance, there are no traces of endogenic activity on it. This is also the reason why the degree of differentiation of the rocks that constitute Callisto is considerably lower than that of Ganymede. This is based on the measured quadrupole moment of the gravitational field to which corresponds a higher value of its dimensionless moment of inertia, $I = C/MR^2 = 0.359 \pm 0.005$, while for Ganymede this is $I = 0.3105 \pm 0.0028$ (and for an ideal sphere of uniformly distributed density $I = 0.4$).

In the absence of an external energy source, it is hard to expect liquid water to be preserved at depth, because the Callisto interiors should have cooled over the geological time by subsolidus convection that is controlled by ice viscosity, though this is less efficient than on Ganymede. However, there is also another scenario, in which an absorption of impact energy by a liquid interior layer is backed by the absence of noticeable seismic traces of the impact from the fall of a large body onto the side opposite to the Valhalla basin. The existence of a magnetic field on Callisto comparable in strength to that on Europa provides evidence for this hypothesis (Schubert, 1997; Anderson et al., 1998; Khurana et al., 1998).

In addition to the Galilean satellites, the Saturnian satellite Titan and the Neptunian satellite Triton attract great attention as well. But also Enceladus, quite a small, almost entirely icy satellite of Saturn with a size of 500 km and a density of 1.120 g/cm^3, deserves to be mentioned. Despite its small size, much of its surface is strongly modified by active geological processes that left ridges and troughs, while the absence of craters and the satellite's albedo close to unity suggest that the surface is very young. Apparently, the observed structures are a consequence of self-organization resulting from the slightly elliptical orbit of Enceladus in a 2:1 resonance with another Saturnian satellite, Dione. The tidal effect from Dione is sufficient to heat Enceladus' interior to a temperature of 176 K, which is the melting temperature of water-ammonia eutectic. This mechanism is probably responsible for the preserved geological activity of this cold body and also for the source of particles that fill Saturn's ring E, within which Enceladus is located (see Kuskov et al., 2008).

Judging by the mean density, Titan and Triton, just like Ganymede and Callisto, consist half of rocks and half of ices, probably mainly water ice, with different types of crystallization. Titan, comparable in size to Mercury and Ganymede, is unique in that it has a thick nitrogen-argon atmosphere of about 400 km in extent with several inversion layers with a pressure near the surface of 1.6 atm at a temperature of 94 K. This temperature is close to the triple point of methane at which phase transitions occur on the surface. Since the acceleration owing to gravity on Titan is approximately one seventh of the terrestrial one, the mass of Titan's atmosphere must be an order of magnitude larger than the terrestrial one to produce a pressure of 1.6 atm (Hunten et al., 1984; Coustenis and Lorenz, 1999). The presence of ^{40}Ar in the atmosphere implies volcanic activity. The morphological features of the surface do not rule out that it may have been affected by tectonic processes.

Another interesting feature of Titan is the circulation of methane, including the formation of methane clouds in the atmosphere and precipitation in the form of methane rains on the surface. The existence of this methane cycle was predicted previously based on computer models and on the investigation of surface properties by radar measurements and measurements made with the Hubble Space Telescope (Meier et al., 2000). These expectations were confirmed by the imaging of surface panoramas from the Huygens lander that separated from the Cassini orbiter when it approached Saturn and landed on Titan in January 2005. The images show boulders of round shapes, probably composed of water and methane ices and likely other organic compounds, dark spots, valleys resembling the beds of rivers flowing down uplands, rows of dunes consisting of "hydrocarbon dust" particles apparently formed by strong winds, and separate filled or dried lakes hundreds of kilometers in size, whose formation can be associated with the precipitation of liquid methane from the atmosphere (Fig. 1.3.22). Methane condenses into clouds at an altitude of several tens of kilometers, weak frost continuously precipitates from them to the surface, and even larger "rain" drops from particularly dense clouds found near the South Pole, compensated for partially by evaporation, which is an analog of the

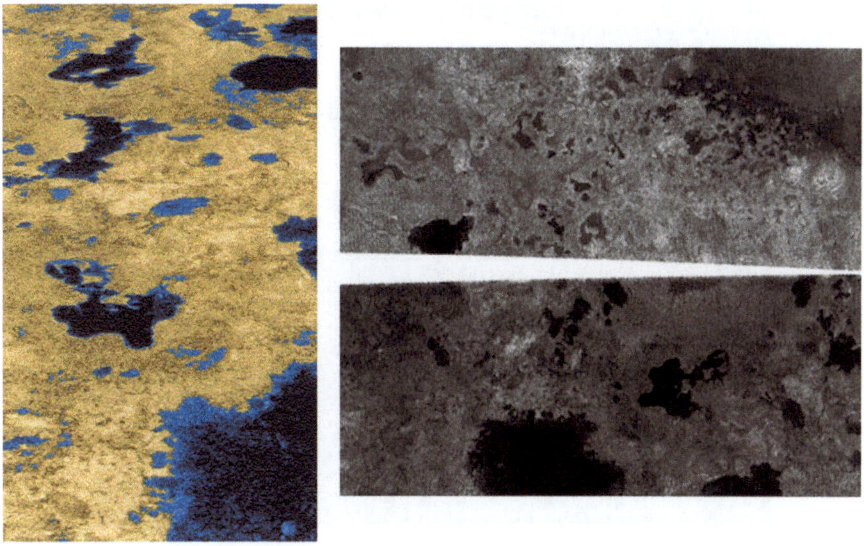

Fig. 1.3.22 The surface of Saturn's satellite Titan. The dark spots on the lighter surface composed of water and hydrocarbon ices are associated with methane lakes, which corroborates the hypothesis about the existence of a methane cycle between the surface and the atmosphere. Two bands on the right are original images. The colors in the left image are distorted to emphasize the contrast of features. The images were obtained by the Huygens lander (ESA courtesy)

hydrological cycle on Earth. There can be no liquid water on the surface because of the very low temperature, but it may well be that it is present in greater depths in the subsurface layer, just as on the Galilean satellites. There is also reason to believe that in addition to methane, there are more complex unsaturated hydrocarbons (ethane, ethylene, acetylene, diacetylene, methylacetylene, cyanoacetylene) on the surface, along with propane, hydrocyanic acid, and other organic compounds, which form in the upper atmosphere and hydrocarbon clouds under ultraviolet radiation through methane photolysis processes. Hydrocarbons impart a characteristic red-orange color to the atmosphere. It is believed that favorable conditions for the initial stages of biogenic synthesis similar to those that existed on the early Earth are created on Titan (Sagan et al., 1984) and this attracts particular interest to Titan.

In contrast to Titan, Triton with a size slightly smaller than that of the Moon is virtually devoid of an atmosphere (the pressure does not exceed 15 microbars) and the temperature of its nitrogen-methane surface is only 38 K (Benner, 1997; McCinnon and Kirk, 1999). Several signatures make Triton resemble Pluto, which until recently was considered the ninth planet of the Solar System and has now been transferred to the category of large Kuiper Belt bodies, plutonoids. But significant differences became obvious after the Voyager flyby of Neptune. First of

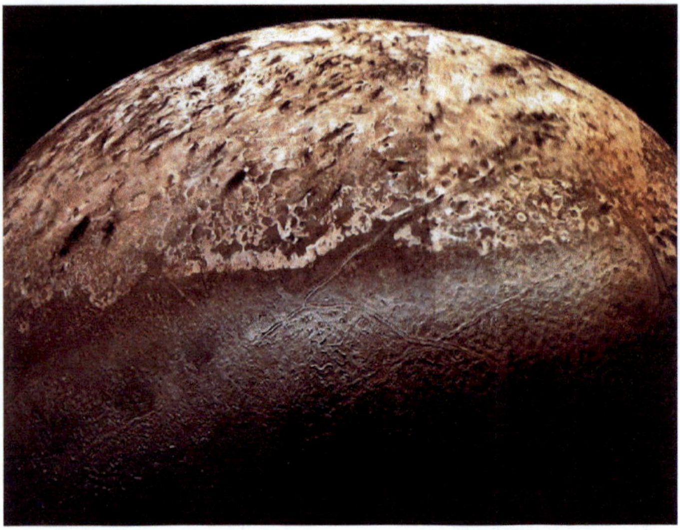

Fig. 1.3.23 Neptune's satellite Triton. Active nitrogen geysers were detected on the methane surface. An image from the Voyager spacecraft. Courtesy of NASA

all, dozens of dark bands were detected on the surface of Triton in the southern polar cap region. Some of these were identified to be geyser-like ejections of liquid nitrogen that reach an altitude of several kilometers (cryovolcanism, Fig 1.3.23). The deposits of dust particles that are transportable by predominant winds even in a highly rarefied medium on frozen methane are probably also associated with geysers. In addition, structures were detected on the surface that resemble frozen lakes with nitrogen-methane coastal terraces up to one kilometer in height, whose formation is probably associated with successive melting-freezing epochs when the insolation conditions changed, but which are more likely caused by the tidal interactions of Triton with Neptune. Just as on the Galilean satellites, the interior heating through the dissipation of tidal energy probably serves as the main source of cryovolcanism on this very cold body. Another energy source of geyser-like ejections could be a rise in insolation level, because these ejections are observed in latitudes at which the Sun is at its zenith. But this mechanism most likely plays a secondary role. The few impact craters on the young surface of Triton provide evidence for its continuing geological activity, however.

Triton has a very unusual orbit, which acutely poses the question about its origin. It is highly inclined to the ecliptic plane and has a nearly zero eccentricity, while the motion of Triton itself, in contrast to all other large planetary satellites, is not prograde (counterclockwise), but retrograde. The peculiarities of Triton's orbital motion suggest that it was initially formed in the Kuiper Belt, just like Pluto, and was subsequently captured by Neptune. However, calculations showed that an ordinary gravitational capture is unlikely. Therefore, it is additionally hypothesized that Triton was a member of a binary system or gradually decelerated in the upper atmosphere of Neptune. This hypothesis is corroborated by the fact that when passing to the orbit around Neptune, Triton would have experienced a strong tidal effect from Neptune and the existing system of its satellites (in particular Nereid), which led to the melting of its predominantly water-ice interiors (the mean density is 2.07 g/cm). It is quite probable that the ongoing tidal effect from Neptune and Nereid at the present epoch heats the planet; as a result, the thermal flux from its interior exceeds the insolation value almost by a factor of 3. Another consequence is that Triton gradually approaches Neptune and will enter inside the Roche limit in the distant future, where it will be torn apart.

It can be concluded from the brief overview of celestial bodies in the outer Solar System that they are all characterized by a surprising order of natural complexes. Obviously, these were formed in the process of self-organization in an open dissipative system of initially chaotic structures in the form of a swarm of planetesimals, out of which the giant planets and the systems of their satellites were born. Self-organization was provided primarily by the gravitational forces that gave rise to commensurabilities and resonances in the motions of planets and satellites and by the tidal interactions that caused interior heating and unique natural phenomena in the form of volcanoes, a hypothetical warm water ocean, and nitrogen geysers under conditions of extremely low temperatures. The formation of volcanic plumes accompanying the outflow of gases into a vacuum is directly related to the turbulent motions of multicomponent media and the phase transitions in these media responsible for the surface morphology and properties, as on Io. Turbulence apparently contributes to the formation of faults and fractures on the icy surface that emerge during convective motions in the water column of Europa's hypothetical ocean. Many features of the cloud formation, photochemical processes, surface morphology, and the formation of complex organic compounds are associated with the circulation of methane in Titan's turbulized atmosphere.

The rings of giant planets are another remarkable example of self-organization. It emerges in a system of particles that are in orbital motion and simultaneously experience chaotic interactions. In this case, order is produced in the configurations of rings, which owe their origin primarily to the emergence of collective processes and the presence of inelastic collisions between macroparticles in the disk system (Fridman and Polyachenko, 1984; Gorkavyi, Fridman, 1994). In other words, self-organization is built into the system itself, while the satellites close to or inside the ring structure, often called shepherds, exert an additional "stimulating" influence. Besides, the particles of the rings, which are an infinite number of small satellites themselves, turn out to be in resonances with larger planetary satellites (Fig. 1.3.24). This disrupts the homogeneous structure of the rings and gives rise to gaps inside them, such as the Cassini and Enke gaps in the rings of Saturn, similar in nature to the Kirkwood gaps in the Main Asteroid Belt (Greenberg and Brahic, 1984; Esposito et al., 1984). The same mechanism is responsible for the generation of density waves, the formation of a hierarchical structure of rings, and their stratification into thousands of thin spiral ringlets through the development of a gravitational-dissipative instability. Another example of self-organization and the interaction of nearby planetary satellites with rings is the formation of a system of

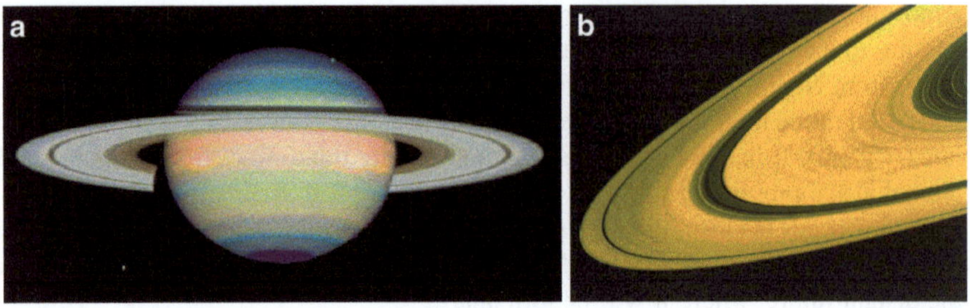

Fig. 1.3.24 **(a)** Saturn in contrast-enhanced artificial colors. The blue, red- and- orange, and green colors correspond to the main clouds, upper-level clouds, and the overcloud haze, respectively. Three main rings are distinguished in the rings surrounding Saturn: C, B, and A. There is the Cassini Division between the A and B rings. A Hubble Space Telescope image. **(b)** Saturn's rings. Their fine structure, the set of isolated ringlets that result from the process of self-organization in the ring structure, is clearly distinguished. The radial dark bands (dark spokes) in the B ring are attributable to the formation of dusty plasma located above the plane of rings and controlled by the planet's magnetic field. An image from the Voyager-2 spacecraft. Courtesy of NASA

very narrow rings around Uranus (Fig. 1.3.25), in which the particles are concentrated by gravitational focusing (Elliot and Nicholson, 1984; Esposito et al., 1991; French et al., 1991). In turn, in the rings of Neptune even a nonuniform distribution of particles in the form of separate "arcs" is observed along the orbit, which drift in the azimuthal direction. The two most characteristics rings with arcs were called Adams and Le Verrier in honor of Neptune's discoverers by prediction ("on the tip of the pen"), which was soon brilliantly confirmed by observations. The arc formation mechanism is not completely understood, although the resonances (42:43) of ring particles with the eccentricity and inclination of the Neptunian satellite Galatea, preventing a uniform distribution of particles along the orbit (Porco et al., 1995; Porco, 1999), serves as one of the explanations. However, this mechanism needs an additional justification.

All planetary rings are probably either a swarm of original particles whose formation into a satellite was hindered by gravitational forces inside the Roche limit or they are a result of the breakup of an asteroid or a comet that entered the Roche limit. Jupiter's ring, filled with very small particles and surrounded from the outside and inside by diffuse nebulae (Ockert-Bell et al., 1999), serves as a characteristic example of such an event. The breakup hypothesis was advanced based on the estimated limited lifetime of the rings, ~ 0.5 Gyr, which is much less than the age of the Solar System (Esposito et al., 1984; 1991). If this is true, the rings should not be considered a relict of the accretion stage of the planet itself, but are periodically appearing and disappearing structures caused by the accidential catastrophic event of gravitational capture of a small body by the planet and its subsequent destruction when it approaches the Roche limit. This seems quite likely, all things considered. This hypothesis is also supported by the fact that, for example, the predominantly icy particles of Saturn's rings have a high albedo, i.e., they were not covered with dark micrometeoric matter, as would happen to the relict rings over the lifetime of the Solar System.

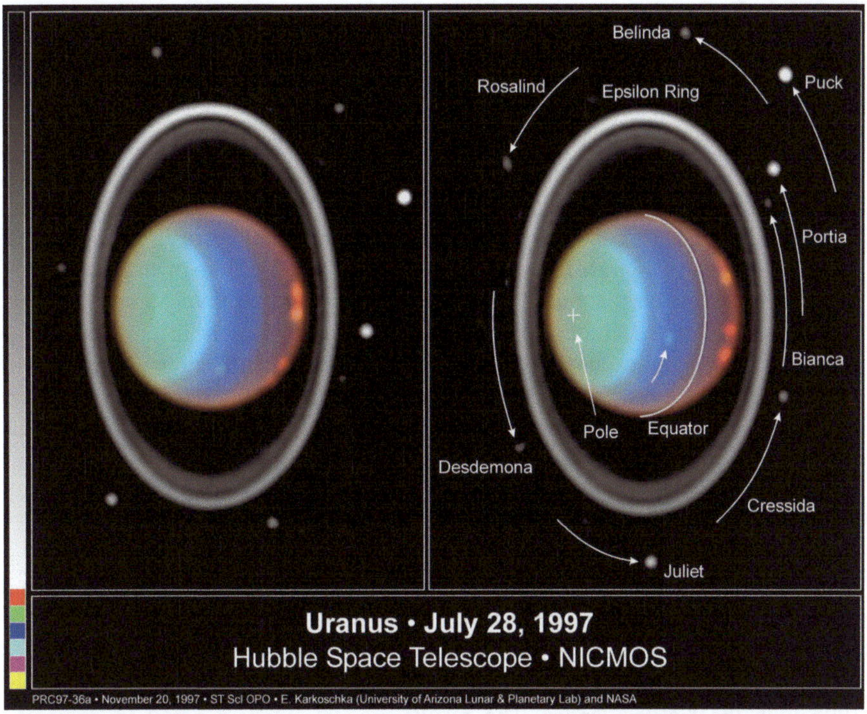

Fig. 1.3.25 The rings of Uranus from Hubble Space Telescope observations. The left picture was taken in 1997; the right one was taken later on and the motions of satellites and atmospheric structures are shown in it. The formation of Uranus's rings is even more pronouncedly than in the case of Saturn affected by the satellites (shown in the right image) located near the rings, along with the collective particle interaction processes. They are responsible for the emergence of additional order in the ring structure. Courtesy of NASA

1.3.4 Atmospheres of the Inner and Outer Planets

We find numerous examples of self-organization in turbulent gas envelopes of planets - the planetary atmospheres.

The atmosphere of a terrestrial planet is a gaseous medium bounded from below by a solid underlying surface, while for a gas-liquid giant planet this lower boundary is just an upper gaseous layer. The atmospheres gradually pass into circumplanetary space. The lower atmosphere adjacent to the surface belongs to the sphere of interests of meteorology, while the upper atmosphere directly exposed to solar short-wavelength electromagnetic and corpuscular radiation belongs to the sphere of interests of planetary aeronomy (Marov and Kolesnichenko, 1987; Izakov, Marov, 1989; Marov, Grinspoon, 1998). Intense heat and mass exchange processes that significantly affect the climate formation and atmospheric evolution take place in the transition regions of the middle atmosphere.

The Main Properties

The properties of the planetary atmospheres differ greatly even at the relatively short distances within the Solar System occupied by the inner (terrestrial type) planets. Mercury, like the Moon, is virtually devoid of an atmosphere – the density of its gaseous envelope near the surface does not exceed 10^{-17} g/cm^3, although a number of interesting dynamical effects arise when the solar wind plasma interacts with the magnetosphere and exosphere of Mercury. The secondary atmospheres of Venus and Mars, which differ in density by almost four orders of magnitude, are oxidizing ones and consist mainly of carbon dioxide with relatively small fractions of nitrogen and trace amounts of oxygen and water. This is in contrast to the primary reducing atmospheres of the giant planets, which are composed of hydrogen, helium, and hydrogen-bearing compounds, the most typical of which are ammonia and methane. A comprehensive overview of the properties for the atmospheres of planets and their satellites considered from the viewpoint of their origin and evolution can be found, for example, in Chamberlen, Hunten (1987), Atreya et al. (1989), Del Genio (1997), Samuelson (1997), and West (1999).

Apart from the chemical composition, several structural and dynamical peculiarities distinguish the atmospheres of giant planets from those of the terrestrial ones. Their tropospheres have a large extent (~120-200 km) and the thermal structure as a whole corresponds to an adiabatic temperature gradient. The atmosphere of Jupiter was directly probed in the pressure range 0.4 - 22 bar in December 1995 with the Galileo lander. The results are presented in detail in dedicated reviews and thematic issues of journals (see JGR, special issues, 1998; ICARUS, special issue, 1998; Marov, Rahe, 1998). Measurements showed that the ratio He/H = 0.24 is close to sola, while the abundances of heavier elements - carbon (in the form of CH_4), nitrogen (in the form of NH_3), and sulfur (in the form of H_2S) are approximately triple their solar values, suggesting that Jupiter's formation is connected with more distant cold regions. Another surprise was the high dryness of the atmosphere and, consequently, its oxygen depletion, although this conclusion most likely does not represent the planet as a whole but is valid only for the local landing site known as a 5-μm spot where thermal emission from deeper atmospheric layers takes place because of a break in the clouds. This is apparently also responsible for a certain anomaly in the formation of a three-layer structure of clouds that, according to model calculations (West et al., 1986; Atreya, 1989), consists of ammonia, ammonia hydrosulfide, and water. Contrary to the condensation of NH_3, H_2S, and H_2O expected in this temperature range, only NH_4SH crystals were detected. At the same time, no ammonia crystals were found, which impart the white color to zones in the dynamical structure of zones and darker belts on Jupiter's disk. At low relative abundances of the cloud-forming compounds, the phase transitions essentially have no effect on the temperature profile, which corresponds to a dry adiabat. The temperature rises from 135 K at the upper cloud boundary to 900 K in the thermosphere. Under conditions of a low solar short-wavelength radiation flux at Jupiter's orbit, this is most likely due to the energy dissipation of internal gravity waves that propagate from deep layers and are heated through energetic particles precipitating from the massive planetary magnetosphere.

The forms in which carbon and nitrogen were present in the protoplanetary disk can be judged from the CH_4 and NH_3 abundances. It follows from the thermochemical equilibrium condition corresponding to the reversible reactions $CH_4+H_2O=CO+3H_2$ and $2NH_3=N_2+3H_2$ that these reactions proceed in the direction of CH_4 and NH_3 formation at low temperatures and in the direction of CO and N_2 formation at high temperatures. However, this condition is by no means always met, as exemplified by the atmosphere of Titan, in which nitrogen dominates at a relatively low methane abundance. The photolysis reactions of NH_3, which was contained in the disk primordial material and initially entered into the composition of Titan's gaseous envelope, may have been competing reactions to the thermochemical reactions.

The formation of CO and N_2 in the regions where the terrestrial planets are located was obviously accompanied by degassing from the interiors, along with the reaction $CO + H_2O \rightarrow CO_2 + H_2$, which explains the carbon dioxide composition of their atmospheres. According to the existing views, the anomalous (from this viewpoint) composition of the Earth's atmosphere with a great abundance of oxygen and nitrogen owes its origin to a biosphere formation. We emphasize that the chemical and isotopic compositions of the atmosphere in many respects reflect the pattern of evolution of the planet itself, which is particularly true for the history of water. The HDO/H_2O ratio, which exceeds the terrestrial one by a factor of 120 on Venus and by a factor of 6 on Mars (see, e.g., Pollack et al., 1980), is most revealing here. The enrichment of the atmosphere with deuterium is a consequence of the loss of water and allows estimating its original reserves and the degree of degassing of the interiors. In turn, water served as the main factor that determined the key features of the evolution of the planet and its atmosphere, primarily the thermal regime and atmospheric dynamics with which the weather and climate formation processes are directly associated.

All planetary atmospheres exhibit a wide spectrum of motions, from micro scales to planetary circulation, through which the nonuniformity of heating by solar radiation is compensated for. Since the Reynolds number **Re**8 in atmospheric motions usually exceeds its critical value $\mathbf{Re}_{cr}$, the flow is turbulent. The atmospheric flows are turbulized either through their significant deformation in a flow around irregularities of the underlying surface or through the loss of hydrodynamic stability by a large-scale flow under the action of enhanced wind velocity and temperature gradients. In a free atmosphere, the loss of stability by internal gravity-shear waves in the middle atmosphere is mainly responsible for the generation of turbulence. These waves can be destroyed by the "primary" or "secondary" instability. The primary (Kelvin-Helmholtz) instability develops in the shear layer between flows with different velocities if the condition $\mathbf{Re}<\mathbf{Re}_{cr}$ is valid for the bulk of the layer. Under the secondary instability, the flow is stable on average, while its unstable parts are localized near the wave crests. This type of instability is characteristic mainly of layers with steeply curved vertical temperature and wind velocity profiles.

Additional problems concerning the stability and energy exchange arise when turning to the upper and middle planetary atmosphere. Peculiar features such as the multicomponent structure and chemical activity of atmospheric gases manifest themselves here and, hence, the aeronomical processes begin to play a crucial role (see Marov, Kolesnichenko, 1987). The patterns of turbulence in a single-component fluid flow and a reacting multicomponent flow differ significantly. Particularly the appearance of local inhomogeneities in the density, temperature, and composition distributions for the atmospheric mixture of gases attributable to photolysis processes and accompanying chemical reactions leads to additional flow turbulization. The emerging gradients in fluid density (which is variable in this case) produce additional vorticity through the interaction with the gradients (or fluctuations) in the pressure field. This is because the source term $\rho^{-1}(\partial\rho^{-1}/\partial r)$ $\times(\partial p/\partial r)$ (equal to zero in barotropic media) appears in the well-known Friedman equation for vorticity

$$\frac{d}{dt}\left(\frac{\Omega_a}{\rho}\right) = \left(\frac{\Omega_a}{\rho}\cdot\frac{\partial}{\partial \mathbf{r}}\mathbf{u}\right) - \rho^{-1}\left(\frac{\partial\rho^{-1}}{\partial \mathbf{r}}\times\frac{\partial p}{\partial \mathbf{r}}\right) + \rho^{-1}\left(\frac{\partial}{\partial \mathbf{r}}\times F\right), \tag{1.3.1}$$

where $\Omega_a = \omega + 2\Omega$ is the curl of the absolute velocity, $\omega = (\partial/\partial\mathbf{r})\times\mathbf{u}$ is the vorticity vector, Ω is the angular velocity vector for the planet, and $\mathbf{F}$ is the sum of the accelerations of the viscous and external mass forces (see, e.g., Monin and Yaglom, 1992). When the mass density ρ is constant, all these dynamical effects, which potentially depend on this "vorticity" term, are absent. Thus, the existence of local inhomogeneities (gradients) in the mass density is a very important property of multicomponent reaching flows that is usually disregarded in classical single-component fluid turbulence models. The local energy sources associated with chemical reactions are another complicating circumstance in the multicomponent turbulence modeling problem. Local heat release in gas flows accelerates the expansion of the medium and can induce Rayleigh-Taylor instability in flows stratified in the gravity field when buoyancy forces are present, thereby realizing a feedback with hydrodynamics. All these problems are considered in detail in our monographs (Marov, Kolesnichenko, 1987; Kolesnichenko, Marov, 1999; Marov and Kolesnichenko, 2001).

From the viewpoint of energy exchange, the solid (liquid) underlying and gas layers constitute a single thermodynamic system whose energy balance is determined either by the incoming solar radiation or predominantly by the internal energy of the interiors, which on Jupiter, Saturn, and Neptune is approximately twice the insolation value. Obviously, the nature of this source is associated with the release of gravitational energy through the ongoing contraction of these planets, in accordance with the virial theorem for a system with a gravitational potential and in hydrostatic equilibrium.

The energy absorbed by the planetary atmosphere and converted into the internal and kinetic energy of the gaseous medium is partially emitted into space in the form of infrared radiation, eventually providing an energy balance. The heat and mass transport takes place against the background of global ($L \gg H(H = p/\rho g$ is the scale height)), mesoscale ($L \approx H$), and local ($L \ll H$) atmospheric dynamics. It is responsible for the diversity of meteorological processes in the atmosphere, the role of molecular transport processes is negligible. Advection driven by planetary in which circulation and convection are important components of regular heat and mass transport. Convection stems from the fact that because the temperature at a given pressure level on rotating terrestrial planets systematically changes with latitude φ, a barotropic atmosphere becomes baroclinic, in which turbulence emerges due to nonlinear evolution of the hydrodynamic Rayleigh-Taylor instability. Strong perturbations in fluid flows also manifest themselves as cyclonic winds accompanied by turbulent heat and mass transport (Chamberlain and Hunten, 1987).

Dynamics of the Atmosphere of the Inner Planets

Specific realizations of dynamical mass transport on different planets in the presence of four types of forces acting on an elementary volume of air (gravity force, Coriolis force ($2\rho\Omega \times \mathbf{u}$), viscous friction, and pressure gradient) depend mainly on the ratios of the planetary rotation periods and thermal atmospheric relaxation. The relative contribution of the Coriolis force to atmospheric dynamics is determined by the Rossby number $\mathbf{Ro} = U/fL$, where U and L are the typical horizontal velocity and length scales for synoptic processes (for the terrestrial atmosphere, $\mathbf{Ro} \approx 10^{-1}$). Balancing the horizontal pressure gradients in the scales of synoptic processes, the Coriolis force imparts geostrophicity to the flows with which the peculiar oscillations in the atmosphere with the inertial frequency$\tau = 2\Omega_z$ are associated, where Ω_z is the vertical component of the planet's angular velocity.

The combined effect of the planet's rotation and sphericity, which is responsible for the change in frequency τ (or f) with latitude ϕ (the so-called "β-effect"), generates Rossby-Blinova planetary waves, in whose troughs and crests the main weather generators are formed, cyclones and anticyclones. In the Earth's atmosphere, these planetary waves propagating horizontally westward have a period exceeding both the planetary rotation period and the periods of three other types of atmospheric wave motions – tidal, acoustic, and internal gravity waves (IGWs). Short-period IGWs with relatively low amplitudes belong to the category of micrometeorological atmospheric oscillations and, along with convection, serve as the source of small-scale turbulence. Thus, they energetically (through dissipation) affect the formation of large-scale weather processes.

For Rossby-Blinova waves, Ertel's conservation condition for the so-called potential vorticity Ω_* of the main flow is valid:

$$d\Omega_*/dt = 0, \quad \Omega_* \equiv \rho^{-1}\Omega_a \cdot (\partial S/\partial \mathbf{r}) \tag{1.3.2}$$

where S is the entropy per unit mass of the medium, the direction of $(\partial S/\partial \mathbf{r})/|\partial S/\partial \mathbf{r}|$ in a stratified atmosphere is in which roughly vertical – this is the so-called thermodynamic vertical; Ω_* is the potential vorticity – an adiabatic Lagrangian invariant that, by the definition of adiabatic processes, the entropy S also is ($dS/dt \cong 0$). Condition (1.3.2) follows from Friedman's equation for vorticity (1.3.1) in adiabatic processes when $F = 0$. Another characteristic feature is the oscillatory exchange between the global vorticity gradient β defined as $\beta = r^{-1}\partial f/\partial\varphi$ and the relative vorticity of the wave itself (Charney and Stern, 1962; Monin, 1988; Ingersoll et al., 1995).

Since the forces from pressure gradients in the Earth's atmosphere are essentially balanced by the Coriolis forces ($Ro \ll 1$), the geostrophic wind is a typical synoptic characteristic and its zonal and meridional components ($u = -(\rho f)^{-1}\partial p/\partial y$, $\nu = (\rho f)^{-1}\partial p/\partial x$) can be determined from the known pressure distribution. In contrast to the Earth, the influence of Coriolis forces on the slowly rotating Venus (one rotation in 243 terrestrial days) is insignificant ($Ro \gg 1$) and, therefore, the cyclostrophic balance condition is valid (Marov and Grinspoon, 1998). It is characterized by a zonal velocity component that increases with altitude and is superimposed on a Hadley circulation cell, which emerges because of the equator-pole temperature gradient. As a result, atmospheric superrotation is produced (also called carrousel circulation), causing the zonal wind velocity in the Venusian atmosphere to increase from ~ 0.5 m/s near the surface to ~100 m/s at an altitude of about 65 km. The latter corresponds to the upper cloud boundary near the tropopause where the characteristic ultraviolet contrasts, which are presumably attributable to the absorption of solar radiation by sulfur allotropes contained in clouds, move with this velocity (see Fig. 1.3.6a).

Interestingly, although the static stability condition is met in the Venusian troposphere,

$$d\theta/dz = (\theta/T)(dT/dz - \gamma_a) > 0, \tag{1.3.3}$$

where $\theta = T(p_0/p)^{(\gamma-1)/\gamma}$ is the potential temperature, p_0 is the standard pressure, $\gamma_a = g/c_p$ is the adiabatic gradient, and the gradient Richardson number satisfies the inequality

$$\mathbf{Ri} = (g/\theta)(\partial\theta/\partial z)(\partial u_h/\partial z)^{-2} > \mathbf{Ri}_{cr} = 1/4, \tag{1.3.4}$$

theoretically no turbulence should emerge in it. The point is that the inequality $\mathbf{Ri} < 0.25$ is a theoretically necessary condition for the instability of small perturbations in an inviscid shear flow, while in the Venusian troposphere $\mathbf{Ri} \approx 2\text{-}6$, just as in the terrestrial atmosphere. Nevertheless, turbulence actually exists, as suggested by the characteristic (for shear turbulence) temperature fluctuations ~ 0.1 K with scales ~100 m measured in the Venusian atmosphere in the range of altitudes 45–50 km (Woo et al., 1982; Seiff, 1983). Turbulence is apparently also present in the lower troposphere at altitudes > 15–20 km, where vertical velocity fluctuations of 0.2–0.3 m/s were detected during the parachute descent of the Venera spacecraft through measurement of the vertical Doppler frequency shift of the onboard master oscillators (Kerzhanovich, Marov, 1983). In turn, the stability of the atmosphere

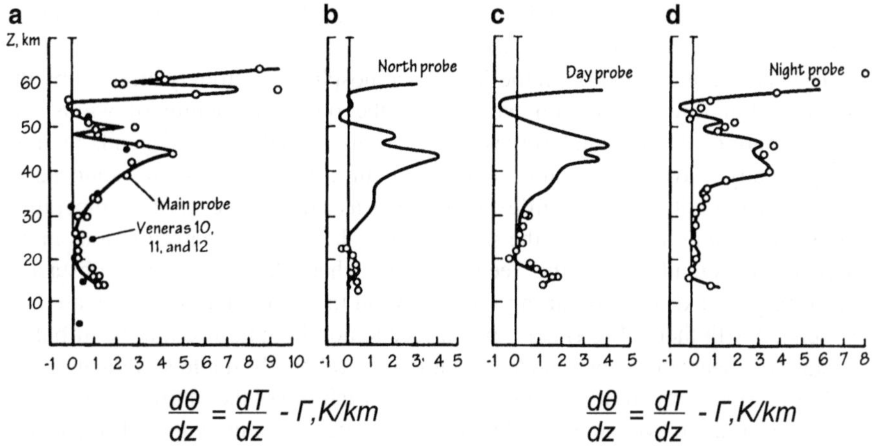

Fig. 1.3.26 Variation of the atmospheric stability parameter d/dz with height z on Venus from the Venera 10, 11, and 12 spacecraft and Pioneer-Venus probes : (**a**) "Veneras" (o) and "Main probe"; (**b**) "North Probe", (**c**) "Day Probe"; (**d**) "Night Probe". The regions of convective instability at altitudes of 52-57 km and shear flows between 45 and 50 km are distinguished. The observed good agreement between the data in various regions of measurements and at different times of the day suggests that the dynamics of the Venusian atmosphere is homogeneous. According to Seiff, 1983

increases (the parameter $\partial\theta/\partial z$ grows) near the upper cloud boundary, but, apart from shear flows, vortex structures of various spatial and temporal scales associated with global circulation processes are also observed. All of this suggests a static stability of at least some regions in the lower atmosphere of Venus for which turbulence is an important element of atmospheric dynamics (Fig. 1.3.26). Turbulence could play an equally important role at early evolutionary stages of Venus, which is associated with the hypothesis about a primordial ocean that was subsequently lost due to the development of a runaway greenhouse effect, as we considered in Sect. 1.3.2.2 (Marov and Grinspoon, 1998).

The dynamics of the Martian tenuous atmosphere with a low thermal inertia differs in many respects from the terrestrial (and Venusian) atmosphere primarily by the absence of oceans, which are heat accumulators and damp the diurnal-seasonal temperature inhomogeneities, and the formation (under conditions of a low moisture content in the atmosphere and conservation of the dry-adiabatic gradient) of massive clouds. The global circulation model (GCM) based on the geostrophic balance condition (**Ro** $\ll$ 1) predicts a similar topology of motions in the troposphere and stratosphere with a predominance of winds blowing eastward at high latitudes in winter and in subtropics in summer, and westward at the remaining latitudes. At the same time, the seasonal carbon dioxide exchange between the atmosphere and polar caps serves as the main driving mechanism of transport in the meridional direction. This gives rise to Hadley cell configurations with upflows and downflows, a rearranging system of winds near the surface and at high altitudes in the summer and winter hemispheres, and seasonal changes of the cloud cover (Kahn, 1984; Zurek et al., 1992; Marov, 1994).

The pattern of circulation in the Martian atmosphere is strongly affected by the surface relief (areography) on which both the observed picture of winds and the generation of horizontal waves with various spatial scales depend. In turn, the planetary waves attributable to the baroclinic instability of the atmosphere and internal gravity waves manifest themselves as irregularities in the temperature and vertical motion profiles in the stratosphere. The observed wave motions in the structure of clouds from the leeward side when flowing around obstacles, suggesting the existence of strong shear flows in the Martian atmosphere, are also associated with them. The entire near-surface atmosphere turns out to be turbulent because of shear flows, even under conditions of relatively stable stratification.

Convection compensates for the high static instability of the Martian atmosphere close to saturation even at a very low relative content of water vapor. The convection excitation efficiency at day hours is approximately an order of magnitude higher than that in the Earth's atmosphere, while during the night it is completely blocked by the formation of an inversion layer with a positive temperature gradient near the surface. Convection also maintains a high content of dust that is constantly present in the Martian troposphere and produces an additional dynamical effect superimposed on the global system of winds. This effect arises from a positive feedback between the dust content and the degree of atmospheric gas heating, which manifests itself as thermally generated diurnal and semidiurnal tides. This peculiar feature of the Martian atmosphere is most pronounced during periodically emerging global dust storms when fine dust rises to an altitude above 30-40 km because of turbulent mixing. Since the troposphere is highly opaque (with an optical depth $\tau \geq 5$), a reverse greenhouse effect that strongly dampens the circulation transport is produced near the surface. The discovery of this unique natural phenomenon during a global dust storm on Mars in 1971 served as impetus for drawing an analogy between it and the possibility of "nuclear winter" on Earth as an inevitable consequence of using atomic weapons (Sagan, 1977).

The turbulence of a dispersed medium whose pattern depends significantly on the dynamical and energetic interaction of the gas and dust phases undoubtedly plays an important role in the complex of processes responsible for the formation, maintenance, and decay of a Martian dust storm, although the details of these mechanisms are not yet completely clear. As is well known, the turbulent momentum and heat fluxes can be estimated from the measured altitude profiles of the mean velocity $u(z)$ and temperature $T(z)$ in the surface boundary layer (Monin and Yaglom, 1992). Assuming that these patterns are also valid for the atmospheres of other planets, the corresponding profiles, which are modified for turbulent flows with a heavy admixture, can be estimated for the boundary layer of Mars by reversing the problem (Barenblat, 1955; Golitsyn, 1973). The mean velocity profile for neutral stratification was shown to be reduced to

$$u(z) = (u_*/\kappa\omega)\ln(z/z_0), \tag{1.3.5}$$

where $u_* = \sqrt{\tau/\rho}$ is the dynamic (friction) velocity; $\tau = -\rho\overline{u''v''}$ is the turbulent momentum flux; κ is the Karman constant; and ρ is the gas-phase density. The dimensionless parameter $\omega = \nu/\alpha\kappa u_*$ is defined via the dust particle settling rate ν, the ratio of the turbulent exchange coefficients for the admixture and momentum $\alpha(\alpha \approx 1)$, and the quantity u_*. It turns out that when relatively small particles ($\nu<\alpha\kappa u_*$, or $\omega<1$) are present in the flow, Eq. (1.3.5) can be interpreted such that the presence of dust in the flow leads to a decrease in the Karman constant κ. In other words, the velocity gradients near the surface increase, which contributes to the saltation effect – the separation and rise of large numbers of dust particles into the atmosphere. This is yet another characteristic example of the nonequilibrium structure in a dynamical dissipative system in which the entropy production tends to its minimum value, corresponding to an equilibrium state as the dust storm decays.

Dynamics of the Atmosphere of the Outer Planets

The main property of the atmospheric circulation on giant planets is alternating prograde (eastward) strong equatorial jets on Jupiter and Saturn and retrograde (westward) jets on Uranus and Neptune. They have different speeds and widths and are responsible for the above mentioned ordered system of zones and belts at low and middle latitudes, which is most clearly exhibited on Jupiter and Saturn (Fig. 1.3.27). Coriolis forces induce zonal winds northward and southward of each zone, giving rise to vortices and shear turbulence. The zonal wind velocity on Saturn reaches 500 m/s compared to its mean value of ~150 m/s on Jupiter; the steepest temperature gradients are also observed here, while there is no noticeable temperature difference between the equator and poles. The light zones and dark belts are the regions of upflows and downflows, respectively. Since $\mathbf{R} \ll 1$ for these rapidly rotating planets, strong zonal flows emerge between zones and belts due to the Coriolis interaction of meridional flows. The strongest winds of a variable direction are observed in the transition regions where shear flows are formed. The most detailed investigation and comprehensive analysis of the mechanism of planetary dynamics on Jupiter was made by the Galileo mission (see Galileo Orbiter Team, 1998a, b).

At the same time, the main mechanism of planetary dynamics along with the disordered motions and numerous coherent vortex structures (convective cells) observed at high latitudes is natural convection caused by a heat source in the interiors. This is apparently responsible for the growth in zonal wind velocity with depth from 70 to 175 m/s between the levels with a pressure of 0.4 and 5 bar measured by the Galileo probe (see Galileo Probe Team, 1996; 1998). This mechanism is similar in many respects to the classical problem of the hydrodynamic Rayleigh-Taylor instability of a horizontal fluid flow heated from below – Benard's problem. However, in the case under consideration, the convective interiors are in a close dynamical interaction with the overlying gas layer where the solar energy is absorbed. In other words, convection generated by the deep

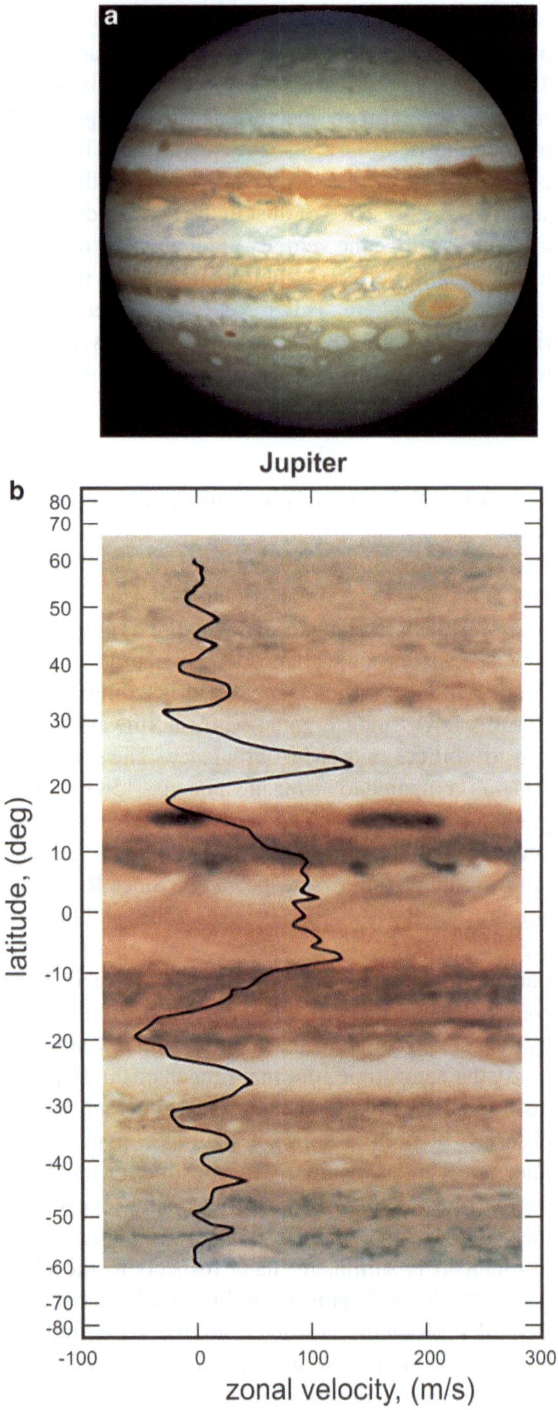

Fig. 1.3.27 (**a**) Jupiter with characteristic light zones and dark belts in its strongly turbulized atmosphere at the upper cloud deck. A Hubble Space Telescope image in October 1995. (**b**) Structure of zones and belts on Jupiter. The lighter regions of upflows are zones and the darker regions of downflows are belts. The solid curve indicates winds of a variable direction (with the velocity both in the direction of the planet's rotation in zones and in the opposite direction in belts); strong shear flows emerge in the transition regions. Disordered motions of various scales observed in the cloud structure are superimposed on this image. The two dark regions in the belt at latitudes 8-18 °N are hot spots recorded in the 5-micron wavelength. An image from the Voyager spacecraft. Courtesy of NASA

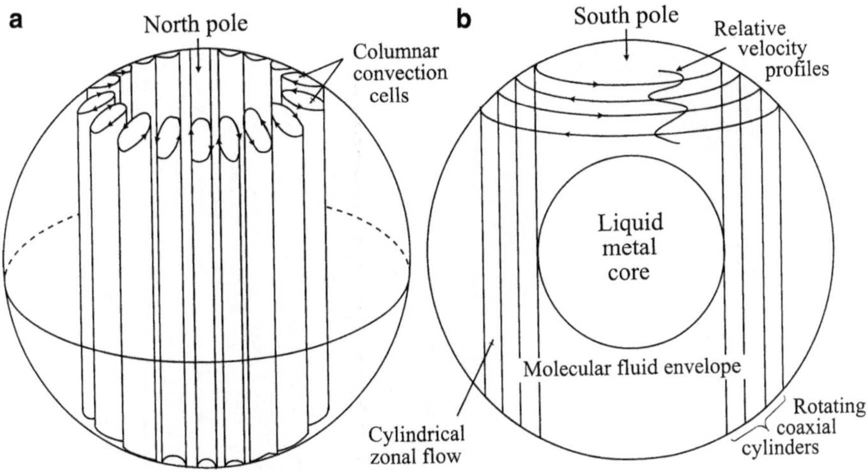

Fig. 1.3.28 Formation of roll-type convective cells (**a**) and a cylindrical zonal flow (**b**) on a rapidly rotating fluid sphere. Roll convection is the most characteristic form of the convective instability of a viscous conducting fluid heated from below in the case of uniform axisymmetric rotation, while coaxial cylindrical surfaces serve as the most general form of the zonal flow of an ideal fluid with an internal adiabatic temperature gradient. The transfer of energy from inclined convective cells to the zonal flow in the horizontal shear layer reflects the interaction between these two forms of motion. According to Ingersoll and Pollard (1982) and Ingersol et al. (1984)

intrinsic heat flux penetrates into the upper troposphere where equatorial Rossby waves arise. This leads to a very complex flow pattern with numerous vortex structures and a high degree of turbulization, as is observed most clearly on the disk of Jupiter. The scheme shown in Fig. 1.3.28 corresponds to this model (Busse, 1976; Ingersoll and Pollard, 1982). In turn, the formation of a zonal flow and the maintenance of planetary circulation on a rotating planet can be explained by a nonlinear mechanism of vortex energy transformation into the kinetic energy of mean motion (Ingersoll et al., 1998). A three-dimensional general circulation modeling and differential rotation seen in laboratory experiments shows that the various flow patterns can be reproduced by mechanisms universal across the giant planets, provided differences in their radiative heating and intrinsic heat fluxes are taken into account (see, e.g., Glatzmaier et al., 2009; Liu, Shneider, 2010). Nonetheless, we cannot yet explain the very complicated dynamics and flow configuration on the giant planets in an energetically consistent manner.

Clouds serve as an indicator of various motions in the strongly turbulized atmospheres of Jupiter and Saturn (West et al., 1986). However, the composition of the main condensates does not explain the observed palette of colors on the disks of these planes, which represent a truly unique natural phenomenon. Therefore, it should be assumed that there are more complex compounds in the atmosphere and clouds, including phosphine (PH_3), hydrocarbons (C_nH_n), and organic polymers, which presumably formed under solar ultraviolet radiation and lightning discharges.

Fig. 1.3.29 Strongly developed turbulence near the Great Red Spot (GRS) in Jupiter's atmosphere with a rich gamut of cloud colors. The GRS on Jupiter (~ 25000Ч12 000 km in size) with strongly turbulized flow regions westward and southward from it. Smaller vortices ("white ovals") are on the right. The maximum image resolution is 95 km. An image from the Voyager spacecraft (courtesy of NASA)

The upper cloud layers on Uranus and Neptune at a lower effective temperature are composed of methane below which the cloud layers of ammonia and sulfur-bearing compounds are probably located. The intense absorption of the red part of the solar spectrum by methane explains the characteristic aquamarine color of these planets. Obviously, the dynamical processes involving chemical transformations are typical of the multicomponent turbulent media of giant planets (Kolesnichenko, Marov, 1999); the order in their chaotic structure observed in the atmosphere at the level of clouds emerges through the energy exchange at the microscopic level. Examples of strongly developed turbulence in the atmospheres of Jupiter, Saturn, and Neptune are shown in Figs. 1.3.29 and 1.3.30, while Fig. 1.3.30 also gives an example of the formation of ordered vortex structures in Saturn's turbulized clouds that is attributable to self-organization processes.

Just as on Jupiter and Saturn, convection acting as the thermal mechanism when energy is transferred from depth must also play an important role in Neptune's atmospheric dynamics. In contrast to Uranus, this planet has an internal heat source that is twice the energy influx from the Sun. At the same time, the most interesting feature determining the thermal regime and atmospheric dynamics of Uranus is an unusual orientation of its rotation axis lying almost in the plane of the circumsolar orbit, so that the change of seasons occurs twice a year (which is 84 years on Uranus). However, despite the large difference in inclinations and energetics, qualitatively identical meridional temperature and zonal wind profiles are observed

Fig. 1.3.30 (**a**) Saturn's image from the Cassini spacecraft. The band structure of the cloud layer attributable to planetary circulation and the turbulized structure of the clouds near the equator are clearly distinguishable. (**b**) Neptune with the system of bands attributable to zonal circulation at the cloud level. The aquamarine color of the planetary disk is explained by the methane absorption of the red spectral region. An image from the Voyager spacecraft. (**c**) A example of the emergence of ordered structures in the clouds of Saturn's turbulized atmosphere. An image from the Voyager 2 spacecraft on August 20, 1981. Courtesy of NASA

on both planets at the level of clouds, although the wind on Uranus is weaker approximately by half. In addition, it is important to emphasize that on Neptune, despite the fact that the power of its energy sources per unit area is approximately a factor of 20 lower than that in the Jovian atmosphere, the wind velocity is almost a factor of 2.5 higher, reaching 400 m/s on the equator, i.e., almost the same as on Saturn. This is because the Neptunian atmosphere has a very low turbulent viscosity and, accordingly, a low level of energy dissipation of wind motions and shear flow turbulization (Ingersoll et al., 1995). Interestingly, in contrast to Neptune, the terrestrial atmosphere has the highest level of dissipation in which the processes related to the hydrologic cycle play a major role along with small-scale convection and surface friction. Therefore, although the Earth derives incomparably more solar energy, the wind velocities on it are almost an order of magnitude lower than those on Neptune. Another interesting feature is that the direction of winds on Neptune and Uranus is opposite to the direction of their rotation. This distinguishes them from Jupiter, Saturn, and Venus (as well as the Sun and Titan), which are characterized by equatorial superrotation.

The numerous large-scale vortex structures whose nature, as has already been said, is associated with variable directions of zonal winds under the Coriolis force, particularly stand out among the remarkable features of the cloud structures on the giant planets. The Great Red Spot (GRS) on Jupiter, whose size varies from 10 000 to 14 000 km in the latitudinal direction and from 24 000 to 40 000 km in the longitudinal one, and the Great Dark Spot (GDS) on Neptune with a size of $\sim$ 15 000 km are the best known of these vortex structures (Figs. 1.3.29, 1.3.31). The GDS reflects intense atmospheric dynamics (a strong zonal wind reaching 400 m/s on the equator) under conditions in which the heat flux from the interiors exceeds the insolation value by a factor of 1.6. The photochemical processes producing complex hydrocarbons in the atmosphere and clouds are triggered in the methane atmosphere of Neptune even at a comparatively low flux of solar ultraviolet photons at a distance of 30 AU. At the same time, the absence of ammonia in the atmosphere of Neptune (and Uranus) was quite unexpected. This can be explained by assuming that concurrently with its formation in the equilibrium thermochemical reactions mentioned above, its evacuation in the reaction with H_2S takes place, giving rise to NH_4SH crystals in clouds. However, this requires assuming that the accretion disk region where Uranus and Neptune were formed contained much more H_2S than NH_3 and that the S/N ratio exceeded the solar one by a factor of 5 (Lunine, 1993; Cruikshank et al., 1995).

The GRS on Jupiter is a giant anticyclone whose lifetime estimated from the similarity parameters is several thousand years, as distinct from terrestrial cyclonic structures with a typical lifetime on the order of one week. It is located in the atmosphere above its surrounding cloud layer and is caused by ascending motions, and the release of latent vaporization heat may serve as an additional energy source. The GRS has a bright gamma of colors and a complex morphology of internal vortex flows. Particularly strong flow turbulization and gas and cloud particle exchange between the vortex and neighboring zones are observed on the periphery where the velocities exceed 100 m/s. The vortex makes a complete turn in seven days,

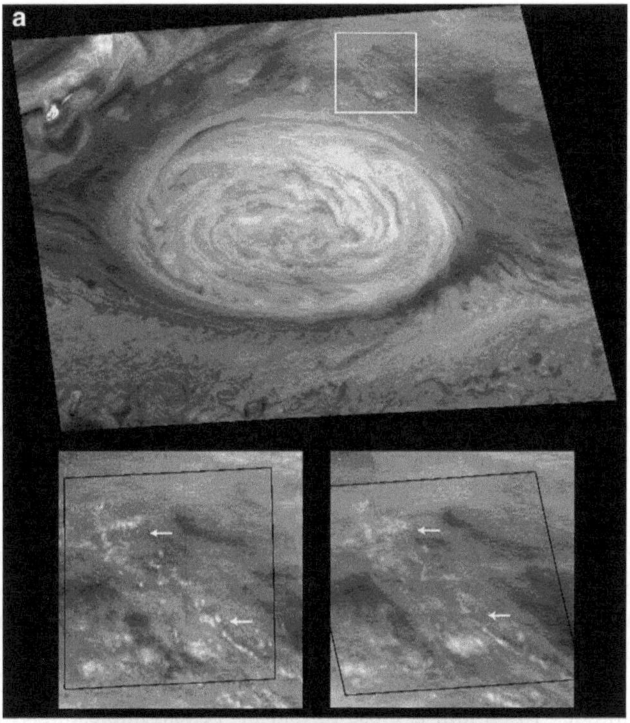

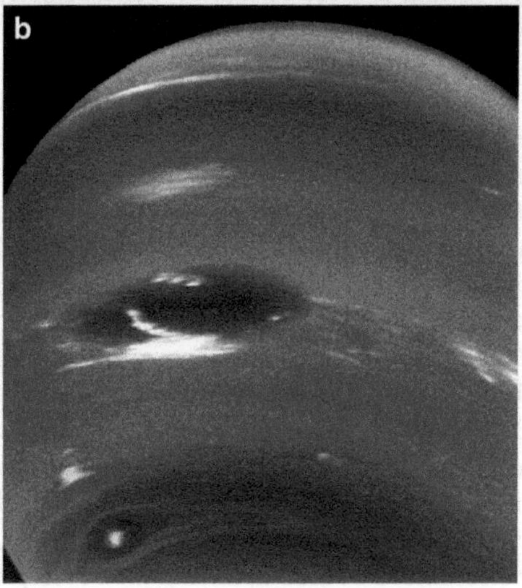

Fig. 1.3.31 Comparison of the Great Red Spot (GRS) on Jupiter and the Great Dark Spot (GDS) on Neptune as examples of self-organization in a planetary atmosphere. **(a)** The GRS is a huge stable vortex with a counterclockwise motion in the Southern Hemisphere (anticyclone) with mean sizes of ~ 12 000 km in latitude and ~ 25 000 km in longitude. The sequence of cloud displacement in Jupiter's turbulent atmosphere northward from the GRS is shown in the bottom part. The size of the highlighted region is 50 km; the time between the images is 70 min. An image from the Galileo spacecraft. **(b)** The GDS is a stable cyclone-type vortex with a size of ~ 15000 km. The bright white spots formed by overlying clouds are seen on the southern side of the GDS and below it. An image from the Voyager spacecraft. Courtesy of NASA

while the GRS itself periodically drifts forward and backward in latitude, having turned three times around the planet in the 20th century. There are no such large structures on Saturn, but numerous vortices (ovals) with smaller sizes and shorter lifetimes exist, just as on Jupiter and Neptune. Additionally, powerful storms occur more frequently in the Saturnian atmosphere. The motions in the ovals are clockwise, while their longitude-latitude oscillations resemble the motion of the upper part of a vortex in a stably stratified shear flow. Just as for ordered zonal flows, it is natural to consider them from the standpoint of the formation of a hydrologic cycle in a stratified gas-liquid medium by taking into account its chemical composition, energetics, and the fulfillment of the stability criterion.

We cannot but mention the unique dynamics of the atmosphere on one more body in the system of giant planets – the Saturnian satellite Titan. The insolation on Titan is too low to provide the development of intense dynamical processes, but, nevertheless, they do exist. The tidal effects from Saturn, which are stronger than the lunar tides on Earth by a factor of 400, can be assumed to serve as the main energy source. The assumption about the tidal mechanism of wind motions is backed by the above-mentioned orientation of the rows of dunes ubiquitously encountered on Titan. As measurements of the Cassini spacecraft showed, the preferential direction of winds is indicative of superrotation in the atmosphere of Titan similar to that in the Venusian atmosphere. The altitude profile of the wind velocity, which in the atmosphere of Titan increases from several to $\sim$ 30m/s in the range of altitudes 10-60 km, is close to that of Venus. Interestingly, the motions weaken greatly above this region, and strong turbulization of the atmosphere is observed, starting from an altitude of $\sim$120 km. Unfortunately, the reason for these unusual dynamical properties of the medium cannot yet be explained. They may be associated with the formation of a layered structure of hydrocarbon clouds in this region of Titan's atmosphere.

We have seen that more or less stable structures periodically emerge in the atmospheres of planets against the background of chaotic (turbulent) gas motions within an open nonlinear system that exchanges energy with the environment. Cyclones and anticyclones on Earth, superrotation of the atmosphere on Venus and Titan produced by various energy sources, powerful dust storms on Mars, stable structures such as the GRS and GDS and other cyclonic ovals in the atmospheres of Jupiter, Saturn, and Neptune are the most characteristic stable structures. They are observed for a more or less long time in a turbulent medium against the background of chaotic small-scale activity in the form of relatively small clouds appearing and disappearing for several hours. These apparently also include the emergence of a reverse greenhouse effect on Mars accompanying the development of a global dust storm and the role of minor atmospheric components that are of great importance in the complex of aeronomical processes and that have an enormous influence on the course of the planet's evolution. Based on the concept of nonlinear dynamics of complex dissipative systems, these phenomena of atmospheric dynamics can be considered from the standpoint of self-organization in the structures of chaotic turbulent media.

1.3.5 Nature and Dynamics of Small Bodies

The small bodies (comets, asteroids, meteoroids, meteoric dust) whose size spectrum extends from several hundred kilometers (large asteroids) to microns (dust grains) are the most dynamic objects in the Solar System. They played and continue to play an important role in its evolution owing to migration processes and numerous collisions with planets. The comets whose nuclei have a complex structure and chemical composition are particularly interesting because of the diversity and complexity of their physical-chemical processes. In addition to the prime interest they trigger in planetary cosmogony, the comets also attract increasing interest as possible carriers of primordial life forms (Marov, 1994; 2005; Gaidos and Selsis, 2007).

It is fundamentally important to investigate the small bodies primarily from a cosmochemical viewpoint, because they are the carriers of the primordial matter from which the Solar System was formed. It is believed that the comets and carbonaceous chondrites, the most primitive asteroids of CI class, have retained particles from the protoplanetary cloud and the gas-dust accretion disk in their composition, because they have undergone the smallest changes in the course of evolution (Weidenschilling, 1997; Wooden et al., 2007). A more detailed discussion of this issue is contained in Sect. 1.3.6. Therefore, it is natural to consider the comets as a kind of "probes" of the Galactic regions nearest to the Solar System. The comets can be associated with the planetesimals ejected by tidal perturbations from the formation regions of the Jovian planets during their evolution to the periphery of the Solar System where the main cometary reservoirs - the Kuiper Belt near the ecliptic plane and the spherical Oort cloud - were formed early in the Solar System's history (Fig. 1.3.32).

The main bulk of asteroids in the inner Solar System is located in the above mentioned Main Belt between the orbits of Mars and Jupiter between ~ 2.7 and 3.2 A.U. The number of bodies larger $\sim$ 1km is estimated to be about 10^5, and there are many more bodies of an even smaller size, while there are also large bodies of a few hundred km across, the largest being the miniplanets Ceres and Vesta. Nonetheless, the total mass of bodies in the Main Belt of asteroids is less than 10^{-3} M_E. They are considered to be a relict of a planet that failed to form because of great tidal perturbations from nearby Jupiter (see, e.g., Marov, 1986).

The Kuiper Belt is located beyond the orbit of Neptune and extends from 40 to approximately 100 AU. The number of trans-Neptunian objects (TNOs) in the Kuiper Belt is estimated to be $\sim 10^8$; more than 10^4 can be larger than 200 km in size (Eneev, 1981; Jewitt, 1999). About a thousand TNOs have been discovered to date. Among them are bodies comparable to Pluto in size (Sedna, Quaoar), which are characteristic representatives of this family of trans-Neptunian icy miniplanets – plutonoids (Jewitt, 1999; Jewitt et al., 1998; Cruikshank et al., 2007). Trans-Neptunian objects are characterized by a great variety of physical processes that manifest themselves in different albedos (from 2.5 to $> 60\%$) and spectral features, through which one can diagnose the properties of their surfaces. In particular, H_2O, CH_4, and N_2 ice deposits were detected in some of the bodies, which makes them akin to Pluto and Triton, although the mean density is appreciably lower and closer to the nuclei of comets (~ 0.5 g/cm^3). Interestingly, about 11% of the investigated objects are binary systems, while Pluto even has three satellites (Cruikshank et al., 2007). It may well be that the Trojan asteroids at the libration points L4 and L5 are the bodies from the Kuiper Belt captured by Jupiter at an early chaotic phase of the Solar System formation (Morbidelli et al., 2005), contrary to the traditional viewpoint that they were formed as the planet itself was growing and were then captured into a 1:1 resonance (Chiang and Lithwick, 2005). The similarity between the color albedos of Trojans and cometary nuclei (the growth in reflectivity with wavelength) and the absence of distinct spectral signatures, as in most trans-Neptunian bodies, provides evidence for the hypothesis of their connection with TNOs. Similar properties are characteristic of the groups of bodies located between the orbits of Jupiter and Neptune - the Centaurs, which are apparently also genetically related to the Kuiper Belt and partially evolve to the orbits of Jupiter-family comets.

The comets in the Oort cloud occupy the very vast region well beyond the Kuiper Belt, whose outer boundary extends to approximately half the distance to the nearest stars (30-60 thousand AU). These comets periodically experience gravitational perturbations from giant interstellar gas-dust clouds and the Galactic disk, as well as during random encounters with stars. Some of the bodies from the cloud can then pass to highly elliptical orbits and can be observed as long-period comets as they approach the Sun. Subsequently, under the influence of gravitational perturbations from planets, they either replenish the well-known families of short-period comets that regularly return to the Sun, or they leave the Solar System forever by passing to parabolic or hyperbolic orbits (Fernandez, 1994; Weissman, 1996). However, the main source of short-period comets is the Kuiper Belt. It experiences gravitational perturbations from Neptune, causing a relatively small fraction of icy bodies to migrate into the inner Solar System, as will be seen from the subsequent analysis. There they replenish the Main Asteroid Belt between the orbits of Mars and Jupiter (Fig. 1.3.32).

The nuclei of comets have small sizes, from a few to several tens of kilometers, and a very low mean density, which usually does not exceed fractions of g/cm^3. This suggests a porous structure of these bodies, which are composed predominantly of water ice and some other low-temperature condensates with an admixture of silicates, graphites, metals, and organic compounds (Marov, 1994; Boice and Huebner, 1999). The water-ice composition of comets is explained by the fact that the water molecule is most abundant in the Solar System and, not coincidentally, H_2O ice also accounts for a significant fraction of the mass of the satellites around the giant planets and other small bodies. The processes involving liquid water at the initial evolutionary stages of the Solar System are particularly clearly traceable by the mineral composition of meteorites and, consequently, their main parent bodies from the Main Belt asteroids (Jewitt et al., 2007).

Based on the measurements made during the encounter of spacecraft with several comets, we can say that, on the whole, the nucleus corresponds in its physical properties to the model of a dirty snowball (or a ball of frozen dirt)

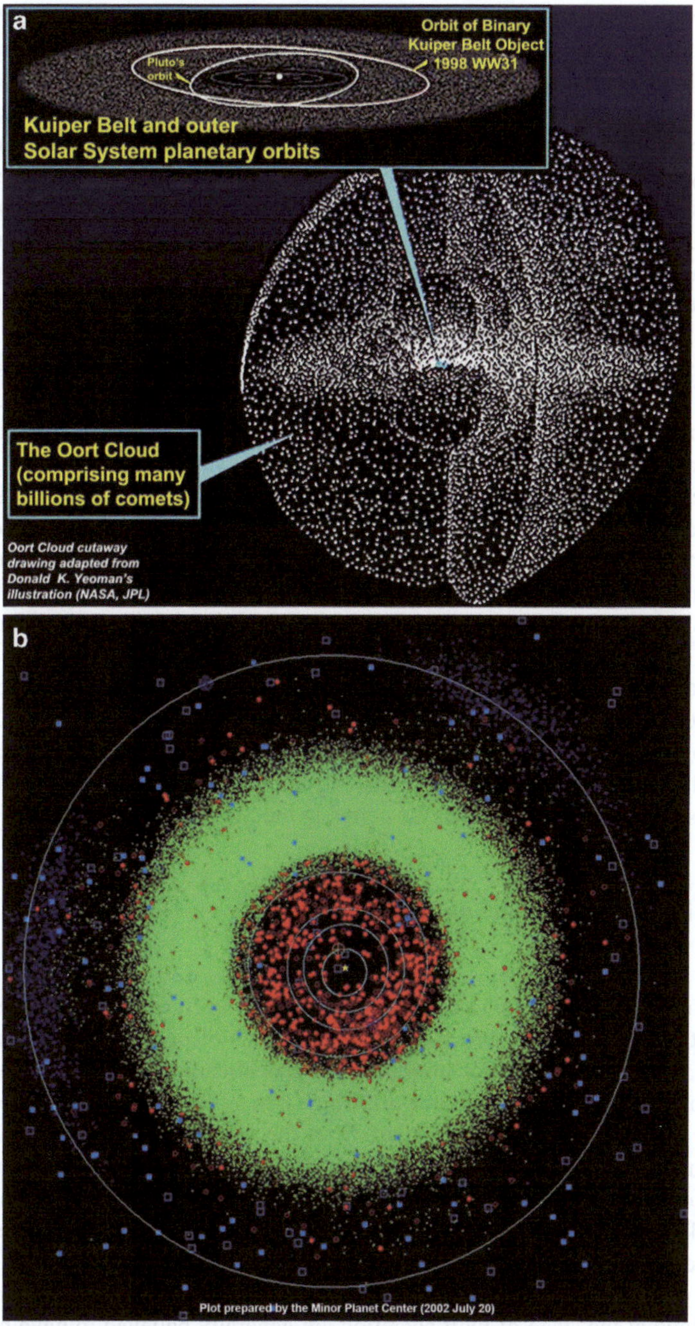

Fig. 1.3.32 **(a)** The Oort cloud and the Kuiper Belt. The Kuiper Belt located at the outskirt of our planetary system (40 – 100 AU) lies deep inside the Oort cloud whose outer boundary is at a distance of $10^4 – 10^5$ AU **(b).** The Main Asteroid Belt between the orbits of Mars and Jupiter (2.8 – 3.2 AU) is indicated by the green dots. There are groups of Earth-approaching asteroids (Amur, Apollo, Aton) also called near-Earth objects (NEOs; red dots) inside the Belt. Courtesy of Yeomans and US Planetary Society

proposed back in the middle of the past century by F. Whipple (Whipple, 1951). An atmosphere (coma) tens of thousands of kilometers in extent, a hydrogen halo hundreds of thousands of kilometers in size, and a tail stretching in the antisolar direction for millions of kilometers are formed through ice sublimation as a comet approaches the Sun and the insolation grows. The coma composed mainly of water, hydroxyl, and hydrogen molecules forms the comet head, which shines because of luminescence processes and partially ionized by short-wavelength solar radiation (Fig. 1.3.33). Dust is carried away into the atmosphere in great amounts during ice sublimation; its submicron particles produce a cometary tail of the first type under light pressure, while the plasma forming in the outer coma produces a tail of the second type when interacting with the solar wind.

Studying the nonstationary heat and mass transport processes in a porous nucleus and the formation of an inhomogeneous structure of the surface from which an icy conglomerate sublimates and a gas-dust coma is formed is of key importance for understanding the nature and evolution of comets at various heliocentric distances. As kinetic modeling results showed (Skorov et al., 2002; Marov et al., 1996), the flow near the nuclei of active comets in the entire daytime hemisphere is close to an equilibrium flow, the gas density decreases rapidly as one recedes from the surface of the nucleus, while the temperature drops to $\sim$ K at a distance of $\sim$ km from the nucleus due to adiabatic expansion into a vacuum (Bisikalo et al., 1989; Marov et al., 1996), and a well-defined jet is formed near the symmetry axis. Several jets attributable to intense removal of gas and dust are seen in the image for the nucleus of Comet Halley obtained by the Giotto spacecraft during its flyby (Fig. 1.3.33). This inhomogeneous sublimation from the surface of the nucleus can be explained by thermal deformations producing faults and fractures in the dust-ice crust formed during successive approaches of the comet to the Sun.

Migration of Small Bodies and Implications

As has already been noted, the orbital properties of small bodies reflect the dynamics of regular and chaotic motions in the Solar System and manifest themselves through migration processes. The collisions of comets and asteroids with planets are a consequence of migration. As has already been noted, they lead to catastrophic events and mass transport, which is very important for the evolution of planets and their atmospheres. Apart from the previously mentioned Tunguska event and Chicxulub, the fall of more than 20 fragments from Comet Shoemaker-Levy torn apart by tidal forces in the immediate vicinity of Jupiter onto this planet in 1994 serves as a convincing confirmation that the collisional processes occurs periodically in the Solar System (Fig. 1.3.34).

The primary atmospheres on the terrestrial planets captured from the protoplanetary cloud at the stage of their accretion are generally believed to have been lost. The fact that the abundances and isotopic composition of inert gases in the existing secondary atmospheres differ sharply from solar abundances serves as convincing evidence for this loss. Undoubtedly, degassing from the interiors during the

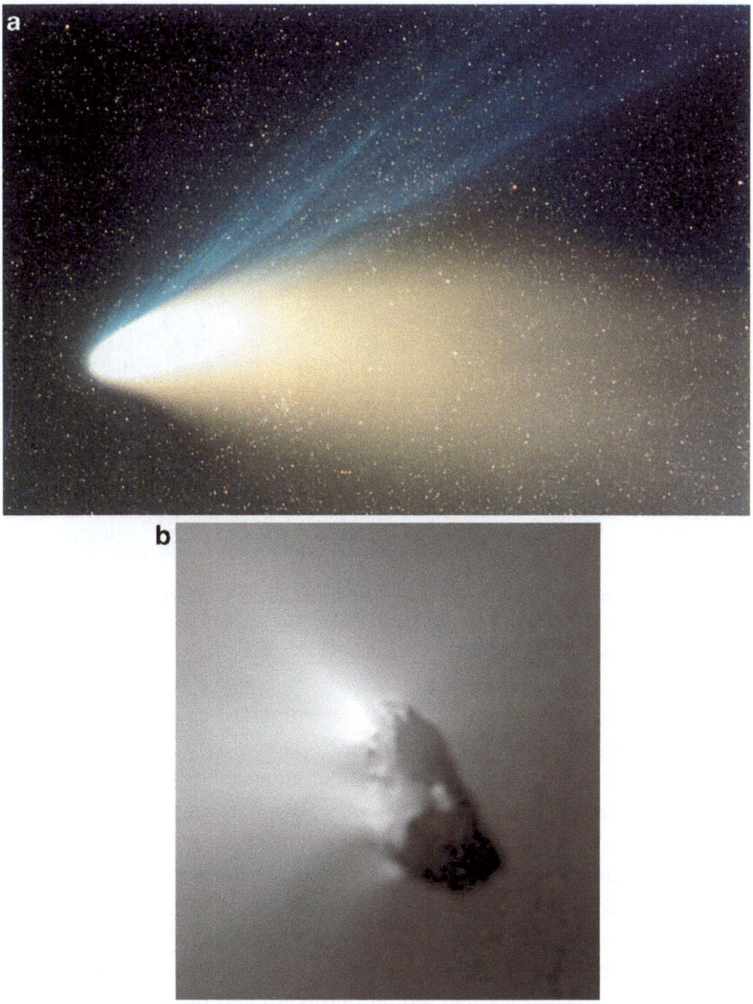

Fig. 1.3.33 (**a**) Image of Haley Bopp comet during its encounter with the Sun. A small nucleus (~10 km) is hidden deep inside a bright region – coma (cometary atmosphere) tens of thousands of kilometers across produced by the sublimation of gas and dust from the icy surface of the nucleus. Extended type I and II tails are clearly seen. (**b**) The nucleus of Halley comet from the distance of 4000 km. The mean size of the nucleus is ~15 km. The streams of gas and dust (jets) from the icy surface of the nucleus as the comet approached the Sun are seen on the left. An image from the Giotto spacecraft (ESA courtesy)

subsequent evolution of these planets contributed significantly to the formation of their hydrosphere and atmosphere of a secondary origin. However, this source apparently was not unique and sufficient to compensate for the loss of low-temperature volatiles (water, nitrogen, carbon, sulfur, etc.) in the formation zone of the terrestrial planets at a temperature > 1000 K in this part of the protoplanetary disk confined by the so-called "snow line" through Jeans dissipation. The low

Fig. 1.3.34 The fall of fragments from Comet Shoemaker-Levy 9 torn asunder by tidal forces when approaching Jupiter in critical distance in 1994. This event clearly manifests the chaotic nature of migration and collision processes in the Solar system

masses of these planets along with their high temperatures also contributed to the escape of the lightest atmophile elements into space.

The existing, nevertheless relatively high abundances of volatiles on the terrestrial planets can be explained by the intense bombardment of these planets by comets and asteroids from the outer Solar System at an early evolutionary stage. This scenario is known as the heterogeneous accretion mechanism (Anders and Owen, 1977; Owen and Bar-Nun, 1995; Chyba et al., 1994). The efficiency of this mechanism was quantitatively estimated by studying the orbital dynamics of small bodies, primarily the comets from the trans-Neptunian belt containing the largest amounts of volatiles. The classical N-body problem and the BULSTO and RMVS computational codes (Ipatov, 2000; Marov, Ipatov, 2001; 2007; Ipatov et al., 2006a, 2006b) were used as a basis for the developed numerical migration model. According to the existing estimates (see Duncan and Levison, 1997), $\sim$ 20000 of $\sim 10^9$ TNOs with sizes $\geq$ 1 km can pass from the above-mentioned accumulation zone of secular perturbations in the Kuiper Belt to Jupiter-crossing orbits (JCOs) during their migration into the Solar System and can stay there for $T \sim 0.13$ Myr. Out of these 20000 JCOs, the evolution of 5500 virtual objects from known families or types of comets toward the Sun was numerically integrated (in time intervals of 10 Myr with a step of 500 yr and a relative error of 10^{-8}-10^{-9}) by taking into account the gravitational effects from planets. Examples of the computations are shown in Fig. 1.3.35. As a result, the probability of their collisions with the Earth, Venus, and Mars was estimated. The integration was performed with initial conditions corresponding to the orbital parameters of three types of comets: the Jupiter family,

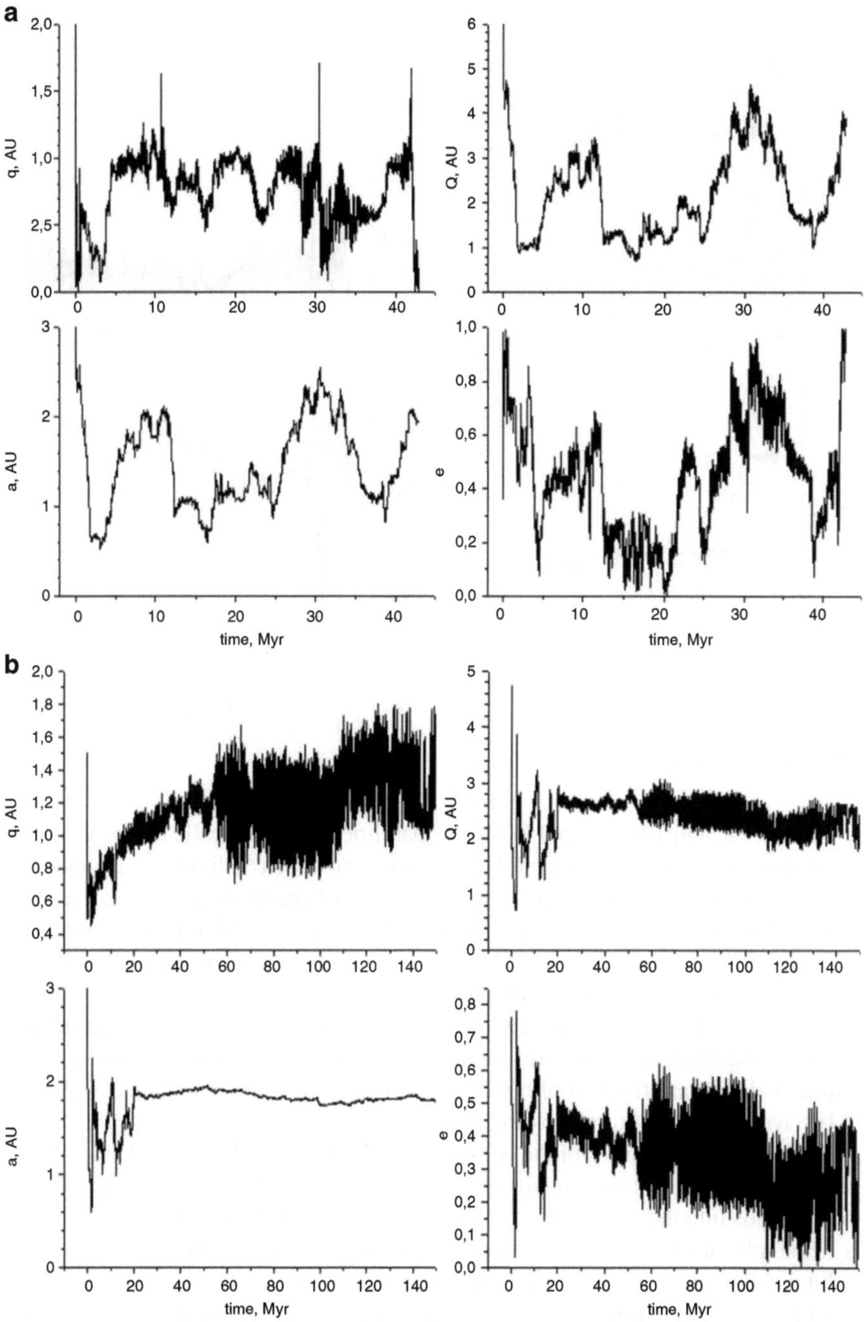

Fig. 1.3.35 Examples of the calculations of the migration of virtual bodies that were initially in orbits like those of Jupiter family comets (**a**), 10P (**b**) and 2P (**c**) further inwards the Solar system. The time variations in perihelion distance q, aphelion distance Q, semimajor axis a, and eccentricity e are shown

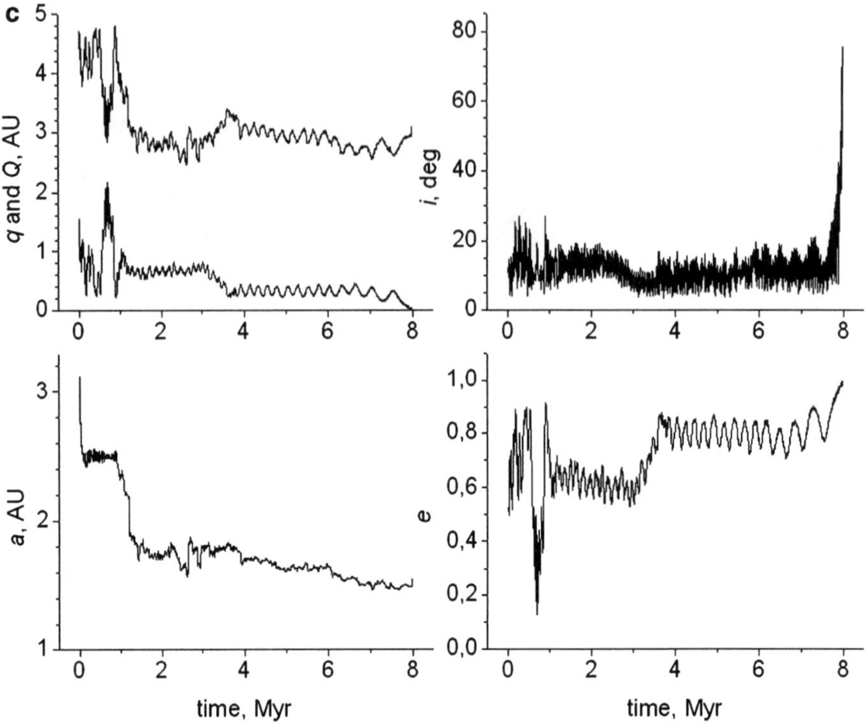

Fig. 1.3.35 (continued)

9P Tempel 1, and 10P Tempel 2. The probabilities P_S of the collision between one ~1-km object and a planet in time T_S were found from the condition $q < a$, where q is the perihelion distance, a is the semimajor axis of the planetary orbit; the mean probabilities and times were defined as $P_r = P_S/N$ и $T_r = T_S/N$. It turned out that for the Earth $P_r = 6.65{\cdot}10^{-6}$ in $T_r > 1$ Myr, i.e., one object can collide with the Earth every ~ 0.5 Myr. Comparable, though slightly lower ($P = 4{\cdot}10^{-7}$ and $P \leq (1-3){\cdot}10^{-6}$), probabilities were obtained in the models by Morbidelli et al. (2000) and Levison and Duncan (1997), respectively. This may be because according to our model a significant fraction of bodies (~ 0.0015) with aphelion distances $Q < 4.7$ AU stayed longer in Earth-crossing orbits.

The mean probabilities of the collision with Venus and Mars turned out to be the same and approximately a factor of 3 lower, respectively. One of every 300 such objects falls into the Sun – these are the so-called *sun-grazers*. Figure 1.3.35 shows an example of the computations for the migration of a virtual TNO with initial orbital parameters of a Jupiter-family comet in the form of evolution of the perihelion and aphelion distances, semimajor axis, and eccentricity *q, Q, a*, and *e* in intervals with widths of 0.005 (a-b) and 0.1 (c-d) AU.

Several important conclusions were drawn from the simulation results regarding the pattern of migration of the bodies from the Kuiper Belt that were initially captured on a Jupiter-crossing orbit into the Solar System and their transition times to the Main Belt and NEO families. The fairly high probability of the collisions of small bodies between themselves and with planets leads to permanent fragmentation/ejection of material and its transport into the interplanetary medium, providing a kind of "exchange processes". This can cause catastrophic consequences on time scales of ~1 Myr.

The key corollary of the model is confirming the important role of comets as a source of volatiles in the evolution of the Earth, Venus, and Mars. Based on the estimated lower limit for the collision probability $P_r = 4 \times 10^{-6}$ and the estimated total mass of the Jupiter-crossing planetesimals, ~100M_E (M_E is the Earth's mass), we obtain the mass of the bodies that collided with the Earth, M ~ 0.0004 M_E. Next, assuming that the content of water ice was ~ 0.5 M, we conclude that the water mass delivered to the Earth from the feeding zone of the giant planets is ~ 2×10^{24} g, which is almost a factor of 1.5 larger than the water mass in the terrestrial oceans (Vinogradov, 1967). Venus and Mars obtained approximately the same volumes of water per unit mass. This important conclusion is consistent with the views about the existence of ancient oceans on the planets adjacent to the Earth that were lost in the course of their subsequent evolution: the runaway greenhouse effect on Venus and the burial in the Martian cryosphere, as was discussed in the preceding section of this chapter.

Naturally, the estimates of the amount of water supplied to the Earth by comets are valid only in order of magnitude and we are far from asserting that this source was unique. Undoubtedly, a certain contribution was made through degassing from the interiors, although, as has been mentioned, so far it is hardly possible to determine the relative importance of the endogenic and exogenic sources. However, it is important to emphasize that an estimate with an error of 50% and not by orders of magnitude is undeniable evidence that the comet model is realistic. It is constrained by the deuterium-to-hydrogen ratio D/H derived from the measurements of H_2O and HDO in six comets, among them Hyakutake and Hale-Bopp, which turned out to be twice that in the terrestrial oceans, namely 2.96×10^{-4} compared to the standard oceanic value (SMOW) D/H=$1{,}57 \times 10^{-4}$ (see Meier and Owen, 1999). However, this hardly rejects the connection of the terrestrial hydrosphere with the cometary source (see Delsemme, 1999). First of all, the measured cometary D/H ratio refers to long-period comets from the Oorth cloud that differ significantly in genesis from the Jupiter-family comets (JFC) with which the delivery of volatiles to the terrestrial planets is associated. Both groups may have been formed in different regions of the protoplanetary disk at different temperatures, which should have affected the D/H ratio[2]. In addition, the estimate of the amount of water supplied by comets has an error within at least a factor of 2 and, as has already been said, we do not rule out the assumption that nearly half of the oceanic water owes its origin to an endogenic source. Finally, we also mention

[2] Indeed, most recent measurements made with the heterodyne detector of the Herschel Space Telescope found the D/H ratio in the comet Hartley-2 to be $1{,}57 \times 10^{-4}$ - similar to that in the Earth's oceans (Hartough et al., Nature 478, pp. 218–220, 2011).

that according to several dynamical models, apart from comets, hydrated asteroids from the Main Belt could also be responsible for a certain fraction of water in the terrestrial oceans (Morbidelli et al., 2000). The interested reader can find a more complete discussion of these problems, including the ratio of isotopes and the abundances of inert gases, in Jewitt et al. (2007).

It can be assumed that the transport of material, including its primordial forms, and the migration of interplanetary dust particles bear the most direct relation not only to the evolution of planets but also to the origin of life as one of the alternative models discussed above. The relative fraction of the cometary and trans-Neptunian particles containing $\sim$10% of volatiles in their composition in the overall balance of the dust that fell onto the terrestrial planets was estimated based on numerical calculations of dust migration by taking into account both gravitational and several other effects (Marov and Ipatov, 2005; 2006,a,b; 2007). Examples of these calculations are shown in Fig. 1.3.36. It turned out that compared to the contribution of small bodies, the contribution of dust to the supply of volatiles was lower by three or four orders of magnitude. However, the dust particles may have been most efficient in delivering organic or even biogenic matter to the Earth because they are considerably less heated when passing through its atmosphere.

Such a coincidental and at first glance insignificant impact may have excited the initially stable system of our Earth and may have triggered the emergence of primitive life forms under favorable conditions. Subsequently, these life forms rapidly developed and hugely gained in biological diversity, with the transition (bifurcation) of the natural medium to a new stable state, which was then the prototype of the Earth's early biosphere. A process like this can still be imagined under similar circumstances on a primitive body.

1.3.6 Protoplanetary Accretion Disks

Studying the birth and evolution of the Solar System and the formation of planetary systems around other stars is among the fundamental problems of modern natural science. By its content, this problem is interdisciplinary and requires elaborating on a generalizing theory that underlies the development of mathematical models for the physical-chemical structure and evolution of a turbulent gas-dust accretion disk around a single solar-type star. Star-forming regions are strongly turbulized chaotic media in which successive ordering processes take place, from the fragmentation of a molecular cloud and the birth of a star to the formation of a planetary system. Examples of these media are shown in Fig. 1.3.37. According to model views, the circumstellar disk is formed through a finite angular momentum of the molecular cloud, and the star recieves much of its mass from it. There are magnetic accretion flows inside the disk that are responsible for the transport of angular momentum that accompanies the growth of the stellar mass and the disk evolution.

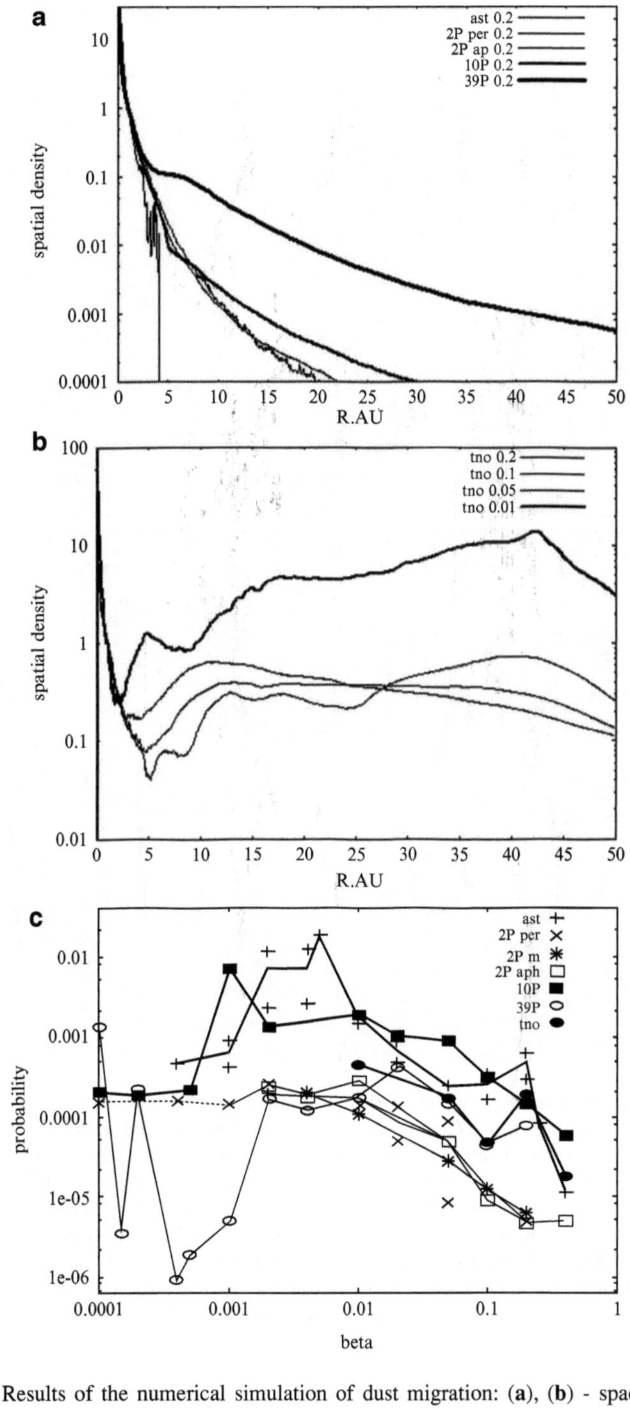

Fig. 1.3.36 Results of the numerical simulation of dust migration: (**a**), (**b**) - space density of interplanetary dust particles of various origin depending on the radial distance (in a.u.); (**c**) - probabilities for the collision of dust particles of various origin with the Earth. ast – asteroids; 2P, 10P, 39P – comets (per – perihelion, aph – aphelion); tno – trans-Neptunian objects

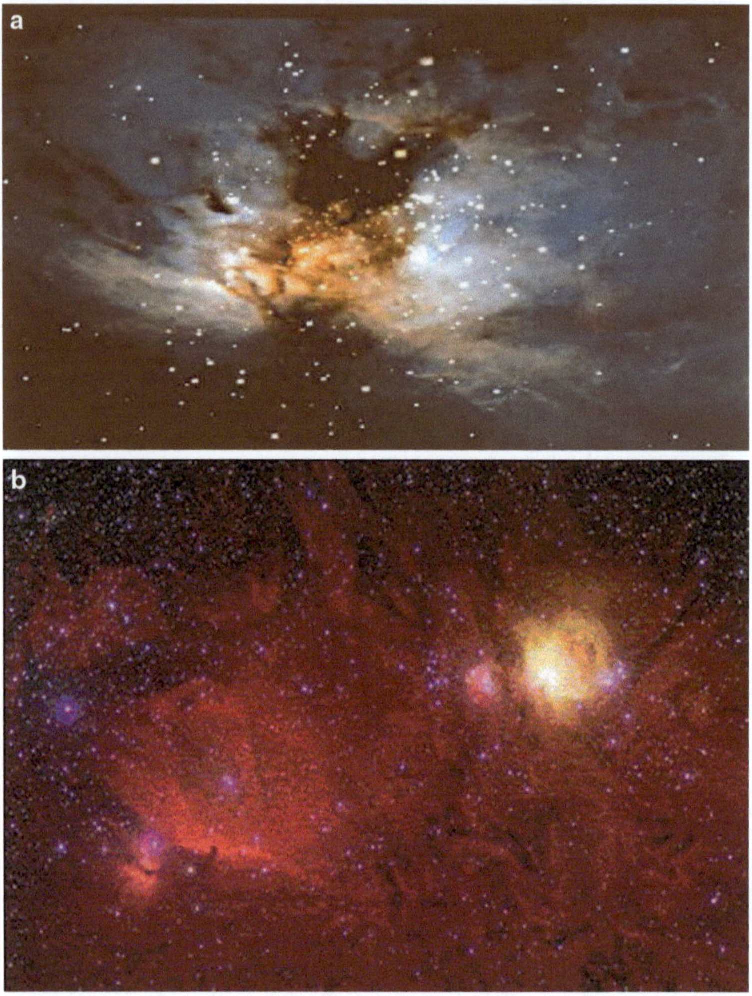

Fig. 1.3.37 **(a)** Star-forming region NGC 2024 in the Orion Nebula. The images were obtained with the infrared camera of the Gemini spacecraft in the three near-infrared bands ($\lambda = 1.2$, 1.65, and 2.2 μm - the blue, green, and red colors in the image, respectively) and subsequently processed by superposition of the images.. Whereas the central part of the nebula in an ordinary photograph in visible light appears dark due to the absorption of light by dust, a dense cluster of young stars is detected in infrared light. Courtesy of UCLA and NASA. **(b)** The star-forming region in Orion (according to M Bessell, RSAA, ANU). **(c)** The star-forming region η Carinae. The image was obtained with the Spitzer Space Infrared Telescope Facility (SIRTF). The same sky region in the visible spectral range is shown on the left for comparison. Courtesy of NASA

c

Fig. 1.3.37 (continued)

The fact that a significant fraction of young stars are surrounded by gas-dust disks similar to the disk around the young Sun had become obvious already by the early 1990s (Strom et al., 1993). Subsequent observations (see, e.g., Lagrange et al., 2000; Schneider et al., 1999) backed the starting concept that a certain part of the material from the parent cloud (nebula) with an appreciable angular momentum remains at the orbit around the central clump and is incorporated into the protoplanetary disk in the process of stellar collapse. Basically, the Kant-Laplace hypotheses about the simultaneous formation of the Sun and the protoplanetary cloud, along with the idea of rotational instability responsible for the successive separation of plane concentric rings (from the material of which the planets were subsequently formed) from the cloud periphery, underlie these views. It is believed that after the isolation of the star, the material from the outer cloud regions continued to accrete onto the disk, causing strong turbulization of the gas-dust medium due to a mismatch between the specific angular momenta of the falling matter and the disk matter involved in Keplerian rotation (Lissauer, 1993; Kwok, 2000; Makalkin, 2003; Dorofeeva and Makalkin, 2004; Russell et al., 2006).

The construction of cosmogonic models requires a detailed study of a turbulized multicomponent gas-dust medium including the kinetic processes in the protoplanetary disk, its dynamical and thermal evolution, and the cosmochemical constraints imposed by the properties of the primordial matter from which the planets were formed (see, e.g., Lewis, 1997; Lagrange et al., 2000). Note in this connection that the thermal bifurcation processes apparently contribute significantly to the initiation of molecular cloud fragmentation and collapse, much as they contribute to the formation of silicate particles in the outer regions of expanding red supergiants or to the autocatalytic cooling of CO regions in the solar chromosphere (Ayres, 1981). This is an example of the structuring of a molecular gas-dust cloud and a forming turbulent disk where the self-organization processes play an important role.

Enormous progress in studying the problems of planetary cosmogony has been achieved in recent years through the use of a wide arsenal of observing (primarily space) facilities and the improvement of mathematical models. Infrared gas and dust emission measurements (Millan-Gabet et al., 2007) and high-angular-resolution submillimeter observations made the most significant contribution to the chemical study of protoplanetary disks. These observations allowed numerous transitions of molecular components, such as CO, and their isotopic composition to be identified (Dutrey et al., 2007). The studies carried out with the Spitzer space telescope (see Meyer et al., 2007), in particular, the discovery of several young stars surrounded by disks in a fairly small star-forming region (Fig. 1.3.38), especially contributed to the understanding of the structure and evolution of circumstellar disks. The extension of some disks is comparable with Neptune's orbit in the Solar System (Fig. 1.3.39).

Numerous molecules were found in protoplanetary disks. Some of these molecules are probably genetically related to the volatiles contained in the frozen granules of primordial accreting material (such as H_2O, CO, N_2, H_2CO, HCN, etc.) and were subsequently subjected to substantial chemical and thermal processing. Some of these molecules are apparently in a nonequilibrium state due to the photolysis processes attributable to hard ultraviolet and X-ray radiation from a young star, causing ionization of the surface layers, among others.

Data on the dust composition and properties are fundamentally important for understanding the evolution of a protoplanetary accretion disk, particularly the structure, thermal regime, and dynamics of its inner regions. This understanding is helped by studies that are also carried out in the optical, near- and thermal infrared spectral ranges, including the measurements of emission spectra with long-baseline interferometry. This provides information about the mineralogy of dust particles (Hildebrand, 1983; Pollack et al., 1994; Watson et al., 2007; Millan-Gabet et al., 2007) and thus serves this purpose. An efficient technique for solving inverse light scattering problems by taking into account the wavelength dependence of particle properties was developed. It allows reconstructing the physical characteristics of dust, with the meteor particles in the terrestrial atmosphere serving as its analog. Light scattering data and millimeter interferometry led to the interesting conclusion that the particles in a disk are much larger than those in a diffuse interstellar medium, reaching millimeter and even centimeter sizes, i.e., they resemble sand and pebbles (Natta et al., 2007). It was found that their sizes obey altitude stratification, so that small micron particles are concentrated near the disk surface; this stratification can persist for millions of years. Naturally, the size spectrum of the solid particles (granules), which affect the medium's opacity and flow turbulence because of the dependence on dust concentration and inertia, should be taken into account when modeling the disk evolution and especially the subdisk formation.

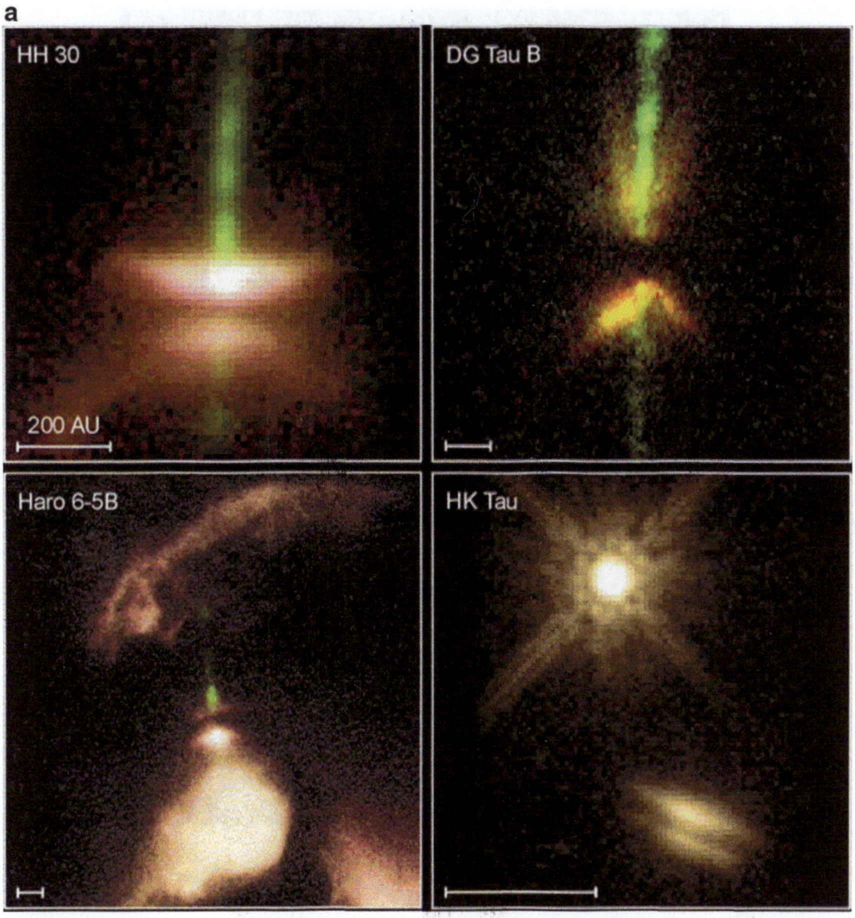

Fig. 1.3.38 **(a)** Gas-dust disks (the dark bands between the bright regions) around young stars from Hubble Space Telescope observations. **(b)** The star-forming region NGC 1333 from Spitzer Space Infrared Telescope Facility (SIRTF) observations. The surrounding disks are clearly seen around several stars. Courtesy of NASA

The content and properties of the granules are directly related to the chemical transformations in a gaseous medium and collectively have the strongest effect on the thermal regime and viscous properties of the disk, especially in the active zones determined by the distance from the star and from the central plane (Bergin et al., 2007). There is reason to believe that the solid particles are roughly similar in origin and nature to those located in the interstellar medium and found in meteorites. They incorporate components such as carbon-bearing insoluble organic matter (IOM) and annealed glass with embedded metal and sulfides (GEMS). These submicron particles can be identified with the crystalline component of the predominantly amorphous interstellar dust silicates enriched with magnesium. However, it may

Fig. 1.3.38 (continued)

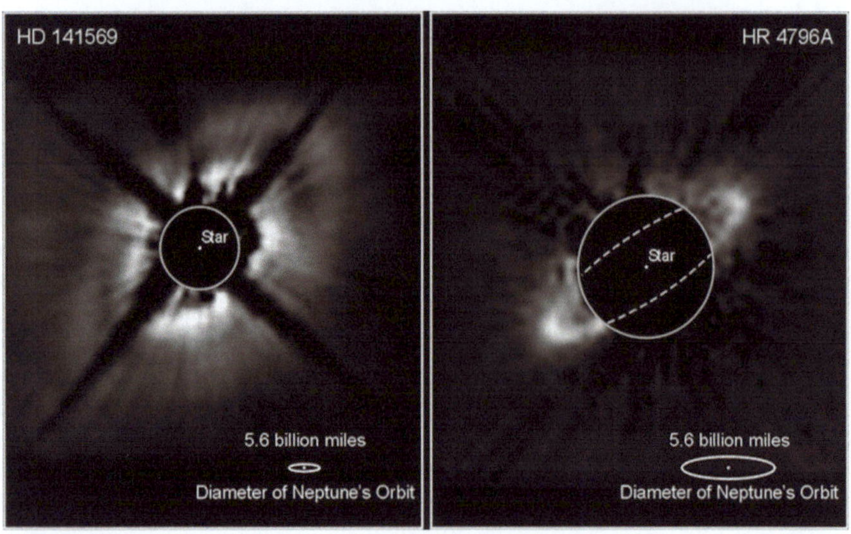

Fig. 1.3.39 Images of gas-dust disks obtained from STS OPO and Hubble Space Telescope observations. Courtesy of NASA

well be that they are the condensates of the early Solar System retained in comets and related meteor streams and also in chondrite porous interplanetary dust particles (CP-IDPs). Consequently, by their genesis, they can correspond to the matter that is produced in the immediate vicinities of circumstellar disks and that may have subsequently undergone evaporation-crystallization during its radial motion in the disk and during its heating by shock waves in the accretion zone and subsequent rapid cooling (see Mathis, 1990; Kemper et al., 2004; Bouwman et al., 2001, 2005; van Boekel et al., 2004; Alexander et al., 2007; Wooden et al., 2007).

The formation time of differentiated iron and stony meteorites (which coincides with the crystallization time of cores and shells of their parent bodies) was defined from the radio isotope study of the meteorite matters based on measurements of ratios of long- and short-lived isotopes and products of their decay, in particular ^{208}Pb-^{207}Pb–^{206}Pb, ^{26}Al–^{26}Mg, ^{53}Mn–^{53}Cr, ^{87}Rb-^{86}Sr, ^{129}I-^{129}Xe, ^{182}Hf – ^{182}W. As we said earlier, the refractory inclusions of millimeter sizes enriched with the refractory elements Al and CA (Calcium **A**luminum **I**nclusions, CAIs) contained in the chondrite meteorites Allende and Efremovka were thoroughly analyzed (Amelin et al., 2002; 2006; 2010; Bouvier et al., 2007; 2008; 2009; 2010). Because they were assumed to belong to the ancient solid material, the most probable value of the absolute age of the Solar System (as a star with disk) was deduced to be 4567.22 $\pm$ 0.21 mil. yrs. (Ma). Concurrently, the absolute age of iron meteorites was defined 4567.5 $\pm$ 0.5 Ma, i.e., accounting for referred margins, their age is within the next $\sim$ 1 Ma. In turn, the absolute age of stony meteorites is 4564.91 $\pm$ 2.58 Ma and therefore, the difference $\Delta = 3.64 \pm 1.52$ Ma can be admitted as an estimate of the total time of accumulation and thermal evolution (differentiation) of the parent bodies of these ancient meteorites. Note that the origin of submillimeter chondrules *(spherules*), which are composed of ferromagnetic silicates and are imbedded in stony meteorites, is also dated to between 1.7 and 2.0 Ma after CAIs origin, close to the formation timeframe for chondrites of different petrological classes (see Amelin et al., 2002; Bouvier et al., 2008; Klein et al., 2008b; 2009). Obviously, the primordial parent bodies of about 100 km in size formed in the very first millions of years since the Solar System began to evolve and experienced differentiation through intense heating by short-lived isotopes, mostly ^{26}Al (half-life time decay only 0.74 Ma, which means that in $\sim$ 5 Ma this isotope has been fully exhausted) with an iron core emergence. The follow-up core and silicon shell fragmentation caused by numerous collisions has probably been responsible for the above iron and stony meteorites. It is also believed that several million years later, when the most efficient heat sources such as the short-lived ^{26}Al and ^{60}Fe nuclides were exhausted, non-differentiated chondrites were formed that underwent no melting and eventually became the parent bodies of planetesimals (Wadhwa et al., 2007).

This time scale agrees with the results of our computer modeling (Marov et al., 2008), which argue for the age of the accretion disk of 0.65 - 1.9 Ma depending on the accretion rate from disk onto the protosun (5×10^{-7} - 5×10^{-8}) $M_{sun..}$, respectively. In turn, the time of accretion termination and maximum mass of the disk was estimated 1 – 2.5 Ma. The time of subdisk formation, where the disk was composed of 1 – 10 cm particles before it reached the critical density/gravity instability at $r \sim$ 1 a.u., was only 0,01 – 0,1 Ma. Evidently, this time was sufficient for the accumulation and thermal evolution of the first solid bodies. Assuming that the mass of the protoplanetary cloud M_{cl} was $\sim 0.1\ M_{sun}$ and that $\sim 0.1\ M_{cl}$ ultimately entered the planets, we estimate that $\sim 10^9$ original bodies with a size of $\sim$ 100 km were born, followed by subsequent collisional evolution.

The idea that the protosolar nebula was produced by a supernova explosion in the vicinity of a compact gas cloud that had initially formed through the fragmentation of a more massive gas cluster gained great popularity in planetary cosmogony. The study of short-lived radionuclides in the Allende meteorite provided a basis for this hypothesis (Wasserburg, 1985; Wasserburg and Papanastassiou, 1982). Particularly elucidating in this respect is the ratio of the isotope ^{26}Al to its daughter isotope ^{26}Mg, which is formed by nucleosynthesis in the star at the closing phase of its life and is implanted into the protosolar nebula's particles when the supernova remnant was injected into it. The general views about the role of short-lived radionuclides with a mean half-life $\geq$ 10 Myr in the evolution of the early Solar System are associated with this idea (see Wadhwa and Russell, 2000; McKeegan and Davis, 2003; Kita et al., 2005; Gounelleu et al., 2006; Wadhwa et al., 2007). It provided evidence for the probable contribution of ^{10}Be and ^{36}Cl to the energetics. In turn, this allowed one to refine the abundances of other short-lived radionuclides, such as ^{60}Fe, ^{41}Ca, and ^{182}Hf, and to reconstruct the chronology of events in the solar protoplanetary disk with a fairly high accuracy.

The modeling results suggesting that an excess pressure is needed to cause the gravitational collapse of a diffuse cloud similar to the parent cloud of the Solar System and the disk separation favor the concept of a supernova explosion. In principle, such an excess pressure, along with the process-accelerating turbulization of the interstellar medium, could be provided by the shock waves generated by a supernova explosion (see the extensive bibliography to the review by Ballesteros-Paredes et al., 2007).

Based on the models being developed, attempts are made to answer with a greater certainty the questions regarding the sequence of changes in the aggregate state of the main protoplanetary matter components, the location of the condensation-evaporation fronts depending on thermodynamic parameters of the disk, the role of particle sublimation and coagulation in a two-phase medium, the relative contribution of radiation and turbulence to the heat and mass transport, and the mechanisms for the development of hydrodynamic and gravitational instabilities with allowance made for the shear stresses in boundary layers and the polydispersity of suspended dust particles. Of particular interest are the poorly resolvable inner disk regions within several astronomical units, where matter actively accretes onto the young star. This causes great changes in the dust/gas ratio, the contribution of photochemical processes, and the thermal regime, whose formation conditions are determined by the two limiting models of optically thin and optically thick disks (Najita et al., 2007; Dullemond et al., 2007). It is very important to understand how much the evolution processes in the circumsolar protoplanetary disk differed from those that are now observed for young solar-type stars. This is especially true for the above-mentioned topical problem of the formation of peculiar configurations of extrasolar planetary systems, especially in the immediate vicinities of the parent star, to which their stability on long time scales is associated.

Planetary System Formation

Whereas the model views about the origin of protoplanetary nebulae are backed by observational data, mainly the mechanical, physical, and cosmochemical characteristics of the Solar System serve as the starting concept of the formation of the planets themselves (see Lewis, 1997; Cassen and Woolum, 1999; Dorofeeva and Makalkin, 2004)). Indeed, the existing patterns in the system of planets and satellites definitely point to a common process of their formation, while the data on the surface properties and matter composition for planets and small bodies when compared with the samples of material from their embryos and "debris", meteorites, provide an insight into the probable sources, paths, and chronology of this process, as has been said above. The idea of the accretion of planets from a gas-dust disk after its separation gained the greatest recognition (Schmidt, 1957; Safronov, 1969; 1982). It incorporates the dynamics of primordial gravitating bodies after developing perturbations in a rotating dense gas-dust subdisk and its breakup because of a gravitational instability as well as the sequence of matter accretion on bodies of intermediate sizes, embryonic clumps, and gradual scooping of smaller bodies by them in the course of swarm evolution (Fig. 1.3.40). Since the braking increases in a gas, their rotation velocity becomes lower than the Keplerian one, which must contribute to the acceleration of this process. The mechanism of original matter exchange in the radial direction, whose efficiency imposes certain constraints on the possibility of realizing various evolution scenarios for the disk and the degree of its chaotization when the embryos of planets were formed, could also play an important role (Levy, Lunine, 1993; Weidenschilling, 2000a, 2000b).

The matter of the protoplanetary gas-dust cloud is a complex multiphase medium with regions of different densities, temperatures, and degrees of ionization. This matter, which is generally a dusty plasma, is magnetized and in a state of strong turbulization. According to present views, the planets are formed after the loss of gravitational stability by the dust subdisk produced by the differential rotation of turbulized protoplanetary matter in an orbit around a solar-type star and the accretion processes when the dust component settled to the equatorial (central) disk plane perpendicular to the rotation axis (Toomre, 1964; Safronov, 1969, 1982; Goldreich, Ward, 1973; Nakagawa et al., 1981; Makalkin, 1994; Youdin and Shu, 2002; Durisen et al., 2007). Note that the flattening of a rotating protoplanetary cloud results from the confrontation of two main dynamical forces – the gravitational and centrifugal ones. When these forces are in balance, weaker factors, such as the thermal and viscous processes, disk self-gravity, and electromagnetic phenomena, become significant for the cloud evolution. It is believed that owing to the viscous forces of friction (arising from the relative displacement of gas

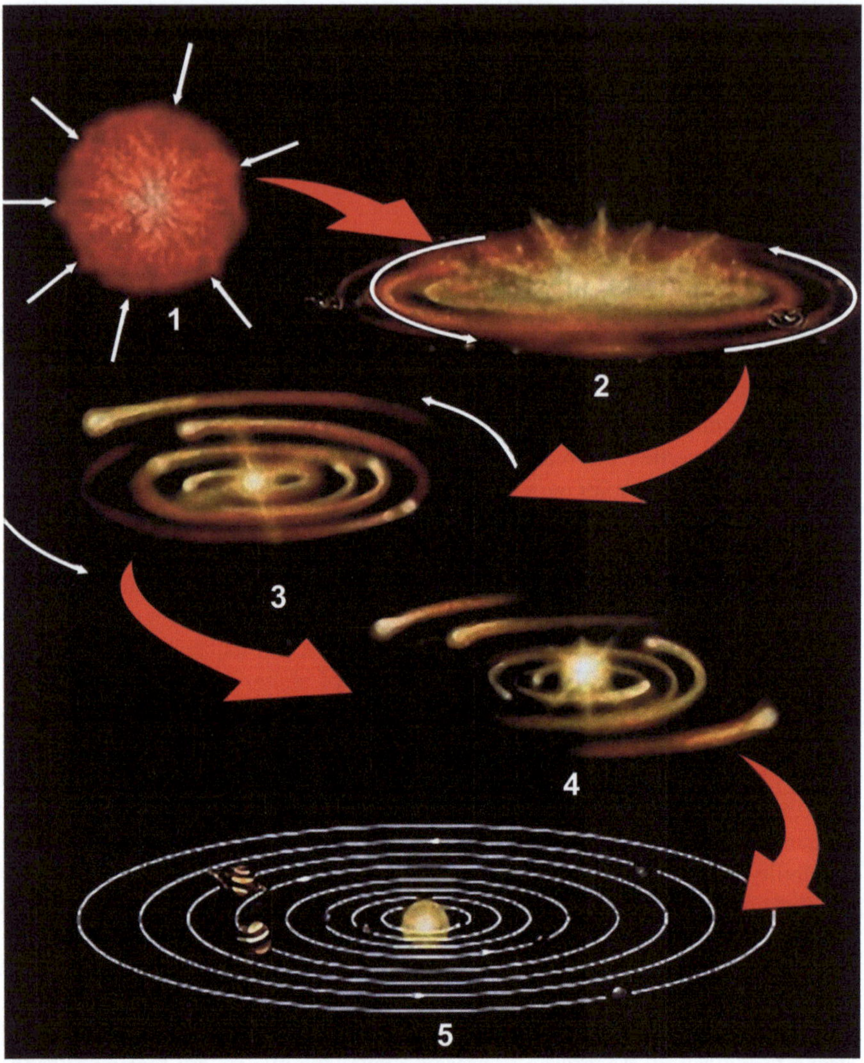

Fig. 1.3.40 Scheme for the formation of the Solar system from the collapse of a molecular cloud fragment through the formation of the proto-Sun and protoplanetary disk (1,2), then its breakup into individual ring clumps of solid particles giving eventually birth to planetesimals (3,4). Continuing collisional interactions of planetesimals ultimately leads to the formation of planets (5)

suspension elements during their orbital motion), the disk matter drifts toward the proto-sun along a very gently sloping spiral trajectory as its angular momentum is transported outward – from the inner disk regions to the outer ones. Since viscosity depends on temperature, which also decisively affects the condensation conditions for volatiles, including water, this has a significant effect on the relative content of solid particles, disk energetics, and angular momentum transport conditions (Cuzzi et al., 2005; Marov et al., 2007; 2008).

Basically, the turbulent nature of any circumstellar gas-dust accretion disk is recognized to be of key importance in their dynamics (see, e.g., Zel'dovich, 1981; Fridman, 1989; Dubrulle, 1993; Balbus and Hawley, 1998; Richard and Zahn, 1999; Bisnovaty-Kogan and Lovelace, 2001). At the same time, the turbulence processes in the cosmogonic models that are developed, which largely determine the disk structure and evolution, are, unfortunately, taken into account very approximately only. This is not coincidental because the disk matter is an inhomogeneous disperse medium composed of a (generally multicomponent) gas and dust particles with various sizes. The role of dust particles, including the dynamics of their drift in the directions radial and orthogonal to the central disk plane and the phase transitions during particle evaporation and/or condensation with allowance made for the temperature stratification in the disk, turns out to be crucial.

From dimensional considerations it can be assumed for a turbulent shear flow of thickness z rotating with an angular velocityΩ that the vortex resulting from the existence of a vertical rotation velocity gradient $dv/dz{=}\Omega = \ell \boldsymbol{Ro}$ (where $\boldsymbol{Ro}$ is the Rossby number) at $\boldsymbol{Ro} = \boldsymbol{const}$ is adapted to the angular velocity of the entire disk. However, this is by no means obvious in the presence of dust particles (Safronov, 1987; Cuzzi, 2004; Cuzzi and Weidenschilling, 2006). The gradient dv/dz leads to the settling of particles near the $z = 0$ plane and the formation of a dust subdisk. This is accompanied by an emerging radial drift in the disk and by a change in the thermal regime due to turbulent energy dissipation. Obviously, turbulence absent under complete mixing conditions at $dv/dz = 0$ should increase as the growing particles settle. The settling rate and the time scale of these competing processes will change as dv/dz increases to $\ell \boldsymbol{Ro}$. In turn, this should affect the emerging conditions for gravitational instability in the subdisk that determine the growth of larger bodies. Invoking the shear turbulence mechanism, in general, backs the views about the possibility of ring-like contraction of a flat protoplanetary cloud and the planet formation from initially "porous" gas-dust clumps that fill the main part of their sphere of attraction (Hill's sphere) and slowly contract owing to internal gravitational forces (Eneev and Kozlov, 1981).

Note that the turbulence generated at the boundaries of the protoplanetary disk layers corresponds in character to the parameters of an Ekman boundary layer with a thickness $\delta{\sim}(\nu^{turb}/\Omega)^{1/2} \sim \Delta v\ /\Omega \mathbf{Re}^{turb}$ (where, as above, ν^{turb} is the turbulent viscosity, $\Delta\nu\, v$ is the difference between the rotation velocity of the gas disk and the Keplerian velocity of dust particles, and $\mathbf{Re}^{turb} = \Delta\nu\delta/\nu^{turb}$ is the turbulent Reynolds number (Safronov, 1969)). In this case, the mixing length is $\sim\delta$ and the turbulent mixing rate is $\nu^{turb}{\sim}\delta^2\ \Omega$. In most evolutionary models for the accretion disk around the young Sun under consideration, the coefficient ν^{turb} corresponds to a higher relative velocity of particles ν_r. Nevertheless, this does not allow one to assert that the cohesion processes are more probable than the fragmentation ones during the collisions of particles, on which depends the solution of the problem of dust cluster formation and gradual growth of planetary embryos in the dust subdisk. Moreover, the available laboratory experiments are inconsistent with the views existing until recently that this scenario is efficient (see, Blum, Wurm, 2000; Sekiya, Takeda, 2003; 2005; Chiang, 2004).

A major problem of a planetary system formation is associated with the mechanism for the transport of angular momentum $J_{\odot}$ from a collapsing solar-type star to the protoplanetary disk. For a uniform-density protosolar nebula, it lies within the range $10^{52} < J_{\odot} < 10^{53}$ g cm^2 s^{-1} and is limited by a value that is approximately an order of magnitude higher for a disk with a mass concentrating on the center. This transport is most likely due to turbulent viscosity in a rotating, convectively unstable gas disk, which determined the time scale of its expansion (Safronov, 1969; 1987; Cassen and Summers, 1983). Subsequently, the turbulent vortices through which the particles accelerated and coalesced more easily into "rings" of matter could play similarly important role in the embryo accretion process.

Alternative possibilities for the loss of angular momentum by the Sun at an early stage of evolution are associated with shear motions during Keplerian disk rotation and, in the presence of a partially ionized medium, with the action of electromagnetic forces or emerging local shear instabilities in a poloidal magnetic field. The transport of angular momentum by small-scale magnetic fields was suggested in Shakura and Sunyaev (1973), along with the assumption about disk Ë-viscosity, while Balbus and Hawley (see the review by Hawley and Balbus, 1991) showed this mechanism to be important for an accretion disk. Under conditions of significant disk matter ionization at a certain ratio of the disk thickness to the radial distance H/r *(aspect ratio)*, this approach allows lifting the restrictions on the possibility of hydrodynamic turbulence generation in a purely Keplerian flow. In other words, the magnetic (Maxwellian) stresses and not the flow dynamics itself turn out to be dominant in this case. The possibility that the excess angular momentum is carried away at the contraction stage to which a magnetic field contributes cannot be ruled out either, although the magnetic fields in stars are relatively weak. Subsequently, the angular momentum of matter can be carried away by the stellar wind, and since the sum of the plasma angular momentum per unit mass and the angular momentum related to magnetic stresses remains constant, the angular momentum is transported to the surface via magnetic stresses. This causes the angular velocity of the star to decrease gradually. However, an adequate mechanism for the appearance of an effective viscosity cannot yet be found in any of the listed models, including turbulent convection, to explain the transport of angular momentum $J_{\odot}$ in the radial direction away from the central clump.

However, the turbulent and electromagnetic mechanisms should be considered to be most physical and probable for explaining this phenomenon. Indeed, the protoplanetary accretion disks have a significant viscosity, which, incidentally, combined with the differential rotation of matter, leads to a constant "intrinsic" source of thermal energy in the disks. According to present views, shear turbulences (Fridman, 1989; Gorkavyi and Fridman, 1994) and chaotic magnetic fields (Armitage et al., 2001) are the most likely causes of the viscosity of differentially rotating disks. The energy of the shear turbulences is often comparable to the energy of hydrodynamic turbulence. Chaotic magnetic fields stretching with accreting plasma, mixed by the differential rotation of the disk, and undergoing reconnection at the boundaries between chaotic cells also probably contribute significantly to the viscosity in the inner disk region and the outer layers of its atmosphere, where a sufficient degree of matter ionization is attained. In turn, large-scale magnetic fields can play an important role in both angular momentum transport and accretion physics (Eardley et al., 1978).

The problem of angular momentum transport is easier to solve in terms of the model of a disk structure that forms during the evolution of a close binary star system. The specific angular momentum of the original gas-dust cloud is known to be the main factor determining the difference between the early evolutionary stages of binary and single stars. The angular momentum for approximately one third of the protostars with gas-dust disks was estimated to be too high for a single star from the distribution of close binary stars in angular momentum, but insufficient for the formation of a close binary system (Masevich and Tutukov, 1988). This led to the conclusion that some planetary systems are probably formed mainly around such protostars, while others are formed in binary star system when one of the components breaks up into a disk near the more massive component (see Protostars and Planets, 2007).

In numerical simulations, the evolution scenario was considered as a function of parameters for the original binary star and it was shown that different configurations are obtained depending on the donor-to-accretor mass ratio and on the extent to which they fill their Roche lobes (see Marov et al., 2007). Constraints were found on the range of parameters at which either binary star coalescence or mass exchange takes place. As a result, either planetary systems like the solar one or systems with superplanets are formed. Moreover, it was shown that for the proto-Solar System to be formed, the mass ratio of the original binary system must be $(0.63/0.47)M_S$ at the binary's semimajor axis $\sim 6R_S$. It turned out that a protoplanetary system with a central star and an expanding excretion disk or with an extended gas arm in the shape of a spiral breaking up into individual clouds (condensations) with masses on the order of those of the giant planets is formed in a time on the order of a hundred orbital periods of the original binary system if the binary star coalesces. The most massive clouds were obtained in orbits with semimajor axes in the range $1\ \mathrm{AU} < a < 3\ \mathrm{AU}$, which can be reduced through tidal dissipation and friction when interacting with the disk, providing the angular momentum conservation condition. Within the framework of a numerical model for a circumstellar gas accretion-decretion disk, it was independently shown that the conservation of angular momentum requires the development of an extended decretion part of the disk that accumulates its "excess" angular momentum on the periphery. Accordingly, accretion builds a ring with a radius of only several stellar radii, while the disk expands to tens or even hundreds of stellar radii (Tutukov and Pavlyuchenkov, 2004).

Irrespective of whether the protoplanetary disk is formed around a single star or in a close binary system, shear flows are present in it and, therefore, the turbulent transport mechanism deserves the greatest attention. Moreover, it seems quite obvious that the conditions for the generation and maintenance of shear turbulence at various evolutionary stages of the disk should be considered to be based on the mathematical description of a two-phase (gas-dust) medium with a differential angular velocity of rotation Ω, different relative contents of dust particles η (so that the gas, ρ_g, and dust, ρ_d, densities are related by the relation$\rho_d = \eta\rho_g$), and their size distribution by taking into account the observational data discussed above and the coagulation processes examined in detail in Ch. 7.

1.3.7 Evolution of Cosmic Objects in the Universe

The processes of self-organization against the background of turbulent motion are among the most important mechanisms that shape the properties of objects in the Universe at different evolutionary stages. These include the fundamental processes of the evolution of primary matter, the formation of galaxies and galaxy clusters, the birth of stars and gas-dust clouds from a diffuse medium, and the formation of accretion disks and planetary systems. The basic ideas of the sequence of matter transformation processes and the developed models of stellar-galactic evolution are a very important element of the cosmology of the Universe (see, e.g., Narlikar, 1993; Peebles, 1993).

The Model of the Universe

The model of a Big Bang, whose echo is the cosmic microwave background (CMB) radiation that almost uniformly fills the space and was serendipitously discovered in 1964 by Penzias and Wilson (1965), is used as the basis for the present views of the Universe. The key question for astrophysics, cosmology, and particle physics is what the Universe mainly consists of. The visible matter in the form of stars is observed through the light emission of their constituent ordinary baryonic matter. On the other hand, determinations of the gravitating masses of galaxies and galaxy clusters from their dynamical properties revealed an enormous invisible mass – dark matter (CDM). This term was suggested by Fritz Zwicky back in 1933, but was underestimated for nearly half a century when in-depth study of this phenomenon was begun (Trimble, 1987; EAA, 2001). The detection of a temperature anisotropy and CMB fluctuations (ripples in space) first by COBE (Cosmic Background Explorer) in 1992. (Mather et al., 1999; Masser, 2006) and then by WMAP (Wilkinson Microwave Anisotropic Probe) in 2001 (Fixsen, Mather, 2002; Smoot, 2007) was certain proof of the existence of CDM. First of all, these measurements convincingly demonstrated the validity of the Big Bang theory, providing the image of a newborn Universe that was in a nearly ideal state of thermodynamic equilibrium, and revealed primordial density perturbations that subsequently formed its large-scale structure. Concurrently, these measurements

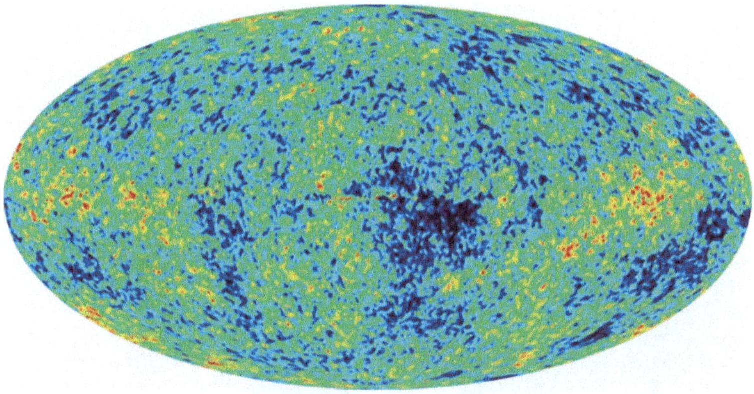

Fig. 1.3.41 Cosmic microwave background radiation from WMAP (**W**ilkinson **M**icrowave **A**nisotropic **P**robe) measurements. The fluctuations and anisotropy are less than 10^{-5} (Maser, 2007). Courtesy of NASA

confirmed the theoretically predicted CMB non-uniformity and anisotropy to be less than 10^{-5} within only a factor of 2, to which the model of a homogeneous and flat Universe corresponds (Holtzman, 1989; Holtzman and Primack, 1993; Liddle, Rand, 1993; Zuckerman and Becklin, 1993; Dalal and Kochanek, 2002). The observed baryonic oscillations (Fig. 1.3.41) may be the imprint of acoustic waves in the early Universe (see Smoot, 2007). This makes it possible to refine the relative contents of visible and dark matter.

The constraints imposed on the physical properties of CDM do not correspond to the views about the baryonic category of elementary particles as the main component of matter and lead to the conclusion that massive galaxies are gravitationally confined by a different, non-baryonic form of matter. This role could, in principle, be fulfilled by weakly interacting neutrinos with a mass on the order of several electron-volts called "hot dark matter" (HDM), which, however, was not confirmed. Nevertheless, models according to which the Universe is composed of cold and hot dark matter were proposed and the CHDM theory was used as their basis (Klypin et al., 1993; Primack et al., 1995; Primack and Gross, 2001; Elgaroy et al., 2002), concurrence between these models is still discussed. Currently, Weak Interacting Massive Particles (WIMP) of non-baryonic nature born at the Big Bang that have a mass exceeding that of protons by about three orders of magnitude were suggested as the most suitable candidate of CDM (see, e.g., Eidelman et al., 2004).

According to the present-day model of the Universe, cold dark matter accounts for about 25% of its critical density. The content of visible matter does not exceed $4.4 \pm 0.4\%$, in agreement with the estimates corresponding to the model of primordial nucleosynthesis. In the standard model, the ratio of several billion photons per one nucleon filling up the Universe (in the form of mainly CMB) per nucleon corresponds to this value. The remaining Universe is filled with vacuum energy, as followes, in particular, from the analysis of the clustering pattern of $\sim$ 250 000 galaxies and supernova observations. About 70% of the matter with

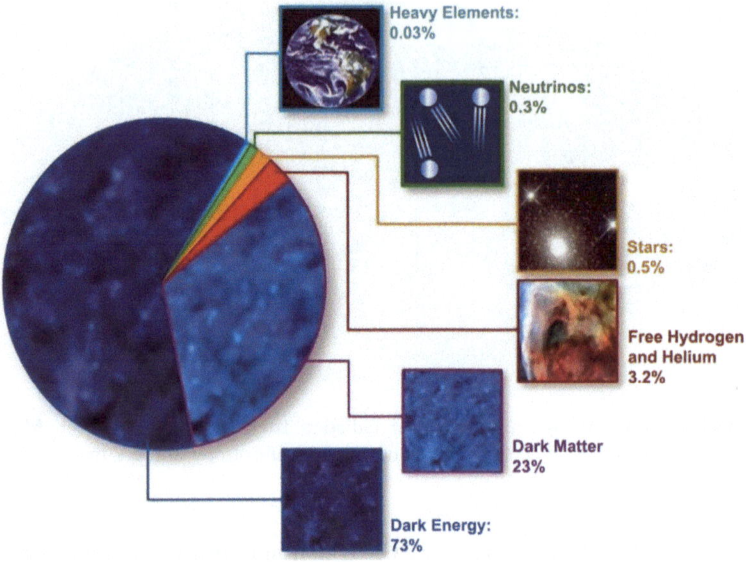

Fig. 1.3.42 Bulk composition of the Universe

which the discovered cosmological acceleration of the Universe is associated and to which the cosmological constant introduced by A. Einstein (and subsequently erroneously rejected by him) corresponds to is in the form of a vacuum energy density. This, in turn, backs the view mentioned above that the Universe has a spatially flat geometry with a nearly zero (at a level $<10^{-41}$) space curvature, which is also independently suggested by the fact that the mean density of the Universe is close (within one order of magnitude) to the critical one. It was hypothesized that the correct value of the term Λ in the mass balance equation could balance the negative space curvature produced by it with the positive curvature created by ordinary and dark matter (Maser, 2007). The bulk composition model of the Universe is shown schematically in Fig. 1.3.42.

As we can see, visible matter is composed mostly of hydrogen and helium, while the huge mass of the stellar matter accessible to observations is only ~ 0.5 percent. Stars are grouped in the great stellar association of the thin disk shape - galaxies - the main “building blocks” of the visible Universe and essentially islands of matter in a "predominantly" empty space. Every galaxy contains hundreds of billions of stars of different age, mass, luminosity, and chemical composition that are at different evolutionary stages, representing the main state of matter in the visible Universe, while the dark matter in a galaxy may exceed nearly two orders of magnitude the visible one. Galaxies are beautiful examples of the matter self-organization in the Universe (see Fig. 1.3.43). Stars in galaxies have a non-uniform distribution - maximum in the center. Our Galaxy (Milky Way) is a thin revolving disk system with a central bulge out of which twist and wind several spiral arms interspersed with clouds of gas and dust. There are about 200 billion stars in our

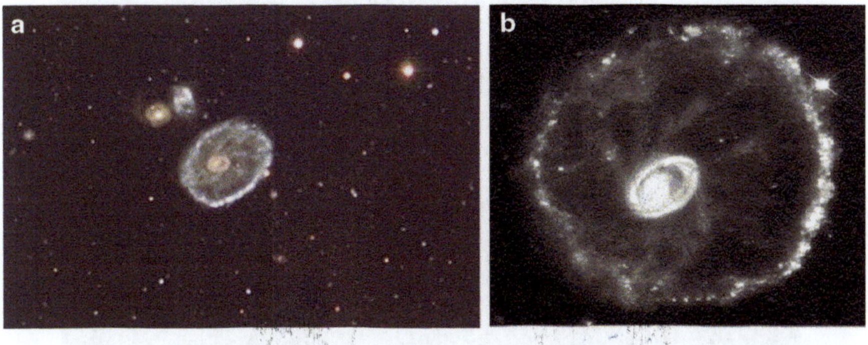

Fig. 1.3.43 **(a)** An example of self-organization of matter – the Cartwheel ring galaxy. Two other galaxies are at the top left. An image of the Anglo-Australian Observatory. **(b)** A high-resolution of the Cartwheel ring galaxy. Hubble Space Telescope image. Courtesy of NASA

Galaxy, and it stretches to $\sim 10^5$ light years (l.y.), or $\sim 30 \text{ x}10^3$ parsec (pc)., 1 pc = 3.26 l.y. The center of our Galaxy is assumed to contain millions of stars of solar mass and it is associated with a massive black hole. Quasars are distinguished as sources of enormous energy release and are thought to be the bright very massive nuclei of distant active galaxies.

Galaxies whose relative velocities reach $\sim$ 1000 km/s form clusters of progressively growing size. Clustering ranges from pairs and loose groups to giant clusters. Our local cluster of galaxies contains about 20 galaxies, while some clusters contain well over 1000 galaxies. Clusters are $\sim$ 3 Mpc in size, or 100 times the size of our Galaxy. As an example, the NGC 3190 (Hickson 44) Local Group of galaxies, which contains more than 30 galaxies, including the Milky Way and the nearest brightest galaxy Andromeda and the Magellan Clouds galaxies, is shown in Fig. 1.3.44a. The central part of the distant galaxy cluster CL 0939+4713, which contains a set of spiral galaxies, is shown in Fig. 1.3.44b. Another galactic cluster in the Virgo constellation is shown in Fig. 1.3.44c.

Clusters of galaxies form superclusters, some with more than 100 clusters in the supercluster. Our cluster of galaxies is a member of a supercluster. The size of the supercluster is about 20 to 30 times the size of a galaxy cluster, or about 300 million l.y. ($\sim$100 Mpc). A supercluster of galaxies as viewed in the radio wavelength is shown in Fig. 1.3.45. The size of the observable Universe, which contains more than 100 billion galaxies, is about 30 times the distance between superclusters, or more than 10 billion light years ($\sim$ 4 Γpc), close to the age of the Universe (13.7 billion light years). This forms the large-scale structure of the Universe *(Cosmic web)*, which consists of *Walls* (strings of superclusters) separated by *Voids*. These walls with the galaxy clusters located between them at the nodes are the largest observable structures (Fig. 1.3.46). They are remnants of the early Universe evolution and were generated by the above-mentioned primordial density perturbations that were revealed by WMAP.

Dark matter may be contained only in very large galaxies with a mass exceeding $\sim 10^{10} M_S$. This follows from the fact that the galaxy mass spectrum $dN \approx dM/M^n$

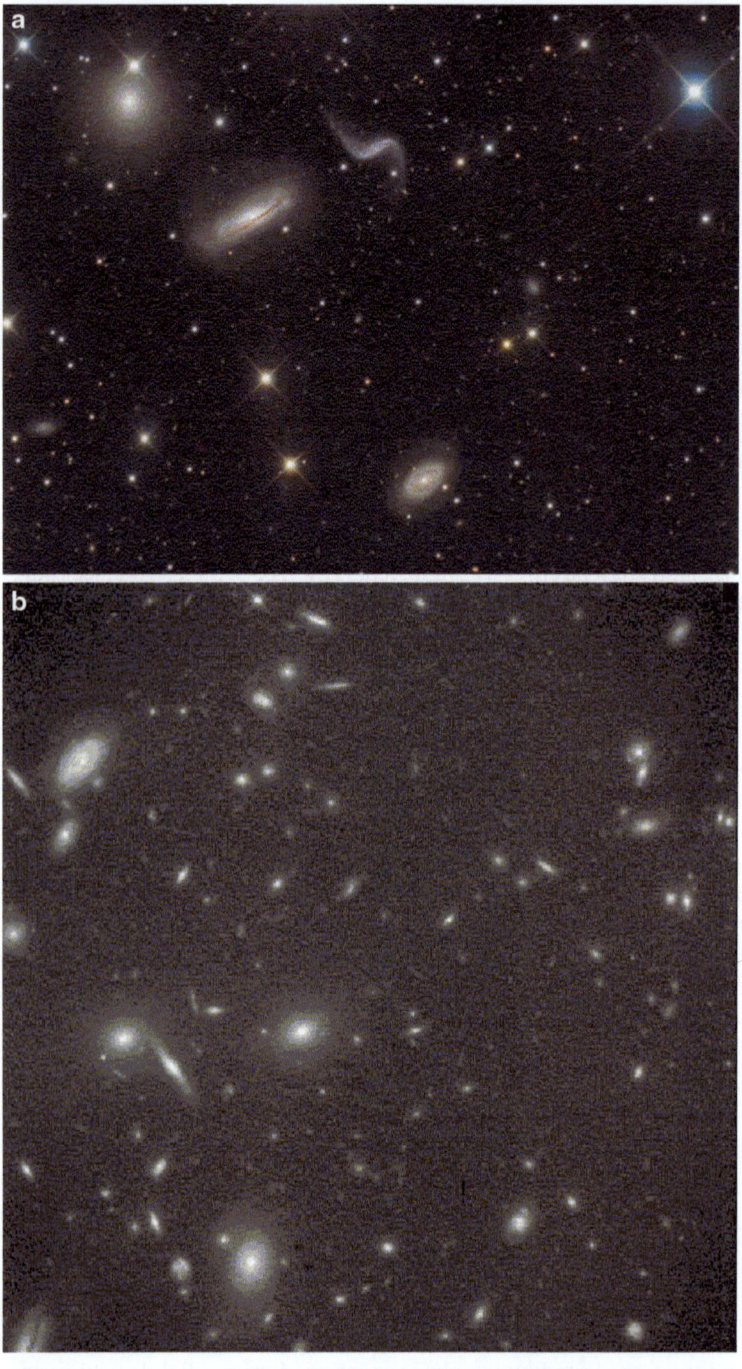

Fig. 1.3.44 Galaxies are brilliant examples of ordering in a chaotic turbulized intergalactic medium: (**a**) the NGC 3190 (Hickson 44) Local Group of galaxies is an example of the structuring of visible matter in the Universe. This group contains more than 30 galaxies, including our Galaxy (Milky Way) and the nearest brightest Andromeda galaxy and the Magellan Clouds (note that a star at the upper right corner and the elliptical galaxy at the upper left corner are comparable in brightness). (**b**) The central part of distant galaxy cluster CL 0939 + 4713 containing a set of spiral galaxies some of which appear perturbed. Hubble Space Telescope images (Courtesy of NASA). (**c**) Virgo galaxy cluster

Fig. 1.3.44 (continued)

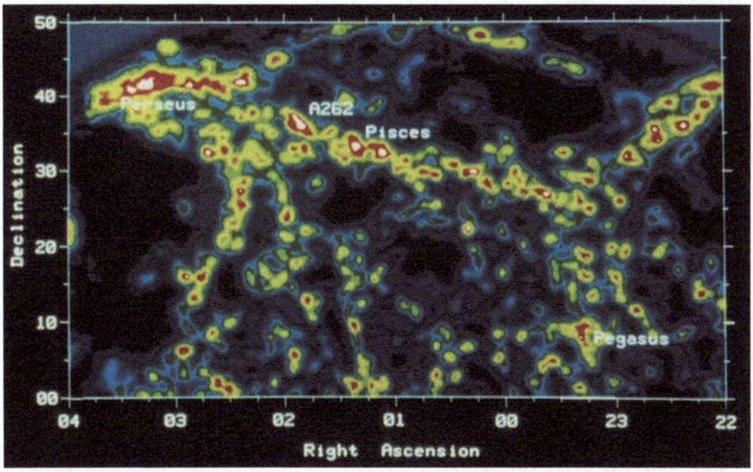

Fig. 1.3.45 Galaxy supercluster observed in the radio wavelength. Courtesy of NASA

has $n = 1$ before this limit (i.e., there is no selected mass) and $n = 2 - 5$ after it, which may be the result of intergalactic collisions. The galaxies and their clusters are remarkable examples of order produced by self-organization processes in a chaotic turbulized intergalactic medium in which dark matter performs structure-forming functions (see Figs. 1.3.43 and 1.3.44). Basically, we are talking about the existence of a generator of stable structures emerging from the chaos of structured elements in the Universe. An example for this is the possibility of the formation of a supermassive CD galaxy at the center of a typical cluster with a mass of about one percent of the cluster mass shown in numerical simulations (Tutukov et al., 2007).

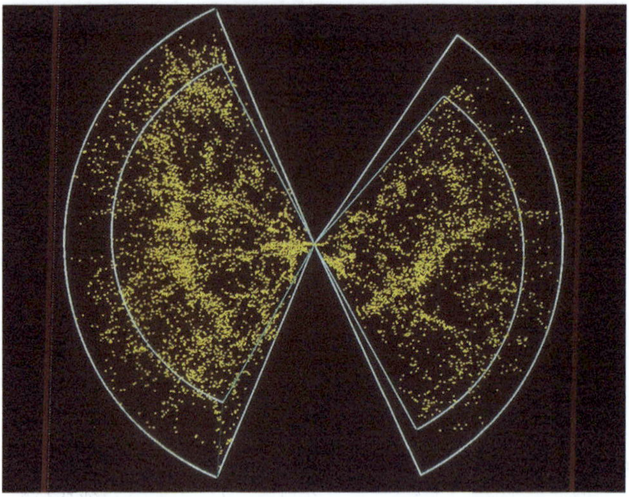

Fig. 1.3.46 Large-scale structure of the Universe. The galaxy clusters and superclusters (*Walls*) are separated by "hollows" (*Voids*)

The concept of self-organization, from the onset of density variations and vortex motions in primordial plasma to the completion of recombination with the formation of dense neutral gas clouds, the appearance of shock waves, vortices, and turbulence in a compressed gas, and the birth of galaxies and stars, was touched on in one way or another in various models for the early evolution of a hot Universe. According to one of these models, which is now only of historical interest, the expanding Universe had anisotropy and dynamical structuring in which a noticeable vortex component was present already at the earliest stages, while the role of turbulence was crucial (Ozernoi, Chibisov, 1970; Ozernoi, 1976). This pregalactic or cosmological turbulence, called "photonic" was supposed to have produced density perturbations that were subsequently enhanced through gravitational instability, which led to self-organization processes with the formation of protogalaxies. Unfortunately, the vortex theory ran into a number of difficulties and was not developed further.

Another example is the adiabatic theory of galaxy formation, in which turbulence emerges in a natural way and plays an important role in the "hydrodynamics" of galactic and intergalactic gases (Sunyaev, Zeldovich, 1972; Doroshkevich et al., 1976). The sequence of processes includes the formation of dense flattened gas clouds characterized by anisotropy in the strain tensor perturbations through the action of tidal forces, the adiabatic compression of gas in the local neighborhood of some point as it moves along one of the directions, and the subsequent accretion of the bulk of matter onto the already compressed gas. As a result, a peculiar kind of matter distribution emerges in the forming disk with a sharp density maximum at the center that rapidly falls off toward the periphery. The set of processes in matter compressed into a "pancake" determines the subsequent evolution of the compressed matter into galaxies and stars. The pancake model suggests the presence of powerful vortex motions and turbulization of both hot and cooling gases. They are maintained

Fig. 1.3.47 Schematic view of the Solar system and planetary nebula to be left behind after the Sun (a G2 star with the lifetime ~10 Gyr) will exhaust its nuclear fuel in ~ 5 Gyr from now. According to EAA (2002)

by the vortex velocity flux through the shock front in the hot gas and by the thermal instability and density inhomogeneity as the gas cools down, which causes the pancake to break up into individual clouds. According to these views, the turbulent motions, especially in its outer regions, have a strong effect on the parameters of the forming galaxies, while the galaxy formation itself may be considered as a self-organization process in a turbulent medium as this dynamical system evolves. The formation of flat structures – pancakes with the scale of superclusters (surrounding quasi-spherical empty spaces, *voids)* that subsequently fragment into galaxies was also assumed based on the HDM model for the dominance of neutrinos in the Universe having a mass with Gaussian primordial fluctuations. However, this model did not receive any confirmation, giving way to the cold dark matter model (Silk et al., 1983; White et al., 1983; Gawiser and Silk, 1998) and is hence, as we said, today only of historical interest, specifically from the viewpoint of important role of turbulence in the Universe's structural formation.

Let us now briefly summarize the main points of the above discussion. The Universe is assumed to start from a subatomic point at an infinitely high temperature and density. At $< 10^{-43}$ sec (T $\sim 10^{32}$ K) it represented a Planck universe (singularity) and is associated with space-time growth. Between 10^{-36} - 10^{-34} sec (T $= 10^{28}$ K) at the stage of "boiling vacuum", superinflation occurred, followed by the Universe's expansion by the factor $\sim 10^{50}$ and symmetry violation. At 10^{-6} sec (T $= 10^{13}$ K) – baryons (protons and neutrons) formed, then within 1 - 1000 sec (T $= 10^{10}$ - 10^{7}K) nucleosynthesis started and nuclear reactions occured as precursor of the origin of the chemical elements. The earliest time that we can look back

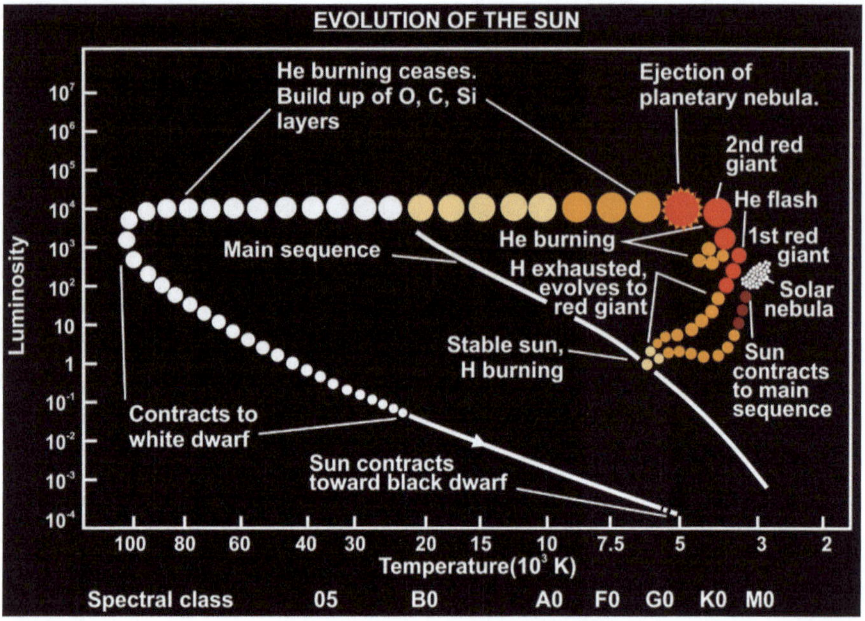

Fig. 1.3.48 Schematic view of our Sun evolution after the nuclear fusion is terminated . The Sun will evolve towards the state of a red giant and then contracts to a white dwarf. The planetary system collapses approximately at the same time

on of the Universe's history is 380,000 years since its origin, when under T $\sim 10^3$ K the atoms of H, D, He, and Li formed. In the following hundreds of millions - 1 billion years (T = 100 K) the stars, galaxies, galaxy clusters, and the cosmic web formed. Today, 13.7 billion years from the beginning, we have cosmic microwave background radiation (CMB) below T = 2.7 K as the Big Bang remnant.

We see that based on this cosmological paradigm, which is backed by COBE and WMAP results, the Universe continued to expand and cool after the original explosion, which was followed by superinflation. Galactic structures resulted from small density fluctuations shortly after the Big Bang and were subsequently enhanced by gravity, although the physical nature of these fluctuations is not yet completely clear. According to the superinflation model, when the temperature reached $\sim 10^{12}$ eV, the strong and electroweak interactions emerged from the initial combined force. At the same time, a huge vacuum energy contained in the scalar field of spacetime was released. At the moment of the "phase change" in the state of a chaotic early Universe, the first neutral hydrogen atoms and CMB were formed from primordial hot plasma and cold dark matter, and a predominant mass density also presumably emerged (Kolb and Turner, 1990; Borner, 1993; Green, 1999; 2004).

The galaxies were probably formed within dark matter by the aggregation of sub-galactic mass haloes, the collapse of self-gravitating objects during the development of gravitational instability that enhanced the quantum fluctuations that emerged at the earliest inflation epoch. Obviously, the internal sources of energy release through dissipative gas-dynamic processes, shock heating during supernova explosions, and supermassive black hole accretion were involved in this process (Binney and Merrifield, 1998; Smoot, 2007; Mayer et al., 2008). Subsequently, the protogalaxies grew by accreting smaller clumps and/or combining with a halo of comparable sizes. Intense radiative cooling of gas accompanied by X-ray emission, recombination, and collisional excitation of emission lines should have taken place at this stage (White et al., 1983). The redistribution of the system's angular momentum, the formation of a gaseous galactic disk, and the onset of star formation should apparently be attributed to the same period. The formation of galaxies and galaxy clusters in an expanding Universe that obeys the present-day gravitation theory and Einstein's general relativity theory can be associated with the presence of Gaussian adiabatic metric fluctuations in the inflation model, which is also consistent with the CMB observations.

Origin and Evolution of Stars

Stars are the main astrophysical objects in the Universe accessible to observation. They go through consecutive cycles of evolution from birth to death (Shklovsky, 1975; 1987). A quite definite order is observed in the processes of stellar evolution, in particular, in the birth of stars with various masses in the Galaxy. This order corresponds in first approximation to the initial mass function (IMF), while its slope in the region of accreting massive stars corresponds to the well-known Salpeter function (see Salpeter, 1955; Andersen, 1991; Bonnell et al., 2007).

The sites of star formation are genetically related to the massive, comparatively flat Galactic disk, in which the visible matter, a mixture of gas and dust with a nonuniform density distribution, is concentrated in its symmetry plane. Such a multiphase multicomponent medium is typical of massive cold dense clouds that are in the process of gravitational contraction and are apparently the early formation stage of star clusters and associations. In disk galaxies, the star formation is active in OB-associations with a mass of $\sim 10^7 M_S$ and with sizes on the order of the gas disk thickness, so that about 90% of the cluster breaks up into stars with a mass of $\sim 10^3 M_S$ in the course of collapse (Pikelner and Kaplan, 1976; Kippenhahn and Weigert, 1990; Hansen and Kawaler, 1994). Gas condensations accompanied by contraction and intense heat release through the conversion of potential energy into kinetic energy and angular momentum transport from the forming clump to peripheral regions emerge from this medium when some critical density is reached due to gravitational instability. When the temperature in the interiors of a protostar reaches several million degrees, thermonuclear fusion begins and the gravitational forces are balanced by the internal gas pressure (gravitational equilibrium). As a result, the contraction ceases and the star occupies a certain position on the Hertzsprung-Russell diagram that depends on the mass-luminosity relation, and the star stays there until the reserves of nuclear fuel are exhausted (see, e.g., Chiosi et al., 1992). The energy release through nuclear reactions in the central region of the star is accompanied by radiative and convective heat transport, which leads to active mixing of interior matter (low-mass stars are entirely convective).

It follows from the theory of stellar evolution that after the completion of hydrogen burning, a star with a mass $M \leq 8M_S$, where M_S is the solar mass, moves off the main sequence, turning initially into a red giant, and subsequently evolves into a white (degenerate) dwarf. The existence of such an object was implicitly predicted back in 1844 by Bessel. Based on the observations of irregularities in the proper motion of Sirius, Bessel hypothesized that Sirius has an invisible companion with a nearly solar mass (see EAA, 2001). As subsequent observations showed, the companion turned out to be the white dwarf Sirius B, and the name itself was given to such degenerate stars by Fowler (1926).

White dwarfs are the most characteristic form of the final evolutionary stage for the bulk of the stellar population (up to 95% of the stars in our Galaxy), with their initial masses lying in the range (0.08 - 8) M_S. The lower threshold corresponds to the condition for the onset of a classical hydrogen-helium thermonuclear reaction when the star is in thermal equilibrium. For solar-type stars, this is the so-called PP chain, which precedes the C-N-O burning cycle of the shells following hydrogen and helium for more massive stars. Accordingly, carbon-oxygen white dwarfs are the final stage of single stars and companions of wide (with $a \geq 300 R_S$) binaries with $M \leq 8 M_S$, while degenerate helium white dwarfs are the final stage of companions of close (with $a \leq 300 R_S$) binary systems with $M \sim 2.5M_S$. They should be distinguished from brown dwarfs, quasi-stars with $0.01M_S < M < 0.08M_S$ whose temperature in the interiors is not sufficiently high to excite the PP cycle, and, therefore, they are intermediate between low-mass stars and planets (Basri, 2000; Luhman et al., 2007; Whitworth et al., 2007). The sequence of the processes of evolution for our Sun is shown in Figs. 1.3.45 and 1.3.46.

In some of the white dwarfs, the final evolutionary phase can be accompanied by a nova explosion caused by the accretion of the hydrogen envelope onto the core. In this envelope, the temperature rises sharply and unstable hydrogen burning takes place in a short C-N-O cycle under conditions of partial electron degeneracy. The final mass of most white dwarfs determined by the "steepness" of the Salpeter spectrum $M_{WO} \approx 0,6M^{0,4}$ is about 0.6 M_S and their size is on the order of the size of a terrestrial planet. For this reason, they have a high mean density. The effective temperatures lie within a wide range, from 150 000 to 4 000 K. Cool degenerate dwarfs oppose their gravitational contraction due to the high density of the degenerate gas of free electrons that conserves the kinetic energy even near absolute zero. The amplitude of motions and the degree of degeneracy increase with density, which prevents the collapse of such a star at the final stage of its evolution.

More massive stars with $8 M_S \leq M \leq 100M_S$ go through all of the successive hydrogen, helium, and heavier-element burning stages up to the production of iron. The star's contraction after the formation of an iron core can no longer be stopped by the energy release through thermonuclear reactions in its central part.

The contraction energy is expended on the disintegration of iron nuclei up to the formation of a neutron core; the central pressure is determined by electron gas degeneracy, while the density is determined by a gas of atomic cores. According to statistical physics, the maximum mass that can be confined by cold electrons is equal to the limiting Chandrasekhar mass $M = 5.75M_S/\mu_e$, where μ_e is the number of nucleons per one electron (see, e.g., Bethe, 1966).

In contrast to low-mass stars, stars with $M \geq 7M_S$ eject their outer envelopes (explode) at the end of their life. This is observed as a supernova explosion, when the total gravitational and kinetic energy (mainly in the form of neutrinos) that is released reaches 10^{53} erg/s. The visible radiation accounts for about 1% of this enormous energy release. This phenomenon can be described in terms of the theory of a strong explosion, while its propagation through the interstellar gas can be described as a detonation wave (Sedov, 1965; Zel'dovich and Raizer, 1966; Zel'dovich, 1983). Two types of supernovae with a threshold mass of (30-40) M_S are distinguished. Stars with a mass below this limit evolve in an ordinary way and shine as red supergiants after the explosion phase, lying on the corresponding branch of the Hertzsprung-Russell diagram. More massive objects with $M \geq 10M_S$ completely lose their hydrogen envelope and the supernova explosion is accompanied by core collapse, following which a supermassive compact object in the form of a neutron star (pulsar) or (at $M \geq 30M_S$) a black hole is left (see, e.g., Masevich and Tutukov, 1988; Lyne and Graham-Smith, 1998). The density inside a neutron star composed mainly of neutrons with a few percent of protons and electrons corresponds to the nuclear density, reaching $\rho_0 = 2.8 \times 10^{14}$ g cm^{-3} at a radius $R \simeq 10$-15 km. Supermassive black holes (quasars) accreting galactic gas and releasing a huge amount of gravitational energy are formed in the nuclei of CD galaxies.

The configurations forming at the final stage of evolution depend fundamentally on the stellar mass. Low-mass stars, in the case of a relatively slow mass outflow from red giants, leave planetary nebulae with various but fairly ordered shapes at the stage of their transition to a white dwarf (Fig. 1.3.49). They contain a strongly turbulized gas and these shapes are believed to have been acquired in the process of self-organization of the nebular matter. This order, including a pronounced cylindrical symmetry, can be partly attributed to the existence of a close companion to the star and to the joint evolution of their envelopes. Extremely complex configurations produced by the interaction of an expanding gas with the interstellar medium, in which certain order can be distinguished against the background of chaotization, also emerge during supernova explosions. The conditions for the formation of first- and subsequent-generation stars are distinguished depending on the extent to which the interstellar galactic medium is enriched with heavy elements that are produced in nucleosynthesis processes in the interiors of stars and during supernova explosions (Alpher et al., 1948). In particular, this determines the metallicity (the ratio of heavier elements to hydrogen) of stellar matter.

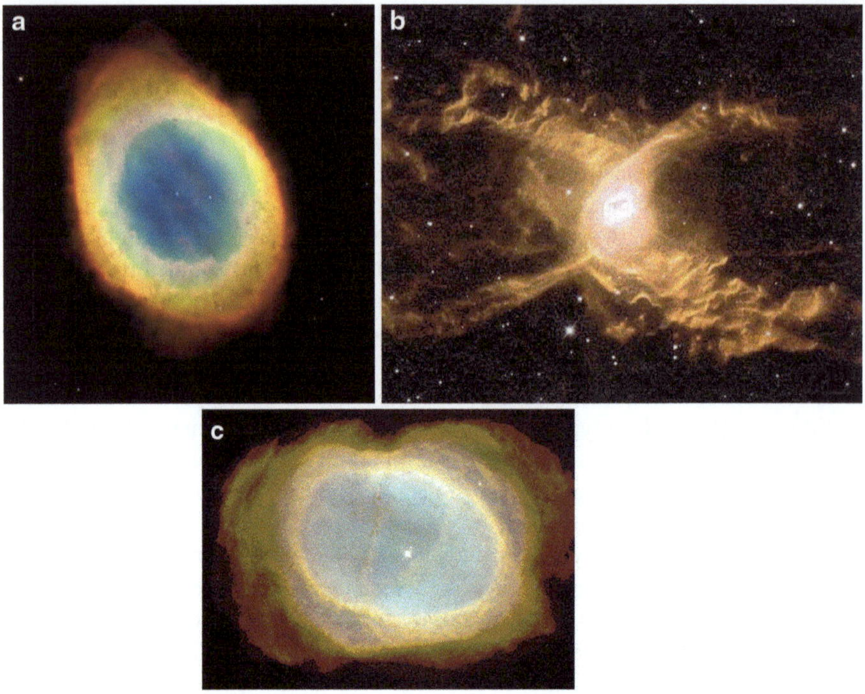

Fig. 1.3.49 Configurations of planetary nebulae. (**a**) The Ring Nebula in Lyra (M57, NGC 6720) in colors close to the real ones. The blue, green, and red correspond, respectively, to the ionized helium, doubly ionized oxygen, and hydrogen and ionized nitrogen. There is a 15 magnitude star at the final stage of its evolution toward a white dwarf with a surface temperature of ~ 150000 K at the center. (**b**) The Tarantula Planetary Nebula. The gas of the ejected envelope is strongly turbulized and has an irregular structure with some ordering. (**c**) A quasi-ring planetary nebula with an inhomogeneous outflow of gas and ordered structures in the inner zone. Hubble Space Telescope images. Courtesy of NASA

The Self-organization of the Universe

As has already been noted, modeling the structure and evolution of galaxies and the entire Universe is closely related to studying the nature of the CMB radiation, which fills the Universe almost uniformly and isotropically. Its discovery served as a major impetus in developing the standard Big Bang theory. According to this, the Universe at the post-inflation stage was an extremely dense hot plasma in a state of complete thermodynamic equilibrium with blackbody radiation. Its gradual cooling as it expanded from singularity corresponds to an ideal blackbody spectrum with a temperature $T = 2.725 \pm 0.001$K at the present epoch. CMB maps with high sensitivity and angular resolution obtained by WMAP brought evidence about relatively small spatial variations in CMB intensity and anisotropy at a level of $\Delta T/T \approx 10^{-5}$ K , or $\Delta T \sim 30$ μK on an angular scale $> 7°$ (Smoot et al., 1992;

Masser, 2006), which can be associated with the relatively small variations in the density of neighboring regions. This ultimately led to the formation of giant galaxy clusters because of gravitational instability. These probably were the processes of initial self-organization against the background of a chaotic structure of primordial atomic matter formed shortly after the Big Bang.

Invoking the views about the dynamical evolution of nonlinear dissipative systems can help in searching for answers to the fundamental questions about the origin of galaxies and galaxy clusters as well as about the nature of dark matter and dark energy. The development of ideas about the topology of the Universe is associated with the solution of these problems, which originate in the inflation model mentioned above with subsequent gradual ordering of matter from radiation that initially chaotically filled the space. These include the attempt to unify the present views about the nature of elementary particles and acting forces in the framework of the advanced Standard Model, which is a key of the unified field theory (which A. Einstein dreamed about) and a step toward the modern view of the Universe, which incorporates the Superstring theory. Basically, this theory is a synergy of micro- and macro- physics (elementary particles/quantum mechanics and cosmology/gravity). It postulates that all elementary particles result from oscillations of an ultramicroscopic loop – a tiny vibrating string of minimal possible Planck length of 10^{-33} cm and enormous tension, of which every particle is just a wave (analogy with musical tone/pitch). Thus string vibrations (resonance modes in the infinite wavelength) are responsible for the nature and mass of all diversity of elementary particles and the four interactions in nature: strong, weak, electromagnetic, and gravitational. String-based space is assumed to be multidimensional (strings exist in 11-dimensional space) and leads us to a new view of the Universe (see, e.g., Green, 1999; 2004).

In the context of super gravity and cosmic strings stretching through huge regions of space, attempts are being made to improve the scenarios of an inflationary Universe by invoking the models of extended, topological, hybrid, and other types of inflation. The same physical principles and the initial chaotic inflation model underlie these models, while the choice between various scenarios with postulated unique properties can obviously be made only on the basis of very complex and sophisticated experiments (Linde, 1990; 1994). The WMAP became an important step on this path. Based on these measurements, highly accurate estimates were made for the densities of baryonic matter, dark matter, and dark energy and the age of the Universe, which is now taken to be 13.7 Gyr (Fixsen and Mather, 2002; Smoot, 2007). Concurrently, predictions of the inflation theory were confirmed and additional arguments were obtained for a nearly flat Universe with the above quoted ratios of baryonic matter, dark matter, and dark energy. Additional precision measurements of the angular power-law spectrum of CMB temperature fluctuations and the polarization induced by gravitational waves since the Big Bang are called for to answer the key question about the nature of the differences between the baryonic composition of the Universe and its dynamical mass density and to give an approach to the possibility of identifying the particles or objects of cold (yet exotic) dark matter.

Let us notice that the Big Bang theory received an additional underpinning in connection with the discovery of an accelerating expansion of the Universe in 1998. As we have already emphasized, this acceleration may have been promoted by antigravity and it is consistent with the cosmological constant Λ introduced by Einstein. Furthermore, it was attributed to exotic dark energy – an even more intriguing entity, which is right now associated with vacuum – the new form of the energy density exceedes by far everything known only a few years ago. The properties of vacuum look really fantastic: it possesses negative pressure, its energy density is independent of the expansion it causes. Because the ratio of routine matter, dark matter, and dark energy is roughly scaled as 1:10:25, vacuum is the most dense media with a fully uniform density over the whole Universe. Theory predicts that antigravity will grow, and in billions of years a new process, the Big Cutting of the Universe, will begin and result in galaxy clusters, galaxies, stars, planetary systems, and then chemical elements and even atom destruction. In other words, antigravity (vacuum) will dominate gravity, giving rise to repulsion rather than the now existing attraction. Therefore, according to the modern views, vacuum fully defines the fate of our Universe.

The modern theory also postulates that there are a multitude of multidimensional parallel universes that permanently grow and decay in different regions and different time. Theses parallel universes form multiverses, and one origin of a universe might be the result of quantum oscillations in vacuum *(creatio ex nihilo)*. Universes floating in space form "temporal-spacious foam" and may experience collisions. Following this scenario (see, e.g., Tegmark, 2006; Novikov et. al., 2006) our Universe's origin might be the result of such a collision that gave rise to the Big Bang effects. Moreover, time-space tunnels resembling black holes could exist in a multiverse to enable an escape to hyperspace. Multidimensional universes are related with the idea of worm holes, which was first discussed by A. Einstein as early as in 1935. Accretion of (phantom) matter onto a worm hole with a regular magnetic field may result in a black hole appearance. The difference is hard to reveal, though the processes may resemble the chaotic inflation in the Big Bang. Thus the worm hole hypothesis assumes interactions between an infinite multitude of parallel universes in the multiverse. These interactions are responsible for numerous mutual collisions of parallel universes (invisible "vesiculars" in space) and occur along a hidden dimension. Again, in the case of our Universe, such a collision resulted in great heating and the effects of the Big Bang scenario.

In this context, there is also another important aspect to which we would like to draw the reader's attention again. According to Masser (2006), for example, "it would be most appropriate to represent the Big Bang not as a single event but as an ongoing process of gradual formation of order out of chaos". In other words, the evolution of the Universe is a continuous self-organization process that has led to its currently observed structure with a host of galaxies, galaxy clusters, and planetary systems. If, alternatively, we proceed from the views of modern cosmology about the existence of parallel universes as small parts of the multiverse, then the initial quantum fluctuations and the birth of post-inflation domains (with different space-time dimensions and different physical constants) postulated by the theory may be considered as a universal paradigm of an infinite process of ordering of an infinite number of initially chaotic structures filled with fields and matter. Here, however, we go far outside the less abstract problems that this monograph deals with.

Chapter 2
Foundations of Mathematical Modeling of Reacting Gas Mixtures

This introductory chapter devoted to the formulation of general mass, momentum, and energy balance laws in a multicomponent chemically active gas mixture is preparatory to the more detailed considerations of turbulence presented in succeeding sections of the book (in this case, references to this chapter are given). For example, the outer gaseous envelopes of celestial bodies (planets, comets) serve as characteristic representatives of such natural media. Their distinctive features are a multicomponent or multiphase composition and a direct impact of solar electromagnetic and corpuscular radiation under which various photochemical and chemical reactions take place. The structure and physical–chemical properties of these media, which lie in the atmosphere-space boundary regions and are commonly called the upper atmosphere, differ significantly, depending on the mass of the celestial body, its distance from the Sun, the presence of a magnetic field, and the formation and evolution conditions. The local temperature, concentration, and pressure gradients emerging in this case lead to the development of different-scale hydrodynamic motions and heat and mass transfer processes whose pattern remains turbulent down to the base of the thermosphere. The atmospheric modeling problem acquires completely new facets in the presence of turbulence, because the classical views of turbulent motions in an incompressible fluid are interwoven with other areas of mechanics that combine multicomponent hydrodynamics, nonequilibrium thermodynamics, and chemical reaction kinetics. In addition to the velocity fluctuations, the density, temperature, and concentration fluctuations of the individual chemical components in a mixture become important. As a result, we run into one of the most complex problems in the mechanics of turbulized media pertaining to semiempirical modeling of interrelated hydrodynamic, physical–chemical, and radiative processes and phenomena in a turbulent flow. In particular, the problem of closing the hydrodynamic equations of mean motion arising in this case is enormously complicated because of the necessity of averaging the strongly nonlinear "source terms" of substance production (consumption) in chemical reactions (which, as a rule, are exponential) and modeling a large number of additional correlation moments related to the fluctuations of various thermohydrodynamic parameters.

M.Y. Marov and A.V. Kolesnichenko, *Turbulence and Self-Organization: Modeling Astrophysical Objects*, Astrophysics and Space Science Library 389,
DOI 10.1007/978-1-4614-5155-6_2, © Springer Science+Business Media New York 2013

Obviously, the most complete and rigorous mathematical description of a multicomponent medium in the case of a regular (laminar) flow can be given in terms of the kinetic theory of multicomponent polyatomic ionized gas mixtures based on the system of generalized integro-differential Boltzmann equations for the distribution functions of particles of each type in the mixture (with the right-hand sides containing the collision and reaction integrals) supplemented by the radiative transfer equation and the Maxwell equations for electromagnetic fields. In particular, such an approach was developed in our monograph (Marov and Kolesnichenko 1987), where we applied a generalized Chapman–Enskog method to obtain the system of differential hydrodynamic equations for a reacting mixture. At the same time, from the viewpoint of macroscopic properties, the upper atmosphere of a planet can be considered as a continuous medium and the phenomenological methods of mixture mechanics can be used for its adequate description. These methods allow the system of hydrodynamic equations with all of the necessary closing relations to be obtained. Such a phenomenological approach also makes it possible to develop (using extended irreversible thermodynamics) semiempirical models for developed turbulent flows of reacting gaseous media.

2.1 Basic Conservation Laws and Balance Equations for Regular Motions of a Gas Mixture

2.1.1 On the Models of Continuous Media and Mathematical Modeling

It is often convenient to study phenomena in the surrounding material world using mathematical models for media, fields, and processes. They allow one to interpret observations and to develop methods for predicting forthcoming events. Each such model is a certain schematization of the physical phenomenon being studied that includes not all of its inherent factors but only some of them characterizing the phenomenon from a particular standpoint. A plausible schematization is often a difficult task that requires considerable experience, intuition, and deep penetration into the essence of the methods of research into the mechanisms of the natural phenomenon being studied from the researcher. One of the general methods for schematizing the motions of liquids, gases, and other deformable bodies is a method based on the construction of new continuum models for a medium with complicated physical–chemical properties (Sedov 1984). The schematization whereby a real natural medium composed of separate molecules is replaced by a continuum turns out to be very convenient for applying the powerful mathematical apparatus of continuous functions and, as practice has shown, is quite efficient for studying a multitude of observed natural phenomena and processes in an overwhelming majority of cases.

There exists a plethora of interesting and topical problems that can and should be solved in terms of the models of continuous media that have already been introduced in the scientific arsenal and have become classical, such as the models for the motion of ideal or viscous incompressible (compressible) liquids and gases, the models for the motion of multicomponent or heterogeneous mixtures, the models for the motion of ionized gases, etc. It seems pertinent to note here an exceptional role of academician L.I. Sedov who was the first to raise the question about the necessity of constructing nonclassical models of continuous media with complicated physical–chemical and thermal properties that allow one to describe new mechanical effects and to solve important problems arising in various mechanical applications. In particular, he was the first to substantiate the fundamental role of the thermodynamic concepts and kinetic laws of physics and chemistry in constructing continuum mechanics models to obtain the defining constitutive equations that, along with the basic conservation laws, describe the motion and physical–chemical phenomena and processes (see Sedov 1962). Note that new models of cosmic natural media have been intensively developed in scientific literature in the last three or four decades to describe the outer gaseous envelopes of celestial bodies and the surrounding space. Such models are characterized by an additional set of physical–mechanical parameters like the concentrations of various substances and phases in a chemically active gas suspension, the electromagnetic characteristics of a turbulized electrically conducting medium, the internal structure parameters, the tensor of correlation quantities, etc.

The problem of deriving the fundamental relations on unified grounds that would contain as special cases the corresponding basic laws whose fruitfulness in studying material media and fields has already been demonstrated and proven in mechanics, physics, and chemistry arises in connection with the construction of up-to-date models for continuous media. Obviously, the universal laws that can serve as a basis for introducing various hypotheses and laws to be tested when modeling various physical phenomena must form a fundamental basis for obtaining closed systems of the corresponding differential equations. As is well known, the universal laws of mechanics on the conservation of mass, momentum, and angular momentum, the universal laws of thermodynamics on the conservation of energy (the first law of thermodynamics) and on the change or conservation of entropy (the second law of thermodynamics), the laws of chemical kinetics, etc. serve as such a reference basis in theoretically constructing the models of continuous media. It is particularly important to note that the laws of nonequilibrium thermodynamics play a primary role in constructing new continuum models, because they are also used to establish the defining constitutive relations that express the peculiar properties of a specific continuous medium. Modern nonequilibrium thermodynamics incorporates several relatively new directions, such as the theories with internal variables, the new formulations of rational thermodynamics (Truesdell 1984), the theories that include fluctuations or are based on the kinetic theory of gases (Keizer 1987), the information theory or nonlinear dynamical systems (Anishchenko et al. 2003). All scientific directions of this kind are based on the common idea that the models of continuum mechanics and conjugate nonequilibrium thermodynamics should be developed simultaneously in close association between themselves. They suggest that the modified thermodynamics being used is predetermined by the hypothesis about local thermodynamic equilibrium, but it can differ from another one, depending on the chosen variables, the definition of entropy and entropy flux, the interpretation of the second law, etc. At the same time, the so-called extended irreversible thermodynamics, which goes beyond the scope of the hypothesis about local equilibrium through a modification of such basic and conceptual notions as the entropy, temperature, pressure, or chemical potential, has gained wide acceptance in the last decade (see Jou et al. 2001). This theory introduces dissipative thermodynamic fluxes (like the heat and diffusion fluxes or the viscous stress tensor) appearing in the mass, momentum, and energy balance equations as additional independent variables. This makes it possible to develop formalism both for deriving the defining constitutive equations for the classical variables specified by ordinary balance laws and for finding the corresponding evolutionary equations (compatible with the second law of thermodynamics) for the dissipative fluxes that describe the reaction of a system to high-frequency (short-wavelength) disturbances—by the introduction of memory functions and generalized transport coefficients dependent on the frequency and wave vector. Thus, among the defining relations there can be both the relations between equilibrium state parameters, various equations of state for the medium, and the relations between the parameters describing the process—kinetic constitutive relations, for example, the evolutionary equations that relate the stress tensor and the strain rate tensor in a viscous fluid. A closed system of differential equations defining the model of a material medium and, in combination with the boundary and initial conditions, the specific mathematical problem can be obtained within the framework of this model using constitutive relations of this kind.

One of the most important questions of modeling a physical phenomenon is how to choose the defining parameters of the model that should be considered as independent arguments varying within a certain range when considering the full set of possible states and processes in the medium for which the model is constructed (Sedov 1962). The defining parameters can be scalars and tensors (specifying, e.g., the corresponding symmetry groups of elementary particles in a continuous medium), dimensional and dimensionless, constant and variable, but, in any case, they should be invariant with respect to the choice of a coordinate system and units of measurement. The appearance of various structural parameters of a mechanical or physical-–chemical nature is related to a quantitative description of the mechanisms of energy exchange, energy dissipation, and other interactions between particles inside material bodies that are of great importance in the phenomena being studied. For instance, in the case of a stochastic turbulized motion, various scalar and tensor correlation quantities should be additionally introduced as characteristics of this kind. The question arises as to how to reveal the system of defining parameters. This can be easily done if, for example, there is an adequate mathematical schematization of the phenomenon or process being studied. Generally, if the closed system of equations has been written, then the space of basic independent variables can be easily identified and enumerated. If the equations of the system are differential and their number is limited, then the number of defining parameters for an infinitesimally small fluid particle can be small and, in any case, finite.

In this chapter and, as a rule, in all the succeeding chapters of the book, emphasis is placed on the construction of new continuum models for natural and cosmic media. They allow various formulations of the formation and evolution problems to be modeled for a number of astrophysical and geophysical objects in space. Models of this kind include the models for the motions of multicomponent or heterogeneous systems with diffusion, heat transfer and viscosity, radiation processes, chemical and physical transformations, the models for various kinds of turbulent motions of chemically active gases and gas suspensions, the models of structured turbulent homogeneous fluid flows, the models of turbulized electrically conducting media interacting with an electromagnetic field, etc.

2.1.2 The General Balance Equation

Let us now concretize the above general remarks by mathematically modeling regular motions of an N-component gas mixture that is considered below as a single continuum with complicated physical–chemical properties. The additional complications related to the influence of an electromagnetic field are disregarded, because gas mixtures of uncharged particles are considered in this chapter. In contrast to the hydrodynamics of a single-fluid continuum, we characterize the mixture by a number of additional parameters defining its chemical composition. The mean mass density $\rho(\boldsymbol{r}, t)$, hydrodynamic velocity $\boldsymbol{u}(\boldsymbol{r}, t)$, thermodynamic pressure $p(\boldsymbol{r}, t)$, temperature $T(\boldsymbol{r}, t)$, and concentrations $n_\alpha(\boldsymbol{r}, t)$ of the individual chemical components are the state variables. Below, we assign the Greek letters $\alpha, \beta, \ldots$ $(\alpha, \beta, \ldots = 1, 2, \ldots, N)$ to the system parameters referring to various mixture components as subscripts. The state variables are assumed to be continuous functions of time t and spatial (Eulerian) coordinates $\boldsymbol{r}$ in a coordinate system fixed with respect to the cosmic object under consideration (here, $\boldsymbol{r}$ is the radius vector with components x_1, x_2, x_3).

The partial differential equations following from integral conservation laws and the defining relations for the diffusion and heat fluxes and the viscous stress tensor are known (see, e.g., de Groot and Mazur 1962) to underlie the hydrodynamic model of a reacting mixture. These equations should be supplemented by the corresponding equations of chemical kinetics and the equations of state for pressure (thermal equation of state) and internal energy (caloric equation of state). In addition, it is necessary to know the algebraic expression for every possible thermodynamic function (such as the internal energy, entropy, or heat capacity of the medium), the formulas for various molecular transport coefficients, and the formulas for chemical reaction rate coefficients. The differential equations of motion require specifying the initial and boundary conditions that, describing the geometry of the thermohydrodynamic system under consideration (a material object with well-defined boundaries) and the mass, momentum, and energy exchange with the environment, must be formulated ad hoc for each specific hydrodynamic problem.

Let us first consider the general balance equation for various thermohydrodynamic parameters characterizing the mixture state and dynamics (e.g., the mass, momentum, internal energy, etc.). The local change in a particular field quantity $\mathit{\Lambda} = \int\limits_W (\mathsf{A}\rho)\mathrm{d}W$ per unit time

$$\frac{\partial \mathit{\Lambda}}{\partial t} = \frac{\partial}{\partial t}\int\limits_W (\mathsf{A}\rho)\mathrm{d}W = \int\limits_W \frac{\partial}{\partial t}(\mathsf{A}\rho)\mathrm{d}W$$

is equal to the difference between the production rate $\int\limits_W \sigma_{(\mathsf{A})}\,\mathrm{d}W$ of the quantity $\mathit{\Lambda}$ in a fixed (in the system of coordinates x_1, x_2, x_3) fluid volume W and the outflow rate $\oint\limits_{\partial W} (\boldsymbol{J}'_{(\mathsf{A})} \cdot \boldsymbol{n})\mathrm{d}\Sigma$ of this quantity through the surface ∂W bounding the volume W. Here, $\partial(\cdot)/\partial t$ is the local Eulerian time derivative; $\mathsf{A}(\boldsymbol{r}, t)$ is the specific value of the extensive parameter $\mathit{\Lambda}$; $\sigma_{(\mathsf{A})}(\boldsymbol{r}, t)$ is the density of the internal source of the attribute A (the production (>0) or destruction (<0) rate of the quantity A per unit volume); $\boldsymbol{J}'_{(\mathsf{A})}(\boldsymbol{r}, t)$ is the molecular flux density of the attribute A (i.e., the sum of all types of transport of A through a unit surface) including the convective part ($\rho\mathsf{A}\boldsymbol{u}$) and the conductive substantial flux $\boldsymbol{J}_{(\mathsf{A})}$ defining the surface impact of the system under consideration on the integral quantity $\mathit{\Lambda}$ ($\boldsymbol{J}'_{(\mathsf{A})} \equiv \rho\mathsf{A}\boldsymbol{u} + \boldsymbol{J}_{(\mathsf{A})}$); $\boldsymbol{n}$ is a normal unit vector on the outer side of the surface ∂W bounding the fluid volume W and $\mathrm{d}\Sigma$ is the area on ∂W. Thus, the integral balance (conservation) equation for the quantity $\mathit{\Lambda}$ is

$$\begin{aligned}\int\limits_W \frac{\partial}{\partial t}(\mathsf{A}\rho)\mathrm{d}W &= \int\limits_W \sigma_{(\mathsf{A})}\mathrm{d}W - \oint\limits_{\partial W} ((\rho\mathsf{A}\boldsymbol{u}) \cdot \boldsymbol{n})\mathrm{d}\Sigma - \oint\limits_{\partial W} (\boldsymbol{J}_{(\mathsf{A})} \cdot \boldsymbol{n})\mathrm{d}\Sigma \\ &= \int\limits_W \left\{\sigma_{(\mathsf{A})} - \left(\frac{\partial}{\partial \boldsymbol{r}} \cdot (\rho\mathsf{A}\boldsymbol{u})\right) - \left(\frac{\partial}{\partial \boldsymbol{r}} \cdot \boldsymbol{J}_{(\mathsf{A})}\right)\right\}\mathrm{d}W,\end{aligned}$$

where the second equality on the right-hand side is written by taking into account the divergence theorem (here, $\partial(\cdot)/\partial\boldsymbol{r}$ denotes the del operator). Since this equation is valid for any volume W at rest relative to the system of coordinates x_1, x_2, x_3, the integral balance equation (if the integrands are continuous functions of coordinates) can be written in differential (local) form:

$$\frac{\partial}{\partial t}(\rho\mathsf{A}) + \left(\frac{\partial}{\partial \boldsymbol{r}} \cdot (\rho\mathsf{A}\boldsymbol{u})\right) = -\left(\frac{\partial}{\partial \boldsymbol{r}} \cdot \boldsymbol{J}_{(\mathsf{A})}\right) + \sigma_{(\mathsf{A})}. \tag{2.1}$$

If the characteristic A in this equation is a vector or tensor quantity, then the flux $\boldsymbol{J}_{(\mathsf{A})}$ is a tensor higher than that of the parameter A by one order, while the production rate $\sigma_{(\mathsf{A})}$ is a tensor of the same rank.

It should be noted that, in general, the quantities $\boldsymbol{J}_{(\mathsf{A})}$ and $\sigma_{(\mathsf{A})}$ are not determined uniquely, because their substitution with $\boldsymbol{J}_{(\mathsf{A})} + \boldsymbol{K}$ and $\sigma_{(\mathsf{A})} + ((\partial/\partial\boldsymbol{r}) \cdot \boldsymbol{K})$, respectively, does not change the balance equation (2.1). However, when the flux $\boldsymbol{J}_{(\mathsf{A})}$ has been ultimately chosen, the source $\sigma_{(\mathsf{A})}$ is determined uniquely. When choosing a specific expression for the flux $\boldsymbol{J}_{(\mathsf{A})}$, the physical meaning of the quantity $\varLambda$ should be taken into account.

Note also that if the fluid region W contains some surface of a strong discontinuity G such that all integrands in the surface integrals over δW have finite values (different on different sides of G) as δW collapses to G, then universal dynamical and thermodynamic conditions at strong discontinuities can be obtained for material media using the integral balance equation:

$$-\left[(\mathsf{A}\rho\boldsymbol{u}) \cdot \boldsymbol{n}\right] + \left[(\boldsymbol{n} \cdot \boldsymbol{J}_{(\mathsf{A})})\right] = \sigma^{G}_{(\mathsf{A})}.$$

Here, the straight brackets denote the jump in a particular characteristic A at a discontinuity and the quantity $\sigma^{G}_{(\mathsf{A})}$ is the surface density of the distribution of the production rate of the attribute A on G (obviously, $\sigma^{G}_{(\mathsf{A})} = 0$ for conserved quantities).

Let us now analyze successively the cases of various parameters A in (2.1) that describe the regular motion of a mixture of reacting gases.

2.1.3 *The Mass Balance Equations for a Reacting Mixture of Gases*

A classical example of the balance equation (2.1) is the well-known fluid continuity equation

$$\frac{\partial\rho}{\partial t} = -\left(\frac{\partial}{\partial\boldsymbol{r}} \cdot \rho\boldsymbol{u}\right) \equiv -\mathrm{div}(\rho\boldsymbol{u}), \tag{2.2}$$

which reflects the conservation of the total mixture mass at an arbitrary internal point of the medium during its motion. In this case, $\mathsf{A} \equiv 1$, $\varLambda \equiv M = \int_W \rho \, \mathrm{d}W$, $\boldsymbol{J}_{(M)} \equiv 0, \sigma_{(M)} \equiv 0$. For a multicomponent medium, the mass density $\rho(\boldsymbol{r}, t)$ and the total hydrodynamic flow velocity $\boldsymbol{u}(\boldsymbol{r}, t)$ are defined by the formulas

$$\rho \equiv \sum_{\alpha=1}^{N} \rho_\alpha = \sum_{\alpha=1}^{N} m_\alpha n_\alpha, \qquad \rho\boldsymbol{u} \equiv \sum_{\alpha=1}^{N} m_\alpha n_\alpha \boldsymbol{u}_\alpha, \tag{2.3}$$

where $n_\alpha(\boldsymbol{r}, t)$, $\rho_\alpha(\boldsymbol{r}, t)$, $\boldsymbol{u}_\alpha(\boldsymbol{r}, t)$, and m_α are, respectively, the number density per unit mixture volume, mass density, hydrodynamic velocity, and molecular mass of particles of type α.

Below, we widely use the operator relation

$$\rho \frac{\mathrm{dA}}{\mathrm{d}t} = \frac{\partial}{\partial t}(\rho \mathrm{A}) + \left(\frac{\partial}{\partial \boldsymbol{r}} \cdot (\rho \mathrm{A} \boldsymbol{u}) \right). \tag{2.4}$$

It relates the substantial and local changes in characteristic A and allows the differential balance equation (2.1) to be written in substantial form:

$$\rho \frac{\mathrm{dA}}{\mathrm{d}t} = -\left(\frac{\partial}{\partial \boldsymbol{r}} \cdot \boldsymbol{J}_{(\mathrm{A})} \right) + \sigma_{(\mathrm{A})}. \tag{2.1*}$$

Relation (2.4) is a corollary of the continuity equation (2.2) and the definition of the substantial (Lagrangian) time derivative

$$\frac{\mathrm{dA}}{\mathrm{d}t} \equiv \frac{\partial \mathrm{A}}{\partial t} + \left(\boldsymbol{u} \cdot \frac{\partial}{\partial \boldsymbol{r}} \mathrm{A} \right) \tag{2.5}$$

which describes the rate of change of A in a comoving reference frame associated with a fluid element moving with velocity $\boldsymbol{u}$. It is often convenient to express the mass balance (2.2) in terms of the specific mixture volume $v(\boldsymbol{r}, t)(\equiv 1/\rho)$. If we set $\mathrm{A} \equiv v$ in (2.4), then we obtain (2.2) in the form

$$\rho \frac{\mathrm{d}}{\mathrm{d}t} \left(\frac{1}{\rho} \right) = \mathrm{div} \mathbf{u}. \tag{2.6}$$

The substantial balance equations for the mixture's individual chemical components are

$$\rho \frac{\mathrm{d}}{\mathrm{d}t}(Z_\alpha) \equiv \frac{\partial}{\partial t}(\rho Z_\alpha) + \mathrm{div}(\rho Z_\alpha \mathbf{u}) = -\mathrm{div} \mathbf{J}_\alpha + \sum_{s=1}^{r} \nu_{\alpha s} \xi_s, \quad (\alpha = 1, 2, \ldots, N). \tag{2.7}$$

Here, $Z_\alpha(\boldsymbol{r},t) \equiv n_\alpha/\rho$ is the specific (per unit mass of the total continuum) number density of the α-component; $\boldsymbol{J}_\alpha(\boldsymbol{r}, t) \equiv n_\alpha(\boldsymbol{u}_\alpha - \boldsymbol{u})$ is the molecular diffusion flux of particles of type α ($\boldsymbol{J}_\alpha \equiv \boldsymbol{J}_{(Z_\alpha)}$); $\sigma_{(Z_\alpha)} \equiv \sigma_\alpha = \sum_{s=1}^{r} \nu_{\alpha s} \xi_s$ is the "source term" representing the production rate of particles (atoms and molecules) of type α through chemical reactions; $\xi_s(\boldsymbol{r}, t)$ is the rate of the sth chemical reaction ($s = 1, 2, \ldots, r$); $\nu_{\alpha s}$ are the stoichiometric coefficients of a substance of type α in the sth reaction, which are positive for the reaction products and negative for the reactants. Note that (2.7) are linearly dependent, because the following relations hold:

$$\sum_{\alpha=1}^{N} m_\alpha Z_\alpha = 1 \quad (^1), \qquad \sum_{\alpha=1}^{N} m_\alpha \boldsymbol{J}_\alpha = 0 \qquad (^2),$$
$$\sum_{\alpha=1}^{N} m_\alpha \sigma_\alpha = \sum_{s=1}^{r} \xi_s \left[\sum_{\alpha=1}^{N} m_\alpha \nu_{\alpha s} \right] = 0 \qquad (^3). \tag{2.8}$$

Here, identities ($2.8^{(1,2)}$) are a corollary of definitions (2.3) for the mass density ρ and hydrodynamic velocity $\boldsymbol{u}$ of the total continuum, while relation ($2.8^{(3)}$) is a corollary of the mass conservation law in chemical reactions, $\sum_{\alpha} m_\alpha \nu_{\alpha s} = 0,\ (s = 1, 2, \ldots, r)$. Therefore, one of the differential equations (2.7) can always be excluded from consideration by replacing it with the algebraic integral ($2.8^{(2)}$).

2.1.4 The Equation of Motion for a Multicomponent Gas Mixture

In astrophysical and geophysical problems, one often has to deal with the relative motions of a gaseous medium being studied in a reference frame associated with the rotating surface of a cosmic body. For this reason, additional terms that allow for the Coriolis force and the centripetal acceleration (often small compared to the acceleration due to gravity) related to the rotation of the cosmic object appear in the corresponding equations. In this case, the complete momentum conservation equation for a multicomponent gas mixture in substantial form is

$$\rho \frac{\mathrm{d}\boldsymbol{u}}{\mathrm{d}t} \equiv \frac{\partial}{\partial t}(\rho\, \boldsymbol{u}) + \left(\frac{\partial}{\partial \boldsymbol{r}} \cdot (\rho\, \boldsymbol{u}\boldsymbol{u})\right)$$
$$= -\frac{\partial p}{\partial \boldsymbol{r}} + 2\rho\, \boldsymbol{u} \times \Omega + \left(\frac{\partial}{\partial \boldsymbol{r}} \cdot \Pi\right) + \rho \sum_{\alpha=1}^{N} Z_\alpha \boldsymbol{F}_\alpha, \tag{2.9}$$

where $\boldsymbol{u}(\boldsymbol{r}, t)$ is the total hydrodynamic velocity of the medium relative to the moving reference frame; $p(\boldsymbol{r}, t)$ is the thermodynamic pressure of the mixture; $\Pi(\boldsymbol{r}, t)$ is the viscous stress tensor related to the molecular momentum transport of all mixture components (written here in the so-called diffusion approximation (Sedov 1984; Nigmatullin 1987; see also Chap. 4)); Ω is the angular velocity vector of rotation of the reference frame in which (2.9) is written (below, the rotation rate is assumed to be $\Omega \equiv$ const); $\boldsymbol{F}_\alpha(\boldsymbol{r}, t)$ is the total external force acting on one particle of type α; and $\boldsymbol{u}\boldsymbol{u}$ is a dyad (a tensor with nine Cartesian coordinates). In what follows, we separate out the gravity force in explicit form:

$$\boldsymbol{F}_\alpha \equiv m_\alpha \boldsymbol{g} + \boldsymbol{F}^*_\alpha, \qquad \rho \sum_{\alpha=1}^{N} Z_\alpha \boldsymbol{F}_\alpha = \rho\, \boldsymbol{g} + \sum_{\alpha=1}^{N} n_\alpha \boldsymbol{F}^*_\alpha, \tag{2.10}$$

where $\boldsymbol{g} = \{0{,}0, -g\}$; $g = |\boldsymbol{g}|$ is the reduced (with allowance made for the centrifugal force) acceleration due to gravity, $\boldsymbol{g} = \frac{\partial}{\partial \boldsymbol{r}}\left(1/2|\Omega \times \boldsymbol{R}|^2 - G\frac{M_{\mathrm{pl}}}{|\boldsymbol{R}|}\right)$; G is the gravitational constant; M_{pl} is the mass of the cosmic body; $\boldsymbol{R}$ is the radius vector drawn from the body's center to the point of space under consideration; and $\boldsymbol{F}^*_\alpha$ is a nongravitational force (e.g., the Lorentz electromagnetic force). When writing (2.10), we took into account the fact that the Oz axis was directed upward along the normal to the equipotential surface of the gravity force (the coordinate $x_3 \equiv z$ is measured from some initial equipotential surface). Comparison of (2.9) and (2.1) written for $\mathsf{A} \equiv \boldsymbol{u}$ gives the following specific expressions for the substantial flux and volume source of the parameter $\boldsymbol{u}$:

$$\boldsymbol{J}_{(\boldsymbol{u})} \equiv -\Pi, \qquad \sigma_{(\boldsymbol{u})} \equiv -\frac{\partial p}{\partial \boldsymbol{r}} + 2\rho \boldsymbol{u} \times \Omega + \rho \sum_{\alpha=1}^{N} Z_\alpha \boldsymbol{F}_\alpha. \tag{2.11}$$

We use these and other similar formulas in Chap. 5 to derive the transport equations for the second correlation moments of thermohydrodynamic parameters fluctuating in a turbulized flow. These additional equations for correlations can be invoked to close the averaged hydrodynamic equations for a mixture, in particular, in the so-called invariant modeling of developed turbulence.

2.1.5 The Energy Balance Equations

In our phenomenological construction of models (with various levels of complexity) for a multicomponent turbulent medium, we need various energy balance equations for a reacting mixture in a laminar flow. In this introductory section, we briefly derive these equations in the case of conservative external fields.

2.1.5.1 Potential Energy Balance

Let us define the specific potential energy of a gas mixture by the relation

$$\Psi(\boldsymbol{r}, t) = \sum_{\alpha=1}^{N} \Psi_\alpha Z_\alpha. \tag{2.12}$$

where $\Psi_\alpha(\boldsymbol{r})$ is the scalar potential (per one particle of type α) satisfying the condition $\boldsymbol{F}_\alpha = -\partial\Psi_\alpha/\partial\boldsymbol{r}, \quad \partial\Psi_\alpha/\partial t = 0$. We obtain the corresponding balance equation from the operator relation (2.4). Supposing for this purpose that $\mathsf{A} \equiv \Psi$, we have

$$\begin{aligned}
\rho\frac{\mathrm{d}\Psi}{\mathrm{d}t} &= \sum_{\alpha=1}^{N} \Psi_\alpha \left[\frac{\partial}{\partial t}(\rho Z_\alpha) + \operatorname{div}(\rho Z_\alpha \boldsymbol{u})\right] + \sum_{\alpha=1}^{N} \rho Z_\alpha \left[\frac{\partial \Psi_\alpha}{\partial t} + \left(\boldsymbol{u} \cdot \frac{\partial \Psi_\alpha}{\partial \boldsymbol{r}}\right)\right] \\
&= \sum_{\alpha=1}^{N} \Psi_\alpha \rho \frac{\mathrm{d}Z_\alpha}{\mathrm{d}t} - \left(\boldsymbol{u} \cdot \sum_{\alpha=1}^{N} n_\alpha \boldsymbol{F}_\alpha\right) \\
&= -\operatorname{div}\left(\sum_{\alpha=1}^{N} \Psi_\alpha \boldsymbol{J}_\alpha\right) + \sum_{\alpha=1}^{N}\left(\boldsymbol{J}_\alpha \cdot \frac{\partial \Psi_\alpha}{\partial \boldsymbol{r}}\right) - \left(\boldsymbol{u} \cdot \sum_{\alpha=1}^{N} n_\alpha \boldsymbol{F}_\alpha\right) + \sum_{\alpha=1}^{N} \Psi_\alpha \sigma_\alpha.
\end{aligned}$$

Assuming next that the density of the internal source of potential energy becomes zero $\left(\sigma_{(\Psi)}{}^{(i)} \equiv \sum_{\alpha=1}^{N} \Psi_\alpha \sigma_\alpha = 0\right)$, which is valid when the potential energy conservation condition is met in chemical reactions (see Gyarmati 1970):

$$\sum_{\alpha=1}^{N} \Psi_\alpha \nu_{\alpha s} = 0, \qquad (s = 1, 2, \ldots, r). \tag{2.13}$$

we ultimately obtain

$$\begin{aligned}\rho \frac{\mathrm{d}\Psi}{\mathrm{d}t} &\equiv \frac{\partial}{\partial t}(\rho \Psi) + \operatorname{div}(\rho \Psi \boldsymbol{u}) \\ &= -\operatorname{div}\left(\sum_{\alpha=1}^{N} \Psi_\alpha \boldsymbol{J}_\alpha\right) + \sum_{\alpha=1}^{N}\left(\boldsymbol{J}_\alpha \cdot \frac{\partial \Psi_\alpha}{\partial \boldsymbol{r}}\right) - \left(\boldsymbol{u} \cdot \sum_{\alpha=1}^{N} n_\alpha \boldsymbol{F}_\alpha\right).\end{aligned} \tag{2.14}$$

Comparing now this equation with (2.1) written for $\mathsf{A} \equiv \Psi$, we get the following expressions for the substantial flux $J_{(\Psi)}$ and volume source $\sigma_{(\Psi)}$ of the potential energy of a multicomponent mixture:

$$\boldsymbol{J}_{(\Psi)} \equiv \sum_{\alpha=1}^{N} \Psi_\alpha \boldsymbol{J}_\alpha, \qquad \sigma_{(\Psi)} \equiv -\sum_{\alpha=1}^{N} (\boldsymbol{J}_\alpha \cdot \boldsymbol{F}_\alpha) - \left(\boldsymbol{u} \cdot \sum_{\alpha=1}^{N} \boldsymbol{n}_\alpha \boldsymbol{F}_\alpha\right). \tag{2.15}$$

2.1.5.2 Total Energy Conservation for a Mixture

The balance equation (conservation law) for the total energy of a multicomponent continuum in substantial form is

$$\rho \frac{\mathrm{d}U_{\text{tot}}}{\mathrm{d}t} \equiv \frac{\partial}{\partial t}(\rho\, U_{\text{tot}}) + \operatorname{div}(\rho\, U_{\text{tot}}\, \boldsymbol{u}) = -\operatorname{div}\boldsymbol{J}_{(U_{\text{tot}})}, \tag{2.16}$$

where

$$U_{\text{tot}}(\boldsymbol{r}, t) = 1/2|\boldsymbol{u}|^2 + \Psi + E \tag{2.17}$$

is the total specific energy of the medium; $E(\boldsymbol{r}, t)$ is the specific internal energy of the gas mixture (which, in fact, is defined by this relation);

$$\boldsymbol{J}_{(U_{\text{tot}})} \equiv \boldsymbol{q} + (p\,\boldsymbol{U} - \boldsymbol{\Pi}) \cdot \boldsymbol{u} + \sum_{\alpha=1}^{N} \Psi_\alpha \boldsymbol{J}_\alpha \tag{2.18}$$

is the substantial total energy flux vector for a moving mixture; $\boldsymbol{q}(\boldsymbol{r},t)$ is the molecular heat flux; and $\boldsymbol{U}$ is a unit vector. Relation (2.18) can actually be considered as a precise definition of the heat flux $\boldsymbol{q}$, which is known to be chosen quite freely in experimental and thermal physics.

It is important to clearly realize that the definition of the internal energy of a mixture $E \equiv \sum_\alpha Z_\alpha e_\alpha$ (where $e_\alpha(\boldsymbol{r},t)$ is the partial (per one particle) internal energy of component α including the chemical energy) introduced by relation (2.17) is generally improper. This is because, apart from the terms related to the thermal motion of molecules and the short-range intermolecular force interaction (in agreement with the conventional understanding of internal energy), the quantity E also implicitly includes the terms dependent on the macroscopic kinetic diffusion energy of the chemical components (in the center-of-mass system). For a multicomponent medium, these additional terms are known to be quantities of the second order of smallness and, hence, can often be discarded (this is the case of the so-called diffusion approximation (Sedov 1984)). However, for a multiphase medium (e.g., a gas–dust one), this additive $1/2\sum_\alpha m_\alpha Z_\alpha |\boldsymbol{u}_\alpha - \boldsymbol{u}|^2$ can be significant and should be taken into account explicitly (see Chap. 7).

2.1.5.3 Mechanical Energy Balance for a Mixture

Let us write the balance equation for the specific kinetic energy of the translational center-of-mass motion. For this purpose, scalarly multiplying the equation of motion (2.9) by the velocity vector $\boldsymbol{u}$ and making standard transformations, we obtain the work-kinetic energy theorem for a multicomponent mixture in substantial form:

$$\rho\,\frac{\mathrm{d}\left(|\boldsymbol{u}|^2/2\right)}{\mathrm{d}t} = p\mathrm{div}\,\boldsymbol{u} - \mathrm{div}(p\boldsymbol{u} - \mathit{\Pi}\cdot\boldsymbol{u}) - \left(\mathit{\Pi} : \frac{\partial}{\partial \boldsymbol{r}}\boldsymbol{u}\right) + \left(\boldsymbol{u}\cdot\sum_{\alpha=1}^{N} n_\alpha \boldsymbol{F}_\alpha\right). \tag{2.19}$$

Here, the product $\mathit{\Pi}\cdot\boldsymbol{u}$ of tensor $\mathit{\Pi}$ by vector $\boldsymbol{u}$ is defined as a vector with components $(\mathit{\Pi}\cdot\boldsymbol{u})_k = \sum_j \mathit{\Pi}_{kj}\boldsymbol{u}_j$, while the double product of two tensors $(\mathit{\Pi} : (\partial/\partial\boldsymbol{r})\boldsymbol{u})$ is defined as a scalar equal to $(\mathit{\Pi} : (\partial/\partial\boldsymbol{r})\boldsymbol{u}) = \sum_{i,j} \mathit{\Pi}_{ij}(\partial \boldsymbol{u}_j/\partial x_i)$ (the formulas of tensor calculus used here are briefly summarized in Appendix).

If we now add (2.14) and (2.19), then we obtain the following substantial balance equation for the specific mechanical energy $e_\mathrm{m}(\boldsymbol{r},t) \equiv |\boldsymbol{u}|^2/2 + \Psi$:

$$\rho\,\frac{\mathrm{d}e_\mathrm{m}}{\mathrm{d}t} \equiv \frac{\partial}{\partial t}(\rho e_\mathrm{m}) + \mathrm{div}(\rho e_\mathrm{m}\,\boldsymbol{u}) = -\mathrm{div}\boldsymbol{J}_{(e_\mathrm{m})} + \sigma_{(e_\mathrm{m})}, \tag{2.20}$$

where

$$\boldsymbol{J}_{(e_{\mathrm{m}})} \equiv (p\boldsymbol{U} - \Pi)\cdot\boldsymbol{u} + \sum_{\alpha=1}^{N}\Psi_\alpha \boldsymbol{J}_\alpha, \qquad (^1)$$
$$\sigma_{(e_{\mathrm{m}})} \equiv p\mathrm{div}\,\boldsymbol{u} - \left(\Pi : \frac{\partial}{\partial \boldsymbol{r}}\boldsymbol{u}\right) - \sum_{\alpha=1}^{N}\boldsymbol{J}_\alpha\cdot\boldsymbol{F}_\alpha^* \qquad (^2) \tag{2.21}$$

are, respectively, the substantial flux density of the mechanical energy e_{m} and the mechanical energy production rate for the mixture.

2.1.6 The Internal Energy Balance Equation for a Medium

We obtain the heat influx equation for a multicomponent gas mixture by subtracting (2.20) from (2.16):

$$\rho\frac{\mathrm{d}E}{\mathrm{d}t} \equiv \frac{\partial(\rho E)}{\partial t} + \mathrm{div}(\rho E\boldsymbol{u}) = -\mathrm{div}\boldsymbol{q} - p\mathrm{div}\boldsymbol{u} + \left(\Pi : \frac{\partial}{\partial \boldsymbol{r}}\boldsymbol{u}\right) + \sum_{\alpha=1}^{N}\boldsymbol{J}_\alpha\boldsymbol{F}_\alpha^*. \tag{2.22}$$

In astrophysical and geophysical applications, it is often convenient to use the heat influx equation written via the total specific enthalpy $H(\boldsymbol{r},t)$ of a gas mixture defined by the relation

$$H \equiv \sum_{\alpha=1}^{N} h_\alpha Z_\alpha = \sum_{\alpha=1}^{N}\left(e_\alpha + \frac{p_\alpha}{n_\alpha}\right)Z_\alpha = E + \frac{p}{\rho}, \tag{2.23}$$

where $h_\alpha(\boldsymbol{r},t)$ and $p_\alpha(\boldsymbol{r},t)$ are, respectively, the partial enthalpy and partial pressure of component α. Given the transformation $\rho\,\mathrm{d}E/\mathrm{d}t + p\mathrm{div}\boldsymbol{u} = \rho\,\mathrm{d}H/\mathrm{d}t - \mathrm{d}p/\mathrm{d}t$, which is a corollary of the continuity equation (2.2) and the definition of H, we then obtain

$$\rho\frac{\mathrm{d}H}{\mathrm{d}t} \equiv \frac{\partial(\rho H)}{\partial t} + \mathrm{div}(\rho H\boldsymbol{u}) = -\mathrm{div}\boldsymbol{q} + \frac{\mathrm{d}p}{\mathrm{d}t} + \left(\Pi : \frac{\partial}{\partial \boldsymbol{r}}\boldsymbol{u}\right) + \sum_{\alpha=1}^{N}\boldsymbol{J}_\alpha\boldsymbol{F}_\alpha^*. \tag{2.24}$$

For most of the goals considered in this book, it suffices to approximate the partial enthalpy $h_\alpha(T)$ with the expression $h_\alpha \cong c_{p\alpha}T + h_\alpha^0$, where $c_{p\alpha}(T)$ is the partial isobaric (at constant pressure) heat capacity per one particle of component α and h_α^0 is the partial enthalpy of component α at zero temperature (which is often called the standard formation heat of particles of type α). Using (2.23), we can then derive the following caloric equation of state:

$$H(Z_\alpha, T) \cong c_pT + \sum_{\alpha=1}^{N} h_\alpha^0 Z_\alpha, \qquad c_p = \sum_{\alpha=1}^{N} c_{p\alpha}Z_\alpha, \tag{2.25}$$

where $c_p(\boldsymbol{r}, t)$ is the specific isobaric heat capacity of the mixture. The quantities $c_{p\alpha}$ are assumed below to be the constants approximating the actual changes in partial heat capacities $c_{p\alpha}(T)$ in a limited temperature range. By comparing (2.24) with (2.1) written for $\mathsf{A} \equiv H$, we can find the following specific representations for the substantial enthalpy flux and "source" of the mixture:

$$\boldsymbol{J}_{(H)} \equiv \boldsymbol{q}, \qquad \sigma_{(H)} \equiv \frac{\mathrm{d}p}{\mathrm{d}t} + \left(\Pi : \frac{\partial}{\partial \boldsymbol{r}} \boldsymbol{u}\right) + \sum_{\alpha=1}^{N} \boldsymbol{J}_\alpha \cdot \boldsymbol{F}_\alpha^*. \tag{2.26}$$

Finally, let us write the heat influx equation (2.24) via the temperature $T(\boldsymbol{r}, t)$. Using (2.7), (2.23), and (2.25), we obtain

$$\rho \frac{\mathrm{d}H}{\mathrm{d}t} = \rho c_p \frac{\mathrm{d}T}{\mathrm{d}t} - \operatorname{div}\left\{\sum_{\alpha=1}^{N} h_\alpha \boldsymbol{J}_\alpha\right\} + \sum_{s=1}^{r} q_s \xi_s + \left(\frac{\partial T}{\partial \boldsymbol{r}} \cdot \sum_{\alpha=1}^{N} c_{p\alpha} \boldsymbol{J}_\alpha\right), \tag{2.27}$$

where the relation

$$q_s(T) = \sum_{\alpha=1}^{N} h_\alpha \nu_{\alpha s} = q_s^0 + \sum_{\alpha=1}^{N} c_{p\alpha} \nu_{s\alpha}, \qquad (s = 1, 2, \ldots, r) \tag{2.28}$$

introduces the so-called heat of a chemical reaction s at pressure p and temperature T. It is equal to the difference between the sum of the products of the partial enthalpies for the reaction products by the corresponding stoichiometric coefficients and the analogous sum for the reactants. The quantities $q_s^0 = \sum_\alpha h_\alpha^0 \nu_{\alpha s}$, which can be interpreted as the chemical reaction heats at zero temperature, are the input parameters for the model of a chemically active medium being developed. From (2.24) and (2.27), we ultimately have

$$\begin{aligned} \rho c_p \frac{\mathrm{d}T}{\mathrm{d}t} = &-\operatorname{div}\left\{\boldsymbol{q}_R + \boldsymbol{q} - \sum_{\alpha=1}^{N} h_\alpha \boldsymbol{J}_\alpha\right\} + \frac{\mathrm{d}p}{\mathrm{d}t} + \left(\Pi : \frac{\partial}{\partial \boldsymbol{r}} \boldsymbol{u}\right) \\ &- \sum_{s=1}^{r} q_s \xi_s + \sum_{\alpha=1}^{N} \boldsymbol{J}_\alpha \cdot \boldsymbol{F}_\alpha^* - \left(\frac{\partial T}{\partial \boldsymbol{r}} \cdot \sum_{\alpha=1}^{N} c_{p\alpha} \boldsymbol{J}_\alpha\right) + Q_d. \end{aligned} \tag{2.29}$$

The next-to-last term on the right-hand side of (2.29) represents the effect of the so-called "diffusing heat capacities." In most of the practically important cases, the specific isobaric heat capacities of the individual components in a gas mixture $c_{p\alpha}^* \equiv c_{p\alpha}/m_\alpha$ may be considered to be very close (so that $c_{p\alpha}^* \approx idem$). That is why, in view of (2.8$^{(2)}$), this term in (2.29) can be discarded.

It is important to keep in mind that the chemical heat sources (sinks) are explicitly present in the energy equation (2.29) for T due to the direct influence of chemical reactions precisely on the temperature field. To completely define the model of a multicomponent medium, the partial heat capacities $c_{p\alpha}$ of all mixture components and the heats $q_s(T)$ of all chemical reactions should be additionally specified.

When writing the energy equation (2.29) in final form, we added, for greater generality, the divergence term div$\boldsymbol{q}_R$ in which the flux

$$\boldsymbol{q}_R = \int_0^\infty \boldsymbol{q}_{R\nu} \mathrm{d}\nu = \int_0^\infty \boldsymbol{\Omega} I_\nu \, \mathrm{d}\Omega \, \mathrm{d}\nu, \tag{2.30}$$

is related to the radiation energy (the so-called Poynting vector) and the term Q_d characterizing the possible additional (local) heating sources of a multicomponent medium. Here, I_ν is the spectral radiation intensity: the amount of radiation energy contained in a unit interval of frequencies ν and in a unit solid angle dΩ transferred per unit time through a unit surface area placed at point $\boldsymbol{r}$ perpendicular to the propagation direction of the radiation energy $\boldsymbol{\Omega}$.

We make one more important remark on (2.29). As we showed in our monograph (Marov and Kolesnichenko 1987), an adequate description of the additional heat influx into the atmospheric region under consideration is the most complex problem, for example, in modeling the upper planetary atmosphere. This heat can result from the direct absorption of solar short-wavelength (and corpuscular) radiation by atmospheric components and its subsequent transformation through photochemical and chemical reactions as well as from individual dissipative processes of a hydrodynamic nature (including, e.g., the dissipation of magnetohydrodynamic and internal gravity waves and other wave motions of various spatial and temporal scales), whose action changes the temperature field in the atmosphere. The local heating sources of this kind for a multicomponent mixture of atmospheric gases that are represented in (2.29) by the term Q_d can be analyzed in detail only for a specific problem.

2.1.7 The Thermal Equation of State

In this book, we use the equation of state for a mixture of perfect gases as the thermal equation of state for a total multicomponent gas continuum (the equation for pressure):

$$p = \sum_{\alpha=1}^{N} p_\alpha = k_\mathrm{B} T \sum_{\alpha=1}^{N} n_\alpha = \mathsf{R}^* \rho T, \tag{2.31}$$

where $p_\alpha = k_\mathrm{B} T n_\alpha$ is the Clapeyron equation for the individual gas component; $\mathsf{R}^* = k_\mathrm{B} \sum_{\alpha=1}^{N} Z_\alpha = k_\mathrm{B}/\mathsf{M}$ is the so-called gas constant of the mixture; k_B is the

Boltzmann constant; $p_\alpha(\boldsymbol{r},t)$ is the partial pressure of component α; and $\mathsf{M} \equiv \rho/n$ is the mean molecular mass of the mixture. Specifying the internal energy as a function of temperature and composition $E = E(Z_\alpha, T)$ redefines the specific model of a gas mixture. For a mixture of perfect gases with constant partial heat capacities $c_{V\alpha}$ (at constant volume), we have

$$e_\alpha(T) = c_{V\alpha}T + h^0_\alpha, \qquad E(Z_\alpha, T) = c_V T + \sum\nolimits_\alpha Z_\alpha h^0_\alpha, \tag{2.32}$$

where $c_V \equiv \sum_\alpha Z_\alpha c_{V\alpha}$ is the specific heat capacity of the mixture at constant volume; it is easy to show that $c_p - c_V = \mathsf{R}^*$. Note that the first equality in (2.31) (the Dalton law) is valid for any gas mixture (irrespective of whether the individual gas components are perfect gases or not), because it is a corollary of the definition of p_α as the product of the total thermodynamic pressure of a medium p by the molar fraction $x_\alpha = n_\alpha/n$ of particles of type α.

$$p_\alpha = x_\alpha p, \qquad \left(n = \sum_{\alpha=1}^{N} n_\alpha\right),$$

where $n(\boldsymbol{r},t)$ is the total number density of the medium.

The system of hydrodynamic equations (2.2), (2.7), (2.9), (2.29), and (2.31) should be supplemented by the equations of chemical kinetics, the corresponding expressions for the chemical reaction rates and heats, and the defining relations for the diffusion and heat fluxes and the viscous stress tensor.

2.2 The Second Law of Thermodynamics

2.2.1 The Emergence of Entropy in Viscous Heat-Conducting Gas Mixtures

The mass, momentum, and energy transfer processes in the above hydrodynamic equations are described by some of their parts, each containing the divergence of the flux of the corresponding quantity related, at least implicitly, to the velocity, temperature, pressure, and component concentration gradients in a medium (the so-called thermodynamic forces). There are two ways of deriving the defining relations between the thermodynamic fluxes and their conjugate thermodynamic forces based on the gas-kinetic and phenomenological approaches. In general, the gas-kinetic approach is limited, because it is associated with the methods of solving the system of generalized Boltzmann equations for a multicomponent gas mixture (in particular, with the Chapman–Enskog method). It has now been completely developed only for moderate-density monoatomic gases and in the case where the molecular interaction force between two colliding particles is known, for example, where the spherically symmetric intermolecular interaction potential or the functional dependence of the collision frequency of particles on their velocity etc. has been specified (see Chapman and Cowling 1952; Ferziger and Kaper 1972). The phenomenological approach, which is based on the application of nonequilibrium thermodynamics to a macroscopic mixture volume, is unrelated to the postulation of a specific molecular model for the interaction between particles and is suitable for a wide class of continuous media with complicated properties (e.g., for electrolytes). However, the thermodynamics of irreversible processes, unfortunately, does not provide a method for determining the so-called transport coefficients – the proportionality coefficients at the gradients of thermodynamic parameters in the defining relations. Explicit algebraic expressions for these coefficients in a generally arbitrary gas–liquid medium must be found experimentally; at the same time, for monoatomic multicomponent gas mixtures at low densities, they can be calculated by considering in detail the dynamics of molecular collisions using the kinetic theory (see, e.g., Ferziger and Kaper 1972).

In this book, we derive phenomenologically the defining equations (for the thermodynamic fluxes) and additional algebraic relations between the transport coefficients.

2.2.2 The Onsager Principle

Before applying the formalism of nonequilibrium thermodynamics for continuous media to the description of heat and mass transfer in a laminar (and subsequently also turbulent) flow of a multicomponent reacting mixture, we briefly discuss here the essence of the basic postulates that underlie the theory and can be practically used in a thermodynamic analysis of any irreversible process. The phenomenological linear equations for irreversible processes (the so-called constitutive Onsager relations) are known (see, e.g., de Groot and Mazur 1962) to be used in nonequilibrium thermodynamics as the defining relations that supplement the system of hydrodynamic equations:

$$\mathrm{J}_k = \sum_{k=1}^{f} L_{kl}\mathrm{X}_l, \quad (k = 1, 2, \ldots, f), \tag{2.33}$$

where $L_{kl}(\boldsymbol{r},t)$ are the scalar kinetic (phenomenological) coefficients (functions of local state parameters of the medium: the temperature, pressure, concentrations, possibly magnetic field strength, etc.) that relate the fluxes $\mathrm{J}_k(\boldsymbol{r},t)$ and the thermodynamic (generalized) forces $X_l(\boldsymbol{r},t)$; f is the number of independent scalar fluxes and forces. Here, the scalars J_k and X_l denote the independent scalar thermodynamic fluxes and forces and, in the case of vector and tensor processes, all Cartesian components of the corresponding tensor and vector quantities in the bilinear expressions for entropy production [see (2.55) below]. The fluxes J_k correspond to the rates of change of extensive parameters (such as the mass, momentum, and energy), for which the conservation laws exist, or transferable quantities (such as the heat), which enter into the divergence terms of the conservation laws; the generalized forces X_l are proportional to the gradients of intensive parameters or external forces disturbing the thermodynamic system from equilibrium. The diagonal coefficients L_{kk} are the coefficients representing direct effects (ordinary diffusion, heat conduction, etc.); the off-diagonal coefficients $L_{kl}(k \neq j)$ are the phenomenological coefficients of cross effects (thermal heat conduction, thermal diffusion (Dufour effect), etc.). The linear relations (2.33) are applicable in states close to local thermodynamic equilibrium (LTE) (and, in fact, define the domain of applicability of the thermodynamics of irreversible processes).

In principle, each Cartesian component of the flux vector J_k can be a linear function of the Cartesian components of all thermodynamic forces. Note, however, that the fluxes and thermodynamic forces in (2.53) generally have different tensor properties, i.e., during rotation and reflection transformations, the Cartesian components of these quantities are transformed differently. As a result, it may turn out that, in view of certain symmetry properties of the material medium under consideration, the Cartesian components of the fluxes will not depend on all Cartesian components of the thermodynamic forces, which limits the number of nonzero kinetic coefficients L_{kl} in (2.33) (the Curie symmetry principle). For example, in the special case of an isotropic system whose properties in an equilibrium state are identical in all directions, physical-chemical and hydrodynamic processes with different tensor dimensions do not interact with one another (see de Groot and Mazur 1962).

In addition, the symmetry relations for kinetic coefficients, the so-called Onsager–Casimir reciprocity principle, are taken as an independent postulate in the axiomatic approach:

$$L_{kl}(\boldsymbol{B}, \Omega) = \varepsilon_k \varepsilon_l \mathrm{L}_{kl}(-\boldsymbol{B}, -\Omega), \quad (k,l = 1, 2, \ldots, f). \tag{2.34}$$

These relations are a corollary of the invariance of microscopic dynamics under time reversal. In writing (2.34), we assumed that the mechanical system could be subjected to an external magnetic induction $\boldsymbol{B}$ or undergo an angular rotation Ω. In the presence of an external magnetic induction $\boldsymbol{B}$, for the particles of the medium to pass along their previously traversed path under time reversal, not only the velocity vector but also the magnetic field vector must be reversed. This is a corollary of the formula representation for the Lorentz electromagnetic force [see Chap. 9]. Similar reasoning is also valid for processes whose evolution is described in noninertial reference frames rotating with an angular velocity Ω. The reciprocity relations (2.34) impose constraints on the configuration of the constitutive relations (2.33), minimizing the number of kinetic coefficients in them.

For an isotropic system (when the external magnetic field and the angular velocity are zero), the reciprocity relations (2.34) take a simpler form:

$$L_{kl} = \varepsilon_k \varepsilon_l L_{lk}, \quad (k,l = 1, 2, \ldots, f). \tag{2.34*}$$

If the coefficients L_{kl} refer to the cross effects that can be described either only by α-parameters (in statistical mechanics, the variables that are even functions of the molecular velocities are commonly called so) or only by β-parameters (the variables that are odd functions of the molecular velocities), then the coefficients $\varepsilon_k = \varepsilon_l = 1$ (a variant of the symmetric Onsager reciprocity relations). However, in the presence of mixed cross effects for α- and β-variables, $\varepsilon_l = 1$, while $\varepsilon_k = -1$ (a variant of the antisymmetric Casimir reciprocity relations).

At present, the symmetry relations for kinetic coefficients are regarded as an empirically stable axiom irrespective of their proof within the framework of statistical mechanics (Miller 1974). According to Mason (1974), the experimental confirmation of the Onsager–Casimir reciprocity principle is as convincing as the confirmation of the first, second, and third laws of thermodynamics. This gives grounds to raise postulate (2.34) to the status of a paradigm and to use it as the basis for describing a wide range of macroscopic transport processes.

A specific representation of the production rate $\sigma_{(S)}(\boldsymbol{r},t)$ (source density) of the entropy $S(\boldsymbol{r},t)$ inside the system under consideration (with a prespecified set of irreversible processes) written as a bilinear form is commonly used to define explicitly the fluxes and their conjugate thermodynamic forces in (2.33):

$$\sigma_{(S)}(\boldsymbol{r},t) = \sum_{k=1}^{f} \mathrm{J}_k X_k \geq 0 \tag{2.35}$$

in this case, each individual flux and each individual force become zero in LTE. It is important to emphasize that the decomposition of $\sigma_{(S)}(\boldsymbol{r},t)$ into the thermodynamic forces and fluxes is fairly arbitrary; nevertheless, this has no direct effect on the interpretation of the final results: once the fluxes J_k have been established, their conjugate thermodynamic forces X_l can be found unambiguously as the coefficients of the corresponding fluxes in (2.35).

Thus, to decipher (2.35) for the entropy source density in terms of the phenomenological theory, we must have the balance equation for the entropy $S(\boldsymbol{r},t)$ of a continuous system in explicit form:

$$\rho\frac{\mathrm{d}S}{\mathrm{d}t} \equiv \frac{\partial}{\partial t}(\rho S) + \mathrm{div}(\rho S\boldsymbol{u}) = -\mathrm{div}\boldsymbol{J}_{(S)} + \sigma_{(S)}, \quad (\sigma_{(S)} \geq 0), \tag{2.36}$$

where the vector $\boldsymbol{J}_{(S)}(\boldsymbol{r},t)$ defines the substantial entropy flux density. Here, $\mathrm{div}\boldsymbol{J}_{(S)}$ describes a reversible heat exchange between the system under consideration and the environment, while the inequality $\sigma_{(S)} \geq 0$ corresponds to the second law of thermodynamics, which states that the entropy of a closed system cannot decrease. It is important to clearly realize that the meaning of the inequality $\sigma_{(S)} \geq 0$ in nonequilibrium thermodynamics goes beyond the scope of the ordinary formulation of the second law of equilibrium thermodynamics, in which only the global increase in entropy between two equilibrium states of an isolated system is considered.

In contrast, here we assume that the condition $\sigma_{(S)} \geq 0$ is met at any point of the coordinate space at any time of the system's evolution.

The specific form of (2.36), which plays a fundamental role in nonequilibrium thermodynamics, can be obtained for a multicomponent thermohydrodynamic system using the balance equations for the specific volume (2.6), the specific number densities of the mixture components (1.7), and the specific internal energy (2.22) from the fundamental Gibbs identity for these quantities. Written along the path of the center of mass of an elementary physical volume, this identity is (see, e.g., Prigogine and Defay 1954):

$$T\frac{\mathrm{d}S}{\mathrm{d}t} = \frac{\mathrm{d}E}{\mathrm{d}t} + p\frac{\mathrm{d}}{\mathrm{d}t}\left(\frac{1}{\rho}\right) - \sum_{\alpha=1}^{N} \mu_\alpha \frac{\mathrm{d}}{\mathrm{d}t}\left(\frac{n_\alpha}{\rho}\right), \tag{2.37}$$

where $\mu_\alpha(\boldsymbol{r}, t)(= h_\alpha - Ts_\alpha)$ and $s_\alpha(\boldsymbol{r},t)$ are, respectively, the partial chemical potential calculated per one particle of substance and the partial entropy of particles of type α.

As regards the applicability of the Gibbs identity for nonequilibrium processes in any continuous thermodynamic system, it is important to take into account the following. The fundamental assumption on which nonequilibrium thermodynamics (in the form in which it was developed by Onsager, Prigogine, and other authors) is based is the assumption about LTE. The essence of LTE is that the entire system can be mentally broken down into sufficiently small (but macroscopic) regions each of which may be considered as an equilibrium thermodynamic system for which the entropy is the same function of state parameters as that in the state of complete thermodynamic equilibrium. In particular, if the specific volume $1/\rho$, the specific density of the internal energy E, and the specific concentrations (n_α/ρ) of the chemical components have been chosen as the mixture state variables when constructing the model of a continuous medium, then the thermodynamic state of an elementary physical volume in the vicinity of point $\boldsymbol{r}$ at an instant of time t is completely described by the fundamental Gibbs integral equation (see Munster 2002)

$$S(\boldsymbol{r}, t) = S(E, (1/\rho), (n_1/\rho), \ldots, (n_N/\rho)),$$

whose differential form is (2.37). Actually, based on the Gibbs identity alone, we can assign the meaning of local temperature $T(\boldsymbol{r}, t)$, local pressure $p(\boldsymbol{r}, t)$, and local chemical potential $\mu_\alpha(\boldsymbol{r}, t)$ for particles of type α to the conjugate quantities

$$\begin{aligned} 1/T &\equiv \left(\frac{\partial S}{\partial E}\right)_{(1/\rho),\{n_\alpha/\rho\}}, \qquad p/T \equiv \left(\frac{\partial S}{\partial(1/\rho)}\right)_{E,\{n_\alpha/\rho\}}, \\ -\mu_\alpha/T &\equiv \left(\frac{\partial}{\partial(n_\alpha/\rho)}\right)_{E,(1/\rho),\{n_\beta \neq n_\alpha\}} \qquad (\alpha = 1, 2, \ldots, N) \end{aligned} \tag{2.38}$$

In other words, these quantities are, in fact, defined by relations (2.38). The entropy S in this case is a thermodynamic potential. The LTE postulate is generally valid when the dissipative processes in a system play a significant role and rule out the appearance of large gradients in state parameters. It should also be kept in mind that the hypothesis about LTE of a medium is equivalent to the assumption about the validity of not only the Gibbs identity but also all the other relations of equilibrium thermodynamics for infinitesimally small regions of nonequilibrium systems.

Let us now summarize the main assertions that are taken as independent postulates when classical nonequilibrium thermodynamics is applied to any irreversible process:

- The principle of quasi-local thermodynamic equilibrium is valid.
- The inequality $\sigma_{(S)} \geq 0$, which expresses the second law of thermodynamics, is valid for entropy production related to irreversible processes in the system itself.
- When restricted to a linear region, the linear phenomenological relations (2.33) for the fluxes and thermodynamic forces appearing in (2.35) for entropy production are valid.
- The symmetry relations (2.34) for the kinetic coefficients appearing in the linear laws (2.33) hold true.

To conclude this section, we additionally note the following. Classical nonequilibrium thermodynamics is not the only thermodynamic theory suitable for studying the nonequilibrium states of a continuous system. There are also other well-developed formalisms, for example, the so-called rational thermodynamics (whose foundations were developed by Coleman (1964), Truesdell (1984), and Noll (1974)), which often allow the problems defying solution in terms of classical irreversible thermodynamics to be overcome. In particular, rational thermodynamics allows the constitutive equations for materials with memory to be derived, because no assumption about a local equilibrium state is made in this approach. At the same time, from a practical point of view, the constitutive equations written as functionals of the entire prehistory of variables are a difficult object of investigation for mathematical modeling and generally require knowing an excessively large volume of information.

As has been noted above, in the last decade significant progress has been achieved in developing the so-called extended irreversible thermodynamics (see Jou et al. 2001). It also goes beyond the scope of the hypothesis about LTE due to a modification of the dissipative fluxes of the corresponding physical quantities that act as additional independent variables of the system under consideration. Of course, the fluxes are not the only possible set of new nonequilibrium variables: in particular, the variables related to the internal degrees of freedom of the molecules or the microstructure of a system can figure as the latter. Both possibilities may be considered, to a certain extent, to complement each other. Naturally, going beyond the scope of LTE suggests a comprehensive rethinking of such basic and conceptual notions as the entropy, temperature, pressure, and chemical potential for more general cases than the multicomponent mixture considered in this chapter, for example, when modeling turbulized continuous media (see Chap. 3).

2.2.3 The Balance Equation for Entropy and Entropy Production in Reacting Gas Mixtures

To study the transport processes in multicomponent mixtures of gases, we use the formalism of classical nonequilibrium thermodynamics, by means of which, as is well known, a wide class of nonequilibrium transport processes can be described in complete agreement with experimental data. To derive the Onsager relations (2.33), it is first necessary to find an explicit form of the entropy balance equation (2.36) for the model of a multicomponent gaseous medium under consideration. For this purpose, we eliminate the time derivatives of the quantities E, $(1/\rho)$, and (n_α/ρ) on the right-hand side of the Gibbs identity (2.37) using the hydrodynamic equations for a mixture; as a result, we obtain a specific form of the substantial balance equation (2.36) for the entropy of a mixture S where the entropy flux density $\boldsymbol{J}_{(S)}$ is specified by the expression

$$\boldsymbol{J}_{(S)} \equiv \frac{1}{T}\left\{\boldsymbol{q} - \sum_{\alpha=1}^{N} \mu_\alpha \boldsymbol{J}_\alpha\right\}, \tag{2.39}$$

while the entropy production $\sigma_{(S)}$ multiplied by T (the so-called energy dissipation) is defined as

$$\begin{aligned} 0 \leq T\sigma_{(S)} \equiv &-\left(\boldsymbol{J}_q \cdot \frac{\partial \ln T}{\partial \boldsymbol{r}}\right) + \left(\Pi : \frac{\partial}{\partial \boldsymbol{r}}\boldsymbol{u}\right) \\ &- \sum_{\alpha=1}^{N} \boldsymbol{J}_\alpha \cdot \left\{T\frac{\partial}{\partial \boldsymbol{r}}\left(\frac{\mu_\alpha}{T}\right) + h_\alpha \frac{\partial \ln T}{\partial \boldsymbol{r}} - \boldsymbol{F}_\alpha\right\} + \sum_{s=1}^{r} A_s \xi_s. \end{aligned} \tag{2.40}$$

According to the second law of thermodynamics, the energy dissipation $T\sigma_{(S)}(\boldsymbol{r},t)$, a quantitative measure of the nonequilibrium processes in a system, is a positive definite quantity. Here,

$$A_s(\boldsymbol{r},t) \equiv -\sum_{\alpha=1}^{N} \mu_\alpha \nu_{\alpha s}, \quad (s = 1, 2, \ldots, r) \tag{2.41}$$

is the so-called chemical affinity of the sth reaction (the system's state function that, according to de Donde, is the driving force of a chemical reaction (see Prigogine and Defay 1954));

$$\boldsymbol{J}_q \equiv \boldsymbol{q} - \sum_{\alpha=1}^{N} h_\alpha \boldsymbol{J}_\alpha \tag{2.42}$$

is the reduced heat flux vector. In contrast to the total heat flux $\boldsymbol{q}$ including the heat $\sum_\alpha h_\alpha \boldsymbol{J}_\alpha$ transferred by diffusion due to the difference in enthalpies (heat contents) of the diffusing substances, the reduced heat flux $\boldsymbol{J}_q$ is the heat transferred only by pure heat conduction.

For the subsequent analysis, it is convenient to decompose the velocity gradient tensor $\partial\boldsymbol{u}/\partial\boldsymbol{r}$ (a second-rank tensor) in (2.40) into symmetric and antisymmetric parts,

$$\partial\boldsymbol{u}/\partial\boldsymbol{r} = (\partial\boldsymbol{u}/\partial\boldsymbol{r})^s + (\partial\boldsymbol{u}/\partial\boldsymbol{r})^a = \overset{0}{(\partial\boldsymbol{u}/\partial\boldsymbol{r})^s} + (\partial\boldsymbol{u}/\partial\boldsymbol{r})^a + 1/3\boldsymbol{U}\,\mathrm{div}\boldsymbol{u}, \quad (2.43)$$

and to represent the symmetric viscous stress tensor Π as

$$\Pi = \pi\boldsymbol{U} + \overset{0}{\Pi^s}\,. \quad (2.44)$$

Here,

$$(\partial\boldsymbol{u}/\partial\boldsymbol{r})^a_{jk} \equiv \frac{1}{2}\left(\frac{\partial u_k}{\partial x_j} - \frac{\partial u_j}{\partial x_k}\right) = \frac{1}{2}(\mathrm{curl}\ \boldsymbol{u})_{jk}, \quad (j,k = 1,2,3) \quad (2.45)$$

is the skew-symmetric velocity gradient tensor;

$$\tilde{D}_{jk} \equiv (\partial\boldsymbol{u}/\partial\boldsymbol{r})^s_{jk} = \frac{1}{2}\left(\frac{\partial u_k}{\partial x_j} + \frac{\partial u_j}{\partial x_k}\right). \quad (2.46)$$

is the strain rate tensor, $\tilde{D} \equiv (\partial\boldsymbol{u}/\partial\boldsymbol{r})^s$;

$$\tilde{S}_{jk} \equiv \overset{0}{(\partial\boldsymbol{u}/\partial\boldsymbol{r})^s_{jk}} = \frac{1}{2}\left(\frac{\partial u_k}{\partial x_j} + \frac{\partial u_j}{\partial x_k}\right) - \frac{1}{3}\delta_{\mathrm{ik}}\mathrm{div}\boldsymbol{u} \quad (2.47)$$

is the shear rate tensor, $\tilde{S} \equiv \tilde{D} - 1/3\boldsymbol{U}\mathrm{div}\boldsymbol{u} \equiv \overset{0}{(\partial\boldsymbol{u}/\partial\boldsymbol{r})^s}$;

$$\pi \equiv 1/3(\Pi : \boldsymbol{U}) = 1/3\Pi_{kk} \quad (2.48)$$

is the so-called viscous pressure. The scalar product of the viscous stress tensor and the velocity gradient tensor can then be represented as

$$\sigma^v_{(S)} \equiv \frac{1}{T}\left(\Pi : \frac{\partial}{\partial\boldsymbol{r}}\boldsymbol{u}\right) = \frac{1}{T}\left(\overset{0}{\Pi^s} : \overset{0}{(\partial\boldsymbol{u}/\partial\boldsymbol{r})^s}\right) + \pi\frac{\mathrm{div}\boldsymbol{u}}{T}, \quad (2.49)$$

here, we use the fact that the scalar product of symmetric and antisymmetric tensors is always equal to zero. Note that the total contribution of viscous phenomena to the entropy production turns out to be split into two parts. The second part, $-T^{-1}\pi\mathrm{div}\boldsymbol{u} = -T^{-1}\pi\rho\,\mathrm{d}(1/\rho)/\mathrm{d}t$, is related to the rate of change of the specific volume and is attributable to the presence of a bulk (second) viscosity in a compressible medium.

Thus, the entropy production $\sigma_{(S)}$ in a moving multicomponent reacting gas mixture is defined by the following set of thermodynamic fluxes $\boldsymbol{J}_q, \xi_s, \boldsymbol{J}_\alpha, \pi, \overset{\circ}{\Pi}{}^s$ and their conjugate thermodynamic forces:

$$\boldsymbol{X}_q \equiv -\frac{1}{T^2}\frac{\partial T}{\partial \boldsymbol{r}} = \frac{\partial}{\partial \boldsymbol{r}}\left(\frac{1}{T}\right), \tag{2.50}$$

$$A_s^* \equiv \frac{A_s}{T} = -\sum_{\beta=1}^{N}\frac{\mu_\beta}{\mathrm{T}}\nu_{\beta s}, \qquad (s = 1, 2, \ldots, r), \tag{2.51}$$

$$\boldsymbol{X}_\alpha^* \equiv -\frac{\partial}{\partial \boldsymbol{r}}\left(\frac{\mu_\alpha}{T}\right) + h_\alpha \frac{\partial}{\partial \boldsymbol{r}}\left(\frac{1}{T}\right) + \frac{1}{T}\boldsymbol{F}_\alpha, \quad (\alpha = 1, 2, \ldots, N), \tag{2.52}$$

$$\boldsymbol{X}_\pi \equiv \frac{1}{T}\mathrm{div}\boldsymbol{u}, \tag{2.53}$$

$$\overset{\circ}{\boldsymbol{X}}{}^s \equiv \frac{1}{T}\left(\partial \overset{\circ}{\boldsymbol{u}}/\partial \boldsymbol{r}\right)^s = \frac{1}{T}\tilde{S}. \tag{2.54}$$

Here, the polar vector $\boldsymbol{X}_q$ is the thermodynamic force responsible for heat conduction; the chemical affinity for the sth reaction A_s is the scalar thermodynamic force conjugate to the chemical reaction rate ξ_s; the thermodynamic diffusion force $\boldsymbol{X}_\alpha^*$ is conjugate to the diffusion flux density $\boldsymbol{J}_\alpha$ (a polar vector); the scalar viscosity force X_π is conjugate to the viscous pressure π (as to the scalar momentum flux responsible for the phenomena related to bulk viscosity); the (second-rank) tensor force $\overset{0}{\boldsymbol{X}}{}^s$, which is conjugate to the part of the viscous stress tensor $\overset{0}{\Pi}{}^s$ with a zero trace, produces a viscous shear (note that the viscous forces are β-variables, because the velocity $\boldsymbol{u}$ is a typical β-parameter). Using the fluxes and thermodynamic forces introduced in this way, we can write the entropy production $\sigma_{(S)}$ in a reacting gas mixture in a bilinear form,

$$\sigma_{(S)} = \sum_{s=1}^{r} A_s^*\xi_s + \pi \boldsymbol{X}_\pi + \boldsymbol{J}_q \cdot \boldsymbol{X}_q + \sum_{\alpha=1}^{N} \boldsymbol{J}_\alpha \cdot \boldsymbol{X}_\alpha^* + \overset{0}{\Pi}{}^s : \overset{0}{\boldsymbol{X}}{}^s \geq 0, \tag{2.55}$$

which corresponds to three (scalar, vector, and tensor) independent sources of nonequilibrium processes in a reacting mixture differing significantly in physical nature. Expression (2.55) allows the constitutive linear relations (2.33) between the fluxes and thermodynamic forces to be specified for the multicomponent continuum considered here.

2.3 Defining Relations for the Diffusion and Heat Fluxes and the Viscous Stress Tensor

In this section, for simplicity, we restrict ourselves to the case of an isotropic medium for which the phenomena described by thermodynamic forces and fluxes with various tensor dimensions do not influence one another (the Curie principle). The anisotropic case is considered in detail, for example, in the well-known monograph by de Groot and Mazur (1962), to which we refer the interested reader.

2.3.1 Linear Kinematic Constitutive Equations

In the isotropic case, (2.55) for entropy production can be split into four different parts each of which, according to the second law of thermodynamics, must be a positively defined quantity. In other words, according to the second law and simultaneously in accordance with the Curie principle, it is necessary that not only the condition

$$\sigma_{(S)} = \overbrace{\sigma_{(S)}}^{\text{(scalar)}} + \overbrace{\sigma_{(S)}}^{\text{(vector)}} + \overbrace{\sigma_{(S)}}^{\text{(tensor)}} \geq 0,$$

be met but also the entropy production be a positively defined quantity for all processes of various tensor natures noninteracting with one another.

The corresponding phenomenological equations (2.33) in rectangular coordinates can then be written as

$$\boldsymbol{J}_q = L_{qq}\boldsymbol{X}_q + \sum_{\beta=1}^{N} L_{q\beta}\boldsymbol{X}^*_\beta, \tag{2.56}$$

$$\boldsymbol{J}_\alpha = L_{\alpha q}\boldsymbol{X}_q + \sum_{\beta=1}^{N} L_{\alpha\beta}\boldsymbol{X}^*_\beta, \quad (\alpha = 1, 2, \ldots, N), \tag{2.57}$$

$$\overset{0}{\Pi}{}^s_{jk} = L\tilde{S}_{jk} = \frac{L}{2T}\left\{\left(\frac{\partial u_k}{\partial x_j} + \frac{\partial u_j}{\partial x_k}\right) - \frac{2}{3}\delta_{jk}\text{div}\boldsymbol{u}\right\}, \quad (j, k = 1, 2, 3) \tag{2.58}$$

$$\pi = l_{\pi\pi}\frac{\text{div}\boldsymbol{u}}{T} + \sum_{s=1}^{r} l_{\pi s}A^*_s, \tag{2.59}$$

$$\xi_s = -l_{s\pi}\frac{\text{div}\boldsymbol{u}}{T} + \sum_{m=1}^{r} l_{sm}A^*_s, \quad (s = 1, 2, \ldots, r) \tag{2.60}$$

with the following Onsager–Casimir symmetry relations existing between the scalar kinetic coefficients $L_{q\alpha}$, $L_{\alpha\beta}$, l_{sv}, and l_{sm} in these relations:

$$L_{q\beta} = L_{\beta q}, \quad (\beta = 1, 2, \ldots, N); \qquad L_{\alpha\beta} = L_{\beta\alpha}, \quad (\alpha, \beta = 1, 2, \ldots, N), \tag{2.61}$$

$$l_{s\pi} = -l_{\pi s}, \quad (s = 1, 2, \ldots, r); \qquad l_{sm} = l_{ms} \quad (s, m = 1, 2, \ldots, r). \tag{2.62}$$

Here, the kinetic coefficients describing the cross effects between the chemical reactions and bulk viscosity satisfy the antisymmetric Casimir relations, because the chemical affinity A_s is an α-force, while the viscous force X_π is a β-one.

It should be noted that, in addition to the reciprocity relations (2.61), the Onsager coefficients $L_{q\alpha}$ and $L_{\alpha\beta}$ satisfy $(N + 1)$ additional relationships that follow from the definition $\boldsymbol{J}_\alpha \equiv n_\alpha(\boldsymbol{u}_\alpha - \boldsymbol{u})$ of the diffusion fluxes and from the linear independence of the thermodynamic forces $\boldsymbol{X}_q$ and $\boldsymbol{X}^*_\beta$. Indeed, substituting relations (2.57) into identity (2.8$^{(2)}$) yields

$$0 = \sum_{\alpha=1}^{N} m_\alpha \boldsymbol{J}_\alpha = \left(\sum_{\alpha=1}^{N} m_\alpha L_{\alpha q}\right) \boldsymbol{X}_q + \sum_{\beta=1}^{N} \left(\sum_{\alpha=1}^{N} m_\alpha L_{\alpha\beta}\right) \boldsymbol{X}^*_\beta,$$

hence, in view of the linear independence of the forces $\boldsymbol{X}_q$ and $\boldsymbol{X}^*_\beta$, it follows that

$$\sum_{\alpha=1}^{N} m_\alpha L_{\alpha q} = 0, \quad (^1) \quad \sum_{\alpha=1}^{N} m_\alpha L_{\alpha\beta} = 0, \quad (\beta = 1, 2, \ldots, N) \quad (^2) \tag{2.63}$$

As to the linear approximation (2.60), it is important to clearly realize the following with regard to the chemical reaction rates: the linear relations (2.60) of irreversible thermodynamics are known to be valid for chemical reactions only when the condition $A_s << RT$ is met [see (4.1.4)]. However, this condition is generally satisfied only at the final stage of a chemical reaction. Thus, the linear approximation of the phenomenological laws is invalid for chemical reactions, although the general expression (2.40) for entropy production through reactions is valid for those reactions for which the mass action law holds (see Sect. 3.3). It is for this reason that we use Arrhenius gas-phase kinetics (see Stiller 1989) to describe turbulized chemically reacting mixtures.

2.3.2 *Viscous Flow of an Isotropic Fluid*

Using (2.58) and (2.59), it is easy to show that the relation between the tensors Π and $(\partial/\partial \boldsymbol{r})\boldsymbol{u}$ for an isotropic medium has the following form (without the terms that allow for the "chemistry," which, as is well known, may be neglected in real cases):

$$\Pi_{jk} = \mu\left(\frac{\partial u_k}{\partial x_j} + \frac{\partial u_j}{\partial x_k} - \frac{2}{3}\delta_{jk}\mathrm{div}\boldsymbol{u}\right) + \mu_\vartheta \delta_{jk}\mathrm{div}\boldsymbol{u}, \tag{2.64}$$

where the molecular shear viscosity $\mu(\boldsymbol{r},t)(\equiv L/2T)$ and bulk viscosity $\mu_\vartheta(\boldsymbol{r},t)$ $(\equiv l_{\pi\pi}/T)$ depend only on the local thermodynamic properties of the medium. The shear viscosity μ plays an important role in flows in which the successive plane fluid layers move with different velocities. The bulk viscosity μ_ϑ turns out to be significant during the expansion of the medium. It is often convenient to introduce the molecular kinematic viscosity $\nu(\boldsymbol{r},t)$ related to the ordinary viscosity (the coefficient of internal friction) μ by the relation $\nu \equiv \mu/\rho$.

It follows from (2.49) and (2.64) that the contribution of viscosity effects to the expression for the dissipation rate of kinetic energy into heat (dissipative function) is

$$
\begin{aligned}
0 \le T\sigma_{(S)}^{\text{vis}} &\equiv 2\mu(\tilde{\boldsymbol{S}}:\tilde{\boldsymbol{S}}) + \mu_\vartheta(\text{div}\boldsymbol{u})^2 \\
&= \frac{\mu}{2}\sum_{k,j}\left(\frac{\partial u_k}{\partial x_j} + \frac{\partial u_j}{\partial x_k} - \frac{2}{3}\delta_{\text{jk}}\text{div}\boldsymbol{u}\right)^2 + \mu_\vartheta(\text{div}\boldsymbol{u})^2,
\end{aligned}
\tag{2.65}
$$

where $\tilde{\boldsymbol{S}}$ is the shear rate tensor defined by (2.47). The factors in front of μ and μ_ϑ are the sums of squares and, consequently, are both positive. It follows from the condition that one of the factors can be zero, while the other is nonzero that $\mu \ge 0$ and $\mu_\vartheta \ge 0$.

2.3.3 Heat Conduction, Diffusion, and Cross Effects

Various definitions of the heat and diffusion fluxes as well as the diffusion, $D_{\alpha\beta}$ $(\boldsymbol{r},t)$, thermal conductivity, $\lambda(\boldsymbol{r},t)$, and thermal diffusion, $D_{T\alpha}(\boldsymbol{r},t)$, coefficients for a multicomponent mixture can be encountered in the literature. The correct definitions of the transport coefficients are those that are consistent with the Onsager–Casimir reciprocity relations in nonequilibrium thermodynamics. For the gas-kinetic theory of multicomponent mixtures, this consistency acquires particular significance when its results are generalized to the case of systems with nonconserved particle densities of individual components, for example, in gas mixtures where there are chemical reactions between their components or in monoatomic gases where transitions are possible between states with different internal degrees of freedom. The most adequate definition of these coefficients consistent with the thermodynamic Onsager–Casimir reciprocity relations was proposed by Ferziger and Kaper (1972). It is preferred to the definition given, for example, in the widely known monograph by Hirschfelder et al. (1954). In this monograph, the multicomponent diffusion coefficients are defined as nonsymmetric functions with respect to the permutation of indices $D_{\alpha\beta} \neq D_{\beta\alpha}$, which is a corollary of the assumption made there that the diagonal coefficients are zero, $D_{\alpha\alpha} = 0$.

To be able to compare the defining relations for the molecular diffusion, $\boldsymbol{J}_\alpha$, and heat, $\boldsymbol{q}$, fluxes (derived below by the methods of nonequilibrium thermodynamics) and the relations obtained by gas-kinetic methods in the book by Ferziger and Kaper (1972), let us express the gradient of the chemical potential μ_α for particles of type α in terms of the gradients of thermohydrodynamic parameters. Assuming the gas mixture under consideration to be an ideal system (note that diluted solutions and mixtures of perfect gases are essentially ideal systems), we write the chemical potential μ_α as (see Prigogine and Defay 1954)

$$\mu_\alpha = \mu_\alpha^0(T,p) + k_{\mathrm{B}}T \ln\left(\frac{n_\alpha}{n}\right), \tag{2.66}$$

where $\mu_\alpha^0(T,p)$ is the chemical potential of a pure component α at fixed temperature T and pressure p. If we now use the well-known thermodynamic relations

$$\left(\frac{\partial}{\partial T}\left(\frac{\mu_\alpha^0(T,p)}{T}\right)\right)_p = -\frac{h_\alpha}{T^2}, \qquad \left(\frac{\partial}{\partial p}\mu_\alpha^0(T,p)\right)_T = v_\alpha = 1/n, \tag{2.67}$$

(where v_α is the partial molar volume), then it is easy to obtain the following expression for the thermodynamic force $\boldsymbol{X}_\alpha^*$:

$$\boldsymbol{X}_\alpha^* \equiv -\frac{1}{nT}\frac{\partial p}{\partial \boldsymbol{r}} - k_B\frac{\partial}{\partial \boldsymbol{r}}\ln(n_\alpha/n) + \frac{\boldsymbol{F}_\alpha}{T}. \tag{2.68}$$

2.3.3.1 Diffusive Thermodynamic Forces

For the subsequent analysis, it is convenient to write the constitutive relations (2.56) and (2.57) for the fluxes $\boldsymbol{J}_q$ и $\boldsymbol{J}_\alpha$ via other linearly dependent vectors $\boldsymbol{d}_\alpha$ closely related to $\boldsymbol{X}_\alpha^*$. This can be done, for example, as follows: let us assume that $\boldsymbol{d}_\alpha \equiv -Tn_\alpha\boldsymbol{X}_\alpha^*/p + n_\alpha m_\alpha \boldsymbol{K}/p$ and determine the vector $\boldsymbol{K}$ common to all components α from the condition

$$\sum_{\alpha=1}^{N} \boldsymbol{d}_\alpha = 0, \tag{2.69}$$

we then obtain

$$\boldsymbol{d}_\alpha = -\frac{Tn_\alpha}{p}\boldsymbol{X}_\alpha^* - \frac{n_\alpha m_\alpha}{p\rho}\frac{\partial p}{\partial \boldsymbol{r}} + \frac{n_\alpha m_\alpha}{p\rho}\sum_{\beta=1}^{N} n_\beta \boldsymbol{F}_\beta,$$

If the vectors $\boldsymbol{X}_\alpha^*$ are now excluded from this formula using (2.68), then we have

$$\boldsymbol{d}_\alpha \equiv \frac{\partial}{\partial \boldsymbol{r}}\left(\frac{n_\alpha}{n}\right) + \left(\frac{n_\alpha}{n} - \frac{\rho_\alpha}{\rho}\right)\frac{\partial \ln p}{\partial \boldsymbol{r}} - \frac{n_\alpha}{p}\left(\boldsymbol{F}_\alpha - m_\alpha \sum_{\beta=1}^{N} Z_\beta \boldsymbol{F}_\beta\right). \tag{2.70}$$

The vector $\boldsymbol{d}_\alpha$, which is also widely used in the gas-kinetic theory, is commonly called the diffusive thermodynamic force of component α (here, $\rho_\alpha(\boldsymbol{r},t)$ $(\equiv m_\alpha n_\alpha)$ is the mass density of particles of type α). Substituting now the vectors $\boldsymbol{X}_\alpha^* = -(k_B n/n_\alpha)\boldsymbol{d}_\alpha + m_\alpha \boldsymbol{K}/T$ into (2.56) and (2.57) and taking into account (2.63), we obtain

$$\boldsymbol{J}_q = L_{qq}\boldsymbol{X}_q - k_B n \sum_{\alpha=1}^{N} L_{q\alpha}\frac{\boldsymbol{d}_\alpha}{n_\alpha}, \tag{2.71}$$

$$\boldsymbol{J}_\alpha = L_{\alpha q}\boldsymbol{X}_q - k_B n \sum_{\beta=1}^{N} L_{\alpha\beta}\frac{\boldsymbol{d}_\beta}{n_\beta}, \quad (\alpha = 1,2,\ldots,N). \tag{2.72}$$

Let us now introduce the symmetric multicomponent diffusion coefficients $D_{\alpha\beta}$, thermal diffusion coefficients $D_{T\alpha}$, and partial thermal conductivity coefficient λ' of a multicomponent gas mixture (in the absence of diffusive forces) using the following definitions for this purpose:

$$D_{\alpha\beta} = D_{\beta\alpha} \equiv pL_{\alpha\beta}/Tn_\alpha n_\beta, \quad (\alpha,\beta = 1,2,\ldots,N), \tag{2.73}$$

$$D_{T\alpha} \equiv L_{q\alpha}/Tn_\alpha \quad (\alpha = 1,2,\ldots,N), \quad \lambda' \equiv L_{qq}/T^2. \tag{2.74}$$

According to (2.63) relating the kinematic coefficients $L_{q\alpha}$ and $L_{\alpha\beta}$, the generalized diffusion and thermal diffusion coefficients we introduced are linearly dependent:

$$\sum_{\alpha=1}^{N} C_\alpha D_{T\alpha} = 0, \quad (^1) \quad \sum_{\alpha=1}^{N} C_\alpha D_{\alpha\beta} = 0, \quad (\beta = 1,2,\ldots,N) \quad (^2), \tag{2.75}$$

(here, $C_\alpha \equiv \rho_\alpha/\rho$ is the mass concentration of particles of type α); therefore, for an N-component medium, there are only $N(N-1)/2$ independent diffusion coefficients and $(N-1)$ independent thermal diffusion coefficients. When using definitions (2.73) and (2.74), the constitutive linear equations (2.71) and (2.72) for the diffusion, $\boldsymbol{J}_\alpha$, and heat, $\boldsymbol{q}$, fluxes take the following final form:

$$\boldsymbol{q} = -\lambda'\frac{\partial T}{\partial \boldsymbol{r}} - p\sum_{\beta=1}^{N} D_{T\beta}\boldsymbol{d}_\beta + \sum_{\beta=1}^{N} h_\beta \boldsymbol{J}_\beta, \tag{2.76}$$

$$\boldsymbol{J}_\alpha = -n_\alpha D_{T\alpha}\frac{\partial \ln T}{\partial \boldsymbol{r}} - n_\alpha \sum_{\beta=1}^{N} D_{\alpha\beta}\boldsymbol{d}_\beta \quad (\alpha = 1, 2, \ldots, N), \tag{2.77}$$

in complete agreement with the present-day kinetic theory of gases (see Ferziger and Kaper 1972).

2.3.3.2 Thermal Diffusion Ratios

Thermal diffusion is known to be an effect of the second order of smallness that is important only in the cases where there is a large difference in masses of the components involved in the transport process. For the subsequent analysis, it is convenient to introduce new transport coefficients to describe the thermal diffusion effects—the so-called thermal diffusion ratios $k_{T\alpha}$:

$$\sum_{\beta=1}^{N} D_{\alpha\beta} k_{T\beta} = D_{T\alpha}, \quad (\alpha = 1, 2, \ldots, N), \tag{2.78}$$

$$\sum_{\alpha=1}^{N} k_{T\alpha} = 0. \tag{2.79}$$

The thermal diffusion ratio $k_{T\alpha}$ is a measure of the relative significance of thermal diffusion with respect to ordinary one. In view of ($2.75^{(2)}$), the determinant of the coefficients for the system of equations (2.78) becomes zero. The only solution of the system of homogeneous equations corresponding to (2.78) is the N -component vector of mass concentrations $C_\alpha \equiv \rho_\alpha/\rho \quad (\alpha = 1,2,\ldots,N)$, but, according to ($2.75^{(1)}$), this vector is orthogonal to the N-component vector of thermal diffusion coefficients. Thus, (2.78) have a solution in the form of some vector $\boldsymbol{k}_{\mathrm{T}}$ with components $k_{T\alpha}$ determined to within some arbitrary constant multiplied by C_α; condition (2.79) ensures the uniqueness of the vector $\boldsymbol{k}_{\mathrm{T}}$. Note that explicit expressions for the quantities $k_{T\alpha}$ in the gas-kinetic theory were derived by assuming that the gases in a mixture are monoatomic or the internal degrees of freedom of the gas molecules are "frozen" (see Ferziger and Kaper 1972).

Using the coefficients $k_{T\alpha}$, (2.77) for the diffusion flux $\boldsymbol{J}_\alpha$ can be written as the following generalized Fick law:

$$\boldsymbol{J}_\alpha = -\sum_{\beta=1}^{N} n_\alpha D_{\alpha\beta}\left(\boldsymbol{d}_\beta + k_{T\beta}\frac{\partial \ln T}{\partial \boldsymbol{r}}\right), \quad (\alpha = 1, 2, \ldots, N). \tag{2.80}$$

Let us now make the following important remark regarding (2.76) for the heat flux $\boldsymbol{q}$. The term $\sum_\beta h_\beta \boldsymbol{J}_\beta$ in (2.76) corresponds to the enthalpy flux attributable to the diffusion of molecules in a reference frame moving with the hydrodynamic velocity $\boldsymbol{u}$. The thermal conductivity is usually measured under conditions when all diffusion fluxes $\boldsymbol{J}_\alpha$ are zero. Therefore, the coefficient λ' cannot be determined through a direct experimental measurement, because a temperature gradient in a gas mixture produces thermal diffusion and, hence, leads to concentration gradients. Consequently, the vectors $\boldsymbol{d}_\beta$ are nonzero even in stationary processes and, hence, the heat flux due to a temperature gradient is always accompanied by the heat flux due to concentration gradients. Using (2.78), (2.73), and (2.77), we can express the vectors $\boldsymbol{d}_\beta$ in terms of the diffusion fluxes and the temperature gradient:

$$\sum_{\alpha=1}^{N} D_{T\alpha}\boldsymbol{d}_\alpha = \sum_{\alpha=1}^{N}\sum_{\beta=1}^{N} D_{\alpha\beta}k_{T\beta}\boldsymbol{d}_\alpha = \sum_{\alpha=1}^{N}\sum_{\beta=1}^{N} D_{\beta\alpha}k_{T\beta}\boldsymbol{d}_\alpha$$
$$= -\sum_{\beta=1}^{N} k_{T\beta}\left(\frac{1}{n_\beta}\boldsymbol{J}_\beta + D_{T\beta}\frac{\partial \ln T}{\partial \boldsymbol{r}}\right).$$

Substituting this expression into (2.76) yields an equation for the heat flux vector in the form

$$\boldsymbol{q} = -\lambda\frac{\partial T}{\partial \boldsymbol{r}} + p\sum_{\beta=1}^{N}\left(\frac{k_{T\beta}}{n_\beta} + h_\beta\right)\boldsymbol{J}_\beta, \tag{2.81}$$

where λ is the true (molecular) thermal conductivity coefficient for a multicomponent mixture,

$$\lambda \equiv \lambda' - nk_B\sum_{\beta=1}^{N} k_{T\beta}D_{T\beta}. \tag{2.82}$$

It is this coefficient that can be directly measured experimentally in a stationary system, because all diffusion fluxes $\boldsymbol{J}_\alpha$ in the stationary case are zero when the gas as a whole is at rest. As can be seen from (2.82), the difference between λ' and λ is of the order of D_T^2. Since the thermal diffusion coefficients are known to be small (recall that all cross effects are of the second order of smallness (see de Groot and Mazur 1964)), this difference is often neglected under normal conditions. The term $p\sum_\beta \frac{k_{T\beta}}{n_\beta}\boldsymbol{J}_\beta$ in (2.81) for the heat flux vector corresponds to the diffusion heat flux.

Finally, it follows from (2.80) and (2.81) that the contribution of the heat conduction and diffusion effects to (2.55) for the entropy production rate is

$$\sigma_{(S)}^{(\mathrm{vect})} = \lambda\left|\frac{\partial \ln T}{\partial \boldsymbol{r}}\right|^2 + kn\sum_{\alpha=1}^{N}\sum_{\beta=1}^{N} D_{\alpha\beta}\left(\boldsymbol{d}_\alpha + k_{T\alpha}\frac{\partial \ln T}{\partial \boldsymbol{r}}\right)\cdot\left(\boldsymbol{d}_\beta + k_{T\beta}\frac{\partial \ln T}{\partial \boldsymbol{r}}\right) \geq 0. \tag{2.83}$$

It follows from the condition that one of the terms can be zero, while the others are nonzero that the thermal conductivity $\lambda>0$ and the inequality

$$\sum_{\alpha=1}^{N}\sum_{\beta=1}^{N} D_{\alpha\beta}\boldsymbol{x}_\alpha\boldsymbol{x}_\beta \geq 0.$$

holds for the multicomponent diffusion coefficients for any vectors $\boldsymbol{x}_\alpha$ satisfying the condition $\sum_\alpha \boldsymbol{x}_\alpha = 0$. The following Sylvester conditions can be pointed out among the properties of the nonnegative definite matrix of coefficients $D_{\alpha\beta}$: $D_{\alpha\alpha} \geq 0$—nonnegativity of all elements of the matrix $D_{\alpha\beta}$ on the main diagonal; $D_{\alpha\beta}D_{\beta\alpha} \leq D_{\alpha\alpha}D_{\beta\beta}$ (each minor of the nonnegative definite matrix $D_{\alpha\beta}$ containing the elements of its main diagonal as its own main diagonal should also be nonnegative), etc. Within these constraints, the diffusion coefficients $D_{\alpha\beta}$ vary over a very wide range in accordance with the degree of coupling between the heat and mass transfer processes.

2.3.4 The Stefan–Maxwell Relations for Multicomponent Diffusion

It is very difficult to use the defining relations (2.77) for the diffusion fluxes $\boldsymbol{J}_\alpha$ in the general, multicomponent case, because, with rare exception, there is no practical information on the generalized multicomponent diffusion coefficients $D_{\alpha\beta}$ (the quantities $D_{\alpha\beta}$ depend on the molar fractions and are usually not tabulated) in the literature, while the available experimental data refer mainly to the diffusion coefficients $\mathsf{D}_{\alpha\beta}$ in binary gas mixtures. At the same time, the gas-kinetic theory of mixtures gives very cumbersome formulas relating the diffusion coefficients $D_{\alpha\beta}$ to the binary coefficients $\mathsf{D}_{\alpha\beta}$ for various pairs of components in a mixture. These formulas are usually difficult to use in solving specific problems. In addition, the system of diffusion equations obtained after the substitution of $\boldsymbol{J}_\alpha$ from (2.77) into the balance equations (2.7) turns out to be unresolvable with respect to the higher-order derivatives. A numerical realization of such systems is known to involve certain difficulties. Therefore, when the diffusion processes in multicomponent gas mixtures are analyzed, a different formulation of the problem where the defining relations (2.77) for the diffusion fluxes $\boldsymbol{J}_\alpha \equiv n_\alpha(\boldsymbol{u}_\alpha - \boldsymbol{u})$ are used in a form resolved with respect to the diffusive thermodynamic forces $\boldsymbol{d}_\alpha$ via the fluxes $\boldsymbol{J}_\alpha$ is often more advantageous. Such an inverse transformation can be written in the form of the so-called generalized Stefan–Maxwell relations

$$\boldsymbol{d}_\beta = \sum_{\substack{\alpha=1\\ \alpha\neq\beta}}^{N} \frac{n_\beta n_\alpha}{n^2 \mathsf{D}_{\alpha\beta}}(\boldsymbol{u}_\alpha - \boldsymbol{u}_\beta) + \frac{\partial \ln T}{\partial \boldsymbol{r}} \sum_{\substack{\alpha=1\\ \alpha\neq\beta}}^{N} \frac{n_\beta n_\alpha}{n^2 \mathsf{D}_{\alpha\beta}}(D_{T\alpha} - D_{T\beta}), \tag{2.84}$$

which include the molecular diffusion coefficients in binary gas mixtures $\mathsf{D}_{\alpha\beta}$ instead of the generalized multicomponent diffusion coefficients $D_{\beta\alpha}$. The advantage of these equations is that the quantity $\mathsf{D}_{\alpha\beta}$ in them is almost independent of the concentrations of the individual components in gas mixtures at low densities.

Historically, relations (2.84) (without thermal diffusion, $D_{T\alpha} = 0$) were derived by Stefan (1871) and Maxwell (1890) phenomenologically by assuming that the force exerted on particles of type α from particles of type β was proportional to the difference of their diffusion velocities and that the total resistance force to the motion of particles of type α in a mixture was equal to the sum of the independent resistance forces from all the remaining particles (of other types). Generalized Stefan–Maxwell relations (including thermal diffusion and the effects of external mass forces) were derived by the methods of the kinetic theory of monoatomic gases in the book by Hirschfelder et al. (1954) in the first approximation of the Chapman–Enskog theory for the multicomponent diffusion coefficients $D_{\alpha\beta}$ and in the second approximation for the thermal diffusion coefficients $D_{T\alpha}$. Muckenfuss and Curtiss (1958) derived these relations in the full second approximation of the Chapman–Enskog. Subsequently, unsuccessful attempts were made in a number of works to obtain relations (2.84) from the gas-kinetic theory in any approximation of the transport coefficients (see, e.g., Muckenfuss and Curtiss 1958). For this reason, Truesdell (1984) suggested that the Stefan–Maxwell relations (2.84) were not universal but were an approximate result of the kinetic theory. However, subsequently, these relations were first derived by Kolesnichenko and Tirsky (1976) by the methods of nonequilibrium thermodynamics, thereby proving their absolute universality (i.e., the legitimacy of applying to any mutually diffusing media).

2.3.4.1 Thermodynamic Derivation of the Stefan–Maxwell Relations

To derive the Stefan–Maxwell relations, let us resolve the defining relations (2.71) and (2.72) with respect to the generalized thermodynamic forces $\boldsymbol{X}_q$ and $\boldsymbol{X}_\alpha \equiv (p/Tn_\alpha)\boldsymbol{d}_\alpha$ $(\alpha = 1, 2, \ldots, N)$ via the fluxes $\boldsymbol{J}_q$ and $\boldsymbol{J}_\alpha (\alpha = 1, 2, \ldots, N)$. Omitting for this purpose the last equation in system (2.72) and using (2.63), we represent relations (2.71) and (2.72) as

$$\boldsymbol{J}_q - L_{qq}\boldsymbol{X}_q = \sum_{\beta=1}^{N-1} L_{q\beta}\left(\boldsymbol{X}_\beta - \frac{m_\beta}{m_N}\boldsymbol{X}_N\right), \tag{2.85}$$

$$\boldsymbol{J}_\alpha - L_{\alpha q}\boldsymbol{X}_q = \sum_{\beta=1}^{N-1} L_{\alpha\beta}\left(\boldsymbol{X}_\beta - \frac{m_\beta}{m_N}\boldsymbol{X}_N\right), \quad (\alpha = 1, 2, \ldots, N-1). \tag{2.86}$$

Resolving now system (2.86) with respect to the differences $\boldsymbol{X}_\beta - (m_\beta/m_N)\boldsymbol{X}_N$, we find

$$\boldsymbol{X}_\beta - (m_\beta/m_N)\boldsymbol{X}_N = \sum_{\alpha=1}^{N-1} \mathsf{M}_{\beta\alpha}\left(\boldsymbol{J}_\alpha - L_{\alpha q}\boldsymbol{X}_q\right), \quad (\beta = 1, 2, \ldots, N-1). \tag{2.87}$$

Here, $\mathsf{M}_{\beta\alpha}$ are the inverse matrix elements that satisfy the relations

$$\sum_{\alpha=1}^{N-1} \mathsf{M}_{\beta\alpha}L_{\alpha\gamma} = \delta_{\beta\gamma} = \begin{cases} 1, & \beta = \gamma; \\ 0, & \beta \neq \gamma. \end{cases} \tag{2.88}$$

where $\delta_{\beta\gamma}$ is the Kronecker symbol. The symmetry of the coefficients $\mathsf{M}_{\beta\alpha}$ follows from the symmetry of the phenomenological coefficients $L_{\alpha\beta}$:

$$\mathsf{M}_{\alpha\beta} = \mathsf{M}_{\beta\alpha}, \quad (\alpha, \beta = 1, 2, \ldots, N-1). \tag{2.89}$$

Using (2.87), we obtain the following relation from (2.85) to define the thermodynamic force $\boldsymbol{X}_q$ via the fluxes $\boldsymbol{J}_q$ and $\boldsymbol{J}_\alpha$

$$\boldsymbol{J}_q = \left(L_{qq} - \sum_{\alpha=1}^{N-1}\sum_{\beta=1}^{N-1} L_{q\beta}\mathsf{M}_{\beta\alpha}L_{\alpha q}\right)\boldsymbol{X}_q + \sum_{\alpha=1}^{N-1}\left(\sum_{\beta=1}^{N-1} L_{q\beta}\mathsf{M}_{\beta\alpha}\right)\boldsymbol{J}_\alpha. \tag{2.90}$$

Let us now multiply each of (2.87) by Z_β (note that $\sum_\beta m_\beta Z_\beta = 1$, $\sum_\beta Z_\beta \boldsymbol{X}_\beta = 0$) and sum them from 1 to $(N-1)$; as a result, we obtain the sought-for relations for the diffusive force vectors $\boldsymbol{d}_\alpha \equiv (Tn_\alpha/p)\boldsymbol{X}_\alpha$ in the form

$$\boldsymbol{X}_N = \left(m_N \sum_{\alpha=1}^{N-1}\sum_{\gamma=1}^{N-1} Z_\gamma \mathsf{M}_{\gamma\alpha}L_{\alpha q}\right)\boldsymbol{X}_q - m_N \sum_{\alpha=1}^{N-1}\left(\sum_{\gamma=1}^{N-1} Z_\gamma \mathsf{M}_{\gamma\alpha}\right)\boldsymbol{J}_\alpha, \tag{2.91}$$

$$\begin{aligned}\boldsymbol{X}_\beta = {} & \sum_{\alpha=1}^{N-1} L_{\alpha q}\left(m_\beta \sum_{\gamma=1}^{N-1} Z_\gamma \mathsf{M}_{\gamma\alpha} - \mathsf{M}_{\beta\alpha}\right)\boldsymbol{X}_q \\ & - \sum_{\alpha=1}^{N-1}\left(\mathsf{M}_{\beta\alpha} - m_\beta \sum_{\gamma=1}^{N-1} Z_\gamma \mathsf{M}_{\gamma\alpha}\right)\boldsymbol{J}_\alpha, \quad (\beta = 1, 2, \ldots, N-1).\end{aligned} \tag{2.92}$$

Thus, the system of equations (2.91) and (2.92) is the inversion of (2.86).

To write (2.91) and (2.92) in the form of the generalized Stefan–Maxwell relations, we add the identity $\sum_{\alpha=1}^{N} m_\alpha \boldsymbol{J}_\alpha = 0$ multiplied, respectively, by the as yet indefinite constants a_q, a_N, and a_β $(\beta = 1, 2, \ldots, N-1)$ to (2.90), (2.91), and (2.92). Then, we obtain

$$\boldsymbol{J}_q = \mathsf{A}_{qq}\boldsymbol{X}_q + \sum_{\alpha=1}^{N} \mathsf{A}_{q\alpha}\boldsymbol{J}_\alpha, \tag{2.93}$$

$$-\boldsymbol{X}_\beta = \mathsf{A}_{\beta q}\boldsymbol{X}_q + \sum_{\alpha=1}^{N} \mathsf{A}_{\beta\alpha}\boldsymbol{J}_\alpha, \quad (\beta = 1, 2, \ldots, N), \tag{2.94}$$

where we find the free parameters a_q and a_β from the requirement for symmetry of the coefficients $\mathsf{A}_{\beta\alpha} = \mathsf{A}_{\alpha\beta}$ $(\alpha,\beta = 1, 2, \ldots, N)$:

$$a_q = -\sum_{\alpha=1}^{N-1}\sum_{\gamma=1}^{N-1} Z_\gamma \mathsf{M}_{\gamma\alpha} L_{\alpha q}, \tag{2.95}$$

$$a_\beta = (m_\beta/m_N)a_N - \sum_{\gamma=1}^{N-1} Z_\gamma \mathsf{M}_{\gamma\beta}, \quad (\beta = 1, 2, \ldots, N-1), \tag{2.96}$$

and the coefficients $\mathsf{A}_{\beta\alpha}$ themselves are

$$\mathsf{A}_{qq} = L_{qq} - \sum_{\beta=1}^{N-1}\sum_{\gamma=1}^{N-1} L_{q\beta}\mathsf{M}_{\beta\gamma}L_{\gamma q}, \tag{2.97}$$

$$\mathsf{A}_{qN} = \mathsf{A}_{Nq} = a_q m_N = -m_N \sum_{\alpha=1}^{N-1}\sum_{\gamma=1}^{N-1} Z_\gamma \mathsf{M}_{\gamma\alpha} L_{\alpha q}, \tag{2.98}$$

$$\begin{aligned}\mathsf{A}_{q\alpha} = \mathsf{A}_{\alpha q} &= a_q m_\alpha + \sum_{\beta=1}^{N-1} L_{q\alpha}\mathsf{M}_{\beta\alpha} \\ &= \sum_{\beta=1}^{N-1} L_{\beta q}\left(\mathsf{M}_{\beta\alpha} - m_\alpha \sum_{\gamma=1}^{N-1} Z_\gamma \mathsf{M}_{\gamma\beta}\right), \quad (\alpha = 1, 2, \ldots, N-1)\end{aligned} \tag{2.99}$$

$$\begin{aligned}\mathsf{A}_{\beta\alpha} = \mathsf{A}_{\alpha\beta} = &-\frac{m_\alpha m_\beta}{m_N} a_N \\ &+ \sum_{\gamma=1}^{N-1} Z_\gamma (m_\alpha \mathsf{M}_{\gamma\beta} + m_\beta \mathsf{M}_{\gamma\alpha}) - \mathsf{M}_{\beta\alpha}, \quad (\alpha,\beta = 1, 2, \ldots, N-1)\end{aligned} \tag{2.100}$$

$$\mathsf{A}_{N\alpha}=\mathsf{A}_{\alpha N}=-a_N m_\alpha+m_N\sum_{\gamma=1}^{N-1}\mathsf{Z}_\gamma\mathsf{M}_{\gamma\alpha}, \tag{2.101}$$

$$\mathsf{A}_{NN}=-a_N m_N, \tag{2.102}$$

with the following identity holding:

$$\sum_{\alpha=1}^{N}\mathsf{Z}_\alpha\mathsf{A}_{\alpha q}=0. \tag{2.103}$$

Thus, the coefficients $\mathsf{A}_{\alpha\beta}$ have been determined to within the constant a_N, which, in general, can be chosen arbitrarily. Let us choose it here from the condition that, apart from identity (2.103), the following identities hold:

$$\sum_{\beta=1}^{N}\mathsf{Z}_\beta\mathsf{A}_{\beta\alpha}=0,\quad(\alpha=1,2,\ldots,N). \tag{2.104}$$

Substituting here (2.100) and (2.101), we find a_N:

$$a_N=-m_N\sum_{\beta=1}^{N-1}\sum_{\gamma=1}^{N-1}\mathsf{Z}_\beta\mathsf{Z}_\gamma\mathsf{M}_{\beta\gamma}. \tag{2.105}$$

Thus, the coefficients $\mathsf{A}_{\beta\alpha}$ have been defined completely via the elements of the symmetric matrix of phenomenological coefficients $L_{\alpha\beta}$ and the elements of the inverse symmetric matrix $\mathsf{M}_{\alpha\beta}$.

Let us now reduce (2.94) to the form of the generalized Stefan–Maxwell relations for multicomponent diffusion. Subtracting for this purpose equalities (2.104) multiplied by $\boldsymbol{J}_\beta/\mathsf{Z}_\beta$ from (2.94), we then find

$$-\boldsymbol{X}_\beta=\mathsf{A}_{\beta q}\boldsymbol{X}_q+\sum_{\substack{\alpha=1\\ \alpha\neq\beta}}^{N}\mathsf{A}_{\beta\alpha}n_\alpha(\boldsymbol{u}_\alpha-\boldsymbol{u}_\beta),\quad(\beta=1,2,\ldots,N), \tag{2.106}$$

or, in more habitual notation,

$$\boldsymbol{d}_\beta=\sum_{\substack{\alpha=1\\ \alpha\neq\beta}}^{N}\frac{n_\beta n_\alpha}{p}\mathsf{A}_{\beta\alpha}(\boldsymbol{u}_\alpha-\boldsymbol{u}_\beta)-\frac{\mathsf{A}_{\beta q}n_\beta}{Tp}\frac{\partial\ln T}{\partial t},\quad(\beta=1,2,\ldots,N). \tag{2.107}$$

It still remains to show that

$$\mathsf{A}_{\beta q} = -\sum_{\alpha=1}^{N} n_\alpha \mathsf{A}_{\beta\alpha}\left(\frac{L_{q\alpha}}{n_\alpha} - \frac{L_{q\beta}}{n_\beta}\right), \quad (\beta = 1, 2, \ldots, N). \tag{2.108}$$

Using for this purpose (2.99), (2.100) and identities (2.103), (2.104), we find

$$\begin{aligned}
\sum_{\alpha=1}^{N} n_\alpha \mathsf{A}_{\beta\alpha}\left(\frac{L_{q\alpha}}{n_\alpha} - \frac{L_{q\beta}}{n_\beta}\right) &= \sum_{\alpha=1}^{N} \mathsf{A}_{\beta\alpha} L_{q\alpha} - \frac{L_{q\beta}}{n_\beta}\sum_{\alpha=1}^{N} n_\alpha \mathsf{A}_{\beta\alpha} \\
&= \mathsf{A}_{\beta N} L_{qN} + \sum_{\alpha=1}^{N-1} \mathsf{A}_{\beta\alpha} L_{q\alpha} \\
&= \sum_{\alpha=1}^{N-1} L_{q\alpha}\left(\mathsf{A}_{\beta\alpha} - \frac{m_\alpha}{m_N}\mathsf{A}_{\beta N}\right) \\
&= \sum_{\alpha=1}^{N-1} L_{q\alpha}\left(-\mathsf{M}_{\beta\alpha} + m_\beta \sum_{\gamma=1}^{N-1} \mathsf{Z}_\gamma \mathsf{M}_{\gamma\alpha}\right) = -\mathsf{A}_{\beta q}.
\end{aligned} \tag{2.109}$$

Finally, substituting (2.108) into (2.107) and using notation (2.74) for the molecular thermal diffusion coefficients ($D_{T\alpha} \equiv L_{q\alpha}/Tn_\alpha$), for the diffusive force vectors we find

$$\begin{aligned}
\boldsymbol{d}_\beta = &\sum_{\substack{\alpha=1 \\ \alpha\neq\beta}}^{N} \frac{n_\beta n_\alpha}{p}\mathsf{A}_{\beta\alpha}(\boldsymbol{u}_\alpha - \boldsymbol{u}_\beta) \\
&+ \frac{\partial \ln T}{\partial t}\sum_{\substack{\alpha=1 \\ \alpha\neq\beta}}^{N} \frac{n_\beta n_\alpha}{p}\mathsf{A}_{\beta\alpha}(D_{T\alpha} - D_{T\beta}), \quad (\beta = 1, 2, \ldots, N-1)
\end{aligned}$$

which closely match the classical Stefan–Maxwell relations (2.84)

$$\begin{aligned}
\boldsymbol{d}_\beta = &\sum_{\substack{\alpha=1 \\ \alpha\neq\beta}}^{N} \frac{n_\beta \boldsymbol{J}_\alpha - n_\alpha \boldsymbol{J}_\beta}{n^2 \mathsf{D}_{\alpha\beta}} \\
&+ \frac{\partial \ln T}{\partial \boldsymbol{r}}\sum_{\substack{\alpha=1 \\ \alpha\neq\beta}}^{N} \frac{n_\beta n_\alpha}{n^2 \mathsf{D}_{\alpha\beta}}(D_{T\alpha} - D_{T\beta}), \quad (\beta = 1, 2, \ldots, N-1)
\end{aligned} \tag{2.110}$$

if the binary diffusion coefficients are defined by the relation

$$\mathsf{D}_{\alpha\beta} = k_B T / n \mathsf{A}_{\alpha\beta}, \quad (\alpha, \beta = 1, 2, \ldots, N). \tag{2.111}$$

As has been noted above, the thermal diffusion ratios $k_{T\beta}$ are very convenient parameters of thermal diffusion processes [see (2.80) and (2.81)]. Here, we define them via the phenomenological coefficients $\mathsf{A}_{q\beta}$ by the formula

$$k_{T\beta} \equiv n_\beta \mathsf{A}_{\beta q} / p, \quad (\beta = 1, 2, \ldots, N). \tag{2.112}$$

In view of (2.103), the equality $\sum\limits_{\beta} k_{T\beta} = 0$ is then valid [cf. (2.79)]. In addition, it is easy to verify that the relations

$$\sum_{\beta=1}^{N} L_{\alpha\beta} \mathsf{A}_{\beta q} = L_{q\alpha}, \quad (\alpha = 1, 2, \ldots, N), \tag{2.113}$$

hold. Given definitions (2.73), (2.74), and (2.112), the system of equations (2.78) $\sum\limits_{\beta} D_{\alpha\beta} k_{T\beta} = D_{T\alpha}$ to find the thermal diffusion ratios $k_{T\beta}$ via the multicomponent diffusion coefficients $D_{\alpha\beta}$ and the thermal diffusion coefficients $D_{T\alpha}$ follows from them. Indeed, using (2.63), (2.88), (2.98), and (2.99), we can successively obtain

$$\sum_{\beta=1}^{N} L_{\alpha\beta} \mathsf{A}_{\beta q} = L_{\alpha N} \mathsf{A}_{Nq} + \sum_{\beta=1}^{N-1} L_{\alpha\beta} \mathsf{A}_{\beta q} = -a_q \sum_{\beta=1}^{N-1} m_\beta L_{\beta\alpha}$$
$$+ \sum_{\beta=1}^{N-1} L_{\alpha\beta} \left(a_q m_\beta + \sum_{\gamma=1}^{N-1} L_{q\gamma} \mathsf{M}_{\gamma\beta} \right) = \sum_{\beta=1}^{N-1} \sum_{\gamma=1}^{N-1} L_{\alpha\beta} \mathsf{M}_{\gamma\beta} L_{q\gamma} = L_{q\beta}. \tag{2.114}$$

Thus, the thermal diffusion ratios $k_{T\beta}$ introduced previously by (2.78) and (2.79) closely coincide with coefficients (2.112).

Using now definitions (2.73), (2.74), and (2.111) and (2.108) for the quantities $k_{T\beta}$ specified by (2.112), we obtain the important algebraic relations

$$k_{T\beta} = \sum_{\substack{\alpha = 1 \\ \alpha \neq \beta}}^{N} \frac{n_\beta n_\alpha}{n^2 \mathsf{D}_{\alpha\beta}} (D_{T\beta} - D_{T\alpha}), \quad (\beta = 1, 2, \ldots, N-1), \tag{2.115}$$

that relate the thermal diffusion ratios $k_{T\beta}$ to the thermal diffusion coefficients $D_{T\beta}$ for a multicomponent mixture and the binary diffusion coefficients $\mathsf{D}_{\alpha\beta}$. Thus, these relations, which also appear in the present-day gas-kinetic theory, are universal, because their derivation has become possible within the framework of the thermodynamic approach developed here.

Finally, using (2.115), let us rewrite the generalized Stefan–Maxwell relations (2.110) as

$$\boldsymbol{d}_\beta = \sum_{\substack{\alpha=1\\ \alpha\neq\beta}}^{N} \frac{n_\beta \boldsymbol{J}_\alpha - n_\alpha \boldsymbol{J}_\beta}{n^2 \mathsf{D}_{\alpha\beta}} + k_{T\beta}\frac{\partial \ln T}{\partial \boldsymbol{r}}, \quad \sum_{\gamma=1}^{N} m_\gamma \boldsymbol{J}_\gamma = 0, \quad (\beta = 1,2,\ldots,N-1) \qquad (2.110^*)$$

where $\boldsymbol{d}_\beta$ is defined by equality (2.70). This form of the Stefan–Maxwell relations is more convenient for the actual solution of hydrodynamic problems, because the expressions for the coefficients $\mathsf{D}_{\alpha\beta}$ and $\mathrm{k}_{T\beta}$ are much simpler for practical calculations than those for the coefficients $D_{\alpha\beta}$ and $D_{T\beta}$ (see Ferziger and Kaper 1972).

2.3.4.2 Approximate Fick Diffusion Law

In the case of a direct numerical solution of (2.110*) for the diffusion fluxes $\boldsymbol{J}_\alpha$, additional difficulties related to the necessity of inverting matrices of dimension $(N-1)^2$ arise when the number of necessary arithmetic operations increases proportionally to $(N-1)^3$ per cell. This essentially rules out the possibility of using (2.110*) without any additional modification for a mixture with more than four or five components. Therefore, when the diffusion fluxes are determined from (2.110*), it is convenient to apply computational procedures (e.g., the method of successive approximations) that allow this inversion of matrices to be avoided (see Oran and Boris 2001). For example, as applied to the problem of calculating the composition of the upper atmosphere, some iterative procedure based on the following form of (2.110*) can be implemented:

$$\boldsymbol{J}_\beta = -\boldsymbol{D}_\beta \left(n\boldsymbol{d}_\beta - nk_{T\beta}\frac{\partial \ln T}{\partial \boldsymbol{r}} - \sum_{\substack{\alpha=1\\ \alpha\neq\beta}}^{N} \frac{x_\beta \boldsymbol{J}_\alpha}{\mathsf{D}_{\alpha\beta}} \right) = -\rho D_\beta \frac{\partial}{\partial \boldsymbol{r}}\left(\frac{n_\beta}{\rho}\right) + \delta \boldsymbol{J}_\beta, \qquad (2.116)$$

where

$$\delta \boldsymbol{J}_\beta \equiv n_\beta D_\beta \left\{ -\frac{\partial \ln \mathsf{M}}{\partial \boldsymbol{r}} - \left(1 - \frac{m_\beta}{\mathsf{M}}\right) \frac{\partial \ln p}{\partial \boldsymbol{r}} + \frac{n}{p} \left(\boldsymbol{F}_\beta - m_\beta \sum_{\alpha=1}^{N} \mathsf{Z}_\alpha \boldsymbol{F}_\alpha \right) \right.$$
$$\left. + \frac{\mathrm{k}_{T\beta}}{x_\beta} \frac{\partial \ln T}{\partial \boldsymbol{r}} + \frac{1}{n} \sum_{\substack{\alpha = 1 \\ \alpha \neq \beta}}^{N} \frac{\boldsymbol{J}_\alpha}{\mathsf{D}_{\alpha\beta}} \right\}, \frac{1}{D_\beta} \equiv \sum_{\substack{\alpha = 1 \\ \alpha \neq \beta}}^{N} \frac{x_\alpha}{\mathsf{D}_{\alpha\beta}},$$

$$\mathsf{M} \equiv \frac{1}{n} \sum_{\alpha=1}^{N} m_\alpha n_\alpha, \qquad n \equiv \sum_{\alpha=1}^{N} n_\alpha, \qquad x_\alpha \equiv \frac{n_\alpha}{n}$$

are, respectively, the mean molecular weight, the total number density, and the molar concentration of particles of type α in the mixture. We derived relations (2.116) by formally resolving each of the equations in system (2.110*) with respect to the flux $\boldsymbol{J}_\beta$ as if the remaining fluxes were known to us. Equations (2.116) are similar in form to the Fick law, but with the significant difference that the effective diffusion coefficient D_β for each of the components β is found by averaging according to Wilke's rule (see Frank-Kamenetskii 1987). The coefficient D_β depends on the mixture composition and, strictly speaking, is a variable quantity under multicomponent diffusion.

According to (2.116), the diffusion flux of particles of type β depends on the composition and diffusion fluxes of other components; however, since the methods of successive approximations are commonly used for a numerical solution of problems, this circumstance is not essential. A numerical calculation based on (2.116) can be implemented, in particular, by the Gauss-Seidel method, in which only one iteration is performed at each time step (of the global iteration when solving a stationary problem by the relaxation method) (see Lapin et al. 1984). It should be noted that from the viewpoint of the expenditure of computer time, this method is essentially equivalent to the method of calculating the diffusion fluxes $\boldsymbol{J}_\beta$ based on the generalized Fick law for multicomponent diffusion

$$\boldsymbol{J}_\beta = -\rho D_\beta \frac{\partial}{\partial \boldsymbol{r}} \left(\frac{n_\beta}{\rho} \right), \qquad \frac{1}{D_\beta} \equiv \sum_{\substack{\alpha = 1 \\ \alpha \neq \beta}}^{N} \frac{x_\alpha}{\mathsf{D}_{\alpha\beta}}, \quad (\beta = 1, 2, \ldots, N-1). \quad (2.116^*)$$

The latter is known to be valid only in the approximation of independent diffusion (i.e., when thermal diffusion is negligible, the molecular masses of the diffusing substances can be assumed to be approximately constant, and the mass force per unit mass is the same for each component, e.g., ($\boldsymbol{F}_\alpha = m_\alpha \boldsymbol{g}$) with Wilke's effective diffusion coefficient D_β. At the same time, it is free from significant shortcomings of the latter related, in particular, to the nonfulfillment of identity ($2.8^{(2)}$), which generally leads to a noticeable violation of the integral mass balance conditions for the individual mixture components. At the same time, a weak point of the Gauss-Seidel method based on (2.116) is the necessity of storing large arrays, the fields of vectors $\boldsymbol{J}_\alpha$, in computer memory.

2.3.4.3 The Heat Flux in a Multicomponent Mixture

It was shown above that the reduced heat flux $\boldsymbol{J}_q$ due to heat conduction and diffusion is defined by (2.76):

$$\boldsymbol{J}_q \equiv \boldsymbol{q} - \sum_{\beta=1}^{N} h_\beta \boldsymbol{J}_\beta = -\lambda' \frac{\partial T}{\partial \boldsymbol{r}} - p \sum_{\beta=1}^{N} D_{T\beta} \boldsymbol{d}_\beta. \tag{2.117}$$

On the other hand, when (2.112) is taken into account, it follows from (2.93) [cf. (2.81)] that

$$\boldsymbol{J}_q = \mathsf{A}_{qq} \boldsymbol{X}_q + \sum_{\alpha=1}^{N} \mathsf{A}_{q\alpha} \boldsymbol{J}_\alpha = -\lambda \frac{\partial T}{\partial \boldsymbol{r}} + p \sum_{\alpha=1}^{N} k_{T\alpha} \frac{\boldsymbol{J}_\alpha}{n_\alpha}, \tag{2.118}$$

where the expression

$$\lambda = \mathsf{A}_{qq} / T^2 \tag{2.119}$$

defines the true thermal conductivity related to the partial thermal conductivity λ' by the relation [cf. (2.82)]

$$\lambda = \lambda' - k_B n \sum_{\alpha=1}^{N} k_{T\alpha} D_{T\alpha} = \lambda' - k_B n \sum_{\alpha=1}^{N} \sum_{\beta=1}^{N} k_{T\alpha} D_{\alpha\beta} k_{T\beta}. \tag{2.120}$$

Indeed, in view of (2.88), (2.97), and (2.114), we have

$$\begin{aligned} \mathsf{A}_{qq} &\equiv L_{qq} - \sum_{\alpha=1}^{N-1} \sum_{\delta=1}^{N-1} L_{q\delta} \mathsf{M}_{\delta\alpha} L_{q\alpha} = L_{qq} - \sum_{\alpha=1}^{N-1} \sum_{\delta=1}^{N-1} L_{q\delta} \mathsf{M}_{\delta\alpha} \left(L_{qN} A_{Nq} + \sum_{\beta=1}^{N-1} L_{\alpha\beta} \mathsf{A}_{\beta q} \right) \\ &= L_{qq} - \sum_{\beta=1}^{N-1} L_{q\beta} \mathsf{A}_{\beta q} - \mathsf{A}_{Nq} L_{qN} = L_{qq} - \sum_{\beta=1}^{N} L_{q\beta} \mathsf{A}_{\beta q}, \end{aligned} \tag{2.121}$$

whence, using notation (2.74), (2.112), and (2.119), follows (2.120). Note that very useful semiempirical formulas to calculate the thermal conductivity λ for a multicomponent polyatomic mixture are given in Mason and Saxena (1958). Estimates of the accuracy of these formulas and other semiempirical relations can be found in the book by Ferziger and Kaper (1972) repeatedly cited above.

The heat influx equation (2.29) written by taking into account (2.65) and (2.117) takes the form

$$\rho c_V \frac{dT}{dt} = -\mathrm{div}\boldsymbol{q}_R + \mathrm{div}\left(\lambda' \frac{\partial T}{\partial \boldsymbol{r}} + p \sum_{\beta=1}^{N} D_{T\beta} \boldsymbol{d}_\beta\right) + \frac{\mu}{2} \sum_{k,j} \left(\frac{\partial u_k}{\partial x_j} + \frac{\partial u_j}{\partial x_k} - \frac{2}{3}\delta_{jk}\mathrm{div}\boldsymbol{u}\right)^2$$
$$+ \mu_\vartheta (\mathrm{div}\boldsymbol{u})^2 - p\mathrm{div}\boldsymbol{u} - \sum_{s=1}^{r} q_s \xi_s + \sum_{\alpha=1}^{N} \boldsymbol{J}_\alpha \cdot \boldsymbol{F}_\alpha^* + Q_d. \tag{2.29*}$$

This differential equation corresponds to the first law of thermodynamics (i.e., the thermal energy conservation law) for the case of conservative external forces.

2.3.5 *Formulas for Defining Multicomponent Diffusion Coefficients via Binary Coefficients*

The method developed above also allows one to derive algebraic equations for defining the symmetric diffusion coefficients $D_{\alpha\beta}$ via the binary coefficients $\mathsf{D}_{\alpha\beta}$. It is easy to verify that the following relation holds:

$$\sum_{\substack{\beta=1 \\ \beta\neq\alpha}}^{N} \mathsf{A}_{\beta\alpha}(n_\beta L_{\gamma\alpha} - n_\alpha L_{\gamma\beta}) = n_\gamma(\delta_{\alpha\gamma} - C_\alpha), \quad (\alpha, \gamma = 1, 2, \ldots, N). \tag{2.122}$$

Indeed, in view of (2.63), (2.100), and (2.104), we have

$$\sum_{\substack{\beta=1 \\ \beta\neq\alpha}}^{N} \mathsf{A}_{\beta\alpha}(n_\beta L_{\gamma\alpha} - n_\alpha L_{\gamma\beta}) = \sum_{\beta=1}^{N} \mathsf{A}_{\beta\alpha}(n_\beta L_{\gamma\alpha} - n_\alpha L_{\gamma\beta}) - n_\alpha \sum_{\beta=1}^{N} \mathsf{A}_{\beta\alpha} L_{\gamma\beta}$$
$$= -n_\alpha \left(\sum_{\beta=1}^{N-1} \mathsf{A}_{\beta\alpha} L_{\gamma\beta} + \mathsf{A}_{\beta N} L_{\gamma N}\right)$$
$$= -n_\alpha \left[m_\alpha \sum_{\delta=1}^{N-1} Z_\delta \left(\sum_{\beta=1}^{N-1} \mathsf{M}_{\delta\beta} L_{\beta\gamma}\right) - \sum_{\beta=1}^{N-1} \mathsf{M}_{\beta\alpha} L_{\gamma\beta}\right]$$
$$= -n_\alpha \left(m_\alpha \sum_{\delta=1}^{N-1} Z_\delta \delta_{\delta\gamma} - \delta_{\alpha\gamma}\right) = n_\gamma(\delta_{\alpha\gamma} - C_\alpha).$$

Using notation (2.73) and (2.111) for the quantities $\mathsf{A}_{\beta\alpha}(=k_BT/n\mathsf{D}_{\alpha\beta})$ and $L_{\gamma\alpha}(=n_\gamma n_\alpha D_{\gamma\alpha}/p)$, (2.122) can be rewritten as the algebraic equations

$$\sum_{\substack{\beta=1\\ \beta\neq\alpha}}^{N}\frac{x_\alpha x_\beta}{\mathsf{D}_{\alpha\beta}}(D_{\gamma\alpha}-D_{\gamma\beta})=\delta_{\gamma\alpha}-C_\alpha,\quad(\gamma,\alpha=1,2,\ldots,N)\tag{2.123}$$

which are suitable for defining the multicomponent diffusion coefficients of a mixture $D_{\gamma\beta}$ via the binary diffusion coefficients $\mathsf{D}_{\gamma\beta}$. Equations (2.123) are linearly dependent (because the summation of (2.122) over α leads to an identity). Therefore, one more equation, namely (2.75$^{(2)}$), should be added to them.

Equations (2.123) and (2.75$^{(2)}$) can be written in the following final form:

$$\sum_{\substack{\beta=1\\ \beta\neq\alpha}}^{N}\left(\frac{n_\alpha n_\beta}{\mathsf{D}_{\alpha\beta}}+\sum_{\substack{\delta=1\\ \delta\neq\alpha}}^{N}\frac{m_\beta}{m_\alpha}\frac{n_\alpha n_\delta}{\mathsf{D}_{\alpha\delta}}\right)D_{\gamma\beta}=n^2(C_\alpha-\delta_{\gamma\alpha}),\quad(\alpha,\gamma=1,2,\ldots,N),\tag{2.124}$$

which is very convenient for practical calculations of the multicomponent diffusion coefficients $D_{\gamma\beta}$. For a binary mixture, the following expressions for the multicomponent diffusion coefficients can be derived from (2.124):

$$D_{11}=\frac{m_1m_2\rho_2n^2}{\rho_1\rho^2}\mathsf{D}_{12},\quad D_{22}=\frac{m_1m_2\rho_1n^2}{\rho_2\rho^2}\mathsf{D}_{12}$$

$$D_{12}=D_{21}=-\frac{m_1m_2n^2}{\rho^2}\mathsf{D}_{12}.\tag{2.125}$$

For a ternary mixture, the typical multicomponent diffusion coefficients are

$$D_{11}=\frac{n^2}{\rho^2n_1^2}\left(\frac{n_1n_3m_3^2\mathsf{D}_{23}\mathsf{D}_{31}+n_1n_2m_2^2\mathsf{D}_{12}\mathsf{D}_{23}+(\rho_2+\rho_3)^2\mathsf{D}_{12}\mathsf{D}_{31}}{n_1\mathsf{D}_{23}+n_2\mathsf{D}_{31}+n_3\mathsf{D}_{12}}\right),$$

$$D_{12}=\frac{n^2}{\rho^2}\left(\frac{n_3m_3^2\mathsf{D}_{23}\mathsf{D}_{31}-m_2(\rho_1+\rho_2)\mathsf{D}_{12}\mathsf{D}_{23}-m_1(\rho_2+\rho_3)\mathsf{D}_{31}\mathsf{D}_{11}}{n_1\mathsf{D}_{32}+n_2\mathsf{D}_{31}+n_3\mathsf{D}_{12}}\right),\tag{2.126}$$

where $\rho_\alpha=m_\alpha n_\alpha$ is the mass density of particles of type α. The expressions for other coefficients $D_{\alpha\beta}$ can be derived from (2.126) by a simple permutation of indices.

Thus, the technique for a thermodynamic derivation of the generalized Stefan–Maxwell equations for multicomponent diffusion proposed in this chapter also allows one to obtain a number of algebraic relations that relate the thermal diffusion ratios to the thermal diffusion and multicomponent diffusion coefficients, the true and partial thermal conductivities, and the multicomponent and binary diffusion coefficients (Kolesnichenko and Marov 1999). All of the relations we derived are in complete agreement with the results of the gas-kinetic theory for multicomponent mixtures of monoatomic gases obtained in the second approximation of the Chapman–Enskog method (see Curtiss 1968; Ferziger and Kaper 1972). However, in contrast to the gas-kinetic approach, the thermodynamic one is not related to the postulation of a specific microscopic model for the interaction of molecules. This proves the universality of all the results obtained here, which can be used to describe a wide class of continuous media, in particular, polyatomic active gas mixtures, dense gases, liquid solutions, etc.

Chapter 3
Closed System of Hydrodynamic Equations to Describe Turbulent Motions of Multicomponent Media

The growing interest in investigating developed turbulent flows of compressible gases and liquids in recent years (see, e.g., van Mieghem 1973; Ievlev 1975, 1990; Kompaniets et al. 1979; Bruyatsky 1986; Kolesnichenko and Marov 1999) has been triggered by the necessity of solving numerous problems of rocket, space, and chemical technologies and problems related to environmental protection. Concurrently, the methods for theoretical modeling of natural media, including the previously inaccessible regions of near-Earth space and the atmospheres of other planets in the Solar system, are improved. In particular, it has become obvious that modeling the upper planetary atmosphere requires developing an appropriate model of turbulent motion that would take into account the multicomponent structure and compressibility of the medium, the heat and mass transfer processes, and chemical reactions (Marov and Kolesnichenko 1987).

We begin this chapter with the derivation of a closed system of averaged hydrodynamic equations designed to describe a wide class of turbulent flows and physical–chemical processes in multicomponent media. We analyze the physical meaning of the individual terms in these equations, including the energy transition rates between various energy balance components. Here, we systematically use the weighted-mean Favre (1969) averaging, which allows the form and analysis of the averaged equations of motion for chemically active gases with variable densities and thermophysical properties to be simplified considerably, along with the traditional probability-theoretic averaging. Special attention is paid to the derivation of closing relations for the turbulent diffusion and heat fluxes and the Reynolds turbulent stress tensor by thermodynamic methods. For the reader's convenience, all calculations are performed comprehensively and can be traced in all details.

Progress in developing and applying semiempirical turbulence models of the first closure order (the so-called gradient models) for a single-fluid medium allows some of them to be generalized to the case of turbulent flows of reacting gas mixtures that is important in astrophysics and geophysics (see, e.g., Libby and Williams 1994). At the same time, assessing the status of the first-order closure problem on the whole, it should be recognized that at present there is actually no general phenomenological theory of turbulent heat conduction and turbulent

M.Y. Marov and A.V. Kolesnichenko, *Turbulence and Self-Organization: Modeling Astrophysical Objects*, Astrophysics and Space Science Library 389,
DOI 10.1007/978-1-4614-5155-6_3, © Springer Science+Business Media New York 2013

diffusion for multicomponent mixtures. The gradient relations used in the literature (see, e.g., Hinze 1963; Monin and Yaglom 1992) are not general enough and were derived mainly for turbulent flows with a well-defined dominant direction under strong and not always justified assumptions, such as, for example, the conservatism of the flow characteristics transferred by turbulent fluctuations or the equality of the mixing lengths for various turbulent transport processes. This necessitates considering other approaches to the closure of averaged hydrodynamic equations for a mixture at the level of first-order turbulence models, in particular, using the methods of extended irreversible thermodynamics. In this case, the Onsager formalism allows the most general structure of the closing gradient relations to be obtained both for the Reynolds stress tensor and for the turbulent heat and diffusion fluxes in a multicomponent mixture, including those in the form of generalized Stefan–Maxwell relations for multicomponent turbulent diffusion. At the closure level under consideration, such relations describe most comprehensively the turbulent heat and mass transport in a multicomponent medium. Both classical models dating back to Prandtl, Taylor, and Karman (see, e.g., Problems of Turbulence 2006) and more recent second-order closure models based, in particular, on the differential balance equations for the turbulent energy and integral turbulence scale, can be used to determine the turbulent exchange coefficients.

3.1 Basic Concepts and Equations of Mechanics of Turbulence for a Mixture of Reacting Gases

One of the main tasks of theoretical geophysics is to numerically calculate the spatial distributions and temporal variations of the density, velocity, temperature, and concentrations of chemical components as well as some other thermohydrodynamic characteristics of a gas mixture in a turbulized planetary atmosphere at large Reynolds numbers $\boldsymbol{Re} = UL/\nu$ (here, U is the characteristic flow velocity in the atmosphere, L is the scale of the main energy-carrying vortices, and ν is the molecular kinematic viscosity). Below, we assume that the system of differential equations for a reacting gas mixture given in the Chap. 2 also describes all details of the true (instantaneous, pulsating) state of the fields of these quantitics under specified initial and boundary conditions in the case of developed turbulence in the atmosphere. However, it is essentially useless without a certain averaging-related modification, because it cannot be solved with present-day computing facilities. The application of numerical computation methods in this case would entail the approximation of an enormous spatiotemporal flow field by a finite number of grid points that should be used when the differential equations are replaced with their finite-difference analogs. At present, there is only one economically justified way out of this situation: to solve the stochastic hydrodynamic equations of a mixture only for large spatiotemporal scales of motion that determine the averaged structural parameters of a turbulized atmosphere and to model small-scale motions (the so-called subgrid turbulence) phenomenologically.

In this case, stochasticity implies the existence of an ensemble of possible realizations of the turbulent flow field for which the concept of a statistical (mathematically expected) average is defined for all fluctuating thermohydrodynamic characteristics. Any flow parameter can then be averaged either over a set of realizations at various times at a given point of coordinate space or over a set of values at various spatial points of some volume at a fixed time. As has already been mentioned in Chap. 1, to eliminate the obvious inconsistency in the averaged hydrodynamic equations (when the flow parameters are defined as time-averaged ones, although they are represented in these equations by time derivatives), the time interval T over which this averaging is performed should be sufficiently long compared to the time scale of individual turbulent fluctuations but, at the same time, short compared to the time scale of a noticeable change in averaged quantities if the averaged motion is nonstationary. Accordingly, the spatial averaging scale should satisfy conditions similar to those imposed on the time interval T. In particular, in atmospheric dynamics it is customary to distinguish the mean zonal motions (with horizontal sizes ~10^4 km) and the deviations from these mean motions (called pulsations, fluctuations, vortices). These fluctuations can have various spatial scales, from several meters to thousands of kilometers. Thus, by the "turbulent fluctuations" we often mean simply the deviations from the mean irrespective of their scales (Brasseur and Solomon 1984).

Thus, the separation of the real stochastic motion of a turbulized medium into slowly varying mean and turbulent (irregular, fluctuating near the means) motions depends entirely on the choice of the spatiotemporal region for which the means are defined. The size of this region fixes the scale of averaged motion. All larger vortices contribute to the averaged motion determined by the mean values of the state parameters $\rho, \boldsymbol{u}, T, Z_\alpha$ $(\alpha = 1, 2, \ldots, N)$. All smaller vortices filtered out in the averaging process contribute to the turbulent motion determined by the corresponding fluctuations of the same parameters. To obtain representative means and the corresponding fluctuations of physical quantities, the spatiotemporal averaging region must include a very large number of vortices with sizes smaller than the averaging region and a very small fraction of vortices with sizes larger than the averaging region (see van Mieghem 1973).

3.1.1 Choosing the Averaging Operator

Averaging is a central problem in the mechanics of continuous media and, in the case of such a complex system as a turbulized fluid, the construction of its macroscopic model itself often depends precisely on the averaging method. In liquid and gas turbulence theories, various methods of averaging physical quantities $\mathsf{A}(\boldsymbol{r}, t)$ are used. For example, these include the temporal averaging

$$\overline{\mathsf{A}(r,t)} = (1/T)\int_0^T \mathsf{A}(r,t+\tau)\mathrm{d}\tau, \tag{3.1}$$

where the averaging interval T is assumed to be sufficiently long compared to the characteristic period of the corresponding fluctuation field but much shorter than the period of variation in the averaged field; the spatial averaging through integration over a spatial volume W; and the probability-theoretic averaging over a statistical ensemble of possible realizations of random hydrodynamic turbulent flow fields. The latter approach is most fundamental. It uses the concept of an ensemble, i.e., an infinite set of hydrodynamic systems of the same physical nature that differ from one another by the state of the field of velocities and/or other thermohydrodynamic parameters at a given time. According to the well-known ergodicity hypothesis (see Monin and Yaglom 1992), the time and ensemble averages are identical for a stationary stochastic process. Without discussing here the advantages and disadvantages of various averaging methods in more detail, we only note that "the practice of constructing phenomenological models to study turbulent motions shows that the techniques for introducing the averaged characteristics of motion are, in general, unimportant for setting up the complete system of averaged hydrodynamic equations if one requires the fulfillment of the following Reynolds postulates in the process of any averaging" (Sedov 1980):

$$\overline{\mathbf{A}+\mathbf{B}} = \overline{\mathbf{A}} + \overline{\mathbf{B}}, \quad \overline{a\mathbf{A}} = a\overline{\mathbf{A}}, \quad \overline{\overline{\mathbf{A}}\mathbf{B}} = \overline{\mathbf{A}}\,\overline{\mathbf{B}}. \quad (^1) \tag{3.2}$$

Here, $\mathbf{A}(\boldsymbol{r},t)$ and $\mathbf{B}(\boldsymbol{r},t)$—are some fluctuating characteristics of the turbulent field, $\overline{\mathbf{A}}(\boldsymbol{r},t)$ and $\overline{\mathbf{B}}(\boldsymbol{r},t)$ are their mean values, and a is a constant (without any fluctuations). Next, we assume that any averaging operator used in (3.2(^{1})) commutes with the differentiation and integration operators both in time and in space:

$$\overline{\partial\mathbf{A}(\boldsymbol{r},t)/\partial t} = \partial\overline{\mathbf{A}(\boldsymbol{r},t)}/\partial t, \quad \overline{\int \mathbf{A}(\boldsymbol{r},t)\mathrm{d}t} = \int \overline{\mathbf{A}(\boldsymbol{r},t)}\mathrm{d}t, \quad (^2)$$

$$\overline{\partial\mathbf{A}(\boldsymbol{r},t)/\partial \boldsymbol{r}} = \partial\overline{\mathbf{A}(\boldsymbol{r},t)}/\partial \boldsymbol{r}, \quad \overline{\int \mathbf{A}(\boldsymbol{r},t)\,d\boldsymbol{r}} = \int \overline{\mathbf{A}(\boldsymbol{r},t)}\mathrm{d}\boldsymbol{r}. \quad (^3) \tag{3.2*}$$

Note that in the case of temporal (and/or spatial) averaging, some of relations (3.2), in general, hold only approximately, although the smaller the change of $\overline{\mathbf{A}(\boldsymbol{r},t)}$ in time and space in the domain of integration under consideration, the more accurate they are. At the same time, for the probability-theoretic averaging of the hydrodynamic equations (over the corresponding statistical ensemble of realizations), the Reynolds postulates (3.2) hold exactly, because they simply follow from ordinary properties of the mathematical expectation in the probability theory. That is why we use them below without any restrictions.

In the classical theories of turbulence for homogeneous incompressible fluids that have been developed by now fairly thoroughly (see, e.g., Townsend 1956; Monin and Yaglom 1992), the averagings are introduced in a similar way and, as a rule, without any weight factors for all thermohydrodynamic parameters without exception. In the case of averaging over time (space) or over an ensemble of possible realizations,

$$\overline{\mathbf{A}(\boldsymbol{r},t)} = \lim_{M\to\infty} \frac{1}{M} \sum_{p=1}^{M} \mathbf{A}^{(p)}, \tag{3.3}$$

(where the summation is over the set of realizations $(p = 1, 2, \ldots, M)$, while the corresponding average field $\overline{\mathbf{A}(\boldsymbol{r},t)}$ is defined as the expected value of $\mathbf{A}$ for an ensemble of identical hydrodynamic systems), the instantaneous value of the parameter $\mathbf{A}$ is represented as the sum of the averaged, $\overline{\mathbf{A}}$, and fluctuation, $\mathbf{A}'$, components:

$$\mathbf{A} = \overline{\mathbf{A}} + \mathbf{A}', \quad (\overline{\mathbf{A}'} = 0). \tag{3.4}$$

However, when applied to a multicomponent continuum with a varying density $\rho(\boldsymbol{r},t)$, such averaging, which is the same for all physical parameters of the medium, not only leads to cumbersome hydrodynamic equations for the scale of mean motion (because it is necessary to retain correlation moments like $\overline{\rho'\boldsymbol{u}'}$, $\overline{\rho'\boldsymbol{u}'\boldsymbol{u}'}$, $\overline{\rho' Z'_\alpha}$, etc. in the equations), but also makes it difficult to physically interpret each individual term of these averaged equations. Bearing in mind the various applications of the phenomenological turbulence model for a reacting mixture being developed in this book, in particular, to some astrophysical phenomena in which the ratio of the characteristic fluid velocity to the averaged speed of sound (a measure of significance of the density fluctuations) is much greater than unity, below we assume the mass density ρ to be variable.

As is well known (see, e.g., Kolesnichenko and Marov 1999), when constructing a model of developed turbulence in a compressible multicomponent medium, apart from the "ordinary" means of physical quantities (such as the density, pressure, molecular mass, momentum, and energy transfer fluxes), it is convenient to use the so-called weighted means (or Favre means (see Favre 1969)) for some other parameters (e.g., the temperature, internal energy, entropy, hydrodynamic velocity, etc.) specified by the relation

$$\langle \mathbf{A} \rangle \equiv \overline{\rho \mathbf{A}}/\overline{\rho} = \left(\lim_{M\to\infty} \frac{1}{M} \sum_{p=1}^{M} \rho^{(p)} \mathbf{A}^{(p)} \right) \Big/ \left(\lim_{M\to\infty} \frac{1}{M} \sum_{p=1}^{M} \rho^{(p)} \right); \tag{3.5}$$

in this case,

$$\mathbf{A} = \langle \mathbf{A} \rangle + \mathbf{A}'', \quad (\overline{\mathbf{A}''} \neq 0), \tag{3.6}$$

where $\mathbf{A}''$ is the corresponding turbulent fluctuation of the field $\mathbf{A}(\boldsymbol{r},t)$. Thus, two symbols are used below in the book to denote the means of physical quantities: the overbar designates averaging over an ensemble of realizations (time and/or space), while the angle brackets designate weighted-mean averaging. The double prime is used below to denote the fluctuations of the same Favre-averaged quantities.

If $\rho \cong \rho$ – const (e.g., in a fluid with Boussinesq properties (Boussinesq 1977)), then both averaging procedures coincide. At the same time, using averaging (3.5) for a number of fluctuating physical quantities that characterize a multicomponent continuum simplifies considerably the form and analysis of the averaged hydrodynamic equations (Kolesnichenko and Marov 1999). In addition, it is also convenient, because precisely these means are probably measured in experimental studies of turbulent flows by conventional methods (see, e.g., Kompaniets et al. 1979).

3.1.1.1 Weighted Means

Some properties of the weighted-mean averaging of physical quantities widely used below can be easily derived from definition (3.5) and the Reynolds postulates (3.2) (see van Mieghem 1973; Kolesnichenko and Marov 1979):

$$\begin{aligned}
&\overline{\langle\mathbf{A}\rangle} = \langle\mathbf{A}\rangle, \quad \langle\overline{\mathbf{A}}\rangle = \overline{\mathbf{A}}, \quad \langle\mathbf{A}\langle\mathbf{B}\rangle\rangle = \langle\mathbf{A}\rangle\langle\mathbf{B}\rangle, \quad \overline{\rho'\mathbf{A}'} = \overline{\rho'\mathbf{A}''},\\
&\overline{\rho\mathbf{A}''} = 0, \quad \overline{\mathbf{A}''} = -\overline{\rho'\mathbf{A}''}/\overline{\rho}, \quad \overline{\rho\mathbf{AB}} = \overline{\rho}\langle\mathbf{A}\rangle\langle\mathbf{B}\rangle + \overline{\rho\mathbf{A}''\mathbf{B}''},\\
&(\mathbf{AB})'' = \langle\mathbf{A}\rangle\mathbf{B}'' + \langle\mathbf{B}\rangle\mathbf{A}'' + \mathbf{A}''\mathbf{B}'' - \overline{\rho\mathbf{A}''\mathbf{B}''}/\overline{\rho}, \quad (\rho\mathbf{A})' = \overline{\rho}\mathbf{A}'' + \rho''\langle\mathbf{A}\rangle,\\
&\overline{\frac{\partial\langle\mathbf{A}\rangle}{\partial r}} = \frac{\partial\langle\mathbf{A}\rangle}{\partial r}, \quad \overline{\rho\mathbf{A}\frac{\partial\mathbf{B}}{\partial r}} = \overline{\rho}\langle\mathbf{A}\rangle\frac{\partial\langle\mathbf{B}\rangle}{\partial r} + \overline{\rho\mathbf{A}\frac{\partial\mathbf{B}''}{\partial r}},\\
&\overline{\rho\frac{\mathrm{d}\mathbf{A}}{\mathrm{d}t}} = \overline{\rho}\frac{D\langle\mathbf{A}\rangle}{Dt} + \left(\frac{\partial}{\partial \boldsymbol{r}}\cdot\overline{\rho\mathbf{A}''\boldsymbol{u}''}\right),
\end{aligned} \tag{3.7}$$

where $D\langle\mathbf{A}\rangle/Dt$ is the substantial time derivative for averaged motion defined by (3.11).

3.1.1.2 Averaged Continuity Equation

It is easy to verify that the average density $\overline{\rho}$ and weighted-mean hydrodynamic velocity of a mixture $\langle\boldsymbol{u}\rangle \equiv \overline{\rho\,\boldsymbol{u}}/\overline{\rho}$ satisfy the continuity equation for mean motion

$$\frac{\partial\overline{\rho}}{\partial t} + \left(\frac{\partial}{\partial r}\cdot\overline{\rho}\langle\boldsymbol{u}\rangle\right) = 0 \tag{3.8}$$

This equation can be obtained by applying the Reynolds averaging operation (3.2) to the continuity equation (2.2), which is assumed to be valid for the instantaneous (true) density and hydrodynamic velocity. Since the turbulent mass flux $\overline{\rho\,\boldsymbol{u}''} = 0 \quad (\overline{\boldsymbol{u}''} \neq 0)$, there is no mass transport through turbulence on average (in the case of Favre averaging). Given the well-known difficulties of modeling the correlations $\overline{\rho' \boldsymbol{u}'}$ that appear in the case of "ordinary" averaging (without any weight) of (2.2), the retention of the standard from of the continuity equation (when formally replacing the true density and velocity by the averaged ones) is a strong argument for using the weighted-mean averaging $\langle \boldsymbol{u} \rangle$ for the hydrodynamic flow velocity (see van Mieghem 1973). Below, when developing the model of multi-component turbulence, we use the stochastic averaging operator (3.3) unless any other averaging method is specified specially.

3.1.1.3 Averaged Operator Relation

Averaging the operator relation (2.4) when using (3.7) and (3.8) leads to the identity

$$\begin{aligned}\overline{\rho \frac{\mathrm{d}\mathbf{A}}{\mathrm{d}t}} &= \frac{\partial}{\partial t}(\overline{\rho}\langle \mathbf{A} \rangle) + \left(\frac{\partial}{\partial r} \cdot (\overline{\rho}\langle \mathbf{A} \rangle \langle u \rangle)\right) + \left(\frac{\partial}{\partial r} \cdot \overline{\rho \mathbf{A}'' \boldsymbol{u}''}\right) \\ &= \overline{\rho}\frac{\partial \langle \mathbf{A} \rangle}{\partial t} + \overline{\rho}\langle \boldsymbol{u} \rangle \frac{\partial \langle \mathbf{A} \rangle}{\partial r} + \left(\frac{\partial}{\partial r} \cdot \overline{\rho \mathbf{A}'' \boldsymbol{u}''}\right).\end{aligned} \tag{3.9}$$

Let us define the turbulent flux of the attribute $\mathbf{A}(\boldsymbol{r}, t)$, which is the second statistical moment (a one-time one-point pair correlation function) representing the transport of some fluctuating characteristic $\mathbf{A}''$ of a turbulent medium by turbulent velocity fluctuations $\boldsymbol{u}''$, by the formula

$$\boldsymbol{J}^{\mathrm{turb}}_{(\mathbf{A})} \equiv \overline{\rho \mathbf{A}'' \boldsymbol{u}''} = \overline{\rho}\langle \mathbf{A}'' \boldsymbol{u}'' \rangle \tag{3.10}$$

and denote the substantial time derivative for an averaged continuum by

$$\frac{D}{Dt}(\cdots) \equiv \frac{\partial}{\partial t}(\cdots) + \langle \boldsymbol{u} \rangle \frac{\partial}{\partial \boldsymbol{r}}(\cdots). \tag{3.11}$$

Identity (3.9) then takes the form

$$\overline{\rho \frac{\mathrm{d}\mathbf{A}}{\mathrm{d}t}} = \overline{\rho}\frac{D\langle \mathbf{A} \rangle}{Dt} + \left(\frac{\partial}{\partial \boldsymbol{r}} \boldsymbol{J}^{\mathrm{turb}}_{(\mathbf{A})}\right). \tag{3.12}$$

In addition, in view of (3.8), the operator relation

$$\overline{\rho}\frac{D\mathbf{A}}{Dt} \equiv \frac{\partial}{\partial t}(\overline{\rho}\mathbf{A}) + \left(\frac{\partial}{\partial \boldsymbol{r}}(\overline{\rho}\mathbf{A}\langle \boldsymbol{u} \rangle)\right) \tag{3.13}$$

between the substantial and local changes in $\mathbf{A}(r, t)$ in an averaged flow is valid. It should be emphasized that the quantity $\mathbf{A}$ in the latter relation can be both the instantaneous value of some specific flow field characteristic (a scalar, a vector, or a tensor) and its averaged value $\langle \mathbf{A} \rangle$ or fluctuation component $\mathbf{A}''$.

3.1.2 *Mass and Momentum Conservation Laws for Averaged Motion*

Below, we consider a turbulized multicomponent gas mixture as a continuous medium whose true (instantaneous) states of motion can be described by the system of hydrodynamic equations (2.2), (2.7), (2.9), (2.29), and (2.31) for a random sample of initial and boundary conditions. This is possible for spatiotemporal scales between the scales of molecular motions and the minimum turbulence scales (determined by the linear sizes and lifetimes of the smallest vortices). The latter generally exceed the scales of molecular motions, i.e., the separation between molecules, let alone the molecular sizes, by several (at least three) orders of magnitude. Highly rarefied gases, which are not considered here, constitute an exception.

3.1.2.1 General Averaged Balance Equation

Using identity (3.12) for the probability-theoretic averaging of the balance equation (2.1), we obtain a general differential form of the substantial balance equation for some structural parameter $\mathbf{A}(\boldsymbol{r},t)$ for an averaged continuum:

$$\overline{\rho}\frac{D\langle\mathbf{A}\rangle}{Dt} \equiv \overline{\rho}\frac{\partial\langle\mathbf{A}\rangle}{\partial t} + \overline{\rho}\langle\boldsymbol{u}\rangle\frac{\partial\langle\mathbf{A}\rangle}{\partial\boldsymbol{r}} = -\left(\frac{\partial}{\partial\boldsymbol{r}}\boldsymbol{J}^{\Sigma}_{(\mathbf{A})}\right) + \overline{\sigma_{(\mathbf{A})}}. \tag{3.14}$$

Here,

$$\boldsymbol{J}^{\Sigma}_{(\mathbf{A})} \equiv \overline{\boldsymbol{J}_{(\mathbf{A})}} + \boldsymbol{J}^{\text{turb}}_{(\mathbf{A})} \tag{3.15}$$

is the substantial total flux density including the averaged molecular, $\overline{\boldsymbol{J}_{(\mathbf{A})}}$, and turbulent, $\boldsymbol{J}^{\text{turb}}_{(\mathbf{A})}$, fluxes of the attribute $\mathbf{A}$; $\overline{\sigma_{(\mathbf{A})}}$ is the averaged volume density of the internal source of $\mathbf{A}$. Note that the main problem of the phenomenological turbulence theory, the so-called closure problem, is related precisely to finding the unknown turbulent fluxes $\boldsymbol{J}^{\text{turb}}_{(\mathbf{A})}$ via the medium's averaged state parameters.

Finally, if we transform the left-hand side of (3.14) using relation (3.10), then we obtain a local form of the differential balance equation for the Favre-averaged field quantity $\mathbf{A}(\boldsymbol{r},t)$:

$$\frac{\partial}{\partial t}(\overline{\rho}\langle\mathbf{A}\rangle) + \frac{\partial}{\partial\boldsymbol{r}}\boldsymbol{J}^{\Sigma 0}_{(\mathbf{A})} = \overline{\sigma_{(\mathbf{A})}}. \tag{3.16}$$

Here,

$$J^{\Sigma 0}_{(\mathbf{A})} \equiv \overline{\rho}\langle\mathbf{A}\rangle\langle\boldsymbol{u}\rangle + \boldsymbol{J}^{\Sigma}_{(\mathbf{A})} = \overline{\rho}\langle\mathbf{A}\rangle\langle\boldsymbol{u}\rangle + \overline{\boldsymbol{J}_{(\mathbf{A})}} + \boldsymbol{J}^{\text{turb}}_{(\mathbf{A})}$$

is the local total flux density of the characteristic $\langle \mathbf{A} \rangle$ in an averaged turbulized continuum including the convective term $\overline{\rho}\langle \mathbf{A} \rangle \langle \boldsymbol{u} \rangle$. The flux density $J^{\Sigma 0}_{\langle \mathbf{A} \rangle}$ is the amount of $\langle \mathbf{A} \rangle$ passing per unit time through a unit surface area ∂W (the position of the surface area is specified by a unit vector n lying on the outer side of the surface ∂W bounding the turbulized fluid volume W).

Let us now turn to the derivation of averaged multicomponent hydrodynamic equations by successively considering the cases of different defining parameters $\mathbf{A}$ that describe the instantaneous thermohydrodynamic state of a turbulized medium in (3.14). In contrast to the ordinary hydrodynamic equations for a mixture that are assumed to describe random fluctuations of all physical parameters, these equations contain only smoothly varying averaged quantities; it is this circumstance that allows the powerful mathematical apparatus of continuous functions and efficient numerical methods to be used for their solution.

3.1.2.2 Specific Volume Balance Equation for Averaged Motion

Let us assume that $\mathbf{A} \equiv 1/\rho$ in (3.14) and use (2.6) for the quantities $J_{(1/\rho)} \equiv -\boldsymbol{u}$ and $\sigma_{(1/\rho)} \equiv 0$. We then obtain

$$\overline{\rho}\frac{D}{Dt}(1/\overline{\rho}) \equiv \overline{\rho}\frac{\partial}{\partial t}(1/\overline{\rho}) + \overline{\rho}\boldsymbol{u} \cdot \frac{\partial}{\partial \boldsymbol{r}}(1/\overline{\rho}) = -\mathrm{div}\,\boldsymbol{J}^{\Sigma}_{(1/\rho)}, \tag{3.17}$$

where

$$\boldsymbol{J}^{\Sigma}_{(1/\rho)}\,(\boldsymbol{r},t) \equiv \overline{\boldsymbol{J}_{(1/\rho)}} + \boldsymbol{J}^{\mathrm{turb}}_{(1/\rho)} \tag{3.18}$$

is the substantial total flux density of the specific volume in a turbulized continuum; the averaged molecular and turbulent fluxes of $(1/\rho)$ are defined, respectively, by the relations [see (3.7)]

$$\overline{\boldsymbol{J}_{(1/\rho)}} = -\overline{\boldsymbol{u}} = -\langle \boldsymbol{u} \rangle - \overline{\boldsymbol{u}''}, \tag{3.19}$$

$$\boldsymbol{J}^{\mathrm{turb}}_{(1/\rho)}(\boldsymbol{r},\,t) \equiv \overline{\rho(1/\rho)''\boldsymbol{u}''} = \overline{\boldsymbol{u}''} = -\overline{\rho'\boldsymbol{u}''}/\overline{\rho}. \tag{3.20}$$

Therefore, for the total flux of the specific volume $J^{\Sigma}_{(1/\rho)}$ we have

$$J^{\Sigma}_{(1/\rho)}\,(\boldsymbol{r},t) = -\langle \boldsymbol{u} \rangle - \overline{\boldsymbol{u}''} + \overline{\boldsymbol{u}''} = -\langle \boldsymbol{u} \rangle. \tag{3.18*}$$

Thus, the substantial averaged specific volume balance equation takes the following ultimate form:

$$\overline{\rho}\frac{D}{Dt}(1/\overline{\rho}) = \text{div}\,\langle \boldsymbol{u}\rangle. \tag{3.21}$$

Finally, below we widely use the relation

$$\overline{\rho}\,(1/\rho)'' = -\rho'/\,\rho, \tag{3.22}$$

between the fluctuations in density ρ' and specific volume $(1/\rho)''$. This relation follows directly from the definition of $(1/\rho)''$:

$$(1/\rho)'' = 1/\rho - <1/\rho> = 1/\rho - 1/\overline{\rho} = (\rho - \overline{\rho})/\rho\,\overline{\rho} = -\rho'/\,\rho\,\overline{\rho}.$$

3.1.2.3 Chemical Component Balance Equations for Averaged Motion

To derive the averaged diffusion equations, we assume in (3.14) that $\mathbf{A} \equiv \mathrm{Z}_\alpha = n_\alpha/\rho$. The quantities $\boldsymbol{J}_{(Z_\alpha)} \equiv \boldsymbol{J}_\alpha$ and $\sigma_{(Z_\alpha)} \equiv \sigma_\alpha = \sum\limits_{s=1}^{r} \nu_{\alpha s}\xi_s$ are then, respectively, the diffusion fluxes of components α and the generation rates of particles of type α in chemical reactions [see Sect. 2.1]. As a result, the sought-for balance equation takes the form

$$\overline{\rho}\frac{D\langle Z_\alpha\rangle}{Dt} \equiv \frac{\partial \overline{n}_\alpha}{\partial t} + \text{div}(\overline{n}_\alpha\langle \boldsymbol{u}\rangle) = -\text{div}\boldsymbol{J}_\alpha^\Sigma + \sum_{s=1}^{r} \nu_{\alpha s}\overline{\xi}_s, \quad (\alpha = 1,2,\ldots,N) \tag{3.23}$$

where

$$\boldsymbol{J}_\alpha^\Sigma(\boldsymbol{r},t) \equiv \overline{\boldsymbol{J}}_\alpha + \boldsymbol{J}_\alpha^{\text{turb}} \tag{3.24}$$

is the total diffusion flux of component α in an averaged turbulized medium;

$$\boldsymbol{J}_\alpha^{\text{turb}}(\boldsymbol{r},t) \equiv \overline{\rho Z_\alpha''\boldsymbol{u}''} = \overline{\rho}\langle Z_\alpha''\boldsymbol{u}''\rangle = \overline{n_\alpha \boldsymbol{u}''} \tag{3.25}$$

is the turbulent diffusion flux of a substance of type α; $\langle Z_\alpha\rangle \equiv \overline{n}_\alpha/\overline{\rho}$.

Using the weighted-mean averaging properties (3.7), it is easy to obtain a different (more traditional) form for the turbulent diffusion flux: $\boldsymbol{J}_\alpha^{\text{turb}} = \overline{n_\alpha'\boldsymbol{u}'} - (\overline{n}_\alpha/\overline{\rho})\,\overline{\rho'\boldsymbol{u}'}$. The cumbersomeness of this expression compared to (3.25) once again suggests that using the weighted Favre averaging for a turbulized mixture with a variable density is efficient.

Applying the averaging operator (3.3) to equalities (2.8) and (2.9) yields their equivalents for averaged motion:

$$\sum_{\beta=1}^{N} m_\beta \langle Z_\beta \rangle = 1 \quad (^1); \quad \sum_{\beta=1}^{N} m_\beta \bar{J}_\beta = 0 \quad (^2); \quad \sum_{\beta=1}^{N} m_\beta \bar{\sigma}_\beta = 0 \quad (^3). \tag{3.26}$$

In addition, the identity

$$\sum_{\beta=1}^{N} m_\beta \boldsymbol{J}_\beta^{\text{turb}} = \sum_{\beta=1}^{N} m_\beta \overline{\rho\, Z_\beta \boldsymbol{u}''} = \overline{\rho \left(\sum_{\beta=1}^{N} m_\beta Z_\beta \right) \boldsymbol{u}''} = \overline{\rho\, \boldsymbol{u}''} = 0$$

is valid for the turbulent diffusion fluxes J_α^{turb} and, hence,

$$\sum_{\beta=1}^{N} m_\beta \boldsymbol{J}_\beta^{\Sigma}(\mathbf{r},t) = 0. \tag{3.27}$$

Thus, the averaged diffusion equations (3.23) for a multicomponent turbulized continuum, just like their regular analogs (2.7), are linearly dependent; for this reason, one of them can be replaced by the algebraic integral (3.27).

3.1.2.4 Averaged Momentum Equation

The equation of averaged motion for a mixture (called the Reynolds equation in the literature) can be derived from (3.14) by assuming that $\mathbf{A} \equiv \boldsymbol{u}$. In this case, the quantity $\boldsymbol{J}_{(\mathbf{u})} \equiv -\Pi$ (viscous stress tensor) corresponds to the surface forces [see (2.11)], as in ordinary hydrodynamics, while the source density

$$\sigma_{(\mathbf{u})} \equiv -\frac{\partial p}{\partial \boldsymbol{r}} + 2\rho\, \boldsymbol{u} \times \Omega + \rho \sum_{\alpha=1}^{N} Z_\alpha \boldsymbol{F}_\alpha$$

is related to the volume forces acting on a unit volume of a multicomponent mixture (below, we neglect the fluctuations in Ω and $\boldsymbol{F}_\alpha$). As a result, the averaged equation of motion can be written in vector form as

$$\bar{\rho} \frac{D\langle \boldsymbol{u} \rangle}{Dt} = -\frac{\partial \bar{p}}{\partial \boldsymbol{r}} + \left(\frac{\partial}{\partial \boldsymbol{r}} \Pi^{\Sigma} \right) + 2\bar{\rho} \langle \boldsymbol{u} \rangle \times \Omega + \bar{\rho} \sum_{\alpha=1}^{N} \langle Z_\alpha \rangle \boldsymbol{F}_\alpha. \tag{3.28}$$

Here,

$$\Pi^{\Sigma}(\boldsymbol{r},t) \equiv -\boldsymbol{J}_{(\mathbf{u})}^{\Sigma} = -\overline{\boldsymbol{J}_{(\mathbf{u})}} - \boldsymbol{J}_{(\mathbf{u})}^{\text{turb}} = \overline{\Pi} + \boldsymbol{R} \tag{3.29}$$

is the total stress tensor in a turbulized flow acting as the viscous stress tensor with respect to the averaged motion; $\overline{\Pi}$ is the averaged viscous stress tensor describing the momentum exchange between fluid particles due to the forces of molecular viscosity; and

$$\boldsymbol{R}(\boldsymbol{r},t) \equiv -\boldsymbol{J}_{(u)}^{\text{turb}} = -\overline{\rho \boldsymbol{u}''\boldsymbol{u}''} = -\overline{\rho}\langle \boldsymbol{u}''\boldsymbol{u}''\rangle \tag{3.30}$$

is the so-called Reynolds tensor having the meaning of additional (turbulent) stresses. The appearance of tensor $\boldsymbol{R}$ in (3.28) is a direct consequence of the nonlinearity of the original (instantaneous) equations of motion (2.9). The Reynolds tensor written in Cartesian coordinates is

$$R_{ij}(\boldsymbol{r},t) \equiv -\overline{\rho u_i'' u_j''} = \begin{pmatrix} -\overline{\rho u_1''^2} & -\overline{\rho\, u_1''\, u_2''} & -\overline{\rho\, u_1'' u_3''} \\ -\overline{\rho\, u_2'' u_1''} & -\overline{\rho\, u_2''^2} & -\overline{\rho\, u_2'' u_3''} \\ -\overline{\rho\, u_3''\, u_1''} & -\overline{\rho\, u_3'' u_2''} & -\overline{\rho\, u_3''2} \end{pmatrix}, \tag{3.31}$$

where u_1'', u_2'', and u_3'' are the velocity fluctuation components relative to the x_1, x_2, and x_3 axes, respectively. It is a symmetric second-rank tensor and describes the turbulent stresses attributable to the interaction of moving turbulent vortices. The turbulent stresses, like the molecular ones, are actually the result of momentum transfer from some fluid volumes to others but through turbulent mixing produced by turbulized fluid velocity fluctuations. When turbulent mixing dominates in a flow (e.g., in the case of developed turbulence emerging at very large Reynolds numbers), the averaged viscous stress tensor $\overline{\Pi}$ can generally be neglected compared to the Reynolds stresses $\boldsymbol{R}$ (except the viscous-sublayer regions bordering the solid surface). The turbulent stress tensor components $R_{ij}(\boldsymbol{r},t)$ are, thus, new unknown quantities. Note once again that the construction of various shear turbulence models is actually associated with the proposed methods of finding the closing relations for these quantities [see Chap. 4].

As has been pointed out above, the choice of $\boldsymbol{J}_{(u)}$ and $\sigma_{(u)}$ is not unique and, in general, can be different. For example, for geophysical applications the total pressure of a mixture is commonly represented as the sum of two terms $p = p^d + p_0$, where p^d is the so-called dynamic pressure and p_0 is the part of the pressure that satisfies the hydrostatic equation

$$\partial p_0/\partial x_j = \rho_0 g_j = -\rho_0 g \delta_{3j}, \quad (j = 1,2,3). \tag{3.32}$$

Here, ρ_0 is some constant mass density typical of the atmosphere (e.g., at the sea level) and $\boldsymbol{g} = \{0,0,-g\}$ is the gravity vector, $g = |\boldsymbol{g}|$. In this case, when the quantities $\boldsymbol{J}_{(u)}$ and $\sigma_{(u)}$ are defined by the relations

$$\boldsymbol{J}_{(u)} \equiv p^d \boldsymbol{U} - \Pi,$$

$$\sigma_{(u)} \equiv \Delta\rho(\boldsymbol{U} \cdot \mathbf{g}) + 2\rho \boldsymbol{u} \times \Omega + \rho \sum_{\alpha=1}^{N} Z_\alpha \boldsymbol{F}_\alpha^*, \quad (\text{where } \Delta\rho = \rho - \rho_0),$$

we can obtain the averaged equation of motion for a mixture in a different form:

$$\overline{\rho}\frac{D\langle \boldsymbol{u}\rangle}{Dt} = (\mathbf{U}\cdot \boldsymbol{g})\Delta\overline{\rho} - \frac{\partial \overline{p}\, d}{\partial \boldsymbol{r}} + \left(\frac{\partial}{\partial \boldsymbol{r}} \cdot \Pi^\Sigma\right) + 2\overline{\rho}\langle \boldsymbol{u}\rangle \times \Omega + \overline{\rho}\sum_{\alpha=1}^{N} \langle Z_\alpha\rangle \boldsymbol{F}_\alpha^*. \quad (3.33)$$

For flows in a free stratified atmosphere, where the buoyancy forces (the first term on the right-hand side of (3.33)) are important, all terms in (3.33) generally have the order $g\Delta\overline{\rho}$ or smaller. Since the total pressure gradient is the sum of the dynamic and hydrostatic pressure gradients, the following approximate equality holds:

$$\partial\overline{p}/\partial x_j = \partial\overline{p}^d/\partial x_j + \partial p_0/\partial x_j \approx -\delta_{j3} g \Delta\overline{\rho} - \delta_{j3}\rho_0 g = -\delta_{j3}\rho_0 g(1 + \Delta\overline{\rho}/\rho_0).$$

Hence it follows that the total pressure gradient in the cases where the estimate $\Delta\overline{\rho}/\rho_0 \ll 1$ is valid can be represented by the approximate relation

$$\partial\overline{p}/\partial x_j \approx -\delta_{j3}\rho_0 g. \quad (3.34)$$

This relation is used in Chap. 4.

3.1.3 *The Energetics of a Turbulent Flow*

In the averaged flow of a turbulized mixture, in contrast to its laminar analog, there are a large number of all possible exchange mechanisms (transition rates) between various forms of energy of the moving elementary fluid volumes that contribute to the conserved total energy of the total material continuum. For the most comprehensive physical interpretation of the individual energy balance components, we analyze here various energy equations for the averaged motion of a multicomponent mixture, including the kinetic energy balance equation for turbulent fluctuations.

3.1.3.1 Balance Equation for the Averaged Potential Energy of a Mixture

In view of identity (3.12), the Reynolds averaging of (2.14) leads to the following balance equation for the averaged specific potential energy of a multicomponent mixture:

$$\overline{\rho}\frac{D\langle \Psi\rangle}{Dt} = -\mathrm{div}\boldsymbol{J}^\Sigma_{(\Psi)} + \overline{\sigma_{(\Psi)}}, \quad (3.35)$$

where

$$\boldsymbol{J}^{\Sigma}_{(\Psi)}(\boldsymbol{r},t) \equiv \overline{\boldsymbol{J}_{(\Psi)}} + \boldsymbol{J}^{\text{turb}}_{(\Psi)} = \sum_{\alpha=1}^{N} \Psi_{\alpha} \boldsymbol{J}^{\Sigma}_{\alpha} \tag{3.36}$$

is the total substantial potential energy flux in a turbulized continuum;

$$\overline{\boldsymbol{J}_{(\Psi)}} = \sum_{\alpha=1}^{N} \Psi_{\alpha} \overline{\boldsymbol{J}}_{\alpha} \tag{3.37}$$

is the averaged molecular potential energy flux in the mixture; and

$$\boldsymbol{J}^{\text{turb}}_{(\Psi)}(\boldsymbol{r},t) \equiv \overline{\rho}\langle \Psi'' \boldsymbol{u}'' \rangle = \sum_{\alpha=1}^{N} \Psi_{\alpha} \overline{\rho Z_{\alpha} \boldsymbol{u}''} = \sum_{\alpha=1}^{N} \Psi_{\alpha} \boldsymbol{J}^{\text{turb}}_{\alpha} \tag{3.38}$$

is the turbulent potential energy flux in the mixture.

The averaged potential energy source for a multicomponent mixture is specified by the relation

$$\overline{\sigma_{(\Psi)}} = -\left(\sum_{\alpha=1}^{N} \boldsymbol{J}^{\Sigma}_{\alpha} \cdot \boldsymbol{F}^{*}_{\alpha} \right) - \overline{\rho} \left(\langle \boldsymbol{u} \rangle \cdot \sum_{\alpha=1}^{N} \langle Z_{\alpha} \rangle \boldsymbol{F}_{\alpha} \right). \tag{3.39}$$

Here, the quantity $\left(\sum_{\alpha=1}^{N} \boldsymbol{J}^{\Sigma}_{\alpha} \cdot \boldsymbol{F}^{*}_{\alpha} \right)$ is the total transformation rate (per unit mixture volume) of the potential energy of mean motion into other forms of energy, which follows from the comparison of (3.39) and (3.54); the quantity $\overline{\rho} \left(\langle \boldsymbol{u} \rangle \cdot \sum_{\alpha=1}^{N} \langle Z_{\alpha} \rangle \boldsymbol{F}_{\alpha} \right)$ is related to the transformation rate of the averaged potential energy into the kinetic energy of mean motion [see (3.40)], with this process being reversible (adiabatic).

3.1.3.2 Balance Equation for the Kinetic Energy of Mean Motion

Scalar multiplication of (3.28) by the velocity vector $\langle \boldsymbol{u} \rangle$ yields an equation for the averaged motion of a multicomponent mixture (the work-kinetic energy theorem) in the following substantial form:

$$\begin{aligned} \overline{\rho} \frac{D}{Dt} \left(|\langle \boldsymbol{u} \rangle|^{2}/2 \right) &= \overline{p} \operatorname{div} \langle \boldsymbol{u} \rangle + \operatorname{div}\left(-\overline{p} \langle \boldsymbol{u} \rangle + \Pi^{\Sigma} \cdot \langle \boldsymbol{u} \rangle \right) \\ &- \left(\Pi^{\Sigma} : \frac{\partial}{\partial \boldsymbol{r}} \langle \boldsymbol{u} \rangle \right) + \left(\langle \boldsymbol{u} \rangle \cdot \sum_{\alpha=1}^{N} \overline{n}_{\alpha} \boldsymbol{F}_{\alpha} \right), \end{aligned} \tag{3.40}$$

where $|\langle \boldsymbol{u} \rangle|^2/2$ is the specific kinetic energy of the mean motion. This equation describes the transformation law of the kinetic energy of the mean motion into the work of external mass and surface forces and into the work of internal forces (and back) without allowance for the irreversible transformation of mechanical energy into thermal one or other forms of energy.

Let us explain the physical meaning of the individual terms in (3.40): $\mathrm{div}\left(\Pi^{\Sigma} \cdot \langle u \rangle\right)$ represents the rate at which the total surface stress Π^{Σ} does the work per unit volume of the averaged moving system; the quantity $\overline{p}\mathrm{div}\langle \boldsymbol{u} \rangle$ is related to the reversible (adiabatic) transformation rate of the averaged internal energy (heat) into mechanical one [see (3.54)] and represents the work done by the moving mixture flow against the averaged pressure $\overline{p}$ per unit time in a unit volume; the sign of $\overline{p}\mathrm{div}\langle \boldsymbol{u} \rangle$ depends on whether the mixture flow expands $(0<\mathrm{div}\langle \boldsymbol{u} \rangle)$ or compresses $(0>\mathrm{div}\langle \boldsymbol{u} \rangle)$; the quantity $\left(\Pi^{\Sigma} : (\partial/\partial \boldsymbol{r})\langle \boldsymbol{u} \rangle\right)$ represents the total irreversible transformation rate of the kinetic energy of the mean motion into other forms of energy per unit volume [see (3.54) and (3.68)], with the energy of the mean motion dissipating under the influence of both molecular viscosity at a rate $\left(\overline{\Pi} : (\partial/\partial \boldsymbol{r})\langle \boldsymbol{u} \rangle\right)>0$ and turbulent viscosity at a rate $(\boldsymbol{R} : (\partial/\partial \boldsymbol{r})\langle \boldsymbol{u} \rangle)$ (generally, this quantity can be different in sign).

Adding (3.35) and (3.40) yields the balance equation for the mechanical energy $\langle E_{\mathrm{m}} \rangle \equiv |\langle \boldsymbol{u} \rangle|^2/2 + \langle \Psi \rangle$ for the averaged motion of a turbulized multicomponent continuum:

$$\overline{\rho}\frac{D}{Dt}\left(|\langle \boldsymbol{u} \rangle|^2/2 + \langle \Psi \rangle\right) = -\mathrm{div}\left(\overline{p}\langle \boldsymbol{u} \rangle - \Pi^{\Sigma} \cdot \langle \boldsymbol{u} \rangle + \sum_{\alpha=1}^{N} \Psi_{\alpha} \boldsymbol{J}_{\alpha}^{\Sigma}\right)$$
$$+\overline{p}\mathrm{div}\langle \mathbf{u} \rangle - \left(\Pi^{\Sigma} : \frac{\partial}{\partial \mathbf{r}}\langle \boldsymbol{u} \rangle\right) - \left(\sum_{\alpha=1}^{N} \boldsymbol{J}_{\alpha}^{\Sigma} \cdot \boldsymbol{F}_{\alpha}\right). \tag{3.41}$$

3.1.3.3 Heat Influx Equation for the Averaged Motion of a Mixture

We derive this equation from the general balance equation (3.14) by assuming that $\mathbf{A} \equiv H$ and using the expressions

$$\mathbf{J}_{(H)} \equiv \mathbf{q}, \quad \sigma_{(H)} \equiv \frac{\mathrm{d}p}{\mathrm{d}t} + \left(\Pi : \frac{\partial}{\partial \mathbf{r}}\mathbf{u}\right) + \left(\sum_{\alpha=1}^{N} \mathbf{J}_{\alpha} \cdot \mathbf{F}_{\alpha}^{*}\right)$$

for the mixture enthalpy flux and source, respectively [see (2.26)]. As a result, we have

$$\bar{\rho}\frac{D\langle H\rangle}{Dt} = -\mathrm{div}\boldsymbol{q}^{\Sigma} + \frac{\overline{\mathrm{d}p}}{\mathrm{d}t} + \overline{\left(\Pi : \frac{\partial}{\partial \boldsymbol{r}}\boldsymbol{u}\right) + \left(\sum_{\alpha=1}^{N}\boldsymbol{J}_{\alpha}\cdot\boldsymbol{F}_{\alpha}^{*}\right)}, \tag{3.42}$$

where

$$\boldsymbol{q}^{\Sigma}(\boldsymbol{r},t) \equiv \overline{\boldsymbol{q}} + \boldsymbol{q}^{\mathrm{turb}} \tag{3.43}$$

is the total heat flux in an averaged turbulized multicomponent continuum;

$$\boldsymbol{q}^{\mathrm{turb}}(\boldsymbol{r},t) \equiv \overline{\rho H''\boldsymbol{u}''} \cong \langle c_p\rangle\overline{\rho T''\boldsymbol{u}''} + \sum_{\alpha=1}^{N}\langle h_{\alpha}\rangle\boldsymbol{J}_{\alpha}^{\mathrm{turb}} \tag{3.44}$$

is the turbulent heat (explicit—the first term and latent—the second term) flux that results from the correlation between the specific enthalpy fluctuations H'' and the hydrodynamic mixture flow velocity fluctuations $\boldsymbol{u}''$. The approximate equality (3.44) is written here to within terms containing triple correlations. It can be easily obtained using the expression

$$H'' = \sum_{\alpha=1}^{N}\left(\langle Z_{\alpha}\rangle h_{\alpha}'' + \langle h_{\alpha}\rangle Z_{\alpha}'' + (Z_{\alpha}''h_{\alpha}'')''\right) \cong \langle c_p\rangle\, T'' + \sum_{\alpha=1}^{N}\langle h_{\alpha}\rangle Z_{\alpha}'' \tag{3.45}$$

for the specific mixture enthalpy fluctuations and the properties of weighted-mean Favre averaging suitable for this case [see (3.7)]. Here, the formulas

$$\begin{aligned} h_{\alpha}'' &= c_{p\alpha}T'', \quad (^{1}) \\ \langle c_p\rangle &= \sum_{\alpha=1}^{N}c_{p\alpha}\langle Z_{\alpha}\rangle \quad (^{2}) \end{aligned} \tag{3.46}$$

define, respectively, the fluctuations in the partial enthalpies of individual components and the averaged specific isobaric heat capacity of a turbulized mixture. Below, we assume the following relation to be valid for the averaged total enthalpy in (3.42):

$$\langle H\rangle \cong \langle c_p\rangle\langle T\rangle + \sum_{\alpha=1}^{N}h_{\alpha}^{0}\langle Z_{\alpha}\rangle = \sum_{\alpha=1}^{N}\langle h_{\alpha}\rangle\langle Z_{\alpha}\rangle. \tag{3.47}$$

This relation can be derived from (2.25) through its Favre averaging and by neglecting the small fluctuations of the heat capacity c_p in a turbulized medium ($c_p'' \cong 0$).

It is convenient to transform the substantial derivative of the total mixture pressure in the expression for the source $\sigma_{(h)}$ to

$$\begin{aligned}\frac{\mathrm{d}p}{\mathrm{d}t} &= \frac{Dp}{Dt} + \left(\boldsymbol{u}''\frac{\partial p}{\partial \boldsymbol{r}}\right) = \frac{Dp}{Dt} + \left(\boldsymbol{u}''\frac{\partial \overline{p}}{\partial \boldsymbol{r}}\right) + \left(\boldsymbol{u}''\frac{\partial p'}{\partial \boldsymbol{r}}\right)\\ &= \frac{Dp}{Dt} + \left(\boldsymbol{u}''\frac{\partial \overline{p}}{\partial r}\right) + \mathrm{div}(p'\boldsymbol{u}'') - p'\mathrm{div}\boldsymbol{u}''.\end{aligned}$$

Hence it follows that

$$\overline{\frac{\mathrm{d}p}{\mathrm{d}t}} = \frac{D\overline{p}}{Dt} + \left(\boldsymbol{J}^{\mathrm{turb}}_{(1/\rho)} \cdot \frac{\partial \overline{p}}{\partial \boldsymbol{r}}\right) + \mathrm{div}\left(\overline{p'\boldsymbol{u}''}\right) - \overline{p'\mathrm{div}\boldsymbol{u}''}. \tag{3.48}$$

In addition, below we use the transformation

$$\overline{\Pi : \frac{\partial \boldsymbol{u}}{\partial \boldsymbol{r}}} = \left(\overline{\Pi} : \frac{\partial \langle \boldsymbol{u}\rangle}{\partial \boldsymbol{r}}\right) + \overline{\Pi : \frac{\partial \boldsymbol{u}''}{\partial \boldsymbol{r}}} = \left(\overline{\Pi} : \frac{\partial \langle \boldsymbol{u}\rangle}{\partial \boldsymbol{r}}\right) + \overline{\rho}\langle \varepsilon_b\rangle, \tag{3.49}$$

where the formula

$$\overline{\rho}\langle \varepsilon_b\rangle \equiv \overline{\left(\Pi : \frac{\partial \boldsymbol{u}''}{\partial \boldsymbol{r}}\right)} \tag{3.50}$$

defines the so-called (specific) dissipation rate of turbulent energy into heat under the influence of molecular viscosity. We note at once that the quantity $\langle \varepsilon_b\rangle$ is among the key statistical characteristics of a turbulized medium.

Substituting now (3.43), (3.48), and (3.49) into (3.42) yields an averaged heat influx equation for a turbulized mixture in the following substantial form [cf. (2.24)]:

$$\begin{aligned}\overline{\rho}\frac{D\langle H\rangle}{Dt} &= -\mathrm{div}\left(\overline{\boldsymbol{q}} + \boldsymbol{q}^{\mathrm{turb}} - \overline{p'\boldsymbol{u}''}\right) + \frac{D\overline{p}}{Dt} + \left(\overline{\Pi} : \frac{\partial \langle \boldsymbol{u}\rangle}{\partial \boldsymbol{r}}\right)\\ &\quad - \overline{p'\mathrm{div}\boldsymbol{u}''} + \left(\boldsymbol{J}^{\mathrm{turb}}_{(1/\rho)} \cdot \frac{\partial \overline{p}}{\partial \boldsymbol{r}}\right) + \left(\sum_{\alpha=1}^{N} \overline{\boldsymbol{J}}_\alpha \cdot \boldsymbol{F}^*_\alpha\right) + \overline{\rho}\langle \varepsilon_b\rangle.\end{aligned} \tag{3.51}$$

For the subsequent analysis, we need (3.51) written via the averaged internal energy $\langle E\rangle$. The quantity $\langle E\rangle$ is defined by the expression

$$\langle E\rangle = \langle H\rangle - \frac{\overline{p}}{\overline{\rho}} \cong \langle c_V\rangle\langle T\rangle + \sum_{\alpha=1}^{N} h^0_\alpha \langle Z_\alpha\rangle, \tag{3.52}$$

which is the result of the Favre averaging of (2.32). Using the transformation

$$\overline{\rho}\frac{D\langle E\rangle}{Dt} + \overline{p}\mathrm{div}\langle \boldsymbol{u}\rangle = \overline{\rho}\frac{D\langle H\rangle}{Dt} - \frac{D\overline{p}}{Dt}, \tag{3.53}$$

which is a corollary of (3.52) and (3.8), we then ultimately obtain

$$\overline{\rho}\frac{D\langle E\rangle}{Dt} = -\mathrm{div}\left(\overline{\boldsymbol{q}} + \boldsymbol{q}^{\mathrm{turb}} - \overline{p'\boldsymbol{u}''}\right) - \overline{p}\mathrm{div}\langle \boldsymbol{u}\rangle + \left(\overline{\Pi} : \frac{\partial\langle \boldsymbol{u}\rangle}{\partial \boldsymbol{r}}\right) + \left(\sum_{\alpha=1}^{N}\overline{\boldsymbol{J}}_{\alpha}\cdot\boldsymbol{F}_{\alpha}^{*}\right) - \overline{p'\mathrm{div}\boldsymbol{u}''} + \left(\boldsymbol{J}_{(1/\rho)}^{\mathrm{turb}}\cdot\frac{\partial\overline{p}}{\partial\boldsymbol{r}}\right) + \overline{\rho}\langle\varepsilon_b\rangle. \tag{3.54}$$

The quantity $\overline{p'\mathrm{div}\boldsymbol{u}''}$ in (3.54) is related to the transformation rate of the kinetic energy of turbulent vortices into the averaged internal energy [see (3.69)] and represents the work done by the environment on the vortices per unit time in a unit volume as a consequence of the pressure fluctuations p' and the expansion ($\mathrm{div}\boldsymbol{u}''>0$) or compression ($\mathrm{div}\mathbf{u}''<0$) of vortices. Comparison of (3.54) and (3.35) shows that the quantity $\left(\sum_{\alpha=1}^{N}\overline{\boldsymbol{J}}_{\alpha}\cdot\boldsymbol{F}_{\alpha}^{*}\right)$ defines the transition rate between the averaged internal and averaged potential energies as a result of the work done by nongravitational external forces. Similarly, comparison of (3.54) and (3.40) shows that the quantities $\overline{p}\mathrm{div}\langle\boldsymbol{u}\rangle$ and $\left(\overline{\Pi} : (\partial/\partial\boldsymbol{r})\langle\boldsymbol{u}\rangle\right)$ are related to the transition rate between the internal and kinetic energies of the mean motion. The correlation $\overline{\rho}\langle\varepsilon_b\rangle \equiv \overline{(\Pi : (\partial/\partial\boldsymbol{r})\boldsymbol{u}'')}$ $\left(\cong \overline{(\Pi' : (\partial/\partial\boldsymbol{r})\boldsymbol{u}')}\right)$ in a developed turbulent flow [see Chap. 4] can be identified with the mean work (per unit time per unit volume) done by the viscous stress fluctuations on turbulent vortices with a velocity shear ($(\partial/\partial\boldsymbol{r})\boldsymbol{u}''\neq 0$). This work is always positive, because $\langle\varepsilon_b\rangle$ represents the dissipation rate of turbulent kinetic energy into heat under the influence of molecular viscosity.

Let us now analyze the transformation rate $\left(\boldsymbol{J}_{(1/\rho)}^{\mathrm{turb}}\cdot\partial\overline{p}/\partial\boldsymbol{r}\right)$. Under the action of buoyancy forces, it is convenient to extrapolate this quantity using (3.34) by the expression

$$\left(\boldsymbol{J}_{(1/\rho)}^{\mathrm{turb}}\cdot\partial\overline{p}/\partial\boldsymbol{r}\right) \approx g(\rho_0/\overline{\rho})\overline{\rho' u_3''}. \tag{3.55}$$

The following two cases are known (see, e.g., van Mieghem 1973) to be generally admissible in a turbulized fluid flow:

- For large vortices, the quantity $g\overline{\rho' u_3''}$ is negative. This is because the large-scale density fluctuation ρ' (of a thermal origin) determines the sign of the vertical vortex displacement under the effect of buoyancy. Indeed, since light ($\rho'<0$) and heavy ($\rho'>0$) vortices are, respectively, warm and cold ones, for example, for warm vortices ($\rho'<0$) rising ($u_3''>0$) in a gravitational field $g\overline{\rho' u_3''}<0$. Thus, large vortices transform the thermal (internal) energy of the flow into the kinetic energy of turbulent motion.

- For small-scale turbulence, the quantity $g\overline{\rho' u_3''}$ is always positive. Indeed, in this case, the approximate relation $(\rho' \approx -(\partial\overline{\rho}/\partial x_3)\xi_3$, where ξ_3 is the vertical vortex displacement or mixing length, holds for the Eulerian density fluctuation ρ'. The mean density distribution in a gravitational field is stable $[-(\partial\overline{\rho}/\partial x_3)>0]$. As a result, the turbulent vortices coming to a given level from below $[u_3''>0,\ x_3>0]$ produce positive density fluctuations $(\rho'>0)$, while those coming from above $[u_3''<0,\ x_3<0]$ produce negative ones $(\rho'<0)$; whence $g\overline{\rho' u_3''}>0$. Thus, in this case, the buoyancy force is a restoring one, i.e., turbulence expends its energy on the work against the buoyancy forces. The quantity $g\overline{\rho' u_3''}$ represents the transformation rate of turbulent energy into averaged internal energy per unit volume of the medium or, in other words, small-scale vortices transform the turbulence energy into heat [см. (3.69)].

Finally, let us write the averaged heat influx equation for a multicomponent turbulized mixture via the temperature. Using (3.46) and (3.47) for the quantities $\langle c_p\rangle$ and $\langle H\rangle$, respectively, and the diffusion equations for mean motion (3.23), it is easy to obtain the expression [cf. (2.27)]

$$\overline{\rho}\frac{D\langle H\rangle}{Dt} = \overline{\rho}\langle c_p\rangle\frac{D\langle T\rangle}{Dt} - \operatorname{div}\left(\sum_{\alpha=1}^{N}\langle h_\alpha\rangle\boldsymbol{J}_\alpha^\Sigma\right) + \left(\frac{\partial\langle T\rangle}{\partial\boldsymbol{r}}\cdot\sum_{\alpha=1}^{N}c_{p\alpha}\boldsymbol{J}_\alpha^\Sigma\right) + \sum_{s=1}^{r}\langle q_s\rangle\overline{\xi}_s, \tag{3.56}$$

where

$$\langle q_s\rangle = \sum_{\alpha=1}^{N}\nu_{\alpha s}\langle h_\alpha\rangle, \quad (s = 1,2,\ldots,r) \tag{3.57}$$

is the averaged heat of the sth reaction. Using this expression, the averaged heat influx equation (3.52) takes the following final form [cf. (2.29)]:

$$\begin{aligned}\overline{\rho}\langle c_p\rangle\frac{D\langle T\rangle}{Dt} = &-\operatorname{div}\left(\boldsymbol{q}^\Sigma - \overline{p'\boldsymbol{u}''} - \sum_{\alpha=1}^{N}\langle h_\alpha\rangle\boldsymbol{J}_\alpha^\Sigma\right)\\ &+\frac{D\overline{p}}{Dt} + \left(\overline{\Pi} : \frac{\partial\langle\boldsymbol{u}\rangle}{\partial\boldsymbol{r}}\right) - \sum_{s=1}^{r}\langle q_s\rangle\overline{\xi}_s + \left(\sum_{\alpha=1}^{N}\overline{\boldsymbol{J}}_\alpha\cdot\boldsymbol{F}_\alpha^*\right) - \overline{p'\operatorname{div}\boldsymbol{u}''}\\ &+\left(\boldsymbol{J}_{(1/\rho)}^{\text{turb}}\cdot\frac{\partial\overline{p}}{\partial\boldsymbol{r}}\right) + \overline{\rho}\langle\varepsilon_b\rangle - \left(\frac{\partial\langle T\rangle}{\partial\boldsymbol{r}}\cdot\sum_{\alpha=1}^{N}c_{p\alpha}\boldsymbol{J}_\alpha^\Sigma\right)\end{aligned} \tag{3.58}$$

(the last term is usually discarded [see Chap. 2]). This is the most general form of the energy equation that can be used in reacting turbulence models of various complexities, in particular, those based on simple gradient closure schemes. It is important to emphasize that the heat influx equation (3.58) written via the averaged temperature $\langle T\rangle$ allows the contribution from the heats of individual chemical reactions to the energetics of a turbulized reacting gaseous medium to be separated out in explicit form, with the chemical source being an averaged quantity in the case of a turbulent flow. The nonlinearity of the algebraic dependence of the reaction rate $\xi_s(T, n_\alpha)$ on the mixture temperature and composition implies that, generally, the quantities $\overline{\xi}_s$ cannot be calculated only from the averaged mixture temperature and composition (i.e., $\overline{\xi}_s \neq \xi s(\langle T\rangle, \overline{n}_\alpha)$), because they depend significantly on the intensity of the turbulent fluctuations in these parameters. We postpone a detailed consideration of this question to the next chapter.

3.1.3.4 Total Energy Conservation Law for the Averaged Motion of a Mixture

Let us now write out the averaged total energy conservation law for a turbulized multicomponent mixture in substantial form. This equation allows us to obtain, in particular, the transfer equation for turbulent energy (the averaged kinetic energy of the turbulent velocity fluctuations), which is fundamental in the turbulence theory. Applying the averaging operator (3.3) to (2.16) and using relations (2.17) and (2.18) for the quantities $E(\boldsymbol{r}, t)$ and $J_{(E)}$, we have

$$\overline{\rho}\frac{D\langle U_{\text{tot}}\rangle}{Dt} + \operatorname{div}\left(\overline{\boldsymbol{J}_{U_{\text{tot}}}} + \boldsymbol{J}^{\text{turb}}_{U_{\text{tot}}}\right) = 0, \tag{3.59}$$

where

$$\langle U_{\text{tot}}\rangle = \left\langle |\boldsymbol{u}|^2/2 \right\rangle + \Psi + \langle E\rangle, \tag{3.60}$$

is the total specific energy of the averaged continuum;

$$\boldsymbol{J}^{\text{turb}}_{U_{\text{tot}}}(\boldsymbol{r},t) \equiv \overline{\rho}\langle U''_{\text{tot}}\boldsymbol{u}''\rangle = \overline{\rho(|\boldsymbol{u}|^2/2 + \Psi + E)\boldsymbol{u}''} \tag{3.61}$$

is the turbulent total energy flux in the mixture; and

$$\begin{aligned}\overline{\boldsymbol{J}_{U_{\text{tot}}}} &\equiv \overline{\boldsymbol{q} + p\,\boldsymbol{u} - \boldsymbol{\Pi}\cdot\boldsymbol{u} + \sum_{\alpha=1}^{N}\Psi_\alpha \boldsymbol{J}_\alpha} \\ &= \overline{\boldsymbol{q}} + \overline{p}\,\langle\boldsymbol{u}\rangle - \overline{\boldsymbol{\Pi}}\cdot\langle\boldsymbol{u}\rangle + \overline{p\boldsymbol{u}''} - \overline{\boldsymbol{\Pi}\cdot\boldsymbol{u}''} + \sum_{\alpha=1}^{N}\Psi_\alpha\overline{\boldsymbol{J}_\alpha}\end{aligned} \tag{3.62}$$

is the averaged molecular total energy flux in the multicomponent medium.

For the subsequent analysis, it is convenient to transform the kinetic energy of the instantaneous motion of the medium as

$$\rho|\boldsymbol{u}|^2/2 \equiv \rho(\langle\boldsymbol{u}\rangle + \boldsymbol{u}'')\cdot(\langle\boldsymbol{u}\rangle + \boldsymbol{u}'')/2 = \rho|\langle\boldsymbol{u}\rangle|^2/2 + \rho(\langle\boldsymbol{u}\rangle\cdot\boldsymbol{u}'') + \rho|\boldsymbol{u}''|^2/2.$$

Performing the (Reynolds) averaging of this expression yields

$$\overline{\rho|\boldsymbol{u}|^2/2} \equiv \overline{\rho}|\langle\boldsymbol{u}\rangle|^2/2 + \overline{\rho|\boldsymbol{u}''|^2/2},$$

or

$$\langle|\boldsymbol{u}|^2/2\rangle \equiv |\langle\boldsymbol{u}\rangle|^2/2 + \langle b\rangle, \tag{3.63}$$

where the formula

$$\langle b\rangle(\boldsymbol{r},t) \equiv \overline{\rho b}/\overline{\rho} = \overline{\rho|\boldsymbol{u}''|^2}/2\overline{\rho} \tag{3.64}$$

defines yet another key statistical characteristic of turbulent motion—the turbulent energy; the quantity $b(\boldsymbol{r},t) \equiv |\boldsymbol{u}''|^2/2$ represents the specific fluctuation kinetic energy of the flow. As a result, (3.60) and (3.61) can be rewritten as

$$\langle U_{\text{tot}}\rangle = |\langle\boldsymbol{u}\rangle|^2/2 + \langle\Psi\rangle + \langle E\rangle + \langle b\rangle, \tag{3.65}$$

$$\begin{aligned}\boldsymbol{J}_{U_{\text{tot}}}^{\text{turb}}(\boldsymbol{r},t) &\equiv \overline{\rho}\langle U_{\text{tot}}''\boldsymbol{u}''\rangle = \overline{\rho(|\boldsymbol{u}|^2/2 + \Psi + E)\boldsymbol{u}''} \\ &= \overline{\rho}\langle b\boldsymbol{u}''\rangle - \boldsymbol{R}\cdot\langle\boldsymbol{u}\rangle + \boldsymbol{J}_{(\Psi)}^{\text{turb}} + \boldsymbol{J}_{(E)}^{\text{turb}},\end{aligned} \tag{3.66}$$

where the correlation function

$$\boldsymbol{J}_E^{\text{turb}}(\boldsymbol{r},t) \equiv \overline{\rho}\langle E''\boldsymbol{u}''\rangle = \overline{\rho(H - p/\rho)\boldsymbol{u}''} = \boldsymbol{q}^{\text{turb}} - \overline{p\boldsymbol{u}''} \tag{3.67}$$

defines the turbulent specific internal energy flux in the mixture.

Finally, combining (3.38), (3.62), (3.66), and (3.67), we rewrite the balance equation (3.59) for the total energy of the mean motion of a turbulized mixture as

$$\begin{aligned}&\overline{\rho}\frac{D}{Dt}\left(\frac{|\langle\boldsymbol{u}\rangle|^2}{2} + \langle\Psi\rangle + \langle E\rangle + \langle b\rangle\right) + \operatorname{div}\left(\boldsymbol{q}^{\Sigma} - \overline{p'\boldsymbol{u}''} + \overline{\rho\left(b + \frac{p'}{\rho}\right)\boldsymbol{u}''}\right. \\ &\left. - \overline{\Pi\cdot\boldsymbol{u}''} + \overline{p}\langle\boldsymbol{u}\rangle - \Pi^{\Sigma}\cdot\langle\boldsymbol{u}\rangle + \sum_{\alpha=1}^{N}\psi_\alpha\boldsymbol{J}_\alpha^{\Sigma}\right) = 0.\end{aligned} \tag{3.68}$$

Here, $\boldsymbol{q}^{\Sigma}(\boldsymbol{r},t) \equiv \overline{\boldsymbol{q}} + \boldsymbol{q}^{\text{turb}}$ is the total heat flux due to the averaged molecular and turbulent transport; $\overline{p}\ \langle\boldsymbol{u}\rangle$ is the mechanical energy flux; $\Pi^{\Sigma}\cdot\langle\boldsymbol{u}\rangle$ is the total energy flux due to the work done by the viscous and turbulent stresses; $(\overline{\rho\, b\, \boldsymbol{u}''} - \overline{\Pi \cdot u''})$ is the turbulent vortex energy flux as a result of turbulent diffusion; and $\sum_{\alpha=1}^{N} \Psi_\alpha \boldsymbol{J}_\alpha^\Sigma$ is the total potential energy flux due to the averaged molecular and turbulent diffusion.

It should be emphasized that the term $\overline{p'\boldsymbol{u}''}$ in (3.68) does not act as the energy flux, because, as is easy to see, it drops out of the complete energy equation and is introduced here and below for convenience.

3.1.3.5 Turbulent Energy Balance Equation

The turbulent energy balance equation (or some of its modifications), which is fundamental in the turbulence theory, is known to underlie many present-day semiempirical turbulence models (see, e.g., Monin and Yaglom 1992). It can be derived by various methods, one of which is presented in Chap. 4. Here, we consider its derivation for a multicomponent mixture based on the above averaged energy equations.

Subtracting (3.41) and (3.54) from (3.59) we obtain the sought-for balance equation for the specific turbulent energy $\langle b\rangle \equiv \overline{\rho|\boldsymbol{u}''|^2}/2\overline{\rho}$ in the following general form:

$$\overline{\rho}\frac{D\langle b\rangle}{Dt} = -\mathrm{div}\boldsymbol{J}_{\langle b\rangle}^{\mathrm{turb}} + \sigma_{\langle b\rangle}, \qquad (^1)$$

$$\boldsymbol{J}_{\langle b\rangle}^{\mathrm{turb}} \equiv \overline{\rho(|\boldsymbol{u}''|^2/2 + p'/\rho)\boldsymbol{u}'' - \Pi \cdot \boldsymbol{u}''}, \qquad (^2)$$

$$\sigma_{\langle b\rangle} \equiv \left(\boldsymbol{R} : \frac{\partial\langle \boldsymbol{u}\rangle}{\partial \boldsymbol{r}}\right) + \overline{p'\mathrm{div}\boldsymbol{u}''} + \left(\sum_{\alpha=1}^{N} \boldsymbol{J}_\alpha^{\mathrm{turb}} \cdot \boldsymbol{F}^*\right) - \left(\boldsymbol{J}_{(1/\rho)}^{\mathrm{turb}} \cdot \frac{\partial \overline{p}}{\partial \boldsymbol{r}}\right) - \overline{\rho}\langle \varepsilon_b\rangle, \qquad (^3)$$

$$(3.69)$$

where $\boldsymbol{J}_{\langle b\rangle}^{\mathrm{turb}}(\boldsymbol{r},t)$ and $\sigma_{\langle b\rangle}(\boldsymbol{r},t)$ are, respectively, the turbulent diffusion flux and the local source (sink) of the averaged kinetic energy of turbulent fluctuations (turbulent energy). The left-hand part of this equation characterizes the change in turbulent energy $\langle b\rangle$ with time and the convective transport of $\langle b\rangle$ by the averaged motion; the second term in the right-hand part of (3.69$^{(3)}$) represents the work done by the pressure forces in the fluctuation motion; the third and fourth terms represent the turbulence energy generation rate under the action of nongravitational forces and buoyancy; finally, the fifth term represents the dissipation rate of turbulent kinetic energy into thermal internal energy due to molecular viscosity. The quantity $\boldsymbol{R} : (\partial\langle \boldsymbol{u}\rangle/\partial \boldsymbol{r})$ on the right-hand sides of (3.40) and (3.69$^{(3)}$) has opposite signs and, hence, it can be interpreted as the transition rate of the kinetic energy of the mean motion into the kinetic energy of turbulent fluctuations. It is important to emphasize once again that this energy transition is a purely kinematic process dependent only on the choice of the turbulent field averaging procedure. Since it is well known that

$\boldsymbol{R} : (\partial\langle\boldsymbol{u}\rangle/\partial\boldsymbol{r}){>}0$ in the case of small-scale turbulence, the latter always transforms the kinetic energy of the mean motion into the kinetic energy of turbulent fluctuations. This is the so-called dissipative effect of small-scale turbulence. However, large-scale turbulence can transform the turbulence kinetic energy into the energy of the mean motion (see van Mieghem 1973).

3.1.3.6 Heat Influx Equation for Quasi-stationary Turbulence

In many practical applications, the heat influx equation (3.54) for a turbulized mixture in its general form defies solution. However, it can be simplified considerably in some special cases. If a stationary-nonequilibrium state in which the turbulent energy $\langle b\rangle \equiv \overline{\rho|\boldsymbol{u}''|^2}/2\overline{\rho}$ is conserved both in time and in space is established in the structure of the fluctuation field in the case of developed turbulence, then $\sigma_{\langle b\rangle} \cong 0$. In this case, an important relation follows from (3.69$^{(3)}$):

$$\left(\boldsymbol{R} : \frac{\partial\langle\boldsymbol{u}\rangle}{\partial\boldsymbol{r}}\right) = -\overline{p'\mathrm{div}\boldsymbol{u}''} - \left(\sum_{\alpha=1}^{N}\boldsymbol{J}_\alpha^{\mathrm{turb}}\cdot\boldsymbol{F}^*\right) + \left(\boldsymbol{J}_{(1/\rho)}^{\mathrm{turb}}\cdot\frac{\partial\overline{p}}{\partial\boldsymbol{r}}\right) + \overline{\rho}\langle\varepsilon_b\rangle \equiv \Im_{E,b}.$$

Using this relation, the heat influx equation for a turbulized mixture (3.54) can be written in an almost "classical" form [cf. (2.22)]:

$$\overline{\rho}\frac{D\langle E\rangle}{Dt} = -\mathrm{div}\left(\boldsymbol{q}^\Sigma - \overline{p'\boldsymbol{u}''}\right) - \overline{p}\mathrm{div}\langle\boldsymbol{u}\rangle + \left(\Pi^\Sigma : \frac{\partial\langle\boldsymbol{u}\rangle}{\partial\boldsymbol{r}}\right) + \left(\sum_{\alpha=1}^{N}\boldsymbol{J}_\alpha^\Sigma\cdot\boldsymbol{F}_\alpha^*\right). \tag{3.54*}$$

Accordingly, the averaged heat influx equation (3.58) written via the temperature takes the form

$$\begin{aligned}\overline{\rho}\langle c_p\rangle\frac{D\langle T\rangle}{Dt} &= -\mathrm{div}\left(\boldsymbol{q}^\Sigma - \overline{p'\boldsymbol{u}''} - \sum_{\alpha=1}^{N}\langle h_\alpha\rangle\boldsymbol{J}_\alpha^\Sigma\right)\\ &\quad + \frac{D\overline{p}}{Dt} + \left(\Pi^\Sigma : \frac{\partial\langle\boldsymbol{u}\rangle}{\partial\boldsymbol{r}}\right) - \sum_{s=1}^{r}\langle q_s\rangle\overline{\xi}_s.\end{aligned} \tag{3.58*}$$

3.1.4 Equation of State for a Turbulized Mixture as a Whole

The averaged equations of motion for a turbulized reacting mixture should be supplemented with the averaged equation of state for pressure. Throughout this book, the multicomponent gas mixture is considered as a compressible baroclinic medium for which the equation of state for pressure is the equation of state for a mixture of perfect gases (2.31). Applying the statistical averaging operator (3.3) to the equation of state (2.31), we obtain the following exact expression for the averaged pressure:

$$\overline{p} = \sum_{\alpha=1}^{N} \overline{p}_\alpha = \overline{\rho} k_B \langle T \rangle \sum_{\alpha=1}^{N} \langle Z_\alpha \rangle + \overline{\rho} k_B \sum_{\alpha=1}^{N} \langle T'' Z''_\alpha \rangle$$

$$= \overline{\rho} k_B \langle T \rangle \sum_{\alpha=1}^{N} \langle Z_\alpha \rangle \left(1 + \frac{\langle T'' Z''_\alpha \rangle}{\langle T \rangle \langle Z_\alpha \rangle} \right). \tag{3.70}$$

Generally, it contains a large number of correlation functions that relate the fluctuations in the temperatures and concentrations of individual components. In the cases where the correlation terms $\langle T'' Z''_\alpha \rangle$ are small compared to the first-order terms $\langle T \rangle \langle Z_\alpha \rangle$ (e.g., when $m_\alpha \cong m$; in this case, we have $Z_\alpha \cong n_\alpha / nm = x_\alpha / m$, $\sum_\alpha \langle T'' Z''_\alpha \rangle = \sum_\alpha \langle T'' Z_\alpha \rangle \approx \langle T'' \rangle / m = 0$), the equation of state for pressure relates the averaged density, temperature, and pressure in a turbulent flow in the same way as in a regular flow:

$$\overline{p} = \overline{\rho} k_{\mathrm{B}} \langle T \rangle \sum_{\alpha=1}^{N} \langle Z_\alpha \rangle = \overline{\rho} \langle \mathbf{R}^* \rangle \langle T \rangle, \tag{3.71}$$

where

$$\langle \mathbf{R}^* \rangle = k_{\mathrm{B}} \sum_{\alpha=1}^{N} \langle Z_\alpha \rangle = k_{\mathrm{B}}\, \overline{n} / \overline{\rho} \tag{3.72}$$

is the Favre-averaged "gas constant" of the mixture. The thermal equation of state (3.71) is usually applied in simple models of multicomponent turbulence based on gradient closure hypotheses.

3.1.5 *The Closure Problem of the Averaged Equations for a Mixture*

Thus, we derived the basic hydrodynamic partial differential equations designed to describe turbulent flows (on the scale of mean motion) of gas-phase reacting mixtures within the continuum model of a multicomponent medium based on the general mass, momentum, and energy conservation laws using weighted-mean Favre averaging[1]. These equations are the same in general form as the

[1] Note that Favre averaging allowed us to obtain exact balance equations for various quantities conserved in a flow, because when deriving them we made no simplifying assumptions as a result of which it would be possible to discard a priori some indefinite terms in the averaged equations.

hydrodynamic equations of a reacting mixture for laminar motion presented in Chap. 2. However, the system of averaged turbulent equations (3.21), (3.23), (3.28), (3.54), (3.69), and (3.72) is not closed, because it contains new indefinite fluxes that emerged when averaging the original nonlinear hydrodynamic equations for a mixture, along with the mean thermohydrodynamic state parameters $\overline{\rho}$, $\langle \boldsymbol{u} \rangle$, $\overline{p}$, $\langle T \rangle$, $\langle Z_\alpha \rangle$ and their derivatives. It can be seen from this system that, apart from the averaged molecular fluxes $\overline{\boldsymbol{q}}, \overline{\boldsymbol{\Pi}}, \overline{\boldsymbol{J}_\alpha}$, and $\overline{\xi}_s$, the averaged motion is also described by the unknown mixed second-order (one-point and one-time) correlation moments. This raises the central problem of the turbulence theory (known as the closure problem) associated with the construction of defining relations for all of the indefinite quantities that appear in the turbulent averaged hydrodynamic equations. This problem for a chemically active multicomponent mixture is also coupled with additional difficulties. The first difficulty is related to the necessity of allowance for the compressibility of the total continuum corresponding to the fluid motion under consideration. The existence of density gradients is one of the most important properties of reacting flows that was barely considered by the classical models of nonreacting turbulence. In particular, turbulent convective flows were considered in meteorology exclusively in the Boussinesq approximation. In this approximation, the density change is known to be taken into account only in the terms describing the influence of the acceleration due to gravity. However, this approach is absolutely inapplicable, for example, to slow (deflagration) turbulent burning, when multiple density changes emerge in the flow. The second difficulty (to be considered in more detail in Chap. 4) is revealed when modeling a large number of additional pair correlations of temperature and concentration fluctuations. These appear (as shown below) when averaging the source terms of substance production $\overline{\sigma}_\alpha$ in the diffusion equations (3.23) describing the change in the composition of a reacting mixture. The evolutionary transfer equations for such correlations in the case of turbulized motion of a compressible reacting mixture are complicated enormously.

Regarding the averaged molecular fluxes, it is important to note the following: since the Favre averaging does not allow their regular analogs given, for example, in the Chap. 2 of this book (in particular, as is easy to verify, the Reynolds averaging of the Navier–Stokes relation (2.64) for the viscous stress tensor Π complicates considerably its form when using the weighted mean value $\langle u \rangle$ for the velocity) to be easily averaged, from the viewpoint of consistently constructing a phenomenological model of compressible turbulence, it seems more appropriate to directly derive the defining relations for these fluxes in terms of an averaged turbulized continuum, for example, by the methods of nonequilibrium thermodynamics, as was done in Sect. 2.3 for their regular analogs.

It is also appropriate to perform this procedure, because the linear algebraic relationships (turbulence models) between the turbulent fluxes appearing in the averaged hydrodynamic equations and the averaged state parameters of the medium (or their derivatives), which are assumed to be known or can be easily calculated, can be obtained simultaneously and by exactly the same thermodynamic method (see Kolesnichenko 1980). We are talking primarily about the turbulent heat flux

$\boldsymbol{q}^{\text{turb}}$ (3.44), the turbulent diffusion fluxes $\boldsymbol{J}_{\alpha}^{\text{turb}}$ $(\alpha = 1, 2, \ldots, N)$ (3.25), the turbulent Reynolds stresses $\boldsymbol{R}$ (3.30), and the large number of pair correlations $\langle Z_{\alpha}'' T'' \rangle$ and $\langle Z_{\alpha}'' Z_{\beta}'' \rangle$ $(\alpha, \beta = 1, 2, \ldots, N)$ that enter explicitly into the averaged equation of state for pressure (3.70) or appear (for a chemically active mixture) when averaging the source terms in the diffusion equations (3.95). In addition, it is required to also model the turbulent specific volume flux $\boldsymbol{J}_{(1/\rho)}^{\text{turb}}$ (3.20) related to the density fluctuations, the averaged source terms of mass production $\overline{\sigma}_{\alpha}$ in chemical reactions, and a number of unknown correlation terms including the pressure fluctuations.

Recall that the simplest closure schemes based on the Boussinesq gradient hypothesis (Boussinesq 1977) initially gained the widest acceptance in the simplest turbulence models for an incompressible single-component fluid (including those with a passive admixture that does not affect the dynamic regime of turbulence). This approach allows the unknown turbulent mass, momentum, and energy fluxes to be related linearly to the gradients of the medium's averaged state parameters via some local proportionality coefficients, the so-called turbulent transport (or exchange) coefficients. For a compressible multicomponent mixture, such relations were first derived in the most general form by the methods of nonequilibrium thermodynamics (Kolesnichenko and Marov 1984) and are given in the next section. Using the gradient closing relations for turbulent flows, we can write the turbulent averaged hydrodynamic equations for a reacting mixture in exactly the same form as that for a regular motion. In particular, this allows the hydrodynamic problems for which the transitions of a laminar reacting gas mixture flow to a turbulent one are very important to be solved numerically. At the same time, it should be noted that the gradient hypothesis by no means solves the closure problem unless some additional assumptions about the turbulent exchange coefficients are made and the methods of their calculation are specified. Moreover, this approach is completely inapplicable when the influence of the turbulization prehistory on the local flow characteristics is significant; in these cases, adequate turbulent exchange coefficients cannot be determined at all (see Ievlev 1990).

An objective assessment of the status of the first-order closure problem shows that, in fact, no general phenomenological theory of turbulent heat conduction and turbulent diffusion for multicomponent mixtures has been developed as yet. As has already been pointed out, the gradient relations widely used in the literature (see, e.g., Monin and Yaglom 1992; van Mieghem 1973; Lapin and Strelets 1989) are not general enough and were derived mainly for a single-fluid medium with a passive admixture. This necessitates considering more general approaches to the closure of the turbulent equations for a mixture at the level of first-order models, for example, through thermodynamic modeling of the turbulence of a compressible continuum. In this case, the Onsager formalism of nonequilibrium thermodynamics allows the most general structure of the defining (rheological) relations to be obtained for turbulent flows, including those in the form of generalized Stefan–Maxwell relations for turbulent multicomponent diffusion and the corresponding expression for the total heat flux. At the closure level under consideration, these defining relations appear to describe adequately the turbulent heat and mass transport in a multicomponent medium. However, since the experimental data on turbulent exchange coefficients are limited, simpler models still have to be often used in practice.

Thus, our subsequent objective is to derive explicit gradient expressions for the averaged molecular and turbulent heat, momentum, and mass transfer fluxes, i.e., to obtain the so-called defining relations for turbulence in a purely phenomenological way using the methods of extended nonequilibrium thermodynamics.

3.2 Rheological Relations for the Turbulent Diffusion and Heat Fluxes and the Reynolds Stress Tensor

This section of the monograph is devoted to developing a thermodynamic model of multicomponent turbulence that describes the relationships between the correlation moments in the averaged hydrodynamic equations for a mixture and the averaged thermohydrodynamic variables that are known or can be easily calculated. Here, within the framework of nonequilibrium thermodynamics, we develop a method of deriving the closing gradient relations for the turbulent diffusion, $\boldsymbol{J}_{\alpha}^{\mathrm{turb}}(\boldsymbol{r},t)$, and heat, $\boldsymbol{q}^{\mathrm{turb}}(\boldsymbol{r},t)$, fluxes and for the Reynolds stress tensor $\boldsymbol{R}(\boldsymbol{r},t)$ that generalize the corresponding results of regular hydrodynamics presented in Chap. 2 to the turbulent motion of a multicomponent mixture. The phenomenological turbulence model developed here is based on the representation of the mixture fluctuation motion by a thermodynamic continuum that consists of two interacting open subsystems (continua): the subsystem of averaged motion obtained by the probability-theoretic averaging of the hydrodynamic equations for an instantaneous mixture flow and the subsystem of turbulent chaos (the so-called turbulent superstructure) related to the fluctuation motion of the medium (Kolesnichenko 1998). We emphasize at once that the proposed "two-fluid turbulence model," just like the model of two fluids in the theory of helium superfluidity, is only a convenient way of phenomenologically describing such a complex phenomenon as hydrodynamic turbulence and does not purport to explain completely the physics of the process. Nevertheless, it allows, in particular, not only the "classical" gradient relations for a single-component turbulized fluid but also the most general structure of such relations for a turbulized multicomponent medium to be obtained using the Onsager formalism of nonequilibrium thermodynamics.

Here, by averaging the fundamental Gibbs identity, which is assumed to be valid for the system's micromotions, we derive the balance equation for the averaged entropy $\langle S\rangle$ of a turbulized medium and find an explicit form for the flux $\mathbf{J}_{\langle S\rangle}^{\Sigma}(\boldsymbol{r},t)$ of entropy $\langle S\rangle$ and its local production $\sigma_{\langle S\rangle}(\boldsymbol{r},t)$ due to irreversible physical processes both within the subsystem of averaged motion and during the interaction with the subsystem of turbulent chaos. Such characteristics as the turbulization entropy $S_{\mathrm{turb}}(\boldsymbol{r},t)$ and temperature $T_{\mathrm{turb}}(\boldsymbol{r},t)$ as well as the pulsation pressure $p_{\mathrm{turb}}(\boldsymbol{r},t)$ can be introduced by postulating the Gibbs thermodynamic identity for the subsystem of turbulent chaos (see Nevzglyadov 1945a, b). These generalized parameters are related to the turbulent fluctuations and dynamical changes in a quasi-stationary state of chaos in exactly the same way as, for example, the local equilibrium entropy $S(\boldsymbol{r},t)$ is related to the molecular fluctuations and dynamical changes in a quasi-equilibrium state. Using the balance equation for the total entropy $S_\Sigma \equiv \langle S\rangle + S_{\rm turb}$ of a turbulized mixture, we obtain linear gradient relations for the turbulent diffusion and heat fluxes and the Reynolds stress tensor. We give a detailed derivation of these relations for isotropic turbulence, when the statistical properties of the turbulent field do not depend on the direction. We derive generalized Stefan–Maxwell relations for turbulent multicomponent diffusion and an expression for the turbulent heat flux that describe most comprehensively the heat and mass exchange in a turbulent mixture flow.

3.2.1 Balance Equation for the Weighted-Mean Entropy of a Mixture

In this chapter, we perform a thermodynamic analysis of the motion of a turbulized multicomponent medium by assuming that the one-point correlations $\langle \mathbf{A}''\mathbf{B}''\rangle$ for any (not equal to the hydrodynamic flow velocity $\mathbf{u}$) fluctuating thermodynamic parameters $\mathbf{A}$ and $\mathbf{B}$ are small compared to the first-order terms $\langle\mathbf{A}\rangle\langle\mathbf{B}\rangle$ and can be omitted, i.e., we assume below that

$$\frac{\langle \mathbf{A}''\mathbf{B}''\rangle}{\langle\mathbf{A}\rangle\langle\mathbf{B}\rangle} \ll 1, \quad (\mathbf{A}\neq \boldsymbol{u},\ \mathbf{B}\neq\boldsymbol{u}). \tag{3.73}$$

We obtain the balance equation for the weighted-mean specific entropy $\langle S\rangle(\boldsymbol{r},t) \equiv \overline{\rho S}/\overline{\rho}$ of a turbulent mixture by the statistical averaging (3.5) of the evolutionary equation (2.36) for the fluctuating entropy S:

$$\overline{\rho}\frac{D\langle S\rangle}{Dt} \equiv \frac{\partial}{\partial t}(\overline{\rho}\langle S\rangle) + \mathrm{div}(\overline{\rho}\langle S\rangle\langle\boldsymbol{u}\rangle) = -\mathrm{div}\left(\overline{\boldsymbol{J}_{(S)}} + \boldsymbol{J}^{\rm turb}_{\langle S\rangle}\right) + \sigma_{\langle S\rangle}. \tag{3.74}$$

Here, $\sigma_{\langle S\rangle}(\boldsymbol{r},t) \equiv \overline{\sigma_{(S)}}$ is the local production of the averaged mixture entropy, i.e., the production of $\langle S\rangle(\boldsymbol{r},t)$ per unit time per unit volume of the medium; $\overline{\boldsymbol{J}_{(S)}}$ and $\boldsymbol{J}^{\rm turb}_{\langle S\rangle}(\boldsymbol{r},t) \equiv \overline{\rho S''\boldsymbol{u}''}$ are the averaged instantaneous molecular entropy flux of the mixture and the turbulent entropy flux of the subsystem of averaged motion, respectively.

There are two possible ways to obtain (decipher) an explicit form of the expressions for $\overline{\boldsymbol{J}_{(S)}}$, $\boldsymbol{J}^{\rm turb}_{\langle S\rangle}$, and $\sigma_{\langle S\rangle}$ in (3.74): either to average (e.g., over an ensemble of possible realizations) their respective instantaneous analogs or to compare the averaged equation (3.74) with the equation derived from the averaged

Gibbs identity (2.37) once the respective substantial derivatives of the averaged state parameters $\langle 1/\overline{\rho}\rangle$, $\langle Z_\alpha\rangle$, and $\langle E\rangle$ have been eliminated from it. Here, we make use of the latter way.

3.2.1.1 Averaged Gibbs Identity

Averaging the fundamental Gibbs identity (2.37) (written along the trajectory of the center of mass of a physical elementary volume), which is valid for mixture micromotions, leads to the following equation for the weighted-mean specific entropy $\langle S\rangle$ and specific internal energy $\langle E\rangle$ of a mixture (Kolesnichenko 1998)

$$\overline{\rho}\langle T\rangle\frac{D\langle S\rangle}{Dt}=\overline{\rho}\frac{D\langle E\rangle}{Dt}+\overline{\rho}\overline{p}\frac{D\langle 1/\overline{\rho}\rangle}{Dt}-\overline{\rho}\sum_{\alpha=1}^{N}\langle\mu_\alpha\rangle\frac{D\langle Z_\alpha\rangle}{Dt}+\varDelta. \tag{3.75}$$

Here, we use the following notation:

$$\begin{aligned}\varDelta\equiv&-\overline{T''\rho\mathrm{d}S/\mathrm{d}t}-\langle T\rangle\mathrm{div}\left(\overline{\rho S''\boldsymbol{u}''}\right)+\mathrm{div}\left(\overline{\rho E''\boldsymbol{u}''}\right)+\overline{p\mathrm{div}\boldsymbol{u}''}\\&-\sum_{\alpha=1}^{N}\overline{\mu''_\alpha\rho\mathrm{d}Z_\alpha/\mathrm{d}t}-\sum_{\alpha=1}^{N}\langle\mu_\alpha\rangle\mathrm{div}\left(\overline{\rho Z''_\alpha\boldsymbol{u}''}\right).\end{aligned} \tag{3.76}$$

It can be shown that if the same thermodynamic relations are valid for the averaged thermodynamic parameters as those for their values in the case of micromotions (and this is true when condition (3.73) is met) and, in particular, if the basic thermodynamic identities

$$\begin{aligned}&\langle G\rangle\equiv\sum_{\alpha=1}^{N}\langle\mu_\alpha\rangle\langle Z_\alpha\rangle=\langle E\rangle+\overline{p}\langle 1/\overline{\rho}\rangle-\langle T\rangle\langle S\rangle, &(^1)\\&\langle S\rangle\delta\langle T\rangle+\sum_{\alpha=1}^{N}\langle\mu_\alpha\rangle\delta\langle Z_\alpha\rangle=\delta\langle E\rangle+\overline{p}\delta\langle 1/\overline{\rho}\rangle, &(^2)\end{aligned} \tag{3.77}$$

are valid, then $\varDelta\equiv 0$ (here, δ denotes an increment of any form), i.e., the fundamental Gibbs identity (3.75) in its substantial form retains its "classical" form for the subsystem of averaged motion as well (Kolesnichenko 1980).

Indeed, averaging the identity

$$\delta(\rho\mathbf{A}\varepsilon)-T\delta(\rho\mathbf{A}S)+p\delta\mathbf{A}-\sum_{\alpha=1}^{N}\mu_\alpha\delta(\rho\mathbf{A}Z_\alpha)\equiv 0,$$

which holds for any field quantity $\mathbf{A}$, over an ensemble of possible realizations, we have

$$
\begin{aligned}
0 = {} & \delta(\overline{\rho}\langle \mathbf{A}\rangle\langle E\rangle) - \langle T\rangle\delta(\overline{\rho}\langle \mathbf{A}\rangle\langle S\rangle) + \overline{p}\delta\langle \mathbf{A}\rangle \\
& - \sum_{\alpha=1}^{N}\langle \mu_\alpha\rangle\delta(\overline{\rho}\langle \mathbf{A}\rangle\langle Z_\alpha\rangle) = -\,\delta\left(\overline{\rho \mathbf{A}'' E''}\right) + \langle T\rangle\delta\left(\overline{\rho \mathbf{A}'' S}\right) \\
& + \overline{T''\delta(\rho S\mathbf{A})} - \overline{p\delta\mathbf{A}''} + \sum_{\alpha=1}^{N}\langle \mu_\alpha\rangle\delta\left(\overline{\rho Z''_\alpha \mathbf{A}''}\right) + \sum_{\alpha=1}^{N}\overline{\mu''_\alpha\delta(\rho Z_\alpha \mathbf{A})};
\end{aligned}
\tag{3.78}
$$

in view of assumption (3.77), the left-hand side of this equality is equal to zero for any $\mathbf{A}$. Setting successively $\mathbf{A} = 1$ and $\mathbf{A} = \boldsymbol{u}$ in (3.78), we obtain, respectively, the following two identities:

$$
\overline{T''\frac{\partial(\rho S)}{\partial t}} + \sum_{\alpha=1}^{N}\overline{\mu''_\alpha\frac{\partial(\rho Z_\alpha)}{\partial t}} = 0, \qquad (^1)
$$

$$
\begin{aligned}
& - \sum_{\alpha=1}^{N}\langle \mu_\alpha\rangle\overline{\mathrm{div}(\rho Z''_\alpha \boldsymbol{u}'')} + \overline{\mathrm{div}(\rho E''\boldsymbol{u}'')} - \langle T\rangle\overline{\mathrm{div}(\rho S''\boldsymbol{u}'')} \\
& - \overline{T''\mathrm{div}(\rho S\boldsymbol{u})} + \overline{p\mathrm{div}\boldsymbol{u}''} - \sum_{\alpha=1}^{N}\overline{\mu''_\alpha\mathrm{div}(\rho Z_\alpha \boldsymbol{u})} = 0, \qquad (^2)
\end{aligned}
\tag{3.79}
$$

from which, as is easy to see, it follows that $\Delta \equiv 0$.

3.2.1.2 Formula for the Production of the Weighted-Mean Entropy of a Mixture

Let us now eliminate the substantial derivatives of the parameters $(1/\overline{\rho})$, $\langle Z_\alpha\rangle$ ($\alpha = 1, 2, \ldots, N$), and $\langle E\rangle$ from the right-hand side of the averaged Gibbs relation (3.75) using the averaged equations (3.21), (3.23), and (3.54). As a result, we obtain a substantial balance equation for the averaged mixture entropy $\langle S\rangle(\boldsymbol{r}, t)$ in the following explicit form [cf. (2.39) and (2.40)]

$$
\overline{\rho}\frac{D\langle S\rangle}{Dt} + \mathrm{div}\left(\frac{\boldsymbol{q}^\Sigma - \sum_{\alpha=1}^{N}\langle \mu_\alpha\rangle \boldsymbol{J}^\Sigma_\alpha}{\langle T\rangle}\right) = \sigma_{\langle S\rangle} = \sigma^{(i)}_{\langle S\rangle} + \sigma^{(e)}_{\langle S\rangle}, \tag{3.80}
$$

where the local production of the averaged entropy is defined by the relation

$$\sigma_{\langle S\rangle} \equiv \frac{1}{\langle T\rangle}\Bigg\{-\left(\tilde{\boldsymbol{J}}_q^{\Sigma}\frac{\partial \ln\langle T\rangle}{\partial \boldsymbol{r}}\right)+\left(\overline{\Pi}:\frac{\partial\langle \boldsymbol{u}\rangle}{\partial \boldsymbol{r}}\right) - \left(\sum_{\alpha=1}^{N}\boldsymbol{J}_\alpha^{\Sigma}\left[\langle T\rangle\frac{\partial}{\partial \boldsymbol{r}}\left(\frac{\langle\mu_\alpha\rangle}{\langle T\rangle}\right)+\langle h_\alpha\rangle\frac{\partial\ln\langle T\rangle}{\partial \boldsymbol{r}}\right]\right)+\sum_{s=1}^{r}\langle A_s\rangle\overline{\xi}_s - \left(\sum_{\alpha=1}^{N}\overline{\boldsymbol{J}}_\alpha\cdot\boldsymbol{F}_\alpha^{*}\right)-\overline{p'\mathrm{div}\boldsymbol{u}''}+\left(\boldsymbol{J}_{(1/\rho)}^{\mathrm{turb}}\cdot\frac{\partial\overline{p}}{\partial\boldsymbol{r}}\right)+\overline{\rho}\langle\varepsilon_b\rangle\Bigg\} \tag{3.81}$$

Here, using the relations

$$\langle A_s\rangle(\boldsymbol{r},t)\equiv-\sum_{\alpha=1}^{N}\nu_{\alpha s}\langle\mu_\alpha\rangle,\quad (s=1,2,\ldots,r) \tag{3.82}$$

we introduced the averaged chemical affinities $\langle A_s\rangle$ for reactions s in a turbulized reacting medium [cf. (2.41)] and use the notation

$$\begin{cases}\tilde{\boldsymbol{J}}_q^{\Sigma}\equiv\overline{\boldsymbol{J}}_q+\tilde{\boldsymbol{J}}_q^{\mathrm{turb}},\quad \overline{\boldsymbol{J}}_q\cong\overline{\boldsymbol{q}}-\sum\limits_{\alpha=1}^{N}\langle h_\alpha\rangle\overline{\boldsymbol{J}_\alpha},\quad \tilde{\boldsymbol{J}}_q^{\mathrm{turb}}\equiv\tilde{\boldsymbol{q}}^{\mathrm{turb}}-\sum\limits_{\alpha=1}^{N}\langle h_\alpha\rangle\boldsymbol{J}_\alpha^{\mathrm{turb}},\\ \tilde{\boldsymbol{J}}_q^{\Sigma}\equiv\tilde{\boldsymbol{q}}^{\Sigma}-\sum\limits_{\alpha=1}^{N}\langle h_\alpha\rangle\boldsymbol{J}_\alpha^{\Sigma},\quad \tilde{\boldsymbol{q}}^{\Sigma}(\boldsymbol{r},t)\equiv\overline{\boldsymbol{q}}+\tilde{\boldsymbol{q}}^{\mathrm{turb}}=\boldsymbol{q}^{\Sigma}-\overline{p'\boldsymbol{u}''},\\ \boldsymbol{J}_\alpha^{\Sigma}\equiv\overline{\boldsymbol{J}_\alpha}+\boldsymbol{J}_\alpha^{\mathrm{turb}},\quad \tilde{\boldsymbol{q}}^{\mathrm{turb}}\equiv\boldsymbol{q}^{\mathrm{turb}}-\overline{p'\boldsymbol{u}''}\quad .\end{cases} \tag{3.83}$$

for the total diffusion and heat fluxes in a multicomponent turbulent continuum. Comparing now (3.80) and (3.81) with (3.74), we obtain the following expressions for the two entropy diffusion fluxes (the averaged molecular, $\overline{\boldsymbol{J}_{(S)}}$, and turbulent, $J_{\langle S\rangle}^{\mathrm{turb}}$, ones) and for the entropy production $\sigma_{\langle S\rangle}$ in the subsystem of averaged motion:

$$\overline{\boldsymbol{J}_{(S)}}\equiv\frac{1}{\langle T\rangle}\left(\overline{\boldsymbol{q}}-\sum_{\alpha=1}^{N}\langle\mu_\alpha\rangle\overline{\boldsymbol{J}}_\alpha\right)=\frac{1}{\langle T\rangle}\overline{\boldsymbol{J}}_q+\sum_{\alpha=1}^{N}\langle S_\alpha\rangle\overline{\boldsymbol{J}}_\alpha, \tag{3.84}$$

$$\boldsymbol{J}_{\langle S\rangle}^{\mathrm{turb}}\equiv\frac{1}{\langle T\rangle}\left[\tilde{\boldsymbol{q}}^{\mathrm{turb}}-\sum_{\alpha=1}^{N}\langle\mu_\alpha\rangle\boldsymbol{J}_\alpha^{\mathrm{turb}}\right]=\frac{1}{\langle T\rangle}\tilde{\boldsymbol{J}}_q^{\mathrm{turb}}+\sum_{\alpha=1}^{N}\langle S_\alpha\rangle\boldsymbol{J}_\alpha^{\mathrm{turb}} \tag{3.85}$$

$$\sigma_{\langle S\rangle}^{(i)}(\boldsymbol{r},t)\equiv\frac{1}{\langle T\rangle}\Bigg\{-\left(\tilde{\boldsymbol{J}}_q^{\Sigma}\cdot\frac{\partial\ln\langle T\rangle}{\partial\boldsymbol{r}}\right)+\left(\overline{\Pi}:\frac{\partial\langle\boldsymbol{u}\rangle}{\partial\boldsymbol{r}}\right)+\sum_{s=1}^{r}\langle A_s\rangle\overline{\xi}_s - \sum_{\alpha=1}^{N}\boldsymbol{J}_\alpha^{\Sigma}\cdot\left[\langle T\rangle\frac{\partial}{\partial\boldsymbol{r}}\left(\frac{\langle\mu_\alpha\rangle}{\langle T\rangle}\right)+\langle h_\alpha\rangle\frac{\partial\ln\langle T\rangle}{\partial\boldsymbol{r}}-\boldsymbol{F}_\alpha\right]\Bigg\}\geq 0, \tag{3.86}$$

$$\sigma_{\langle S\rangle}^{(e)}(\boldsymbol{r},t) \equiv \frac{1}{\langle T\rangle}\left\{\left(-\sum_{\alpha=1}^{N}\boldsymbol{J}_{\alpha}^{\text{turb}}\cdot\boldsymbol{F}_{\alpha}\right)-\overline{p'\text{div}\boldsymbol{u}''}+\left(\boldsymbol{J}_{(1/\rho)}^{\text{turb}}\cdot\frac{\partial\overline{p}}{\partial\boldsymbol{r}}\right)+\overline{\rho}\langle\varepsilon_b\rangle\right\}$$
$$\equiv\frac{\Im_{E,b}}{\langle T\rangle} \tag{3.87}$$

Here, $\langle\mu_{\alpha}\rangle \cong \langle \mathrm{h}_{\alpha}\rangle - \langle T\rangle\langle S_{\alpha}\rangle$ is the averaged partial chemical potential; the positive quantity $\sigma_{\langle S\rangle}^{(i)}(\boldsymbol{r},t)$ defines the local production rate of the averaged mixture entropy $\langle S\rangle$ due to irreversible transport processes and chemical reactions within the subsystem of averaged motion; as will be clear from the subsequent analysis, the quantity $\sigma_{\langle S\rangle}^{(e)}(\boldsymbol{r},t)$ (the sink or source of entropy) reflects the entropy exchange between the subsystems of turbulent chaos and averaged motion.

It should be noted that the quantity $\sigma_{\langle S\rangle}^{(e)}(\boldsymbol{r},t)$ can be different in sign, depending on the specific regime of turbulent flow. Indeed, the dissipation rate of turbulent energy $\langle\varepsilon_b\rangle(\boldsymbol{r},t)$ is always positive. However, the energy transition rate $\overline{p'\text{div}\boldsymbol{u}''}$ (representing the work done on turbulent vortices per unit time per unit volume by the environment due to the pressure fluctuations p' and the expansion ($\text{div}\boldsymbol{u}''>0$) or compression ($\text{div}\boldsymbol{u}''<0$) of vortices) can be different in sign. The quantity $\boldsymbol{J}_{(1/\rho)}^{\text{turb}}\cdot(\partial/\partial\boldsymbol{r})\overline{p}\approx g\overline{\rho' u_3}$, which represents the turbulence energy generation rate under the action of buoyancy forces, is positive in the case of small-scale turbulence, but it can be both positive and negative for large vortices (see van Mieghem 1973). Thus, it follows from (3.81) that, generally, the entropy $\langle S\rangle$ for the subsystem of averaged motion can both increase and decrease, which is a characteristic feature of thermodynamically open systems.

Note also that attributing the individual terms in (3.80) to the turbulent flux or to the production of averaged entropy is to some extent ambiguous: a number of alternative formulations using various definitions of the turbulent heat flux different from (3.80) are possible. Considerations of this kind are expounded in de Groot and Mazur (1962) and Gyarmati (1970).

3.2.2 Entropy Balance Equations and Entropy Production for the Subsystem of Turbulent Chaos

Thus, we have made sure that the Favre-averaged entropy $\langle S\rangle$ alone is not enough for an adequate description of all features of a turbulized continuum, because it is not related to any parameters characterizing the internal structure of the subsystem of turbulent chaos and, in particular, to such a paramount parameter as the turbulence energy (the averaged fluctuation kinetic energy per unit mass of the medium)

$$\langle b\rangle(\boldsymbol{r},t) \equiv \overline{\rho|\boldsymbol{u}''|^2}/2\overline{\rho}. \tag{3.88}$$

Therefore, when a phenomenological model of turbulence is constructed, a thermodynamic consideration of the subsystem of turbulent chaos also seems necessary. This goal can be achieved by increasing the number of independent variables in the thermodynamic description of this subsystem, which is in a non-equilibrium stationary state in the case of strongly developed turbulence. Below, we characterize the physically elementary volume $\mathrm{d}\boldsymbol{r}$ of turbulent chaos (as a rule, when a continuum model is constructed, infinitely small particles are considered as thermodynamic systems for which the physical concepts of an internal state are defined) by the following structural parameters: the extensive state variables $E_{\mathrm{turb}}(\boldsymbol{r},t)$ (internal turbulization energy density) and $S_{\mathrm{turb}}(\boldsymbol{r},t)$ (generalized local turbulization entropy) and the intensive state variables $T_{\mathrm{turb}}(\boldsymbol{r},t)$ (generalized turbulization temperature characterizing the intensity of turbulent fluctuations) and $p_{\mathrm{turb}}(\boldsymbol{r},t)$ (turbulization pressure) (Blackadar 1955). It is important to note that such generalized parameters of the state of chaos as the turbulization entropy S_{turb} and energy E_{turb} (considered below as primary concepts) are introduced here a priori to ensure coherence of the theory and, in general, have no precise physical interpretation (see Jou et al. 2001). Nevertheless, we assume below that the general thermodynamic relations holding in a quasi-equilibrium state also remain valid for a quasi-stationary state of turbulent chaos. In particular, an important point is the formulation of the second law of thermodynamics that serves exclusively as a constraint on the form of the corresponding constitutive equations. By admissible physically real processes (i.e., processes in which a sequence of states can be realized in the course of time within the framework of the applied model of turbulent motion) we mean the solution of the balance conservation equations supplemented by defining relations (obtained in a standard way) when the Clausius principle holds: the changes in the total entropy $S_\Sigma = \langle S\rangle + S_{\mathrm{turb}}$ of a turbulized system caused by internal irreversible processes can be only positive or (in the extreme case) equal to zero.

Let us now turn to corollaries of this formalism. Following the elegant Gibbs method (see, e.g., Münster 2002), we choose the following fundamental Gibbs equation (in integral form) for the generalized entropy as a local characteristic function (containing all thermodynamic information about the subsystem of turbulent chaos in a stationary state):

$$S_{\mathrm{turb}}(\boldsymbol{r},t) = S_{\mathrm{turb}}(E_{\mathrm{turb}}(\boldsymbol{r},t), 1/\overline{\rho}(\boldsymbol{r},t)); \tag{3.89}$$

this functional relation is assumed to be specified a priori. Let us now take, as is usually done in the formalized construction of classical locally equilibrium thermodynamics, the following definitions of the conjugate variables $T_{\mathrm{turb}}(\boldsymbol{r},t)$ and $p_{\mathrm{turb}}(\boldsymbol{r},t)$ (by assuming all these derivatives to be positive):

$$1/T_{\mathrm{turb}} \equiv \{\partial S_{\mathrm{turb}}/\partial E_{\mathrm{turb}}\}_{1/\overline{\rho}}, \quad p_{\mathrm{turb}}/T_{\mathrm{turb}} \equiv \{\partial S_{\mathrm{turb}}/\partial(1/\overline{\rho})\}_{E_{\mathrm{turb}}}.$$

The meaning of generalized (turbulization) temperature and pressure can then be assigned to the intensive variables $T_{\text{turb}}(\boldsymbol{r},t)$ and $p_{\text{turb}}(\boldsymbol{r},t)$, respectively. The corresponding differential form of the fundamental Gibbs equation (3.89) written along the trajectory of the center of mass of a physically elementary volume is

$$T_{\text{turb}}(\boldsymbol{r},t)\frac{D}{Dt}(S_{\text{turb}}(\boldsymbol{r},t)) = \frac{D}{Dt}(E_{\text{turb}}(\boldsymbol{r},t)) + p_{\text{turb}}(\boldsymbol{r},t)\frac{D}{Dt}\left(\frac{1}{\overline{\rho}(\boldsymbol{r},t)}\right). \tag{3.90}$$

Obviously, it is admissible to interpret the various kinds of functional relations between the variables E_{turb}, T_{turb}, p_{turb}, and S_{turb}, which can be derived by a standard (for thermodynamics) method from (3.90), as the "equations of state" for the subsystem under consideration. Below, we identify the quantity $E_{\text{turb}}(\boldsymbol{r},t)$ with the turbulence energy

$$E_{\text{turb}}(\boldsymbol{r},t) \equiv \langle b\rangle(\boldsymbol{r},t) + \text{const} = \overline{\rho|\boldsymbol{u}''|^2}/2\overline{\rho} + \text{const} \tag{3.91}$$

and assume that the subsystem of turbulent chaos in the thermodynamic sense is a perfect classical gas with three degrees of freedom in which the energy is distributed uniformly (the key hypotheses of the model). In particular, we then have

$$\begin{gathered}\langle b\rangle = c_V^{\text{turb}}T_{\text{turb}} = \tfrac{3}{2}\mathbf{R}^*T_{\text{turb}} = \tfrac{3}{2}p_{\text{turb}}/\overline{\rho}, \quad p_{\text{turb}} = \mathbf{R}^*T_{\text{turb}}\overline{\rho}, \\ S_{\text{turb}} = \tfrac{3}{2}\mathbf{R}^*\ln\left(p_{\text{turb}}/\overline{\rho}^{\frac{5}{3}}\right) + \text{const}.\end{gathered} \tag{3.92}$$

We derive the corresponding balance equation for the turbulization entropy S_{turb} from (3.90) by the above method [see Sect. 2.2] using (3.21) for the specific volume $(1/\overline{\rho})$ and the balance equation (3.69) for the turbulent energy $\langle b\rangle$; as a result, we obtain

$$\overline{\rho}\frac{DS_{\text{turb}}}{Dt} + \text{div}\boldsymbol{J}_{(S_{\text{turb}})} = \sigma_{(S_{\text{turb}})} \equiv \sigma^{(i)}_{(S_{\text{turb}})} + \sigma^{(e)}_{(S_{\text{turb}})}, \tag{3.93}$$

where

$$\boldsymbol{J}_{(S_{\text{turb}})} \equiv \frac{1}{T_{\text{turb}}}\left(\overline{\rho(|\boldsymbol{u}''|^2/2 + p'/\rho)\boldsymbol{u}'' - \Pi\cdot\boldsymbol{u}''}\right) = \frac{1}{T_{\text{turb}}}\boldsymbol{J}^{\text{turb}}_{\langle b\rangle}, \tag{3.94}$$

$$0 \le \sigma^{(i)}_{(S_{\text{turb}})} = \frac{1}{T_{\text{turb}}}\left\{-\left(\boldsymbol{J}^{\text{turb}}_{\langle b\rangle}\cdot\frac{\partial \ln T_{\text{turb}}}{\partial \boldsymbol{r}}\right) + \left(\boldsymbol{R} : \frac{\partial\langle\boldsymbol{u}\rangle}{\partial\boldsymbol{r}}\right) + p_{\text{turb}}\text{div}\langle\boldsymbol{u}\rangle\right\}, \tag{3.95}$$

$$\sigma^{(e)}_{(S_{\text{turb}})} \equiv \frac{1}{T_{\text{turb}}}\left\{\left(\sum_{\alpha=1}^{N}\boldsymbol{J}^{\text{turb}}_{\alpha}\cdot\boldsymbol{F}^*\right) + \overline{p'\text{div}\boldsymbol{u}''} - \left(\boldsymbol{J}^{\text{turb}}_{(1/\rho)}\cdot\frac{\partial\overline{p}}{\partial\boldsymbol{r}}\right) - \overline{\rho}\langle\varepsilon_b\rangle\right\} \equiv -\frac{\Im_{E,b}}{T_{\text{turb}}} \tag{3.96}$$

Here, $\boldsymbol{J}_{(S_{\text{turb}})}(\boldsymbol{r},t)$ is the substantial flux of the entropy S_{turb} for the subsystem of turbulent chaos; the quantities $\sigma^{(i)}_{(S_{\text{turb}})}$ and $\sigma^{(e)}_{(S_{\text{turb}})}$ mean the local production and sink rates of the fluctuation entropy S_{turb}, respectively.

For the subsequent analysis, it is convenient to decompose the gradient of the averaged velocity $\partial\langle\boldsymbol{u}\rangle/\partial\boldsymbol{r}$ (a second-rank tensor) in (3.86) and (3.95) into symmetric and antisymmetric parts [see (2.43)],

$$\begin{aligned}\partial\langle\boldsymbol{u}\rangle/\partial\boldsymbol{r} &= (\partial\langle\boldsymbol{u}\rangle/\partial\boldsymbol{r})^s + (\partial\langle\boldsymbol{u}\rangle/\partial\boldsymbol{r})^a \\ &= \boldsymbol{S} + \frac{1}{3}\boldsymbol{U}\mathrm{div}\langle\boldsymbol{u}\rangle + (\partial\langle\boldsymbol{u}\rangle/\partial\boldsymbol{r})^a,\end{aligned} \tag{3.97}$$

and represent the symmetric Reynolds stress tensor $\boldsymbol{R}$ (given the equation of state (3.92)) as

$$\overset{0}{\boldsymbol{R}} \equiv \boldsymbol{R} - \frac{1}{3}(\boldsymbol{R}{:}\boldsymbol{U})\boldsymbol{U} = \boldsymbol{R} + p_{\text{turb}}\boldsymbol{U} = \boldsymbol{R} + \frac{2}{3}\overline{\rho}\langle b\rangle\boldsymbol{U}, \tag{3.98}$$

where

$$\begin{aligned} p_{\text{turb}} &= -\frac{1}{3}(\boldsymbol{R}:\boldsymbol{U}), \quad \boldsymbol{D} \equiv (\partial\langle\boldsymbol{u}\rangle/\partial\boldsymbol{r})^s, \\ \boldsymbol{S} \equiv \overset{0}{\boldsymbol{D}} &\equiv \overset{0}{(\partial\langle\boldsymbol{u}\rangle/\partial\boldsymbol{r})^s} = \boldsymbol{D} - \frac{1}{3}\boldsymbol{U}\mathrm{div}\langle\boldsymbol{u}\rangle \end{aligned} \tag{3.99}$$

are, respectively, the turbulization pressure, the strain rate tensor, and the shear rate tensor for a turbulized continuum. The scalar product of the Reynolds tensor and the velocity gradient can then be written as $(\boldsymbol{R} : (\partial/\partial\boldsymbol{r})\langle\boldsymbol{u}\rangle) = \overset{0}{\boldsymbol{R}} : \overset{0}{\boldsymbol{D}} - p_{\text{turb}}\mathrm{div}\langle\boldsymbol{u}\rangle$ and the balance equation for the turbulization entropy S_{turb} (3.93) takes the form

$$\overline{\rho}\frac{DS_{\text{turb}}}{Dt} + \mathbf{div}\left\{\frac{1}{T_{\text{turb}}}\boldsymbol{J}^{\text{turb}}_{\langle b\rangle}\right\} = \frac{1}{T_{\text{turb}}}\left\{-\left(\boldsymbol{J}^{\text{turb}}_{\langle b\rangle}\cdot\frac{\partial \ln T_{\text{turb}}}{\partial\boldsymbol{r}}\right) + \overset{0}{\boldsymbol{R}} : \overset{0}{\boldsymbol{D}} - \Im_{E,b}\right\}. \tag{3.100}$$

In writing (3.100), we used the fact that the scalar product of symmetric and antisymmetric tensors is always equal to zero.

3.2.3 *Balance Equation for the Total Entropy of a Turbulized Continuum*

The introduction of two entropies, $\langle S\rangle$ and S_{turb}, concretizes our view of the initial turbulized continuum as a thermodynamic complex that consists of two mutually open subsystems—the subsystems of averaged motion and turbulent chaos.

The balance equation for the total entropy $S_\Sigma = (\langle S\rangle + S_{\text{turb}})$ of a multicomponent system follows from (3.80) and (3.100):

$$\overline{\rho}\frac{DS_\Sigma}{Dt} + \operatorname{div}\left\{\frac{\boldsymbol{J}^{\text{turb}}_{\langle b\rangle}}{T_{\text{turb}}} + \frac{\left(\boldsymbol{q}^\Sigma - \sum_{\alpha=1}^{N}\langle\mu_\alpha\rangle\boldsymbol{J}^\Sigma_\alpha\right)}{\langle T\rangle}\right\} = \sigma_\Sigma \equiv \sigma^{(i)}_{\langle S\rangle} + \sigma^{(i)}_{S_{\text{turb}}} + \sigma_{\langle S\rangle,\,S_{\text{turb}}}, \tag{3.101}$$

where

$$\begin{aligned}0 \le \sigma_\Sigma \equiv \frac{1}{\langle T\rangle}\Bigg\{&-\left(\tilde{\boldsymbol{J}}^\Sigma_q\cdot\frac{\partial\ln\langle T\rangle}{\partial\boldsymbol{r}}\right) + \overline{\pi}\operatorname{div}\langle\boldsymbol{u}\rangle + \overset{0}{\overline{\Pi}} : \overset{0}{D}\\&+\sum_{s=1}^{r}\langle A_s\rangle\overline{\xi}_s - \sum_{\alpha=1}^{N}\boldsymbol{J}^\Sigma_\alpha\cdot\left[\langle T\rangle\frac{\partial}{\partial\boldsymbol{r}}\left(\frac{\langle\mu_\alpha\rangle}{\langle T\rangle}\right) + \langle h_\alpha\rangle\frac{\partial\ln\langle T\rangle}{\partial\boldsymbol{r}} - \boldsymbol{F}_\alpha\right]\Bigg\}\\&+\frac{1}{T_{\text{turb}}}\left\{-\left(\boldsymbol{J}^{\text{turb}}_{\langle b\rangle}\cdot\frac{\partial\ln T_{\text{turb}}}{\partial\boldsymbol{r}}\right) + \overset{\circ}{R} : \overset{\circ}{\boldsymbol{D}}\right\} + \Im_{E,b}\left(\frac{T_{\text{turb}} - \langle T\rangle}{T_{\text{turb}}\langle T\rangle}\right),\end{aligned} \tag{3.102}$$

$$\Im \equiv -\left(\sum_{\alpha=1}^{N}\boldsymbol{J}^{\text{turb}}_\alpha\cdot\boldsymbol{F}_\alpha\right) - \overline{p'\operatorname{div}\boldsymbol{u}''} + \left(\boldsymbol{J}^{\text{turb}}_{(1/\rho)}\cdot\frac{\partial\overline{p}}{\partial\boldsymbol{r}}\right) + \overline{\rho}\langle\varepsilon_b\rangle, \tag{3.103}$$

$$\sigma_{\langle S\rangle,\,S_{\text{turb}}} \equiv \sigma^{(e)}_{\langle S\rangle} + \sigma^{(e)}_{S_{\text{turb}}}. \tag{3.104}$$

The local production of the total entropy σ_Σ related to irreversible processes within a turbulized continuum is thus seen to be defined by the set of thermodynamic fluxes $\tilde{\boldsymbol{J}}^\Sigma_q$, $\overline{\xi}_s$, $\boldsymbol{J}^\Sigma_\alpha$, $\overline{\pi}$, $\overline{\Pi}$, $J^{\text{turb}}_{\langle b\rangle}$, p_{turb}, $\boldsymbol{R}$, $\Im_{E,b}$ and their conjugate thermodynamic forces [cf. (2.50), (2.51), (2.52), (2.53) and (2.54)]

$$Y^\Sigma_q \equiv -\frac{1}{\langle T\rangle^2}\frac{\partial\langle T\rangle}{\partial\boldsymbol{r}} = \frac{\partial}{\partial\boldsymbol{r}}\left(\frac{1}{\langle T\rangle}\right), \quad Y_{\langle b\rangle} \equiv -\frac{1}{T^2_{\text{turb}}}\frac{\partial T_{\text{turb}}}{\partial\boldsymbol{r}} \tag{3.105}$$

$$Y_{A_s} \equiv \frac{\langle A_s\rangle}{\langle T\rangle} = -\sum_{\beta=1}^{N}\frac{\langle\mu_\beta\rangle}{\langle T\rangle}\nu_{\beta s}, \quad (s = 1, 2, \ldots, r), \tag{3.106}$$

$$\boldsymbol{Y}^*_\alpha \equiv -\frac{\partial}{\partial\boldsymbol{r}}\left(\frac{\langle\mu_\alpha\rangle}{\langle T\rangle}\right) + \langle h_\alpha\rangle\frac{\partial}{\partial\boldsymbol{r}}\left(\frac{1}{\langle T\rangle}\right) + \frac{\boldsymbol{F}_\alpha}{\langle T\rangle}, \tag{3.107}$$

$$Y_\pi \equiv \frac{\operatorname{div}\langle\boldsymbol{u}\rangle}{\langle T\rangle}, \tag{3.108}$$

$$\boldsymbol{Y}_D \equiv \frac{1}{\langle T \rangle} \overset{0}{\boldsymbol{D}}, \quad \boldsymbol{Y}_R \equiv \frac{1}{T_{\text{turb}}} \overset{0}{\boldsymbol{D}}, \tag{3.109}$$

$$Y_{E,b} \equiv \left(\frac{T_{\text{turb}} - \langle T \rangle}{T_{\text{turb}} \langle T \rangle} \right). \tag{3.110}$$

Using these definitions, the entropy production σ_Σ can be written in the following bilinear form:

$$\begin{aligned} 0 \le \sigma_\Sigma = & \overbrace{\sum_{s=1}^{r} \overline{\xi}_s Y_{A_s} + \tilde{\boldsymbol{J}}_q^\Sigma \cdot \boldsymbol{Y}_q^\Sigma + \sum_{\alpha=1}^{N} \boldsymbol{J}_\alpha^\Sigma \cdot \boldsymbol{Y}_\alpha^* + \overline{\pi} Y_\pi + \overset{0}{\overline{\Pi}} : \boldsymbol{Y}_D}^{\sigma_{\langle S \rangle}^{(i)}} \\ & + \overbrace{\boldsymbol{J}_{\langle b \rangle}^{\text{turb}} \cdot \boldsymbol{Y}_{\langle b \rangle} + \overset{0}{\boldsymbol{R}} : \boldsymbol{Y}_R}^{\sigma_{S_{\text{turb}}}^{(i)}} + \overbrace{\Im_{E,b} Y_{E,b}}^{\sigma_{\langle S \rangle, S_{\text{turb}}}}, \end{aligned} \tag{3.111}$$

which corresponds to three independent sources of nonequilibrium processes in a turbulized mixture with a distinctly different physical nature.

According to the main postulate of generalized nonequilibrium thermodynamics [see Sect. 2.2], when the thermodynamic system is near local equilibrium or near a stable stationary-nonequilibrium state, the thermodynamic fluxes can be represented as linear functions of their conjugate macroscopic forces: $J_{\gamma i} = \sum_{\delta} L_{\gamma\delta}^{ij} X_{\delta j}$ $(\gamma, \delta = 1, 2, \ldots f)$. It is important to note that (3.111) allows the defining relations to be obtained for three main regimes of a turbulized mixture flow—for an averaged laminar flow, for developed turbulence when the turbulent fluxes are much more efficient than the corresponding averaged molecular fluxes ($T_{\text{turb}} \gg \langle T \rangle$, $R \gg \overline{\Pi}$, $\boldsymbol{q}^{\text{turb}} \gg \overline{\boldsymbol{q}}$, etc.), and finally, in the general case where the processes of averaged molecular and turbulent transport are comparable in efficiency. As can be seen from (3.111), the spectrum of possible cross effects for a turbulent flow is extended considerably compared to a laminar one. Thus, for example, the reduced heat flux $\tilde{\boldsymbol{J}}_q^\Sigma \equiv \boldsymbol{q}^\Sigma - \overline{p' \boldsymbol{u}''} - \sum_{\alpha=1}^{N} \langle h_\alpha \rangle \boldsymbol{J}_\alpha^\Sigma$ in a turbulized continuum can emerge not only under the influence of its conjugate thermodynamic force $\boldsymbol{Y}_q^\Sigma$ but also through the action of the force $\boldsymbol{Y}_{\langle b \rangle}$ conjugate to the flux $\boldsymbol{J}_{\langle b \rangle}^{\text{turb}}$ (which describes the "diffusion" transfer of turbulent energy). However, unfortunately, there are no reliable experimental data at present that quantitatively describe such cross effects in a turbulized medium. In addition, the contribution from any cross effects to the total transfer rate is generally an order of magnitude smaller than that from direct effects (see de Groot and Mazur 1962). Taking these circumstances into account, we use below the requirements that the production rates of the total entropy $\sigma_{\langle S \rangle}^{(i)}$, $\sigma_{S_{\text{turb}}}^{(i)}$ $\sigma_{\langle S \rangle, S_{\text{turb}}}$ be positive independent of one another, i.e., by assuming that any linear relations referring, for example, to the subsystem of averaged motion (in particular, between the symmetric part of the averaged viscous stress tensor $\overset{\circ}{\overline{\Pi}}$ with a zero trace and the tensor viscous force $\boldsymbol{Y_D}$) are not affected noticeably by the subsystem of turbulent chaos (the tensor force $\boldsymbol{Y_R}$). We also omit a number of cross effects in the linear constitutive relations without any special stipulations.

To conclude this section, we make two remarks:

- The quantity $\sigma_{\langle S \rangle, S_{\text{turb}}}$ describing the entropy production within the full system through irreversible entropy exchange between the subsystems of turbulent chaos and averaged motion is also always positive in view of the second law of thermodynamics. Therefore, the "direction" of the thermodynamic flux $\Im_{E,b}(\boldsymbol{r}, t)$ is specified by the sign of the state function $Y_{E,b} \equiv (1/\langle T \rangle - 1/T_{\text{turb}})$, which should be considered as the conjugate thermodynamic force (macroscopic factor) producing this entropy flux. Such entropy exchange between two mutually open subsystems is known to be an indispensable condition for a structured collective behavior, i.e., it can be a source of self-organization in one of them (see Chap. 5).
- Generally, the matrix of phenomenological coefficients $L^{ij}_{\gamma\delta}$ for a turbulized continuum depends not only on averaged state parameters (temperature, density, etc.) but also on characteristics of the turbulent superstructure itself, for example, on the parameters $\overline{\rho}$, $\langle \varepsilon_b \rangle$, and T_{turb} (or $\langle b \rangle$). Such a situation, in which there is a functional dependence of the tensor of kinematic coefficients $L^{ij}_{\gamma\delta}$ on the thermodynamic fluxes themselves (e.g., on the turbulent energy dissipation rate $\langle \varepsilon_b \rangle$), is known to be typical for self-organizing systems (see Haken 1983, 1988). In general, it can lead to the individual terms in the sum σ_Σ being not positive definite, although the sum itself $\sigma_\Sigma \geq 0$. In this case, a superposition of various fluxes, in principle, can lead to negative values of individual diagonal elements in the matrix $L^{ij}_{\gamma\delta}$. This probably explains the effect of negative viscosity in some turbulent flows (see Chaps. 5 and 8).

3.2.4 Linear Closing Relations for a Turbulized Multicomponent Mixture of Gases

To concretize the gradient closing relations (constitutive Onsager laws) relating the averaged molecular and turbulent thermodynamic fluxes to the corresponding thermodynamic forces, we now use the formalism of nonequilibrium thermodynamics presented in Sect. 2.2. We consider here the general case where the averaged molecular and turbulent transport processes are comparable in significance and restrict ourselves to the derivation of such relations for meso- and small-scale turbulence. For the latter, as is well known, there is a tendency for local statistical isotropy of its characteristics to be established (the statistical properties of a turbulent flow in this case do not depend on direction). This approach can be easily generalized to the case of nonisotropic (large-scale) turbulence.

As is well known from the general theory of tensor functions (see Sedov 1984), the symmetry properties of isotropic media are completely characterized by a metric tensor g^{ij}: all tensors will be tensor functions of only the metric tensor, in particular, $L^{ij}_{\gamma\delta} = L_{\gamma\delta} g^{ij} \quad (\gamma, \delta = 1, 2, \ldots f)$, where $L_{\gamma\delta}$ are scalar coefficients. In addition, since there is no interference between the fluxes and thermodynamic forces of various tensor dimensions in an isotropic system (the Curie principle), we may consider, for example, phenomena described by polar vectors (heat conduction or diffusion) independently of scalar and tensor phenomena (see de Groot and Mazur 1962). Adopting the additional hypothesis that the system is Markovian (when the fluxes at a given time depend on the generalized forces taken at the same time), we then obtain the following phenomenological relations (written in rectangular coordinates, $g^{ij} \equiv \delta_{ij}$) (Kolesnichenko 1998) from (3.111):

$$\tilde{\boldsymbol{J}}^{\Sigma}_q \equiv \boldsymbol{q}^{\Sigma} - \overline{p'\boldsymbol{u}''} - \sum_{\alpha=1}^{N} \langle h_\alpha \rangle \boldsymbol{J}^{\Sigma}_\alpha = L^{\Sigma}_{qq} \frac{\partial}{\partial \boldsymbol{r}} \left(\frac{1}{\langle T \rangle} \right) + \sum_{\beta=1}^{N} L^{\Sigma}_{q\beta} \boldsymbol{Y}^*_\beta, \tag{3.112}$$

$$\boldsymbol{J}^{\Sigma}_\alpha = L^{\Sigma}_{\alpha q} \frac{\partial}{\partial \boldsymbol{r}} \left(\frac{1}{\langle T \rangle} \right) + \sum_{\beta=1}^{N} L^{\Sigma}_{\alpha\beta} \boldsymbol{Y}^*_\beta, \quad (\alpha = 1, 2, \ldots, N), \tag{3.113}$$

$$(\overset{0}{\overline{\Pi}})_{jk} = L(\boldsymbol{Y_D})_{jk} = \overline{\mu} \left\{ \left(\frac{\partial \langle u_k \rangle}{\partial x_j} + \frac{\partial \langle u_j \rangle}{\partial x_k} \right) - \frac{2}{3} \delta_{jk} \mathrm{div} \langle \boldsymbol{u} \rangle \right\}, \tag{3.114}$$

$$\overline{\pi} = \frac{l_{\overline{\pi}\overline{\pi}}}{\langle T \rangle} \left(\mathrm{div} \langle \boldsymbol{u} \rangle + \sum_{s=1}^{r} l_{\overline{\pi}s} \langle A_s \rangle \right) \cong \overline{\mu}_\vartheta \mathrm{div} \langle \boldsymbol{u} \rangle, \tag{3.115}$$

$$\overline{\xi}_s = -\mathrm{l}_{s\overline{\pi}} \frac{\mathrm{div} \langle \boldsymbol{u} \rangle}{\langle T \rangle} + \sum_{m=1}^{r} \mathrm{l}_{sm} \frac{\langle A_s \rangle}{\langle T \rangle}, \quad (s = 1, 2, \ldots, r), \tag{3.116}$$

$$\begin{aligned} (\boldsymbol{R})_{\mathrm{jk}} &= -\frac{2}{3} \overline{\rho} \langle b \rangle \delta_{jk} + L_{\mathrm{turb}} (\boldsymbol{Y_R})_{jk} \\ &= -\frac{2}{3} \overline{\rho} \langle b \rangle \delta_{jk} + \mu^{\mathrm{turb}} \left\{ \left(\frac{\partial \langle u_k \rangle}{\partial x_j} + \frac{\partial \langle u_j \rangle}{\partial x_k} \right) - \frac{2}{3} \delta_{jk} \mathrm{div} \langle \boldsymbol{u} \rangle \right\}, \end{aligned} \tag{3.117}$$

$$\boldsymbol{J}^{\mathrm{turb}}_{\langle b \rangle} = -\frac{l_b}{T^2_{\mathrm{turb}}} \frac{\partial T_{\mathrm{turb}}}{\partial \boldsymbol{r}} = -\frac{\mu^{\mathrm{turb}}}{\sigma_b} \frac{\partial \langle b \rangle}{\partial \boldsymbol{r}}, \tag{3.118}$$

$$\Im_{E,b} = l_{E,b} \left(\frac{T_{\mathrm{turb}} - \langle T \rangle}{T_{\mathrm{turb}} \langle T \rangle} \right). \tag{3.119}$$

Here, the formulas

$$\overline{\mu} \equiv L/2\langle T\rangle, \quad \overline{\mu}_{\vartheta} \equiv l_{\overline{\pi\pi}}/\langle \mathrm{T}\rangle, \quad \mu^{\mathrm{turb}} \equiv L_{\mathrm{turb}}/2T_{\mathrm{turb}}, \quad \nu^{\mathrm{turb}} \equiv \mu^{\mathrm{turb}}/\overline{\rho} \qquad (3.120)$$

introduce the averaged molecular viscosity, $\overline{\mu}(\boldsymbol{r},\mathrm{t})$, and second viscosity, $\overline{\mu}_{\vartheta}(\boldsymbol{r},t)$, coefficients needed to define the averaged viscous stress tensor $\overline{\Pi}$ as well as the turbulent viscosity, $\mu^{\mathrm{turb}}(\boldsymbol{r},t)$, and kinematic turbulent viscosity, $\nu^{\mathrm{turb}}(\boldsymbol{r},t)$, coefficients defining the turbulent stress tensor $\boldsymbol{R}$. The coefficient σ_b is the "Prandtl number" for the turbulent energy, whose value is usually assumed to be constant. The scalar kinematic coefficients $L^{\Sigma}_{q\beta}$ and $L^{\Sigma}_{\alpha\beta}$, as in the laminar case [see (2.61) and (2.63)], satisfy the Onsager-Casimir symmetry conditions $L^{\Sigma}_{\alpha\beta} = L^{\Sigma}_{\beta\alpha}$ $(\alpha,\beta = 1,2,\ldots N)$ and the conditions

$$\sum_{\alpha=1}^{N} m_{\alpha} L^{\Sigma}_{q\alpha} = 0, \quad (^1) \quad \sum_{\alpha=1}^{N} m_{\alpha} L^{\Sigma}_{\alpha\beta} = 0, \quad (\beta = 1,2,\ldots N). \qquad (^2) \qquad (3.121)$$

It should be kept in mind that, in contrast to the ordinary molecular viscosity coefficients $\overline{\mu}$ and $\overline{\mu}_{\vartheta}$, the turbulent viscosity coefficient μ^{turb} characterizes not the physical properties of a fluid but the statistical properties of its fluctuation motion; that is why it can take on negative values in some cases. In addition, the well-known increase in turbulent viscosity compared to its molecular analog once again suggests that a turbulent motion is more ordered (organized) than a laminar one. Indeed, the viscosity in a laminar motion is determined by the momentum transfer at a chaotic molecular level. In contrast, in a turbulent motion, momentum is transferred from layer to layer by collective degrees of freedom and this is an indubitable indication of its greater order.

Regarding the defining relation (3.117) for the tensor $\boldsymbol{R}$, we note the following: when the turbulent field anisotropy is taken into account, this relation becomes considerably more complicated, because it requires replacing the scalar turbulent viscosity coefficient μ^{turb} by a (fourth-rank) tensor [see Chap. 7 and the monograph by Monin and Yaglom (1992)]. Note also that we managed to derive here the defining relation (in standard form)

$$\overline{\Pi}_{jk} = \overline{\mu}\left\{\left(\frac{\partial\langle u_k\rangle}{\partial x_j} + \frac{\partial\langle u_j\rangle}{\partial x_k}\right) - \frac{2}{3}\delta_{jk}\mathrm{div}\langle \boldsymbol{u}\rangle\right\} + \overline{\mu}_{\vartheta}\mathrm{div}\langle \boldsymbol{u}\rangle \qquad (3.114^*)$$

for the averaged viscous stress tensor directly, i.e., without invoking the corresponding regular analog [see (2.64)] for a laminar motion and its subsequent averaging.

As we see, the linear law (3.116) can also be used to obtain the limiting form of the expressions for the averaged chemical reaction rates near a chemical equilibrium state. However, since this result has a limited domain of applicability, here we not dwell on it, deferring a more detailed consideration to Chap. 4.

3.2.4.1 Heat Conduction and Diffusion in a Turbulized Mixture

Using the formal similarity of the defining relations for the vector turbulent diffusion and heat processes specified by (3.112) and (3.113) to those for a laminar flow [see (2.56) and (2.57)], we rewrite (using the approach developed in Sect. 2.3) (3.112) and (3.113) as

$$\boldsymbol{J}_{\alpha}^{\Sigma} = -\overline{n}_{\alpha} D_{T\alpha}^{\Sigma} \frac{\partial \ln\langle T\rangle}{\partial \boldsymbol{r}} - \overline{n}_{\alpha} \sum_{\beta=1}^{N} D_{\alpha\beta}^{\Sigma} \boldsymbol{d}_{\beta}^{\text{turb}}, \quad (\alpha = 1, 2, \ldots, N), \tag{3.122}$$

$$\tilde{\boldsymbol{J}}_{q}^{\Sigma} = -\hat{\lambda}^{\Sigma} \frac{\partial\langle T\rangle}{\partial \boldsymbol{r}} - \overline{p} \sum_{\beta=1}^{N} D_{T\beta}^{\Sigma} \boldsymbol{d}_{\beta}^{\text{turb}}, \tag{3.123}$$

where

$$\boldsymbol{d}_{\beta}^{\text{turb}} \equiv \frac{\partial}{\partial \boldsymbol{r}}\left(\frac{\overline{n}_{\beta}}{\overline{n}}\right) + \left(\frac{\overline{n}_{\beta}}{\overline{n}} - \langle C_{\beta}\rangle\right) \frac{\partial \ln \overline{p}}{\partial \boldsymbol{r}} - \frac{\overline{n}_{\beta}}{\overline{p}}\left(\boldsymbol{F}_{\beta} - m_{\beta} \sum_{\alpha=1}^{N} \langle Z_{\alpha}\rangle \boldsymbol{F}_{\alpha}\right) \tag{3.124}$$

are the generalized thermodynamic forces for a turbulent mixture motion. These are similar to the corresponding expressions (2.70) for a regular motion and can be introduced for a turbulized mixture using the relations

$$\boldsymbol{d}_{\beta}^{\text{turb}} \equiv -\frac{\langle T\rangle \overline{n}_{\beta}}{\overline{p}} \boldsymbol{Y}_{\beta}^{*} - \langle C_{\beta}\rangle \frac{\partial \ln \overline{p}}{\partial \boldsymbol{r}} + \frac{\overline{\rho}_{\beta}}{\overline{p}} \sum_{\alpha=1}^{N} \langle Z_{\alpha}\rangle \boldsymbol{F}_{\alpha}, \quad (^{1})$$

$$\sum_{\alpha=1}^{N} \boldsymbol{d}_{\alpha}^{\text{turb}} = 0, \; (^{2}) \tag{3.125}$$

i.e., in exactly the same way as was done in Sect. 2.3.3 (here, $\langle C_{\beta}\rangle = m_{\beta}\overline{n}_{\beta}/\overline{\rho}$ is the Favre-averaged mass concentration of particles of type β).

In relations (3.122) and (3.123), by analogy with the formulas for a laminar fluid flow, we introduced the symmetric multicomponent turbulent diffusion coefficients $D_{\alpha\beta}^{\Sigma}$ $(\alpha, \beta = 1, 2, \ldots, N)$, turbulent thermal diffusion coefficients $D_{T\beta}^{\Sigma}$ $(\beta = 1, 2, \ldots, N)$, and turbulent thermal conductivity coefficients $\hat{\lambda}^{\Sigma}$ for a multicomponent gas using the definitions

$$\hat{\lambda}^{\Sigma} \equiv \frac{L_{qq}^{\Sigma}}{\langle T\rangle^{2}}, \quad D_{T\beta}^{\Sigma} = \frac{L_{q\beta}^{\Sigma}}{\langle T\rangle \overline{n}_{\beta}}, \quad D_{\alpha\beta}^{\Sigma} = D_{\beta\alpha}^{\Sigma} = \frac{\overline{p}}{\langle T\rangle \overline{n}_{\alpha}\overline{n}_{\beta}} L_{\alpha q}^{\Sigma}. \tag{3.125b}$$

In view of (3.121), the scalar turbulent transport coefficients $D_{T\beta}^{\Sigma}$ and $D_{\alpha\beta}^{\Sigma}$ satisfy the conditions

$$\sum_{\alpha=1}^{N} \langle C_\alpha \rangle D^{\Sigma}_{T\alpha\beta} = 0, \quad \sum_{\alpha=1}^{N} \langle C_\alpha \rangle D^{\Sigma}_{\alpha\beta} = 0, \quad (\alpha, \beta = 1, 2, \ldots, N). \tag{3.126}$$

The coefficients defined by (3.125) are the effective transport coefficients attributable not only to the molecular mass and heat transfer from some fluid volumes to other ones but also to the turbulent mixing produced by turbulized fluid velocity fluctuations; therefore, it can be assumed that $D^{\Sigma}_{\alpha\beta} \equiv \overline{D}_{\alpha\beta} + D^{\mathrm{tyrb}}_{\alpha\beta}$ and $\hat{\lambda}^{\Sigma} \equiv \overline{\hat{\lambda}} + \hat{\lambda}^{\mathrm{turb}}$. Since the cross processes related to thermal diffusion and diffusive heat conduction for turbulized mixtures are completely unstudied at present, below we neglect them by assuming that $D^{\Sigma}_{T\alpha\beta} \cong 0$.

Thus, the defining relations for the turbulent diffusion and heat fluxes can be written in the following final form:

$$\boldsymbol{J}^{\Sigma}_{\alpha} = -\overline{n}_\alpha \sum_{\beta=1}^{N} D^{\Sigma}_{\alpha\beta} \boldsymbol{d}^{\mathrm{turb}}_{\beta}, \quad (\alpha = 1, 2, \ldots, N), \tag{3.127}$$

$$\boldsymbol{q}^{\Sigma} - \overline{p'\boldsymbol{u}''} = -\hat{\lambda}^{\Sigma} \frac{\partial \langle T \rangle}{\partial \boldsymbol{r}} + \sum_{\beta=1}^{N} \langle h_\beta \rangle \boldsymbol{J}^{\Sigma}_{\beta}. \tag{3.128}$$

These relations describe most completely the heat and mass transfer processes in a developed isotropic turbulent flow of a multicomponent gas mixture. Unfortunately, since the experimental data on multicomponent turbulent diffusion coefficients are limited at the current stage of development of the phenomenological turbulence theory, more simplified models have to be used in practice. It should also be added that the turbulent exchange coefficients introduced here, in particular, the coefficients $D^{\Sigma}_{\alpha\beta}$, can be defined in terms of the so-called K-theory of developed turbulence by invoking additional transfer equations for the pair correlations of fluctuating thermohydrodynamic mixture parameters [see Chap. 4].

3.2.4.2 Generalized Stefan–Maxwell Relations for a Turbulized Mixture

Just as in the case of laminar mass and heat transfer in a mixture, it is convenient to reduce the defining relations (3.127) and (3.128) for the turbulent diffusion and heat fluxes (in particular, when multicomponent flows are simulated numerically) to the form of generalized Stefan–Maxwell relations including the binary (for a binary mixture) turbulent diffusion coefficients $\mathbf{D}^{\Sigma}_{\alpha\beta}$. This is because, in contrast to the multicomponent diffusion coefficients $D^{\Sigma}_{\alpha\beta}$, empirical data are, in general, easier to use for the coefficients $\mathbf{D}^{\Sigma}_{\alpha\beta}$.

The procedure for deriving the generalized Stefan–Maxwell relations for multicomponent diffusion in a turbulent flow does not differ in any way from that performed in Sect. 2.3.4 when deriving these relations for a laminar mixture flow. Using this analogy, we immediately present the final result (Kolesnichenko 1998):

$$\sum_{\substack{\alpha=1\\ \alpha\neq\beta}}^{N} \frac{\overline{n}_\beta \boldsymbol{J}^\Sigma_\alpha - \overline{n}_\alpha \boldsymbol{J}^\Sigma_\beta}{\overline{n}^2 \mathbf{D}^\Sigma_{\alpha\beta}} = \boldsymbol{d}^{\text{turb}}_\beta, \quad (\beta = 1,2,\dots,N-1); \quad \sum_{\alpha=1}^{N} m_\beta \boldsymbol{J}^\Sigma_\alpha = 0, \quad (3.129)$$

where

$$\boldsymbol{d}^{\text{turb}}_\beta \equiv \frac{\partial}{\partial \boldsymbol{r}}\left(\frac{\overline{n}_\beta}{\overline{n}}\right) + \left(\frac{\overline{n}_\beta}{\overline{n}} - \langle C_\beta\rangle\right)\frac{\partial \ln \overline{p}}{\partial \boldsymbol{r}} - \frac{\overline{n}_\beta}{\overline{p}}\left(\boldsymbol{F}_\beta - m_\beta \sum_{\alpha=1}^{N} \langle Z_\alpha\rangle \boldsymbol{F}_\alpha\right).$$

In the case of a direct numerical solution of these relations for the turbulent diffusion fluxes $\boldsymbol{J}^\Sigma_\alpha$, it is convenient to reduce them, by analogy with a laminar mixture flow, to the form of a generalized Fick law [see (2.116)]. As a result, we obtain

$$\boldsymbol{J}^\Sigma_\beta = -D^\Sigma_\beta \left(\overline{n}\boldsymbol{d}^{\text{turb}}_\beta - \frac{1}{\overline{n}} \sum_{\substack{\alpha=1\\ \alpha\neq\beta}}^{N} \frac{\overline{n}_\beta}{\mathbf{D}^\Sigma_{\alpha\beta}} \boldsymbol{J}^\Sigma_\alpha\right) = -\overline{\rho}\, D^\Sigma_\beta \frac{\partial}{\partial \boldsymbol{r}}\left(\frac{\overline{n}_\beta}{\overline{\rho}}\right) + \delta \boldsymbol{J}^\Sigma_\beta, \quad (3.130)$$

where

$$\delta \boldsymbol{J}^\Sigma_\beta \equiv \overline{n}_\beta D^\Sigma_\beta \times \left\{ -\frac{\partial \ln \mathbf{M}}{\partial \boldsymbol{r}} - \left(1 - \frac{m_\beta}{\mathbf{M}}\right)\frac{\partial \ln \overline{p}}{\partial \boldsymbol{r}} + \frac{\overline{n}}{\overline{p}}\left(\boldsymbol{F}_\beta - m_\beta \sum_{\alpha=1}^{N} \langle Z_\alpha\rangle \boldsymbol{F}_\alpha\right) + \frac{1}{\overline{n}} \sum_{\substack{\alpha=1\\ \alpha\neq\beta}}^{N} \frac{\boldsymbol{J}^\Sigma_\alpha}{\mathbf{D}^\Sigma_{\alpha\beta}} \right\}; \quad (3.131)$$

$$D^\Sigma_\beta \equiv \left(\frac{1}{\overline{n}} \sum_{\substack{\alpha=1\\ \alpha\neq\beta}}^{N} \frac{\overline{n}_\alpha}{\mathbf{D}^\Sigma_{\alpha\beta}}\right)^{-1}; \quad \mathbf{M} \equiv \sum_{\alpha=1}^{N} m_\alpha \overline{n}_\alpha \Big/ \sum_{\alpha=1}^{N} \overline{n}_\alpha = \frac{\overline{\rho}}{\overline{n}}. \quad (3.132)$$

By introducing the effective diffusion coefficient D^Σ_β, we can simplify considerably the numerical solution of the problem despite the fact that the generalized Fick law in form (3.130) generally does not allow each diffusion equation (3.23) to be considered separately from the other ones. However, since the methods of successive approximations are commonly used for the numerical solution of problems, the presence of the term $\delta \boldsymbol{J}_\beta^\Sigma$ in (3.130) is often not important.

We see from relations (3.130), (3.131) and (3.132) that the ordinary Fick diffusion law strictly holds for a turbulized mixture in the following cases: (a) the thermal diffusion is negligible; (b) the mixture is binary; (c) the mass force per unit mass is the same for each component ($\boldsymbol{F}_\alpha/m_\alpha = \boldsymbol{F}_\beta/m_\beta$); and (d) either the pressure gradients are zero or the molecular weights of both substances are identical (if $m_\alpha \equiv m_\beta = m$, then $\mathbf{M} = m$). These conditions are rather stringent and it is often difficult to justify them when modeling real turbulent transport processes. Nevertheless, since the generalized Stefan–Maxwell equations for multicomponent diffusion are complex and since the turbulent coefficients $\mathbf{D}_{\alpha\beta}^\Sigma$ have been studied inadequately, for simplicity, the generalized Fick diffusion law (3.130) (without the second term on the right-hand side) can be used in many analytical applications. For the integral mass balance condition $\sum_\alpha m_\beta \boldsymbol{J}_\alpha^\Sigma = 0$ to be retained, all Wilkey coefficients must be assumed to be equal, $D_\beta^{\text{turb}} \equiv D^{\text{turb}}$.

3.2.5 Formulas to Determine the Correlations Including Density Fluctuations

Let us now consider the derivation of a defining relation for the turbulent specific volume flux $\boldsymbol{J}_{(1/\rho)}^{\text{turb}}$ that so far remains unknown. In contrast to a single-fluid turbulized continuum, where the compressibility effects are often negligible, the total mass density $\rho(\boldsymbol{r}, t)$ in a multicomponent chemically active turbulent medium generally changes significantly from point to point, for example, due to the formation of new components and local heat release in chemical reactions. As we have already seen, when the compressibility of the mass density is taken into account (in the turbulence model), one more unknown correlation function $\boldsymbol{J}_{(1/\rho)}^{\text{turb}} \equiv \overline{\rho(1/\rho)''\boldsymbol{u}''} = \overline{\boldsymbol{u}''} = -\overline{\rho'\boldsymbol{u}''}/\overline{\rho}$, the turbulent specific volume flux, enters into the heat influx equation for mean motion (3.54) and the turbulent energy balance equation (3.69). Correlation moments of this type (e.g., $\overline{\rho' Z_\alpha''}/\overline{\rho}$, $\overline{\rho' T''}/\overline{\rho}$, etc.) also appear in other transfer equations for the second moments of the local turbulent field characteristics that are invoked below when developing complicated models of multicomponent turbulence in the second approximation [see Chap. 4].

It should be noted that in the case of so-called developed turbulent flows, where the turbulence energy production and dissipation rates are approximately equal, these additional balance equations for the second correlation moments transform from differential ones into a system of algebraic relations between the sought-for second-order correlation moments (like $\overline{\rho \mathbf{A}''\mathbf{B}''}$ and $\overline{\rho \mathbf{A}''\mathbf{B}''}$) and the gradients of the averaged velocities, temperatures, and concentrations for the individual mixture components. It is these relations that can also be used to establish various forms of complicated algebraic dependences for the turbulent exchange coefficients on the gradients of the averaged defining parameters for a medium (in particular, for stratified flows).

To determine the correlations $\overline{\mathbf{A}''} = -\overline{\rho'\mathbf{A}''}/\overline{\rho}$, it is generally necessary to invoke special differential equations for them. These, in turn, contain a number of new correlation terms that are poorly amenable to modeling. Nevertheless, this approach was investigated in the literature (see, e.g., Methods of Turbulent Flow Calculation 1984; Kolesnichenko and Marov 1999). At the same time, a simpler way of determining the correlations $\overline{\rho'\mathbf{A}''}/\overline{\rho}$ is possible, which allows them to be related algebraically to the turbulent diffusion and heat fluxes. It is based on the fact that the relative density fluctuations caused by the pressure fluctuations are often negligible compared to their variations caused by the temperature and concentration fluctuations of the individual components in a multicomponent medium.

To derive such algebraic relations, let us first find the expression for the density fluctuations ρ' in a gas mixture. For this purpose, we rewrite the equations of state for a multicomponent mixture of perfect gases

$$p = \mathbf{R}^*\rho T, \quad \mathbf{R}^* = k_{\mathrm{B}}n/\rho = k_{\mathrm{B}}\sum_{\alpha=1}^{N} Z_\alpha \tag{3.133}$$

as

$$\begin{aligned} p/\rho &= \langle \mathbf{R}^*\rangle\,\langle T\rangle + (\mathbf{R}^*)''\langle T\rangle + \langle \mathbf{R}^*\rangle T'' + (\mathbf{R}^*)''T'' \\ &= \langle \mathbf{R}^*\rangle\langle T\rangle + k_B\langle T\rangle\sum_{\alpha=1}^{N} Z''_\alpha + \langle \mathbf{R}^*\rangle T'' + k_{\mathrm{B}}\sum_{\alpha=1}^{N}(Z''_\alpha T''). \end{aligned} \tag{3.134}$$

Here, we write the true values of the quantities T and $\mathbf{R}^*$ as the sum of averaged and fluctuation values, $T = \langle T\rangle + T''$ and $(\mathbf{R}^* = \langle \mathbf{R}^*\rangle + (\mathbf{R}^*)'')$, and use the easily derivable formulas

$$\langle \mathbf{R}^*\rangle = k_{\mathrm{B}}\sum_{\alpha=1}^{N}\langle Z_\alpha\rangle = k_{\mathrm{B}}\,\overline{n}/\overline{\rho}, \quad (\mathbf{R}^*)'' = k_{\mathrm{B}}\sum_{\alpha=1}^{N} Z''_\alpha. \tag{3.135}$$

Let us now Favre average (3.134). As a result, we have

$$\langle \mathbf{R}^*\rangle\langle T\rangle = \overline{p}/\overline{\rho} - k_{\mathrm{B}}\sum_{\alpha=1}^{N}\langle Z''_\alpha T''\rangle. \tag{3.136}$$

If we eliminate $\langle \mathbf{R}^* \rangle \langle T \rangle$ from (3.136) rewritten as

$$p = \overline{p}\ \rho/\overline{\rho} + \langle \mathbf{R}^* \rangle \rho T'' + k_{\mathrm{B}} \rho \langle T \rangle \sum_{\alpha=1}^{N} Z''_\alpha + k_{\mathrm{B}} \rho \sum_{\alpha=1}^{N} (Z''_\alpha T'') - k_{\mathrm{B}} \rho \sum_{\alpha=1}^{N} \langle Z''_\alpha T'' \rangle$$
$$= \overline{p} + \overline{p} \rho'/\overline{\rho} + \langle \mathbf{R}^* \rangle \rho T'' + k_{\mathrm{B}} \rho \langle T \rangle \sum_{\alpha=1}^{N} Z''_\alpha + k_{\mathrm{B}} \rho \sum_{\alpha=1}^{N} (Z''_\alpha T'')'',$$

using this expression, then we obtain the sought-for exact relation

$$\frac{\rho'}{\overline{\rho}} = \frac{1}{\overline{p}} \left(p' - \langle \mathbf{R}^* \rangle \rho T'' - k_{\mathrm{B}} \langle T \rangle \rho \sum_{\alpha=1}^{N} Z''_\alpha - k_{\mathrm{B}} \rho \sum_{\alpha=1}^{N} (Z''_\alpha T'')'' \right) \tag{3.137}$$

relating the turbulent mass density fluctuations ρ' to the pressure, temperature, and concentration fluctuations of the individual gas mixture components.

When the turbulence models in the second approximation are developed, it is convenient to express the temperature fluctuations T'' in (3.137) in terms of the enthalpy fluctuations H''. This is because the transfer equations of the second correlation moments for a multicomponent mixture containing temperature fluctuations are much more complex than the transfer equations of the moments containing total mixture enthalpy fluctuations (see Chap. 4). In particular, the balance equation for the mean square of the enthalpy fluctuations $\langle H''^2 \rangle$, in contrast to that for the variance $\langle T''^2 \rangle$, does not contain a large number of pair correlations $\langle Z''_\alpha T'' \rangle$ whose presence in the equation for $\langle T''^2 \rangle$ is ultimately determined by the presence of a chemical source of thermal energy in the instantaneous equation for the temperature [cf. (2.24) and (2.29)]. To eliminate the fluctuations T'' from (3.137), we use relation (3.45) written as

$$\langle c_p \rangle T'' = H'' - \sum\nolimits_\alpha \langle h_\alpha \rangle Z''_\alpha. \tag{3.45*}$$

Finally, to calculate the sought-for correlation $\overline{\mathbf{A}''} \equiv -\overline{\rho' \mathbf{A}''}/\overline{\rho}$, we multiply (3.137) by $\mathbf{A}''$ and average the derived expression over an ensemble of possible realizations. As a result, after discarding the third-order correlation moments, we obtain

$$\overline{\mathbf{A}''} = -\frac{\overline{p' \mathbf{A}''}}{\overline{p}} + \frac{\langle \mathbf{R}^* \rangle}{\overline{p} \langle c_p \rangle} \left(\overline{\rho H'' \mathbf{A}''} - \sum_{\alpha=1}^{N} \langle h_\alpha \rangle \overline{\rho Z''_\alpha \mathbf{A}''} \right) + \frac{k_B \langle T \rangle}{\overline{p}} \sum_{\alpha=1}^{N} \overline{\rho Z''_\alpha \mathbf{A}''}. \tag{3.138}$$

Relation (3.138) is used below both in constructing relatively simple models of multicomponent turbulence based on the gradient closure schemes and in constructing more complex turbulence models based on the differential transfer equations for various second correlation moments of fluctuating thermohydrodynamic quantities [see Chap. 4].

Let us now successively identify the parameter **A** in the general formula (3.138) with the hydrodynamic flow velocity, the specific number density Z_α, and the mixture enthalpy H; as a result, we obtain a number of key relations for semiempirical modeling of multicomponent turbulence (when using the weighted Favre averaging):

$$\boldsymbol{J}^{\text{turb}}_{(1/\rho)} \equiv -\frac{\overline{\rho' \boldsymbol{u}''}}{\overline{\rho}} = -\frac{\overline{p' \boldsymbol{u}''}}{\overline{p}} + \frac{1}{\overline{\rho}\langle c_p\rangle\langle T\rangle}\boldsymbol{J}^{\text{turb}}_q + \frac{1}{\overline{n}}\sum_{\alpha=1}^{N}\boldsymbol{J}^T_\alpha, \tag{3.139}$$

$$\overline{H''} \equiv -\frac{\overline{\rho' H''}}{\overline{\rho}} = -\frac{\overline{p' H''}}{\overline{p}} + \frac{1}{\overline{\rho}\langle c_p\rangle\langle T\rangle}\left(\overline{\rho H''^2} - \sum_{\alpha=1}^{N}\langle h_\alpha\rangle\overline{\rho Z''_\alpha H''}\right) + \frac{1}{\overline{n}}\sum_{\alpha=1}^{N}\overline{\rho Z''_\alpha H''}, \tag{3.140}$$

$$\overline{Z''_\beta} \equiv -\frac{\overline{\rho' Z''_\beta}}{\overline{\rho}} = -\frac{\overline{p' Z''_\beta}}{\overline{p}} + \frac{1}{\overline{\rho}\langle c_p\rangle\langle T\rangle}\left(\overline{\rho H'' Z''_\beta} - \sum_{\alpha=1}^{N}\langle h_\alpha\rangle\overline{\rho Z''_\alpha Z''_\beta}\right) + \frac{1}{\overline{n}}\sum_{\alpha=1}^{N}\overline{\rho Z''_\alpha Z''_\beta}, \tag{3.141}$$

It should be kept in mind that in the case of flows with small Mach numbers ***Ma***, the relative turbulent pressure fluctuations in (3.137) may be neglected compared to the relative density and/or temperature fluctuations. This also implies that the first terms in (3.139), (3.140) and (3.141) may be discarded. This principle is known as Morkovin's hypothesis, whose validity for flows without chemical reactions was confirmed in the literature up to $\boldsymbol{Ma} = 5$ (Morkovin 1961). In particular, for forced convection in the terrestrial atmosphere, which is pronounced only in jet flows where the wind velocity gradients reach fairly high values, the ratios of the fluctuations ρ', p', and T'' to the corresponding means $\overline{\rho}$, $\overline{p}$, and $\langle T\rangle$ have the following order of magnitude (see van Mieghem 1973):

$$|\rho'|/\overline{\rho} \approx |T''|/\langle T\rangle \approx 10^{-4}, \quad |p'|/\overline{p} \approx 10^{-5}. \tag{3.142}$$

At the same time, the scale of turbulent pressure fluctuations for free convection in the terrestrial atmosphere is generally much larger than that for forced convection. Therefore, in general, the terms with the pressure fluctuations p' in (3.137) cannot be discarded.

3.2.6 *Rheological Relations for the Turbulent Diffusion and Heat Fluxes in the Case of Strongly Developed Turbulence*

In Sect. 3.2.4, we derived the defining relations (3.127) and (3.128) for the turbulent diffusion, $\boldsymbol{J}^{\text{turb}}_\alpha$, and heat, $\boldsymbol{q}^{\text{turb}}$, fluxes in the case where the molecular and turbulent mass and energy transport processes are equivalent in their contribution to the flow dynamics and energetics. Let us now obtain the analogous relations for strongly developed turbulence when turbulent mixing is much more efficient than molecular one, i.e., when $\boldsymbol{R} \gg \overline{\Pi}, \boldsymbol{q}^{\text{turb}} \gg \overline{\boldsymbol{q}}$, etc. In this case, the turbulization temperature T_{turb}, which is known to be a measure of the intensity of turbulent mixing, is much higher than the averaged mixture temperature $\langle T\rangle$, a characteristic of the mean energy in the chaotic thermal motion of atoms. Given the inequality $T_{\text{turb}} \gg \langle T\rangle$, (3.102) for the production rate of the system's total entropy σ_Σ can be represented as

$$\begin{aligned} 0 \le \sigma_\Sigma \equiv \tilde{\boldsymbol{J}}_q^{\text{turb}} \left\{ \frac{\partial}{\partial \boldsymbol{r}}\left(\frac{1}{\langle T\rangle}\right) + \frac{1}{\overline{\rho}\langle c_p\rangle \langle T\rangle^2}\frac{\partial \overline{p}}{\partial \boldsymbol{r}} \right\} + \boldsymbol{J}_{\langle b\rangle}^{\text{turb}} \cdot \frac{\partial}{\partial \boldsymbol{r}}\left(\frac{1}{T_{\text{turb}}}\right) \\ + \sum_{\alpha=1}^{N} \boldsymbol{J}_\alpha^{\text{turb}} \cdot \left\{ -\frac{\partial}{\partial \boldsymbol{r}}\left(\frac{\langle \mu_\alpha\rangle}{\langle T\rangle}\right) + \langle h_\alpha\rangle \frac{\partial}{\partial \boldsymbol{r}}\left(\frac{1}{\langle T\rangle}\right) + \frac{1}{\langle T\rangle \overline{n}}\frac{\partial \overline{p}}{\partial \boldsymbol{r}} \right\} \\ + \frac{1}{T_{\text{turb}}} \overset{0}{\boldsymbol{R}} : \overset{0}{\boldsymbol{D}} + \sum_{s=1}^{r} \overline{\xi}_s \frac{\langle A_s\rangle}{\langle T\rangle} + \frac{1}{\langle T\rangle}\overline{\rho}\langle \varepsilon_b\rangle. \end{aligned} \tag{3.143}$$

In writing (3.143), we discarded the small terms referring to the averaged molecular motion and used (3.139) for the turbulent flux $\boldsymbol{J}_{(1/\rho)}^{\text{turb}}$ without the small term that allows for the pressure fluctuations (the approximation of free convection).

Thus, in the case of strongly developed turbulence, the entropy production σ_Σ is determined by the set of thermodynamic fluxes $\tilde{\boldsymbol{J}}_q^{\text{turb}}$, $\boldsymbol{J}_{\langle b\rangle}^{\text{turb}}$, $\boldsymbol{J}_\alpha^{\text{turb}}$, $\overset{0}{\boldsymbol{R}}$, $\overline{\xi}$ and the corresponding conjugate thermodynamic forces

$$\boldsymbol{Y}_q^{\text{turb}} \equiv -\frac{1}{\langle T\rangle^2}\left(\frac{\partial \langle T\rangle}{\partial \boldsymbol{r}} - \frac{\mathbf{1}}{\overline{\rho}\langle c_p\rangle}\frac{\partial \overline{p}}{\partial \boldsymbol{r}}\right), \quad (^1) \quad \boldsymbol{Y}_{\langle b\rangle} \equiv -\frac{\mathbf{1}}{T_{\text{turb}}^2}\frac{\partial T_{\text{turb}}}{\partial \boldsymbol{r}}, \quad (^2) \tag{3.144}$$

$$\boldsymbol{Y}_\alpha^{**} \equiv -\frac{\partial}{\partial \boldsymbol{r}}\left(\frac{\langle \mu_\alpha\rangle}{\langle T\rangle}\right) + \langle h_\alpha\rangle \frac{\partial}{\partial \boldsymbol{r}}\left(\frac{1}{\langle T\rangle}\right) + \frac{1}{\overline{n}\langle T\rangle}\frac{\partial \overline{p}}{\partial \boldsymbol{r}} = -\frac{\overline{p}}{\overline{n}_\alpha \langle T\rangle}\frac{\partial}{\partial \boldsymbol{r}}\left(\frac{\overline{n}_\alpha}{\overline{n}}\right) \tag{3.145}$$

$$\boldsymbol{Y}_{\boldsymbol{R}} \equiv \frac{\mathbf{1}}{T_{\text{turb}}}\overset{0}{\boldsymbol{D}}, \quad (^1) \quad Y_{A_s} \equiv \frac{\langle A_s\rangle}{\langle T\rangle} = -\sum_{\beta=1}^{N} \frac{\langle \mu_\beta\rangle}{\langle T\rangle}\nu_{\beta s}, \quad (s = 1, 2, \ldots, r), \quad (^2) \tag{3.146}$$

which correspond to the scalar, vector, and tensor sources of nonequilibrium in the system. Using these thermodynamic fluxes and forces, the production of the total entropy $S_\Sigma = \langle S\rangle + S_{\text{turb}}$ in a turbulized mixture can be written in the following bilinear form:

$$0 \le \sigma_\Sigma \equiv \tilde{\boldsymbol{J}}_q^{\text{turb}} \cdot \boldsymbol{Y}_q^{\text{turb}} + \boldsymbol{J}_{\langle b\rangle}^{\text{turb}} \cdot \boldsymbol{Y}_{\langle b\rangle} + \sum_{\alpha=1}^{N} \boldsymbol{J}_\alpha^{\text{turb}} \cdot \boldsymbol{Y}_\alpha^{**} + \overset{0}{\boldsymbol{R}} : \boldsymbol{Y}_{\boldsymbol{R}} + \sum_{s=1}^{r} \overline{\xi}_s Y_{A_s} + \frac{1}{\langle T\rangle}\overline{\rho}\langle \varepsilon_b\rangle. \tag{3.147}$$

In particular, (3.147) allows us to obtain defining relations corrected for the case of strongly developed turbulence for the vector thermodynamic fluxes that differ insignificantly from relations (3.127) and (3.128). For an isotropic medium, the corresponding relations when the small cross effects are neglected have the form

$$\tilde{\boldsymbol{J}}_q^{\mathrm{turb}} = -\frac{L_{qq}^{\mathrm{turb}}}{\langle T\rangle^2}\left(\frac{\partial\langle T\rangle}{\partial \boldsymbol{r}} - \frac{1}{\overline{\rho}\langle c_p\rangle}\frac{\partial\overline{p}}{\partial \boldsymbol{r}}\right), \tag{3.148}$$

$$\boldsymbol{J}_\alpha^{\mathrm{turb}} = -L_{\alpha\alpha}^{\mathrm{turb}}\frac{\overline{p}\overline{\rho}}{\overline{n}_\alpha\overline{n}\langle T\rangle}\left\{\frac{\partial}{\partial \boldsymbol{r}}\left(\frac{\overline{n}_\alpha}{\overline{\rho}}\right) + \frac{\overline{n}_\alpha}{\overline{\rho}}\frac{\partial \ln \mathbf{M}}{\partial \boldsymbol{r}}\right\}. \tag{3.149}$$

If we now define the turbulent thermal conductivity and diffusion coefficients by the formulas

$$\lambda^{\mathrm{turb}} \equiv \frac{1}{\langle T\rangle^2}L_{qq}^{\mathrm{turb}}, \quad (^1) \quad D_\alpha \equiv \frac{\overline{p}}{\langle T\rangle\overline{n}_\alpha\overline{n}}L_{\alpha\alpha}^{\mathrm{turb}} \quad (^2) \tag{3.150}$$

then the closing relations (3.148) and (3.149) can be written in the following final form:

$$\boldsymbol{q}^{\mathrm{turb}} = \overline{p'\boldsymbol{u}''} - \lambda^{\mathrm{turb}}\left(\frac{\partial\langle T\rangle}{\partial \boldsymbol{r}} - \frac{1}{\overline{\rho}\langle c_p\rangle}\frac{\partial\overline{p}}{\partial \boldsymbol{r}}\right) + \sum_{\alpha=1}^{N}\langle h_\alpha\rangle\boldsymbol{J}_\alpha^{\mathrm{turb}}, \tag{3.151}$$

$$\boldsymbol{J}_\alpha^{\mathrm{turb}} \cong -\overline{\rho}D_\alpha\frac{\partial}{\partial \boldsymbol{r}}\left(\frac{\overline{n}_\alpha}{\overline{\rho}}\right). \tag{3.152}$$

Expression (3.151) generalizes the analogous expression for the heat flux, which is commonly used in meteorology for atmospheric turbulence with a passive admixture and in a number of other important cases, for example, when modeling free turbulent convection (see Monin and Yaglom 1992), to multicomponent mixtures.

Note that, as a rule, the so-called potential temperature

$$\theta \equiv \langle T\rangle\left(\frac{p_0}{\overline{p}}\right)^{\frac{\gamma-1}{\gamma}} = \langle T\rangle\left(\frac{p_0}{\overline{p}}\right)^{\frac{\mathbf{R}^*}{c_p}}. \tag{3.153}$$

(here, p_0 is some standard pressure), which is related to the entropy S of a gas mixture by the relation $S = c_p \ln\theta + \mathrm{const}$, is used in meteorology instead of the "ordinary" temperature $\langle T\rangle$. Hence it is clear that, for example, during adiabatic vertical motions of small elements of a turbulized medium, the potential temperature θ does not change (i.e., it is a strictly conservative flow characteristic), while the ordinary temperature $\langle T \rangle$ changes with altitude. It is easy to see that the following approximate equality holds:

$$\frac{\partial \theta}{\partial \boldsymbol{r}} \cong \left(\frac{\partial \langle T \rangle}{\partial \boldsymbol{r}} - \frac{\mathbf{1}}{\overline{\rho} \langle c_p \rangle} \frac{\partial \overline{p}}{\partial \boldsymbol{r}} \right),$$

using which (3.151) takes a habitual form for meteorological studies.

To conclude this section, let us summarize the basic (reference) differential equations and defining relations that characterize the relatively simple gradient model of multicomponent reacting turbulence. It corresponds to the case where the second-order correlation moments $\langle \mathbf{A}''\mathbf{B}'' \rangle$ for the fluctuating thermohydrodynamic parameters of a mixture $\mathbf{A}$ and $\mathbf{B}$ (different from the hydrodynamic flow velocity $\boldsymbol{u}$) in all of the applied formulas are small compared to the first-order terms $\langle \mathbf{A} \rangle \langle \mathbf{B} \rangle$ and can be discarded. The gradient model of multicomponent turbulence is defined primarily by the hydrodynamic equations for mean motion (3.21), (3.23), (3.28), (3.58), the averaged equation of state for pressure (3.71), relation (3.139) defining the turbulent specific volume flux for a mixture $\boldsymbol{J}^{\text{turb}}_{(1/\rho)}$, and the rheological relations (3.127), (3.128), and (3.117) for the turbulent diffusion, $\boldsymbol{J}^{\Sigma}_{\alpha}$, and heat, $\boldsymbol{q}^{\Sigma}$, fluxes and the Reynolds stress tensor $\boldsymbol{R}$. Data on the kinetics of reactions in a turbulized flow and at its boundaries are also needed to completely describe the flow and the heat exchange between chemically reacting gases. In the simple turbulence model under consideration, the quantities $\overline{\xi_s}$ can be assumed to be calculated only from the averaged temperature and composition of the mixture, $\overline{\xi_s} = \xi_s(\langle T \rangle, \overline{n}_\alpha)$ [cf. (4.1.4)]. In addition, the system of differential equations and final relations should be supplemented by the set of chemical components with allowance made for their gasdynamic, thermophysical, and chemical properties; the universal laws of kinetics and thermodynamics including the equations of state and the expressions for various thermodynamic functions retaining its ordinary form in the approximation under consideration; the formulas for the molecular and turbulent transport coefficients; and the initial and boundary conditions. Although the continuum model of reacting multicomponent turbulence constructed in this way is a "simplified" one, it can, nevertheless, be used to formulate and plausibly solve various astrophysical and geophysical problems.

3.3 Modeling the Turbulent Transport Coefficients

3.3.1 The Turbulence Scale

The gradient model or the so-called algebraic model of turbulence in the first approximation constructed in the preceding section works efficiently for simple quasi-stationary flows with well-chosen turbulent exchange coefficients. However, as has been noted above, this model is completely unsuitable for flows in which the local average conditions of motion change abruptly or when there is a significant flow region affected by the turbulent flow as a whole. This is because when such models are derived, it is commonly assumed that some "local equilibrium" state is established in the structure of developed turbulence in which the turbulence characteristics at each point of the flow are entirely determined only by the local characteristics of the averaged flow field (and the field of volume forces) near the same point, i.e., by the local values of the dissipation rate and the turbulence scale as well as the local averaged state parameters of the medium itself. If, alternatively, the convective and diffusive terms are significant in the balance equation for turbulent energy (i.e., the flow parameters at a point depend on the characteristics of the turbulent flow as a whole), then these local formulas are, in general, invalid (see Ievlev 1975).

Thus, the specific application of gradient models to the calculation of turbulent flows requires a preliminary development of methods for determining the turbulent exchange coefficients. The formulas for calculating the turbulent diffusion, D^{turb}, turbulent thermal conductivity, λ^{turb}, and turbulent viscosity, ν^{turb}, coefficients can be obtained by various methods differing from one another by the degree of complexity. One of the possible methods for a semiempirical determination of these coefficients based on the differential transfer equations for the second correlation moments (simplified to algebraic relations) is analyzed below in Chap. 4. In this section, we briefly discuss the traditional approach to modeling turbulence based on the concept of mixing length and analyze some simple semiempirical methods for modeling the turbulent exchange coefficients.

3.3.2 Gradient Hypothesis

The concept of mixing length introduced by Prandtl (1925) in the theory of turbulence allows one not only to simply express the turbulent exchange coefficients in the defining relations for turbulent fluxes in terms of the mixing length but also to derive these relations themselves for some special cases. We now demonstrate this for the turbulent transport of a conservative (passive) admixture as an example. But first we once again draw the reader's attention to the fact that the defining relations for turbulent fluxes derived thermodynamically in Sect. 3.2 are not related to the key assumption of this approach, namely, the assumption that the flow characteristics transported by turbulent fluctuations are conservative, and hence, have a wider domain of applicability.

Thus, we assume that the transport of a particular scalar flow characteristic by turbulent fluctuations is a diffusion process and that the existence of some mixing length scale can be admitted—the distance traversed by an elementary volume of gas in a turbulized flow before it irreversibly mixes with the environment as a result of velocity fluctuations. Denote the Lagrangian turbulent fluctuation of a field quantity $\mathbf{A}(\boldsymbol{r},t)$ corresponding to the Eulerian fluctuation $\mathbf{A}''$ by $\mathbf{A}''_L$ and the effective mixing length of the attribute $\mathbf{A}$ traversed by the turbulent vortices in the flow before they are destroyed through their interaction with other perturbations by $\xi_{(\mathbf{A})}(\boldsymbol{r},t)$. We then have

$$\mathbf{A}''_L = \mathbf{A}'' + \xi_{(\mathbf{A})} \cdot (\partial/\partial \boldsymbol{r})\langle \mathbf{A} \rangle. \tag{3.154}$$

3.3.2.1 Turbulent Diffusion Flux

Assuming now the component composition of a turbulized mixture to be conservative (i.e., assuming that, being displaced by the distance $\xi_{(\alpha)}$, the vortices retain the same specific number density $(Z_\alpha \equiv n_\alpha/\rho)$ of component α in the Lagrangian volume that they possessed at the initial level), we obtain

$$(Z_\alpha)''_L = 0, \qquad Z''_\alpha = -\xi_{(\alpha)} \cdot (\partial/\partial \boldsymbol{r})\langle Z_\alpha \rangle. \tag{3.155}$$

Hence follows a simple rheological relation for the turbulent flux $\boldsymbol{J}_\alpha^{\text{turb}}$ $(\alpha = 1, 2, \ldots, N)$ [cf. (3.130)]:

$$\boldsymbol{J}_\alpha^{\text{turb}}(\boldsymbol{r},t) \equiv \overline{\rho}\langle \boldsymbol{u}'' Z''_\alpha \rangle = -\overline{\rho}\langle \boldsymbol{u}''\xi_{(\alpha)}\rangle \cdot \frac{\partial}{\partial \boldsymbol{r}}\left(\frac{\overline{n}_\alpha}{\overline{\rho}}\right) = -\overline{\rho}\, D_\alpha^{\text{turb}} \cdot \frac{\partial}{\partial \boldsymbol{r}}\left(\frac{\overline{n}_\alpha}{\overline{\rho}}\right), \tag{3.156}$$

where the formula $D_\alpha^{\text{turb}}(\boldsymbol{r},t) \equiv \langle \boldsymbol{u}''\xi_{(\alpha)}\rangle$ defines the nonsymmetric tensor of turbulent diffusion coefficients that, in the general anisotropic case, allows for the differences in the intensities of the turbulent velocity and composition fluctuations along different coordinate axes. Obviously, the length $\xi_{(\alpha)}$ here is, in a sense, analogous to the mean free path in the kinetic theory of gases. Thus, the turbulent diffusion coefficient D_α^{turb} is expressed very simply in terms of the mixing length $\xi_{(\alpha)}$ (which is a random variable). Relation (3.156) is equivalent to the assertion that the turbulent flux of a substance of type α is proportional to the gradient of the mean concentration $\langle Z_\alpha \rangle \equiv \overline{n}_\alpha/\overline{\rho}$ and has the opposite direction with respect to it. The coefficients $D_\alpha^{\text{turb}}(\boldsymbol{r},t)$ are functions of the coordinates and time and usually exceed their laminar analogs by many times.

Generally, for different mixture components, it is, of course, necessary to introduce different effective mixing lengths $\xi_{(\alpha)}$ for the transport of a substance. In addition, the conservatism (i.e., Lagrangian invariance) hypothesis for the concentrations Z_α of the individual components in a chemically active flow is not justified enough, because the averaged Arrhenius chemical kinetics affects significantly the turbulent mass transport processes at finite chemical reaction rates (see

Ievlev 1975). Nevertheless, since this question has not been studied adequately in the literature, below, for simplicity, we assume that $\xi_{(\alpha)} \equiv \xi$.

Note once again that, in contrast to the molecular diffusion coefficients, the turbulent diffusion coefficients D^{turb} describe not the physical properties of a gas mixture but the properties of a specific turbulent fluid motion and, therefore, directly depend both on the turbulent field intensity and on the method (scale) of averaging the fluctuating characteristics of this motion. For this reason, the way of introducing the averaged turbulence parameters is the fundamental basis on which both the approaches to developing the methods of experimental measurements for the turbulent exchange coefficients themselves and the results of comparing the theoretical and experimental data depend. Of course, this remark also pertains to all of the other turbulent exchange coefficients considered below.

3.3.2.2 Turbulent Heat Flux

Let us now derive the rheological relation for the turbulent heat flux in a multicomponent mixture defined by the formula $\boldsymbol{q}^{\text{turb}}(\boldsymbol{r},t) \equiv \overline{\rho H''\boldsymbol{u}''}$ [see (3.44)] in exactly the same way. Since the mixture entropy S rather than enthalpy H is a conservative flow characteristic, to obtain the needed relation we use (3.83) and (3.85) rewritten as

$$\boldsymbol{q}^{\text{turb}} = \tilde{\boldsymbol{J}}_q^{\text{turb}} + \overline{p'\boldsymbol{u}''} + \sum_{\alpha=1}^{N} \langle h_\alpha \rangle \boldsymbol{J}_\alpha^{\text{turb}}, \tag{3.157}$$

$$\tilde{\boldsymbol{J}}_q^{\text{turb}} = \langle T \rangle \boldsymbol{J}_{\langle S \rangle}^{\text{turb}} - \langle T \rangle \sum_{\alpha=1}^{N} \langle S_\alpha \rangle \boldsymbol{J}_\alpha^{\text{turb}}. \tag{3.158}$$

Using definition (3.10) for the turbulent flux $\boldsymbol{J}_{(\mathbf{A})}^{\text{turb}} \equiv \overline{\rho \mathbf{A}''\boldsymbol{u}''}$ of an attribute $\mathbf{A}$, we rewrite (3.158) as

$$\tilde{\boldsymbol{J}}_q^{\text{turb}} = \langle T \rangle \boldsymbol{J}_{\langle S \rangle}^{\text{turb}} - \langle T \rangle \sum_{\alpha=1}^{N} \langle S_\alpha \rangle \, \boldsymbol{J}_\alpha^{\text{turb}} = \langle T \rangle \overline{\rho \boldsymbol{u}'' \left(S'' - \sum_{\alpha=1}^{N} \langle S_\alpha \rangle Z_\alpha'' \right)}$$

and substitute here the expressions

$$S_L'' = S'' + \xi.(\partial/\partial \boldsymbol{r})\langle S \rangle \cong 0 \quad \text{and} \quad Z_{\alpha L}'' = Z_\alpha'' + \xi(\partial/\partial \boldsymbol{r})\langle Z_\alpha \rangle \cong 0, \tag{3.159}$$

which relate the Lagrangian and Eulerian turbulent fluctuations for the entropy S and concentration Z_α (the characteristics of a multicomponent medium that are assumed to be conservative). As a result, we obtain

$$\tilde{\boldsymbol{J}}_q^{\text{turb}} = \langle T\rangle \overline{\rho \boldsymbol{u}''\left(S_{\text{L}}'' - \sum_{\alpha=1}^{N}\langle S_\alpha\rangle Z_{L\alpha}''\right)} - \langle T\rangle \overline{\rho \boldsymbol{u}''\xi}.\left(\frac{\partial\langle S\rangle}{\partial \boldsymbol{r}} - \sum_{\alpha=1}^{N}\langle S_\alpha\rangle \frac{\partial\langle Z_\alpha\rangle}{\partial \boldsymbol{r}}\right)$$
$$\cong -\overline{\rho}\langle T\rangle \boldsymbol{D}^{\text{turb}}.\left(\frac{\partial\langle S\rangle}{\partial \boldsymbol{r}} - \sum_{\alpha=1}^{N}\langle S_\alpha\rangle \frac{\partial\langle Z_\alpha\rangle}{\partial \boldsymbol{r}}\right) = -\boldsymbol{\lambda}^{\text{turb}}.\left(\frac{\partial\langle T\rangle}{\partial \boldsymbol{r}} - \frac{1}{\overline{\rho}\langle c_p\rangle}\frac{\partial\overline{p}}{\partial \boldsymbol{r}}\right). \tag{3.160}$$

In writing the last equality in (3.160), we used the transformation

$$\langle T\rangle\delta\langle S\rangle - \langle T\rangle\sum_{\alpha=1}^{N}\langle S_\alpha\rangle\delta\langle Z_\alpha\rangle \equiv \langle T\rangle\sum_{\alpha=1}^{N}\langle Z_\alpha\rangle\delta\langle S_\alpha\rangle = \sum_{\alpha=1}^{N}\langle Z_\alpha\rangle(\delta\langle h_\alpha\rangle - \delta\,\overline{p}_\alpha/\overline{n}_\alpha)$$
$$= \langle c_p\rangle\delta\langle T\rangle - \delta\overline{p}/\overline{\rho}, \tag{3.161}$$

which is a corollary of the Gibbs–Duhem identity, and the relation

$$\boldsymbol{\lambda}^{\text{turb}}(\boldsymbol{r},t) = \overline{\rho}\langle c_p\rangle \boldsymbol{D}^{\text{turb}}, \tag{3.162}$$

which defines the tensor of turbulent thermal conductivity coefficients. Thus, for the turbulent heat flux we ultimately have [cf. (3.151)]

$$\boldsymbol{q}^{\text{turb}}(\boldsymbol{r},t) = \overline{p'\boldsymbol{u}''} - \boldsymbol{\lambda}^{\text{turb}}\cdot\left(\frac{\partial\langle T\rangle}{\partial \boldsymbol{r}} - \frac{1}{\overline{\rho}\langle c_p\rangle}\frac{\partial\overline{p}}{\partial \boldsymbol{r}}\right) + \sum_{\alpha=1}^{N}\langle h_\alpha\rangle \boldsymbol{J}_\alpha^{\text{turb}}. \tag{3.163}$$

We make several more remarks regarding the derivation of (3.163). First, it was obtained by assuming the parameters S and Z_α to be conservative characteristics of the medium, i.e., by assuming that the turbulent motion of a Lagrangian vortex particle in the mixture from the level $\boldsymbol{r}(x_j,t)$ where it separated from the overall flow to the level $\boldsymbol{r}(x_j,t)+\xi_j$ occurs not only isentropically but also with an invariable spatial distribution of chemical gas components. However, as has been noted above, these flow characteristics are generally not the Lagrangian invariants ($S_L'' \neq 0$, $Z_{\alpha L}'' \neq 0$) of the turbulent field, because the motion of vortices can be accompanied by various thermal effects (e.g., local heat release through chemical reactions or small-scale turbulent heating through viscous dissipation) and by changes in chemical composition, which leads to the inverse effect of heat transfer on the development of turbulence [see Chap. 7]. Second, when deriving (3.163), we assumed that the so-called turbulent Lewis number is equal to unity, $\boldsymbol{Le}^{\text{turb}} \equiv \chi^{\text{turb}}/D^{\text{turb}} = 1$, where χ^{turb} is the turbulent thermal diffusivity related to the ordinary thermal conductivity by the relation 9b

$$\chi^{\text{turb}} \equiv \lambda^{\text{turb}}/\overline{\rho}\langle c_p\rangle. \tag{3.162*}$$

This ordinary assumption in the theory of turbulence (see, e.g., Lapin and Strelets 1989) is equivalent to the mixing lengths for the mixture substance and entropy being equal, $\xi = \xi_{(S)}$ [see (3.159)]. However, of course, it is generally necessary to distinguish these scales, because the turbulent vortices can be involved in heat transfer more actively than in mass transfer (and vice versa).

For a stratified atmosphere, the rheological relation (3.163) for the vertical component of the turbulent heat flux can be written by taking into account the hydrostatic equation (3.34) as

$$q_z^{\text{turb}} = \overline{p'u_z''} + \sum_{\alpha=1}^{N} \langle h_\alpha \rangle J_{\alpha z}^{\text{turb}} - \lambda^{\text{turb}} \left(\frac{\partial \langle T \rangle}{\partial z} + \frac{g}{\langle c_p \rangle} \right). \tag{3.164}$$

Recall that the quantity $\gamma_a \equiv g/<c_p>$ in meteorological literature is called the dry adiabatic lapse rate ($\gamma_a = 0.98°/100$ m in the terrestrial troposphere) and the expression $\theta \equiv \langle T \rangle + \gamma_a z$ roughly coincides with the potential temperature used in meteorology [see (3.153)].

Thus, according to relation (3.164), there exist two mechanisms of thermal energy transfer through a turbulized multicomponent gas: under the action of an averaged temperature gradient (the potential temperature θ in a stratified medium) and by the turbulent diffusion fluxes J_α^{turb} when each particle of substance α transfers, on average, $\langle h_\alpha \rangle$ of thermal energy. It is important to recall once again that the first term in (3.164) does not act as the energy flux—the quantity $\overline{p'\boldsymbol{u}''}$ drops out of the averaged heat influx equation (3.58) when (3.164) for the heat flux is substituted into it.

3.3.2.3 Reynolds Tensor

The turbulent stresses, along with the molecular ones, are actually the result of momentum transfer, but through the turbulent velocity fluctuations. In the simplest case of a plane shear (along the z axis) flow, the horizontal component of the Reynolds stress (3.185) takes the form

$$R_{xz} \equiv -\overline{\rho\, u_x'' u_z''} = \bar{\rho} \nu^{\text{turb}} \partial \langle u_x \rangle / \partial z, \tag{3.165}$$

where the formula $\nu^T = \langle \xi_z\, u_z'' \rangle$ introduces the vertical turbulent viscosity coefficient defining the ratio of the apparent internal stress to the corresponding averaged strain rate. In writing (3.165), we assumed that $u_x'' = -\xi_z \partial \langle u_x \rangle / \partial z$, i.e., the vortices being displaced vertically by the distance ξ_z retain the momentum at the level $z + \xi_z$ that they had at the initial level z (Prandtl's hypothesis).

However, the Reynolds tensor R_{ij} is generally related to the strain rate tensor by the more complex linear relation (3.117), which, in projections onto the Cartesian coordinate axes, takes the form,

$$R_{xx} \equiv -\overline{\rho(u_x'')^2} = -\frac{2}{3}\overline{\rho}\langle b\rangle + 2\overline{\rho}\nu^{\text{turb}}\left(\frac{\partial\langle u_x\rangle}{\partial x} - \frac{1}{3}\text{div}\langle \boldsymbol{u}\rangle\right),$$

$$R_{yy} \equiv -\overline{\rho(u_y'')^2} = -\frac{2}{3}\overline{\rho}\langle b\rangle + 2\overline{\rho}\nu^{\text{turb}}\left(\frac{\partial\langle u_y\rangle}{\partial y} - \frac{1}{3}\text{div}\langle \boldsymbol{u}\rangle\right),$$

$$R_{zz} \equiv -\overline{\rho(u_z'')^2} = -\frac{2}{3}\overline{\rho}\langle b\rangle + 2\overline{\rho}\nu^{\text{turb}}\left(\frac{\partial\langle u_z\rangle}{\partial z} - \frac{1}{3}\text{div}\langle \boldsymbol{u}\rangle\right),$$

$$R_{xy} = R_{yx} \equiv -\overline{\rho u_x'' u_y''} = -\overline{\rho u_y'' u_x''} = \overline{\rho}\nu^{\text{turb}}\left(\frac{\partial\langle u_x\rangle}{\partial y} + \frac{\partial\langle u_y\rangle}{\partial x}\right),$$

$$R_{yz} = R_{zy} \equiv -\overline{\rho\, u_z'' u_y''} = -\overline{\rho\, u_y'' u_z''} = \overline{\rho}\nu^{\text{turb}}\left(\frac{\partial\langle u_z\rangle}{\partial y} + \frac{\partial\langle u_y\rangle}{\partial z}\right),$$

$$R_{zx} = R_{xz} \equiv -\overline{\rho u_z'' u_x''} = -\overline{\rho u_x'' u_z''} = \overline{\rho}\nu^{\text{turb}}\left(\frac{\partial\langle u_z\rangle}{\partial x} + \frac{\partial\langle u_x\rangle}{\partial z}\right). \tag{3.166}$$

3.3.3 First Approximation Modeling for the Turbulent Transport Coefficients

Obviously, the rheological relations for turbulent fluxes derived above using the concept of mixing length do not solve the turbulence modeling problem either: formulas of the type $D^{\text{turb}} \equiv \langle \boldsymbol{u}''\xi\rangle$ do not allow the turbulent transport coefficients to be determined experimentally, because the local mixing length ξ is a quantity too uncertain to be measured. Thus, the problem of closing the averaged hydrodynamic equations for a mixture is again reduced to the problem of finding approximating algebraic formulas for the turbulent exchange coefficients. This approach is called a semiempirical first-order (approximation) turbulence theory.

Let us initially analyze the simplest method of modeling the turbulent viscosity coefficient ν^{turb}. But first we note that the assumption commonly used to find the turbulent thermal conductivity, λ^{turb}, and turbulent diffusion, D^{tyrb}, coefficients is that the turbulent Prandtl and Schmidt numbers defined by the formulas

$$\boldsymbol{Pr}^{\text{tyrb}} \equiv \overline{\rho}\langle c_p\rangle \nu^{\text{turb}}/\lambda^{\text{turb}}, \quad \boldsymbol{Sc}^{\text{tyrb}} \equiv \nu^{\text{turb}}/D^{\text{turb}} \tag{3.167}$$

are approximately constant in a fluctuating multicomponent flow. This assumption stems from the fact that, in contrast to the turbulent transport coefficients themselves, these ratios change only slightly both within some turbulized flow and when passing from flow to flow. It is also commonly assumed that the turbulent thermal diffusivity $\chi^{\text{turb}} \equiv \lambda^{\text{turb}}/\overline{\rho}\langle c_p\rangle$ and the turbulent diffusion coefficient D^{turb} coincide, i.e., that the Lewis number is $\boldsymbol{Le}^{\text{turb}} \equiv \chi^{\text{turb}}/D^{\text{turb}} = 1$. In this case, the turbulent Prandtl and Schmidt numbers are equal, $\boldsymbol{Pr}^{\text{tyrb}} = \boldsymbol{Sc}^{\text{tyrb}}$. According to the currently available experimental data, $\boldsymbol{Pr}^{\text{tyrb}} = 0.86/0.90$ for flows near a solid wall and $\boldsymbol{Pr}^{\text{tyrb}} = 0.5$ for plane jets and in mixing layers.

3.3.3.1 Prandtl Algebraic Model

As an example, let us consider an averaged flow where the gravitational forces produce a preferential direction in coordinate space. The following principle of local similarity (which is consistent with the principle of local internal equilibrium in the structure of a turbulent flow) is commonly used in the theory of turbulent transport: the turbulent transport coefficients at each point depend only on the properties of the medium at the same point, on the local turbulence scale, and on some characteristics of the averaged flow and volume force fields at this point. Therefore, the expression for the turbulent viscosity coefficient ν^{turb} is then

$$\nu^{\text{turb}} = \nu^{\text{turb}}(\overline{\rho}, \nu, L, f^i, \partial\langle u_x\rangle/\partial z), \tag{3.168}$$

where $L(\boldsymbol{r})$ is the external turbulence scale at a given point of the flow, $f^i(\boldsymbol{r})$ are the local characteristics of the fields of volume forces (in particular, the forces of inertia related to the longitudinal fluid acceleration; then, $f \propto \partial\langle u_x\rangle/\partial x$). When the influence of the first and second derivatives of the averaged velocity on the local properties of the fluid flow is taken into account, this dependence can be different. The scale $L(\boldsymbol{r})$ characterizes the geometric structure of the turbulent field or the characteristic size (and it is then the integral turbulence scale L) of the large vortices involved in turbulent transport that carry a major fraction of the flow kinetic energy. Occasionally, $L(\boldsymbol{r})$ can be treated as the mean mixing length $\Lambda(z) = \sqrt{\overline{\xi^2}}$ (as was originally done by Prandtl (1925, 1942)). In this case, it coincides in order of magnitude with the correlation length of the velocity field. The external turbulence scale $L(\boldsymbol{r})$ should be determined from additional considerations. Because of this uncertainty, it remains possible to take into account the integrated properties of the flow and its prehistory in the purely local formulas for the turbulent exchange coefficients. In particular, for free shear layers, the parameter Λ can be assumed to be equal to a length proportional to the layer thickness over the entire layer. However, the proportionality coefficient depends on the pattern of the free flow. For example, for a flow around an infinite flat wall, it has been established that the mean mixing length Λ is proportional to the distance to the wall: $\Lambda(z) = \kappa z$, where κ is the Karman constant that can be set equal to ~0.4.

Far from a solid wall, turbulence depends weakly on the molecular viscosity of the medium ν and, hence, the quantity ν can be excluded from the set of arguments in (3.168). For the turbulent viscosity coefficient, we can then establish (at $f^i = 0$)

$$\nu^{\text{turb}} = \nu^{\text{turb}}(\overline{\rho}, L, \partial\langle u_x\rangle/\partial z);$$

whence follows the famous Prandtl formula, when using the dimensional theory:

$$\nu^{\text{turb}}(z) = \alpha L^2 |\partial\langle u_x\rangle/\partial z)|,$$

or

$$\nu^{\text{turb}}(z) = \Lambda^2 |\partial\langle u_x\rangle/\partial z)|, \quad (\Lambda(z) = \kappa z). \tag{3.169}$$

The constant factor α is determined for each specific type of motion based on experimental data; in some cases, it is convenient to drop this factor by appropriately redefining the turbulence scale L.

Near a solid wall, where the influence of molecular viscosity ν is significant, the well-known functional dependence

$$\nu^{\text{turb}} = \nu\varphi\left(\frac{\Lambda^2|\partial\langle u_x\rangle/\partial z)|}{\nu}\right) \tag{3.170}$$

also follows from (3.168) (at $f^i = 0$). This expression can be further improved both using theoretical (to be more precise, semiempirical) considerations and purely experimentally (see, e.g., Lapin and Strelets 1989; Monin and Yaglom 1992). If the forces of inertia in the longitudinal direction are included in the set of arguments, then relations (3.169) and (3.170) are modified and take the form

$$\nu^{\text{turb}} = L^2\left|\frac{\partial\langle u_x\rangle}{\partial z}\right|\left(\frac{\partial\langle u_x\rangle/\partial x}{\partial\langle u_x\rangle/\partial z}\right), \tag{3.169*}$$

$$\nu^{\text{turb}} = \nu\varphi\left(\frac{\Lambda^2}{\nu}\left|\frac{\partial\langle u_x\rangle}{\partial z}\right|, \frac{\partial\langle u_x\rangle/\partial x}{\partial\langle u_x\rangle/\partial z}\right). \tag{3.170*}$$

Relation (3.165) by no means always describes correctly the turbulent momentum transport. In particular, there can exist regions in a turbulent flow behind the grid where the averaged flow velocity is constant and the gradient $\partial\langle u_x\rangle/\partial z) = 0$, while the correlator $\langle u_x'' u_z''\rangle \neq 0$, because turbulence is generated immediately behind the grid and is subsequently transported downstream by the averaged flow. However, the mixing length hypothesis (3.169) requires zero ν^{turb} and, according to the Prandtl model (Prandtl 1942), turbulence is absent. This circumstance reveals the main shortcoming of such models: the mixing length hypothesis suggests local equilibrium of the turbulent field. Fortunately, the displacement of the points in space at which $\langle u_x'' u_z''\rangle \neq 0$ and $\partial\langle u_x\rangle/\partial z) = 0$ relative to each other is often small and, therefore, applying (3.165) does not lead to significant errors in numerical flow simulations.

Now, using the rheological relations (3.117), (3.156), and (3.164), we give convenient (for practical purposes) expressions for the Reynolds stresses and the turbulent diffusion and heat fluxes that describe the vertical momentum, mass, and thermal energy transfer for a turbulent motion of a multicomponent mixture:

$$R_{xz} = \overline{\rho}L^2 \left| \frac{\partial\langle u_x \rangle}{\partial z} \right| \frac{\partial\langle u_x \rangle}{\partial z}, \qquad (^1)$$

$$J_{\alpha z}^{\text{turb}} = -\frac{1}{\boldsymbol{Pr}^{\text{turb}}} \overline{\rho}L^2 \left| \frac{\partial\langle u_x \rangle}{\partial z} \right| \frac{\partial}{\partial z}\left(\frac{\overline{n}_\alpha}{\overline{\rho}} \right), \qquad (^2)$$

$$q_z^{\text{turb}} = -\frac{1}{\boldsymbol{Pr}^T} \overline{\rho}\langle c_p \rangle L^2 \left| \frac{\partial\langle u_x \rangle}{\partial z} \right| \left(\frac{\partial\langle T \rangle}{\partial z} + \frac{g}{\langle c_p \rangle} \right) + \sum_{\alpha=1}^{N} \langle h_\alpha \rangle J_{\alpha z}^{\text{turb}}. \qquad (^3) \qquad (3.171)$$

It is important to keep in mind that in the general case of a temperature-inhomogeneous medium, some correction to the linear scale L that allows for the inverse effect of a nonuniform temperature distribution (determining the degree of flow stability) on the turbulent mixing efficiency should be introduced into these relations. This correction is necessary, because the hypothesis about Lagrangian invariance of any transferable substance is invalid for a chemically active gas mixture stratified in a gravitational field. In a homogeneous stratified medium (for example, in a well-mixed lower planetary atmosphere), a correction of this kind can emerge only due to the existing vertical temperature gradients in certain regions of space; as a result, the buoyancy forces facilitating or preventing the generation of turbulence energy appear. To take this fact into account, Richardson proposed a dimensionless criterion—the Richardson gradient number

$$\boldsymbol{Ri} = \frac{g}{\langle T \rangle}\left(\frac{\partial\langle T \rangle}{\partial z} + \frac{g}{\langle c_p \rangle} \right) \Big/ \left(\frac{\partial\langle u_x \rangle}{\partial z} \right)^2. \qquad (3.172)$$

Based on the similarity theory, it is natural to assume that all dimensionless characteristics of a turbulent flow are certain functions of the dimensionless number $\boldsymbol{Ri}$. To take into account the buoyancy forces in relations (3.171), the following corrections to the scale L can be applied:

- in the case of stable stratification ($\boldsymbol{Ri}>0$) hampering the development of turbulence: $L = L^*(1 - \beta_1 \boldsymbol{Ri})$, $5<\beta_1<10$ (typically $\beta_1 \cong 7$) (Monin and Yaglom 1992);
- in the case of unstable stratification ($\boldsymbol{Ri}<0$) increasing the turbulence energy through the instability energy: $L = L^*(1 - \beta_2 \boldsymbol{Ri})^{-1/4}$ ($\beta_2 \cong 14$) (Lumley and Panofsky 1964); the formula $L = L^*(1 - c\boldsymbol{Ri})^{0,25}$, where c is an empirical coefficient (Bradshaw 1969), is also recommended for the same case;
- in the limiting case ($\boldsymbol{Ri} = 0$) where the adiabatic height distribution of the temperature ($\partial\langle T \rangle/\partial z = -g/\langle c_p \rangle \equiv -\gamma_a$) does not affect the development of turbulence: $L = L^*$, the mixing length in the absence of buoyancy forces.

At the same time, the buoyancy forces change the Prandtl–Schmidt number (Munk and Anderson 1948):

$$\boldsymbol{Pr}^{\text{turb}*} = \boldsymbol{Pr}^{\text{turb}}(1 + 3.33\boldsymbol{Ri})^{1.5}/(1 + 10\boldsymbol{Ri})^{0.5}. \qquad (3.173)$$

Apart from the Prandtl first-order turbulence model, other mixing length models based on different concepts of a conserved quantity when passing from one layer of a turbulent flow to another are also known. These modes include, for example, the Taylor (1932) turbulence model, which is based on the assumption that the vorticity rather than the momentum has the property of a conserved transferrable substance in a turbulent flow, and the Karman model, in which the mixing length Λ in (3.169) depends on two derivatives, $\Lambda = \mathrm{const}(\partial\langle u_x\rangle/\partial z)/(\partial^2\langle u_x\rangle/\partial z^2)$. However, the shortcomings of the Taylor model are the same as those of the Prandtl one—the impossibility to describe the turbulent transport processes at the points where $\partial\langle u_x\rangle/\partial z) = 0$, while the Karman formula is difficult to apply, for example, for free turbulent flows where the velocity profile has a point of inflection at which $\partial^2\langle u_x\rangle/\partial z^2 = 0$ and, hence, ν^{turb} becomes infinite, which does not correspond to reality (see Loitsyanskii 1978).

3.3.4 *Differential Kolmogorov–Prandtl Model [b − L Model]*

To overcome the limitation of the mixing length hypothesis noted above, it was necessary to construct turbulence models that would, nevertheless, allow the absence of a real internal equilibrium between the turbulence field and the field of averaged flow parameters to be somehow taken into account. In equilibrium turbulence, where the production of turbulence energy at each point of the flow is compensated for by its dissipation, there is no need to include the turbulence energy $\langle b\rangle \equiv \overline{\rho|\boldsymbol{u}''|^2}/2\overline{\rho}$ in the set of arguments defining the turbulent viscosity coefficient ν^{turb}, because it is defined by the same set of local parameters. To approximately take into account the real "nonequilibrium" of the turbulent field, Kolmogorov (1942) and Prandtl (1945) proposed introducing the parameter $\langle b\rangle$ into the set of arguments defining the coefficient ν^{turb} and using also the turbulence energy balance equation (3.69), along with the hydrodynamic equations for averaged motion, to solve specific problems.

In the field of gravitational forces, i.e., when $\left(\sum_{\alpha=1}^{N}\boldsymbol{J}_{\alpha}^{\mathrm{turb}}\cdot\boldsymbol{F}_{\alpha}^{*}\right) = 0$, (3.69) to determine the quantity $\langle b\rangle$ takes the form

$$\overline{\rho}\frac{\partial\langle b\rangle}{\partial t} + \overline{\rho}\left(\langle\boldsymbol{u}\rangle\cdot\frac{\partial\langle b\rangle}{\partial\boldsymbol{r}}\right) = -\mathbf{div}\boldsymbol{J}_{\langle b\rangle}^{\mathrm{turb}} + \left(\boldsymbol{R}:\frac{\partial\langle\boldsymbol{u}\rangle}{\partial\boldsymbol{r}}\right) - \left(\boldsymbol{J}_{(1/\rho)}^{\mathrm{turb}}\cdot\frac{\partial\overline{p}}{\partial\boldsymbol{r}}\right) - \overline{\rho}\langle\varepsilon_b\rangle, \tag{3.174}$$

where

$$\boldsymbol{J}_{\langle b\rangle}^{\mathrm{turb}} \equiv \overline{\rho(|\boldsymbol{u}''|^2/2 + p'/\rho)\boldsymbol{u}'' - \Pi'\cdot\boldsymbol{u}''} \tag{3.175}$$

is the turbulent energy flux. According to the (second-order) Kolmogorov–Prandtl closure model, the turbulent viscosity coefficient ν^{turb} and the turbulent energy dissipation rate $\langle \varepsilon_e \rangle$ [see (3.50) and (3.88)] are related to the turbulent flow energy $\langle b \rangle$ by the fundamental relations

$$\nu^{\text{turb}} = c_\mu L \sqrt{\langle b \rangle}, \tag{3.176}$$

$$\langle \varepsilon_b \rangle = c_\varepsilon \frac{\langle b \rangle^{\frac{3}{2}}}{L} \tag{3.177}$$

(where c_μ and c_ε are empirical constants). These relations basically follow from the dimensional theory and are a generalization of the well-known Kolmogorov (1941, 1942) hypothesis that the turbulent energy dissipation rate $\langle \varepsilon_b \rangle(\boldsymbol{r}, t)$ at a given point of a developed turbulent flow is determined only by the local values of the turbulent energy $\langle b \rangle(\boldsymbol{r}, t)$ and the turbulence scale $L(\boldsymbol{r}, t)$.

To determine the other indefinite terms in (3.174), we use relations (3.117), (3.118), and (3.139). The diffusion term

$$\boldsymbol{J}^{\text{turb}}_{\langle b \rangle} \equiv \overline{\rho(b + p'/\rho)\boldsymbol{u}'' - \Pi' \cdot \boldsymbol{u}''} = -\frac{\overline{\rho}\, \nu^{\text{turb}}}{\sigma_b} \frac{\partial \langle b \rangle}{\partial \boldsymbol{r}} \tag{3.178}$$

describes the total substantial turbulence energy flux related to various transport mechanisms in space. In particular, the quantity $\boldsymbol{J}^{\text{tyrb}}_{(b)} \equiv \overline{\rho\, b\boldsymbol{u}''}$ is interpreted as the kinetic energy flux of the fluctuation (vortex) motion, so that $\mathbf{div} J^{\text{tyrb}}_{(\text{b})}$ describes the mean rate of decrease in the vortex kinetic energy per unit volume due to "turbulent diffusion." In turn, the quantity $-\operatorname{div}\left(\overline{\Pi' \cdot \boldsymbol{u}''}\right)$ characterizes the mean rate of increase in the vertex kinetic energy due to the work done by the viscous stress tensor fluctuations at the boundary of an elementary volume.

The expression for the rate of turbulence production by a shear flow appearing in (3.174) can be written as

$$\begin{aligned}(\boldsymbol{R} : (\partial/\partial \boldsymbol{r})\langle \boldsymbol{u} \rangle) &= \mu^{\text{turb}} \overset{0}{\boldsymbol{D}} : \overset{0}{D} - p_{\text{turb}} \operatorname{div}\langle \boldsymbol{u} \rangle \\ &= \mu^{\text{turb}} \left(\frac{\partial \langle u_k \rangle}{\partial x_j} + \frac{\partial \langle u_j \rangle}{\partial x_k} - \frac{2}{3} \delta_{jk} \operatorname{div}\langle \boldsymbol{u} \rangle \right)^2 - \frac{2}{3} \overline{\rho} \langle b \rangle \operatorname{div}\langle \mathbf{u} \rangle \\ &\equiv \Phi_\nu - \frac{2}{3} \overline{\rho} \langle b \rangle \operatorname{div}\langle \boldsymbol{u} \rangle,\end{aligned} \tag{3.179}$$

where the positive function

$$\Phi_\nu = \overline{\rho} \nu^{\text{turb}} \overset{0}{\boldsymbol{D}} : \overset{0}{\boldsymbol{D}} = \overline{\rho} \nu^{\text{turb}} \left(\frac{\partial \langle u_k \rangle}{\partial x_j} + \frac{\partial \langle u_j \rangle}{\partial x_k} - \frac{2}{3} \delta_{jk} \operatorname{div}\langle \boldsymbol{u} \rangle \right)^2 \tag{3.180}$$

is the rate at which heat is generated by viscous turbulent friction per unit volume per unit time and, accordingly, is called the dissipative function. According to (3.40), the quantity $(\boldsymbol{R} : (\partial/\partial \boldsymbol{r})\langle \boldsymbol{u}\rangle)$ can be treated as the rate of exchange between the turbulent energy and the kinetic energy of mean fluid motion (in this case, it should be kept in mind that this energy exchange is a purely kinematic process depending significantly on the choice of the averaging operation).

The generation of turbulent energy in the gravity field $\overline{\rho}\,\mathrm{G}$ due to a nonuniform temperature and/or mixture composition distribution present in (3.174) can be represented as

$$\begin{aligned}\overline{\rho}G &\equiv \left(\boldsymbol{J}^{\text{turb}}_{(1/\rho)} \cdot \frac{\partial \overline{p}}{\partial \boldsymbol{r}}\right)\\ &= -\frac{\nu^{\text{turb}}}{\boldsymbol{Pr}^{\text{tyrb}}}\frac{1}{\langle T\rangle}\left(\frac{\partial \langle T\rangle}{\partial \boldsymbol{r}} - \frac{1}{\overline{\rho}\langle c_p\rangle}\frac{\partial \overline{p}}{\partial \boldsymbol{r}}\right)\cdot\frac{\partial \overline{p}}{\partial \boldsymbol{r}} + \frac{\nu^{\text{turb}}}{\boldsymbol{Pr}^{\text{tyrb}}}\frac{\partial \ln \mathbf{M}}{\partial \boldsymbol{r}}\cdot\frac{\partial \overline{p}}{\partial \boldsymbol{r}}\end{aligned} \tag{3.181}$$

using the formulas

$$\boldsymbol{J}^{\text{turb}}_{\alpha}(\boldsymbol{r},t) = -\frac{\overline{\rho}\,\nu^{\text{turb}}}{\boldsymbol{Pr}^{\text{turb}}}\frac{\partial}{\partial \boldsymbol{r}}\left(\frac{\overline{n}_\alpha}{\overline{\rho}}\right), \tag{3.182}$$

$$\begin{aligned}\boldsymbol{J}^{\text{turb}}_{q}(\boldsymbol{r},t) &\equiv \boldsymbol{q}^{\text{turb}}(\boldsymbol{r},t) - \sum_{\alpha=1}^{N}\langle h_\alpha\rangle \boldsymbol{J}^{\text{turb}}_{\alpha}\\ &= -\frac{\overline{\rho}\langle c_p\rangle \nu^{\text{turb}}}{\boldsymbol{Pr}^{\text{tyrb}}}\left(\frac{\partial \langle T\rangle}{\partial \boldsymbol{r}} - \frac{1}{\overline{\rho}\langle c_p\rangle}\frac{\partial \overline{p}}{\partial \boldsymbol{r}}\right).\end{aligned} \tag{3.183}$$

to determine the turbulent flux $\boldsymbol{J}^{\text{turb}}_{(1/\rho)}$ (the term with pressure fluctuations in (3.139) was discarded).

3.3.4.1 Richardson and Kolmogorov Numbers

As can be seen from (3.181), two additional turbulence generation mechanisms are possible in turbulized multicomponent mixture flows. Whereas the first mechanism is thermal in nature, the second turbulence generation mechanism is diffusive in nature and emerges when there is a nonuniform distribution of individual components in the region of space under consideration. This eventually stems from the fact that the spatiotemporal nonuniformity (fluctuations) of the mass density is attributable to two factors: spatial temperature nonuniformity and concentration nonuniformity [see (3.137)]. As is well known, if a local region with a density lower than that of the environment appears in a liquid, then the so-called buoyancy force will act on it in the gravity field. Under certain conditions, the equilibrium stability is lost and this force sets the liquid in convective motion.

The appearance (disappearance) of turbulence of two types, thermal turbulence and "concentration" turbulence produced by spatial composition nonuniformity, is also eventually related to the dual nature of the buoyancy force.

It is generally accepted to allow for the influence of thermal stratification of a medium on the evolution of a turbulent flow using the Richardson number $\boldsymbol{Ri}$. In the model being developed, it can be written in the following general form:

$$\boldsymbol{Ri} = -\frac{\left\{\frac{\partial \ln\langle T\rangle}{\partial \boldsymbol{r}} - \frac{1}{\langle c_p\rangle\langle T\rangle}\left(\frac{1}{\overline{\rho}}\frac{\partial \overline{p}}{\partial \boldsymbol{r}}\right)\right\}\cdot\left(\frac{1}{\overline{\rho}}\frac{\partial \overline{p}}{\partial \boldsymbol{r}}\right)}{\overset{0}{D}:\overset{0}{D}} \cong \frac{\frac{g}{\langle T\rangle}\left(\frac{\partial\langle T\rangle}{\partial z} + \frac{g}{\langle c_p\rangle}\right)}{\overset{0}{\boldsymbol{D}}:\overset{0}{\boldsymbol{D}}} \tag{3.184}$$

(the second form of the number $\boldsymbol{Ri}$ was obtained by taking into account the hydrostatic equation (3.34))

To allow for the influence of a nonuniform chemical composition distribution in a multicomponent mixture on the generation (disappearance) of turbulence under the action of buoyancy forces, below we use the Kolmogorov number

$$\boldsymbol{Ko} = \frac{\left\{\frac{\partial \ln \mathbf{M}}{\partial \boldsymbol{r}}\cdot\left(\frac{1}{\overline{\rho}}\frac{\partial \overline{p}}{\partial \boldsymbol{r}}\right)\right\}}{\overset{0}{\boldsymbol{D}}:\overset{0}{\boldsymbol{D}}} \cong \frac{g\frac{\partial \ln M}{\partial z}}{\overset{0}{\boldsymbol{D}}:\overset{0}{\boldsymbol{D}}}, \qquad \mathbf{M} \equiv \overline{\rho}/\overline{n}, \tag{3.185}$$

which was introduced into the turbulence theory by Barenblatt (1978).

Using these definitions, the mixture turbulent energy transfer equation to determine the parameter $\langle b\rangle$ can be written in the following general form:

$$\begin{aligned}\overline{\rho}\frac{\partial\langle b\rangle}{\partial t} + \overline{\rho}\left(\langle \boldsymbol{u}\rangle\cdot\frac{\partial\langle b\rangle}{\partial \boldsymbol{r}}\right) = \frac{\partial}{\partial \boldsymbol{r}}\left(\frac{\overline{\rho}\nu^{\mathrm{turb}}}{\sigma_b}\cdot\frac{\partial\langle b\rangle}{\partial \boldsymbol{r}}\right) = -\frac{2}{3}\overline{\rho}\langle b\rangle\mathbf{div}\langle \boldsymbol{u}\rangle \\ + \overline{\rho}\nu^{\mathrm{turb}}\overset{0}{\boldsymbol{D}}:\overset{0}{\boldsymbol{D}}\left\{1 - \frac{\boldsymbol{Ri}+\boldsymbol{Ko}}{\boldsymbol{Pr}^{\mathrm{turb}}}\right\} - \overline{\rho}\,c_\varepsilon\frac{\langle b\rangle^{\frac{3}{2}}}{L}.\end{aligned} \tag{3.186}$$

It can be seen from (3.186) that if $\boldsymbol{Ri}+\boldsymbol{Ko}<0$, then the turbulent energy is generated by both wind shear and buoyancy forces. When $\boldsymbol{Ri}+\boldsymbol{Ko}\to\boldsymbol{Pr}^{\mathrm{turb}}$, the corresponding sum of the terms in the turbulent energy balance equation becomes zero, implying that no turbulent motion is maintained. In those practically important cases where one of the above two generation mechanisms of buoyancy forces is inefficient, one can talk about critical Richardson numbers $\boldsymbol{Ri}_c$ (Kolmogorov numbers $\boldsymbol{Ko}_c$). These numbers can be determined from the condition that the turbulent motion exists only at $\boldsymbol{Ri}<\boldsymbol{Ri}_c$ (at constant composition) or only at $\boldsymbol{Ko}<\boldsymbol{Ko}_c$ (at constant temperature). When the two generation mechanisms of buoyancy forces are switched on, they can generally act both in the same and opposite directions. If the thicknesses of the thermal and diffusive mixing layers differ significantly, then this different directionality of the action of the turbulent energy source can lead to a reversal in some (external for the thinner layer) region of the jet flow.

There exists a whole group of semiempirical Kolmogorov–Prandtl turbulence models in which (3.186) is used to determine the turbulent viscosity coefficient in free shear flows (plane or axisymmetric ones) by its numerical solution together with the hydrodynamic equations of mean motion and the differential equation (or algebraic expression) for the external turbulence scale L (see, e.g., Turbulence: Principles and Applications 1980).

3.3.5 Equations for the Turbulence Scale: A Model with Two Transfer Equations

To close (3.186), it is necessary to have some algebraic relations or an additional differential equation to determine the turbulence scale L. Deriving the differential equation for the scale L is one of the most complex problems in the semiempirical theory of developed turbulence. The fact is that the parameter L cannot be defined only via the one-point moments of the fluctuating characteristics for a moving flow. Being a measure of the distance between two points r_1 and r_2 in a turbulized flow at which the two-point correlation moments $\langle \mathbf{A}''(\boldsymbol{r}_1)\mathbf{A}''(\boldsymbol{r}_2)\rangle$ still differ noticeably from zero, the scale L is determined from the complex differential equations for these moments by their integration over the distance between points $\boldsymbol{r}_1$ and $\boldsymbol{r}_2$ (see, e.g., Lumley and Panofsky 1964; Lewellen 1977). The differential equations for L derived in this way, which describe its convection, generation, and dissipation, contain a large number of proportionality coefficients poorly established from experiments, i.e., they are generally much less reliable than, for example, the balance equation for the Reynolds stress tensor where many terms are determined almost exactly [see Chap. 4].

For this reason, to ensure the efficiency of practical calculations, the turbulence scale L is often specified in the form of purely empirically determined functions or is found using an algebraic formula (and, occasionally, a simplified differential equation) that takes into account only the flow geometry (the distance to the wall z, the boundary layer thickness δ, the channel shape, etc.) and that does not depend on the peculiarities of the fluid flow. For example, the empirical formula derived by Nikuradze when investigating a fluid flow in smooth tubes can be used for L/δ. Replacing the tube radius R by δ, we have the following formula to determine L in the boundary layer:

$$L/\delta = \left(0.14 - 0.08\,(1 - z/R)^2 - 0.06(1 - z/R)^4\right), \tag{3.187}$$

where z is the distance along the normal from the wall. In the case of free convection in stratified shear layers, the following simple differential equation (see Laikhtman 1970) can be used to determine the scale L:

$$L = -\kappa c^{\frac{1}{4}}\left(\frac{\langle b\rangle}{\nu_{\mathrm{turb}}}\right) \Big/ \left[\frac{\partial}{\partial z}\left(\frac{\langle b\rangle}{\nu_{\mathrm{turb}}}\right)\right], \tag{3.188}$$

which allows L to be calculated via the mean characteristics of a moving flow. Here, $\kappa = 0,4$ is the Karman constant; c is an empirical constant. In the limiting cases of free convection and strong stability, this formula transforms into the well-known asymptotic relations characteristic of these limiting stratification regimes.

At the same time, the above formulas are distinguished by relatively small universality and, being suitable for one class of flows, should be modified when passing to the description of another type of flows. In addition, formulas of type (3.187) can be used only in the case of "local equilibrium" turbulence, whose characteristics are determined by the local conditions at each point. For "nonequilibrium" turbulence, where the influence of the flow prehistory on the flow characteristics at a point is very important, the quantity L should still be determined using a dynamical equation including all forms of energy transformations in a turbulent flow and with coefficients independent of the flow geometry.

In the case of invariant modeling of the turbulent transport processes, one seeks to use a "universal" evolutionary equation for the integral turbulence scale L that to some extent removes the above shortcomings in its derivation. Lewellen (1977) provided such an equation for the turbulence macroscale L defined by the formula

$$L = \frac{\text{const}}{\langle b \rangle} \iiint\limits_{w} \langle u_k''(r) u_k''(r+l) \rangle \frac{dw}{l^2},$$

which was obtained by integrating the transfer equation for the two-point velocity correlations $\langle u_k''(r) u_k''(r+l) \rangle$ over the volume. This equation is

$$\begin{aligned} \frac{\partial L}{\partial t} + \langle \boldsymbol{u} \rangle \frac{\partial L}{\partial \boldsymbol{r}} &= 0.3 \frac{\partial}{\partial \boldsymbol{r}} \left(L \langle b \rangle^{\frac{1}{2}} \frac{\partial L}{\partial \boldsymbol{r}} \right) - 0.35 \frac{L}{\bar{\rho} \langle b \rangle} \boldsymbol{R} : \frac{\partial \langle \boldsymbol{u} \rangle}{\partial \boldsymbol{r}} \\ &+ 0.6 \frac{\nu L}{\lambda^2} - \frac{0.375}{\langle b \rangle^{\frac{1}{2}}} \left[\frac{\partial}{\partial \boldsymbol{r}} \left(\langle b \rangle^{\frac{1}{2}} L \right) \right]^2 + \frac{0.8 L}{\langle b \rangle} \frac{g}{\langle T \rangle} q_z^{\text{turb}}, \end{aligned} \tag{3.189}$$

where $\lambda = L/(3 + 0.125 \boldsymbol{Re}^{\text{turb}})^{\frac{1}{2}}$ is the so-called Kolmogorov–Taylor microscale. The term including the scale λ to some extent allows for the relationship between the velocity fluctuations at some distance from the wall and the pressure fluctuations at the wall as well as the difference between the vortex sizes in the transverse direction and along the wall. The difficulties of deriving this and similar equations for L stem from the fact that none of the terms in the original equation for the correlations $\langle u_k''(r) u_k''(r+l) \rangle$ can be integrated and, hence, they must all be modeled. On the other hand, the difficult problem of boundary conditions at the free boundary of the turbulent flow region where the scale L does not approach zero arises for the differential equation (3.189).

For this reason, for multicomponent turbulence it is often convenient to invoke some transfer equation for the combination $\langle \mathbf{A}''^2 \rangle^m L^n$ that, when used jointly with the transfer equations for the moment $\langle \mathbf{A}''^2 \rangle$, defines the scale L instead of (3.189) (see the collective monograph Turbulence: Principles and Applications 1980). One of such equations is the transfer equation for the dissipation rate of turbulent energy $\langle \varepsilon_b \rangle$, which in combination with the empirical relation (3.177) allows the system of averaged hydrodynamic equations to be completely closed at the level of second-order moments. The question about the boundary conditions in this case is simplified considerably, because the quantity $\langle \varepsilon_b \rangle$ tends to zero at the outer boundary.

The evolutionary transfer equation for the dissipation rate of turbulent energy $\langle \varepsilon_b \rangle$ in the case of a homogeneous incompressible fluid flow was first obtained by Davydov (1959, 1961). For developed turbulence, this equation is

$$\overline{\rho}\left(\frac{\partial\langle \varepsilon_b \rangle}{\partial t} + \langle \boldsymbol{u} \rangle \frac{\partial\langle \varepsilon_b \rangle}{\partial \boldsymbol{r}}\right) = 0.15\overline{\rho}\frac{\partial}{\partial \boldsymbol{r}} \cdot \left(\frac{\langle b \rangle^2}{\langle \varepsilon_b \rangle}\frac{\partial\langle \varepsilon_b \rangle}{\partial \boldsymbol{r}}\right) + 1.45\frac{\langle \varepsilon_b \rangle}{\langle b \rangle}\left(\boldsymbol{R} : \frac{\partial\langle \boldsymbol{u} \rangle}{\partial \boldsymbol{r}} + 0.48\,\overline{\rho}\,G\right) - 1.92\frac{\overline{\rho}\langle \varepsilon_b \rangle^2}{\langle b \rangle}, \tag{3.190}$$

where the quantities $\boldsymbol{R} : (\partial\langle \boldsymbol{u} \rangle/\partial \boldsymbol{r})$ and $\overline{\rho}G$ are defined by (3.179) and (3.181), respectively. The terms modeled in the simplest way, which remain at large numbers $\boldsymbol{Re}^{\mathrm{turb}}$, appear on the right-hand side of (3.190): the generation of $\langle \varepsilon_b \rangle$ under the action of a velocity gradient, the production of turbulent energy dissipation by buoyancy forces and nongravitational forces, and the molecular destruction of $\langle \varepsilon_b \rangle$. This equation considered simultaneously with the empirical relations

$$L = \langle b \rangle^{\frac{3}{2}}/\langle \varepsilon_b \rangle,\ \nu^{\mathrm{turb}} = 0.09\langle b \rangle/\langle \varepsilon_b \rangle \tag{3.191}$$

following from dimensional considerations allows the turbulent energy transfer equation (3.186) to be completely closed. Thus, the $b - \varepsilon$-model proposed by (Launder 1975) for the calculation of free shear flows is basically a version of the Kolmogorov–Prandtl model. In the calculations of near-wall flows, when the calculation should be performed up to the wall (where the local Reynolds numbers are small), the empirical coefficients in (3.190) obtained at large Reynolds numbers do not work. Lander and Morse (1979) proposed modifications for this case as well.

To conclude this section, we emphasize once again that the constants in the $b - \varepsilon$ model depend both on the flow geometry and on the physical–chemical nature of the medium being modeled. Another shortcoming of such models is the assumption about a gradient nature of the turbulent transport processes. Nevertheless, these models are widely used for practical purposes, for example, in such programs as PHOENICS™, FLUENT™, FIRE™, NUMECA™, STAR-CD™, and KIVA™ developed for numerical simulations of chemically reacting turbulent flows in engines, turbines, burners, and chemical reactors (see, e.g., Rosten and Spalding 1987).

Chapter 4
Differential Models for the Closure of Turbulently Averaged Hydrodynamic Equations for a Chemically Active Continuous Medium

In the preceding chapter, we constructed several different (in complexity and accuracy) semiempirical gradient models of developed turbulence in the first approximation based on the hypothesis about a linear dependence of the indefinite second-order correlation moments (entering into the system of averaged hydrodynamic equations for a multicomponent mixture) on the gradients of the averaged velocities, temperatures and concentrations of variations chemical substances. The linear defining relations for the turbulent diffusion and heat fluxes and the Reynolds stress tensor obtained there allowed this system of equations to be closed. Basically, this enabled its successful application for numerical simulations of comparatively simple and slowly changing turbulent multicomponent reacting gas flows. At the same time, as has been repeatedly noted above, the first-approximation gradient models do not always properly reflect some important features of the actual behavior of turbulent flows with a complex geometry (in particular, spatial flows, separated flows, etc.). Moreover, the structure of the turbulent exchange coefficients and the set of empirical constants (appearing in the closing gradient relations), which are designed, for example, to describe flows in tubes, channels, boundary layers, jets, and wakes, are not universal enough and they should be chosen every time again for turbulent flows of different intensities or with different geometries.

For a single-component fluid, the second-order turbulence models, which, in the opinion of several authors, have an optimal level of complexity (see, e.g., Turbulence: Principles and Applications 1980), have gain the widest acceptance in recent years. In these schemes, all of the second-order correlation moments entering into the averaged hydrodynamic equations are described by a system of differential transfer equations similar to (3.186) for the turbulent energy whose closure is based on the gradient hypothesis. The latter allows the correlation moments of the third or higher orders entering into them to be approximated by relations including only the second- and first order moments. Such differential equations underlie the so-called invariant method of modeling developed turbulence. The term "invariant modeling" is understood here in the sense that, while remaining semiempirical, the constructed turbulence model, nevertheless, does not contain any greatly changing

M.Y. Marov and A.V. Kolesnichenko, *Turbulence and Self-Organization: Modeling Astrophysical Objects*, Astrophysics and Space Science Library 389, DOI 10.1007/978-1-4614-5155-6_4, © Springer Science+Business Media New York 2013

empirical coefficients whose values should be chosen again for each new flow. This approach is widely used, in particular, in numerical simulations of real turbulized fluid flows with a significant influence of the flow prehistory on the turbulence characteristics at a point (see Turbulence: Principles and Applications 1980).

However, this comparatively efficient way of an engineering solution of the developed turbulence modeling problem runs into the fundamental difficulties of constructing a closed system of differential equations. This is because, for example, the model transfer equations for the second correlation moments include the third moments that are also unknown, the equations for the third moments include the fourth moments, etc. Such an infinite chain of transfer equations, the so-called Friedman–Keller chain (see Keller 1930), can be broken only through the establishment of a new closure hypothesis by means of which the hierarchy of equations and unknowns is broken by the introduction of certain relations between higher-order and lower-order moments. In particular, the same gradient relations (but now for high-order moments) containing various empirical constants can act as such relations. It is important to note that, given such a form of approximating relations for high-order statistical correlation moments, the existence of some internal "equilibrium" in the turbulent field structure is assumed explicitly or implicitly (as in the case of first-approximation turbulence models). This naturally leads to a reduction in the generality of such an approach (Ievlev 1975, 1990).

At present, considerable progress in invariant modeling of developed turbulent multicomponent reacting fluid flows is also related to the use of additional differential transfer equations for various one-point correlation moments of different orders. These equations, along with the turbulent energy transfer equations (3.186) and the turbulence macroscale transfer equation (3.189), close the system of turbulently averaged hydrodynamic equations for a mixture (see, e.g., Turbulence: Principles and Applications 1980; Kolesnichenko and Marov 1999). This chapter is just devoted to the construction of such semiempirical second-approximation turbulence models for a reacting mixture of gases. Here, we propose the derivation of differential transfer equations for the various one-point second correlation moments of the turbulent fluctuations in thermohydrodynamic parameters of a medium entering into the averaged equations of multicomponent hydrodynamics with a variable density and variable thermophysical properties. Although these additional equations (in which approximate expressions containing empirical constants are used in approximating the unknown correlations) are semiempirical, it should be recognized that the turbulence models based on them are fairly flexible. In particular, they allow the effect of the mechanisms of convection, diffusion, formation, redistribution, and dissipation of the listed statistical turbulent field characteristics on the spatiotemporal distribution of averaged thermohydrodynamic parameters for a medium to be taken into account. On the other hand, they can also be used to derive more reliable algebraic relations for the turbulent exchange coefficients in shear mixture flows (including the modeling of natural and cosmic media) than those that we considered in Chap. 3 of this book (see also Marov and Kolesnichenko 1987).

Let us make several more introductory remarks concerning the distinctive features of the invariant modeling method as applied to multicomponent turbulence. As has been noted above, one of the peculiar properties of chemically reacting flows that, as a rule, was not considered by classical turbulence models is the presence of density gradients. However, the density, temperature, and concentration gradients resulting, in particular, from local heat and mass release in chemical reactions can change greatly the hydrodynamic velocity field, thereby realizing a feedback between chemical kinetics and hydrodynamics. At the same time, the chemical reactions themselves in a turbulized medium proceed differently than they do in a laminar flow. In other words, the mutual influence of turbulence and Arrhenius nonequilibrium kinetics should be taken into account. If the motion of a multicomponent mixture is turbulized, then, in addition to the velocity fluctuations, the fluctuations in the mass density, temperature, and concentration of individual components emerge. It is actually for this reason that a large number of additional differential transfer equations for the one-point pair correlations including these fluctuations should generally be invoked to close the turbulently averaged hydrodynamic equations for a mixture.

Furthermore, the closure problem for a reacting medium is generally also greatly complicated due to the necessity of averaging the nonlinear "source terms" of substance production in chemical reactions with an exponential behavior. In this chapter, we propose an original procedure for averaging the rates of chemical reactions of any order and outline a scheme for semiempirical modeling of the additional statistical correlations appearing in this case. Note also that, while not being fundamentally new, the approach being developed here, nevertheless, contains a systematic presentation (when using weighted-mean Favre averaging) of the method for invariant modeling of developed turbulence as applied to the motion of multicomponent chemically active natural media. We did not intend to give a comprehensive overview of the literature on the question touched on here. Such overviews were published extensively in world literature and their selection can be found, for example, in the collective monograph of renowned American scientists cited above.

4.1 Nonequilibrium Arrhenius Kinetics in a Turbulized Flow

We begin this chapter by presenting the procedure for averaging the chemical reaction rates $\xi_s(T, Z_\alpha)$ in the averaged diffusion equations (3.24) for a multicomponent turbulized medium. In the simple gradient turbulence models for a reacting mixture described in the preceding section, when using (3.24) we adopted the approximation according to which the function $\overline{\xi}_s \equiv \xi_s(\langle T\rangle, \langle Z_\alpha\rangle)$ where all arguments are averaged ones is meant by the averaged rate $\overline{\xi_s(T, Z_\alpha)}$. This condition is consistent with the main assumption of our approach (3.73), under which we constructed our simple gradient models—the one-point correlations $\langle \mathsf{A}''\mathsf{B}''\rangle$ for any (not equal to the hydrodynamic flow velocity $\boldsymbol{u}$) fluctuating thermodynamic parameters A and B are small compared to the first-order terms $\langle \mathsf{A}\rangle\langle \mathsf{B}\rangle$ and can be omitted. In more accurate models of reacting turbulence, the second correlation moments related to the temperature and composition fluctuations, of course, cannot be discarded and should be taken into account in the structure of the hydrodynamic equations for mean motion. This is because, in general, the averaged chemical reaction rates in a turbulent flow are not determined by Arrhenius kinetics for averaged parameters but depend significantly on their fluctuations (see, e.g., Kuznetsov 1969; Ievlev 1990). The thermal energy release by chemical reactions triggers an expansion of the medium and can induce an additional buoyancy force if the flow is essentially exothermic. It is in this way that chemical kinetics realizes a feedback with hydrodynamics.

Thus, additional difficulties not only in finding an explicit form of the sources (sinks) $\overline{\sigma}_\alpha$ of chemical substances in the averaged continuity equations (3.24) for particles of type α and determining the exact heat release (heat absorption) $\sum_{s=1}^{r}\langle q_s\rangle$ $\overline{\xi}_s$ in the averaged heat influx equation (3.59) but also in modeling a number of stochastic correlation terms like $\langle H''Z''_\alpha\rangle$ and $\langle Z''_\alpha Z''_\beta\rangle$ entering into the corresponding transport equations for the second correlation moments emerge when chemical turbulence is modeled. Eventually, the effects of the interaction of chemical kinetics and various transport processes in a turbulized multicomponent medium are actually related to the presence of such correlation terms.

The procedure for averaging the quantities $\overline{\xi_s(T, Z_\alpha)}$ is nontrivial and requires a special consideration. The averaging problems stem from the fact that the algebraic expressions relating the instantaneous reaction rates to the fluctuating flow characteristics are essentially nonlinear, with the level of complication, obviously, depending on the total order of the chemical reaction and on the existence of a nonlinear temperature dependence of the reaction rate constant.

4.1.1 Elements of Nonequilibrium Arrhenius Kinetics

Consider some of the information from the formal kinetic theory of homogeneous (volume) chemical reactions necessary for the subsequent analysis. The chemical processes in a reacting mixture can be described by reaction equations that are symbolically written as

$$\beta_{1s}[1] + \beta_{2s}[2] + \ldots \underset{\mathsf{K}_{rs}}{\overset{\mathsf{K}_{fs}}{\Leftrightarrow}} \eta_{1s}[1] + \eta_{2s}[2] + \ldots \qquad (s = 1, 2, \ldots, r); \tag{4.1}$$

here, $\beta_{\alpha s}$ and $\eta_{\alpha s}$ are the stoichiometric coefficients (integers) of component α with respect to the sth chemical reaction, $[\alpha]$ are the chemical symbols of the reacting substances, and $\Leftrightarrow$ signifies that the reaction can proceed in both directions. If we denote the reaction rate constants for the sth forward and reverse reactions by K_{fs} and K_{rs}, respectively, then the forward reaction rate can be written as

$$\omega_{\mathrm{fs}} \equiv \mathsf{K}_{\mathrm{fs}} n_1^{\beta_{1s}} n_2^{\beta_{2s}} \ldots, \tag{4.2}$$

where n_α is the current number density of component α. A similar expression can also be written for the reverse reaction. The total production rate of molecules of type α through all chemical reactions is

$$\begin{aligned} \sigma_\alpha &= \sum_{s=1}^{r} (\eta_{\alpha s} - \beta_{\alpha s})\, \xi_s \\ &= \sum_{s=1}^{r} (\eta_{\alpha s} - \beta_{\alpha s}) \Big[\big(\mathsf{K}_{\mathrm{fs}} n_1^{\beta_{1s}} n_2^{\beta_{2s}} \ldots \big) - \big(\mathsf{K}_{\mathrm{rs}} n_1^{\eta_{1s}} n_2^{\eta_{2s}} \ldots \big) \Big]. \end{aligned} \tag{4.3}$$

The resulting sth reaction rate for ideal thermodynamic mixtures (see Sect. 2.3.3) is defined via the forward reaction rate parameters ω_{fs} and the equilibrium constant K_s by the formula (see, e.g., de Groot and Mazur 1962):

$$\xi_s(T, n_\alpha) = \mathsf{K}_{fs} \prod_{\alpha=1}^{N} n_\alpha^{\beta_{\alpha s}} \left\{ 1 - \frac{1}{\mathsf{K}_s} \prod_{\alpha=1}^{N} n_\alpha^{\nu_{\alpha s}} \right\} = \omega_{fs} \left[1 - \exp\left(-\frac{A_s}{k_B T} \right) \right], \tag{4.4}$$

where

$$\mathsf{K}_s \equiv \mathsf{K}_{fs}/\mathsf{K}_{rs}, \qquad \nu_{\alpha s} \equiv \eta_{\alpha s} - \beta_{\alpha s} \tag{4.5}$$

$A_s \equiv -\sum_{\alpha=1}^{N} \nu_{\alpha s}\, \mu_\alpha$ is the chemical affinity of the s th reaction [see (2.41)]. Note that the total order of the forward reaction ω_{fs} is defined as $\Delta\beta_s \equiv \sum_{\alpha=1}^{N} \beta_{\alpha s}$.

In the approximation of Arrhenius kinetics, the rate constant K_{fs} is known to be approximated by the expression (see, e.g., Williams 1965)

$$\mathsf{K}_{fs}(T) = \mathsf{K}_{fs}^0 T^{a_{\mathrm{fs}}} \exp\left(-E_{fs}/k_{\mathrm{B}} T\right), \tag{4.6}$$

where K_{fs}^0 and a_{fs} denote, respectively, the constant of the frequency factor (the so-called pre-exponential factor) and the temperature exponent of the frequency factor for the chemical reaction, while E_{fs} denotes the Arrhenius activation energy.

Analyzing below ideal multicomponent systems, we write the chemical potential (2.66) in the following general form (see Prigogine and Defay 1954)

$$\mu_\alpha = \mu_\alpha^0(T, p) + k_{\mathrm{B}} T \ln\left(\frac{n_\alpha}{n} \right), \quad (\alpha = 1, 2, \ldots, N), \tag{4.7}$$

where

$$\mu_\alpha^0(T,p) = \mathrm{k_B}T \ln p + h_\alpha^0 - Tc_{p\alpha}^0 \ln T - T\int_0^T \frac{\mathrm{d}T}{T}\int_0^T c_{p\alpha}^*(T)\,\mathrm{d}T - \mathrm{k_B}T\gamma_\alpha \tag{4.8}$$

is the chemical potential of a pure component α at given mixture temperature T and pressure p; γ_α is the chemical constant for particles of type α; $c_{p\alpha}^0$, $c_{p\alpha}^*(T)$ are, respectively, the translational and oscillatory components of the partial isobaric heat capacity $c_{p\alpha}$ defined by the relation $c_{p\alpha} \equiv (\partial h_\alpha/\partial T)_{p,\{n_\beta\}}$; and h_α^0 is the partial enthalpy of component α extrapolated to zero temperature. The equilibrium constant K_s and chemical affinity A_s of the sth reaction can then be written as (Prigogine and Defay 1954):

$$A_s \equiv -\sum_{\alpha=1}^{N} \nu_{\alpha s}\mu_\alpha = k_\mathrm{B}T\ln\left[\mathsf{K}_s(T)/\prod_{\alpha=1}^{N} n_\alpha^{\nu_{\alpha s}}\right], \quad (s = 1,2,\ldots,r) \tag{4.9}$$

$$\begin{aligned} \mathsf{K}_s(T) &= n^{\Delta\nu}\exp\left[-\sum_{\alpha=1}^{N}\nu_{\alpha s}\frac{\mu_\alpha^0}{k_\mathrm{B}T}\right] \\ &= \Delta\nu_s\ln(k_\mathrm{B}T) - \frac{q_s^0}{k_\mathrm{B}T} + \frac{1}{k_\mathrm{B}}\sum_{\alpha=1}^{N}\nu_{\alpha s}n_{p\alpha}^{\prime 0}\ln T \\ &\quad + \frac{1}{k_\mathrm{B}}\int_0^T\frac{\mathrm{d}T}{T}\int_0^T\sum_{\alpha=1}^{N}\nu_{\alpha s}n_{p\alpha}^{\prime *}(T)\,\mathrm{d}T + \sum_{\alpha=1}^{N}\nu_{\alpha s}\gamma_\alpha. \end{aligned} \tag{4.10}$$

Here, $\Delta\nu_s = \sum_{\alpha=1}^{N}\nu_{\alpha s}$ is the algebraic sum of the stoichiometric coefficients for the s-th chemical reaction and $q_s^0 = \sum_{\alpha=1}^{N}\nu_{\alpha s}h_\alpha^0$ is the so-called s-th reaction heat extrapolated to zero temperature.

The expression in square brackets on the right-hand side of (4.9) is of the utmost importance in chemical kinetics. Since both ξ_s and A_s become zero under chemical equilibrium, which follows from purely thermodynamic considerations [see (2.40)], the so-called mass action law follows from (4.9) in this case:

$$\mathsf{K}_s(T) = \prod_{\alpha=1}^{N}\left\{(n_\alpha)_\mathrm{eq}\right\}^{\eta_{\alpha s}-\beta_{\alpha s}}, \quad (s = 1,2,\ldots,r). \tag{4.11}$$

It allows the composition of reaction productions under chemical equilibrium to be calculated (here, the subscript (eq) denotes the equilibrium number density).

4.1.2 Averaging the Nonequilibrium Chemical Reaction Rates

Averaging the production rate (4.3) of the α-component through all chemical reaction (4.1) and, subsequently, discarding the terms containing the third and higher-order terms is generally a nontrivial problem. The emerging difficulties are associated, first, with the strong nonlinearity of (4.4) for the resulting sth reaction rate, with the degree of complication, obviously, depending both on the total reaction order $\Delta\beta_s$ and on the existence of a nonlinear temperature dependence of the reaction rate constant K_{fs}, and, second, with the necessity of subsequently finding a large number of second-order correlation terms containing turbulent temperature and composition fluctuations. These correlation terms should generally be determined from the corresponding evolutionary transfer equations containing, in turn, the third and higher-order correlations. Such equations can be closed only under additional assumptions about the relationship between higher-order statistical correlations and lower-order correlation moments.

Before obtaining the averaged chemical reaction rate written in the most general form, consider a typical second-order chemical reaction $[1] + [2] \xrightarrow{\mathsf{K}_f} [3]$ proceeding only in the forward direction as an example showing the characteristic complexity of the problem. The averaged component destruction rate $[1]$ can be written as

$$-\overline{\sigma}_1 = \overline{\mathsf{K}_f n_1 n_2} = \overline{\rho^2 \mathsf{K}_f Z_1 Z_2} = \langle \rho \mathsf{K}_f \rangle \left(\overline{\rho} \langle Z_1 \rangle \langle Z_2 \rangle + \overline{\rho Z_1'' Z_2''} \right) + \langle Z_1 \rangle \overline{\rho^2 \mathsf{K}_f Z_2''} + \langle Z_2 \rangle \overline{\rho^2 \mathsf{K}_f Z_1''} + \overline{\rho (\rho \mathsf{K}_f)'' Z_1'' Z_2''} \,. \qquad (^*)$$

The first term on the right-hand side of this expression contains the product of averaged quantities. It can be obtained by substituting the corresponding mean flow parameters into the expression for the source σ_1. The second term describes the nonuniformity of the reactant concentration distribution at an effective reaction rate constant corresponding to the adopted averaging method. The remaining terms are attributable to the influence of concentration and reaction rate constant fluctuations. Since the chemical reaction rate constant depends on temperature (4.6), the rate constant fluctuations in the case of a non-isothermal process depends on temperature fluctuations in an extremely complex way, which, in turn, introduces considerable complications into formula (*). However, even in the case of an isothermal process in strongly diluted systems, where the density and the reaction rate constant may be considered, expression (*) contains an additional term leading to a change in the chemical reaction rate compared to its value calculated from the mean values (see Turbulent Reacting Flows 1980)

Let us now consider the general case. We assume below that (4.2)–(4.10) refer to the instantaneous state of a chemically turbulized mixture flow. According to the Reynolds assumption, the physical parameters T, p, and n_α can then be represented as the sum of their averaged values $\langle T \rangle, \overline{p}$, and $\overline{n}_\alpha$ and their turbulent fluctuations $\delta T,\ \delta p,\ \delta n_\alpha$ (in this section, for uniformity, we use the same notation for the fluctuation components of parameters irrespective of the method of their averaging, i.e., $\delta T \equiv \rho T''/\overline{\rho}$, $\delta\, n_\alpha \equiv n'_\alpha$). Since $\xi_s = \xi_s(T, n_\alpha)$, restricting ourselves only to the second-order approximation, we can adopt

$$\xi_s(T, n_\alpha) \cong \xi_s(\langle T\rangle, \overline{n}_\alpha) + \delta\xi_s + \frac{1}{2}\delta^2\xi_s, \tag{4.12}$$

where

$$\delta\xi_s = \left(\frac{\partial\xi_s}{\partial T}\right)_0 \delta T + \sum_{\alpha=1}^{N}\left(\frac{\partial\xi_s}{\partial n_\alpha}\right)_0 \delta\, n_\alpha, \tag{4.13}$$

$$\delta^2\xi_s = \left(\frac{\partial^2\xi_s}{\partial T^2}\right)_0 (\delta T)^2 + \sum_{\beta=1}^{N}\left(\frac{\partial^2\xi_s}{\partial T\,\partial n_\beta}\right)_0 \delta T\,\delta\, n_\beta + \sum_{\alpha=1}^{N}\sum_{\beta=1}^{N}\left(\frac{\partial^2\xi_s}{\partial n_\alpha \partial n_\beta}\right)_0 \delta\, n_\alpha \delta n_\beta. \tag{4.14}$$

Here, the index "0" marks the values of the derivatives calculated at $T = \langle T\rangle$ and $n_\alpha = \overline{n}_\alpha$.

By varying the independent variables T and n_β, it is easy to obtain the first variation of the resulting rate $\xi_s = \xi_s(T, n_\alpha)$ from (4.4). We write it as

$$\delta\xi_s = \left[1 - \exp\left(-\frac{A_s}{k_{\mathrm{B}}T}\right)\right]\delta\omega_{\mathrm{fs}} + \omega_{\mathrm{fs}} \exp\left(-\frac{A_s}{k_{\mathrm{B}}T}\right)\,\delta\left(\frac{A_s}{k_{\mathrm{B}}T}\right), \tag{4.15}$$

where

$$\delta\omega_{fs} = \omega_{fs}\left(\frac{E_{fs} + k_{\mathrm{B}}Ta_{fs}}{k_{\mathrm{B}}T^2}\delta\, T + \sum_{\alpha=1}^{N}\frac{\beta_{\alpha s}}{n_\alpha}\right). \tag{4.16}$$

Next, to obtain the variation $\delta(A_s/k_{\mathrm{B}}T)$, we make use of the well-known thermodynamic relations (de Groot and Mazur 1962)

$$\left(\frac{\partial(A_s/T)}{\partial T}\right)_p = \sum_{\alpha=1}^{N}\nu_{\alpha s}\left(\frac{\partial(\mu_\alpha/T)}{\partial T}\right)_p = \frac{q_s(T,p)}{T^2}, \tag{4.17}$$

$$\left(\frac{\partial A_s}{\partial p}\right)_T = -\sum_{\alpha=1}^{N}\nu_{\alpha\, s}\left(\frac{\partial\,\mu_\alpha}{\partial p}\right)_T = V_s(T,p), \tag{4.18}$$

$$\left(\frac{\partial A_s}{\partial\, n_\alpha}\right)_{T,p} = -\sum_{\beta=1}^{N}\nu_{\beta\, s}\left(\frac{\partial\,\mu_\beta}{\partial\, n_\alpha}\right)_{T,p} = -\sum_{\beta=1}^{N}\nu_{\beta\, s}\mu_{\beta\alpha} = \sum_{\beta=1}^{N}\mu_{\beta\alpha}n_\beta\left(\frac{\nu_{\beta\, s}}{n_\beta} - \frac{\nu_{\alpha\, s}}{n_\alpha}\right), \tag{4.19}$$

in which the following designations are introduced:

$$q_s(T,p) = \sum_{\alpha=1}^{N} h_\alpha(\eta_{\alpha s} - \beta_{\alpha s}) = q_s^0 + \sum_{\alpha=1}^{N} \nu_{\alpha s} \int_0^T c_{p\alpha}(T)\, \mathrm{d}T, \tag{4.20}$$

$$V_s(T,p) = \sum_{\alpha=1}^{N} (\eta_{\alpha s} - \beta_{\alpha s}) v_\alpha, \tag{4.21}$$

$$\mu_{\beta\alpha} = \left(\frac{\partial \mu_\beta}{\partial n_\alpha}\right)_{T,p,\,\{n_\beta\}\,(\beta\neq\alpha)}. \tag{4.22}$$

Here, v_α is the partial molar volume of the α-component [see (2.67)]; q_s is the heat of reaction s at constant T and p, which is equal to the difference between the sum of the partial enthalpies of the reaction products multiplied by the corresponding stoichiometric coefficients and the analogous sum for the reactants [see (2.28)]; $V_s(T,p)$ is the so-called change in specific volume during the sth reaction at constant temperature and pressure; $\mu_{\beta\alpha}$ is the partial derivative of the chemical potential μ_β with respect to the number density n_α at constant temperature T and pressure p and all other number densities $n_\beta \neq n_\alpha$, (note that $\sum_{\alpha=1}^{N} n_\alpha \mu_{\beta\alpha} = 0$).

For a mixture of perfect gases, we have [see (2.67)]

$$v_\beta = \frac{1}{n}, \qquad \mu_{\beta\alpha} = k_{\mathrm{B}} T \left(\frac{\delta_{\beta\alpha}}{n_\alpha} - \frac{1}{n}\right) \tag{4.23}$$

where $\delta_{\beta\alpha}$ is the Kronecker symbol. Then,

$$\begin{aligned} \delta\left(\frac{A_s}{k_{\mathrm{B}}T}\right) &= \frac{q_s(T,p)}{k_{\mathrm{B}}T^2}\delta T - \frac{V_s(T,p)}{k_{\mathrm{B}}T}\delta p + \sum_{\alpha=1}^{N}\left(\frac{\Delta \nu_s}{n} - \frac{\nu_{\alpha s}}{n_\alpha}\right)\delta n_\alpha \\ &= \frac{q_s(T,p) - \Delta\nu_s k_{\mathrm{B}} T}{k_{\mathrm{B}}T^2}\delta T - \sum_{\alpha=1}^{N}\frac{\nu_{\alpha s}}{n_\alpha}\delta n_\alpha. \end{aligned} \tag{4.24}$$

Substituting now relations (4.16) and (4.24) into (4.15), we obtain the following expression for the first variation $\delta\xi_s$ of the sth chemical reaction rate:

$$\delta\xi_s = \Lambda_0^s\, \delta T + \sum_{\alpha=1}^{N} \Lambda_\alpha^s\, \delta n_\alpha, \tag{4.25}$$

where

$$\Lambda_0^s \equiv \omega_{fs}\left\{\frac{a_{fs}}{T}+\frac{E_{fs}}{k_B T^2}+\frac{q_s - E_{fs} - (\Delta v_s + a_{fs})k_B T}{k_B T^2}\left(\frac{1}{\mathsf{K}_s}\prod_{\beta=1}^{N}(n_\beta)^{\nu_{\beta s}}\right)\right\}, \tag{4.26}$$

$$\Lambda_\alpha^s \equiv \omega_{fs}\left\{\frac{\beta_{\alpha s}}{n_\alpha}-\frac{\eta_{\alpha s}}{n_\alpha}\left(\frac{1}{\mathsf{K}_s}\prod_{\beta=1}^{N}(n_\beta)^{\nu_{\beta s}}\right)\right\}. \tag{4.27}$$

Let us now obtain the second variation $\delta^2\xi_s$ of the resulting *s*th chemical reaction rate $\xi_s \equiv \omega_{fs} - \omega_{rs}$ (here, $\omega_{rs} = \omega_{fs}\exp(-A_s/k_B T)$ is the reverse reaction rate). For this purpose, we first represent it as

$$\begin{aligned}\delta^2\xi_s &= \delta(\delta\omega_{fs} - \delta\omega_{rs}) = \delta[\omega_{fs}\delta(\ln\omega_{fs}) - \omega_{rs}\delta(\ln\omega_{rs})]\\ &= \omega_{fs}\left[(\delta\ln\omega_{fs})^2 + \delta^2(\ln\omega_{fs})\right] - \omega_{rs}\left[(\delta\ln\omega_{rs})^2 + \delta^2(\ln\omega_{rs})\right].\end{aligned} \tag{4.28}$$

Using (4.16) and the easily deducible (given (4.16) and (4.24)) relations

$$\delta(\ln\omega_{rs}) = \frac{E_{fs} + a_{fs}k_B T - q_s + \Delta v_s k_B T}{k_B T^2}\delta T + \sum_{\alpha=1}^{N}\frac{\eta_{\alpha s}}{n_\alpha}\delta n_\alpha, \tag{4.29}$$

$$\delta^2(\ln\omega_{fs}) = -\frac{2E_{fs} + a_{fs}k_B T}{k_B T^3}(\delta T)^2 - \sum_{\alpha=1}^{N}\frac{\beta_{\alpha s}}{n_\alpha^2}(\delta n_\alpha)^2, \tag{4.30}$$

$$\delta^2\left(\frac{A_s}{k_B T}\right) = \frac{\Delta v_s k_B T - q_s - q_s^0}{k_B T^3}(\delta T)^2 + \sum_{\alpha=1}^{N}\frac{\nu_{\alpha s}}{n_\alpha^2}(\delta n_\alpha)^2, \tag{4.31}$$

$$\delta^2(\ln\omega_{rs}) = \delta^2(\ln\omega_{fs}) - \delta^2\left(\frac{A_s}{k_B T}\right), \tag{4.32}$$

we then ultimately obtain (Kolesnichenko and Marov 1984)

$$\frac{1}{2}\delta^2\xi_s = \mathsf{B}_0^s(\delta T)^2 + \sum_{\alpha=1}^{N}\mathsf{B}_\alpha^s(\delta T\,\delta n_\alpha) + \sum_{\alpha=1}^{N}\sum_{\gamma=1}^{N}\mathsf{B}_{\alpha\gamma}^s(\delta n_\alpha\delta n_\gamma), \tag{4.33}$$

where

$$\mathsf{B}_\alpha^s \equiv \omega_{fs}\frac{E_{fs}^*}{k_B T^2 n_\alpha}\left\{\beta_{\alpha s} - \eta_{\alpha s}\left(1 - \frac{q_s^*}{E_{fs}^*}\right)\left(\frac{1}{\mathsf{K}_s}\prod_{\beta=1}^{N}(n_\beta)^{\nu_{\beta s}}\right)\right\}, \tag{4.34}$$

$$\mathsf{B}^s_{\alpha\gamma} \equiv \frac{\omega_{\mathrm{fs}}}{2n_\alpha n_\gamma}\left\{\beta^2_{\alpha s} - \beta_{\gamma s} - (\eta^2_{\alpha s} - \eta_{\gamma s})\left(\frac{1}{\mathsf{K}_s}\prod_{\beta=1}^{N}(n_\beta)^{\nu_{\beta s}}\right)\right\}, \tag{4.35}$$

$$\begin{aligned}\mathsf{B}^s_0 \equiv \omega_{\mathrm{fs}}\Bigg\{&\left(1 - \frac{1}{\mathsf{K}_s}\prod_{\beta=1}^{N}(n_\beta)^{\nu_{\beta s}}\right)\frac{E^{*2}_{\mathrm{fs}} - k_{\mathrm{B}}T(2E^*_{\mathrm{fs}} - a_{\mathrm{fs}}k_{\mathrm{B}}T)}{2k^2_{\mathrm{B}}T^4}\\ &+\frac{q^*_s(q^*_s - 2E^*_{\mathrm{fs}} + 2k_{\mathrm{B}}T)}{2k^2_{\mathrm{B}}T^4}\left(\frac{1}{\mathsf{K}_s}\prod_{\beta=1}^{N}(n_\beta)^{\nu_{\beta s}}\right)\Bigg\}\end{aligned} \tag{4.36}$$

Here, the following notation is used:

$$E^*_{fs} \equiv E_{\mathrm{fs}} + a_{\mathrm{fs}}k_{\mathrm{B}}, \quad q^*_s \equiv q_s(T,p) - \Delta\nu_s k_{\mathrm{B}}T. \tag{4.37}$$

All of the physical parameters p, T, n_α in (4.26), (4.27), (4.34), (4.35), and (4.36) are averaged in the above way, but the averaging symbol is omitted to simplify the formulas.

Averaging now the approximate equality (4.12) over an ensemble of possible realizations and using relations (4.25) and (4.33), we ultimately obtain the expression for the averaged chemical reaction rate in a turbulized flow

$$\overline{\xi_s(T,n_\alpha)} = \xi_s(\langle T\rangle, \overline{n}_\alpha) + \mathsf{B}^s_0\overline{(\delta T)^2} + \sum_{\alpha=1}^{N}\mathsf{B}^s_\alpha\overline{\delta T\delta n_\alpha} + \sum_{\alpha=1}^{N}\sum_{\beta=1}^{N}\mathsf{B}^s_{\alpha\beta}\overline{\delta n_\alpha\delta n_\beta}, \tag{4.38}$$

where

$$\overline{(\delta T)^2} \equiv \overline{\rho^2 T''^2}/\overline{\rho}^2 \cong \overline{\rho T''^2}/\overline{\rho} = \langle T''^2\rangle, \tag{4.39}$$

$$\overline{\delta T\delta n_\alpha} \equiv \overline{\rho T'' n'_\alpha}/\overline{\rho} = \overline{\rho^2 T'' Z_\alpha}/\overline{\rho} \cong \overline{\rho T'' Z_\alpha} = \overline{\rho}\langle T'' Z''_\alpha\rangle, \tag{4.40}$$

$$\overline{\delta n_\alpha\delta n_\beta} \equiv \overline{n'_\alpha n'_\beta} \cong \overline{\rho}^2\left(\langle Z''_\alpha Z''_\beta\rangle - \langle Z_\alpha\rangle\overline{Z''_\beta} - \langle Z_\beta\rangle\overline{Z''_\alpha}\right). \tag{4.41}$$

Relation (4.41) follows from the obvious transformation

$$n'_\alpha = n_\alpha - \overline{n}_\alpha = \rho Z_\alpha - \overline{\rho}\langle Z_\alpha\rangle = \rho Z''_\alpha + \rho'\langle Z_\alpha\rangle. \tag{4.42}$$

It follows from (4.38) that when the mutual influence of averaged hydrodynamics and chemical kinetics on the pattern of chemical reactions and the heat and mass transfer processes in a turbulized flow is taken into account, it is necessary to invoke a fairly large number of additional transfer equations for the pair correlation functions of temperature and composition fluctuations in the general case of a multicomponent mixture.

Regarding (4.38) widely used below, we note the following. In view of the exponential form of (4.4) and the slow convergence of the corresponding power series in expansion (4.12), more detailed information about the structure of the turbulent field in a multicomponent reacting mixture flow is generally needed. In other words, the knowledge of the second correlation moments for the fluctuating temperature and composition alone may turn out to be quite insufficient for a satisfactory approximation of the averaged source term $\overline{\sigma}_\alpha$. Naturally, the most complete information about chemical turbulence could be obtained by introducing the joint probability density function for the hydrodynamic velocity $\boldsymbol{u}$ and basic thermodynamic parameters of a chemically active mixture flow at all points of coordinate space. However, this approach is still too complex for practical applications and, therefore, can play only a limited role in reacting turbulence studies.

In recent years, the methods based on a one-point probability density function for the velocity, concentration, and other quantities in turbulent flows, $P(\rho,\ \boldsymbol{u},\ T,\ Z_\alpha;\ \boldsymbol{r}, t)$, (see, e.g., Borgi 1980; O'Brien 1980), have gain wide acceptance (in particular, in connection with turbulent burning studies). However, it is not yet possible to fully assess the prospects for this considerably simplified approach either, because there is no complete theory that would allow the function P to be directly found from its defining integro-differential equation and because the number of numerically solved test turbulent burning problems is still small. Yet, the approximation of the averaged chemical reaction rates $\overline{\xi_s}$ at the level of second-order moments that we proposed, to some extent, allows one to take into account the influence of chemical kinetics on the formation and evolution of reacting turbulence. There is no question that even such a limited approach to solving the complex problem of the interaction between turbulence and chemical kinetics is still considerably better than ignoring it completely.

4.1.3 *Formula for the Correlation Moments Including the Fluctuations in Substance Source Through Chemical Reactions*

For the subsequent analysis, we need an expression that allows the quantities $\overline{\mathcal{A}''\sigma_\alpha}$ containing the fluctuations σ'_α in the source of a substance of type α through chemical reactions to be written via the correlation moments $\langle H''Z''_\alpha\rangle$ and $\langle Z''_\alpha Z''_\beta\rangle$. Using the properties of the weighted-mean averaging (3.7) and (4.3) for the source σ_α, we initially represent the quantity $\overline{\mathcal{A}''\sigma_\alpha}$ as

$$\overline{\mathcal{A}''\sigma_\alpha} = \overline{\sigma}_\alpha\overline{\mathcal{A}''} + \overline{\sigma'_\alpha\mathcal{A}''} = \overline{\mathcal{A}''}\left(\sum_{s=1}^{r}\nu_{\alpha s}\xi_s(\langle T\rangle, \overline{n}_\alpha)\right) + \sum_{s=1}^{r}\nu_{\alpha s}\overline{\xi'_s\mathcal{A}''}, \tag{4.43}$$

where the velocity fluctuation ξ'_s, according to (4.25), is defined by the expression

$$\xi'_s = \Lambda^s_0 (\rho T''/\overline{\rho}) + \sum_{\beta=1}^{N} \Lambda^s_\beta n'_\beta. \tag{4.44}$$

Since the transfer equations for the second-order correlation moments used below (in constructing the phenomenological model of reacting turbulence) contain correlations of the form $\langle H''Z''_\alpha\rangle$ and $\langle Z''_\alpha Z''_\beta\rangle$, it is convenient to express the fluctuations ξ'_s of the chemical reaction rate precisely in terms of them. This can be done using (3.46) rewritten as

$$\langle c_p\rangle T'' = H'' - \sum\nolimits_\alpha \langle h_\alpha\rangle Z''_\alpha.$$

As a result, we obtain

$$\begin{aligned}\overline{\xi'_s \mathcal{A}''} &= \Lambda^s_0 \langle T''\mathcal{A}''\rangle + \sum_{\beta=1}^{N} \Lambda^s_\beta \overline{n'_\beta \mathcal{A}''}\\ &\cong \frac{\Lambda^s_0}{\langle c_p\rangle}\langle H''\mathcal{A}''\rangle + \sum_{\beta=1}^{N}\left(\overline{\rho}\Lambda^s_\beta - \frac{\langle h_\beta\rangle}{\langle c_p\rangle}\Lambda^s_0\right)\langle Z''_\beta \mathcal{A}''\rangle - \overline{\mathcal{A}''}\sum_{\beta=1}^{N}\Lambda^s_\beta \overline{n}_\beta.\end{aligned} \tag{4.45}$$

Hence, in view of (3.138) written as

$$\overline{\mathcal{A}''} \cong \frac{1}{\langle T\rangle\langle c_p\rangle}\langle H''\mathcal{A}''\rangle + \frac{1}{\langle c_p\rangle}\sum_{\alpha=1}^{N}\left(\frac{\overline{\rho}\langle c_p\rangle}{\overline{n}} - c_{p\alpha}\right)\langle Z''_\alpha \mathcal{A}''\rangle.$$

we ultimately derive the sought-for expression for $\overline{\mathcal{A}''\sigma_\alpha}$:

$$\begin{aligned}\overline{\mathcal{A}''\sigma_\alpha} &= \overline{\mathcal{A}''}\left(\sum_{s=1}^{r}\nu_{\alpha s}\xi_s(\langle T\rangle, \overline{n}_\alpha)\right) + \sum_{s=1}^{r}\nu_{\alpha s}\overline{\xi'_s \mathcal{A}''}\\ &= \overline{\rho}L^{\mathrm{Ch}}_{0\alpha}\langle H''\mathcal{A}''\rangle + \overline{\rho}\sum_{\beta=1}^{N}L^{\mathrm{Ch}}_{\alpha\beta}\langle Z''_\alpha \mathcal{A}''\rangle.\end{aligned} \tag{4.46}$$

Here, the following notation is introduced:

$$\overline{\rho}L^{\mathrm{Ch}}_{0\alpha} \equiv \frac{1}{\langle T\rangle\langle c_p\rangle}\sum_{s=1}^{r}\nu_{\alpha s}\left(\xi_s(\langle T\rangle, \overline{n}_\alpha) + \Lambda^s_0\langle T\rangle - \sum_{\beta=1}^{N}\Lambda^s_\beta \overline{n}_\beta\right), \tag{4.47}$$

$$\overline{\rho}L^{\mathrm{Ch}}_{\alpha\beta} \equiv \overline{\rho}\sum_{s=1}^{r}\nu_{\alpha s}\left(\Lambda^s_\beta - \frac{\langle T\rangle}{\overline{n}}\Lambda^s_0\right) + \langle T\rangle\left(\frac{\overline{\rho}\langle c_p\rangle}{\overline{n}} - c_{p\beta}\right)L^{\mathrm{Ch}}_{0\alpha}. \tag{4.48}$$

Below, we consider the cases of various physical quantities $\mathcal{A}$ fluctuating in a turbulized flow.

In conclusion, note the following. The often necessary preliminary analysis of the influence of gaseous mixture flow turbulization on the chemical reaction rates (and vice versa) is possible when the following characteristic times are intercompared: the characteristic time of hydrodynamic processes, t_{hydro}; the turbulence time scale, t_{turb} (e.g., the dissipative time scale $t_{\text{turb}} \sim \langle b \rangle / \langle \varepsilon_e \rangle$); and the time scale of chemical reactions, t_{chem}. These characteristic times can be in various relationships between one another (usually, $t_{\text{turb}} \ll t_{\text{hydro}}$; in particular, for flows in a boundary layer, $t_{\text{turb}}/t_{\text{hydro}} \sim 10^{-2}$).

In geophysical applications, when analyzing the relationship between turbulence and chemical kinetics, the analysis can be restricted to the following possible cases:

- $t_{\text{chem}} > t_{\text{hydro}}$ (in this case, $t_{\text{chem}} \gg t_{\text{turb}}$)—the chemical reactions are nonequilibrium ones with respect to the averaged fluid motion and are "frozen" with respect to the pulsating motion; in this case, the chemical transformation processes do not affect the gasdynamic flow characteristics, but the mean gas composition should be determined from the averaged continuity equations for the individual chemical components in which the source terms are calculated by taking into account the temperature and composition fluctuations (see Ievlev 1990);
- $t_{\text{chem}} < t_{\text{hydro}}$—the chemical reactions are independent with respect to the averaged motion, but the mutual influence of chemical kinetics processes and heat and mass transfer processes in a turbulized flow is significant;
- $t_{\text{chem}} \ll t_{\text{hydro}}$—the averaged flow may be considered a chemically equilibrium one, but the chemical reactions themselves can be nonequilibrium ones with respect to the turbulent pulsations; in this case, the mixture composition is determined from the mass action law (4.11) but will allowance made for the possible influence of temperature and composition fluctuations on the flow characteristics, while the chemical transformation processes affect the turbulent exchange coefficients (Ievlev 1990; Marov and Kolesnichenko 1987).

4.2 Model Transfer Equations for the Second Moments for a Multicomponent Gas Mixture

In this section, the closure problem for the system of averaged multicomponent hydrodynamic equations for a reacting mixture of gases with a variable mass density $\rho(\boldsymbol{r}, t)$ is solved at the level of second-order correlation moments, when model differential transfer equations for their determination are additionally invoked. Here, based on the general balance equation for the second correlation moments $\langle \mathcal{A}'' \mathcal{B}'' \rangle$ of the thermohydrodynamic mixture parameters fluctuating in a turbulized flow, we derive the following model transfer equations in a uniform way: the transfer equations for the components of the Reynolds turbulent stress tensor R_{ij}, the transfer equation for the turbulent energy $\langle b \rangle$, the transfer equation for the turbulent heat flux vector $\boldsymbol{q}^{\text{turb}}$, the transfer equation for the mean-square total mixture enthalpy fluctuation (variance) $\langle H''^2 \rangle$, the transfer equations for the turbulent diffusion vectors $\boldsymbol{J}_\alpha^{\text{turb}}$ $(\alpha = 1, 2, \ldots, N)$, the transfer equations for the second mixed moments of the mixture enthalpy and composition fluctuations $\langle H'' Z_\alpha'' \rangle$ $(\alpha = 1, 2, \ldots, N)$ and, finally, the transfer equations for the second mixed moments of the concentration fluctuations for various pairs of substances in the mixture $\langle Z_\alpha'' Z_\beta'' \rangle$ $(\alpha, \beta = 1, 2, \ldots, N)$. We analyze the physical meaning of all statistical high-order correlations in these equations and propose their simplest approximations. Substantial progress achieved in recent years in the development and application of second-order turbulence models for a single-component fluid with a constant density (see, e.g., Donaldson 1972; Deardorff 1973; Andre et al. 1976; Turbulence: Principles and Applications 1980) gives hope for the efficiency of the generalization of some of them to a compressible multicomponent fluid flow considered here. Bearing in mind that the quality of any constructed multicomponent turbulence model is ultimately determined by the consistency of the model results with the experimental data, below we use the simplest approximation expressions for the unknown moments in the above transfer equations, being guided by the analogy with present-day semiempirical models of developed turbulence for a single-component medium with a constant density.

4.2.1 General Form of the Transfer Equation for the One-Point Second Moments for a Turbulized Mixture

The statistical approach to describing the turbulent transport processes in a single-component fluid is known to be based on the analysis of a chain of engaging prognostic equations for the correlation moments of coupling of an increasing order (see Keller and Friedman 1924). Let us briefly consider a general scheme for setting up these equations using an incompressible fluid with a constant density as an example. The kth-order multi-point and multi-time correlator of the velocity fluctuations is determined by statistically averaging the product of k fluctuation velocities:

$$\mathcal{B}_{\mathrm{A}_1 \ldots \mathrm{A}_k}(\boldsymbol{r}_1, t_1; \ldots; \boldsymbol{r}_k, t_k) = \overline{u'_{\mathrm{A}_1}(\boldsymbol{r}_1, t_1) \ldots u'_{\mathrm{A}_k}(\boldsymbol{r}_k, t_k)},$$

where $\boldsymbol{r}_k$ is the radius vector of a point in space, t_k is the instant of time, and A_k is the index running over x, y, z. The complete statistical description of a random velocity field is equivalent to determining all arbitrary-order correlators of the velocity fluctuations, which is virtually unfeasible. Therefore, only the lowest-order correlators, for example, the pair correlator $\mathcal{B}_{ij}(\boldsymbol{r}, \boldsymbol{r}_1, t) = \overline{u_i'(\boldsymbol{r}, t)\, u_j'(\boldsymbol{r}_1, t)}$, i.e., the multi-point single-time second-order moment of the velocity fluctuations, are actually used. When studying shear turbulence in an incompressible single-component fluid, it is often convenient to use only the one-point pair correlators of the velocity fluctuations $\mathsf{B}_{ij}(\boldsymbol{r},t) = \overline{u_i'(\boldsymbol{r},t)\, u_j'(\boldsymbol{r},t)}$, i.e., the Reynolds tensor.

Since the averaging and differentiation operations are commutative [the Reynolds postulate, see (3.2)], the time derivative of the correlator $\mathcal{B}_{ij}(\boldsymbol{r},t)$ can be represented as

$$\frac{\partial}{\partial t}\mathsf{B}_{ij}(\boldsymbol{r},t) = \overline{u_i'(\boldsymbol{r},t)\frac{\partial u_j'(\boldsymbol{r},t)}{\partial t}} + \overline{u_j'(\boldsymbol{r},t)\frac{\partial u_i'(\boldsymbol{r},t)}{\partial t}}.$$

To derive the evolutionary equations for the second moments $\mathsf{B}_{ij}(\boldsymbol{r},t)$, it is necessary to eliminate the time derivatives on the right-hand side of the latter equality using the appropriate hydrodynamic equations for the velocity fluctuations. The third-order correlation functions for the velocity fluctuations will then appear in the derived equations. More complex evolutionary equations, for example, for the third-order correlators in which the fourth-order correlation functions will appear, etc., can also be derived in a similar way. Breaking this chain at any step leads to an unclosed system of equations, which is the main problem of the Keller–Friedman method.

Additional difficulties arise when this approach is generalized to multicomponent chemically reacting compressible media. Note in this connection that even for a single-component fluid the system of equations for the coupling moments is so complicated that the equations themselves were not written out in the classic work by Keller and Friedman cited above, but only the basic idea of their derivation was pointed out and it was specified how many and what equations are obtained in this case. For this reason, the actual derivation of the chain of equations describing the dynamics of the correlation moments of an increasing order through weighted Favre averaging for a turbulized flow of a multicomponent mixture with chemical reactions is of fundamental importance both for constructing semiempirical models of chemical turbulence with an enhanced level of complexity and for a deeper understanding of the interaction between kinetics and turbulent transport in such media (see Marov and Kolesnichenko 1987).

Let us now derive the transfer equation for the second moments for a turbulized medium in general form. Let $\mathsf{A}\,(\boldsymbol{r},t)$ be the specific value of any scalar quantity (in particular, this can be tensor components) whose substantial balance equation appears as (2.1). In view of (3.14), we can then write

$$-\left\langle\frac{\partial}{\partial \boldsymbol{r}}\boldsymbol{J}_{(\mathsf{A})}\right\rangle + \sigma_{(\mathsf{A})} = \rho\frac{\mathrm{d}\mathsf{A}}{\mathrm{d}t} \equiv \rho\frac{D\langle\mathsf{A}\rangle}{Dt} + \rho\left\langle \boldsymbol{u}''\frac{\partial\langle\mathsf{A}\rangle}{\partial\boldsymbol{r}}\right\rangle + \rho\frac{\mathrm{d}\,\mathsf{A}''}{\mathrm{d}t},$$

whence it follows that

$$\rho\frac{\mathrm{d}\,\mathsf{A}''}{\mathrm{d}t} = -\rho\left\langle \boldsymbol{u}''\frac{\partial\langle\mathsf{A}\rangle}{\partial\boldsymbol{r}}\right\rangle - \rho\frac{D\langle\mathsf{A}\rangle}{Dt} - \left\langle\frac{\partial}{\partial\boldsymbol{r}}\boldsymbol{J}_{(\mathsf{A})}\right\rangle + \sigma_{(\mathsf{A})}. \qquad (4.49)$$

Multiplying (4.49) by the fluctuation B'' of another parameter $\mathsf{B}(\boldsymbol{r},t)$ (for which a balance equation like (2.1) can also be written) and performing turbulent averaging of the result yields

$$\overline{\rho\, \mathsf{B}'' \frac{\mathrm{d}\,\mathsf{A}''}{\mathrm{d}t}} = -\left(\boldsymbol{J}^{\mathrm{turb}}_{(\mathsf{B})} \frac{\partial \langle \mathsf{A} \rangle}{\partial \boldsymbol{r}}\right) + \overline{\mathsf{B}''\left\{\sigma_{(\mathsf{A})} - \left(\frac{\partial}{\partial \boldsymbol{r}} \boldsymbol{J}_{(\mathsf{A})}\right)\right\}}. \tag{4.50}$$

Interchanging now the parameters A and B in this relation, we have

$$\overline{\rho\, \mathsf{A}'' \frac{\mathrm{d}\mathsf{B}''}{\mathrm{d}t}} = -\left(\boldsymbol{J}^{\mathrm{turb}}_{(\mathsf{B})} \frac{\partial \langle \mathsf{B} \rangle}{\partial \boldsymbol{r}}\right) + \overline{\mathsf{A}''\left\{\sigma_{(\mathsf{B})} - \left(\frac{\partial}{\partial \boldsymbol{r}} \boldsymbol{J}_{(\mathsf{B})}\right)\right\}} \tag{4.51}$$

Finally, adding (4.50) and (4.51), we obtain the following balance equation for the one-point second moment $\langle \mathsf{A}''\mathsf{B}'' \rangle$ in substantial form when identity (3.12) is taken into account:

$$\begin{aligned} \overline{\rho \frac{\mathrm{d}(\mathsf{A}''\mathsf{B}'')}{\mathrm{d}t}} &\equiv \overline{\rho} \frac{D}{Dt} \langle \mathsf{A}''\mathsf{B}'' \rangle + \left(\frac{\partial}{\partial \boldsymbol{r}} \overline{\rho\, \mathsf{A}''\mathsf{B}''\boldsymbol{u}''}\right) \\ &= -\left(\boldsymbol{J}^{\mathrm{turb}}_{(\mathsf{A})} \frac{\partial \langle \mathsf{B} \rangle}{\partial \boldsymbol{r}}\right) - \left(\boldsymbol{J}^{\mathrm{turb}}_{(\mathsf{B})} \frac{\partial \langle \mathsf{A} \rangle}{\partial \boldsymbol{r}}\right) + \overline{\mathsf{A}''\left\{\sigma_{(\mathsf{B})} - \left(\frac{\partial}{\partial \boldsymbol{r}} \boldsymbol{J}_{(\mathsf{B})}\right)\right\}} \\ &\quad + \overline{\mathsf{B}''\left\{\sigma_{(\mathsf{A})} - \left(\frac{\partial}{\partial \boldsymbol{r}} \boldsymbol{J}_{(\mathsf{A})}\right)\right\}} \end{aligned} \tag{4.52}$$

When using the notation

$$\boldsymbol{J}^{\Sigma}_{(\mathcal{AB})} \equiv \boldsymbol{J}^{\mathrm{turb}}_{(\mathcal{AB})} + \overline{\mathsf{A}''(\partial/\partial \boldsymbol{r})\boldsymbol{J}_{(\mathsf{B})}} + \overline{\mathsf{B}''(\partial/\partial \boldsymbol{r})\boldsymbol{J}_{(\mathsf{A})}} \tag{4.53}$$

$$\boldsymbol{J}^{\mathrm{turb}}_{(\mathcal{AB})} \equiv \overline{\rho\, (\mathsf{A}''\mathsf{B}'')\boldsymbol{u}''} \tag{4.54}$$

$$\overline{\rho} \langle \varepsilon_{(\mathcal{AB})} \equiv -\overline{\boldsymbol{J}_{(\mathsf{A})}(\partial/\partial \boldsymbol{r})\mathsf{B}''} - \overline{\boldsymbol{J}_{(\mathsf{B})}(\partial/\partial \boldsymbol{r})\mathsf{A}''} \tag{4.55}$$

it takes the following final form:

$$\begin{aligned} \overline{\rho} \frac{\partial \langle \mathsf{A}''\mathsf{B}'' \rangle}{\partial t} + \overbrace{\left(\overline{\rho} \langle \boldsymbol{u} \rangle \frac{\partial \langle \mathsf{A}''\mathsf{B}'' \rangle}{\partial \boldsymbol{r}}\right)}^{\text{Convection}} &= -\overbrace{\left(\frac{\partial}{\partial \boldsymbol{r}} \boldsymbol{J}^{\Sigma}_{(\mathcal{AB})}\right)}^{\text{Diffusion}} \overbrace{- \left(\boldsymbol{J}^{\mathrm{turb}}_{(\mathsf{A})} \frac{\partial \langle \mathsf{B} \rangle}{\partial \boldsymbol{r}}\right) - \left(\boldsymbol{J}^{\mathrm{turb}}_{(\mathsf{B})} \frac{\partial \langle \mathsf{A} \rangle}{\partial \boldsymbol{r}}\right)}^{\text{Reproduction}} \\ &\quad + \overbrace{\overline{\mathsf{A}''\,\sigma_{(\mathsf{B})}} + \overline{\mathsf{B}''\sigma_{(\mathsf{A})}}}^{\text{Redistribution}} - \overbrace{\overline{\rho} \langle \varepsilon_{(\mathcal{AB})} \rangle}^{\text{Dissipation}} \end{aligned} \tag{4.56}$$

Here, $\boldsymbol{J}^{\text{turb}}_{(\mathsf{AB})}$ and $\boldsymbol{J}^{\Sigma}_{(\mathsf{AB})}$ are, respectively, the turbulent flux and total turbulent flux of the mixed correlation $\langle \mathsf{A}''\mathsf{B}''\rangle$ and $\overline{\rho}\langle \varepsilon_{(\mathsf{AB})}\rangle$ is the dissipation rate of the correlation $\langle \mathsf{A}''\mathsf{B}''\rangle$ under the action of molecular transport processes.

In the special case of coincident fluctuating flow characteristics where $\mathsf{A}(\boldsymbol{r},t) \equiv \mathsf{B}(\boldsymbol{r},t)$, (4.55) transforms into the transfer equation for the mean-square fluctuation (variance) $\langle \mathsf{A}''^2\rangle$ of the thermohydrodynamic parameter $\mathsf{A}(\boldsymbol{r},t)$ (Kolesnichenko 1981):

$$\overline{\rho}\frac{\partial\langle \mathsf{A}''^2/2\rangle}{\partial t} + \left(\overline{\rho}\langle \boldsymbol{u}\rangle\frac{\partial\langle \mathsf{A}''^2/2\rangle}{\partial \boldsymbol{r}}\right) = -\frac{\partial}{\partial \boldsymbol{r}}\left(\overline{\rho\,\mathsf{A}''^2\boldsymbol{u}''/2} + \overline{\mathsf{A}''\boldsymbol{J}_{(\mathsf{A})}}\right) - \boldsymbol{J}^{\text{turb}}_{(\mathsf{A})}\frac{\partial\langle \mathsf{A}\rangle}{\partial \boldsymbol{r}} + \overline{\mathsf{A}''\sigma_{(\mathsf{A})}} - \overline{\rho}\langle\varepsilon_{\mathsf{A}}\rangle, \tag{4.57}$$

where

$$\overline{\rho}\langle\varepsilon_{\mathsf{A}}\rangle \equiv \overline{\rho}\langle\varepsilon_{(\mathcal{A}\mathcal{A})}\rangle/2 = -\overline{\boldsymbol{J}_{(\mathsf{A})}(\partial/\partial\boldsymbol{r})\mathsf{A}''} \tag{4.58}$$

is the scalar dissipation rate of the variance $\langle \mathsf{A}''^2\rangle$.

Thus, the general balance equation for mixed pair correlations (4.56), along with the equation for variances (4.57), contains the terms reflecting the influence of the following dynamical processes on the spatiotemporal distribution of the corresponding statistical characteristics: convective transport, diffusion, formation through energy exchange between the averaged and fluctuation motions, redistribution of turbulent energy between the fluctuation motions in various directions, and dissipation of the fluctuating flow characteristic due to molecular transport processes. Let us now successively consider the cases of various quantities A and B.

4.2.2 Transfer Equations for the Turbulent Stress Tensor for a Multicomponent Medium with a Variable Density

The transfer equation for the Reynolds tensor $R_{ki} \equiv -\overline{\rho}\langle u''_k u''_i\rangle$ can be obtained from (4.56) by assuming that $\mathsf{A} \equiv u_k(\boldsymbol{r},t)$ and $\mathsf{B} \equiv u_i(\boldsymbol{r},t)$ and using (2.11) for the corresponding fluxes and volume sources of the velocities u_k and u_i:

$$J_{(u_i)j} \equiv -\pi_{ij}, \quad \sigma_{(u_i)} \equiv -\frac{\partial p}{\partial x_i} + 2\rho\varepsilon_{kji}u_k\Omega_j + \rho\sum_{\alpha=1}^{N} Z_\alpha F_{\alpha i}, \quad (^1)$$
$$J_{(u_k)j} \equiv -\pi_{kj}, \quad \sigma_{(u_k)} \equiv -\frac{\partial p}{\partial x_k} + 2\rho\varepsilon_{jlk}u_j\Omega_l + \rho\sum_{\alpha=1}^{N} Z_\alpha F_{\alpha k}, \quad (^2) \tag{4.59}$$

where ε_{ilk} is the alternating Levi-Civita tensor:

$$\varepsilon_{ilk} = \begin{cases} 1, & i,l,k = 1,2,3; \quad 3,1,2; \quad 2,3,1; \\ 0, & i = l \qquad i = k \qquad l = k; \\ -1, & i,l,k = 2,1,3; \quad 3,2,1; \quad 1,3,2; \end{cases}$$

Ω_j are the components of the angular velocity vector of the reference frame rotation (relative to the absolute reference frame) in which the equation of motion for a mixture (2.9) is written. As a result, we obtain the exact (without discarding any terms) transfer equation

$$-\overline{\rho}\frac{D}{Dt}\left(\frac{R_{ki}}{\overline{\rho}}\right) + \left(\frac{\partial}{\partial \boldsymbol{r}}\boldsymbol{J}^{\Sigma}_{(u_k u_i)}\right) = -\overline{\rho}\left(\langle u''_k \boldsymbol{u}''\rangle \frac{\partial \langle u_i\rangle}{\partial \boldsymbol{r}}\right) - \overline{\rho}\left(\langle u''_i \boldsymbol{u}''\rangle \frac{\partial \langle u_k\rangle}{\partial \boldsymbol{r}}\right) + \overline{u''_k\,\sigma_{(u_i)}} + \overline{u''_i\,\sigma_{(u_k)}} - \overline{\rho}\langle \varepsilon_{(u_k u_i)}\rangle. \tag{4.60}$$

Here,

$$\overline{u''_k\,\sigma_{(u_i)}} \equiv -\overline{u''_k(\partial p/\partial x_i)} - 2\varepsilon_{lji}R_{kl}\Omega_j + \sum_{\alpha=1}^{N} \overline{J^{\mathrm{turb}}_{\alpha k}F_{\alpha i}}, \; (^1)$$
$$\overline{u''_i\,\sigma_{(u_k)}} \equiv -\overline{u''_i(\partial p/\partial x_k)} - 2\varepsilon_{ljk}R_{il}\Omega_j + \sum_{\alpha=1}^{N} \overline{J^{\mathrm{turb}}_{\alpha i}F_{\alpha k}}; \; (^2) \tag{4.61}$$

$$\left(\boldsymbol{J}^{\Sigma}_{(u_k u_i)}\right)_j \equiv \overline{\rho(u''_k u''_i)\,u''_j} - \overline{\pi_{ij}u''_k} - \overline{\pi_{kj}u''_i} \tag{4.62}$$

is the diffusion flux (a third-rank tensor) of the turbulent stress tensor;

$$\overline{\rho}\langle \varepsilon_{(u_k u_i)}\rangle \equiv \overline{\pi_{ij}(\partial/\partial x_j)u''_k} + \overline{\pi_{kj}(\partial/\partial x_j)u''_i} \tag{4.63}$$

is the dissipation rate (a second-rank tensor) of the tensor R_{ki} due to molecular viscosity.

Let us now apply the same technique as that used in deriving (3.48) for the terms with the pressure in (4.61); the following transformation is then valid for any attribute A fluctuating in a flow:

$$\overline{\mathsf{A}''\frac{\partial p}{\partial x_i}} = \overline{\mathsf{A}''}\,\frac{\partial \overline{p}}{\partial x_i} + \overline{\mathsf{A}''\frac{\partial p'}{\partial x_i}} = \overline{\rho(1/\rho)''\mathsf{A}''}\frac{\partial \overline{p}}{\partial x_i} - \overline{p'\frac{\partial \mathsf{A}''}{\partial x_i}} + \frac{\partial}{\partial x_j}\left(\delta_{ij}\overline{p'\mathsf{A}''}\right) \tag{4.64}$$

whence, in the case of $\mathsf{A} \equiv u_k(\boldsymbol{r}, t)$ and $\mathsf{A} \equiv u_i(\boldsymbol{r}, t)$, we have, respectively,

$$\begin{aligned} &\overline{u_k'' \frac{\partial p}{\partial x_i}} = J^{\mathrm{turb}}_{(1/\rho)k} \frac{\partial \overline{p}}{\partial x_i} - \overline{p' \frac{\partial u_k''}{\partial x_i}} + \frac{\partial}{\partial x_j}\left(\delta_{ij}\overline{p' u_k''}\right), \qquad (^1) \\ &\overline{u_i'' \frac{\partial p}{\partial x_k}} = J^{\mathrm{turb}}_{(1/\rho)i} \frac{\partial \overline{p}}{\partial x_k} - \overline{p' \frac{\partial u_i''}{\partial x_k}} + \frac{\partial}{\partial x_j}\left(\delta_{kj}\overline{p' u_i''}\right), \qquad (^2) \end{aligned} \tag{4.65}$$

(here, δ_{ij} is the Kronecker delta).

Substituting now relations (4.61)–(4.65) into the transfer equation for the turbulent stress tensor (4.60), we rewrite it in the following compact final form:

$$-\frac{\partial R_{ki}}{\partial t} - \frac{\partial}{\partial x_j}\left(R_{ki}\langle u_j\rangle\right) + \frac{\partial}{\partial x_j}\left(\boldsymbol{J}^{\Sigma}_{ki}\right)_j = P^*_{ki} + \Phi_{ki} + \overline{\rho}\, G_{ki} - \overline{\rho}\varepsilon_{ki}, \tag{4.66}$$

where

$$\begin{aligned} \left(\boldsymbol{J}^{\Sigma}_{ki}\right)_j &\equiv \left(\boldsymbol{J}^{\Sigma}_{(u_k u_i)}\right)_j + \overline{p'(\delta_{ij}u_k'' + \delta_{kj}u_i'')} \\ &= \overline{\rho(u_k'' u_i'')\, u_j''} + \overline{(\delta_{ij}p' - \pi_{ij})u_k''} + \overline{(\delta_{kj}p' - \pi_{kj})u_i''}, \end{aligned} \tag{4.67}$$

$$P^*_{ki} \equiv R_{kj}\left(\frac{\partial\langle u_i\rangle}{\partial x_j} - 2\varepsilon_{jli}\Omega_l\right) + R_{ij}\left(\frac{\partial\langle u_k\rangle}{\partial x_j} - 2\varepsilon_{jlk}\Omega_l\right), \tag{4.68}$$

$$\Phi_{ki} \equiv \overline{p'(\partial u_k''/\partial x_i + \partial u_i''/\partial x_k)}, \tag{4.69}$$

$$\overline{\rho}\varepsilon_{ki} \equiv \overline{\rho}\langle\varepsilon_{(u_k u_i)}\rangle \equiv \overline{\pi_{ij}(\partial/\partial x_j)u_k''} + \overline{\pi_{kj}(\partial/\partial x_j)u_i''}, \tag{4.63*}$$

$$\overline{\rho}G_{ki} \equiv -J^{\mathrm{turb}}_{(1/\rho)k}\frac{\partial\overline{p}}{\partial x_i} - J^{\mathrm{turb}}_{(1/\rho)i}\frac{\partial\overline{p}}{\partial x_k} + \sum_{\alpha=1}^{N}\left(J^{\mathrm{turb}}_{\alpha k}F_{\alpha i} + J^{\mathrm{turb}}_{\alpha i}F_{\alpha k}\right). \tag{4.70}$$

Let us explain the meaning of the individual terms in (4.66):

- $\boldsymbol{J}^{\Sigma}_{ki}$ is the total "diffusion flux" of the Reynolds stress tensor R_{ki} including the various turbulence transport mechanisms in space, namely the turbulent diffusion tensor (the term $\overline{\rho(u_k'' u_i'')\, u_j''}$), the tensor transport due to molecular viscosity when the adjacent, hitherto non-fluctuating, gas layers are drawn into motion by the fluctuating hydrodynamic velocity (the terms $-\overline{\pi_{ij}u_k''} - \overline{\pi_{kj}u_i''}$); and the tensor transport due to the interaction between the fluctuating velocity and pressure fields (the terms $\overline{\delta_{ij}p'u_k''} + \overline{(\delta_{kj}p'u_i'')}$);
- P^*_{ki} is a tensor quantity related to the generation of the stress tensor R_{ki} due to the interaction of turbulent velocity fluctuations with a nonuniform field of averaged

velocities (attributable to the wind shear or rotation of the medium with a common angular velocity Ω_j);

- Φ_{ki} is the correlation tensor of the turbulent velocity and pressure fluctuations describing the redistribution of turbulent energy between various components of the Reynolds tensor R_{ki};
- $\overline{\rho}G_{ki}$ is a tensor related to the appearance (or disappearance) of turbulence when the medium moves in a nonuniform hydrodynamic pressure field or under the action of mass forces;
- $\overline{\rho}\varepsilon_{ki}$ is a tensor quantity related to the dissipation rate of the turbulent stress tensor R_{ki} due to molecular viscosity.

The derived transfer equation for the tensor R_{ki} (4.66) for a multicomponent mixture differs from the well-known analogous equation for a flow with a constant density (see Monin and Yaglom 1992) mainly by the structure of the tensor quantity G_{ki}. In view of (3.139) for the flux $\boldsymbol{J}^{\text{turb}}_{(1/\rho)}$, (4.70) takes the form

$$\begin{aligned}\overline{\rho}G_{ki} &= \left(\overline{p'u''_k}\frac{\partial \ln\overline{p}}{\partial x_i}+\overline{p'u''_i}\frac{\partial \ln\overline{p}}{\partial x_k}\right)-\frac{1}{\overline{\rho}\langle n'_p\rangle\langle T\rangle}\left(J^{\text{turb}}_{qk}\frac{\partial \overline{p}}{\partial x_i}+J^{\text{turb}}_{qi}\frac{\partial \overline{p}}{\partial x_k}\right)\\ &\quad-\sum_{\alpha=1}^{N}\left\{\left(\frac{1}{\overline{n}}\frac{\partial \overline{p}}{\partial x_i}-F_{\alpha i}\right)J^{\text{turb}}_{\alpha k}+\left(\frac{1}{\overline{n}}\frac{\partial \overline{p}}{\partial x_k}-F_{\alpha k}\right)J^{\text{turb}}_{\alpha i}\right\}.\end{aligned} \tag{4.71}$$

It should be kept in mind that in many practically important cases, for example, for flows with small Mach numbers, the relative pressure fluctuations can be neglected compared to the density and temperature ones; the first term in (4.71) can then be discarded [see the note to (3.139)]. In this case, the expression for G_{ki} written for a stratified flow in the gravity field takes a simpler form:

$$\overline{\rho}G_{ki} \cong \frac{g}{\langle c_p\rangle\langle T\rangle}\left(\delta_{k3}J^{T}_{qi}+\delta_{i3}J^{T}_{qk}\right)+\frac{\overline{\rho}}{\overline{n}}g\sum_{\alpha=1}^{N}\left(\delta_{i3}J^{\text{turb}}_{\alpha k}+\delta_{k3}J^{\text{turb}}_{\alpha i}\right). \tag{4.72}$$

Formula (4.72) represents a similar term of the transfer equation for the tensor R_{ki} obtained previously for single-component fluid flows (see Turbulence: Principles and Applications 1980) modified to the case of a compressible multicomponent medium.

The equations for the Reynolds turbulent stress tensor written in form (4.66) (six equations for the components of the symmetric tensor R_{ki}) cannot be directed used to close the averaged hydrodynamic equations for a mixture, because they contain a large number of new indefinite quantities that represent the statistical correlations with pressure fluctuations, dissipative terms, and third-order moments. We approximate these indefinite quantities for a turbulized medium with a variable density by analogy with the more or less satisfactory and well-tested closure hypotheses used in the "classical" theory of turbulence with a constant density (see Turbulent Shear Flows-I 1979). Note in this connection that no sufficient grounds for a significant modification of such approximations for mixture flows with a variable density have been revealed at present.

Below, we restrict our analysis to simplified second-order closure schemes that use a minimum number of arbitrary empirical constants. More complex parametric relations for the modeled correlation moments for a single-component fluid are given, for example, in Turbulence: Principles and Applications (1980), to which we refer the interested reader. Thus, we assume that the approximation relations used in semiempirical turbulence theories for single-component incompressible fluids are also valid in the case of a multicomponent compressible medium considered here, where the weighted-mean Favre averaging is systematically used; in particular, the following simplified closure scheme is then valid:

$$\overline{\rho}\varepsilon_{ki} = \frac{2}{3}K_{b1}\,\delta_{ki}\frac{\overline{\rho}\langle b\rangle^{\frac{3}{2}}}{L} - K_{b2}\frac{\nu R_{ki}}{L^2}, \tag{4.73}$$

$$\Phi_{ki} = K_{p1}\frac{\langle b\rangle^{\frac{1}{2}}}{L}\left(R_{ki} + \frac{2}{3}\delta_{ki}\overline{\rho}\langle b\rangle\right) - K_{p2}\left(P_{ki} - \frac{2}{3}\delta_{ki}P\right), \tag{4.74}$$

$$(\boldsymbol{J}^{\Sigma}_{ki})_j = \left(c_1 L\langle b\rangle^{\frac{1}{2}} + c_2\nu\right)\frac{\partial R_{ki}}{\partial x_j}. \tag{4.75}$$

Here, $K_e,\ K_p,\ c -$ are free constants determined from experimental data, with the empirical coefficients K_{b2} and c_2 being significant only for small turbulent Reynolds numbers $\boldsymbol{Re}^{\rm turb} \equiv L\langle b\rangle^{1/2}/\nu$, i.e., when $K_{b1}\boldsymbol{Re}^{\rm turb}/K_{b2} \le 0(1)$ and $c_1\boldsymbol{Re}^{\rm turb}/c_2 \le 0(1)$; ν is the molecular kinematic viscosity for a mixture; $L(\boldsymbol{r},t)$ is the external scale of the turbulent velocity field at a point interrelated with, though not coincident with, the integral turbulence scale L (or the mixing length Λ) discussed in the preceding chapter. The following notation is used in (4.74): $P_{ki} = P^*_{ki} + \overline{\rho}G_{ki}$ is the total appearance (disappearance) rate of the Reynolds turbulent stresses R_{ki};

$$P = \frac{1}{2}\delta_{kj}P_{kj} = R_{kj}\partial\langle u_k\rangle/\partial x_j + \overline{\rho}G \tag{4.76}$$

is the total generation rate of the turbulent energy $\langle b\rangle$ under the action of a mean shear (the first term) and buoyancy effects and mutual transformations of the turbulent and potential energies in a stratified multicomponent mixture (the second term). In view of (4.71), the expression for $\overline{\rho}G$ can be transformed to

$$\begin{aligned}\overline{\rho}G &\equiv \frac{1}{2}\overline{\rho}\,\delta_{kj}G_{kj} = -\boldsymbol{J}^{\rm turb}_{(1/\rho)}\frac{\partial\overline{p}}{\partial\boldsymbol{r}} + \sum_{\alpha=1}^{N}\boldsymbol{J}^{\rm turb}_{\alpha}\boldsymbol{F}_{\alpha}\\ &= \frac{\overline{p'\boldsymbol{u}''}}{\overline{p}}\frac{\partial\overline{p}}{\partial\boldsymbol{r}} - \frac{\boldsymbol{J}^{\rm turb}_q}{\overline{\rho}\langle n'_p\rangle\langle T\rangle}\frac{\partial\overline{p}}{\partial\boldsymbol{r}} - \sum_{\alpha=1}^{N}\boldsymbol{J}^{\rm turb}_{\alpha}\left(\frac{1}{\overline{n}}\frac{\partial\overline{p}}{\partial\boldsymbol{r}} - \boldsymbol{F}_{\alpha}\right).\end{aligned} \tag{4.77}$$

Obviously, the closing relations (4.73)–(4.75) ensure the fulfillment of the minimal requirements that any second-order closure schemes should satisfy. Expression (4.73) for the tensor quantity $\overline{\rho}\varepsilon_{ki}$ describes the effect of viscous dissipation on the structure of the Reynolds tensor. The presence of small-scale vortices is the main cause of turbulent energy dissipation. For flows with large Reynolds numbers $\boldsymbol{Re}^T$, this small-scale turbulence is isotropic. The combination of terms in (4.73) ensures isotropy of the viscous dissipation process at large Reynolds numbers (the first term) and allows for the possibility of dissipation anisotropy for each component of the tensor R_{kj} at small Reynolds numbers (the second term). Since the scale of large vortices $L(\boldsymbol{r}, t)$ does not affect the dissipation process, their destruction rate in (4.73) does not depend on the molecular viscosity ν.

The first term in the approximation relation (4.74) for the correlations of pressure and velocity fluctuations describes the redistribution of turbulent energy from the main (averaged) flow between the individual fluctuation velocity components and the tendency for a fluctuating flow to become isotropic, which agrees with the Rotta (1951a, b) model. The sum in parentheses $(R_{ki} + \frac{2}{3}\overline{\rho}\langle b\rangle\delta_{ki})$ (or, more precisely, the difference, because $\overline{\rho}\langle b\rangle = -\frac{1}{2}R_{kj}\delta_{kj}$) characterizes the degree of flow anisotropy and has the necessary tensor symmetry property. The second term in (4.74) corresponding to the Launder relation (Launder 1975; Launder and Morse 1979) for many turbulent flows balances the generation of turbulent stresses in the transfer equation for the tensor R_{kj} (4.66) and describes the influence of turbulent energy production mechanisms on the turbulent velocity and, hence, pressure fluctuations. The proportionality of the tensors Φ_{ki} and $\left(P_{ik} - \frac{2}{3}\delta_{ik}P\right)$ is equivalent to the assumption that the correlations of the pressure and hydrodynamic velocity fluctuations not only tend to make the turbulence isotropic (at a rate linearly dependent on the deviation from isotropy) but also redistribute the generation of turbulence at a rate proportional to the anisotropy of this generation (Turbulence: Principles and Applications 1980).

The flux $\boldsymbol{J}^{\Sigma}_{ki}$ specified by (4.75) determines the diffusive transport of the Reynolds stresses R_{kj} from one flow region into another without their generation or decay. This transport prevents the formation of large gradients in the spatial distribution of R_{kj} (see Turbulent Shear Flows-I 1979). Although both parts of this gradient relation are third-rank tensors, it, nevertheless, is inconsistent (by symmetry conditions) with the tensor $\overline{\rho(u''_k u''_i)\, u''_j}$. In addition, it violates the so-called realizability constraints, which consist in the necessity of satisfying the Schwarz inequality for the third-order moment (see Monin and Yaglom 1992). Nonetheless, we restrict ourselves to this simplest approximation of the diffusion term, because below we use only a locally equilibrium version of the turbulence model being developed here.

Success in applying the transfer equations for the second-order correlation moments depends largely on how well the empirical constants have been chosen. Usually, they are experimentally determined by studying special turbulent flows dependent only on one (sought-for) coefficient. Ideally, for each closed turbulence model, once the method of approximating the unknown terms in the transfer equations has been chosen, all of the introduced free constants should be constant. The following numerical values for these constants can be taken in (4.73)–(4.75):

$$K_{b1} = 0.125; \quad K_{b2} = 6; \quad K_{p1} = 0.275; \quad K_{p2} = 0.55; \quad c_1 = 0.3 \pm 0.05, \tag{4.78}$$

which are given, for example, in the widely cited monograph Turbulence: Principles and Applications (1980).

At the same time, as was justly pointed out by Ievlev (1975), the free coefficients can be assumed to be constant only if some "equilibrium" turbulence spectrum exists for the flow conditions under consideration. For a different flow regime, their values can change greatly. Given this circumstance, some authors consider the empirical coefficients to be actually some single-valued functions of characteristic dimensionless flow parameters (e.g., the Reynolds number, the Richardson number, etc.) and, possibly, some other dimensionless turbulence characteristics. However, it should be kept in mind that, in this case, the efficient invariant modeling method completely loses its advantage over the first-order closure schemes.

4.2.2.1 Transfer Equation for the Turbulent Energy

Convolving (4.60) in indices k and i $(-\frac{1}{2}R_{ki}\delta_{ki} = \overline{\rho}\langle b \rangle)$ leads to an exact equation for the averaged kinetic energy of the turbulent fluctuations in a compressible mixture [cf. (3.174)]:

$$\overline{\rho}\frac{\partial \langle b \rangle}{\partial t} + \overline{\rho}\left(\langle \boldsymbol{u} \rangle \frac{\partial \langle b \rangle}{\partial \boldsymbol{r}}\right) = -\mathrm{div}\, \boldsymbol{J}^{\mathrm{turb}}_{\langle b \rangle} + \left(R : \frac{\partial \langle \boldsymbol{u} \rangle}{\partial \boldsymbol{r}}\right) + \overline{p' \mathrm{div}\, \boldsymbol{u}''} + \overline{\rho}\, G - \overline{\rho}\langle \varepsilon_b \rangle. \tag{4.79}$$

In contrast to the equation for the Reynolds tensor, this equation contains only two unknown correlation terms. The first (diffusion) term

$$\boldsymbol{J}^{\mathrm{turb}}_{\langle b \rangle} \equiv \overline{\rho(b + p'/\rho)\boldsymbol{u}''} - \overline{\Pi\, \boldsymbol{u}''} = \overline{\rho\, b \boldsymbol{u}''} + \overline{(p' U - \Pi')\boldsymbol{u}''} - \overline{\Pi}\, \boldsymbol{J}^{\mathrm{turb}}_{(1/\rho)} \tag{4.80}$$

describes the total substantial turbulence energy flux related to the various transport mechanisms in coordinate space. In particular, $\boldsymbol{J}^{\mathrm{turb}}_{(b)} \equiv \overline{\rho\, b \boldsymbol{u}''}$ is often interpreted as the kinetic energy flux of the fluctuation (vortex) motion, so that div $\boldsymbol{J}^{\mathrm{turb}}_{(b)}$ describes the mean rate of decrease in vortex kinetic energy per unit volume through "turbulent diffusion." In turn, – div $(\overline{\Pi' \boldsymbol{u}''})$ characterizes the mean rate of increase in vortex kinetic energy through the work done by the viscous stress tensor fluctuations at the boundary of an elementary volume.

The second (correlation) term

$$\overline{\rho}\langle\varepsilon_b\rangle \equiv \overline{\Pi : (\partial/\partial \boldsymbol{r})\boldsymbol{u}''} = \overline{\rho}\varepsilon_b + \overline{\Pi} : (\partial/\partial \boldsymbol{r})\,\boldsymbol{J}^{\text{turb}}_{(1/\rho)}, \tag{4.81}$$

where

$$\overline{\rho}\varepsilon_b \equiv \overline{\Pi' : (\partial/\partial \mathbf{r})\ \mathbf{u}'} > 0, \tag{4.82}$$

describes the dissipation of turbulent kinetic energy into heat due to molecular viscosity. It is the mean work per unit time per unit volume done by the viscous stress tensor fluctuations on turbulent vortices with a velocity shear. For a self-sustaining turbulent field, the dissipation rate ε_b has the same order of magnitude as the turbulence generation rate by a shear flow $\boldsymbol{R} : (\partial/\partial \boldsymbol{r})\langle \boldsymbol{u}\rangle$.

The penultimate term on the right-hand side of (4.79), the quantity $\overline{\rho}\,G$ defined by (4.77), describes the generation of turbulent energy in the gravity field due to temperature and/or stratified mixture composition nonuniformity and the mutual transformations of the turbulent and potential energies under the action of nongravitational mass forces. The term with a pressure fluctuation in (4.77) can often be discarded.

For developed turbulence, when $\boldsymbol{Re} \equiv LV/\nu >> 1$, the inequality $\langle b\rangle^{\frac{1}{2}} \geq LV/L_{\text{hydro}}$ holds (see Turbulence: Principles and Applications 1980). Here, the Reynolds number is determined from the integral turbulence scale L corresponding to typical sizes of large vortices and the characteristic flow velocity V; $L_{\text{hydro}} = |\partial \ln\theta/\partial \boldsymbol{r}|^{-1}$ is the distance at which significant variations of some averaged thermohydrodynamic parameters θ occur. It is easy to show that at large Reynolds numbers ($\boldsymbol{Re} >> 1$) the last terms in (4.80) and (4.81) can be discarded. Indeed, estimating the order of magnitude of the individual terms in (4.81), we have

$$\varepsilon_b \propto \langle b\rangle^{\frac{3}{2}}/L, \quad \overline{\Pi} \propto \overline{\rho}\nu V/L_{\text{hydro}}, \quad (\partial/\partial \boldsymbol{r})\ \boldsymbol{J}^{\text{turb}}_{(1/\rho)} \propto \langle b\rangle^{\frac{1}{2}}/L_{\text{hydro}}.$$

Therefore, the followings estimates are valid:

$$\frac{\overline{\Pi} : (\partial/\partial \boldsymbol{r})\boldsymbol{J}^{\text{turb}}_{(1/\rho)}}{\overline{\rho}\varepsilon_b} \propto \frac{\overline{\rho}\nu V}{L_{\text{hydro}}}\frac{\langle b\rangle^{\frac{1}{2}}}{L_{\text{hydro}}}\frac{L}{\overline{\rho}\langle b\rangle^{\frac{3}{2}}} = \frac{\nu VL}{\langle b\rangle L^2_{\text{hydro}}} = \frac{\nu}{LV}\frac{V^2L^2}{\langle b\rangle L^2_{\text{hydro}}} \leq \frac{1}{\boldsymbol{Re}},$$

$$\frac{\left|\overline{\Pi}\,\boldsymbol{J}^{\text{turb}}_{(1/\rho)}\right|}{\left|\overline{\Pi'\,\boldsymbol{u}''}\right|} \propto \frac{\overline{\rho}\nu V}{L_{\text{hydro}}}\langle b\rangle^{\frac{1}{2}}\frac{L}{\overline{\rho}\langle b\rangle^{\frac{3}{2}}L_{\text{hydro}}} = \frac{\nu VL}{\langle b\rangle L^2_{\text{hydro}}} \leq \frac{1}{\boldsymbol{Re}}.$$

Thus, in an extremely developed turbulent flow of a multicomponent mixture, the transfer equation for the turbulent energy takes the form

$$\overline{\rho}\frac{D\langle b\rangle}{Dt} + \operatorname{div}\left(\overline{\rho\, b\boldsymbol{u}''} + \overline{(p'\boldsymbol{U} - \Pi')\,\boldsymbol{u}''}\right) = \overline{p'\operatorname{div}\,\boldsymbol{u}''} + P - \overline{\rho}\varepsilon_b. \tag{4.83}$$

Convolving the approximating relations (4.73)–(4.75) leads to the following models for the unknown correlation terms in the turbulent energy balance equation (4.76):

$$\overline{\rho}\varepsilon_b = \frac{1}{2}<\varepsilon_{ki}>\delta_{ki} = K_{b1}\frac{\overline{\rho}\langle b\rangle^{\frac{3}{2}}}{L} + K_{b2}\frac{\overline{\rho}\nu\langle b\rangle}{L^2}, \tag{4.84}$$

$$\overline{p'\operatorname{div}\,\boldsymbol{u}''} = 0, \tag{4.85}$$

$$\boldsymbol{J}_{\langle b\rangle} = -\left(c_1 L\langle b\rangle^{\frac{1}{2}} + c_2\nu\right)\frac{\partial}{\partial \boldsymbol{r}}(\overline{\rho}\langle b\rangle). \tag{4.86}$$

These relations basically follow from dimensional considerations and are a generalization of the well-known Kolmogorov (1941, 1942) hypothesis. According to the latter, the energy dissipation rate ε_b at a given point in a developed turbulent flow is determined only by the local values of the mean turbulent energy per unit mass $\langle b\rangle$ and by the turbulence scale $L(\boldsymbol{r}, t)$, while the turbulent transport of momentum and fluctuation energy is effected by gradient-type diffusion terms.

As has been mentioned in Chap. 3, the fundamental turbulent energy transfer equation (4.83) underlies many present-day semiempirical turbulence models, for example, it is used in specific calculations of turbulent motions based on the Kolmogorov-Launder and other models. In particular, invoking it, along with the formula for the turbulence scale L, in the case of a locally equilibrium state of the R_{ki} field (when the left-hand part of (4.66) is zero) allows one, to some extent, to take into account the existence (in a real turbulent flow) of a partial equilibrium between the turbulent velocity field and the averaged flow (see Lewellen 1977).

To conclude this section, we note the following. The partial differential equations (4.66) for R_{ki} are fairly complex from a computational viewpoint and, hence, are often unsuitable without any additional simplifications for numerical calculations. At the same time, they may turn out to be very useful as a tool for improving simpler gradient turbulence models. In particular, in the locally equilibrium case where the convective and diffusion terms almost balance each other, the differential equations (4.66) for R_{ki} turn into algebraic equations while retaining some fundamental properties of the original equations. Solving these algebraic equations allows one to derive, in certain cases, a more general rheological relation for the turbulent stress tensor R_{ki} than (3.166), because it basically includes the anisotropy of the turbulent viscosity coefficients. The algebraic formulas to calculate the turbulent viscosity coefficients obtained in this case contain the turbulent energy $\langle b\rangle$, the turbulence scale L (see Sect. 4.3), and a number of empirical constants appearing in the equation for R_{ki}.

4.2.3 Transfer Equations for the Turbulent Diffusion and Heat Fluxes for a Multicomponent Medium with a Variable Density

As we have repeatedly noted above, to generalize the invariant turbulence modeling method (developed in the literature for a single-component medium based on (4.66)) to multicomponent reacting gas mixtures, apart from the transfer equations (4.66) for the spatial components of the Reynolds stress tensor, a number of model transfer equations for second-order correlation moments, such as $\langle H''\boldsymbol{u}''\rangle$, $\langle Z''_\alpha \boldsymbol{u}''\rangle$, $\langle H''^2\rangle$, $\langle H''Z''_\alpha\rangle$, and $\langle Z''_\alpha Z''_\beta\rangle$, must be additionally invoked. Although the approach to deriving these additional equations used below is of the same type, its consistent implementation for these correlations seems absolutely necessary in constructing a semiempirical model of reacting turbulence, because it allows us not only to obtain well-grounded exact transfer equations but also to reveal the limitations inherent in these equations associated with their closure.

4.2.3.1 Transfer Equation for the Turbulent Heat Flux

Let us identify the parameters A and B in the general transfer equation (4.56) with the mixture enthalpy H and the hydrodynamic flow velocity $\boldsymbol{u}$, respectively, and use (2.11) and (2.26) for the fluxes and generation rates of these quantities:

$$\boldsymbol{J}_{(H)} \equiv \boldsymbol{q}, \quad \sigma_{(H)} \equiv \frac{\mathrm{d}p}{\mathrm{d}t} + \left(\Pi : \frac{\partial}{\partial \boldsymbol{r}}\boldsymbol{u}\right) + \sum_{\alpha=1}^{N} \boldsymbol{J}_\alpha \boldsymbol{F}_\alpha, \quad (^1)$$
$$J_{(\boldsymbol{u})} \equiv -\Pi, \quad \sigma_{(\boldsymbol{u})} \equiv -\frac{\partial p}{\partial \boldsymbol{r}} + 2\rho\, \boldsymbol{u} \times \Omega + \rho \sum_{\alpha=1}^{N} Z_\alpha \boldsymbol{F}_\alpha. \quad (^2) \tag{4.87}$$

As a result, we obtain the following exact transfer equation for the turbulent heat flux $\boldsymbol{q}^{\mathrm{turb}} \equiv \overline{\rho H''\boldsymbol{u}''} = \bar{\rho}\langle H''\boldsymbol{u}''\rangle$:

$$\bar{\rho}\frac{D(\boldsymbol{q}^{\mathrm{turb}}/\bar{\rho})}{Dt} + \left(\frac{\partial}{\partial \boldsymbol{r}}\boldsymbol{J}^{\Sigma}_{(H\boldsymbol{u})}\right) = -\left(\boldsymbol{q}^{\mathrm{turb}}\frac{\partial\langle\boldsymbol{u}\rangle}{\partial \boldsymbol{r}}\right) - 2\boldsymbol{q}^{\mathrm{turb}} \times \Omega + \boldsymbol{R}\left(\frac{\partial\langle H\rangle}{\partial \boldsymbol{r}} - \frac{1}{\bar{\rho}}\frac{\partial\bar{p}}{\partial \boldsymbol{r}}\right)$$
$$+\bar{\rho}\,\boldsymbol{G}_H + \overline{p'\frac{\partial H''}{\partial \boldsymbol{r}}} + \overline{\boldsymbol{u}''\left(\frac{\mathrm{d}p'}{\mathrm{d}t} + \Pi : \frac{\partial \boldsymbol{u}''}{\partial \boldsymbol{r}}\right)} - \bar{\rho}\varepsilon_H \tag{4.88}$$

which describes the spatiotemporal distributions of the correlations $\langle H''\boldsymbol{u}''\rangle$ for a turbulent shear flow. Here,

$$\boldsymbol{J}^{\Sigma}_{(H\boldsymbol{u})} \equiv \overline{(\rho\, \boldsymbol{u}''\boldsymbol{u}'' + p'\boldsymbol{U} - \Pi)H''} + \overline{\boldsymbol{q}\boldsymbol{u}''} \tag{4.89}$$

is the diffusion flux (a second-rank tensor) of the correlation $\langle H''\boldsymbol{u}''\rangle$;

$$\overline{\rho}\varepsilon_H \equiv \overline{\rho}\varepsilon_{(H\boldsymbol{u})} \equiv \overline{\Pi \frac{\partial H''}{\partial \boldsymbol{r}}} - \overline{\boldsymbol{q}\frac{\partial \boldsymbol{u}''}{\partial \boldsymbol{r}}} \tag{4.90}$$

is the destruction rate (a vector) of the correlation $\langle H''\boldsymbol{u}''\rangle$ under the action of molecular viscosity and thermal diffusivity;

$$\overline{\rho}\,\boldsymbol{G}_H \equiv -\overline{H''}\,\frac{\partial \overline{p}}{\partial \boldsymbol{r}} = -\left(\frac{\langle H''^2\rangle}{\langle c_p\rangle\langle T\rangle} - \sum_{\alpha=1}^{N}\sum_{\beta=1}^{N}\frac{\overline{n}_\beta(c_{p\alpha}-c_{p\beta})}{\overline{n}\langle c_p\rangle}\langle Z''_\alpha H''\rangle\right)\frac{\partial \overline{p}}{\partial \boldsymbol{r}} \tag{4.91}$$

is the generation (dissipation) rate (a vector) of the turbulent heat flux $\boldsymbol{q}^{\mathrm{turb}}$ under the averaged pressure gradient (the correlation $\overline{H''}$ in this formula is defined by (3.140)). The tensor notation used in (4.88) is presented in Appendix A. We use the term "exact equation" in the sense that no approximate transformations were carried out when deriving (4.88).

By analogy with the transfer equation for the stress tensor (4.60), it can be said that the convective and "diffusive" terms are on the left-hand side of the transfer equation (4.88), while the quantities describing the generation of the turbulent heat flux $\boldsymbol{q}^{\mathrm{turb}}$ under the buoyancy forces and the gradients of the averaged hydrodynamic velocity, enthalpy, and pressure for the mixture, the quantities related to the redistribution of $\langle H''\boldsymbol{u}''\rangle$ by the mixed moment of the pressure and enthalpy gradient fluctuations in the mixture (this term counteracts the generation of $\langle H''\boldsymbol{u}''\rangle$, limiting its growth), and the quantity characterizing the dissipation of the flux $\boldsymbol{q}^{\mathrm{turb}}$ are on the right-hand side. In his monograph, Ievlev (1975) estimated the penultimate term in (4.88) for developed turbulence. It follows from this estimate that it is small compared to the dissipative term $\overline{\rho}\langle\varepsilon_H\rangle$ in all cases where the turbulent heat transport is significant. Therefore, below we omit this quantity.

The transfer equation (4.88) includes a number of indefinite correlation moments creating the closure problem. For the reasons given above, we model these additional correlations using the following simple approximations (Kolesnichenko and Marov 1984):

$$\boldsymbol{J}^{\Sigma}_{(H\boldsymbol{u})} = -\left(c_3 L\langle b\rangle^{\frac{1}{2}} + c_4(\nu+\chi)\right)\frac{\partial}{\partial \boldsymbol{r}}\boldsymbol{q}^{\mathrm{turb}}, \tag{4.92}$$

$$\overline{p'\partial H''/\partial \boldsymbol{r}} = -K_{s1}\frac{\langle b\rangle^{\frac{1}{2}}}{L}\boldsymbol{q}^{\mathrm{turb}} - K_{s2}\boldsymbol{P}_H, \tag{4.93}$$

$$\overline{\rho}\varepsilon_H = K_{b2}\frac{\nu+\chi}{2L^2}\boldsymbol{q}^{\mathrm{turb}}, \tag{4.94}$$

in which K_{s1}, K_{s2}, c_3, and c_4 are universal empirical constants. With regard to these relations, we can say the following: for locally isotropic turbulence at large

Reynolds numbers, the correlation moments $\overline{\Pi(\partial H''/\partial \boldsymbol{r})}$ and $\overline{\boldsymbol{q}\,(\partial \boldsymbol{u}''/\partial \boldsymbol{r})}$ and, thus, the term with viscous dissipation $\langle \varepsilon_H \rangle$ are zero. The dissipative correlation $\langle \varepsilon_H \rangle$ is also negligible in non-isotropic turbulence, provided that the turbulent Reynolds number is large. Consequently, model approximations of the quantity $\langle \varepsilon_H \rangle$ only at small numbers $\boldsymbol{Re}^{\text{turb}}$, for example, in a near-surface layer, are needed for practical applications. The pressure–enthalpy gradient correlation $\overline{p'(\partial H''/\partial \boldsymbol{r})}$ is the equivalent of the pressure–strain correlation (4.63) in the equation for the Reynolds tensor. When the direct dissipation is negligible, this term provides a mechanism that limits the growth of the fluxes. The expression proposed by (Monin and Yaglom 1971), the first term in (4.93), serves as the most commonly used approximation for the pressure–enthalpy gradient correlation $\overline{p'(\partial H''/\partial \boldsymbol{r})}$. This expression is a direct analog of the "tendency-to-isotropy" approximation introduced by Rotta for the correlation quantities Φ_{ki} related to the velocity and pressure fluctuations [see (4.74)]. The second approximating term related to the quantity

$$\boldsymbol{P}_H \equiv -\left(\boldsymbol{q}^{\text{turb}} \frac{\partial \langle \boldsymbol{u} \rangle}{\partial \boldsymbol{r}}\right) - 2\boldsymbol{q}^{\text{turb}} \times \boldsymbol{\Omega} + \overline{\rho}\, \boldsymbol{G}_H \tag{4.94*}$$

defines the generation rate of the turbulent heat flux $\boldsymbol{q}^{\text{turb}}$ under the influence of mean strain and due to the averaged pressure gradient (see Launder and Spalding 1972). Although the correlation $\overline{p'(\partial H''/\partial \boldsymbol{r})}$ in this case partially compensates for the term with the direct generation of $\langle H''\boldsymbol{u}'' \rangle$ by buoyancy forces, these forces cause an increase in the vertical turbulent heat flux for unstable stratification of the medium and its decrease for stable one. The diffusion term in (4.88), which is the spatial transfer rate of the correlation $\langle H''\boldsymbol{u}'' \rangle$ under the action of hydrodynamic velocity and pressure fluctuations, is modeled just as in the case of the equation for the Reynolds tensor R using a gradient-type hypothesis with a scalar diffusion coefficient.

Based on the available information about the modeling of plane mixing layers, we can adopt the following values for the free constants in relations (4.92)–(4.94):

$$K_{b2} = 6; \quad K_{s1} = 0.4; \quad K_{s2} = 0.5; \quad c_3 = 0.8. \tag{4.95}$$

For other types of flows, these constants are to be refined further.

4.2.3.2 Transfer Equations for the Turbulent Diffusion Fluxes

Suppose that $\mathsf{A} \equiv n_\alpha/\rho$ and $\mathsf{B} \equiv \boldsymbol{u}$ in the general transfer equation (4.56). Using (2.7) and (2.11) for the corresponding fluxes and generation rates of these quantities, we then obtain the following transfer equations for the turbulent diffusion fluxes $\boldsymbol{J}_\alpha^{\text{turb}} \equiv \overline{\rho} \langle Z_\alpha'' \boldsymbol{u}'' \rangle \quad (\alpha = 1, 2, \ldots, N)$:

$$\overline{\rho}\frac{D(\boldsymbol{J}_\alpha^{\text{turb}}/\overline{\rho})}{Dt}+\frac{\partial}{\partial\boldsymbol{r}}\boldsymbol{J}^{\Sigma}_{(Z_\alpha\boldsymbol{u})}=\boldsymbol{R}\frac{\partial\langle Z_\alpha\rangle}{\partial\boldsymbol{r}}-\left(\boldsymbol{J}_\alpha^{\text{turb}}\cdot\frac{\partial\langle\boldsymbol{u}\rangle}{\partial\boldsymbol{r}}\right)-2\boldsymbol{J}_\alpha^{\text{turb}}\times\boldsymbol{\Omega}$$
$$+\overline{p'\frac{\partial Z''_\alpha}{\partial\boldsymbol{r}}}+\overline{\rho}\boldsymbol{G}_\alpha+\sum_{s=1}^{r}\nu_{\alpha s}\overline{\xi_s\boldsymbol{u}''}-\overline{\rho}\varepsilon_\alpha,\quad \alpha=1,2,\ldots,N \tag{4.96}$$

Here,

$$\boldsymbol{J}^{\Sigma}_{(Z_\alpha\boldsymbol{u})}\equiv\overline{(\rho\,\boldsymbol{u}''\boldsymbol{u}''+p'\boldsymbol{U}-\boldsymbol{\Pi})Z''_\alpha}+\overline{\boldsymbol{J}_\alpha\boldsymbol{u}''} \tag{4.97}$$

is the diffusion flux (a second-rank tensor) of the correlation $\langle Z''_\alpha\boldsymbol{u}''\rangle$;

$$\overline{\rho}\varepsilon_\alpha\equiv\overline{\rho}\langle\varepsilon_{Z_\alpha\boldsymbol{u}}\rangle\equiv\overline{\boldsymbol{\Pi}\frac{\partial Z''_\alpha}{\partial\boldsymbol{r}}-\boldsymbol{J}_\alpha\frac{\partial\boldsymbol{u}''}{\partial\boldsymbol{r}}} \tag{4.98}$$

is the destruction rate (a vector) of the correlation $\langle Z''_\alpha\boldsymbol{u}''\rangle$ due to molecular viscosity and diffusion;

$$\rho\,\boldsymbol{G}_\alpha\equiv-\overline{Z''_\alpha}\frac{\partial\overline{p}}{\partial\boldsymbol{r}}=-\left(\frac{\langle H''Z''_\alpha\rangle}{\langle c_p\rangle\langle T\rangle}-\sum_{\alpha=1}^{N}\sum_{\beta=1}^{N}\frac{\overline{n}_\beta(c_{p\alpha}-c_{p\beta})}{\overline{n}\langle c_p\rangle}\langle Z''_\alpha Z''_\beta\rangle\right)\frac{\partial\overline{p}}{\partial\boldsymbol{r}} \tag{4.99}$$

is the generation rate of the turbulent diffusion flux $\boldsymbol{J}_\alpha^{\text{turb}}$ due to the averaged pressure gradient [the correlation $\overline{Z''_\alpha}$ in this formula is defined by (3.141)]. According to (4.46), the terms in (4.96) including the correlations with the fluctuations in hydrodynamic velocity and chemical sources of a substance of type α take the form

$$\overline{\sigma_\alpha\boldsymbol{u}''}=\sum_{s=1}^{r}\nu_{\alpha s}\overline{\xi_s\boldsymbol{u}''}=\overline{\boldsymbol{u}''}\left(\sum_{s=1}^{r}\nu_{\alpha s}\xi_s(\langle T\rangle,\overline{n}_\alpha)\right)+\sum_{s=1}^{r}\nu_{\alpha s}\overline{\xi'_s\,\boldsymbol{u}''}$$
$$=\overline{\rho}\,L^{Ch}_{0\alpha}\langle H''\boldsymbol{u}''\rangle+\overline{\rho}\sum_{\beta=1}^{N}L^{Ch}_{\alpha\beta}\langle Z''_\alpha\boldsymbol{u}''\rangle. \tag{4.100}$$

To approximate the dissipative and diffusive terms in the transfer equations for the diffusion fluxes $\boldsymbol{J}_\alpha^{\text{turb}}$, we use the following simple expressions similar to (4.91)–(4.93) for the turbulent heat flux:

$$\boldsymbol{J}^{\Sigma}_{(Z_\alpha\boldsymbol{u})}=-\left(c_{\alpha1}L\langle b\rangle^{\frac{1}{2}}+c_{\alpha2}(\chi+\nu)\right)\frac{\partial}{\partial\boldsymbol{r}}\boldsymbol{J}_\alpha^{\text{turb}}, \tag{4.101}$$

$$\overline{p'(\partial Z''_\alpha/\partial\boldsymbol{r})}=-K_{\alpha1}\frac{\langle b\rangle^{\frac{1}{2}}}{L}\boldsymbol{J}_\alpha^{\text{turb}}-K_{\alpha2}\boldsymbol{P}_\alpha, \tag{4.102}$$

$$\overline{\rho}\varepsilon_\alpha = K_{b2}\frac{\nu+\chi}{2L^2}\boldsymbol{J}_\alpha^{\text{turb}}. \tag{4.103}$$

Here, the coefficients $K_{\alpha1}, \quad K_{\alpha2}, \quad c_{\alpha1}, \quad c_{\alpha2} \quad (\alpha = 1, 2, \ldots, N)$ are empirical constants, the quantity

$$\boldsymbol{P}_\alpha \equiv -\left(\boldsymbol{J}_\alpha^{\text{turb}}\frac{\partial\langle\boldsymbol{u}\rangle}{\partial\boldsymbol{r}}\right) - \boldsymbol{J}_\alpha^{\text{turb}} \times \boldsymbol{\Omega} + \overline{\rho}\,\boldsymbol{G}_\alpha \tag{4.104}$$

characterizes the generation rate of the turbulent diffusion flux $\boldsymbol{J}_\alpha^{\text{turb}}$ due to the mean strain of the medium and buoyancy effects. Since the necessary ad hoc data are completely absent in the literature, we use numerical values similar to (4.95) for the universal constants of the model:

$$K_{b2} = 6; \qquad K_{\alpha1} = 0.4; \qquad K_{\alpha2} = 0.5; \qquad c_{\alpha1} = 0.8. \tag{4.105}$$

This is probably possible for a passive (non-reacting) mixture. In contrast, for a chemically active mixture, additional special experiments that would allow the special cases of reacting turbulence to be modeled are required for an accurate determination of these constants.

It should be kept in mind that the N equations (4.96) for the turbulent diffusion fluxes are not independent; it is easy to show that their addition with prior multiplication by m_α leads to an identity. Therefore, the algebraic integral $\sum_{\alpha=1}^{N} m_\alpha \boldsymbol{J}_\alpha^{\text{turb}} = 0$ should be used as an additional relation to determine the fluxes $\boldsymbol{J}_\alpha^{\text{turb}}$.

We have shown above that the correlations $\langle H''^2\rangle$, $\langle Z''_\alpha H''\rangle$, and $\langle Z''_\alpha Z''_\beta\rangle$ directly enter into the transfer equations for the turbulent diffusion and heat fluxes or, more specifically, into the terms describing the generations of these fluxes under buoyancy forces (see (4.91) and (4.99)). Therefore, below we also consider the transfer equations for these correlations and the questions of their closure.

4.2.4 *Transfer Equations and Dissipation of the Scalar Second Moments for a Multicomponent Medium with a Variable Density*

4.2.4.1 Transfer Equation for the Mean-Square Mixture Enthalpy Fluctuations

Let us identify the parameter A in the general transfer equation for the mean-square fluctuation (4.57) with the enthalpy of a multicomponent mixture H and use (4.87$^{(1)}$) for the flux $\boldsymbol{J}_{(H)}$ and the generation rate $\sigma_{(H)}$. As a result, we obtain the following transfer equation for the enthalpy variance for developed turbulence:

$$\overline{\rho}\frac{D}{Dt}\langle H''^2/2\rangle + \operatorname{div} \boldsymbol{J}_H^\Sigma = -\boldsymbol{q}^{\text{turb}}\left(\frac{\partial\langle H\rangle}{\partial \boldsymbol{r}} - \frac{1}{\overline{\rho}}\frac{\partial \overline{p}}{\partial \boldsymbol{r}}\right) - \overline{H''(\mathrm{d}p'/\mathrm{d}t) + \boldsymbol{\Pi} : (\partial \boldsymbol{u}''/\partial \boldsymbol{r})} - \overline{\rho}\varepsilon_H, \tag{4.106}$$

where

$$\boldsymbol{J}_H^\Sigma \equiv \overline{\rho H''\boldsymbol{u}''/2} + \overline{H''\boldsymbol{q}} = \tfrac{1}{2}\boldsymbol{q}^{\text{turb}} + \overline{H''\boldsymbol{q}} \tag{4.107}$$

is the diffusion flux (a vector) of the variance $\langle H''^2\rangle$;

$$\overline{\rho}\varepsilon_H \equiv \overline{\rho}\langle\varepsilon_{HH}\rangle \equiv -\overline{\boldsymbol{q}(\partial H''/\partial \boldsymbol{r})} \tag{4.108}$$

is the scalar dissipation rate of the mean square of the enthalpy fluctuations $\langle H''^2\rangle$ due to molecular thermal diffusivity. According to the available estimates (Ievlev 1975), the penultimate term in (4.106) is small compared to the last dissipative term and may be discarded.

Equation (4.106) for the variance of the mixture enthalpy $\langle H''^2\rangle$ is similar in structure to (4.83) for turbulent energy transfer. However, it is much simpler than the latter, because it does not contain the correlation moments with pressure fluctuations and the terms attributable to buoyancy forces. In addition, in the case of a multicomponent chemically active medium, (4.106) is also much simpler than that for the temperature variance $\langle T''^2\rangle$, because, in contrast to the latter, it does not contain a large number of mixed second moments of the temperature and composition fluctuations. The presence of such correlations in the equation for the variance $\langle T''^2\rangle$ is ultimately associated with the chemical source of heat in the heat influx equation (2.29) for the temperature. The relative contribution from various processes to the balance of $\langle H''^2\rangle$ is similar to that from the corresponding terms in the equation for the turbulence energy, i.e., the terms in near-wall layers exceed the transfer terms in both generation and dissipation, while advection and diffusion have a stronger effect in free shear flows.

A gradient representation of the flux is adopted in most of the known models closing the diffusion terms in (4.106):

$$\boldsymbol{J}_H^\Sigma = -\left(c_5 L\langle b\rangle^{\frac{1}{2}} + c_6\chi\right)\frac{\partial}{\partial \boldsymbol{r}}\left(\overline{\rho}\langle H''^2/2\rangle\right). \tag{4.109}$$

Another important parameter in (4.106) requiring an approximation is the dissipation rate of $\langle H''^2\rangle$. Below, it is defined by the expression

$$\overline{\rho}\varepsilon_H = K_{H1}\frac{\overline{\rho}\langle b\rangle^{\frac{1}{2}}\langle H''^2\rangle}{L} + K_{H2}\frac{\chi\overline{\rho H''^2}}{L^2}. \tag{4.110}$$

Here, K_{H1}, K_{H2}, c_5, c_6 are universal constants, with K_{H2} and c_6 being significant only for small turbulent Reynolds numbers $\boldsymbol{Re}^{\text{turb}}$, i.e., when $K_{H1}\boldsymbol{Pr}\,\boldsymbol{R}$

$\boldsymbol{e}^{\text{turb}}/K_{H2} \leq 0(1)$ and $c_5 \boldsymbol{Pr}\, \boldsymbol{Re}^{\text{turb}}/c_6 \leq 0(1)$; here, $\boldsymbol{Pr} = \nu/\chi$ is the Prandtl molecular number. Based on data from Turbulence: Principles and Applications (1980), we can recommend the following typical values for these coefficients:

$$K_{H1} = 0.45; \quad K_{H2} = 3; \quad c_5 = 0.3 \pm 0.05. \tag{4.111}$$

It should be noted that, alternatively, the quantity ε_H can be found from its own transfer equation (see, e.g., Launder 1976). Since, according to approximation (4.109), the dissipation rate ε_H of the mean-square mixture enthalpy fluctuations, along with the quantities $\langle H''^2 \rangle$ and $\langle b \rangle$, determines a linear turbulent transport scale for the enthalpy fluctuations, $L_H \propto \langle b \rangle^{\frac{1}{2}} \langle H''^2 \rangle / \varepsilon_H$, this additional transfer equation for ε_H under the assumption of $L = L_H$ can also be used to determine the turbulence scale $L(\boldsymbol{r}, t)$ (see Sect. 3.3.4).

4.2.4.2 Transfer Equation for the Correlations Associated with Mixture Enthalpy and Composition Fluctuations

Let us assume that $\mathsf{A} \equiv Z_\alpha$ and $\mathsf{B} \equiv H$ in the general transfer equation (4.56) and use (2.7) and (4.87$^{(1)}$) for the corresponding fluxes and generation rates of these quantities. As a result, we obtain the following transfer equation for the correlations $\langle H'' Z''_\alpha \rangle$:

$$\bar{\rho} \frac{D\langle H'' Z''_\alpha \rangle}{Dt} + \operatorname{div} \boldsymbol{J}^{\Sigma}_{(HZ_\alpha)} = -\boldsymbol{q}^{\text{turb}} \frac{\partial \langle Z_\alpha \rangle}{\partial \boldsymbol{r}} - \boldsymbol{J}^{\text{turb}}_\alpha \left(\frac{\partial \langle H \rangle}{\partial \boldsymbol{r}} - \frac{1}{\bar{\rho}} \frac{\partial \bar{p}}{\partial \boldsymbol{r}} \right) + \sum_{s=1}^{r} \nu_{\alpha s} \overline{\xi_s H''}$$
$$+ \overline{Z''_\alpha \left(\frac{\mathrm{d}p'}{\mathrm{d}t} + \Pi : \frac{\partial \boldsymbol{u}''}{\partial \boldsymbol{r}} \right)} - \bar{\rho} \varepsilon_{H\alpha}, \quad (\alpha = 1, 2, \ldots, N) \tag{4.112}$$

where

$$\boldsymbol{J}^{\Sigma}_{(HZ_\alpha)} \equiv \overline{\rho H'' Z''_\alpha \boldsymbol{u}'' + H'' \boldsymbol{J}_\alpha + Z''_\alpha \boldsymbol{q}} \tag{4.113}$$

is the diffusion flux (a vector) of the correlation $\langle H'' Z''_\alpha \rangle$;

$$\bar{\rho} \varepsilon_{H\alpha} \equiv \bar{\rho} \langle \varepsilon_{HZ_\alpha} \rangle \equiv -\overline{\boldsymbol{q}(\partial Z''_\alpha / \partial \boldsymbol{r})} - \overline{\boldsymbol{J}_\alpha (\partial H'' / \partial \boldsymbol{r})}, \tag{4.113*}$$

is the scalar dissipation rate of the correlation $\langle H'' Z''_\alpha \rangle$ under the action of molecular processes and thermal diffusivity. Below, we discard the penultimate term on the right-hand side of (4.112), which is small compared to the dissipative term (Ievlev 1975).

According to (4.46), the terms including the correlations with the fluctuations in enthalpy and chemical sources of a substance of type α take the form

$$\sum_{s=1}^{r} \nu_{\alpha s}\overline{\xi_s H''} = \overline{H''}\left(\sum_{s=1}^{r} \nu_{\alpha s}\xi_s(\langle T\rangle, \overline{n}_\alpha)\right) + \sum_{s=1}^{r} \nu_{\alpha s}\overline{\xi_s' H''}$$
$$= \overline{\rho} L_{0\alpha}^{\mathrm{Ch}} \langle H''^2\rangle + \overline{\rho}\sum_{\beta=1}^{N} L_{\alpha\beta}^{\mathrm{Ch}} \langle Z''_\alpha H''\rangle. \tag{4.114}$$

The dissipative and diffusive terms in the transfer equation (4.112) can be approximated with the relations

$$\overline{\rho}\varepsilon_{H\alpha} = K_{H\alpha 1}\frac{\overline{\rho}\langle b\rangle^{\frac{1}{2}}}{L}\langle H''Z''_\alpha\rangle + K_{H\alpha 2}\frac{\overline{\rho}\chi}{L^2}\langle H''Z''_\alpha\rangle, \tag{4.115}$$

$$\boldsymbol{J}^{\Sigma}_{(HZ_\alpha)} = -\left(c_{H\alpha 1}L\langle b\rangle^{\frac{1}{2}} + c_{H\alpha 2}\chi\right)\frac{\partial}{\partial \boldsymbol{r}}\left(\overline{\rho}\langle H''Z''_\alpha\rangle\right). \tag{4.116}$$

The empirical constants $K_{H\alpha 2}$ and $c_{H\alpha 2}$ are significant only for small turbulent Reynolds numbers $\boldsymbol{Re}^{\mathrm{turb}}$, i.e., when $K_{H\alpha 1}\boldsymbol{Pr}\,\boldsymbol{Re}^{\mathrm{turb}}/K_{H\alpha 2} \le 0(1)$ and $c_{H\alpha 1}\boldsymbol{Pr}\,\boldsymbol{Re}^{T}/c_{H\alpha 2} \le 0(1)$. By analogy with (4.111), we can adopt the following values for the empirical coefficients:

$$K_{H\alpha 1} = 0.45; \quad K_{H\alpha 2} = 3; \quad c_{H\alpha 1} = 0.3 \pm 0.05.$$

The transfer equations (4.112) are also linearly dependent. For this reason, one of them should be discarded. In this case, the equality $\sum_{\alpha=1}^{N} m_\alpha\langle H''Z''_\alpha\rangle = 0$ is an additional relation to determine the correlations $\langle H''Z''_\alpha\rangle$.

4.2.4.3 Transfer Equations for the Correlation Moments of the Mixture Composition Fluctuations

Finally, assuming that $\mathsf{A} \equiv Z_\alpha$ and $\mathsf{B} \equiv Z_\beta$ in the general transfer equation (4.56), we obtain $N(N+1)/2$ transfer equations for the mixed pair correlations $\langle Z''_\alpha Z''_\beta\rangle$ symmetric in indices α and β as a result of simple manipulations:

$$\overline{\rho}\frac{D\langle Z''_\alpha Z''_\beta\rangle}{Dt} + \mathrm{div}\,\boldsymbol{J}^{\Sigma}_{\alpha\beta} = -\boldsymbol{J}^{\mathrm{turb}}_{\alpha}\frac{\partial\langle Z_\beta\rangle}{\partial \boldsymbol{r}} - \boldsymbol{J}^{\mathrm{turb}}_{\beta}\frac{\partial\langle Z_\alpha\rangle}{\partial \boldsymbol{r}}$$
$$= \sum_{s=1}^{r}\left(\nu_{\alpha s}\overline{\xi_s Z''_\beta} + \nu_{\beta s}\overline{\xi_s Z''_\alpha}\right) - \overline{\rho}\,\varepsilon_{\alpha\beta}, \quad (\alpha, \beta = 1, 2, \ldots N) \tag{4.117}$$

where

$$\boldsymbol{J}^{\Sigma}_{\alpha\beta} \equiv \overline{\rho Z''_\alpha Z''_\beta \boldsymbol{u}''} + \overline{Z''_\alpha \boldsymbol{J}_\beta + Z''_\beta \boldsymbol{J}_\alpha} \tag{4.118}$$

is the diffusion flux vector of the correlation $\langle Z''_\alpha Z''_\beta \rangle$;

$$\overline{\rho}\varepsilon_{\alpha\beta} \equiv \overline{\rho}\langle \varepsilon_{(Z_\alpha Z_\beta)} \rangle \equiv -\overline{\boldsymbol{J}_\beta(\partial Z''_\alpha/\partial \boldsymbol{r}) + \boldsymbol{J}_\alpha(\partial Z''_\beta/\partial \boldsymbol{r})} \tag{4.119}$$

is the scalar destruction rate of the correlation $\langle Z''_\alpha Z''_\beta \rangle$ due to molecular diffusion. According to (4.46), the terms containing the pair correlation moments of the composition and chemical reaction rate fluctuations $\sum_{s=1}^{r} \nu_{\beta s}\overline{\xi_s Z''_\alpha}$ and $\sum_{s=1}^{r} \nu_{\alpha s}\overline{\xi_s Z''_\beta}$ take the form

$$\sum_{s=1}^{r} \nu_{\beta s}\overline{\xi_s Z''_\alpha} = \overline{\rho}\, L^{(\mathrm{Ch})}_{0\beta}\, \langle H'' Z''_\alpha \rangle + \overline{\rho} \sum_{\gamma=1}^{N} L^{(\mathrm{Ch})}_{\gamma\beta}\, \langle Z''_\gamma Z''_\alpha \rangle, \tag{4.120}$$

$$\sum_{s=1}^{r} \nu_{\alpha s}\overline{\xi_s Z''_\beta} = \overline{\rho}\, L^{(\mathrm{Ch})}_{0\alpha}\, \langle H'' Z''_\beta \rangle + \overline{\rho} \sum_{\gamma=1}^{N} L^{(\mathrm{Ch})}_{\gamma\alpha}\, \langle Z''_\gamma Z''_\beta \rangle. \tag{4.121}$$

The following simple expressions can also be used to approximate the diffusive and dissipative terms in the transfer equation for the correlation $\langle Z''_\alpha Z''_\beta \rangle$:

$$\overline{\rho}\varepsilon_{\alpha\beta} = K_{\alpha\beta 1}\frac{\overline{\rho}\langle b \rangle^{\frac{1}{2}}}{L}\langle Z''_\alpha Z''_\beta \rangle + K_{\alpha\beta 2}\frac{\overline{\rho}\chi}{L^2}\langle Z''_\alpha Z''_\beta \rangle \tag{4.122}$$

$$\boldsymbol{J}^{\Sigma}_{\alpha\beta} = -\left(c_{\alpha\beta 1} L \langle b \rangle^{\frac{1}{2}} + c_{\alpha\beta 2}\chi\right)\frac{\partial}{\partial \boldsymbol{r}}\left(\overline{\rho}\langle Z''_\alpha Z''_\beta \rangle\right). \tag{4.123}$$

Note that (4.117) are also linearly dependent. The equalities

$$\sum_{\alpha=1}^{N} m_\alpha \langle Z''_\alpha Z''_\beta \rangle = 0, \quad \sum_{\beta=1}^{N} m_\beta \langle Z''_\alpha Z''_\beta \rangle = 0 \quad (\alpha, \beta = 1, 2, \ldots, N) \tag{4.124}$$

are additional relations to determine the correlations $\langle Z''_\alpha Z''_\beta \rangle$.

Thus, the various transfer equations for second-order correlation moments derived in this section close the system of turbulently averaged hydrodynamic equations for a compressible multicomponent mixture when the turbulence macroscale L is modeled in one way or another. In combination with the averaged hydrodynamic equations, they form a fairly complex semiempirical model of multicomponent turbulence in the second approximation based on which various, including anisotropic, turbulent flows of a reactive gas mixture can be computed. When the weighted-mean Favre averaging is used systematically, the proposed derivation of these equations makes it possible not only to trace all of the simplifying assumptions that are commonly made to write them but also allows their modifications to be easily obtained in the case of other turbulized flows, in particular, for heterogeneous and electrically conducting media.

4.3 Algebraic Closure Models for a Multicomponent Chemically Active Medium

The complicated phenomenological model of multicomponent turbulence considered in the preceding section includes the differential transfer equations for second-order one-point correlation moments as the closing ones. It potentially surpasses the simpler models based on the gradient relations for turbulent flows. However, the model transfer equations are fairly complex from a computational viewpoint and, hence, are often not very convenient for practical applications at the current stage of development of the mixture turbulence theory. At the same time, they can be used to improve the simpler gradient models. In particular, the differential transfer equations for the second correlation moments can be simplified to algebraic relations for these quantities and can be used subsequently to model the turbulent transport coefficients appearing in the gradient relations.

4.3.1 Local Equilibrium Approximation (the K-Theory of Turbulence for a Chemically Reacting Gas Mixture)

Thus, if we assume that there is an internal equilibrium with the field of averaged thermohydrodynamic parameters in the turbulent field structure at which the convective and diffusive terms in the transfer equations (4.66), (4.88), (4.96), (4.106), (4.112), and (4.117) balance each other, then the second-order correlation moments $R_{ij}, q_j^{\text{turb}}$, $\langle H''^2\rangle, J_{\alpha j}^{\text{turb}}$, $\langle H''Z''_\alpha\rangle, \langle Z''_\alpha Z''_\beta\rangle$ is in local equilibrium with one another. In other words, they do not change in time or coordinate space. This approximation (for a single-fluid medium) was called by Donaldson (1972) the "super-equilibrium" one. The relations between the second-order correlation moments and gradient characteristics of the main flow defined by (4.66), (4.88), (4.96), (4.106), (4.112), and (4.117) form a second-order closure scheme or the so-called K-theory of turbulence. It is valid if: (1) any changes of the averaged flow are very slow compared to the characteristic time for turbulent motion $t_{\text{turb}} \sim L/\langle b\rangle$ and (2) the spatial change in turbulence characteristics is small at distances of the order of the scale L. Generally, these two conditions are satisfied simultaneously very rarely, because the turbulence scale L is usually determined by the spatial changes of the flow. The boundary layer region with a constant momentum flux in which the ratio $L/\langle b\rangle$ is close to zero is the peculiar region where both conditions are satisfied.

Thus, in the local equilibrium approximation, the above transfer equations degenerate and we obtain the following algebraic equations to determine the second correlation moments:

1. *The equations to determine the correlations* $\langle u_k'' u_j'' \rangle$

$$(1-K_{p2})\left\{-\langle u_k'' u_i''\rangle\left(\frac{\partial\langle u_j\rangle}{\partial x_i}-2\varepsilon_{ilj}\Omega_l\right)-\langle u_j'' u_i''\rangle\left(\frac{\partial\langle u_k\rangle}{\partial x_i}-2\varepsilon_{ilk}\Omega_l\right)+G_{kj}\right\}$$
$$+K_{p1}\frac{\sqrt{\langle b\rangle}}{L}\left(-\langle u_k'' u_j''\rangle+\frac{2}{3}\delta_{kj}\langle b\rangle\right)+\frac{2}{3}K_{p2}\delta_{kj}\left(-\langle u_l'' u_m''\rangle\frac{\partial\langle u_l\rangle}{\partial x_m}+G\right)$$
$$-\frac{2}{3}K_{b1}\,\delta_{kj}\frac{\langle b\rangle^{\frac{3}{2}}}{L}-K_{b2}\frac{\nu\,\langle u_k'' u_j''\rangle}{L^2}=0,\qquad (k,j=1,2,3).\tag{4.125}$$

2. *The equations to determine the correlations* $\langle H'' u_j'' \rangle$

$$\left(K_{s1}\frac{\sqrt{\langle b\rangle}}{L}+K_{b2}\frac{\nu+\chi}{2L^2}\right)\langle H'' u_j''\rangle+\langle u_j'' u_k''\rangle\left(\frac{\partial\langle H\rangle}{\partial x_k}-\frac{1}{\bar\rho}\frac{\partial\bar p}{\partial x_k}\right)$$
$$+(1-K_{s2})\left\{\left(\frac{\partial\langle u_j\rangle}{\partial x_k}+2\varepsilon_{ljk}\Omega_l\right)\langle H'' u_k''\rangle-G_{Hj}\right\}=0,\qquad (j=1,2,3).\tag{4.126}$$

3. *The equation to determine the correlations* $\langle H''^2 \rangle$

$$\left(K_{H1}\frac{\sqrt{\langle b\rangle}}{L}+K_{H2}\frac{\chi}{L^2}\right)\langle H''^2\rangle+\langle H'' u_j''\rangle\left(\frac{\partial\langle H\rangle}{\partial x_j}-\frac{1}{\bar\rho}\frac{\partial\bar p}{\partial x_j}\right)=0;\tag{4.127}$$

4. *The equation to determine the correlations* $\langle Z_\alpha'' u_j'' \rangle$

$$\left(K_{\alpha1}\frac{\sqrt{\langle b\rangle}}{L}+K_{b2}\frac{\nu+\chi}{2L^2}\right)\langle Z_\alpha'' u_j''\rangle+(1-K_{\alpha2})\left\{\left(\frac{\partial\langle u_j\rangle}{\partial x_k}+2\varepsilon_{ljk}\Omega_l\right)\langle Z_\alpha'' u_k''\rangle-G_{\alpha j}\right\}$$
$$-\langle u_k'' u_j''\rangle\frac{\partial\langle Z_\alpha\rangle}{\partial x_k}-L_{0\alpha}^{\mathrm{Ch}}\langle H'' u_j''\rangle-\sum_{\beta=1}^{N}L_{\alpha\beta}^{\mathrm{Ch}}\langle Z_\alpha'' u_j''\rangle=0,\quad (j=1,2,3;\ \alpha=1,2,\ldots,N);\tag{4.128}$$

5. *The equations to determine the correlations* $\langle H'' Z_\alpha'' \rangle$

$$\left(K_{H\alpha1}\frac{\sqrt{\langle b\rangle}}{L}+K_{H\alpha2}\frac{\chi}{L^2}\right)\langle H'' Z_\alpha''\rangle+\langle H'' u_k''\rangle\frac{\partial\langle Z_\alpha\rangle}{\partial x_k}+\langle Z_\alpha'' u_k''\rangle\left(\frac{\partial\langle H\rangle}{\partial x_k}-\frac{1}{\bar\rho}\frac{\partial\bar p}{\partial x_k}\right)$$
$$-L_{0\alpha}^{\mathrm{Ch}}\langle H''^2\rangle-\sum_{\beta=1}^{N}L_{\alpha\beta}^{\mathrm{Ch}}\langle Z_\alpha'' H''\rangle=0,\quad (\alpha=1,2,\ldots,N)\tag{4.129}$$

6. *The equations to determine the correlations* $\langle Z''_\alpha Z''_\beta \rangle$

$$\left(K_{\alpha\beta 1} \frac{\sqrt{\langle b \rangle}}{L} + K_{\alpha\beta 2} \frac{\chi}{L^2} \right) \langle Z''_\alpha Z''_\beta \rangle + \langle Z''_\alpha u''_k \rangle \frac{\partial \langle Z_\beta \rangle}{\partial x_k} + \langle Z''_\beta u''_k \rangle \frac{\partial \langle Z_\alpha \rangle}{\partial x_k}$$
$$- L^{(\mathrm{Ch})}_{0\beta} \langle H'' Z''_\alpha \rangle - L^{(\mathrm{Ch})}_{0\alpha} \langle H'' Z''_\beta \rangle - \sum_{\gamma=1}^{N} L^{(\mathrm{Ch})}_{\gamma\beta} \langle Z''_\gamma Z''_\alpha \rangle$$
$$- \sum_{\gamma=1}^{N} L^{(\mathrm{Ch})}_{\gamma\alpha} \langle Z''_\gamma Z''_\beta \rangle = 0, \quad (\alpha, \beta = 1, 2, \ldots, N) \tag{4.130}$$

Here, the quantities G_{ki}, G, G_{Hk}, and $G_{\alpha k}$ describing the generation rate of the corresponding correlation due to the averaged pressure gradient are defined by the relations

$$\overline{\rho}\, G_{ki} = -\frac{1}{\langle n'_p \rangle \langle T \rangle} \left(\langle H'' u''_k \rangle \frac{\partial \overline{p}}{\partial x_i} + \langle H'' u''_i \rangle \frac{\partial \overline{p}}{\partial x_k} \right)$$
$$- \overline{\rho} \sum_{\alpha=1}^{N} \left\{ \left(\frac{1}{\overline{n}} \frac{\partial \overline{p}}{\partial x_i} - F_{\alpha i} \right) \langle Z''_\alpha u''_k \rangle + \left(\frac{1}{\overline{n}} \frac{\partial \overline{p}}{\partial x_k} - F_{\alpha k} \right) \langle Z''_\alpha u''_i \rangle \right\}, \tag{4.131}$$

$$\overline{\rho}\, G = -\frac{1}{\langle c_p \rangle \langle T \rangle} \langle H'' u''_k \rangle \frac{\partial \overline{p}}{\partial x_k}$$
$$+ \sum_{\alpha=1}^{N} \left(\sum_{\beta=1}^{N} \frac{\overline{n}_\beta (c_{p\alpha} - c_{p\beta})}{\overline{n} \langle c_p \rangle} \frac{\partial \overline{p}}{\partial x_k} + F_{\alpha k} \right) \langle Z''_\alpha u''_k \rangle, \tag{4.132}$$

$$\overline{\rho}\, G_{Hk} \equiv \left(-\frac{\langle H''^2 \rangle}{\langle c_p \rangle \langle T \rangle} + \sum_{\alpha=1}^{N} \sum_{\beta=1}^{N} \frac{\overline{n}_\beta (c_{p\alpha} - c_{p\beta})}{\overline{n} \langle c_p \rangle} \langle Z''_\alpha H'' \rangle \right), \tag{4.133}$$

$$\rho\, G_{\alpha k} = -\left(\frac{\langle H'' Z''_\alpha \rangle}{\langle c_p \rangle \langle T \rangle} - \sum_{\alpha=1}^{N} \sum_{\beta=1}^{N} \frac{\overline{n}_\beta (c_{p\alpha} - c_{p\beta})}{\overline{n} \langle c_p \rangle} \langle Z''_\alpha Z''_\beta \rangle \right) \frac{\partial \overline{p}}{\partial x_k}. \tag{4.134}$$

Equations (4.125)–(4.134) form the system of algebraic equations that relate the second-order correlation moments $\langle u''_k u''_j \rangle$, $\langle H'' u''_j \rangle$, $\langle Z''_\alpha u''_j \rangle$, $\langle H''^2 \rangle$, $\langle H'' Z''_\alpha \rangle$, $\langle Z''_\alpha Z''_\beta \rangle$ to the averaged flow parameters, the turbulence scale L, and the turbulent energy $\langle b \rangle$. The quantity $\langle b \rangle$ is defined by the formula $\langle b \rangle = \frac{1}{2} \langle u''_k u''_i \rangle \delta_{ki}$. Some transfer equation for the turbulence macroscale (3.189) (or some algebraic expression for the scale L) should be used to close this system.

Thus, determining the correlations $\langle u''_k u''_j \rangle$, $\langle H'' u''_j \rangle$, $\langle Z''_\alpha u''_j \rangle$, $\langle H''^2 \rangle$, $\langle H'' Z''_\alpha \rangle$, $\langle Z''_\alpha Z''_\beta \rangle$ from the system of algebraic equations (4.125)–(4.134) and assuming the gradient relations (see Monin and Yaglom 1992)

$$J_{\alpha k}^{\text{turb}}/\overline{\rho} \equiv \langle Z_\alpha'' u_k'' \rangle = -D_{\alpha kj}^{\text{turb}} \frac{\partial \langle Z_\alpha \rangle}{\partial x_j}, \tag{4.135}$$

$$q_k^{\text{turb}}/\overline{\rho} \equiv \langle H'' u_k'' \rangle = -\chi_{kj}^{\text{turb}} \left(\frac{\partial \langle H \rangle}{\partial x_j} - \frac{1}{\overline{\rho}} \frac{\partial \overline{p}}{\partial x_j} \right), \tag{4.136}$$

$$\begin{aligned} R_{ij}/\overline{\rho} &\equiv -\langle u_i'' u_j'' \rangle \\ &= -\frac{2}{3} \langle b \rangle \delta_{ij} + \frac{1}{2} \left(\nu_{is}^{\text{turb}} \delta_{jl} + \nu_{js}^{\text{turb}} \delta_{il} \right) \left(\frac{\partial \langle u_s \rangle}{\partial x_l} + \frac{\partial \langle u_l \rangle}{\partial x_s} \right), \end{aligned} \tag{4.137}$$

to be valid, we can, in general, derive the expressions for the turbulent viscosity, ν_{is}^{turb}, thermal diffusivity, χ_{kj}^{turb}, and diffusion, $D_{\alpha kj}^{\text{turb}}$, tensors. It should be specially emphasized that this approach takes into account the effect of chemical reactions on the intensity of the enthalpy and composition fluctuations and on the turbulent momentum, heat and mass transport coefficients. For example, the turbulent heat flux is determined not only by the temperature and concentration gradients but also by the dependence of the chemical source on mixture temperature and composition. The specific formulas for the turbulent exchange coefficients in the case of a horizontal turbulent shear flow in a stratified medium obtained in the local equilibrium approximation are considered in our monograph (Kolesnichenko and Marov 1999), to which we refer the reader.

Thus, we have completed the general description of the method for a "theoretical" (semiempirical) determination of the turbulent exchange coefficients (tensors), which is the objective of this section. Note that the empirical constants in various special cases can be modified significantly.

4.3.2 *Quasi-Equilibrium Approximation*

Generally speaking, it would be more appropriate to neglect the terms in (4.125)–(4.134) and (3.189) that describe the effects inherent in a laminar flow, because small Reynolds numbers are incompatible with the assumption about locally equilibrium turbulence. However, since this approach is approximate, they are often retained, but, at the same time, the complete turbulence energy transfer equation (4.83) with both convective and diffusive terms is invoked for the analysis. Of course, one of (4.125) should be omitted in this case. This approach (the so-called quasi-equilibrium approximation), which was first proposed by Lewellen (see Turbulence: Principles and Applications 1980) for a single-component fluid, appears to be more accurate than the "super-equilibrium approximation" (in which these terms are discarded in the equations), because it allows one, to some extent, to take into account the "nonequilibrium" effects of the turbulent field of fluctuating thermohydrodynamic parameters. In this case, the correlation characteristics of turbulence at each spatial point are related to the field of defining parameters in various flow regions. Applying the complete equation (4.83) to determine the turbulent energy can also be justified by the fact that the relaxation time for the "locally equilibrium" structure of the turbulent field is much shorter than the time it takes for the turbulent velocity field to reach a level at which the production and dissipation of turbulent energy become equal (see Ievlev 1975).

In conclusion, note that the flows commonly encountered in nature in which some conservative flow characteristic is transferred in the direction opposite to its gradient cannot be accurately calculated using the invariant modeling method in the second closure order (Mellor and Yamada 1974, 1982). Such a phenomenon is observed, for example, in the neutrally stratified (in temperature) boundary layer of the Earth's atmosphere in the case of developed convection where the heat flux is directed upward against the potential temperature gradient. As a result, the turbulent thermal conductivity coefficient in (4.114) turns out to be negative—the effect of a negative thermal conductivity. Accordingly, an adequate theory of counter-gradient transfer can probably be developed only on the basis of third-order closure models (see Lykosov 1991).

Chapter 5
Stochastic-Thermodynamic Modeling of Developed Structured Turbulence

The goal of this chapter is to bring the reader up to date on the currently rapidly developing stochastic-thermodynamic theory of irreversible processes using the modeling of structured turbulence as an example. Here, we consider the synergetic approach to developing a phenomenological model of extremely developed turbulence in a compressible homogeneous fluid by taking into account the nonlinear cooperative processes. Until recently, turbulence, which is without exaggeration the most common type of fluid motion in nature, was traditionally represented mainly as a fine-grained fluctuating continuum in a state of total stochastic chaos. However, a different viewpoint on turbulence that was probably first put forward explicitly by Prigogine (Prigogine and Stengers 1984) is also admissible. According to this approach, a real turbulent fluid flow is a less-random and macroscopically more organized process than it seems at first glance; the transition from a laminar flow to a turbulent one is the process of self-organization whereby part of the energy of turbulent chaos corresponding to small-scale fluctuations of thermohydrodynamic fluid parameters passes into a macroscopically organized motion of vortex coherent structures (CSs). This increases the internal order of a turbulized hydrodynamic system compared to molecular chaos (a laminar fluid motion). In particular, the cascade fragmentation of vortices in fully developed turbulence can also be interpreted as an unbounded sequence of self-organization processes. In this case, the set of spatiotemporal scales on which such a process unfolds corresponds to a coherent behavior of an enormous number of particles manifested in the formation of relatively stable mesoscale supermolecular structures (when the molecules are involved in collective, coordinated, and interrelated motions corresponding to different-scale vortices continuously distributed in a real fluid flow). This change of the view on turbulence is clearly expressed in Prigogine's utterance (see Prigogine and Stengers 1994), "Who could predict thirty years ago that nonequilibrium leads to self-organization in the form in which we observe it in hydrodynamic instabilities like Benard cells."

As has now become clear, the presence of relatively large coherent vortex structures (turbulent filaments, rings, vortex spirals, etc.) randomly distributed in space and time is a characteristic feature of many, if not all, developed turbulent

M.Y. Marov and A.V. Kolesnichenko, *Turbulence and Self-Organization: Modeling Astrophysical Objects*, Astrophysics and Space Science Library 389, DOI 10.1007/978-1-4614-5155-6_5, © Springer Science+Business Media New York 2013

flows (see Crow and Champagne 1971; Brown and Roshko 1974; Rabinovich and Sushchik 1990; Monin and Yaglom 1975). According to the latest views (see, e.g., Klimontovich 2002), hydrodynamic turbulence, which is among the most complex dynamical phenomena, is related, in particular, to the formation and development of an enormous number of organized dissipative vortex structures with various spatio-temporal scales under certain fluid flow conditions in an essentially nonequilibrium open system. For instance, the self-organization processes against the background of a chaotic fluctuating motion of cosmic matter appear to be the most important mechanism that form the peculiar features of astro- and geophysical objects at various stages of their evolution, including the formation of galaxies and galaxy clusters, the birth of stars from the diffuse medium of gas–dust clouds, the formation of protoplanetary disks and the subsequent accumulation of planetary systems, the formation of planetary and cometary gaseous envelopes (atmospheres), and the different-scale flows in atmospheres and circumplanetary plasmas.

According to the currently available experimental data (for a comprehensive overview of the corresponding publications, see, e.g., the monograph by Khlopkov et al. (2002)), a coherent vortex structure can be defined as a connected, liquid mass with phase-correlated (i.e., coherent) vorticity in the entire region of coordinate space occupied by the structure (see Sect. 6.2). In the last decade, a large number of various CSs have been discovered and their topological properties have been firmly established owing to progress achieved in developing the technique for a visual observation of turbulized fluid flows. As examples, we can name such CSs as vortex filaments, Taylor vortices, turbulent spots, vortex balls, hairpin-like vortices, burstings, vortex spirals, streaks, Brawn–Thomas structures, and mushroom-shaped vortices. The frequency of occurrence of a particular structure depends on the type of flow (a boundary layer, a mixing layer, a jet, etc.) and on the geometry and regime of turbulized fluid motion. As a rule, such vortex structures are localized in space, do not overlap (therefore, they can often be considered as lumped objects—clusters), and have mean free paths much larger than their own sizes. By definition, the characteristic CS size is the largest spatial scale l on which coherent vorticity exists; the latest results show that l can be much smaller than the characteristic hydrodynamic flow scale L but larger than the Kolmogorov scale η, i.e., it can lie in the inertial range of scales, $\eta < l << L$. Due to the interaction of individual CSs (e.g., according to the Biot–Savart law in the case of two simple vortices), which, generally, is essentially nonlinear, they combine or break up, i.e., a new single structure (e.g., a spiral vortex) or adjacent similar structures (rings, balls, etc.) appear, which can be connected by ropes—the regions of low coherent vorticity. Below, we will proceed from the fact that the mesoscale spatiotemporal CSs are random in all of the parameters used to describe them.

Unfortunately, a direct numerical simulation of developed turbulent motions based on exact (instantaneous) hydrodynamic equations generally involves great mathematical difficulties, while constructing a full-blown theory of structured turbulence is hardly possible at the current stage of development of hydromechanics, because the mechanisms for the formation, interaction, and evolution of different-scale vortex structures are extremely complex. Actually, for this reason, the development of new model approaches (including phenomenological ones) to describing fully developed turbulence, the introduction of additional internal parameters of a medium (characterizing small-scale flow structures), the establishment of universal laws and special relations supplementing already known equations such as the mass, energy, and momentum conservation laws for their calculation are required. Of course, such semiempirical modeling of structured turbulence, despite the idealization of the description of a real fluid flow inevitable in this case, must nevertheless reflect the main features of the most important hydrodynamic effects with minimum necessary computational efforts. Consequently, to construct an adequate model for a "highly organized" turbulent flow, it is generally necessary to also consider relatively large dissipative coherent structures, along with a fine-grained fluctuating field of thermohydrodynamic flow parameters (such an approach was considered in Chap. 3). They should be taken into account both when modeling the process of drawing an as yet non-turbulized surrounding fluid into turbulized motion and when describing the fully developed turbulent mass, momentum, and heat transport processes. In other words, any efficient continuum model of turbulence cannot be constructed without explicitly including spatiotemporal coherent structures and without describing them by some internal state parameters of a fluctuating flow. According to Frisch (1995), such regular vortex structures are, in a sense, the "sinews" of turbulence. Consequently, the true possibilities for efficiently overcoming the mechanical and mathematical problems with which the promising formulations and numerical implementations of various turbulent problems, in particular, in astrophysics and cosmogony, are associated open up only when turbulence is modeled by taking into account its internal structurization (see Sedov 1980; Marov and Kolesnichenko 2001).

In view of the aforesaid, it becomes clear why the phenomenon of structured turbulence provides the researchers with rich information for the development of new ideas concerning the order–chaos relationship in turbulized fluid flows and the simplicity and complexity in the behavior of open fluctuating hydrodynamic systems, which can form spatiotemporal vortex structures without any specific impact from outside through self-organization, i.e., establish "order through fluctuations" far from local thermodynamic equilibrium. Unfortunately, it should be noted that, although more than 30 years has elapsed since the understanding of the synergetic nature of turbulence as a self-organization process, the views of dissipative coherent structures emerging in a flow have not yet been materialized into the development of model approaches aimed at creating practical (engineering) methods of turbulence calculation based, as a rule, on hydrodynamic equations. At the same time, extending the formalism of nonequilibrium thermodynamics to media with excited internal degrees of freedom of macromolecules (which can be described by additional parameters characterizing the internal microstructure of the medium) probably allows this approach to be also extended to modeling the cascade transport of kinetic energy by vortices of various sizes formed by their successive fragmentation in a developed turbulent flow.

Thus, the goal of this chapter is to try to construct a phenomenologically hydrodynamic model of stationary-nonequilibrium turbulence using the methods of extended irreversible thermodynamics with internal variables (see de Groot and Mazur 1962; Jou et al. 2001) and statistical thermodynamics of irreversible processes (Stratonovich 1985; Keizer 1987), which have proven in recent years to be so efficient in studying a wide class of physical problems, by taking into account the nonlinear cooperative processes in them that give rise to various nonequilibrium dissipative vortex structures far from thermodynamic equilibrium (see, e.g., Ebeling et al. 2001). In other words, we will attempt to phenomenologically obtain a closed system of averaged hydrodynamic equations self-consistently describing the various turbulent transport and self-organization processes. In this approach, the vortex CSs belonging to the strongly localized regions of small-scale motion of the subsystem of turbulent chaos should be taken into account both at the stage of laminar-flow turbulization due to the development of a hierarchy of instabilities of some type (e.g., the Kelvin–Helmholtz instability in free shear flows—in mixing layers, jets, wakes, etc., or the Taylor–Gertler instability in the sublayer of a turbulent boundary layer where longitudinal vortices can appear in near-wall layers) and at the stage of developed turbulence given all possible resonant situations associated, for example, with the synchronization of small-scale vorticity in a strongly turbulized fluid flow. Here and below, by the synchronization of regular and chaotic (stochastic) oscillations we mean the establishment of some relations between the characteristic frequencies and phases of self-oscillating systems as a result of their mutual influence (see Sect. 6.2).

In Sect. 5.1, we will discuss the synergetic approach to constructing a phenomenological model of developed turbulence in a compressible homogeneous fluid by taking into account the nonlinear cooperative processes. Including a set of random variables in the model as the internal parameters of the subsystem of turbulent chaos related to its microstructure allows the Fokker–Planck–Kolmogorov (FPK) kinetic equations in configuration space to be derived in this case by thermodynamic methods. These equations are designed to determine the temporal evolution of the conditional probability density for the random structural parameters of chaos (referring, in particular, to the cascade fragmentation of large-scale vortices or temperature inhomogeneities). They will also be used below to analyze the Markovian stochastic processes of the transition from one quasi-nonequilibrium state of turbulent chaos to another through a successive loss of fluid flow stability when varying the corresponding control parameters. Stabilization of the subsystem of chaos near another stationary-nonequilibrium state in configuration space corresponds to the transition of a turbulized system to a new state appropriate to the formation of complex spatiotemporal CSs in a turbulized flow.

In Sect. 5.2, we will consider an alternative method for investigating the mechanisms of such a transition based on stochastic Langevin equations closely related to the FPK kinetic equations (this relation is traced in this section). Here, we analyze the cardinal problem of the approach being developed—the possibility of the existence of asymptotically stable stationary-nonequilibrium states in the subsystem of turbulent chaos. For this purpose, we propose a nonequilibrium thermodynamic potential for the stochastic internal coordinates of turbulent chaos that generalizes the well-known Boltzmann–Planck relation for equilibrium thermodynamic states to stationary-nonequilibrium states for the ensemble of subsystems (representing chaos) and show that this potential is the Lyapunov function for such states.

Finally, the last third section of this chapter is devoted to the thermodynamic derivation of generalized fractional FPK equations describing the evolution of the internal coordinates of the subsystem of turbulent chaos based on fractional dynamics that allows for the structure and metric of the fractal time. Such equations are designed to describe the transition to new quasi-equilibrium states with the topology of a fractal set. The structural stability of CSs in this case is maintained through multiscale correlations, which play a key role in the regime of strong nonlinearity. It can be said that the phase space fractality and the long-range correlation effects are mutually consistent characteristics of the system's self-organization process during its transition to a nonequilibrium turbulent state. Introducing fractional time derivatives into the FPK kinetic equation allows the memory effects (i.e., the non-Markovian nature of the evolution processes in some regions of phase space) and the intermittency in time, with which the presence of turbulent bursts against the background of less intense low-frequency oscillations of fine-grained turbulence is usually associated, to be taken into account in the context of unified mathematical formalism.

5.1 Synergetic Approach to Describing Stationary-Nonequilibrium Turbulence

Developed turbulence realized at a finite but fairly large Reynolds number is known to be characterized by continuous Fourier spectra (both temporal and spatial ones), suggesting the existence of multiscale structure in the field of thermohydrodynamic parameters. Actually, multiscale structure of a fluid flow when an enormous number of degrees of freedom are excited is a key signature of developed turbulence. Therefore, any model approach to describing fully developed turbulence is basically one or another way of limiting the degrees of freedom.

When phenomenologically constructing the model of structured turbulence (aimed at describing "regular" fields of hydrothermodynamic parameters), we will represent a moving fluctuating fluid as a thermohydrodynamic system consisting of two interrelated continua (subsystems) that simultaneously fill the same volume of coordinate space continuously—the subsystem of averaged motion and the subsystem of turbulent spatiotemporal chaos (Marov and Kolesnichenko 1999; Kolesnichenko 2002, 2004). The continuum of averaged motion obtained by the probability-theoretic averaging of the instantaneous hydrodynamic equations for a fluid is designed mainly to investigate the evolution of the averaged fields of thermohydrodynamic parameters (including the description of large vortex structures). In turn, the subsystem of turbulent chaos (turbulent "superstructure") consists of two components (phases): turbulent chaos proper (the so-called incoherent turbulence) associated with the stochastic small-scale fluctuating motion of a turbulized fluid and a coherent component, generally, with the topology of a fractal set and associated with mesoscale vortex structures in a turbulized fluid flow embedded in this almost uniform (fine-grained) fluctuating field of thermohydrodynamic parameters. In the thermodynamic description of the subsystem of turbulent chaos, we will include the set of internal coordinates that ultimately correspond to excited macroscopic degrees of freedom of a turbulized fluid. This makes it possible to use the generalized Onsager theory (see Chap. 2) when modeling the processes of turbulent transport and kinetics in the total continuum. In the case under consideration, this theory describes not only the linear relaxation of the averaged extensive thermodynamic parameters to their stationary values but also the behavior of the turbulent fluctuations in the neighborhood of stationary-nonequilibrium states of chaos (see Keizer 1987). By applying the well-known extension of the formalism of nonequilibrium thermodynamics to systems with internal degrees of freedom (see Prigogine 1967), we can obtain the FPK kinetic equations for the distribution functions of the characteristic parameters of small-scale turbulence and model the Richardson–Kolmogorov cascade process on their basis.

It should be noted that our separation of a real fluctuating fluid flow into imaginary (averaged and turbulent) ones, in general, depends on the choice of the spatiotemporal region for which the mean values of the local physical variables that are continuous functions of the coordinates $\boldsymbol{r}$ and time t have been established, i.e., to some extent, it is arbitrary. Below, we will assume the hydrodynamic scale of averaged motion Λ (the Obukhov scale of observation (Obukhov 1962) or the resolution step of the finite-difference grid) lying in the inertial range $\eta<\Lambda<$ and determined by the size $d\boldsymbol{r} \sim \Lambda^3$ of the averaging region G to be such that the subsystem of turbulent chaos contains the entire set of mesoscale CSs whose size is smaller than the averaging region, $l<\Lambda$. Here, $\eta = (\nu^3/\varepsilon_b)^{1/4}$ is the Kolmogorov length scale that characterizes the effect of viscous dissipation on the structure of small-scale turbulence; L is the external or integral scale of turbulence characterizing its generation mechanism; ε_b is a key parameter of the Richardson–Kolmogorov cascade vortex fragmentation process that is the mean dissipation rate of turbulent energy per unit fluid mass per unit time and simultaneously (under quasi-equilibrium conditions) the transfer rate of the kinetic energy of fluctuating motion over the hierarchy of vortices. According to the existing estimates, for the averaged flow to contain the bulk (80 or 90 %) of the total energy of a turbulized flow, the averaging scale Λ must be smaller than the integral turbulence scale L by a factor of 10–20. Note that the specificity of the two-phase structure of turbulent chaos manifests itself in the additional turbulent momentum and energy transport by vortex coherent structures. This slightly modifies the known closure models (defining relations) and necessitates refining the effective (given the presence of CSs in the flow) turbulent viscosity and thermal conductivity coefficients.

5.1.1 System of Hydrodynamic Equations of Mean Motion for a Single-Component Compressible Fluid

Below, we will follow the classical approach to the phenomenology of fully developed turbulence. It is based on Reynolds' idea of averaging the instantaneous equations of fluid motion for fluctuating thermohydrodynamic parameters over space and/or time or through a different equivalent procedure, for example, through the probability-theoretic averaging over an ensemble of statistically similar hydrodynamic systems under identical external conditions adopted in statistical hydrodynamics (see Monin and Yaglom 1992). Under the standard (for statistical physics) assumption about the system's ergodicity, when the time (space) average of any physical variable can be identified with its probability-theoretic average, these averagings filter out the modes of motion whose scale is smaller than the spatiotemporal averaging interval. These small-scale fluctuating motions excluded in the averaging process are assumed to contribute to the turbulent fluid motion determined by the fluctuations of hydrodynamic parameters with respect to the corresponding average values. Actually, it is this small-scale turbulent motion that is modeled below by the subsystem of turbulent chaos.

It is important to note that the averaging problem is a central one in the mechanics of natural media. In the case of such a complex system as structured turbulence, the construction of its macroscopic model itself often depends precisely on the averaging method. Bearing in mind the various applications of the turbulence model being developed, in particular, to astrophysical phenomena in which the ratio of the characteristic fluid velocity to the average speed of sound (a measure of significance of the fluid density fluctuations) is much greater than unity, in this chapter we will assume the system's mass density $\rho(r,t)$ to be variable. We have already noted above (see. Sect. 3.1) that the averagings in classical theories of turbulence with a constant mass density usually for all physical parameters of the medium without exception are introduced in some identical way, as a rule, without any weight coefficients. At the same time, such an identical averaging for all physical parameters in the case of a fluid with a variable mass density leads not only to cumbersome hydrodynamic equations of mean motion but also to difficulties in the physical interpretation of some individual terms in them. Therefore, when constructing a phenomenological model of structured turbulence in a compressible medium, in addition to the "ordinary" means $\bar{f}(r,t)$ of some thermohydrodynamic variables $f(r,t)$ (e.g., the density or pressure), we will use, as in the two preceding chapters of this book, the Favre means for some other parameters $g(r,t)$ (e.g., the temperature or hydrodynamic velocity) specified by the relation

$$\langle g(r,t)\rangle = \overline{\rho g(r,t)}\,/\,\bar{\rho} = \lim_{M\to\infty}\frac{1}{M}\sum_{p=1}^{M}\rho^{[p]}g^{[p]}\Big/\lim_{M\to\infty}\frac{1}{M}\sum_{p=1}^{M}\rho^{[p]}. \tag{5.1}$$

Here, the summation is over the set of possible realizations $[p]$ ($1 \leq [p] \leq M$) for a stochastic hydrodynamic system; $g = \langle g \rangle + g''$, where g'' is the corresponding turbulent fluctuation, with $\overline{g''} \neq 0$. Below in this chapter, unless stated otherwise, the letter in angle brackets will denote the weighted-mean averaging of the corresponding physical quantity. Let us write out here the properties of the weighted-mean averaging of physical quantities that are often used in this chapter [see (3.7)]:

$$\overline{\rho g f} = \overline{\rho}\langle g \rangle \langle f \rangle + \overline{\rho g'' f''}, \quad \overline{\rho g''} = 0, \quad \overline{g''} = -\overline{\rho' g''} \, / \, \overline{\rho},$$

$$\overline{\rho\left\{\frac{\partial f}{\partial t} + u \cdot \frac{\partial}{\partial r} f\right\}} = \overline{\rho}\left\{\frac{\partial \langle f \rangle}{\partial t} + \langle u \rangle \cdot \frac{\partial}{\partial r} \langle f \rangle\right\} + \frac{\partial}{\partial r} \cdot \overline{(\rho f'' u'')}$$

$$\equiv \overline{\rho}\,\frac{D\langle f \rangle}{Dt} + \frac{\partial}{\partial r} \cdot J_f^{\text{turb}}, \; D(..)\,/\,Dt \equiv \partial(..)/\partial t + \langle u \rangle \cdot \partial(..)/\partial r.$$

Here, the turbulent flux of an attribute f is denoted by

$$J_f^{\text{turb}} \equiv \overline{\rho f'' u''} = \overline{\rho}\,\langle f'' u'' \rangle.$$

These relations can be easily derived from definition (5.1) and the Reynolds averaging postulates (3.2).

The system of exact hydrodynamic equations of mean motion for a single-component compressible fluid obtained by the probability-theoretic averaging of the corresponding instantaneous hydrodynamic equations valid on the microscale are [cf. (3.22), (3.29), (3.55), (3.72)]

$$\overline{\rho}\,\frac{D\langle v \rangle}{Dt} = \text{div}\,\langle u \rangle, \quad (\langle v \rangle \equiv 1/\overline{\rho}), \tag{5.2}$$

$$\overline{\rho}\,\frac{D\langle u \rangle}{Dt} = -\frac{\partial \overline{p}}{\partial r} + \left(\frac{\partial}{\partial r} \cdot \Pi^{\Sigma}\right) + \overline{\rho} F, \tag{5.3}$$

$$\begin{aligned}\overline{\rho}\,\frac{D\langle E \rangle}{Dt} &= -\text{div}\,\left(\overline{q} + q^{\text{turb}} - \overline{p' u''}\right) - \overline{p}\,div\langle u \rangle + \left(\overline{\Pi} : \frac{\partial \langle u \rangle}{\partial r}\right) \\ &\quad - \overline{p'\,div\,u''} + \left(J_v^{\text{turb}} \cdot \frac{\partial \overline{p}}{\partial r}\right) + \overline{\rho}\,\langle \varepsilon_b \rangle,\end{aligned} \tag{5.4}$$

$$\overline{p}(r,t) = \mathcal{R}\overline{\rho}\,\langle T \rangle\,. \tag{5.5}$$

Here, $u(r,t), p(r,t), v(r,t), \; E(r,t)$ are, respectively, the instantaneous values of the hydrodynamic velocity, pressure, specific volume ($v = 1/\rho$), and specific internal energy for a fluid particle; $\langle u(r,t) \rangle \equiv \overline{\rho u} \, / \, \overline{\rho}$ is the Favre-averaged hydrodynamic velocity of the medium; $D(..)\,/\,Dt \equiv \partial(..)/\partial t + \langle u \rangle \cdot \partial(..)/\partial r$ is the total time derivative relative to the averaged velocity field; $\overline{p}(r,t)$ is the average pressure;

R is the gas constant; $F(r,t)$ is the external force acting on a unit mass (in this section, we neglect the fluctuations in mass force); $\Pi^{\Sigma}(r,t) \equiv \overline{\Pi}(r,t) + R(r,t)$ is the total stress tensor in a turbulent flow; $\overline{\Pi}(r,t)$ is the average molecular viscous stress tensor; $R(r,t) \equiv -\overline{\rho u'' u''}$ is the turbulent (Reynolds) stress tensor [see (3.31)]; $\overline{q}(r,t)$ is the average molecular heat flux; $q^{\text{turb}}(r,t) \equiv \overline{\rho H'' u''}$ and $J_{v}^{\text{turb}}(r,t) \equiv -\overline{\rho' u''} / \overline{\rho}$ are, respectively, the turbulent heat and specific volume fluxes (here, we use the relation $\rho v'' = -\rho'/\overline{\rho}$); $H(r,t) \equiv E(r,t) + p(r,t)/\rho(r,t)$ is the instantaneous value of the specific fluid enthalpy; $\langle \varepsilon_b(r,t) \rangle = \overline{\rho \varepsilon_b}/\overline{\rho} \equiv \left(\overline{\Pi : \partial u''/\partial r} \right) / \overline{\rho}$ is the weighted-mean value of the specific dissipation rate of turbulent kinetic energy into heat due to molecular viscosity v [see (3.51)].

It can be seen from the system of equations (5.2)–(5.5) that the averaged motion of a turbulized homogeneous fluid is characterized by: (1) the average molecular thermodynamic fluxes $\overline{q}(r,t)$ and $\overline{\Pi}(r,t)$, for which the corresponding defining relations are needed (they are derived for a turbulized medium using the thermodynamic approach, for example, in Kolesnichenko (1998)); and (2) the as yet indefinite mixed one-point one-time correlations (second-order moments) $R(r,t)$, $q^{\text{turb}}(r,t)$, and $J_{v}^{\text{turb}}(r,t)$, which represent the so-called turbulent fluxes of the fluid characteristics related to the fluctuations of thermohydrodynamic parameters. The correlation terms including pressure fluctuations $\overline{p'\, div\ u''}$ and $\overline{\text{div}(p'u'')}$ as well as the weighted-mean viscous dissipation rate of turbulent energy $\langle \varepsilon_b \rangle$ should also be determined. When the model of developed turbulence is constructed phenomenologically, the defining (constitutive) relations closing system (5.2)–(5.5) can be established by the same method as in the laminar case (see Sect. 2.1), i.e., in accordance with the thermodynamic rules of continuum mechanics by the Onsager method. However, the additional idea that the corresponding thermodynamic forces are also responsible for the linear relaxation of the fluctuating characteristics of turbulized chaos to its stable stationary-nonequilibrium state (see, e.g., Keizer 1987) will be used below as a basis for the latter.

5.1.2 *Thermodynamics of Structured Turbulence: The Internal Fluctuating Coordinates of the Subsystem of Turbulent Chaos*

Thus, our immediate task is to suggest a concept that would allow us to go beyond the classical formalism of irreversible thermodynamics considered in the second chapter. This goal can be achieved by expanding the space of independent basic variables by introducing the internal coordinates defining the microstructure of the subsystem of turbulent chaos. The subsequent step involves finding the evolutionary equations for these additional state parameters.

Within the framework of the complete model of structured turbulence, the system of equations (5.2)–(5.5) obtained by the probability-theoretic averaging of the instantaneous hydrodynamic equations for a single-component fluid is designed to investigate the spatiotemporal evolution of the averaged fields of hydrodynamic quantities, including various large vortex structures. In accordance with Prigogine's viewpoint on hydrodynamic turbulence as a macroscopically highly organized flow we adopted, below we will consider the subsystem of turbulent (vortex) chaos as a continuum with a certain internal microstructure. Moreover, we will assume that the vortex continuum consists of two components: turbulent chaos proper (the so-called incoherent turbulence) associated with the stochastic small-scale fluctuating motion of a turbulized fluid and a coherent component embedded in this almost uniform fluctuating field of hydrodynamic parameters. The latter component is an ensemble of mesoscale vortex structures (multimolecular structures) whose images in the phase space of an equivalent dynamical system are classical attractors (e.g., limit cycles) or strange attractors (with a fractal structure). Each of these two subsystems is thermodynamically open, i.e., capable of exchanging energy and entropy (but not mass) with the adjacent subsystem. Below, we will assume the hydrodynamic velocity fields for the subsystems of averaged motion and turbulent chaos to be coincident, because no separation of the corresponding Lagrangian volumes (diffusion effect) occurs in the process of a real turbulent fluid motion the subsystem of turbulent chaos has no macroscopic hydrodynamic velocity relative to the subsystem of averaged motion. Note that a different approach to modeling structured turbulence associated, in particular, with the triple decomposition of the instantaneous motion of a fluid into an averaged motion and coherent and incoherent fluctuations is also known in the literature. The hydrodynamic equations with double averaging, over time and over a specially chosen ensemble determined by some features characteristic of CSs, serve as a basis for such models. However, this procedure is not free from many contradictions (see, e.g., Khlopkov et al. 2002).

We emphasize once again that our artificial separation of a real turbulized fluid flow into imaginary averaged and turbulent (vortex, fluctuating) ones is a convenient way of pictorially describing the phenomenon, i.e., to a certain extent, it has a model character. For each of these two subsystems for any elementary volume $d\boldsymbol{r}$ of the medium, we will determine local "coarse-grained" thermodynamic parameters (which are continuous functions of the coordinates $\boldsymbol{r}$ and time t), such as the density, pressure, temperature, internal energy, and entropy[1] Furthermore, we will also additionally characterize the subsystem of turbulent chaos by a number of internal coordinates that are ultimately related to its microstructure (see below). We will emphasize at once that the generalized temperature and entropy of the subsystem of turbulent chaos considered as primary concepts have no precise physical interpretation and are introduced only to ensure coherence of the theory (see Jou et al. 2001).

[1] In this connection, it is pertinent to recall the following: according to Onsager (1949), the methods of statistical mechanics can be used to describe turbulent chaos in which different-scale vortices are well mixed and, hence, the methods of statistical thermodynamics of irreversible processes are also applicable.

In addition, we will assume that the generalized thermodynamic state parameters characterizing the stationary-nonequilibrium vortex structure of turbulent chaos are related by ordinary relations for local equilibrium thermodynamics such as the Gibbs and Gibbs–Dugham relations, which serve exclusively as a constraint on the form of the derived defining (constitutive) relations. In other words, we will assume that such identities also remain valid far from local thermodynamic equilibrium of the subsystem of chaos if, however, chaos is in a stable stationary-nonequilibrium state that we will choose as a reference one. This fundamental assumption is a kind of a new postulate (see Prigogine 1967) on which the thermodynamic approach to the model description of fully developed turbulence will be based. It should be noted in this connection that since the energy of turbulent motions continuously dissipates due to molecular viscosity, the situation in which the subsystem of turbulent chaos reaches a local statistically equilibrium state turns out to be impossible in principle. At the same time, for a stationary turbulized fluid flow, where the viscous energy dissipation is, on average, compensated for by the energy from the external source on a long time scale, the stationary-nonequilibrium transport processes in the subsystem of turbulent chaos are quite admissible and do not differ much in their physical essence from the local equilibrium processes in a dissipative system (note that, in the case considered here, the so-called H-theorem is valid; according to the latter, any initial state of turbulent chaos approaches a stationary-nonequilibrium state after a sufficiently long time [see Chap. 6]).

When constructing the turbulence model, we will use one more key concept of the Kolmogorov (1941) theory: in the limit of very large Reynolds, $Re = Lu_0/\nu$, and Peclet, $Pe = L_T u_{T0}/\chi$, numbers corresponding to large-scale motions in a flow, despite the anisotropy, inhomogeneity, and nonstationarity of the averaged flow, the randomness of the fragmentation of vortices (or macrostructural temperature inhomogeneities) and the chaoticity of their energy transfer over the cascade downward cause the statistical regime of turbulent fluctuations within the small spatiotemporal averaging region G of the instantaneous values of hydrodynamic parameters to be almost locally isotropic—homogeneous, isotropic, and quasi-stationary, i.e., varying only with control parameters and primarily with the Reynolds number $\boldsymbol{Re}$ (which eventually determines the number of cascades in the hierarchy of vortices of various orders). In this case, it is important to emphasize that, naturally, there can be no complete local isotropy due to the presence of mesoscale vortex structures. Here, u_0 and u_{T0} are typical changes in the mean velocity at distances L and L_T, respectively; χ and ν are the molecular thermal diffusivity and kinematic viscosity, respectively; L_T is the distance at which the mean temperature changes noticeably. Below, we will assume, for simplicity, that the Prandtl number $Pr = \nu/\chi$ is of the order of unity and $L \approx L_T$; in this case, the boundaries of the inertial and convective ranges within which the molecular viscosity and molecular heat conduction effects are significant may be considered to be coincident.

Let us now use the methods of extended irreversible thermodynamics (see Jou et al. 2001) and nonequilibrium statistical thermodynamics (see, e.g., Keizer 1987) to obtain the defining (closing) relations for the thermodynamic fluxes and forces that describe the various turbulent transport processes in coordinate space and self-organization processes in phase space with an efficiency sufficient for practical purposes.

5.1.2.1 Thermodynamics of the Subsystem of Averaged Motion

We will begin by analyzing the balance equations for the average entropy $\langle S(r,t)\rangle$ of a turbulized homogeneous fluid. The probability-theoretic averaging of the Gibbs identity, which is assumed to be valid for micromotions,

$$T\,\delta S = \delta E + p\delta(1/\rho),$$

where $T(r,t)$, $S(r,t)$ are the instantaneous values of the absolute temperature and specific entropy in a fluid particle, respectively, leads to the fundamental Gibbs identity for the subsystem of averaged motion (see Chap. 3). This identity written along the averaged trajectory of the center of mass of a physically elementary volume $\boldsymbol{dr}$ takes the form [cf. (3.75)]

$$\frac{D\langle S\rangle}{Dt} = \frac{1}{\langle T\rangle}\frac{D\langle E\rangle}{Dt} + \frac{\overline{p}}{\langle T\rangle}\frac{D\langle 1/\rho\rangle}{Dt}. \tag{5.6}$$

Identity (5.6) can be written in the form of a local balance equation for the system's average entropy $\langle S\rangle$ if we eliminate the substantial time derivatives of the parameters $\langle E\rangle$ and $\langle 1/\rho\rangle$ from it using the averaged hydrodynamic equations (5.2) and (5.4). as a result, we obtain (see (3.86), (3.87), and (3.98))

$$\frac{\partial}{\partial t}\left(\overline{\rho}\langle S\rangle\right) + \mathrm{div}\left(\overline{\rho}\langle S\rangle\langle u\rangle + \frac{q^{\Sigma}}{\langle T\rangle}\right) = \sigma^{i}_{\langle S\rangle} + \sigma^{e}_{\langle S\rangle}, \tag{5.7}$$

where

$$0 \le \sigma^{i}_{\langle S\rangle} \equiv \frac{1}{\langle T\rangle}\left\{-\left(q^{\Sigma}\cdot\frac{\partial \ln}{\langle T\rangle}\right) + \overline{\pi}\, div\,\langle u\rangle + \left(\overset{\circ}{\overline{\Pi}}{}^{s} : \overset{\circ}{D}\right)\right\}, \tag{5.8}$$

$$\sigma^{e}_{\langle S\rangle} \equiv \frac{1}{\langle T\rangle}\left\{-\overline{p'\, div\, u''} + \left(J^{\mathrm{turb}}_{v}\cdot\frac{\partial\overline{p}}{\partial r}\right) + \overline{\rho}\,\langle\varepsilon_b\rangle\right\} \equiv \frac{\Im_{E,b}}{\langle T\rangle}. \tag{5.9}$$

Here, $\overset{\circ}{\overline{\Pi}}{}^{s}$ is the (average) shear rate tensor and $\overline{\pi} \equiv \frac{1}{3}\left(\overline{\Pi} : U\right) = \frac{1}{3}\,\overline{\Pi}_{kk}$ is the average viscous pressure [see (2.47) and (2.48)]. The positive quantity $\sigma^{i}_{\langle S\rangle}(r,t)$ defines the local production rate of the system's average entropy $\langle S\rangle$ due to dissipative transport processes inside the subsystem of averaged fluid motion; as will be seen from the subsequent analysis, the quantity $\sigma^{e}_{\langle S\rangle}(r,t) \equiv \Im_{E,b}/\langle T\rangle$ (the entropy sink or source) reflects the entropy exchange between the subsystems of turbulent chaos and averaged motion. It is important to note that the quantity $\sigma^e_{\langle S \rangle}(r,t)$ can be different in sign depending on the specific regime of fluid motion. Indeed, the turbulent energy dissipation rate $\langle \varepsilon_b(r,t) \rangle$ is always a positive quantity. However, the energy transition rate $\overline{p' \mathrm{div}\, u''}$ (the work done on turbulent vortices by the environment per unit time per unit volume due to the pressure fluctuations and the expansion ($\mathrm{div}\, u'' > 0$) or compression ($\mathrm{div}\, u'' < 0$) of turbulent vortices) can be different in sign. The quantity $\left(J_v^{\mathrm{turb}} \cdot (\partial \overline{p} / \partial r) \right)$ is positive for small-scale turbulence, but it can be both positive and negative for large-scale and mesoscale vortices (see, e.g., Marov and Kolesnichenko 1999). Thus, it follows from (5.7) that, generally, the average entropy of a turbulized medium $\langle S \rangle$ can both increase and decrease. This is a characteristic feature of any thermodynamically open systems.

This implies that there is some internal openness in an externally closed turbulized system modeled by two continua. It stems from the fact that the averaged motion of a turbulized fluid is described only by one of the two continua. At the same time, each physically infinitesimal volume element $dr \sim \Lambda^3$ (where Λ is the averaging scale) of the medium is still assumed to be so large that the additional information about the pattern of the fluctuating motion (turbulent superstructure) on scales smaller than or equal to the size of a "mathematical point" can be taken into account in the model. Hence it follows, in particular, that the average entropy $\langle S \rangle$ alone is clearly not enough for an adequate description of all features of structured turbulence, because this quantity is not related to any parameters characterizing the internal structure and thermodynamics of the subsystem of turbulent chaos and, in particular, to such a key parameter of the theory as the turbulence energy (the average fluctuation kinetic energy per unit mass of the medium)

$$\langle b(r,t) \rangle \equiv \overline{\rho (u'')^2} / 2\overline{\rho}\,. \tag{5.10}$$

For this reason, to macroscopically describe structured turbulence and, in particular, the cascade transport of turbulent energy by vortices of various scales (downward over the range of sizes), below we will use a well-known generalization of the formalism of irreversible thermodynamics to media with an internal structure by introducing for this purpose generalized extensive thermodynamic parameters (internal energy $E_{\mathrm{turb}}(r,t)$, generalized chemical potentials $\mu_{\mathrm{turb}}(q,r,t)$, etc.) of the subsystem of turbulent chaos related to the fluctuating fluid motion (Blackadar 1955) and the so-called internal coordinates corresponding to the system's microstructure. In other words, we will proceed just as has long been done in nonequilibrium thermodynamics, for example, in order to take into account various transformations in the internal degrees of freedom of molecules, in particular, to take into account the orientation of polar molecules relative to the external electric field (as is well known, this can be done by introducing an internal coordinate θ, the angle between the field and dipole directions), when the formalism of a generalized chemical potential equal to the standard chemical potential and the field term dependent on the internal coordinate θ is used (see Prigogine 1967).

5.1.2.2 Internal Coordinates of the Subsystem of Turbulent Chaos

Thus, when modeling the stochastic system corresponding to the subsystem of turbulent chaos, we will use the formalism of generalized statistical thermodynamics (Stratonovich 1985; Keizer 1987). It suggests the study of a statistical ensemble of macroscopically identical subsystems of chaos with the same generalized extensive state parameters such as the average specific volume $1/\overline{\rho}\,(r,t)$, internal energy $E_{\rm turb}(r,t)$, and entropy $S_{\rm turb}(r,t)$ of chaos and some infinite sequence of internal variables $n(q_k, r, t)$. The internal variables $n(q_k, r, t)$ can be the number densities of small-scale vortices or temperature inhomogeneities in states characterized by specified values of the parameters q_k—the internal coordinates defining the system's microstructure. We will assume that the vortex structures of chaos are somehow localized in both coordinate space $\boldsymbol{r}$ and configuration space $\boldsymbol{q}$. The internal coordinates $q_k(r,t)$ $\left(k=\overline{1,n}\right)$, which are some characteristics of the ensemble of vortex (or temperature) chaos corresponding to small-scale turbulent fluctuations, are generally random (stochastic) variables fluctuating relative to their mean (stationary) values q_k^{st}. They serve as a measure of the differences in any set of thermodynamically identical subsystems of the vortex ensemble. Allowance for the fluctuations in the internal coordinates describing the state of chaos in purely dynamical modeling (see Sect. 5.2) refines its statistical description and leads to a more adequate reflection of reality.

The continuously changing local random parameters that adequately characterize the evolution of a turbulized fluid, including the spatiotemporal evolution of various mesoscale coherent structures, can be attributed to the stochastic internal coordinates describing the macroscopic state of turbulent chaos. Thus, some of the internal coordinates q_k can refer to the incoherent component of the subsystem of turbulent chaos, while others can characterize individual CSs. In particular, the following positive definite quantities (or their logarithms) can be chosen as the stochastic coordinates $q_k(r,t)$: the kinetic energy of vortices, $b=|u''|^2/2$; the dissipation rate of turbulent energy into heat,

$$\varepsilon=\frac{1}{2}\,\nu\sum\nolimits_{i,j}\left(\partial u''_i\,/\,\partial x_j+\partial u''_j\,/\,\partial x_i\right)^2;$$

the scalar dissipation of temperature inhomogeneities, $\varepsilon_T=\chi\sum_j\left(\partial T''/\,\partial x_j\right)^2$; the mixing rate of a substance with a concentration $\theta(r,t)$ to the molecular level (which does not affect the flow dynamics), $\varepsilon_\theta=\chi\sum_j\left(\partial\theta''/\,\partial x_j\right)^2$ (the quantity ε_θ is a measure of the concentration field inhomogeneity disappearing per unit time through molecular diffusion $D\approx\chi$); the system's enstrophy (in the case of two-dimensional turbulence); the mean vorticities of the field of velocity fluctuations referring to kth type mesoscale vortex structures (the fundamental quantities to characterize CSs), etc.

As we will see below, using the internal coordinates as additional macroscopic parameters of turbulent chaos allows the evolutionary Fokker–Planck–Kolmogorov

(FPK) equations to be thermodynamically obtained in the space of configurations q_k (given Prigogine's central (for this approach) postulate concerning the direction of the irreversible processes in any local volume of the space of internal coordinates (see Prigogine 1967; Chap. 3, Sect. 11)). These kinetic equations are designed to determine the temporal evolution of the probability density function for various stochastic small-scale turbulence characteristics. They also allow one to analyze the conditions for the transition from one stable stationary-nonequilibrium state of turbulent chaos to another that are eventually caused by a successive loss of fluid flow stability when changing the parameters controlling the regime of turbulent motion as a whole.

The purely classical (i.e., without any introduction of internal stochastic coordinates) thermodynamic approach to modeling turbulence that we used in Chap. 3 for the subsystem of turbulent chaos seems not quite adequate in the case of a structured vortex continuum, because any two realizations of the ensemble (the set of subsystems of chaos with the same set of extensive thermodynamic state parameters) will be identical in all respects when applying it, which does not correspond to the real state of things. This is attributable to turbulent fluctuations in the internal coordinates of the state of chaos, which ultimately serve as a measure of the differences in any ensemble of thermodynamically identical subsystems of chaos. Actually, it is these turbulent fluctuations, which are not suppressed under strongly nonequilibrium conditions but, on the contrary, are enhanced in certain situations by internal irreversible processes within the vortex subsystem at the so-called bifurcation points (at which the subsystem "can choose" between various states), that lead to various mechanical manifestations of a real turbulized flow. In particular, the subsystem of turbulent chaos in some stable stationary state (far from local thermodynamic equilibrium) at certain values of control parameters can shift to a new stationary state with a neutral stability (associated with the so-called critical point of stability loss) and, subsequently, abruptly pass to another asymptotically stable stationary state corresponding to one or another form of the supramolecular coherent behavior of a large number of particles (e.g., the oscillations of different-scale vortices). Recall once again that, according to Prigogine (Prigogine and Stengers 1994), this ability to bring order via fluctuations is a fundamental property of any open strongly nonequilibrium thermodynamic systems. Thus, because of the continuous turbulent fluctuations, it is convenient to imagine any quasi-stationary state of turbulent chaos as a state of not one individual subsystem but the whole physical ensemble of subsystems. That is why it is necessary to refine the thermodynamic description of a turbulized flow so as to be able to take into account the effects of stochasticity of the vortex continuum modeling it (Kolesnichenko 2002).

In accordance with the stochastic-thermodynamic approach being developed here, below we will assume that for a complete statistical description of the random vector process $q(r,t)$ in a turbulized continuum (the set of structural small-scale characteristics of chaos $q_k(r,t)$, where $k = \overline{1,n}$, which is often convenient to gather into one column vector q in n-dimensional configuration space), it will suffice to know the one-point probability density $W_1(q,t)$ and the joint two-point probability density $W_2(q_0, t_0; q, t)$. The random processes that are completely described only by these two distribution functions are known to be Markovian ones. It can be said that this cardinal assumption determines the class of random processes (turbulent fluctuations) to which the stochastic-thermodynamic model of developed turbulence being analyzed here is applicable. Below, we will also use the two-point conditional probability density, $P_2(q_0, t_0| q, t)$, which allows the probable value of the parameter q at time t to be found if $q = q_0$ at time t_0 with a probability equal to unity. We will use these probability densities to obtain the mean values of various functions $f(q(t))$ of the random state vector $q(t)$; in particular, the formula

$$\overline{f(q(t))} \equiv \int f(q(t)) W_1(q, t) \mathrm{d}q$$

defines the unconditional mean of $f(q(t))$ and the formula

$$\overline{f(q(t))}^0 \equiv \int f(q) P_2(q_0, |q, t) \mathrm{d}q$$

introduces the mean of $f(q(t))$ at time t provided that $f(q(t_0)) = f(q_0)$ (conditional mean). The relationship between the means over the conditional subensemble $\overline{f(q(t))}^0$ and over the entire physical ensemble $\overline{f(q(t))}$ is implicitly contained in the relation

$$P_2(q_0, t_0|q, t) = W_2(q_0, t_0; q, t)/W_1(q_0, t_0),$$

which actually defines the so-called transition probability function P_2.

In addition, we will restrict our analysis to the so-called stationary physical ensemble of turbulent chaos consisting of adequate (in the above sense) subsystems maintained by continuously acting external sources of turbulence in a state in which the random variables $q(t)$ are invariant with respect to a shift along the time axis, i.e., $q(t_p + \tau) = q(t_p)$ for all p and τ. Clearly, in this case, the one-time probability density $W_1(q)$ will not depend on time, while the joint probability densities will be determined only by the difference $(t - t_0)$, for example, $P_2(q_0, t_0|q, t) = P_2(q_0, 0|q, t - t_0)$. Bearing this in mind, below we will omit the initial time in the expressions for W_2 and P_2 and write them in an abridged form: $P_2(q_0|q, t)$ etc. Note also that, in the case under consideration, the positive function P_2 has the following properties (see, e.g., Keizer 1987):

$$\int P_2(q_0|q, t) \mathrm{d}q = 1, \quad \int W_1(q_0) P_2(q_0|q, t) \mathrm{d}q_0 = W_1(q),$$
$$\lim_{t \to \infty} P_2(q_0|q, t) = W_1(q); \tag{5.11}$$

the latter relation implies that the conditional probability density for stationary processes ceases to depend on the initial condition asymptotically with time.

5.1.2.3 Basic Kinetic Equation

Recall that the key proposition of the Kolmogorov (1941) theory for the generation of small-scale turbulence is that the excitation of vortex structures, their nonlinear interactions, and the viscous dissipation of turbulent energy are strictly separated in the space of wave numbers, when the energy influx into a turbulized flow occurs near the wave number k_L corresponding to the turbulence macroscale L, while the energy dissipation becomes efficient near the wave number k_η (where η is the turbulence microscale often called the Kolmogorov scale). In other words, the existence of an inertial range of scales ($k_L <<< k_\eta$) is a characteristic signature of developed turbulence. The energy transfer from large-scale turbulent vortices to small-scale ones can be imagined as a cascade process of their fragmentation (the idea of an energy cascade was first put forward by L. Richardson in 1922).

For the subsequent analysis, we will need the so-called basic kinetic equation for the rate of change in the number of vortex moles $n(q)$ in the cascade interaction between turbulent motions of different scales or for the function $P_2(q,t) \equiv n(q)/n_\Sigma$. The latter is the (conditional) probability density to detect the system in the interval $(q, q + \mathrm{d}q)$ at time t if it was in a state q^{st} at the initial time (at $t = 0$) with a probability equal to unity. Here,

$$n_\Sigma(r,t) = \int n(q,r,t)\mathrm{d}q \tag{5.12}$$

is the total number of vortex structures with an attribute q in the volume dr.

When deriving the equation for the rate of change in the number of vortex moles $n(q)$ in the cascade interaction between turbulent vortices of different scales, we will assume that the cascade vortex fragmentation mechanism associated with the transition of kinetic energy from large vortices to increasingly small ones is such that the medium "retains memory" only about the last transition (a Markovian process). If the number density of vortices $n(q_i)$ in a state q_i can be changed only through their transition from the neighboring states q_{k-1} (the decay of large vortices with attributes q_{i-1} into smaller vortices with attributes q_i) or to the neighboring states q_{i+1} (the interaction locality principle), then we have $\partial n/\partial t + (J_i - J_{i-1}) = 0$, where J_i is the rate of the transition $i \to (i+1)$ and J_{i-1} is the rate of the transition $(i-1) \to i$. Since the distribution of vorticity in a real turbulized fluid flow is continuous (as is well known, there is a continuum of excited degrees of freedom in the case of developed turbulence (see, e.g., Landau and Lifshitz 1988)), below we will assume that this transfer of the kinetic energy of averaged motion over the Richardson–Kolmogorov cascade downward can be adequately described in terms of the change in random parameters q taking on continuous values. In particular, this corresponds to the cascade destruction of large vortices and the formation of small ones whereby only infinitesimal changes in quantities q occur in a single interaction, while finite changes arise from the cumulative action of a large number of vortex interactions. In the language of chemistry, this process can be considered as a series of consecutive chemical reactions expressed by the scheme $s_1 \to s_2 \to s_3 \to \cdot s \ldots$

Note that the principle of detailed balance can generally break down for thermodynamic systems far from complete thermal equilibrium; this is also true, in particular, for the stationary-nonequilibrium cascade formation of vortex structures. Consequently, for a real cascade process, the equation for the rate of change in the number density of turbulent vortices $n(q,t)$ should be rewritten as the standard continuity equation

$$\partial n(q,t)/\partial t + \partial J(q,t)/\partial q = 0$$

in the space of configurations. Here, $J(q,t)$ is the vortex flux in the inertial range corresponding to the number of "vortex moles" that pass from a state q to a state $q + \mathrm{d}q$ per unit time. In the more general case where the number density of turbulent vortices also depends on the spatial coordinate r, the kinetic equation for the rate of change in $n(q,r,t)$ can be written as the continuity equation (both in the coordinate space r and in the space of internal coordinates q)

$$\frac{\partial n(q,r,t)}{\partial t} + \frac{\partial}{\partial r}\cdot(n(q,r,t)\langle u\rangle) + \frac{\partial}{\partial q}\cdot J(q,r,t) = 0, \tag{5.13}$$

or, in view of (5.2), as

$$\rho\frac{D}{Dt}\left(\frac{n(q,r,t)}{\rho}\right) + \frac{\partial}{\partial q}\cdot J(q,r,t) = 0. \tag{5.13*}$$

Here, $J(q,r,t)$ is the thermodynamic flux in the space of internal coordinates q to be determined.

5.1.2.4 Thermodynamics of the Subsystem of Structured Chaos

Thus, the transfer of turbulent energy over the cascade of vortices in the case of developed turbulence can be considered as a kind of chemical transformations with the corresponding chemical potential $\mu_{\text{turb}}(q)$ for the internal degrees of freedom q and the de Donder chemical affinity

$$A_{\text{turb}}(q) = -\partial\mu_{\text{turb}}/\partial q\,.$$

The latter can be treated as the driving force of the cascade process corresponding to one equivalent $n(q) \to n(q+\partial q)$ of the vortex fragmentation process. Note that the concept of chemical potential is distinguished by great generality: it is applicable to almost any model of a continuous medium if the concept of thermodynamic temperature (which is generally not the absolute temperature of the medium) can be introduced for it in one way or another.

Extending the formalism of chemical potential to stationary-nonequilibrium turbulent chaos, we will determine intensive thermodynamic parameters, such as the generalized turbulization temperature $T_{turb}(r,t)$ and pressure $p_{turb}(r,t)$ (which are not related in any way to the molecular temperature and pressure of the underlying flow) as well as the turbulent chemical potential $\mu_{turb}(q,r,t)$ for the internal degrees of freedom q, from the fundamental Gibbs relation for the generalized entropy (specified a priori as the characteristic function)

$$S_{turb} = S_{turb}\left(E_{turb},\ 1/\overline{\rho},\ n(q)/\overline{\rho}\right) \tag{5.14}$$

(see, e.g., Munster 2002) using the relations

$$\begin{aligned} &\frac{1}{T_{turb}} = \left(\frac{\partial S_{turb}}{\partial E_{turb}}\right)_{1/\overline{\rho},n/\overline{\rho}}; \qquad \frac{p_{turb}}{T_{turb}} = \left(\frac{\partial S_{turb}}{\partial(1/\overline{\rho})}\right)_{E_{turb},n/\overline{\rho}}; \\ &\frac{\mu_{turb}(q)}{T_{turb}} = -\left(\frac{\partial S_{turb}}{\partial(n(q)/\overline{\rho})}\right)_{1/\overline{\rho},E_{turb}}. \end{aligned} \tag{5.15}$$

The chemical potential $\mu_{turb}(q)$ for the internal degrees of freedom is generally defined as a functional derivative. The turbulization entropy S_{turb} used here is assumed to contain all thermodynamic information about the subsystem of stationary-nonequilibrium turbulent chaos, i.e., it is related to the stability, fluctuations, and dynamical changes in a quasi-stationary state in exactly the same way as the equilibrium entropy of some thermodynamic system in a local equilibrium state (Keizer 1987). A remarkable feature of the entropy S_{turb} is that many statistical properties of turbulent chaos in quasi-stationary states can be deduced from this quantity. In particular, it is admissible to interpret the various algebraic relations for the intensive variables $E_{turb}(r,t)$, $T_{turb}(r,t)$, $p_{turb}(r,t)$, and $\mu_{turb}(r,t)$ that can be derived in a standard (for thermodynamics) way from (5.15) as specific "equations of state" for the subsystem of turbulent chaos.

It is important to clearly realize that the applicability of the concepts of turbulization temperature and entropy is not restricted only to the situations that arise near local thermodynamic equilibrium of the subsystem of turbulent chaos. On the contrary, these quantities are introduced precisely with the goal of thermodynamically describing chaos in stationary-nonequilibrium states. Note once again that, in contrast to the local equilibrium entropy of the subsystem of averaged motion $\langle S\rangle$, the generalized turbulence entropy S_{turb} in this approach, in general, remains an indefinite quantity regarding which there are no experimental or theoretical (based on the knowledge of the physics of turbulent chaos) methods for establishing its true functional dependence on state parameters. This quantity is introduced into the theory exclusively with the goal of ensuring its coherence, while the explicit form of the functional equation (5.14) is postulated depending on the goals of the model.

The differential form of the fundamental Gibbs relation for the turbulization entropy S_{turb} written along the trajectory of the center of mass of a physically elementary volume $\mathrm{d}r$ takes the following form (see Kondepudi and Prigogine 1999)

$$\frac{DS_{\text{turb}}}{Dt} = \frac{1}{T_{\text{turb}}} \frac{DE_{\text{turb}}}{Dt} + \frac{p_{\text{turb}}}{T_{\text{turb}}} \frac{D}{Dt}(1/\overline{\rho}) - \frac{1}{T_{\text{turb}}} \int_q \mu_{\text{turb}}(q) \frac{\mathrm{d}(n(q)/\overline{\rho})}{\mathrm{d}t} \mathrm{d}q. \quad (5.16)$$

Below, we identify the internal energy of chaos E_{turb} with the turbulence energy (Kolesnichenko 2002)

$$E_{\text{turb}} \equiv \langle b \rangle + \text{const} = \overline{\rho (u'')^2} / 2\overline{\rho} + \text{const} \quad (5.17)$$

and assume that, thermodynamically, the subsystem of turbulent chaos is an ideal classical gas (with three degrees of freedom in which the energy is distributed uniformly)—these are the key hypotheses of the approach considered here. In particular, the following equations of state can then be written:

$$\begin{gathered} \overline{\rho} \langle b \rangle = \frac{3}{2} R \overline{\rho} T_{\text{turb}}, \quad p_{\text{turb}} = R \overline{\rho} T_{\text{turb}}, \\ \mu_{\text{turb}}(q, T_{\text{turb}}) = k_{\text{B}} T_{\text{turb}} \ln n(q) + \Phi(q, T_{\text{turb}}), \end{gathered} \quad (5.18)$$

where $R(= n_\Sigma k_{\text{B}}/\overline{\rho})$ is the "gas constant" for the vortex continuum, k_{B} is the Boltzmann constant, and $\Phi(q, T_{\text{turb}})$ is the so-called potential energy in internal coordinate q, which also generally depends on the turbulization temperature T_{turb}.

We will note at once that, in some cases, the potential energy $\Phi(q)$ can be eliminated from (5.18) using for this purpose the equilibrium distribution $P_2^{st}(q)$ of internal coordinates q corresponding to some asymptotically stable stationary state of turbulent chaos specified a priori. We have already pointed out that since the energy of turbulent motions dissipates continuously due to viscosity, the situation where a statistically equilibrium state is reached turns out to be impossible here. The stationary state, in contrast to the equilibrium one, is usually dissipative. Indeed, as it well known from the course of thermodynamics (see, e.g., Kondepudi and Prigogine 1999), as some chemically active molecular continuum passes to a stable stationary (but fairly close to equilibrium) state characterized by minimal entropy production, the entropy itself also decreases. By analogy with molecular chemically reacting systems, for a stationary state of the subsystem of turbulent chaos when the internal coordinates q can fluctuate near some stable stationary value q^{st} at certain (for a given elementary volume dr) internal energy and specific volume, the turbulization entropy S_{turb} must also be minimal among all of the other states with the same values of E_{turb} and $1/\overline{\rho}$. Thus, the following condition is valid:

$$\delta S_{\text{turb}} = -\frac{\delta t}{T_{\text{turb}}} \int_q \mu_{\text{turb}}^{st}(q) \frac{D(n(q)/\overline{\rho})}{Dt} \mathrm{d}q = -\frac{1}{\overline{\rho} T_{\text{turb}}} \int \mu_{\text{turb}}^{st} \, \delta n(q) \mathrm{d}q = 0$$

where $\delta n \equiv n(q) - n(q^{st})$. Since the total number of vortex moles n_Σ is constant, we also have $\int \delta n(q) \mathrm{d}q = 0$. It follows from these two conditions that the chemical potential $\mu_{turb}^{st}(q)$ in internal coordinates for a stable stationary state of turbulent chaos does not depend on the state vector q ($\mu_{turb}^{st} = \text{const}$). Using this fact, we can obtain the expression

$$\mu_{turb}(q,t) = k_B T_{turb} \ln\left[n(q)/n(q^{st})\right] + \mu_{turb}^{st}, \tag{5.19}$$

which allows the chemical potential to be determined from the stationary distribution $n(q^{st})$ of an attribute q known in advance. Relation (5.19) can be rewritten as

$$\mu_{turb}(q,t) = k_B T_{turb} \ln\left(\frac{P_2(q_0|q,t)}{W_1^{st}(q)}\right) + \mu_{turb}^{st}. \tag{5.19*}$$

The latter is a generalization of the well-known Einstein formula (for a statistically equilibrium state) to quasi-stationary states in configuration space q; in writing (5.19*), we used the expression

$$f(q,T_{turb}) \equiv -\partial\Phi(q,T_{turb})/\partial q = k_B T_{turb} \partial \ln W_1^{st}(q)/\partial q \tag{5.19**}$$

for the friction force (in configuration space q) generated by the potential field $\Phi(q, T_{turb})$. The function

$$W_1^{st}(q) = \text{const} \cdot \exp\left(-\frac{\Phi(q,T_{turb})}{k_B T_{turb}}\right)$$

defines the (maximum) probability of a stable stationary state q^{st} when the fluctuating internal coordinates q remain constant, while the function $\Phi(q, T_{turb})$ acts as the thermodynamic potential for a stationary state. Recall that no distinction is made between the two concepts of equilibrium in equilibrium thermodynamics—the equilibrium state corresponding to maximum entropy and the equilibrium distribution in possible states that are physically almost equivalent (see, e.g., de Groot and Mazur 1962). A similar situation also holds for stationary states in the thermodynamics of nonequilibrium processes (Keizer 1987). This is because the asymptotic probability densities of the states are concentrated in an extremely narrow region and these Gaussian quantities (see below) transform into delta functions concentrated in q^{st} in the thermodynamic limit.

Using (5.13*), let us transform the Gibbs identity (5.16) by integration by parts and assuming that the flux $J(q, r, t)$ becomes zero at both boundaries q_1 and q_2 of the domain of definition of the variable q (a corollary of the condition $\int_{q_1}^{q_2} \delta n(q) dq = 0$) to

$$\frac{DS_{turb}}{Dt} = \frac{1}{T_{turb}}\frac{DE_{turb}}{Dt} + \frac{p_{turb}}{T_{turb}}\frac{D(1/\bar{\rho})}{Dt} + \frac{1}{T_{turb}}\int_q J(q) \cdot \frac{\partial \mu_{turb}(q,r,t)}{\partial q} dq. \tag{5.20}$$

The last term in this relation

$$\frac{D_i S_{\text{turb}}(r,t)}{Dt} \equiv \int_q \sigma_q(S_{\text{turb}})\mathrm{d}q = -\frac{1}{T_{\text{turb}}}\int_q J(q,r,t)\frac{\partial \mu_{\text{turb}}(q,r,t)}{\partial q}\mathrm{d}q \geq 0 \quad (5.21)$$

describes the total growth of the turbulization entropy S_{turb} due to the irreversible processes of the formation of various vortex structures characterized by the complete set of internal coordinates q. It can be seen from (5.21) that the local fluctuation entropy production $\sigma_q(S_{\text{turb}})$ corresponding to each part of the space of internal coordinates q has an ordinary thermodynamic form:

$$\sigma_q(S_{\text{turb}}) = J(q)\cdot A_{\text{turb}}(q)/T_{\text{turb}},$$

where

$$\begin{aligned} A_{\text{turb}}(q,r,t) &\equiv -\frac{\partial \mu_{\text{turb}}(q,r,t)}{\partial q} = -\frac{k_{\text{B}}T_{\text{turb}}}{n(q,r,t)}\frac{\partial n(q,r,t)}{\partial q} - \frac{\partial \Phi(q)}{\partial q} \\ &= \frac{k_{\text{B}}T_{\text{turb}}}{n(q,r,t)}\exp\left(\frac{\Phi(q)}{k_{\text{B}}T_{\text{turb}}}\right) - \frac{\partial}{\partial q}\left[\exp\left(\frac{\partial \mu_{\text{turb}}(q,r,t)}{k_{\text{B}}T_{\text{turb}}}\right)\right] \end{aligned} \quad (5.22)$$

is the generalized de Donder chemical affinity for configuration q (the state function of the subsystem of turbulent chaos) written here by taking into account (5.15) for the generalized chemical potential $\mu_{\text{turb}}(q,r,t)$.

5.1.2.5 Transfer Equation for the Turbulization Entropy

Let us now derive the transfer equation for the turbulization entropy S_{turb} by applying the same procedure that led to (5.7). For this purpose, from (5.20) we will eliminate the substantial derivatives of the specific volume $\langle v\rangle$ and turbulent energy $\langle b\rangle(\equiv E_{\text{turb}})$, the differential equation for which is [see (3.69)]

$$\overline{\rho}\frac{D\langle b\rangle}{Dt} = -\text{div}\, J^{\text{turb}}_{\langle b\rangle} + \left(R : \frac{\partial\langle u\rangle}{\partial r}\right) + \overline{p'\,\text{div}\,u''} - \left(J^{\text{turb}}_{v}\cdot\frac{\partial\overline{p}}{\partial r}\right) - \overline{\rho}\langle\varepsilon_b\rangle.$$

As a result, we will have

$$\frac{\partial}{\partial t}(\overline{\rho}S_{\text{turb}}) + \text{div}\left(\overline{\rho}S_{\text{turb}}\langle u\rangle + \frac{J^{\text{turb}}_{b}}{T_{\text{turb}}}\right) = \sigma_{S_{\text{turb}}} \equiv \sigma^{i}_{S_{\text{turb}}} + \sigma^{e}_{S_{\text{turb}}}, \quad (5.23)$$

where

$$\sigma^{e}_{S_{\text{turb}}} \equiv \frac{1}{T_{\text{turb}}}\left\{\overline{p'\,div\,u''} - \left(J^{\text{turb}}_{v}\cdot\frac{\partial}{\partial r}\,\overline{p}\right) - \overline{\rho}\langle\varepsilon_b\rangle\right\} = -\frac{\Im_{E,b}}{T_{\text{turb}}}, \quad (5.24)$$

$$
\begin{aligned}
0 \leq \sigma^{i}_{S_{\text{turb}}} &\equiv \frac{1}{T_{\text{turb}}}\left\{ -\left(J^{\text{turb}}_{b} \cdot \frac{\partial \ln T_{\text{turb}}}{\partial r} \right) + \left(R : \frac{\partial}{\partial r} \langle u \rangle \right) \right. \\
&\quad \left. + \, p_{\text{turb}}\, div \,\langle u \rangle - \overline{\rho} \int\limits_{q} J(q) \cdot \frac{\partial \mu_{\text{turb}}}{\partial q}\, \mathrm{d}q \right\} \\
&= \frac{1}{T_{\text{turb}}}\left\{ -\left(J^{\text{turb}}_{b} . \frac{\partial \ln T_{\text{turb}}}{\partial r} \right) + \left(\overset{\circ}{R} : \overset{\circ}{D} \right) - p_{\text{turb}}\, div\,\langle u \rangle \right. \\
&\quad \left. + \overline{\rho} \int\limits_{q} J(q) \cdot A_{\text{turb}}(q) \mathrm{d}q \right\}.
\end{aligned}
\tag{5.25}
$$

Here, $J^{\text{turb}}_{\langle b \rangle} \equiv \overline{\rho(|u''|^2/2 + p'/\rho)u'' - \Pi \cdot u''}$ is the so-called turbulent energy diffusion flux; $\overset{\circ}{R}(r, t)$ is the part with a zero trace of the Reynolds stress tensor defined by (3.98);

$$
p_{\text{turb}} = \frac{1}{3}(R : U) \tag{5.26}
$$

is the turbulization pressure. The quantities $\sigma^{i}_{S_{\text{turb}}}(r, t)$ and $\sigma^{e}_{S_{\text{turb}}}(r, t)$ are, respectively, the local production and sink of the entropy S_{turb} for the subsystem of turbulent chaos. Note that the work of turbulent stresses $R : (\partial \langle u \rangle / \partial r)$ causes the entropy of chaos to grow, while the viscous dissipation $\overline{\rho}\langle \varepsilon_b \rangle$ reduces the turbulization entropy S_{turb}.

5.1.3 Balance Equation for the Total Entropy of the Subsystems of Averaged Motion and Structured Turbulent Chaos

Adding (5.7) and (5.23) yields the balance equation for the total entropy $S_{\Sigma} = \langle S \rangle + S_{\text{turb}}$ of a turbulized fluid system

$$
\frac{\partial}{\partial t}(\overline{\rho} S_{\Sigma}) + \text{div}\left(\overline{\rho} S_{\Sigma} \langle u \rangle + \frac{q^{\Sigma}}{\langle T \rangle} + \frac{J^{\text{turb}}_{b}}{T_{\text{turb}}} \right) = \sigma_{\Sigma}, \tag{5.27}
$$

where

$$
0 \leq \sigma_{\Sigma} = \sigma^{i}_{\langle S \rangle} + \sigma^{e}_{\langle S \rangle} + \sigma^{i}_{S_{\text{turb}}} + \sigma^{e}_{S_{\text{turb}}} = \sigma^{i}_{\langle S \rangle} + \sigma^{i}_{S_{\text{turb}}} + \frac{T_{\text{turb}} - \langle T \rangle}{T_{\text{turb}} \langle T \rangle} \Im_{E,b} \tag{5.28}
$$

is the local entropy production related to irreversible processes inside the total turbulized continuum. The quantity σ_{Σ} written by taking into account (5.8), (5.9) and (5.24), (5.25) has the structure of a bilinear form,

$$\sigma_\Sigma = \sum_\alpha \Im_\alpha(r, t) X_\alpha(r, t)$$

$$\begin{aligned} 0 \le \sigma_\Sigma \equiv q^\Sigma \cdot \frac{\partial}{\partial r}\left(\frac{1}{\langle T\rangle}\right) + \frac{1}{\langle T\rangle}\pi\, div\langle u\rangle + \frac{1}{\langle T\rangle}\left(\overset{\circ}{\overline{\Pi}} : \overset{\circ}{D}\right) \\ + J_b^{\mathrm{turb}} \cdot \frac{\partial}{\partial r}\left(\frac{1}{T_{\mathrm{turb}}}\right) + \frac{1}{T_{\mathrm{turb}}}\left(\overset{\circ}{R} : \overset{\circ}{D}\right)\frac{\overline{\rho}}{T_{\mathrm{turb}}}\int_q J(q)\cdot A_{\mathrm{turb}}(q)\mathrm{d}q \\ + \frac{T_{\mathrm{turb}} - \langle T\rangle}{T_{\mathrm{turb}}\langle T\rangle}\Im_{E,b}. \end{aligned} \quad (5.29)$$

According to the basic postulate of the nonlinear thermodynamics of nonequilibrium processes, if the system is near a relatively stable quasi-stationary state, then the thermodynamic fluxes can be represented as linear functions of the conjugate macroscopic forces:

$$\Im_{\alpha i}(r, t) = \sum_\beta \Lambda^{ij}_{\alpha\beta} X_{\beta j}(r, t) \quad (\alpha, \beta = 1, 2, \ldots) \quad (5.30)$$

(see, e.g., Keizer 1987). It is important to emphasize that the matrix of phenomenological coefficients $\Lambda^{ij}_{\alpha\beta}$ for a turbulized fluid continuum depends not only on the system's average state parameters (i.e., on its average temperature $\langle T\rangle$, density $\overline{\rho}$, etc.) but also on turbulent superstructure characteristics, i.e., on parameters such as ε_b, T_{turb}, and$\langle b\rangle$. Such a situation, where there is a functional dependence of the coefficients $\Lambda^{ij}_{\alpha\beta}$ on the thermodynamic fluxes $\Im_{\alpha i}$ themselves (e.g., on the dissipation rate ε_b, which is also the energy flux over the cascade of vortices in the stationary case), is known to be typical of self-organizing systems (see, e.g., Haken 1981, 1988). Generally, this can lead to the individual terms $\Im_\alpha(r, t)X_\alpha(r, t)$ in the sum σ_Σ being not positive definite, although the entire sum is always greater than or equal to zero, $\sigma_\Sigma \ge 0$. The superposition of various fluxes can then, in principle, lead to negative individual diagonal elements of the matrix $\Lambda_{\alpha\beta}$, which probably explains the negative viscosity effect in some turbulent flows [see the text below (5.63)].

As can be seen from (5.29), the spectrum of possible cross effects for a turbulent flow generally widens compared to a laminar one. Thus, for example, the total heat flux q^Σ in a turbulized continuum can emerge under the influence of not only its conjugate thermodynamic force $\partial(1/\langle T\rangle)/\partial r)$ but also the force $\partial(1/T_{\mathrm{turb}})/\partial r)$ conjugate to the flux J_b^{turb} (describing the "diffusive" transport of turbulent kinetic energy). However, at present there are no reliable experimental data quantitatively describing such cross effects. In addition, the contribution from any cross effects to the total rate of some process is usually an order of magnitude smaller than that from direct ones (see de Groot and Mazur 1962). Taking this circumstance into account, below we will use the requirement that the intensities σ_Σ, $\sigma^i_{\langle S\rangle}$, $\sigma^i_{S_{\mathrm{turb}}}$ are positive independently of one another and will omit a number of cross effects in the linear constitutive relations (5.30) without special stipulations.

We will conclude this section by the following remark. In view of the second law of thermodynamics, the last term on the right-hand side of (5.29) describing the entropy production within the total continuum through irreversible entropy exchange between the subsystems of turbulent chaos and averaged motion is always positive,

$$\sigma_{\langle S\rangle, S_{\mathrm{turb}}}(r,\, t) \equiv \Im_{E,b}\left(\frac{1}{\langle T\rangle} - \frac{1}{T_{\mathrm{turb}}}\right) \geq 0. \tag{5.31}$$

Therefore, the "direction" of the thermodynamic flux $\Im_{E,b}(r,\, t)$ is determined by the sign of the state function $X_{\Im} \equiv (1/\langle T\rangle - 1/T_{\mathrm{turb}})$, which should be considered as the conjugate thermodynamic force (a macroscopic factor) producing the entropy flux $\Im_{E,b}$. Such entropy exchange between two mutually open subsystems is known (see Prigogine and Stengers 1994) to be an indispensable condition for a structured collective behavior, i.e., it can be a source of self-organization in one of them. Let us now show that the dissipative activity of the subsystem of turbulent chaos in the case of stationary-nonequilibrium turbulence is determined by the influx of negative entropy ($\sigma^{e}_{S_{\mathrm{turb}}} \equiv -\Im_{E,b}/T_{\mathrm{turb}}<0$) from the subsystem of averaged motion.

5.1.4 Stationary-Nonequilibrium State of the Turbulent Field: Defining Relations for Structured Turbulence

Since turbulence is accompanied by the dissipation of kinetic energy, a continuously acting external (with respect to the medium under consideration) source is needed to maintain its quasi-stationary regime (when the energy pumping and dissipation almost balance each other). For example, this energy source can be a turbulence-producing wire grid placed perpendicular to the forced fluid flow, stationary boundary conditions causing a large-scale flow velocity shear or a thermal-convective large-scale instability, etc. Such a source should be powerful enough to compensate for the expenditure of turbulent energy dissipated through molecular viscosity. For quasi-stationary turbulence, almost the entire energy being expended will be transferred without any significant (but, in general, existing) losses through the inertial range from the energy range to the viscous one, where it dissipates into heat. The energy transfer from large-scale vortices to small-scale ones can be visualized as a random Richardson–Kolmogorov cascade turbulent vortex fragmentation process.

In the model approach being developed here, we will assume that a continuous process of energy transfer from the subsystem of averaged fluid motion to the subsystem of turbulent chaos corresponds to such quasi-stationary turbulence. Obviously, a stationary-nonequilibrium regime between the influx of energy from the "external source" (attributable to the averaged fluid flow) and its dissipation (due to irreversible processes within the subsystem of turbulent chaos itself) is then established in the vortex continuum associated with small-scale turbulence in which $\mathrm{d}S_{\mathrm{turb}}/\mathrm{d}t \cong 0$. Note, incidentally, that the quasi-stationary state in which the entropy production is minimal is an attractor for the open subsystem of turbulent chaos, while the state corresponding to the total entropy maximum serves as an attractor for the turbulized system as a whole. The condition $\mathrm{d}S_{\mathrm{turb}}/\mathrm{d}t \cong 0$ implies that the production $\sigma^i_{S_{\mathrm{turb}}}$ of turbulization entropy S_{turb} is compensated for by its efflux σ_Σ, $\sigma^i_{\langle S\rangle}$, $\sigma^i_{S_{\mathrm{turb}}}(r,\,t) \equiv \Im_{E,b}\left(\frac{1}{\langle T\rangle} - \frac{1}{T_{\mathrm{turb}}}\right) \geq 0.$ to such an extent that the total generation of entropy S_{turb} is almost absent,

$$\sigma_{S_{\mathrm{turb}}} \equiv \sigma^e_{S_{\mathrm{turb}}} + \sigma^i_{S_{\mathrm{turb}}} \cong 0.$$

It should also be kept in mind that the turbulization entropy flux in the stationary case is constant, $J_{S_{\mathrm{turb}}} \equiv J_b^{\mathrm{turb}}/T_{\mathrm{turb}} \cong \mathrm{const}$ $(\mathrm{div} J_{S_{\mathrm{turb}}} \approx 0)$. Since $\sigma^i_{S_{\mathrm{turb}}} \geq 0$, the inequality $0 > \sigma^e_{S_{\mathrm{turb}}} \cong -\sigma^i_{S_{\mathrm{turb}}}$ holds, i.e., the subsystem of turbulent chaos must export entropy into the "external medium" (give it up to the subsystem of averaged motion) in order to compensate for the entropy production through irreversible internal processes within itself. In other words, an influx of negative entropy (negentropy) from the "external medium" is needed to maintain the stationary-nonequilibrium state within the subsystem of turbulent chaos,

$$\sigma^e_{S_{\mathrm{turb}}} \equiv -\Im_{E,b}/T_{\mathrm{turb}} = -\langle T\rangle \sigma^e_{\langle S\rangle}/T_{\mathrm{turb}} < 0.$$

Such a condition is known to be sufficient for the formation of dissipative coherent structures in a vortex continuum (see Prigogine and Stengers 1986, 1994). Indeed, since the entropy efflux from the subsystem of averaged motion in the stationary-nonequilibrium state of chaos is positive ($0 < \sigma^e_{\langle S\rangle} \Im_{E,b}/\langle T\rangle$), the rate $\Im_{E,b}$ of entropy (heat) exchange between the averaged and turbulent motions is also positive, $\Im_{E,b} \geq 0$. It then follows from inequality (5.31) that the turbulization temperature T_{turb} is higher than the average turbulized fluid temperature ($T_{\mathrm{turb}} > \langle T\rangle$), which is in complete agreement with the basic synergetic principle about the self-organization of a dissipative system. According to this principle, the formation of coherent structures (in our case, the formation of different-scale coherent vortex structures in the subsystem of turbulent chaos) when heat is removed from the system, i.e., when passing to lower temperatures, is a universal property of matter (Ebeling 2004).

5.1.4.1 Defining Relations for Structured Turbulence

Thus, the negentropy entering the subsystem of turbulent chaos is expended on maintaining and improving its internal structure. The relation $0 \leq \sigma^e_{\langle S\rangle} = -T_{\mathrm{turb}}$

$\sigma^e_{S_{\text{turb}}}/\langle T\rangle \cong T_{\text{turb}}\sigma^i_{S_{\text{turb}}}/\langle T\rangle$ is then valid and the balance equation (5.7) for the averaged entropy $\langle S\rangle$ of a turbulized fluid system takes the form

$$\overline{\rho}\frac{D\langle S\rangle}{Dt}+\operatorname{div}\left(\frac{q^{\Sigma}}{\langle T\rangle}\right)=\sigma^i_{\langle S\rangle}+\sigma^e_{\langle S\rangle}\cong\sigma^i_{\langle S\rangle}+\frac{T_{\text{turb}}}{\langle T\rangle}\sigma^i_{S_{\text{turb}}}\cong\sigma_{\Sigma}, \tag{5.32}$$

where the following expression holds for the local energy dissipation $\langle T\rangle\sigma_{\Sigma}$:

$$\langle T\rangle\sigma_{\Sigma}\equiv-\left(q^{\Sigma}\cdot\frac{\partial\ln\langle T\rangle}{\partial r}\right)+\left(\overset{\circ}{R}:\overset{\circ}{D}\right)+\overline{\rho}\int\limits_{q}J(q)\cdot A_{\text{turb}}(q)\mathrm{d}q\geq 0. \tag{5.33}$$

Here, $q^{\Sigma}(r,t)\equiv q^{\text{turb}}-\overline{p'u''}$ is the total heat flux in the subsystem of averaged motion for developed turbulence. Based on (5.33), we can write the following defining (gradient) relations for the turbulent fluxes and their conjugate thermodynamic forces in the linear approximation and using the Curie–Prigogine principle (according to which no relation is possible between tensors of different ranks in a locally isotropic medium (see, e.g., de Groot and Mazur 1962)):

$$q^{\Sigma}(r,t)=-\lambda^{\text{turb}}\frac{\partial\ln\langle T\rangle}{\partial r}, \tag{5.34}$$

$$R(r,t)=\frac{2}{3}\overline{\rho}\langle b\rangle U+\overline{\rho}\nu^{\text{turb}}\left\{\frac{1}{2}\left(\frac{\partial\langle u\rangle}{\partial r}+\left(\frac{\partial\langle u\rangle}{\partial r}\right)^{\text{turb}}\right)-\frac{1}{3}(\operatorname{div}\langle u\rangle)U\right\}, \tag{5.35}$$

$$J(q,r,t)=\int\limits_{\tilde{q}}L(q,\tilde{q},r,t)\cdot A(\tilde{q},r,t)\mathrm{d}\tilde{q}. \tag{5.36}$$

They correspond to the stationary state of the turbulent field. Note that the linearity condition is not strong enough to deprive the case under consideration of practical significance.

Assessing the status of the problem of closing the averaged hydrodynamic equations as a whole, it should be recognized that almost all semiempirical turbulence models are currently based to a certain extent on the gradient relations (see Chaps. 3 and 4). The phenomenological coefficients (turbulent exchange coefficients) $\lambda^{\text{turb}}(r,t)$, $\nu^{\text{turb}}(r,t)$ in these relations are scalar quantities, because, as has been emphasized above, strong turbulence is locally homogeneous and isotropic. In contrast to the molecular exchange coefficients, these quantities are not material constants. This is because in a turbulized continuum the various processes of (mass, momentum, and energy) transfer from one region of the system to another are determined by the collective motions of molecules (vortex structures) and, hence, must depend on turbulence intensity parameters, in particular, on the parameters $\langle\varepsilon_r\rangle$ and L (or $\langle b\rangle$). For example, in the inertial range of vortex scales (η), the turbulent viscosity ν^{turb} corresponding to the Richardson–Obukhov empirical "law of four thirds" (this law follows, in particular, from dimensional and similarity considerations) is $\nu^{\text{turb}} \approx \langle \varepsilon_r \rangle^{1/3} L^{4/3} \approx \langle b \rangle^2 / \langle \varepsilon_r \rangle$.

Thus, when stationary-inhomogeneous turbulence is modeled in the applications where the energy dissipation processes in the system are important, a heat transfer equation for averaged motion in form (5.32) should be invoked; this equation should be supplemented by the linear defining relations (5.34)–(5.36).

5.1.5 *Prigogine's Principle: Thermodynamic Derivation of the Fokker–Planck–Kolmogorov Equations*

According to (5.36), the phenomenological relation for the thermodynamic flux $J(q)$ in the space of internal coordinates q and the corresponding "instantaneous" affinity $A_{\text{turb}}(q)$ generally has an integral form. Following Prigogine (see Prigogine 1967; Chap. 3, Sect. 11), we will now assume that the irreversible processes in each part of the internal coordinate space q proceed in such a direction that only a positive entropy increment occurs. This implies that not only integral (5.19) but also the quantity

$$T_{\text{turb}}\, \sigma_q(S_{\text{turb}}) = J(q,\, r,\, t) \cdot A_{\text{turb}}(q,\, r,\, t) \geq 0, \tag{5.37}$$

which is the energy dissipation per unit volume of configuration space q, will be positive. The defining relation between the flux $J(q)$ and affinity $A_{\text{turb}}(q)$ of a state q (corresponding to one equivalent of the vortex decay process) is then

$$\begin{aligned} J(q) = L(q) \cdot A_{\text{turb}}(q) &= -L(q) \cdot \frac{\partial \mu_{\text{turb}}(q,\, r,\, t)}{\partial q} \\ &= -L(q) \cdot \left(\frac{k_{\text{B}}\, T_{\text{turb}}}{n(q,\, t)} \frac{\partial n(q,\, t)}{\partial q} + \frac{\partial \Phi(q,\, T_{\text{turb}})}{\partial q} \right) \\ &= -\frac{L(q)}{n(q,\, t)} \cdot \left(-f(q,\, T_{\text{turb}}) n(q,\, t) + k_{\text{B}}\, T_{\text{turb}} \frac{\partial n(q,\, t)}{\partial q} \right). \end{aligned} \tag{5.38}$$

Here, $L(q)$ is the positive definite local matrix of transport coefficients satisfying the Onsager–Casimir reciprocity relation $L^{\text{transp}} = L$; $n(q)$ are the numbers of vortex moles in the cascade interaction between turbulent motions of different scales.

Relation (5.38) considered together with (5.13) allows the following evolutionary Fokker–Planck–Kolmogorov equation (in the so-called kinetic form (see Klimontovich 1995)) to be obtained in the space of stochastic variable q for the distribution functions of various statistical small-scale turbulence characteristics:

$$\frac{\partial P_2(q,r,t)}{\partial t} + \frac{\partial}{\partial r} \cdot (\langle u \rangle P_2(q,r,t)) = \frac{\partial}{\partial q} \cdot \left\{ -K(q) P_2(q,r,t) + \frac{\varepsilon^2}{2}\, Q(q) \cdot \frac{\partial P_2(q,r,t)}{\partial q} \right\}. \tag{5.39}$$

Here, the probability flux $J = J_{dr} + J_{dif}$ is the sum of the drift, J_{dr}, and diffusion, J_{dif}, components:

$$J_{dr} = KP_2, \quad J_{dif} = -\frac{\varepsilon^2}{2} Q \cdot \frac{\partial P_2}{\partial q},$$

where the following notation is used for the drift vector K and the matrix of generalized diffusion coefficients D in the space of stochastic variable q:

$$K(q) \equiv L(q) \cdot f(q), \quad D = \varepsilon^2 L(q) = \frac{1}{2}\varepsilon^2 Q, \quad Q(q) \equiv 2L(q). \tag{5.40}$$

The function $P_2(q, t) \equiv n(q)/n_\Sigma$ is the (conditional) probability density to detect the system in an interval $(q, q + \mathrm{d}q)$ at time t if it was in a state q^{st} at the initial time (at $t = 0$) with a probability equal to unity. The parameter $\varepsilon \equiv \sqrt{k_B T_{turb}} = \sqrt{\rho\,\overline{(u'')^2}/3\overline{p}}$ characterizes the total intensity of the action of internal noise in the subsystem of turbulent chaos (generated by its "thermal" structure) on the random process $q(t)$. In writing (5.39), we assumed the mobility parameter $L(q) \equiv L(q)/n(q)$ in internal coordinate q to be independent, in the first approximation, of the density $n(q)$. It should be noted that, in general, the matrix K does not form a vector unless we restrict ourselves only to the linear transformations of coordinates in our modeling.

Apart from the kinetic form of the FPK equation (5.39), other representations of this equation are also used in the literature. Ito's and Stratonovich's representations are most popular among them. The latter are based on a different treatment of the so-called stochastic integrals (see, e.g., Tikhonov and Mironov 1977) that emerge when the corresponding nonlinear stochastic Langevin equations are solved. Although these representations are based on stochastic equations identical in form [see (5.86)], they lead to FPK equations differing in form:

$$\begin{aligned}\frac{\partial P_2(q,r,t)}{\partial t} + \frac{\partial}{\partial r}\cdot(\langle u\rangle P_2(q,r,t)) = \frac{\partial}{\partial q}\cdot\Big\{&-(K(q)+H(q))P_2(q,r,t)\\ &+\frac{1}{2}\varepsilon^2\frac{\partial}{\partial q}\cdot(Q(q)P_2(q,r,t))\Big\},\end{aligned} \tag{5.39*}$$

where $H(q) \equiv \lambda\,\frac{\varepsilon^2}{2}\frac{\partial Q(q)}{\partial q}$ is a "fictitious force" dependent on the choice of calculus. For $\lambda = 0,\ 1/2,\ 1$, we have to do with Ito's, Stratonovich's, and Klimontovich's representations of the FPK equation, respectively.

If the response of the subsystem of turbulent chaos to the impact of the external medium (the subsystem of averaged motion) does not depend on its internal state specified by the stochastic variable q, then the diffusion coefficient does not change with coordinate q either (i.e., $Q(q) = \mathrm{const} = Q$). The stochastic system can then be assumed to possess additive noise that, in our case, is just reduced to the intensity $\varepsilon^2 = k_B T_{turb}$ of the internal noise in the subsystem of turbulent chaos. The fictitious force $H(q)$ for such systems is identically equal to zero, i.e., they are completely insensitive to the choice of calculus. In other words, all three forms of the FPK equation take the same simplest form (5.39) for them. However, a feedback is generally possible between the internal states q of the stochastic subsystem of chaos and the subsystem of averaged motion. It turns out that not only the external fluctuations (associated, for example, with the degrees of freedom of the turbulent field that are not described by the selected coordinates q) affect the stochastic subsystem of turbulent chaos but also the latter has the reverse effect on their intensity. As applied to the case under consideration, this implies that the diffusion coefficient becomes dependent on the random coordinate q, i.e., the external fluctuations are multiplicative. In the presence of multiplicative noise, when $Q(q) \neq$ const, the simplest form of the FPK equation (5.39) holds only in Klimontovich's representation ($\lambda = 1$).

Thus, the problem of choosing calculus arises for systems with multiplicative noise, because the force entering into the FPK equation is determined ambiguously. In Ito's calculus, it is reduced to a real force acting on the selected degree of freedom. When passing to Stratonovich's calculus, an addition proportional to the derivative of the effective diffusion coefficient emerges. The value of this addition doubles in Klimontovich's kinetic representation. It is important that it affects significantly the behavior of the stochastic system. As a result, the question about the physical nature of this addition and the choice of calculus seems topical. We will return to this problem in the next chapter.

5.1.6 *Examples of the Fokker–Planck–Kolmogorov Equations Describing the Evolution of the Fluctuating Characteristics of Turbulent Chaos*

Let us now show (with simple examples) that Prigogine's principle (5.37) can serve as a basis for deriving the evolutionary partial differential equations (Fokker–Planck–Kolmogorov equations) in the space of stochastic variable q for the distribution functions of various stochastic small-scale turbulence characteristics if appropriate hypotheses about the distribution in a stationary-nonequilibrium state are adopted in advance for the latter. It should be kept in mind, however, that such hypotheses are almost always not quite rigorous and are a strong idealization related to the simplification of a real turbulent motion under natural conditions (see Monin and Yaglom 1996).

5.1.6.1 Evolution of Vortices in the Space of Fluctuating Velocities

Let us first use assumption (5.37) to derive the kinetic equation describing the change in the probability density function of vortex velocities with time,

$P_2(u''r,t)$ $\quad (\equiv n(u'',r,t)/n_\Sigma$; n is the number density of turbulent vortices; n_Σ is the total number of vortex moles [see (5.12)]. Below we consider this function as the internal variable of the subsystem of turbulent chaos and the fluctuating velocity u'' as the internal coordinate q. The probability distribution function for the fluctuating velocity is known to be not universal in the case of developed turbulence, because it depends on the mechanism generating the turbulent field. Nevertheless, following Millionshchikov (1941), below we use the hypothesis about the normal distribution of fluctuating velocities (for a locally isotropic turbulent field) in the stationary case

$$W_1(u'') \equiv n^{\text{stat}}(u'')/n_\Sigma = \left(\beta/\sqrt{\pi}\right) \cdot \exp\left(-\beta^{\,2} u''^2\right). \tag{5.41}$$

Millionshchikov applied such a distribution in the turbulence theory for special purposes that will not be discussed here. Note only that there are many examples where the velocity distribution function is approximately Gaussian, for instance, in turbulence generated by grids in wind tunnels or for turbulence in an atmospheric boundary layer (see, e.g., Batchelor 1953; Frisch 1998). For the subsystem of turbulent chaos we consider, the constant β in (5.41) can be related to the turbulization temperature T_{turb} in exactly the same way as is done in the kinetic theory of gases (see Ferziger and Kaper 1972). Using (5.17) and (5.41), it is easy to find that $\beta^2 = (2\,\mathcal{R}T_{\text{turb}})^{-1}$, whence we obtain a different equivalent expression for the function $n^{\text{stat}}(u'')$:

$$\begin{aligned} n^{\text{stat}}(u'') &= n_\Sigma \left(\frac{1}{2\pi \mathcal{R}T_{\text{turb}}}\right)^{1/2} \exp\left(-\frac{u''^2}{2\,\mathcal{R}T_{\text{turb}}}\right) \\ &= \text{const} \cdot \exp\left(-\frac{u''^2}{2\,\mathcal{R}T_{\text{turb}}}\right). \end{aligned} \tag{5.41*}$$

Substituting this distribution into (5.16) yields the following representation of the chemical potential $\mu_{\text{turb}}(u'',r,t)$ for the configuration u'':

$$\mu_{\text{turb}}(u'',r,t) = \frac{k_{\text{B}}}{\mathcal{R}}(u''^2/2) + k_{\text{B}}T_{\text{turb}}(r,t)\,ln\,[n\,(u'',r,t)] + \text{const}. \tag{5.42}$$

Given this formula, the phenomenological relation (5.36) for the probability flux $J(u'',r,t)$ takes the form

$$\begin{aligned} J(u'') &= -\frac{\overline{\rho}}{n_\Sigma} L_{u''}\left(u'' + \frac{\mathcal{R}T_{\text{turb}}}{n(u'')}\frac{\partial n(u'')}{\partial u''}\right) \\ &= -\alpha\left(u''\,n(u'') + \mathcal{R}T_{\text{turb}}\frac{\partial n(u'')}{\partial u''}\right), \end{aligned} \tag{5.43}$$

where $\mathcal{R}\,(= n_\Sigma k_{\text{B}}/\overline{\rho})$. Here, we introduce the coefficient $\alpha \equiv k_{\text{B}}L_{u''}/\mathcal{R}n(u'')$ that can be interpreted as the "mobility" in the space of internal coordinate u'' per unit volume; in the first approximation, this coefficient does not depend on $n(u'')$ and is assumed to be independent of u''. Substituting (5.43) into (5.19) yields the sought-for kinetic FPK equation for the conditional probability density function of the fluctuating velocity u'':

$$\frac{\partial P_2}{\partial t}+\frac{\partial}{\partial r}\cdot\left(P_2\langle u(r,t)\rangle\right)-\alpha\frac{\partial}{\partial u''}\left(u''P_2+\mathcal{R}T_{\mathrm{turb}}(r,t)\frac{\partial P_2}{\partial u''}\right)=0 \tag{5.44}$$

This dynamical equation supplemented by the initial condition $P_2=\delta(u''-0)$ (the δ-function concentrated at point 0 appears on the right-hand side) describes the temporal evolution of the probability density function P for the fluctuating velocity u'', in particular, for decaying (so-called degenerating) turbulence. It should be noted that the quantity $K\equiv-\alpha\,u''$ (the friction coefficient in the corresponding Langevin equation) acts as the drift coefficient in the FPK equation written in standard form and the quantity

$$D\equiv 2L_{u''}k_{\mathrm{B}}T_{\mathrm{turb}}/n(u'')=2\alpha\,\mathcal{R}T_{\mathrm{turb}}=\alpha\,\beta^{-2}$$

is the diffusion coefficient.

The normal distribution (5.41), which is a stationary solution of the one-dimensional (in parameter u'') FPK equation (5.44), can be taken as the initial statistical state of the fluctuating velocity field for the whole class of various motions of degenerating turbulence. A nonstationary solution of this equation can then be constructed in an analytically closed form:

$$P_2(u'',r,t)=\{\pi\,a(r,t)\}^{-1/2}\exp\left\{-[u''-b(r,t)]^2/a(r,t)\right\}, \tag{5.45}$$

where

$$\begin{aligned}a(r,t)&=\frac{D}{\alpha}\{1-\exp(-2\alpha\,t)\}+a_0\exp(-2\alpha\,t),\\ b(r,t)&=b_0(r)\exp(-\alpha\,t);\end{aligned} \tag{5.46}$$

$a_0(r)$ and $b_0(r)$ are the initial conditions. This solution allows the various n-point moments (correlation functions) of the mth order describing the statistical relation between the random velocities at various points of spacetime to be calculated. In particular, for the quantities $\overline{u''}^0(r,t)$ (the conditional mean velocity of the ensemble of vortices at time t) and $\overline{u''(r,t)u''(r,t_1)}$ (the two-time one-point correlation function), we have

$$\overline{u''}^0(r,t)=\int u''P_2(0|u'',t)\,\mathrm{d}u''=b_0(r)\exp(-\alpha\,t), \tag{5.47}$$

whence $b(r,t) \equiv \overline{u''}^{0}(r,t)$;

$$\begin{aligned}\overline{u''(r,t)u''_1(r,t_1)} &= \int u''\,\mathrm{d}u'' \int u''_1\,\mathrm{d}u''\,W_2(u'',t;u''_1,t_1)\\ &= \overline{u''^2(r,t)}\,exp\{-\alpha\,|t-t_1|\} = \frac{1}{2}\frac{D}{\alpha}\exp\{-\alpha\,|t-t_1|\},\end{aligned} \tag{5.48}$$

where the average is taken over the stochastic process (see, e.g., Haken 1981). Here, $W_2(u'',t;u''_1,t_1)$ is the joint probability density; since the generation of new modes of fluctuating motion (the fragmentation of vortex structures) is a Markovian process, it is represented as the product of the probability density at time t_1, $W_1(u''_1,t_1)$, and the conditional probability $P_2(u'',t\,|\,u''_1,t_1)$ (which is reduced to the δ-function at $t = t_1$):

$$\delta(u'' - u''_1))W_2(u'',t;u''_1,t_1) = W_1(u''_1,t_1)P_2(u'',t\,|\,u''_1,t_1).$$

Since $D = 2\alpha\,\mathcal{R}T_{\mathrm{turb}} = \frac{4}{3}\alpha\langle b\rangle$, first, the correct relation

$$\overline{u''^2(r,t)} = \frac{2}{3}\langle b\rangle \cong \frac{1}{3}\overline{|u''|^2},$$

consistent with the assumption about local isotropy of the vortex velocity field in the case of developed turbulence and, second, the effective formula for one of the most important correlation quantities in the theory of statistical turbulence

$$\overline{u''(r,t)u''(r,t_1)} = \frac{2}{3}\langle b\rangle\exp(-\alpha\,|t-t_1|) \tag{5.49}$$

defining the speed with which the fluctuating velocity "forgets its past" (according to this formula, this occurs in a time $t \cong 1/\alpha$) follow from (5.48).

Solution (5.45) at zero values of the parameters a_0 and b_0 takes the form

$$P_2(u'',t) = \{2\pi\,\mathcal{R}T_{\mathrm{turb}}[1-\exp(-2\alpha\,t)]\}^{-1/2}\exp\left\{-\frac{u''^2}{2\mathcal{R}T_{\mathrm{turb}}[1-\exp(-2\alpha\,t)]}\right\}. \tag{5.50}$$

It allows the temporal evolution of the conditional probability distribution function for the fluctuating velocity to be traced if the velocity distribution was Gaussian in the case of stationary turbulence.

It should be kept in mind that choosing the fluctuating velocity u'' as a suitable characteristic of turbulent vortices (the internal coordinate of the subsystem of turbulent chaos) generally does not justify itself, because the Gaussian probability distribution of the fluctuating velocity u'' has been confirmed with a sufficient degree of reliability neither experimentally (it was established, for example, that the deviation from normality behind the grid increases considerably with increasing Reynolds number *Re*) nor theoretically (the classical "two and five thirds" laws of turbulence are known to break down for this distribution). We have already pointed out above that the most acceptable characteristics of small-scale turbulence for the role of an internal coordinate are nonnegative macroscopic variables—even functions of rates such as the dissipation rate of turbulent energy (see, e.g., Monin and Yaglom 1996). According to Kolmogorov's hypothesis, such random characteristics asymptotically satisfy a lognormal probability distribution. This is because the successive fragmentation of vortex structures is similar to the coagulation of solid particles (the latter is known to lead to a lognormal particle size distribution). In Chap. 6, we will analyze the FPK equation describing the temporal evolution of the probability distribution function for the turbulent energy dissipation rate in the corresponding configuration space. However, we will note already here that the lognormal distribution does not accurately describe the edges of the true distribution of a random variable and, hence, can be used to calculate the high-order moments only with great caution.

5.1.6.2 Cascade Process (Thermodynamic Treatment Corresponding to Kolmogorov's Original Similarity Hypotheses)

Let us now apply Prigogine's principle (5.37) to the derivation of the kinetic equation describing the temporal evolution of the vortex distribution function in the space of kinetic energy. We will describe the Richardson–Kolmogorov cascade process (large vortices → small vortices → heat) using an analogy with the process of consecutive chemical reactions. In this case, the original kinetic equation (5.13) for the distribution function $P(0|q;r,t)$ of turbulent vortices in the space of fluctuating energy ($q = \rho|u''|^2/2$) takes the form

$$\frac{\partial P_2(0|q;r,t)}{\partial t} + \frac{\partial}{\partial r} \cdot (P_2(0|q;r,t)\langle u \rangle) = \frac{\partial}{\partial q}(P_2(0|q;r,t)\,\varepsilon(q,r,t)), \tag{5.51}$$

where

$$\varepsilon(q,r,t) \equiv -J(q,r,t)/n(q,r,t) \tag{5.52}$$

is the transition reaction rate from state q to state $q + \mathrm{d}q$ corresponding to the probability flux $J(q)$ in state q; n is the number density of turbulent vortices. In other words, relation (5.51) defines the parameter $\varepsilon(q,r,t)$ that can be interpreted in the case under consideration as the transfer rate of kinetic energy $\rho|u''|^2/2$ over the hierarchy of turbulent vortices along the coordinate q; concurrently, this quantity also defines the kinetic energy dissipation in vortices of type q. Indeed, the equation for the first moment

$$\int q P_2(0|q;t)\mathrm{d}q = \overline{\rho|u''|^2/2}^{\,0} = \overline{\rho}\langle b\rangle$$

derived from (5.51) as a result of integration by parts (and by assuming the flux $J(q, r, t)$ at the boundaries of the domain of integration to be zero) takes the classical form (see, e.g., Landau and Lifshitz 1988)

$$\overline{\rho}\frac{D\langle b\rangle}{Dt} \cong -\int \varepsilon(q, r, t)P_2(0|q; r, t)\mathrm{d}q = -\overline{\varepsilon(r, t)}^{\,0} \cong -\overline{\rho}\langle\varepsilon(r, t)\rangle \tag{5.53}$$

(here, the conditional average is taken over the stochastic process q), in which the parameter $\langle\varepsilon\rangle$ defines the mean turbulent energy dissipation rate at point (r, t). It is for this reason that the quantity $\varepsilon(q)$ can be interpreted as the energy dissipation rate in vortices of type q (at the point $q = \rho|u''|^2/2$ of configuration space). But then for the part of the dissipation energy $\langle T\rangle\sigma_\Sigma$ [see (5.33)] attributable to the transfer of turbulent energy over the cascade, we have

$$(\langle T\rangle\sigma_\Sigma)^{Ch} \equiv -\overline{\rho}\int_q \varepsilon(q, r, t)\, n(q, r, t)\, A_{\mathrm{turb}}(q, r, t)\mathrm{d}q \geq 0. \tag{5.54}$$

Hence follows a local phenomenological Prigogine-type equation:

$$\varepsilon(q, r, t) = L_q A_{\mathrm{turb}}(q, r, t)/n(q, r, t) = -\alpha' A_{\mathrm{turb}}(q, r, t), \tag{5.55}$$

in which

$$A_{\mathrm{turb}}(q) = -k_{\mathrm{B}}T_{\mathrm{turb}}\frac{\partial \ln n(q)}{\partial q} + f(q) \tag{5.56}$$

is the chemical affinity of the vortex fragmentation process (the state function of the subsystem of turbulent chaos); $f(q) = -\partial\Phi/\partial q$ is the so-called friction force; $\alpha' = -L_q/n(q)$ is the mobility coefficient, which is assumed to be independent of q.

In the case where a stationary-nonequilibrium flow is established in a turbulized medium in which the energy transfer rate over the cascade is constant, $\varepsilon(q, r, t) \cong \langle\varepsilon(r, t)\rangle$ (this assumption adopted in the original formulation of the well-known Kolmogorov (1941) hypotheses), inequality (5.54) takes the form

$$\langle T\rangle\sigma_\Sigma = -\overline{\rho}\,\langle\varepsilon(r, t)\rangle\, A^{gl}_{\mathrm{turb}}(r, t) \geq 0. \tag{5.57}$$

Here,

$$A^{gl}_{\mathrm{turb}} \equiv \int_q n(q)A_{\mathrm{turb}}(q)\mathrm{d}q = n_\Sigma\overline{A_{\mathrm{turb}}}$$

is the so-called global affinity of the formation of turbulent structures. Given (5.56), it can be rewritten as

$$\begin{aligned}\overline{A_{\mathrm{turb}}}(r,t) &= -k_{\mathrm{B}}T_{\mathrm{turb}}\int\limits_{q}\frac{\partial P_2(0|q;t)}{\partial q}\mathrm{d}q + \int\limits_{q} P_2(0|q;t)f(q)\mathrm{d}q \\ &= -k_{\mathrm{B}}T_{\mathrm{turb}}[P_2(q_{L_1}) - P_2(q_\eta)] + \overline{f(r,t)} \cong \overline{f(r,t)},\end{aligned} \tag{5.58}$$

because P_2 becomes zero at the boundaries of the domain of integration (here, η is the local value of the Kolmogorov microscale; n_Σ is the total number of vortex moles). Thus, the production σ_Σ of averaged entropy in such a stationary process is the product of the total energy transfer rate over the cascade $\langle\varepsilon\rangle$ and the global affinity A^{gl}_{turb} referring to the entire cascade fragmentation of large vortices into small ones. In this case, the linear phenomenological relation

$$\langle\varepsilon(r,t)\rangle = -\alpha' A^{gl}_{\mathrm{turb}}(r,t) \tag{5.59}$$

is valid, in complete agreement with the results of irreversible thermodynamics for consecutive chemical reactions.

On the other hand, we can adopt the more realistic condition

$$J(q,r,t) \cong J(r) \equiv -n_\Sigma(r)\langle\varepsilon(r)\rangle$$

for the thermodynamic kinetic energy flux $J(q,r,t)$ over the cascade of vortices being quasi-stationary, i.e., for the flux J on various scales of motion being independent of the parameter $q = \rho|u''|^2/2$. This assumption considered together with the linear relation (5.55) leads to a more general (than (5.59)) nonlinear relation between ε and the chemical affinity

$$\tilde{A}(r,t) \equiv \int_q A(q)\mathrm{d}q = \mu(q_{L_1},r) - \mu(q_\eta,r). \tag{5.60}$$

for the cascade process as a whole. The nonlinear defining relation for $\langle\varepsilon\rangle$ can be easily obtained by applying the second formula in (5.22) for the "local affinity" $A_{\mathrm{turb}}(q,r,t)$; as a result, we have [cf. (4.4) for the chemical reaction rate]

$$\langle\varepsilon\rangle \cong \gamma\left[1 - \exp\left(-\frac{\tilde{A}}{k_{\mathrm{B}}T_{\mathrm{turb}}}\right)\right], \tag{5.61}$$

where

$$\gamma = \frac{\left(\alpha'\frac{k_{\mathrm{B}}T_{\mathrm{turb}}}{n_\Sigma}\right)\exp\left(\frac{\mu(q_\eta)}{k_{\mathrm{B}}T_{\mathrm{turb}}}\right)}{\int_{q_\eta}^{q_{L_1}}\exp\left(\frac{\Phi(q)}{k_B T_{\mathrm{turb}}}\right)\mathrm{d}q}. \tag{5.62}$$

Thus, the existing deep analogy between the consecutive chemical reactions ($A \to B \to C \to$ etc.) and the Richardson–Kolmogorov cascade fragmentation of vortices with the corresponding chemical potential and chemical affinity allows structured turbulence to be macroscopically described by the methods of extended irreversible thermodynamics as a self-organization process in an open system. Using the two interpretations of Kolmogorov's parameter ε as the quantity describing the dissipation rate of energy into heat and, simultaneously, as the transfer rate of turbulent energy over the cascade of vortices in the stationary-equilibrium case, we have been able to obtain the defining relations for a key characteristic of the turbulent field, the turbulent energy dissipation rate $\langle\varepsilon\rangle$, when thermodynamically modeling structured turbulence. Recall that in the Kolmogorov (1941) theory this quantity is constant and is called Kolmogorov's parameter. Relations (5.59) and (5.60) closing the system of hydrodynamic equations (5.2)–(5.5) make the thermodynamic approach to modeling developed turbulence to a certain extent self-sufficient. When subsequently concretizing the synergetic approach to describing stationary-nonequilibrium turbulence developed above in the problems of numerical simulations of various astro- and geophysical phenomena, it should obviously be refined and developed further.

To conclude this section, we will make a brief but important remark: at first glance, by analogy with a laminar fluid motion, the condition for the increase in total continuum entropy (5.33) should have imposed certain constraints on the turbulent transport coefficients in the defining relations (5.34), (5.35), and (5.59). The positive definiteness of the direct molecular exchange coefficients is known to follow precisely from such conditions; in this case, the cross coefficients can be different in sign (see, e.g., Landau and Lifshitz 1988). Substituting relations (5.34), (5.35), and (5.59) into (5.33) for the total entropy production of a turbulized system, we obtain

$$0 \le \sigma_\Sigma \equiv \frac{1}{\langle T\rangle}\left\{\lambda^{\mathrm{turb}}\left(\frac{\partial \ln\langle T\rangle}{\partial r}\right)^2 + \overline{\rho}\nu^{\mathrm{turb}}\left(D - \frac{1}{3}(\mathrm{div}\langle u\rangle)U\right)^2 + \overline{\rho}\alpha'\left(A^{gl}_{\mathrm{turb}}\right)^2\right\}. \tag{5.63}$$

As we have already noted, the specificity of the interaction (associated with the functional dependence of the turbulent transport coefficients on parameters $\langle\varepsilon\rangle$ and $\langle b\rangle$) between various dissipative processes in a turbulized continuum is such that "switching off" one of the thermodynamic forces (e.g., the affinity A^{gl}_{turb}) can change (or even "switch off") other processes (e.g., the viscous ones). This implies that the second law of thermodynamics, which requires that the entire sum (5.63) is positive, in general, cannot be applied to its individual terms. Thus, it can happen that, for example, the quantity

$$\overline{\rho}\nu^{\mathrm{turb}}\left(D - \frac{1}{3}(\mathrm{div}\langle u\rangle)U\right)^2 < 0,$$

provided that the sum $\sigma_\Sigma \geq 0$. And this indicates that turbulent flows with a negative turbulent viscosity, $\nu^{\text{turb}}<0$, can appear under certain peculiar conditions. The above considerations are a thermodynamic justification for the possibility that a negative viscosity can appear in turbulent fluid flows.

Let us now summarize the most important results of this section. In elaborating our previous works (Kolesnichenko and Marov 1985; Kolesnichenko 1998; Marov and Kolesnichenko 2001), we performed a stochastic-thermodynamic analysis of developed turbulence in a homogeneous fluid and constructed a phenomenological model of structured turbulence as a self-organization process in an open system. We represented a turbulized continuum as a thermodynamic complex consisting of two mutually open subsystems—the subsystem of averaged motion and the subsystem of turbulent chaos, which, in turn, is considered as an ensemble of vortices with various spatiotemporal scales. This representation allowed us to obtain the defining relations for the turbulent fluxes and forces in the subsystem of turbulent chaos in a nonequilibrium-stationary state by the methods of thermodynamics with internal variables. By introducing a number of additional random parameters for the medium characterizing the excited macroscopic degrees of freedom of a strongly turbulized continuum, we were able to obtain various Fokker–Planck–Kolmogorov equations for the distribution functions of small-scale turbulence characteristics using Prigogine's postulate (concerning the direction of the irreversible processes localized in the space of configurations) and to thermodynamically describe Kolmogorov's cascade process. At the same time, a deeper understanding of the phenomenology of Kolmogorov's cascade is possible only by taking into account a large number of statistically correlated stochastic processes q that comprehensively characterize vortex spatiotemporal structures. The simplified thermodynamic analysis of quasi-stationary turbulence performed here and the idealized macroscopic model constructed on its basis, nevertheless, allow our views of the properties of open dissipative hydrodynamic systems to be extended. The latter are the object of investigation in one of the most important and rapidly developing branches of nonlinear dynamics including the evolution of chaotic motions and the formation of ordered dissipative structures.

In conclusion, note that the dual nature of the irreversible processes leading to disorder near equilibrium and to order far from equilibrium also clearly manifests itself when analyzing the present-day problems of turbulence of the macroworld in the entire variety of spatiotemporal scales. In particular, the latter correspond to the origin and evolution of the Universe, stellar and planetary cosmogony as well as the processes in the gaseous envelopes of celestial bodies and the formation of ecosystems in which the cascades of spatiotemporal configurations are possible. From our viewpoint, the concept of entropy itself becomes much more substantive and deeper owing to the approach to modeling structured turbulent chaos being developed here. In this connection, we will quote another small fragment from the book by Prigogine and Stengers (1994), "Clearly, the aforesaid is no more than a beginning, but it is very important, because it illustrates how wide the field of studies the concept of chaos, i.e., the constructive role of irreversibility, opens before us. We are convinced that these studies will lead to a new face of science with the problem of becoming at its center. However, no matter what the future of science will be, one conclusion is clear: the surrounding world cannot be described without irreversible processes." One of the main objectives of this section was to find an additional confirmation for this assertion by investigating the stationary-nonequilibrium state of turbulent chaos (using stochastic-thermodynamic methods) by theoretically showing the possibility of self-organization in such open systems.

5.2 Investigation of the Self-Organization of Turbulent Chaos Based on Stochastic Langevin Equations

In the preceding section, we constructed a macroscopic model of developed turbulence in a homogeneous compressible fluid based on the stochastic theory of irreversible processes by taking into account the presence of various temporal (or spatiotemporal) dissipative structures in the flow. In particular, these can include the sets of disordered vortex filaments, vortex rings, vortex tubes, and other compact mesoscale structures in the physical (accordingly, configuration) space. When Prigogine's key (for this model) postulate concerning the direction of irreversible processes in some local volume of the space of internal coordinates is taken into account, the approach we use has allowed the kinetic Fokker–Planck–-Kolmogorov (FKP) equation to be derived thermodynamically in the space of stochastic coordinates characterizing turbulent chaos. This equation is designed to determine the temporal evolution of the conditional probability density for some statistical characteristics of small-scale turbulence. In addition, it allows the transitions from one quasi-stationary state of vortex chaos to another caused by a successive loss of stability by the hydrodynamic system when changing a parameter controlling the regime of turbulent motion as a whole, for example, the global Reynolds number, to be investigated.

In this section developing the above concept, we analyze an alternative approach to investigating such transitions near the critical points of a self-organizing subsystem of vortex chaos. This approach based on stochastic Langevin equations and deterministic transfer equations for the conditional means is closely related to the approach based on FPK equations. However, in contrast to the latter, it has an advantage when studying stochastic processes in the neighborhood of quasi-stationary states. This advantage manifests itself in the possibility of using the efficient mathematical methods of the theory of stability of motion, the nonlinear dynamics of chaotic and stochastic systems, etc. It will be pertinent to recall in this connection that a rapprochement between such sciences as the theory of hydrodynamic stability, the statistical thermodynamics of nonequilibrium processes, the theory of bifurcations in nonlinear dynamical systems, etc., has occurred in recent years within the framework of the synergetic approach to modeling turbulent (dynamical) chaos. Moreover, it has become clear that any adequate macroscopic theory of developed turbulence cannot be constructed without explicitly modeling coherent dissipative structures and describing them by some structural parameters (Druden 1948; Crow and Champagne 1971; Brown and Roshko 1974), because it is the small-scale spatiotemporal dissipative structures that are, in a sense, the "building blocks" of this theory.

Four "scenarios" for the initial stage of turbulence generation in hydrodynamic systems that refer almost entirely to the analysis of nonequilibrium nonstationary states with nonlinear transition regimes, periodic trajectories, and restricted aperiodic motions are known in the literature (see Chap. 1). Three of these scenarios refer entirely to the initial stage of turbulent generation, when the number of excited macroscopic (collective) degrees of freedom is still small (Landau and Lifshitz 1988; Monin et al. 1989). In this chapter, when analyzing the model of coherent turbulence, we will adhere mainly to the Landau–Hopf scenario. According to the latter, a continuous transition to fully developed turbulence is accomplished through the cascade of bifurcations associated with successive excitation of newer and newer degrees of freedom.

It is common practice to study the temporal evolution of nonequilibrium dynamical systems using nonlinear (stochastic) differential equations—ordinary or partial ones—or evolution equations with discrete time intervals (e.g., the logistic equation). Such dynamical equations in explicit form generally cannot be solved. At the same time, the behavior of their numerical implementations is much more complex than that of the individual solutions for stationary systems. In particular, isolated closed trajectories (which are the images of periodic motions in the phase space) can exist here—the so-called limit cycles to which the trajectories starting in the region of their attraction tend, the quasi-periodic motions on tori with bifurcations, the trajectories that are chaotic attractors continuously densely filling a compact phase space and corresponding to nonperiodic solutions, the various transitions between such structures at critical points with their own local bifurcations where the stability is lost, etc. This entire huge set of numerical solutions with various types of behavior (from deterministic to chaotic) is also extremely important for a deeper qualitative understanding of the turbulence generation mechanisms. At the same time, it should be noted that, unfortunately, the theory of dynamical systems has not yet exerted a significant influence on the quantitative aspect of studying developed turbulent flows at large Reynolds numbers.

For this reason, we will not touch here on all these important problems but will focus mainly on the macroscopic description of developed structured turbulence, including some aspects of the theory of Markovian fluctuations as applied to the stationary-nonequilibrium states of the subsystem of turbulent chaos using statistical methods in nonequilibrium thermodynamics for this purpose. In the preceding section, the following important question remained open: Are the stationary states of turbulent chaos actually stable like the thermodynamic equilibrium states in classical thermodynamics? To answer this key question for the thermodynamic approach being developed, first of all it is necessary to be able to "prepare" in principle the initial stationary statistical ensemble corresponding to the macroscopic subsystem of turbulent chaos. Clearly, such an ensemble can be formed, for example, when there are asymptotically stable stationary states with finite domains of attraction. In the case of a medium with molecular fluctuations where the stationary state is a statistically equilibrium one, the existence of the corresponding thermodynamic Lyapunov function ensures that the equilibrium state is asymptotically stable. In this section, we show that a similar connection with thermodynamics also exists for the stationary states of a continuum with turbulent fluctuations. In particular, we propose an explicit nonequilibrium thermodynamic potential that generalizes the well-known Boltzmann–Planck relation for the equilibrium-state distribution function to the stationary states of the ensemble corresponding to the subsystem of turbulent chaos and show that it is the Lyapunov function for asymptotically stable stationary states. We qualitatively analyze the behavior of large-scale fluctuations in the neighborhood of critical transition points that gives rise to the simplest types of attractors—attracting periodic motions (limit cycles). We consider the relationship between internal intermittency in the inertial range of scales and dissipation energy fluctuations.

5.2.1 Stochastic Approach to Studying the Evolution of Turbulent Chaos: A Gaussian Process

In Sect. 5.1, we applied the stochastic-thermodynamic formalism based on the inclusion of some number of internal coordinates q_k corresponding to small-scale turbulent fluctuations and describing vortex and temperature structures in the model of turbulent chaos, apart from the generalized "classical" thermodynamic variables,. In general, these variables are random functions fluctuating near their stationary (mean) values q_k^{st}. We emphasized that positive quantities that are even functions of the fluctuating velocities, temperatures, or number densities or their logarithms can figure as the internal coordinates of chaos.

Let us now discuss in more detail the justification for a number of postulates as well as physical and mathematical assumptions on which the approach to modeling structured turbulence being developed in the monograph is based. Let us introduce a Gaussian stochastic process $q(t)$ in the space of configurations widely used below for which all joint and conditional probability densities are Gaussian in form. As will be clear from the subsequent analysis, stationary, Gaussian, and Markovian processes (the so-called Ornstein–Uhlenbeck processes) are a good approximation in describing the asymptotically stable stationary state of the model macroscopic subsystem of turbulent chaos. It is important to bear in mind in this connection that the statistics of strongly nonlinear random velocity and temperature fields in a real turbulized fluid is generally neither Gaussian nor Markovian, especially on large scales (see, e.g., Monin and Yaglom 1996). Nevertheless, it seems appropriate to also consider this restricted approximation. In particular, its advantage is that we can investigate the properties and behavior of those stationary-irreversible dissipative processes in the subsystem of chaos that are associated with the fluctuations, stability, and bifurcation changes of individual stationary states at the level of the well-developed theory of random functions.

The multivariate generalization of the Gaussian curve

$$G(q) = \left((2\pi)^n \det \sigma\right)^{-1/2} \lfloor -\sigma^{-1} : (q - \overline{q})(q - \overline{q})/2 \rfloor \tag{5.64}$$

is defined by two matrix quantities: the mean $\overline{q} = \int q\, G(q) \mathrm{d}q$ and tensor variance $\sigma = \int (q - \overline{q})(q - \overline{q}) G(q) \mathrm{d}q$ (here, σ^{-1} is the matrix inverse to the positive definite nth order matrix σ). For a stationary Gaussian process, only the conditional means $\overline{f(q(t))}^0$ and the corresponding variances depend on time and the deterministic transfer equations for the macroscopic quantities describing the linear relaxation of the means to their stationary values are associated precisely with these quantities (see Keizer 1987).

Let us now explain in more detail why it is occasionally convenient to choose the logarithms of positive stochastic turbulence characteristics such as the turbulent energy dissipation rate as the internal coordinates of chaos. Let $\varepsilon^*(r,t)$ be some positive local characteristic of the turbulent field determined by turbulent fluctuations (e.g., the turbulent energy dissipation rate $\varepsilon(r,t)$ or the degeneracy rate of the temperature variance $\varepsilon_T(r,t)$ etc.). In the case of fully developed turbulence, according to Kolmogorov's refined similarity hypotheses (Kolmogorov 1962), the dissipation $\varepsilon(r,t)$ (or related quantities) then asymptotically satisfies a lognormal probability distribution, i.e., the random variable $\ln \varepsilon^*$ has a Gaussian distribution:

$$W_1(\ln \varepsilon^*) = \left(\sqrt{2\pi}\, \sigma_{\ln \varepsilon^*}\right)^{-1} \exp[-(\ln \varepsilon^* - m^*)^2 / 2\sigma^2_{\ln \varepsilon^*}] \tag{5.65}$$

In addition, the variance $\sigma^2_{\ln \varepsilon^*}$ is assumed in Kolmogorov's theory to depend linearly on $\ln(L/\eta)$ (see Sect. 6.1)

$$\sigma^2_{\ln \varepsilon^*} = \mu^* \ln \frac{L}{\eta} + B(r,t), \quad m^* \equiv \overline{\ln \varepsilon^*} = -\sigma^2_{\ln \varepsilon^*}/2 + \ln \overline{\varepsilon^*}, \tag{5.66}$$

where $\sigma^2_{\ln \varepsilon^*}$ and m^* are the variance and mean (stationary) value of the random variable $\ln \varepsilon^*$; $B(r,t)$ is the term dependent on the characteristics of large-scale motion; μ^* is a universal constant that has different values for different variables ε^*; $L/\eta \approx Re^{3/4}$.

Distribution (5.65) adopted in Kolmogorov's (1961) improved theory can be justified by the fact that the model of a random cascade is similar to the fragmentation of dust particles; the lognormal size distribution is known to correspond asymptotically to the latter process. In addition, there are numerous experimental confirmations for the lognormal probability distribution of the turbulent energy dissipation and related positive small-scale turbulence characteristics in a wide range of moderate values of the argument. A comprehensive overview of the corresponding works is given, for example, in the book by Monin and Yaglom (1996), to which we refer the interested reader. It should be noted, however, that, in all cases, the empirical probability distribution was found to deviate from the lognormal one in the "tails," i.e., at very low or very high values of the argument. In particular, as a result of this fact, the high-order moments ε^* can no longer be accurately calculated using the lognormal distribution. In addition, the validity of the lognormal distribution for such quantities was questioned in a number of publications, because it implicitly suggests the appearance of supersonic speeds at very large Reynolds numbers (see, e.g., Frisch 1998).

The second argument for using the variables $q^* \equiv \ln \varepsilon^*$ is that the quantity $\ln \varepsilon^*$ fluctuates much more weakly than ε^* and, hence, its mean is physically more meaningful: large fluctuations of the random variable ε^*, which emerge, for example, at a critical point, distort the statistics of $\ln \varepsilon^*$ much less than the statistics of ε^*. In addition, in the case of large fluctuations in ε^*, the exponent of the mean $\ln \varepsilon^*$ gives the most probable value of the quantity ε^* itself.

Substituting the multivariate Gaussian distribution $W_1{}^{ss}(q^*)$ for such small-scale characteristics $q^* = \{\ln \varepsilon_1^*, \ln \varepsilon_2^*, \ldots\}$ into (5.19**) for the friction force

$$f(q, T_{\text{turb}}) \equiv -\partial \Phi(q, T_{\text{turb}})/\partial q = k_{\text{B}} T_{\text{turb}} \partial \ln W_1^{st}(q)/\partial q \tag{5.67}$$

yields

$$f(q^*, T_{\text{turb}}) = -k_{\text{B}} T_{\text{turb}} (\sigma^*)^{-1} \cdot (q^* - q^{*ss}), \tag{5.68}$$

where the variance of the process is defined by the relation

$$\sigma^* \cong k_{\text{B}} T_{\text{turb}}(r) + \frac{3}{4} \mu^* \ \ln Re.$$

Thus, the "friction force" appearing in the FPK equation (5.39) in the space of internal variables of turbulent chaos depends, for example, on the global Reynolds number Re controlling the regime of turbulent motion as a whole. This ultimately determines the possibility of a rearrangement of the vortex structure of the ensemble of hydrodynamic systems associated with turbulent chaos, allowing the critical values Re_{cr} at which its abrupt transition from some quasi-stable stationary state to a new stable state occurs to be revealed.

5.2.1.1 Fundamental Solution of the Fokker–Planck–Kolmogorov Equation

Thus, the conditional probability density function $P_2 \equiv P_2(q_0|q, t)$ for the continuous Markovian stationary-nonequilibrium stochastic process $q(r, t)$ describing the evolution of the stochastic internal parameters of turbulent chaos (including, in particular, the temporal evolution of the stochastic characteristics of vortex structures in a random Richardson–Kolmogorov cascade) satisfies the kinetic FPK equation (5.39). Below, we will use the FPK equation in Ito's form (5.39*), which allows the connection between the deterministic and stochastic descriptions of the transition process to be traced particularly easily. In matrix notation, (5.39*) in the spatially homogeneous case can be rewritten as

$$\frac{\partial P_2}{\partial t} = -\frac{\partial}{\partial q} \cdot [K(q, T_{\text{turb}})P_2] + \frac{1}{2}\varepsilon^2 \frac{\partial^2}{\partial q \partial q} : [Q(q)P_2]. \tag{5.69}$$

Here, the vector $K(q,T_{\text{turb}})$ and the second-rank tensor $\frac{1}{2}\varepsilon^2 Q(q)$ on the right-hand side define, respectively, the drift and diffusion parts of the probability flux, with P_2 being a positive function with the following properties:

$$\int P_2(q_0|q,t)\mathrm{d}q = 1, \quad \int W_1(q_0)P_2(q_0|q,t)\mathrm{d}q_0 = W_1(q,t),$$
$$\lim_{t\to\infty} P_2(q_0|q,t) = W_1(q). \tag{5.70}$$

The latter relation implies that the conditional probability density for stationary processes ceases to depend on the initial condition asymptotically with time. The parameter $\varepsilon^2 \equiv k_B T_{\text{turb}}$ reflects the intensity of the natural (internal) source of fluctuations in the variables q associated with the intrinsic nonlinear perturbing mechanism of the subsystem of turbulent chaos—with its "thermal" structure. The deterministic behavior of the state vector $q(t)$ satisfying the deterministic differential equation

$$\partial q/\partial t = K(q, 0) \tag{5.71}$$

corresponds to the limiting case of $\varepsilon \to 0$. Equation (5.69) is convenient for studying the stochastic processes associated with initial conditions of the form $P_2(q_0|q,0) = \delta(q - q_0)$ that correspond to a delta-like probability density concentrated at point $q = q_0$. In other words, we assume that a large number (ensemble) of independently moving trajectories emerge from point q_0 of the configuration space at time $t = 0$ and seek for their probability density in some region of the q-space at time t. Furthermore, particular boundary conditions in q that should be specially formulated to analyze specific problems can be imposed on the function $P_2(q_0|q,t)$. If not the initial state $q(0) = q_0$ but the initial distribution $W_1(q_0)$ is specified at the initial time $t = 0$, then multiplying (5.69) by $W_1(q_0)$ and integrating over q_0 we will obtain (using the second relation in (5.70)) the so-called direct Fokker–Planck–Kolmogorov equation for the one-point probability density $W_1(q, t)$ for which the distribution $W_1(q)$ is a stationary solution.

It is well known that an explicit analytical solution of (5.69) can be found only in exceptional cases, for example, when the drift matrix $K(q) \equiv L \cdot f(q)$ is linear in variables q,

$$K(q) \equiv H \cdot q, \tag{5.72}$$

and the matrix of diffusion coefficients $Q = 2L$ does not depend on q (additive noise). It is this case, which is of greatest interest for thermodynamic ensembles (see, e.g., Haken 1981) that we will consider in this section. The stochastic process $q(t)$ will then be definitely bounded if we additionally accept (which is assumed below) that the nth order (by the number of internal coordinates of chaos) quadratic matrix H is strongly coupled. Recall that strong coupling implies that the matrix H has one zero eigenvalue, while the real parts of all the remaining eigenvalues $\lambda_p(H)$ $(p = 1, \ldots, n-1)$ are negative, $Re\lambda_p(H)<0$.

The fundamental solution of the kinetic equation (5.69) satisfying the initial condition $P_2(q_0|q,0) = \delta(q-q_0)$ and rewritten by taking into account the assumptions made above as

$$\frac{\partial P_2(q_0|q,t)}{\partial t} = -H^{\text{transp}} : \frac{\partial}{\partial q}(q P_2(q_0|q,t)) + k_{\text{B}} T_{\text{turb}} \frac{Q}{2} : \frac{\partial^2}{\partial q \partial q} P_2(q_0|q,t) \tag{5.73}$$

will be a Gaussian distribution:

$$P_2(q_0|q,t) = ((2\pi)^n \det \sigma(t))^{-1/2} \exp\left[-\sigma^{-1}(t) : (q - \overline{q}^0(q_0,t))\,(q - \overline{q}^0(q_0,t))/2\right], \tag{5.74}$$

which can be easily verified by a direct substitution. Here,

$$\overline{q}^0(q_0,t) \equiv \int q(t) P_2(q_0|q,t) \mathrm{d}q \tag{5.75}$$

is the conditional mean for the stochastic process $q(t)$ (where $q(0) = q_0$) that can be written as

$$\overline{q}^0(q_0,t) = \exp(Ht) \cdot q_0.$$

Here, by the exponential $\exp(Ht)$ of the square matrix H we mean the absolutely converging matrix series

$$\exp(Ht) \equiv \sum_{p=0}^{\infty} \frac{1}{p!}(Ht)^p;$$

$$\sigma(t) \equiv \overline{\delta q(t) \delta q^{\text{transp}}(t)}^0 = k_{\text{B}} T_{\text{turb}} \int_0^t \exp(Hs) \cdot Q \cdot \exp(H^{\text{transp}} s) \mathrm{d}s \tag{5.76}$$

is the matrix variance of the conditional fluctuation $\delta q \equiv q(t) - q^0(q_0,t)$ satisfying the differential equation

$$\partial\sigma(t)/\partial t = H\cdot\sigma(t) + \sigma(t)\cdot H^{\text{transp}} + k_{\text{B}}T_{\text{turb}}Q \tag{5.77}$$

with the initial condition $\sigma(0) = 0$ (see, e.g., de Groot and Mazur 1962).

Let us now show that a stationary Gaussian Markovian process corresponds to the asymptotically stable stationary state q^{st}, the attractor state, for which

$$\lim_{t\to\infty} \overline{q}^0(q_0, t) = q^{st} \tag{5.78}$$

if the initial values of the vector $q(0) = q_0$ represented by the physical ensemble are concentrated in the domain of its attraction. Indeed, if we use the third relation in (5.70), which establishes the asymptotic equality of P_2 and W_1, then (5.74) for $t \to \infty$ transforms into a Gaussian distribution for the one-time distribution function,

$$\begin{aligned}\lim_{t\to\infty} P_2(q_0|q,t) &= W_1(q)\\ &= \left((2\pi)^n \det\sigma(q^{st})\right)^{-1/2} \exp\left[-\sigma^{-1}(q^{st}):(q-q^{st})(q-q^{st})/2\right],\end{aligned} \tag{5.79}$$

which characterizes the one-time means in a stationary ensemble. In this case, the stationary variance matrix

$$\sigma(q^{st}) \equiv \overline{(q-q^{st})(q-q^{st})^{\text{transp}}}^0,$$

in view of the existence of the limit $\lim_{t\to\infty}\sigma(t) = \sigma$ and, hence, $\lim_{t\to\infty} \mathrm{d}\sigma/\mathrm{d}t = 0$, satisfies the algebraic equation

$$H(q^{st})\cdot\sigma(q^{st}) + \sigma(q^{st})\cdot H^{\text{transp}}(q^{st}) = -k_{\text{B}}T_{\text{turb}}Q. \tag{5.80}$$

The latter expresses the generalized fluctuation-dissipation theorem for a stationary state q^{st} (recall that we assume that $Re\lambda_p(H)<0$). This equation generalizes the well-known relation between the random force and friction coefficient in the FPK equation (see Klimontovich 2002).

At fixed Q and H, (5.80) can be considered as a linear equation whose solution is the sought-for stationary variance $\sigma(q^{st})$. Since the matrix σ^{st} is symmetric, we can find $n(n+1)/2$ its independent elements $\sigma_{ij}^{st}(1 \le i \le j \le n)$. For example, in the case of two variables, the quantities σ_{11}^{st}, σ_{12}^{st}, and σ_{22}^{st} are the independent elements of the matrix σ . As follows from (5.80), they satisfy the linear equations

$$\begin{aligned}2H_{11}\sigma_{11}^{st} + 2H_{12}\sigma_{12}^{st} &= -k_{\text{B}}T_{\text{turb}}Q_{11}, \quad 2H_{21}\sigma_{12}^{st} + 2H_{22}\sigma_{22}^{st}\\ &= -k_{\text{B}}T_{\text{turb}}Q_{22},\\ H_{21}\sigma_{12}^{ss} + (H_{11}+H_{22})\sigma_{12}^{st} + H_{12}\sigma_{22}^{st} &= -k_{\text{B}}T_{\text{turb}}Q_{12},\end{aligned} \tag{5.81}$$

with a unique solution when the matrix H has eigenvalues with negative real parts.

The opposite case where an explicit form of the diffusion matrix Q (the intensity of the random Langevin forces, see (5.85)) is determined from (5.80) based on the known variance, $\sigma(q^{st})$, and relaxation, $H(q^{st})$, matrices is also important. This procedure is used when an explicit form of the matrix σ for a stationary state is known, for example, expressed in terms of the proper thermodynamic potential.

5.2.2 *Stochastic Langevin Equations in the Space of Internal Coordinates*

If (5.73) is multiplied by q^p and integrated over q from $-\infty$ to $+\infty$, then it is easy to obtain the following infinite chain of engaging equations containing conditional moments of different orders under the assumption that the expressions with H and Q integrated by parts disappear at the boundaries:

$$\frac{\partial \overline{q^p}^0}{\partial t} = pH\cdot\overline{q^p}^0 + \frac{p(p-1)}{2} k_B T_{\text{turb}} Q \cdot \overline{q^{p-2}}^0, \quad p = 1, 2, \ldots \tag{5.82}$$

Only the first deterministic equation of the system (at $p = 1$), which detaches and is closed, constitutes an exception:

$$\partial \overline{q}^0 / \partial t = H \cdot \overline{q}^0. \tag{5.83}$$

Here, $H(\alpha)$ is the constant relaxation matrix that determines the return of the vector of conditional means of the internal variables $\overline{q}^0$ to some stationary value and that generally depends parametrically on the external control parameters of the problem $\alpha = \{Re, Pe, \ldots\}$. The solution $\overline{q}^0(q_0, \alpha, t)$ of (5.83) also depends on the initial condition $q(0) = q_0$.

Equation (5.83) describes the idealized deterministic behavior of the subsystem of turbulent chaos. The simplest method of allowance for the turbulent fluctuations consists in introducing some random source into the deterministic model (5.83). Following the formal structure of Onsager's theory regarding the statistical fluctuations (see Keizer 1987), we will assume that the conditional turbulent fluctuations $\delta q = q(t) - \overline{q}^0(q_0, \alpha, t)$ of the stochastic vector process $q(t)$ satisfy a dynamical equation similar to (5.83) for the conditional mean but supplemented with the term describing the influence of weak "natural noise" in the subsystem of turbulent chaos:

$$\partial\, \delta q / \partial t = H \cdot \delta q + \tilde{F}(q, t)$$

(an analog of Onsager's regression hypothesis for molecular fluctuations). Hence it follows that $\partial q/\partial t$ and $H \cdot q$ can also differ only by some random (Langevin) term $\tilde{F}(q,t)$ with a zero conditional mean:

$$\partial q/\partial t = H \cdot q + \tilde{F}(q,t), \quad \overline{\tilde{F}(q,t)}^{0} = 0. \tag{5.84}$$

The stochastic differential equation (5.84) is the Langevin equation corresponding to an FPK equation of form (5.73). Note that the condition $\overline{\tilde{F}(q,t)}^{0} = 0$ specifies only partially the properties of the so-called random Langevin force (source) $\tilde{F}(q,t)$ acting in the space of configurations q. However, the statistical properties of the stochastic process $\tilde{F}(q,t)$ will be determined completely if we specify the form of the correlation tensor $\overline{\tilde{F}(q_1,t_1)\tilde{F}^{\text{transp}}(q,t)}^{0}$. Usually, when practical problems are solved, the random variable $\tilde{F}(q,t)$ is additionally assumed to be a stationary δ-correlated (in time) Gaussian process (multidimensional white noise)

$$\overline{\tilde{F}(q,t_1)\tilde{F}^{\text{transp}}(q,t)}^{0} = k_{\text{B}}T_{\text{turb}}Q\delta(t-t_1) \tag{5.85}$$

with the matrix of intensities of the random force Q coinciding with the tensor of diffusion coefficients $Q = 2\hat{L}$ in the kinetic FPK equation (5.73).

Given the key assumption (5.85), both approaches, those based on the FPK equation (5.73) and on the system of stochastic first-order differential equations (5.84) for the random function $q(t)$, are known to be completely equivalent (see, e.g., Tikhonov and Mironov 1977). It is important to keep in mind that the Ornstein–Uhlenbeck process is defined precisely by the linear stochastic differential equation with white noise as the random source. In particular, solving (5.84) and (5.85), we can obtain relations (5.74) for the conditional probability density and (5.77) for the variance of the conditional fluctuation as well as the corresponding asymptotic (for $t \to \infty$ expressions (5.79) and (5.80)). For our purposes, this implies that the evolutionary processes in the subsystem of turbulent chaos (these include, e.g., the relaxation processes in the neighborhood of individual stable stationary states, the simplest spatiotemporal structures) can also be analyzed based on such stochastic equations or their corollaries using the powerful methods of the qualitative theory of differential equations and a linear analysis of stability. Note also that the equivalence of the general FPK equation (5.69) and the corresponding nonlinear Langevin equation

$$\partial q/\partial t = K(q, T_{\text{turb}}) + \tilde{F}(q,t),$$
$$\overline{\tilde{F}(q,t_1)\tilde{F}^{\text{transp}}(q,t)}^{0} = k_{\text{B}}T_{\text{turb}}Q(q)\delta(t-t_1)$$

for a random vector process $q(t)$ (where $\tilde{F}(q,t)$ is a normally distributed random variable in the space of parameters (q,t), which requires invoking very sophisticated mathematical methods, is proven, for example, in Klyatskin and Tatarskii (1973)).

Let us now use (5.86) to analyze a cardinal problem of the thermodynamic approach to modeling a stationary cascade process being developed—the possibility of the existence of asymptotically stable stationary-nonequilibrium states q^{st} and limit cycles in the subsystem of turbulent chaos.

5.2.3 Nonequilibrium Stationary States of Turbulent Chaos

Thus, let us first consider the deterministic behavior of the subsystem of turbulent chaos whereby the internal state parameters $q(t)$ satisfy the deterministic differential equation (5.71)

$$\partial q/\partial t = K(q, 0).$$

It corresponds to the limiting case of $\varepsilon \to 0$ when there is no influence of the "internal noise" associated with the "thermal" structure of chaos on the motion. The following can be noted regarding this equation: although the motion of individual vortex moles along their trajectories is deterministic, the motion of individual regions of configuration space, i.e., bundles of trajectories, acquires stochastic properties. In other words, in the evolution of such a dynamical system, internal stochasticity coexists with completely deterministic laws of dynamics. Note also that a similar (coincident literally to details when q is replaced by $\overline{q}^0$) analysis can be performed based on the averaged equation (5.83), which is also valid when the "thermal" fluctuations are taken into account. The internal noise of the subsystem of turbulent chaos introducing some complication into our analysis can be taken into account once the behavior of the deterministic system (5.71) has been investigated comprehensively (see Keizer 1987). The dynamical equation (5.71) is an autonomous first-order (in time) ordinary differential equation whose solutions depend parametrically on the initial conditions $q(0) = q_0$ and on some set of control parameters $\alpha = \{Re, Pe, \ldots\}$ (reproducing the impact of the "external world" on the subsystem of chaos). A change in the latter generally causes a change in the pattern of motion itself and, hence, in the course of evolution of the vortex subsystem.

The establishment of some ordered regime in the subsystem of turbulent chaos can be investigated as follows. Since the "medium" is assumed to be constant, it can be assumed that there exists at least one stationary solution q^{st} of (5.71), i.e., a time-independent solution for which

$$\partial q^{st}/\partial t = K(q^{st}) = 0.$$

This solution corresponds to the trivial unorganized state of turbulent chaos (for the subsequent analysis, we will take it as a reference state). Note that a singular (limiting) point of a certain type (a center, a stable focus, a saddle, an unstable node, etc.) serves as the image of a stationary state in the q-space; in the case of a configuration space with a large dimension n, the number of types of singular points increases. The transition to ordered vortex chaos is associated with the idea of instability and the breaking of its initial symmetry: self-organization emerges when the trivial solution q^{st} becomes unstable and gives way to a new solution of (5.71) with a lower symmetry. The simplest method of investigating this possibility consists primarily in verifying the stability of the stationary state against small perturbations.

The stability of the limiting point implies that the trajectories in its neighborhood describing the relaxation process approach it when $t \to \infty$. In other words, the limiting point with a certain domain of attraction in the space of configurations is an attractor. As has already been mentioned above, the attractor may not be unique at some control parameters. The situations where several attractors exist in the space of internal coordinates, each having its own domain of attraction, are possible. This is because the system of equations $K(q^{st}) = 0$ can have more than one solution even if the K-linear vector function

$$K(q, 0) = H \cdot q;$$

in this special case, $H \cdot q^{st} = 0$, i.e., the states q^{st} are the eigenvectors (modes) of the matrix H with zero eigenvalues.

When the probabilistic aspects of possible scenarios for the evolution of initial stationary states are studied, it is necessary to have an initial physical ensemble corresponding to the subsystem of turbulent chaos. If quite a specific stationary state q^{st} is investigated, then the initial conditions q_0 represented by the ensemble should be concentrated precisely in the domain of its attraction. The question arises as to whether a given stationary state q^{st} is actually stable according to Lyapunov under the initial condition q_0 (or orbitally stable according to Poincare) just like the thermodynamic equilibrium states in classical thermodynamics. Otherwise, the trajectories $q(t)$ that start near q^{st} will not reach the needed stationary state. This implies with a high probability that unstable stationary states without any domain of attraction cannot be described by a quasi-stationary statistical ensemble, because it even cannot be properly organized.

However, this possibility is undoubtedly guaranteed by the so-called asymptotic stability. This is a special case of the Lyapunov stability whereby there exists such a neighborhood of the stationary state q^{st} that all of the trajectories emerging from it remain close to this attractor state and asymptotically approach it with time. In other words, when the state q^{st} is Lyapunov stable and, in addition, such $\eta > 0$ can always be found that at $|q^{st} - q_0| < \eta$,

$$\lim_{t \to \infty} q(q_0, t) = q^{st} \quad \left(or \ \lim_{t \to \infty} \overline{q}^0(q_0, t) = q^{st} \right). \tag{5.87}$$

The Lyapunov asymptotic stability (and generally the asymptotic orbital stability) is characteristic of any systems with dissipation; in particular, it is also inherent in the subsystem of turbulent chaos. In the case of stochastic fluctuations associated with the atomic structure of the medium, it follows from the second law of thermodynamics that all of the thermodynamic equilibrium states that are not critical points (see below) are asymptotically stable. The objective of this section is to show that a similar situation is possible for the stationary-nonequilibrium states of turbulent chaos. Below, we will discuss possible criteria guaranteeing the asymptotic stability of individual stationary states of fluctuating chaos. One of them is based on analysis of the stability of the solution for (5.71) in the first (linear) approximation and another is based on the existence of the so-called "Lyapunov function."

5.2.3.1 Criteria for the Asymptotic Stability of Stationary States

Let us first consider the simple case where the subsystem of turbulent chaos has only one stationary state q^{st},

$$\partial q^{st}/\partial t = K(q^{st}) = 0, \qquad \left(K \equiv -L \cdot \frac{\partial \Phi(q, Re)}{\partial q}\bigg|^{st} \right), \tag{5.88}$$

corresponding to the current Reynolds number Re. We will perform an analysis of this isolated stationary state q^{st} for asymptotic stability in the first approximation. This analysis is known to consist in checking the following linear system for asymptotic stability:

$$\partial \delta q/\partial t = H \cdot \delta q, \tag{5.89}$$

obtained by discarding the nonlinear terms in the Taylor expansion of the right-hand part of (5.71):

$$\begin{aligned} \partial(q^{st} + \delta q)/\partial t &= K(q^{st} + \delta q) = K(q^{st}) + \partial K/\partial q|^{st} \cdot \delta q + \mathcal{O}(\delta q) \\ &= K(q^{st}) + H(q^{st}) \cdot \delta q + \mathcal{O}(\delta q). \end{aligned}$$

Here, $H(q^{st}) \equiv \partial K/\partial q|^{st}$ is the Jacobi matrix. Using now the formal solution $\delta q(t) = \exp(Ht)\delta q(0)$ of (5.89), let us ascertain the cases in which its solutions are asymptotically stable. For this purpose, we will write the normal form of the exponential of the matrix $H(q^{st})$ as

$$\exp(Ht) = S^{-1}\mathrm{diag}[\exp(tJ_1(\lambda_1)), \ldots, \exp(tJ_m(\lambda_m)]\, S, \tag{5.90}$$

where

$$\exp(tJ_p(\lambda_p)) = \exp(\lambda_p t) \sum_{k=0}^{m_p - 1} \frac{t^k}{k!} I_k^{(p)}, \quad (p = 1, \ldots, m).$$

Here, S is some nonsingular matrix reducing the matrix H to the Jordan form ($\det S \neq 0$); $\lambda_1, \ldots, \lambda_m$ are $m (m \leq n)$ eigenvalues of the matrix H corresponding to the various cells $J_p(\lambda_p) \quad (\equiv \lambda_p I_p + I_1^{(p)})$ of its canonical Jordan form and m_p are the corresponding orders of these cells $\left(\sum_{p=1}^{m} m_p = n\right)$; I_p is a unit matrix of order p and $I_1^{(p)}$ is its first unit skew series; $I_k^{(p)} = [I_1^{(p)}]^k \ (k = 1, \ldots, m_p - 1)$ are the corresponding unit skew series, $I_0^{(p)} = I_p$.

The well-known Lyapunov theorems about stability in the first approximation directly follow from the general solution (5.90) of the linear equation (5.89). As applied to the fluctuating motion of turbulent chaos in the space of configurations considered here, they can be reformulated as follows:

- The stationary state of turbulent chaos is asymptotically stable if and only if all characteristic roots $\lambda_p(H) \quad (p = 1, \ldots, n)$ of the relaxation matrix $H(q^{st})$ (at the current Reynolds number) have negative real parts, i.e., all λ_p lie in the left half-plane.
- If there is at least one root with a positive real part among the roots of the secular equation $\det(H(q^{st}) - \lambda I_n) = 0$, then the stationary state q^{st} is unstable.
- If some roots of the secular equation have zero real parts and the remaining roots have negative real parts, then the stationary state of turbulent chaos (called a critical state in this case) will be:
  (a) stable (according to Lyapunov) if simple elementary divisors (i.e., $m_p = 1$) correspond to the roots with zero real parts;
  (b) unstable if at least one root with a zero real part is a multiple root of the corresponding elementary divisor ($m_p > 1$).

$Re\lambda_p(H)$ specifies the rate with which the perturbations in the internal state parameters q of chaos evolve. As a rule, the lifetime of the perturbations in the system is defined by a quantity of the order of $\tau_{\text{macro}} = |1/Re\lambda_p|$. This implies that τ_{macro} is the macroscopic evolution time scale of turbulent chaos. Thus, the emergence of a transition to some ordered regime in an evolving subsystem of turbulent chaos can be easily determined by investigating the behavior of the characteristic roots $\lambda_p(H) \quad (p = 1, \ldots, n)$ of the relaxation matrix $H(q^{st})$ as a function of the current Reynolds number. At $Re < Re_{\text{crit}}$, there exists a unique stable asymptotic solution q^{st} corresponding to the state of the subsystem of turbulent chaos with the highest symmetry that is only compatible with the boundary conditions imposed on the system. At $Re = Re_{\text{crit}}$, when at least one of the quantities $Re\lambda_p$ ceases to be negative and becomes positive, this solution becomes unstable and new branches of the solution with a lower symmetry simultaneously emerge in the supercritical region. For the critical Reynolds number Re_{crit} (which is a bifurcation point) when the relaxation time of the fluctuations in the first approximations tends to infinity, one or more new (stable or unstable) asymptotic solutions of the vector equation (5.71) appear. It is such a bifurcation that appears in Benard's classical problem when the temperature gradient imposed on the system reaches some threshold value. It corresponds to the formation of a coherent spatial structure of convective cells in an initially disordered liquid phase.

Information regarding the stability of the original nonlinear deterministic equation (5.71) can also be obtained by analyzing the stability of the linearized problem (5.89) near the stationary state q^{st}: the asymptotic stability, Lyapunov stability, or instability of the solutions of the original system (5.71) follow, respectively, from the asymptotic stability, Lyapunov stability, or instability of the solutions of the linear system (5.89). At the same time, analysis of the stability of the system near the stationary state in the first approximation gives very limited information regarding the behavior of the solutions of the nonlinear equation (5.71), because the influence of the nonlinear terms should generally be considered. In other words, the above analysis can reveal only the simplest self-organization mechanism of vortex chaos when the bifurcation occurs while the stability of the trivial reference state q^{st} is lost. The strongly nonlinear stochastic equations (5.86) should be used to describe the set of more complex scenarios for the behavior of turbulent chaos (in the space of control parameters (Re, Pe)) associated with the cascades of instabilities.

5.2.3.2 Proving the Asymptotic Stability of Stationary States: The Lyapunov Function

Let us now answer the question posed at the beginning of this section: Are the stationary states q^{st} of turbulent chaos actually asymptotically stable like the thermodynamic equilibrium states in classical thermodynamics? It is well known from the classical theory of stability of motion that a sufficient condition for the asymptotic stability of a stationary state q^{st} is the existence of a Lyapunov function for the differential equation of perturbed motion (5.89)

$$\mathcal{V}(\delta q) \equiv \delta q^{\text{transp}} \cdot V \cdot \delta q \geq 0, \tag{5.91}$$

a real-valued positive definite quadratic form specified in the space of states q and satisfying the condition $\partial \mathcal{V}(\delta q)/\partial t \leq 0$; the time derivative is taken along the trajectory defined by (5.89). In relation (5.91), the equality is achieved only at $q(t) = q^{st}$; $\delta q = q(t) - q^{st}$ is the deviation of the state vector $q(t)$ from the stationary state q^{st} , V is some real, symmetric, and positive definite matrix. For the statistical thermodynamics of nonequilibrium processes, the case where the Lyapunov function is a quadratic form is of particular interest. In addition, the existence of a positive definite quadratic form with a negative definite time derivative follows from the well-known Lyapunov theorem about asymptotic stability. It follows from these two results that the existence of the Lyapunov function (5.91) is a necessary and sufficient condition for the stability of the stationary solution of (5.71) (see, e.g., Demidovich 1998).

We will provide here one of the possible proofs for the existence of a Lyapunov function for the asymptotically stable stationary state q^{st} of turbulent chaos to which the stationary Gaussian Markovian process (5.79) corresponds (details of this proof will be used in the next subsection). But first let us find the necessary condition under which the positive definite quadratic form

$$\mathcal{V}(\Delta q) \equiv \delta\, q^{\text{transp}} \cdot V \cdot \delta q \geq 0$$

has a negative definite time derivative. Let $\mathcal{V}(\delta\, q)$ be the Lyapunov function. For the time derivative along the trajectory defined by (5.89), we then have

$$\begin{aligned} \mathrm{d}\mathcal{V}(\delta q)/\mathrm{d}t &\equiv (\mathrm{d}\,\delta q^{\text{transp}}/\mathrm{d}t) \cdot V \cdot \delta q + \delta q^{\text{transp}} \cdot V \cdot (\mathrm{d}\,\delta q/\mathrm{d}t) \\ &= \delta q^{\text{transp}} \cdot H^{\text{transp}} \cdot V \cdot \delta q + \delta q^{\text{transp}} \cdot V \cdot H \cdot \delta q = \delta q^{\text{transp}} \cdot B \cdot \delta q \leq 0, \end{aligned}$$

where H is a stable matrix (recall once again that, according to our assumption, $Re\lambda_p(H)<0$) and $B \equiv H^{\text{transp}} \cdot V + V \cdot H$ is a negative definite matrix. Since V is a symmetric positive definite matrix, it has an inverse matrix V^{-1} that is also symmetric and positive definite; therefore, $V^{-1} = V^{-1\text{transp}}$. Consequently, performing the orthogonal transformation V^{-1} on the matrix B, we will also obtain a symmetric negative definite matrix

$$G \equiv V^{-1} \cdot B \cdot V^{-1\text{transp}} = H \cdot V^{-1} + V^{-1} \cdot H^{\text{transp}} \leq 0, \tag{5.92}$$

because the matrix B has this property. Note that the sought-for necessary condition (5.92) is identical in form to (5.80) expressing the generalized fluctuation-dissipation theorem. The latter is obviously a manifestation of the deep connection existing between the dynamical properties of the fluctuations in stationary states and their stability.

Let us now turn directly to the proof for the existence of a Lyapunov function, a positive definite quadratic form with a negative definite time derivative, for the asymptotically stable stationary state q^{st} (5.79). This proof is based on the fact that, since the matrix $H(q)$ is stable, (5.80)

$$H(q) \cdot \sigma(q) + \sigma(q) \cdot H^{\text{transp}}(q) = -k_{\mathrm{B}} T_{\text{turb}} Q$$

designed to find the variance (in which Q is a positive definite matrix corresponding to the constant tensor of diffusion coefficients) has a unique positive definite solution:

$$\sigma(q) = k_{\mathrm{B}} T_{\text{turb}} \int_0^{\infty} \exp(Hs) \cdot Q \cdot \exp(H^{\text{transp}} s) \mathrm{d}s.$$

On the other hand, (5.80) is equivalent to relation (5.92) for

$$G \equiv -k_{\mathrm{B}} T_{\text{turb}} Q, \quad V^{-1} \equiv \sigma,$$

in view of which the matrix σ^{-1} (inverse to the variance matrix σ) generates a function $\mathcal{V}(\delta q) \equiv \delta q^{\text{transp}} \cdot \sigma^{-1} \cdot \delta q$ with a negative definite time derivative. Thus, for all stationary states of turbulent chaos with a stationary Gaussian one-time probability density of form (5.79), there exists a Lyapunov function, i.e., these states are asymptotically stable.

To conclude this subsection, recall that the asymptotic stability of stationary states for the subsystem of turbulent chaos can be determined using the unitary Routh–Hurwitz criterion, which gives necessary and sufficient conditions for the asymptotic stability of the stationary solution of system (5.86) (see, e.g., Demidovich 1998). If the secular equation for the constant Jacobi matrix

$$H(q^{st}) = \partial K/\partial q|^{st} = [h_{jk}]$$

is written in the expanded form

$$\lambda^n - H_1\lambda^{n-1} + H_2\lambda^{n-2} + \ldots + (-1)^n H_n = 0, \tag{5.93}$$

where

$$H_1 = SpH, \quad H_2 = \sum_{\alpha<\beta} \begin{vmatrix} h_{\alpha\alpha} & h_{\alpha\beta} \\ h_{\beta\alpha} & h_{\beta\beta} \end{vmatrix}, \quad H_n = \det H \text{ are real coefficients,}$$

then the fulfillment of the inequalities

$$\begin{aligned} &\Delta_1 = -H_1 > 0, \quad \Delta_2 = \begin{vmatrix} -H_1 & 1 \\ -H_3 & H_2 \end{vmatrix} = -H_1H_2 + H_3 > 0, \ldots, \\ &\Delta_n = (-1)^n H_n \Delta_{n-1} > 0, \end{aligned} \tag{5.94}$$

provides the necessary and sufficient conditions under which the algebraic equation has roots only with negative real parts. It is convenient to use this criterion to investigate the loss of stability by the stationary states of the subsystem of turbulent chaos (depending on the current Reynolds number) for $n \le 3$; however, for large n, it is nevertheless appropriate to pass to numerical methods of solving the secular equation, because the determinant $\det(H - \lambda I_n)$ is difficult to expand in this case.

5.2.4 Thermodynamic Stability of Stationary States and Critical Stationary States

Above, we established the connection existing between the dynamical properties of the fluctuations in stationary states and their stability and showed that the matrix σ^{-1} (inverse to the one-time variance matrix) generates a Lyapunov function in the case of an asymptotically stable stationary-nonequilibrium state of turbulent chaos. On the other hand, the existence of a Lyapunov function is a sufficient condition for the asymptotic stability of the stationary state q^{st}. As is well known, in the case of molecular fluctuations where the stationary state is a thermodynamic equilibrium one, there exists one more important interrelationship between the stability and fluctuations of the system's parameters that is established by the second law of thermodynamics and leads to a quadratic Lyapunov function,

$$\mathcal{V}^e(\delta q) \equiv \delta q^{\text{transp}} \cdot (\sigma^e)^{-1} \cdot \delta q,$$

for the relaxation processes near equilibrium. In particular, for closed systems with molecular fluctuations, the second differential of the entropy is a negative definite Lyapunov function, while the inverse variance matrix $(\sigma^e)^{-1}$ in (5.79) for the stationary Gaussian one-time probability density $W_1{}^e(q)$ is related to the matrix of second derivatives of the entropy by Einstein's formula $(\sigma^e)^{-1} = -(1/k_B)\,\partial^2 S/\partial q \partial q|^e$ (see, e.g., de Groot and Mazur 1962). Let us show that a similar relationship to thermodynamics also exists for the physical ensemble corresponding to the stationary-nonequilibrium state of turbulent chaos, a medium undergoing a constant impact from the subsystem of averaged motion and, hence, prevented from approaching the state of complete thermodynamic equilibrium.

The natural approach to generalizing the ideas that explain the formation of equilibrium structures to nonequilibrium situations consists in extending the concept of a potential function describing the dynamical properties of macroscopic systems located at a finite distance from thermodynamic equilibrium. For this purpose, we will make the following generalizing assumption regarding the form of the one-time probability density $W_1(q, T_{\text{turb}})$ for the stationary-nonequilibrium states of the ensemble referring to the subsystem of turbulent chaos: we will assume that the formula

$$W_1(q,\varepsilon)\tilde{}\ \exp\left\{-\frac{\Phi_0(q)}{\varepsilon}+\cdots\right\},\qquad \Phi(q)=\Phi_0(q)+\varepsilon\Phi_1(q)+\cdots \tag{5.95}$$

specifies the probability $W_1(q, T_{\text{turb}})$ as a function of the nonequilibrium thermodynamic global potential $\Phi(q)$ for the system's stationary-nonequilibrium states (Graham 1981). Only the principal part of the function $W_1(q, T_{\text{turb}})$ corresponding to the limiting case of weak noise $\varepsilon \to 0$, i.e., when the variables q satisfy the deterministic version of the stochastic equation (5.84), (5.71), is specified in exponential (5.95). Thus, the potential

$$\Phi_0(q) = -\lim_{\varepsilon\to 0} \varepsilon \ln W_1(q,\varepsilon). \tag{5.96}$$

determines the asymptotic expression for the stationary-nonequilibrium distribution function. We will assume the function $\Phi_0(q)$ to be single-valued and continuous and at least to have piecewise first-order differentiability. Relation (5.95) is to some extent a generalization of the Boltzmann–Planck postulate (for the equilibrium distribution of fluctuations in thermodynamic parameters of a closed system when $\Phi = -S$) to the stationary-nonequilibrium states of the ensemble corresponding to the subsystem of turbulent chaos.

If we substitute (5.95) into (5.69) and pass to the deterministic limit $\omega \to 0$, then we will obtain the equation

$$0 = K(q,0) \cdot \frac{\partial \Phi_0(q)}{\partial q} + \frac{1}{2} Q(q) : \frac{\partial \Phi_0(q)}{\partial q} \frac{\partial \Phi_0(q)}{\partial q} \tag{5.97}$$

that relates the potential $\Phi_0(q)$ to $K(q,0)$ and $Q(q)$. The function $\Phi_0(q)$ is determined as the solution of (5.97) if the boundary conditions correspond to infinitely large values of $\Phi_0(q)$ when $q \to \infty$ (for a stable system).

Let us now point out the three most important properties of the global potential $\Phi_0(q)$:

- The function $\Phi_0(q)$ decreases with time along the trajectory defined by the deterministic equation (5.71), because

$$\frac{\partial \Phi_0(q)}{\partial t} = K(q,0) \cdot \frac{\partial \Phi_0(q)}{\partial q} = -\frac{1}{2} Q(q) : \frac{\partial \Phi_0(q)}{\partial q} \frac{\partial \Phi_0(q)}{\partial q} \leq 0 \tag{5.98}$$

  (recall that the matrix $Q(q)$ is positive definite); consequently, $\Phi_0(q)$ is the Lyapunov function for the deterministic equation (5.71).
- It follows from property (5.98) that $\Phi_0(q)$ has a local minimum for any stable attractor of (5.96) in the space of configurations. On any connected subset of the attractor, the potential $\Phi_0(q)$ is constant. The unstable stationary solutions of (5.96) correspond to the maxima or saddle points of the function $\Phi_0(q)$.
- The potential $\Phi_0(q)$, when it has been found, contains very rich information regarding the probability distribution function in stationary states and the attractors of the deterministic equations.

Let us consider here the special case of a monostable stationary state q^{st} where (5.95) is simplified considerably: the relaxation of the subsystem of fluctuating chaos turns out to be a Gaussian process. Indeed, expanding the function $\Phi_0(q)$ in a Taylor series near this state, we will have

$$\begin{aligned}\Phi_0(q) &= \Phi_0(q|^{st}) + \left.\frac{\partial \Phi_0(q)}{\partial q}\right|^{st} \cdot \delta q + \frac{1}{2} \left.\frac{\partial^2 \Phi_0(q)}{\partial q \partial q}\right|^{st} : \delta q \delta q + \cdots \\ &\cong \Phi_0(q|^{st}) + \frac{1}{2} \left.\frac{\partial^2 \Phi_0(q)}{\partial q \partial q}\right|^{st} : \delta q \delta q \equiv \Phi_0(q|^{st}) + \frac{k_B T_{turb}}{2} (\sigma^*)^{-1} : \delta q \delta q,\end{aligned} \tag{5.99}$$

where $(\sigma^*)^{-1}_{ij} = (\sigma^*)^{-1}_{ji} \equiv (k_B T_{turb})^{-1} \partial^2 \Phi_0 / \partial q_i \partial q_j|^{st}$ is a symmetric positive definite matrix. At the same time, the deterministic equation (5.71) near q^{st} can be rewritten as

$$\partial q/\partial t = K(q|^{st}+\delta q) = K(q|^{st}) + \partial K/\partial q|^{st}\cdot\delta q + \cdots \cong H\cdot\delta q = H\cdot q,$$

where $H \equiv \partial K/\partial q|^{st}$; (5.97) then takes the form

$$0 = (H\cdot q)\cdot\frac{\partial\Phi_0(q)}{\partial q} + \frac{1}{2}Q(q) : \frac{\partial\Phi_0(q)}{\partial q}\frac{\partial\Phi_0(q)}{\partial q}. \tag{5.100}$$

Substituting (5.99) into (5.100) yields the already known algebraic equation (5.80),

$$H(q^{st})\cdot\sigma^*(q^{st}) + \sigma^*(q^{st})\cdot H^{\text{transp}}(q^{st}) = -k_{\text{B}}T_{\text{turb}}Q, \tag{5.80*}$$

to determine $\sigma^*(q^{st})$. Hence, irrespective of the global condition (5.98), we can conclude using the fluctuation-dissipation theorem (5.80) that the thermodynamic potential (5.99) for the internal degrees of freedom q is the Lyapunov function for the nonequilibrium stationary states of fluctuating chaos. Since the transfer equation was linearized near the stationary state, the potential $\Phi_0(q)$ of form (5.99) can be called a local one; it can be used only in the local neighborhood of the stationary state q^{st}. It follows from (5.80*) that σ^* will be a positive definite matrix if the real parts of all eigenvalues of the matrix H are negative, i.e., if the stationary state, as was assumed, is stable.

On the other hand, irrespective of condition (5.80*), we have

$$\frac{\partial\Phi_0(q)}{\partial t} = -\frac{1}{2}(\sigma^*)^{-1}\cdot Q(q)\cdot(\sigma^*)^{-1} : \delta q\delta q<0, \tag{5.101}$$

i.e., in any case, the system moves in the direction of decreasing potential $\Phi_0(q)$. In this case, it tends to the stationary state if the latter is stable and to infinity if it is unstable. The critical points emerge when the second-order terms become zero at low nonzero values of δq. At the same time, the principle of minimum potential guarantees that the matrix $(\sigma^*)^{-1} = (k_{\text{B}}T_{\text{turb}})^{-1}\partial^2\Phi_0/\partial q\partial q|^{st}$ is strictly positive in the neighborhood of a stable stationary state. Thus, for a stable stationary state q^{st} not too close to the critical points, we have the following Gaussian probability distribution from postulate (5.95):

$$W_1^{st}(q) = \sqrt{1/(2\pi)^n\det\sigma^*}\,\exp\left\{-(\sigma^*)^{-1} : \Delta q\Delta q/2\right\}, \tag{5.102}$$

where the normalization factor follows from the normalization condition (5.70).

5.2.4.1 Multiple Stationary States of Turbulent Chaos

No fluctuations were included in the chaos self-organization mechanisms described above. The analysis was quite deterministic: the turbulent fluctuations were disregarded. Meanwhile, the turbulent fluctuations (internal noise) constitute an integral part of the kinetic processes through which the subsystem of turbulent chaos evolves. Being under the influence of internal fluctuations, the system ceases to be in a certain state. It begins to make random walks in the space of states generating some distribution of state parameters, i.e., the FPK equations for the probability density function $W_1(q)$ or the corresponding stochastic Langevin equations should be invoked to describe the system. At the same time, it turns out that the extrema of the probability density with distinct peaks are generally located in the immediate neighborhood of the solution for the deterministic system and coincide with it in the limit of a large system. Consequently, the views of the pattern of bifurcation transitions within the framework of the deterministic description remain valid in the sense that they correspond mainly to the behavior of the probability density extrema corresponding to the system's stationary-nonequilibrium states.

Let us now analyze some statistical aspects of the finite number of asymptotically stable stationary states for the subsystem of turbulent chaos whose domains of attraction fill almost the entire space of internal coordinates. Eliminating, for the time being, the states near the critical points (see below), we will assume that the dynamical equations (5.86) for the conditional means of the ensemble have p attractors $q_1^{st}, q_2^{st}, \ldots, q_p^{st}$, while the remaining stationary states are unstable. Let us first find the form of the one-time stationary probability density $W_1{}^{st}(q)$ associated with one of these attractors. Let q_0 belong to the domain of attraction of quite a definite state q_{j}^{st}, then $\lim_{t\to\infty} \overline{q}^0(q_0, t) = q_{\mathrm{j}}^{st}$. According to Onsager's regression hypothesis, the conditional fluctuations $\delta q = \overline{q}^0(t) - q^{st}$ are defined by the stochastic differential equation

$$\partial \delta q / \partial t = H(q_0, t) \cdot \delta q + \tilde{F}(q, t),$$

$$\overline{\tilde{F}(q,t)}^0 = 0, \quad \overline{\tilde{F}(q,t_1)\tilde{F}^{\mathrm{T}}(q,t)}^0 = k_{\mathrm{B}} T_{\mathrm{turb}} Q(q) \delta(t - t_1), \tag{5.103}$$

in which

$$H = \partial K / \partial q|^{st}, \quad Q(q) = 2L(q). \tag{5.104}$$

Equations (5.103) describe the stationary ensemble locally, i.e., in the neighborhood of an asymptotically stable stationary state. Based on the equivalence of (5.103) to the FPK equation (5.73) pointed out above, we can conclude that a Gaussian probability density of the following form is associated with the state q_j^{st}:

$$\begin{aligned} \lim_{t\to\infty} P_2(q_0|q,t) &= W_1^{(j)}(q) \\ &= \left((2\pi)^n \det \sigma(q_j^{ss})\right)^{-1/2} \exp\left[-\sigma^{-1}(q_j^{st}) : (q - q_j^{st})(q - q_j^{st})/2\right], \end{aligned} \tag{5.105}$$

where the stationary variance matrix $\sigma(q_j^{st})$ satisfies relation (5.80). Thus, the one-time stationary probability density $W_1^{(j)}(q)$ to which all of the conditional probability densities in the domain of attraction of a given stationary state tend is associated with each of the asymptotically stable stationary states q_j^{st}. Since the function $W_1^{(j)}(q)$ is concentrated in a very narrow region around the attractor q_j^{st}, some unique Ornstein–Uhlenbeck process corresponds to each such state. Hence it follows that if the distribution $W_1^{(j)}(q)$ describes the initial ensemble, then the probability that the solution $q(t)$ differs significantly from the solution q_j^{st} or lies outside the domain of its attraction is negligible. Consequently, $W_1^{(j)}(q)$ is also the one-time probability density for the entire ensemble corresponding to the subsystem of turbulent chaos.

This assertion admits the following generalization: the temporal evolution of an ensemble with an arbitrarily prepared (i.e., different from its stationary form) initial distribution $W_1(q_0, t_0)$ and concentrated at the initial time within the domain of attraction of p attractors q_1^{st}, q_2^{st},..., q_p^{st} leads to a superposition of local stationary ensembles. Indeed, it follows from relation (5.105) that

$$W_1(q) \equiv \lim_{t\to\infty} W_1(q,t) = \lim_{t\to\infty} \int W_1(q_0,t_0)P_2(q_0,t_0|q,t)\mathrm{d}q_0$$
$$= \sum_{j=1}^{p} f_j^0 W_1^{(j)}(q), \tag{5.106}$$

where $f_j^0 \equiv \int_{D_j} W_1(q_0, t_0)\mathrm{d}q_0$ is the fraction of the initial ensemble per domain of attraction D_j of the attractor q_j^{st}. Thus, in the course of time, the one-time probability density breaks up into the sum of Gaussian variables concentrated in stable stationary states with weights equal to the fraction of the initial probability density per each domain of attraction.

5.2.4.2 Critical States with Neutral Stability

The reason why the local stationary subensembles behave autonomously is that the fluctuations in the neighborhood of an asymptotically stable stationary state q_j^{st} are relatively small. At the same time, the vanishing of the real part of one or more eigenvalues of the relaxation matrix H (the corresponding secular equation has n roots, among which there can be both real and complex ones; the latter appear as complex conjugate pairs, because all coefficients of the secular equation are real) when the real parts of the remaining roots are negative leads to the so-called critical stationary state $(q_j^{st})_{\mathrm{cr}}$ with a neutral stability (critical point) of the system under consideration. Such a critical state of the macroscopic subsystem of turbulent chaos can be associated with the growth of the Reynolds number. According to the third of the stability theorems listed above, the critical state is not asymptotically stable (though it can be Lyapunov stable). At least one relaxation mode exists at the critical point. It is the solution that is the eigenvector of the matrix H, which either is constant or oscillates with a frequency equal to the imaginary part of the eigenvalue of this mode. In the other possible case where a pair of complex conjugate eigenvalues crosses the imaginary axis (with a zero velocity) and, thus, acquires a positive real part, one of the two typical situations at the critical point is a normal Andronov–Hopf bifurcation whereby the stationary state at the instant of stability loss transforms into a stable periodic trajectory (with a periodic oscillation frequency ω_1) to which all of the neighboring trajectories begin to be attracted. As a result, a limit cycle appears, which is initially very small but gradually increases in size as the control parameters exceed their bifurcation values. A different situation (known as the reverse Andronov–Hopf bifurcation) when an unstable limiting point emerges is also possible. Thus, the statistical law of change in turbulent fluctuations (in the q-space) near the critical points of fluctuating chaos cannot provide their smallness and, hence, (5.106) is inapplicable for such stationary states. A non-Gaussian distribution function is most likely associated with them. To ascertain how the probability density actually appears in this instability region, it is generally necessary to numerically solve the Fokker–Planck–Kolmogorov equation (5.69) or to try to determine the nonequilibrium thermodynamic potential $\Phi_0(q)$ by solving (5.97).

It should also be borne in mind that the fluctuations near the critical point $\left(q_j^{st}\right)_{\rm cr}$ affect unpredictably the "choice" of the subsequent path of evolution by the system (the critical point does not need to be a bifurcation one). In principle, we can also analyze the loss of stability for periodic motions based on (5.86), just like the stationary states, and find the corresponding bifurcations. The rigorous results of such an analysis refer mainly to the bifurcations of some stationary solutions into other stationary or periodic solutions (see, e.g., Marsden and McCracken 1976). In one of the typical situations similar to the normal Andronov–Hopf bifurcation, when the limit cycle ceases to be attractive at some bifurcation values of the control parameters, two new modes with oscillation frequencies ω_1 and ω_2 are excited instead of one mode, a quasi-periodic attractor is generated—an attractive two-dimensional smooth torus T^2 on which all trajectories are located. Thus, one can come to a sequence of normal Andronov–Hopf bifurcations, i.e., to the motion on the surface of some torus T^k with an increasing dimension k. The motion appears increasingly complex and does not differ from a chaotic one at an infinite number of bifurcations (deterministic chaos).

It is in this way, according to Landau's hypothesis (see, e.g., Landau and Lifshitz 1988; Monin et al. 1989), that the turbulence itself emerges when the set of fundamental fluid flow velocity oscillation frequencies is supplemented by a new incommensurable frequency after each normal Andronov–Hopf bifurcation, the solution passes to a new torus (whose topological dimension coincides with the number of independent frequencies), and the flow becomes even more irregular and, when the set of modes ($Re^{9/4}$) is excited, turbulent. This overestimation of the number of modes follows from Kolmogorov's (1941) theory, which disregards the presence of coherent structures in the inertial range. As has already been repeatedly noted above, the most recent studies show that bundles of thin and very intense vortex filaments and other CSs are present in a turbulent flow on small scales. Therefore, the number of degrees of freedom can be much smaller than that actually suggested by this scenario, which corresponds to the motion over an infinite-dimensional torus (quasi-periodic attractor). At the same time, it should be noted that the Landau–Hopf scenario does not give rise to the regime of truly chaotic motions, i.e., motions with statistical properties of a random process, for example, a sensitive dependence of the temporal evolution of the system on initial conditions (the model for the formation of different-scale vortex structures that is not based on the Landau–Hopf scenario will be considered in Sect. 6.2). In addition, the Landau–Hopf scenario for any dissipative dynamical systems is structurally unstable,[2] i.e., atypical: as a rule, the chain of normal bifurcations breaks already after the first three steps—either the open winding on the three-dimensional torus is closed, turning into a limit cycle (mutual synchronization of modes) or the torus is destroyed and the so-called strange attractor emerges as a result of this destruction. True dynamical chaos with a continuum of frequencies, a sensitive dependence on initial conditions, mixing, etc. is realized on such a manifold, which is no longer a smooth torus but has a complex topology. From a geometrical point of view, the strange attractor is, as a rule, a fractal set with a noninteger Hausdorff–Besicovitch dimension (Mandelbrot 1983).

5.2.5 *Intermittency Effects*

Let us now turn to one more reason why it is important to investigate the behavior of the fluctuations in turbulent energy dissipation rate (see Sect. 6.1) or other positive definite small-scale characteristics when constructing a phenomenological model of developed turbulence—to the intermittency problem. Note that the regime of temporal intermittency associated with sharp large-amplitude fluctuations is inherent in any dynamical dissipative systems as they pass from quasi-periodic motion to chaos.

The instability of ordered spatiotemporal structures of the turbulent field (intense velocity and temperature fluctuations localized in the coordinate (phase) space (see Crow and Champagne 1971; Brown and Roshko 1974)) and the related random cascade generation of increasingly small velocity vortices and microstructural temperature inhomogeneities form the basis for hydrodynamic intermittency.

The experimental law of finite energy dissipation stating that the mean rates of turbulent energy dissipation $\overline{\varepsilon}$ and scalar dissipation $\overline{\varepsilon}_T$ in the limit of an infinitely large Reynolds number ($\nu \to 0$) tend to finite and positive limits, $\lim_{\nu \to 0} \overline{\varepsilon}(\nu) = \overline{\varepsilon} > 0$,

[2] The dynamical systems for which small perturbations of the evolution operator lead to topologically equivalent dynamical systems are called structurally stable ones.

$\lim_{\chi \to 0} \bar{\varepsilon}_T(v) = \bar{\varepsilon}_T > 0$, plays a fundamental role in the theory of developed turbulence. It is this circumstance that was used as a basis for the original hypotheses of Kolmogorov's (1941) theory for locally homogeneous and isotropic turbulence. At the same time, numerous experiments suggest that highly irregular distributions of the velocity and temperature gradients, when regions with very small gradients (a nonturbulent fluid for which $\varepsilon \approx 0$, $\varepsilon_T \approx 0$) alternate irregularly with regions where the gradients are very large ($\varepsilon > 0$, $\varepsilon_T > 0$), exist in a turbulent flow. This phenomenon called internal intermittency was first investigated by Batchelor and Townsend (1949) (see also Batchelor 1953). Internal intermittency is particularly pronounced in flows at very large Reynolds numbers, i.e., when the external turbulence scale L is large enough with respect to the internal one η (e.g., in stars, oceans, and atmospheres). The peculiarity of a turbulent flow also manifests itself in the fact that these regions move chaotically in space, while the values of $\varepsilon(r,t)$ and $\varepsilon_T(r,t)$ in them, in general, depend on Re. In addition, there is one more characteristic property of the asymptotic behavior of ε and ε_T. Since the averaged quantities $\bar{\varepsilon}$ and $\bar{\varepsilon}_T$ do not depend on the Reynolds number and the kurtosis coefficients of these quantities estimated by various authors (see Monin and Yaglom 1996) probably increase indefinitely with Re, the following anomalous singularity holds: the scalar dissipation and energy dissipation take place in a volume that approaches zero as the Reynolds number increases indefinitely, $Re \to \infty$. The question arises as to how this result can be interpreted. In the qualitative global theory of three-dimensional turbulence, it is commonly assumed (see, e.g., Mandelbrot 1974) that the motion of a fluid forces the vortex tubes to stretch, while the vortex being stretched flattens and turns into a vortex sheet in order to retain a fixed volume as the length increases. Subsequently, the vortex sheets curl into infinitely complex fractal structures. Thus, when $Re \to \infty$, the energy dissipation is concentrated on a set of points that is topologically a surface. This surface has an infinite area and is so "entangled" that it may be considered as a fractal for which the Hausdorff–Besicovitch dimension lies somewhere near 2.5–2.6 but probably does not exceed 2.66 (Mandelbrot 1974).

Bearing in mind the macroscopic modeling of structured turbulence, it is convenient to distinguish two types of intermittency—external and internal intermittencies referring to the subsystems of averaged motion and turbulent chaos, respectively. In the case of external intermittency, the spatiotemporal distribution of the parameter $\langle \varepsilon(r,t) \rangle$ is very nonuniform—"turbulent spots" where $\langle \varepsilon(r,t) \rangle > 0$ alternate with laminar regions where $\langle \varepsilon(r,t) \rangle = 0$ (the parameter $\langle \varepsilon_T(r,t) \rangle$ behaves similarly). In particular, the oscillations of the boundaries of turbulent wakes, turbulent jets, and boundary layers are associated with the manifestation of external intermittency of a turbulent flow. It is convenient to model this intermittency in terms of an invariant turbulence theory by invoking the evolutionary equations for the average dissipation rate (see, e.g., Marov and Kolesnichenko 2001). On the other hand, the elementary volume dr of the subsystem of turbulent chaos is also characterized by a highly nonuniform spatiotemporal distribution of the turbulent energy dissipation ε or scalar dissipation ε_T: a high degree of dissipation of energy and temperature inhomogeneities is inherent in some subregions of the volume dr, while there is virtually no dissipation in other subregions. In addition, there exist subregions where the energy dissipation and scalar dissipation are much greater than their means.

In view of the aforesaid, it is the dissipation of turbulent energy ε or related quantities (quadratic in velocity and temperature gradients) that can serve as an indicator of intermittency (Obukhov 1962). It should be noted that the statistical models of a random cascade in which the intermittency of turbulence in the inertial range is described in terms of ε fluctuations have now been developed in many works of domestic and foreign scientists (see, e.g., Novikov and Stewart 1964; Frisch et al. 1978; Frisch 1998; Monin and Yaglom 1996). Other approaches to modeling hydrodynamic intermittency based, in particular, on the concept of strange attractors have also been developed (see, e.g., Strange Attractors 1981). There exist even theories in which an attempt was made to explain the observed interrelationship of the intermittency of "cosmic fluid" turbulence and the distribution (origin) of galaxies or stellar intermittency.

The fluctuations of ε and ε_T cased by the variations in hydrodynamic characteristics of averaged motion, in particular, by the variations in numbers *Re* and *Pe*, obviously exert an influence on the statistical properties of arbitrary-order structure functions for the spatial velocity or temperature differences (and their spectra), which are known to be the main mathematical tool for quantitatively studying a random cascade (see Chap. 1). For this reason, the explicit expressions for them derived in terms of Kolmogorov's (1941) original theory cannot be quite universal. Landau's famous "Kazan remark" (see, e.g., Landau and Lifshitz 1988) about the hypotheses of this theory concerned precisely this influence. Actually, Kolmogorov's original concept of a random cascade was refined by Obukhov (1962) in connection with this criticism. He suggested abandoning the condition $\bar{\varepsilon}=$ const in region G (with the center at point r and a characteristic scale $\Lambda<$) and proceeding from the fact that the statistical properties of turbulence are not determined by the probability-theoretic mean $\bar{\varepsilon}\,(r,t)$ of the random variable ε but depend on the dissipation $\varepsilon_r(r,t)$ averaged over some volume V_r with a characteristic size r (the results obtained depend weakly on the shape of the averaging region) small compared to the typical averaged flow inhomogeneity scale, $r<<\Lambda$ (see Chap. 6).

In general, the Kolmogorov–Obukhov theory allows one to improve the semiempirical approaches to describing developed turbulence, for example, the gradient methods used to close the averaged hydrodynamic equations at the level of first-order moments (see, e.g., Kolesnichenko 1998; Marov and Kolesnichenko 1999, 2001). Until recently, the defining relations in such semiempirical models were purely local. In particular, the components of the turbulent viscosity tensor in the gradient relations for the Reynolds stress tensor $R(r,t)$ are usually specified by expressions including the dissipation and kinetic energy (or the local external scale) of turbulence that are determined at the same spatiotemporal point (r,t), i.e., without allowance for the influence of the global Reynolds number *Re* or other characteristics of averaged motion on the pattern of this functional relation (Monin and Yaglom 1992). In this case, the empirical constants in the relations for the turbulent exchange coefficients are usually varied greatly to efficiently describe turbulent flows of various types; it is for this reason that such theories do not require including intermittency characteristics. At the same time, as is clear from the aforesaid, intermittency is a characteristic signature of the nonlocal large-small-scale interaction between turbulent vortices that is disregarded in semiempirical theories for one-point moments.

Let us now summarize the most important results of this section. Using the kinetic FPK equations, we proposed an alternative approach to describing the transition from one quasi-stationary state of turbulent chaos to another based on stochastic Langevin equations. Using the deterministic dynamical equations, we analyzed a fundamental problem of the approach being developed, the possibility of the existence of asymptotically stable stationary-nonequilibrium states of turbulent chaos, and proposed a generalized nonequilibrium thermodynamic potential for the internal variables. This potential was shown to be the Lyapunov function for the nonequilibrium-stationary states of the ensemble corresponding to the subsystem of turbulent chaos.

Although the problem of hydrodynamic turbulence formally refers to the theory of dissipative dynamical systems, one of the rapidly developing branches of nonlinear dynamics including the evolution of chaotic motions and the formation of ordered dissipative structures, unfortunately, the potentialities of this theory (bifurcations, Lyapunov exponents, strange attractors, the fractal dimension of attractors, etc.) being developed mainly for systems with a small number of degrees of freedom still remain virtually untapped when semiempirically modeling developed turbulence with a very large number of excited degrees of freedom. To somewhat advance in this direction, we considered here the stochastic-thermodynamic approach to modeling structured turbulence where not the infinite-dimensional space of states for a turbulized fluid, with each of its points corresponding to a certain spatiotemporal distribution of the random velocity and temperature fields, but the space of configurations of several dimensions corresponding to the internal structure of the subsystem of turbulent chaos serves as the phase space of the physical ensemble. In some cases, this approach corresponds to a more efficient application of the methods of investigating dynamical systems. At the same time, it should be recognized that it will be unrealistic to construct an efficient quantitative mathematical model of developed turbulence including coherent structures by the proposed method until the dynamical and topological properties of CSs (which, of course, are different, for example, for a crooked branched curve, a vortex tube or a corrugated layered surface—a vortex sheet, or a Burgers sheet, etc.) are taken into account and described quantitatively. Actually, it is such characteristics that should be described by an appropriate set of internal parameters for turbulent chaos.

Note also that Kolmogorov's (1961) theory has not been applied in any way in semiempirical turbulence models until the present time. The stochastic-thermodynamic approach proposed in this section allows the dissipation of turbulent energy to be used in the averaged hydrodynamic equations of motion as an indicator of intermittency. It provides a good basis for developing an appropriate model of hydrodynamic turbulence with intermittency (associated with intense velocity and temperature fluctuations). However, a detailed analysis of the phenomenological modeling of intermittency in the inertial range and, thus, the development of a technique for determining the turbulent transport coefficients as a function of the parameters controlling the fluid flow as a whole is outside the scope of this analysis and requires a special consideration.

In conclusion, note that, since turbulent motions are common in nature, one might expect the synergetic approach to modeling developed turbulence proposed here to be used in various astro- and geophysical applications. The dual nature of irreversible processes, which leads to the absence of order near equilibrium and to order far from equilibrium, clearly manifests itself when analyzing the present-day problems of turbulence of the macroworld in the entire variety of spatiotemporal scales, from the origin and evolution of the Universe, stellar and planetary cosmogony, to the processes in the gaseous envelopes of celestial bodies and the formation of ecosystems where the random cascade formation of different-scale turbulent vortices is possible. In view of the latest numerous achievements in the field of experimental astro- and geophysics (see Sect. 1.3), we believe that the time of the construction of more sophisticated (than classical) turbulence models corresponding to real dynamical processes in these natural objects has come. The stochastic-thermodynamic approach to modeling structured turbulence considered in this chapter serves precisely this purpose.

5.3 Fractional-Order FPK Equation to Describe Turbulent Chaos with Memory

The stochastic diffusion process $q(t)$ (corresponding to the preferential excited macroscopic degrees of freedom of a fluctuating flow) described by the FPK diffusion equation (5.39) does not remember its past, i.e., it has no memory. At the same time, as has been noted above, the statistics of the strongly nonlinear random fields of hydrothermodynamic quantities in a real turbulized fluid is generally non-Markovian, especially on large scales (see Monin and Yaglom 1996). In the language of physics, the non-Markovian nature of the flow turbulization process, which is possible in some regions of phase space through multiscale spatiotemporal correlations of fluctuating parameters (generally nondecreasing exponentially), precisely implies the availability of memory. For this reason, it is necessary to generalize the approximate model for the evolution of small-scale turbulent field characteristics based on (5.39) considered above to the more realistic case of turbulence with memory.

In connection with the development of methods for describing the structured field of a turbulized fluid flow, we are interested mainly in the states of turbulent chaos that can excite a cascade of multiplicative-noise-induced phase transitions to new (quasi-) stationary states of the vortex subsystem as the intensity of external hydrodynamic fluctuations (associated, in particular, with the dynamical manifestation of the degrees of freedom disregarded in the model) grows. As was shown in Sect. 5.1, such transitions, to which a set of different-scale stable coherent structures corresponds in a fluctuating fluid flow, are possible under the condition of entropy and energy exchange with the subsystem of averaged motion. It is well known that if the multiplicative noise in a system is modeled by a power function (which, as a rule, is done), then the domain of definition of a stochastic medium in the configuration space q generally acquires a fractal character (see, e.g., Olemskii 1998). A fractal medium is, as a rule, a medium with memory. The diffusion processes on fractal structures are essentially non-Gaussian (in the classical understanding of a Gaussian process) and are determined by a complex interaction of the correlations acting on arbitrarily large spatiotemporal scales. A consistent allowance for these correlations in an appropriate theory of structured turbulence generally requires abandoning the Gaussian kinetic equation (5.69) (describing the motion of "ideal" chaos in a continuous configuration space) and the traditional view of the diffusive transfer of the probability density $P_2(q,t)$ as the classical Brownian motion of vortex particles in a monotonic medium. This is because in the general case of chaos of fractal structures, there is a fractal Brownian motion (a random process with memory (see Mandelbrot and Van Ness 1968)) and a nonlinear dependence of the probability flux $J(q,t)$ on the fractal topology.

Whereas the laminar and ideal chaotic motions of a fluid have a certain degree of monotonicity, the real turbulent motion is intermittent, because it exhibits intermediate properties between those of regular and chaotic motions. One of the characteristic properties of the phase space corresponding to the real turbulent motion of a fluid is its incompleteness. The latter manifests itself in the presence of numerous "islands" in the phase volume of a stochastic "sea" where no chaotic trajectories can penetrate (see Zaslavsky 2004). In the geometric space, the presence of laminar spots, bursts, and intermittency corresponds to this state of the subsystem of vortex chaos. The laminar spots manifest themselves as localized spatial structures of a regular fluid motion against the background of a turbulent flow; the turbulent bursts are associated with intense broad-spectrum fluctuations against a fine-grained background of less intense low-frequency oscillations in hydrothermodynamic parameters of a regular flow; the intermittency in time (in space) is the alternation of laminar and turbulent regions in time or space. Thus, the specificity of structured turbulence itself provides a physical basis for the replacement of homogeneous spacetime by fractal spacetime. In view of the aforesaid, the kinetic description of the fractal properties of the evolution of the subsystem of turbulent chaos requires some generalization.

The necessary generalizations can be obtained by invoking the methods of fractional integro-differential calculus (see, e.g., Samko et al. 1987; Nakhushev 2003) appropriate to complex nonlinear systems with multiscale correlations in space and time. The fractional differential equations take into account the memory, spatiotemporal intermittency, and nonlocality effects, which go far beyond the traditional Gaussian statistics. In this section, we propose a thermodynamic approach to deriving the generalized fractional Fokker–Planck–Kolmogorov (FFPK) equations describing the turbulent transport processes in the subsystem of turbulent chaos based on fractional dynamics that takes into account the structure and metric of the fractal time. The introduction of fractional time derivatives $\partial^{\omega}/\partial t^{\omega}$ (the parameter ω is related to the fractal dimension of the active time) into the kinetic FPK equation allows the effects of memory and intermittency in time (the presence of turbulent bursts) to be taken into account in the context of unified mathematical formalism. Below, for simplicity, we will restrict our analysis to the spatially homogeneous case of a stochastic subsystem of turbulent chaos described by a set of internal coordinates $q(t)$ that vary randomly with time t.

5.3.1 Causality Principle for Non-Markovian Processes in the Subsystem of Turbulent Chaos

Let a random process $q(t)$ occur in an N-dimensional configuration space $0 \leq q_k < \infty$ $(k = 1, 2, \ldots N)$ and in time t from the initial $t = 0$ to the estimated one. Below, we will assume that the temporal domain of definition for the stochastic subsystem of turbulent chaos is a self-similar fractal set. It is quite admissible that the linear nonhereditary law (5.36) is also violated in media with a fractal geometry. The availability of memory in the subsystem of turbulent chaos implies that if the force (5.22)

$$A_{\text{turb}}(q, t') = -\partial \mu_{\text{turb}}(q, t')/\partial q \tag{5.107}$$

of the cascade process corresponding to one equivalent $n(q) \to n(q + \partial q)$ of the vortex fragmentation process (the generalized de Donde chemical affinity for state q) acts on the subsystem of chaos at time t', then the probability flux $J(q, t)$ emerges whose value at a succeeding time $t > t'$ is specified by the hereditary integral equation (linear hereditarity)

$$J(q, t) = \int_0^t L(q, t - t') \cdot A_{\text{turb}}(q, t') \mathrm{d}t' = \int_0^t L(q, \tau) \cdot A_{\text{turb}}(q, t - \tau) \mathrm{d}\tau, \tag{5.108}$$

(cf. the phenomenological relation (5.36) valid for a Markovian turbulized medium). Here, the kernel of the integral operator $L(q, t - t')$ reflects the hereditarity property, i.e., the influence of the prehistory of the process on its state at a given time t. Below, we consider only positive times by assuming that $P_2(q, t) = 0$ at $t < 0$. Note that the term hereditarity (according to the terminology of Vita Volterra) is equivalent to the concepts of memory, consequence, heredity, delay, etc.

Since the response of the subsystem cannot precede in time the effect causing it, we have the following conditions for the matrix L:

$$L(t - t') = 0 \text{ for } t < t', \tag{5.109}$$

or

$$L(\tau) = 0 \text{ for } \tau < 0. \tag{5.110}$$

These relations express the causality principle for the case under consideration. Let us also require that a constant finite driving force cause a finite response of the system. This implies that

$$\int_0^\infty L(q,\tau)\mathrm{d}\tau < \infty, \tag{5.111}$$

i.e., these integrals must exist and be finite.

For a subsystem of turbulent chaos without memory, the time dependence of the memory function is

$$L(q, t - t') = L(q)\delta(t - t') \tag{5.112}$$

where $L(q)$ is the matrix of positive coefficients and $\delta(t - t')$ is the Dirac delta function. Substituting (5.112) into (5.108) yields the relation used in Sect. 5.1:

$$\begin{aligned} J(q,t) &= \int_0^t L(q) \cdot A_{\text{turb}}(q,t')\delta(t - t')\mathrm{d}t' \\ &= L(q) \cdot A_{\text{turb}}(q,t) = -L(q) \cdot \partial\mu_{\text{turb}}(q,t)/\partial t. \end{aligned} \tag{5.113}$$

According to this relation, if there is no memory, the probability flux $J(q,t)$ at time t depends only on the chemical affinity $A_{\text{turb}}(q,t)$ for the state q of a vortex continuum taken at the same time t. When memory is available, the delta function in (5.112) is smeared into a bell-shaped dependence whose width determines the time interval T during which the action of the chemical affinity affects the flux. For systems with so-called ideal memory, we have $T \to \infty$, i.e., the flux $J(q,t)$ is formed over the entire interval of action of the affinity $A_{\text{turb}}(q,t)$ until the time t. Formally, this is expressed by specifying the kernel of the integral relation (5.108) in the form

$$L(q, t - t') = L(q)/t. \tag{5.114}$$

Here, there is no dependence on the time t' of action of the affinity $A_{\text{turb}}(q,t)$ and the dependence of the memory function on the measurement time t of the flux $J(q,t)$ is taken in a form to satisfy the normalization condition

$$\int_0^t L(q, t - t')\,\mathrm{d}t' = L(q). \tag{5.115}$$

Thus, including memory causes the constant kernel (5.112) to be modified into the hyperbolic dependence (5.114).

The phenomenological hereditary relation (5.108) written in time representation is inconvenient due to the presence of convolution (i.e., integration over t'). It can be eliminated by using the integral Laplace transform for the functions $J(q,t)$, $A_{\text{turb}}(q,t)$, and $L(q,t)$

$$J(q,t) = \frac{1}{2\pi i} \int_{-i\infty}^{+\infty} J(q,p)\exp(pt)\mathrm{d}p, \tag{5.116}$$

$$A_{\text{turb}}(q,t) = \frac{1}{2\pi i} \int_{-i\infty}^{+\infty} A_{\text{turb}}(q,p)\exp(pt)\mathrm{d}p, \tag{5.117}$$

$$L(q,t) = \frac{1}{2\pi i} \int_{-i\infty}^{+\infty} L(q,p)\exp(pt)\mathrm{d}p, \tag{5.118}$$

where their Laplace transforms $J(q,p)$, $A_{\text{turb}}(q,p)$, and $L(q,p)$ are defined by the formulas

$$J(q,p) = \int_0^{\infty} J(q,t)\exp(-pt)\mathrm{d}t, \tag{5.119}$$

$$A_{\text{turb}}(q,p) = \int_0^{\infty} A_{\text{turb}}(q,t)\exp(-pt)\mathrm{d}t, \tag{5.120}$$

$$L(q,p) = \int_0^{\infty} L(q,t)\exp(-pt)\mathrm{d}t. \tag{5.121}$$

Given these relations, (5.108) can be written as

$$J(q,p) = L(q,p) \cdot A_{\text{turb}}(q,p). \tag{5.122}$$

It is easy to see that the Laplace transform of kernel (5.112) corresponding to the absence of memory (an equivalent of "one point," $t = t'$) is reduced to a time-independent function,

$$L(q,p) = L(q).$$

For ideal total memory (an equivalent of a "straight line," $0<t'$) in the limit $|p|$ $t>>1$, from (5.114) it is easy to obtain

$$L(q,p) = L(q)/pt.$$

The following question can be posed: Is the existence of a geometric object that is intermediate between a "straight line" and a "point" possible? Fractal geometry answers this question affirmatively, because such an object exists and is a Cantor set or so-called Cantor dust. We will assume below that memory in the system is retained only at the points of a (generally nonsymmetric) Cantor set whose fractal dimension defined by (1.1) is

$$D = \ln j / \ln(1/\xi).$$

Here, j is the number of blocks involved in constructing the elementary figure of the fractal (for a triad Cantor set, $j = 2$); ξ is the self-similarity index determining the extent to which the block size decreases at each set construction step (for a symmetric triad Cantor set, for which the middle link is removed at the kth construction step, $l_k/3$, $\xi = 1/3$ [see (1.2)]). Since the size of the resulting fragment for a Cantor set should not exceed the size of the initial block, the similarity parameter is limited by the condition $\xi j \leq 1$, which, in turn, leads to $D \leq 1$ (for triad Cantor dust, $D = \ln 2/\ln 3 \cong 0.6309$). The algorithm for constructing a Cantor set can be found in almost all books on the theory of fractals (see, e.g., Cronover 2000).

Thus, one may expect the fractal dimension of a temporal Cantor set D to be related to the measure of memory retention (see, e.g., Olemskii and Flat 1993). The question arises as to how the transform $L(q,p)$ will appear for the memory acting at the points of a Cantor set. Whereas the Laplace transforms of the simplest time dependences (5.112) and (5.114) are easy to find, the calculations for the memory acting at the points of a Cantor set are more cumbersome. The corresponding calculations performed by Nigmatullin (1986, 1992) yield the following result:

$$L(q,p) \propto 2^{-D/2}(1-\xi)^{-D}(pt)^{-D}. \tag{5.123}$$

Thus, the physical systems with residual memory at points from a Cantor set are described by the Laplace transform (5.123), where $0 \leq D \leq 1$ specifies a measure of the system's residual memory. Using the inverse Laplace transform (5.118), for the time dependence of the memory function we find

$$L(q,\tau) = L(q)\frac{1}{\Gamma(D)}\left[\sqrt{2}(1-\xi)t\right]^{-D}\tau^{D-1}, \quad \tau \equiv t - t' > 0 \tag{5.124}$$

where $\Gamma(z) = \int_0^\infty \exp(-\eta)\eta^{z-1}\mathrm{d}\eta$ is the Euler gamma function, $Re\, z>0$ (the following relations are valid: $\Gamma(z+1) = z\Gamma(z)$, $\Gamma(z)\Gamma(1-z) = \pi/\sin \pi z$, $\Gamma(n-1)= n!$ $(n \in N)$; $1/\Gamma(z)$ is an integer function and has first-order zeros at points $z = k,\ k \in N \cup \{0\}$). From a mathematical point of view, equality (5.124) implies that the hereditary kernel, which is a power function, satisfies the homogeneity condition: $L(q, m\tau) = m^{D-1}L(q,\tau)$. The homogeneity of (5.126) precisely reflects the self-similarity of the phase space for the stochastic subsystem of turbulent chaos. As has already been said, fractal objects have this property.

The representation of the flux $J(q,t)$ as the fractional integral

$$
\begin{aligned}
J(q,t) &= \int_0^t L(q,\tau)\cdot A_{\text{turb}}(q,t-\tau)\mathrm{d}\tau \\
&= L(q)\cdot\left[\sqrt{2}(1-\xi)t\right]^{-D}\left(\frac{1}{\Gamma(D)}\int_0^t \tau^{D-1}A_{\text{turb}}(q,t-\tau)\mathrm{d}\tau\right) \\
&= \left[\sqrt{2}(1-\xi)\right]^{-D}L(q)\cdot D_{0t}^{-D}(A_{\text{turb}}(q,t)),
\end{aligned}
\tag{5.125}
$$

where the relation

$$
\begin{aligned}
D_{0t}^{\alpha}(A_{\text{turb}}(q,t)) &\equiv \frac{t^{\alpha}}{\Gamma(-\alpha)}\int_0^t \tau^{-\alpha-1}A_{\text{turb}}(q,t-\tau)\mathrm{d}\tau = \frac{t^{\alpha}}{\Gamma(-\alpha)}\int_0^t \frac{A_{\text{turb}}(q,t')}{(t-t')^{1+\alpha}}\mathrm{d}t' \\
&= \frac{1}{\Gamma(-\alpha)}\int_0^1 (1-u)^{-\alpha-1}A_{\text{turb}}(q,ut)\mathrm{d}u \\
&= -\frac{1}{\alpha\Gamma(-\alpha)}\int_0^1 A_{\text{turb}}(q,(1-u)t)\mathrm{d}u^{-\alpha} \\
&= \frac{1}{\alpha\Gamma(-\alpha)}\int_0^1 \frac{\partial\mu_{\text{turb}}(q,(1-u)t)}{\partial q}\mathrm{d}u^{-\alpha}, \quad \alpha<0.
\end{aligned}
\tag{5.126}
$$

Here, $u = t'/t$ is a dimensionless time bounded by the condition $u \le 1$. The fractional nature of this integral is reflected by the presence of the index $-\alpha$ in the differential of the argument u. The negative index $\alpha = -D$ of the fractional integral coincides with the fractal dimension of the Cantor set and points to the fraction of the "channels" constituting some branching fractal structure of a temporal cluster open for the evolution of a physical system (Nigmatullin 1986).

5.3.2 *Fractional Integral and Fractional Derivative (Introductory Information)*

Here, we provide general information from the theory of fractional integration and differentiation (see, e.g., Nakhushev 2003). Let α, c, d, and $a \in [c,d]$ be real numbers; $[\alpha]$ is the integer part of the number α (entier) satisfying the inequality $[\alpha] \le \alpha < [\alpha] + 1$; $L[c,d]$ is the space of real functions $\varphi(t)$ with a finite norm $||\varphi|| = \int_c^d \varphi(t)\mathrm{d}t$; $\Gamma(z)$ is the Euler gamma function. The operator

$$D_{at}^{\alpha}(\varphi(t)) = \frac{\mathrm{sign}(t-a)}{\Gamma(-\alpha)} \int_a^t \frac{\varphi(\xi)}{(t-\xi)^{\alpha+1}}\mathrm{d}\xi, \quad \alpha \neq 0, 1, 2, \ldots, \tag{5.127}$$

(here, the integral is understood in the sense of the Hadamard finite part) acting on the function $\varphi(t)$ from the domain of its definition $D(D_{at}^{\alpha}) \subset L[c,d]$ is then called an α-order fractional integro-differentiation operator with the beginning at point a and the end at point t. The lower limit of integration $t = a$ in (5.127) fixes the time t from which the reading of dynamical phenomena in the medium begins (this choice is largely arbitrary). By definition,

$$D_{at}^{0}(\varphi(t)) = \varphi(t); \quad D_{at}^{n}(\varphi(t)) = \partial^n \varphi(t)/\partial t^n, \quad n = 1, 2, \ldots \tag{5.128}$$

For $\alpha < 0$, the operator D_{at}^{α} coincides with the α-order fractional (in the Riemann–Liouville sense) integration operator. For $\alpha > 0$ and under certain assumptions regarding the smoothness of the function $\varphi(t) \in D(D_{at}^{\alpha})$, we can assume that the fractional differentiation operator

$$D_{at}^{\alpha}(\varphi(t)) = \mathrm{sign}^{[\alpha]+1}(t-a) \frac{\partial^{[\alpha]+1}}{\partial t^{[\alpha]+1}} D_{at}^{\alpha-[\alpha]-1}(\varphi(t)), \tag{5.129}$$

or

$$D_{at}^{\alpha}(\varphi(t)) = \frac{1}{\Gamma(1+[\alpha]-\alpha)} \lim_{\varepsilon \to +0} \frac{\partial^{[\alpha]+1}}{\partial t^{[\alpha]+1}} \int_a^{t_\varepsilon} \frac{\varphi(\xi)\mathrm{d}\xi}{|t-\xi|^{\alpha-[\alpha]}}, \tag{5.129*}$$

where $t_\varepsilon = t + \varepsilon\, sign(a-t)$.

To explain the structure of the fractional integration operator ((5.127) for $\alpha < 0$), recall that it is easy to prove the formula for an n-fold integral of the form (Cauchy's formula)

$$\int_a^t \mathrm{d}t_n \int_a^{t_n} \mathrm{d}t_{n-1} \ldots \int_a^{t_2} \varphi(t_1)\mathrm{d}t_1 = \frac{1}{(n-1)!} \int_a^t \frac{\varphi(\xi)}{(t-\xi)^{1-n}}\mathrm{d}\xi \tag{5.130}$$

by the method of mathematical induction. Since $(n-1)! = \Gamma(n)$, a meaning can be assigned to the right-hand part of (5.130) even at noninteger values of n, i.e., the replacement of the factorial by the Euler gamma function allows one to continue this expression into the domain of noninteger (and even complex) indices n and, thus, to obtain the right-hand part of (5.127). For any function $\varphi(t) \in L[c,d])$, the law of composition of fractional integration operators with identical beginnings is valid:

$$D^{\alpha}_{at}\left(D^{-\beta}_{at}(\varphi)\right) = D^{\alpha-\beta}_{at}(\varphi), \quad \forall\alpha<0. \tag{5.131}$$

In a sense, the fractional differentiation and integration operations are mutually inverse, $D^{\alpha}_{at}\left(D^{-\alpha}_{at}(\varphi(t))\right) = D^{0}_{at}(\varphi(t)) = \varphi(t)$.

Note now that applying the differential operator $\partial/\partial t$ to integral (5.129) reduces its multiplicity by one; applying it n times gives the integrand $\varphi(t)$ and m times gives its $(m-n)$ derivative $\partial^{(m-n)}\varphi(t)/\partial t^{(m-n)}$. Extending this rule to the fractional integral (i.e., just by replacing n by $\alpha>0$), we obtain the expression

$$\frac{\partial^{(m-\alpha)}}{\partial t^{(m-\alpha)}}\varphi(t) \equiv D^{m-\alpha}_{at}(\varphi(t)) = \frac{1}{\Gamma(\alpha)}\frac{\partial^m}{\partial t^m}\int\limits_a^t \frac{\varphi(\xi)}{(t-\xi)^{1-\alpha}}d\xi, \tag{5.132}$$

$$m = 1, 2, 3\ldots,$$

which is the $m-\alpha$-order fractional Riemann–Liouville derivative. In order that the integral converge at the upper limit, we will limit α by the interval $0<\alpha\leq 1$. In the special case of $m = 1$, (5.129) transforms into

$$D^{1-\alpha}_{at}(\varphi(t)) = \frac{1}{\Gamma(\alpha)}\frac{\partial}{\partial t}\int\limits_a^t \frac{\varphi(\xi)}{(t-\xi)^{1-\alpha}}\mathrm{d}\xi \tag{5.132*}$$

in view of Abel's well-known identity, the fractional derivative (5.132*) transforms into an ordinary "integer" derivative $\partial\varphi/\partial t$ when $\alpha \to 0$.

The fractional derivatives (5.131) retain a number of properties of ordinary derivatives: in particular, they are the left inverse operations with respect to the fractional integration (5.129) and have the linearity and α-homogeneity properties

$$D^{\alpha}_{at}\left(D^{-\alpha}_{at}\varphi(t)\right) = \varphi(t), \tag{5.133}$$

$$D^{\alpha}_{at}(c_1\varphi_1(t) + c_2\varphi_2(t)) = c_1 D^{\alpha}_{at}(\varphi_1(t)) + c_2 D^{\alpha}_{at}(\varphi_2(t)), \tag{5.134}$$

$$D^{\alpha}_{at}(\varphi(kt+l)) = k^{-\alpha}D^{\alpha}_{ak+l,t}(\varphi(t)), \quad (k>0). \tag{5.135}$$

At the same time, they have peculiar properties, for example, the dependence on the limit a (nonlocality). As an illustration, we will give several examples of the fractional Riemann–Liouville derivatives of simple functions:

$$D_{at}^{\alpha}\left((t-a)^{\lambda}\right)=\frac{\Gamma(\lambda+1)}{\Gamma(\lambda+1-\nu)}(t-a)^{\lambda-\alpha},\quad D_{at}^{\alpha}\left(t^{\lambda}\right)=\frac{\Gamma(\lambda+1)}{\Gamma(\lambda-\nu+1)}t^{\lambda-\alpha}$$

$$D_{at}^{\alpha}(C)=\frac{Ct^{-\alpha}}{\Gamma(1-\nu)},\quad D_{0t}^{\alpha}(\exp(at))=\frac{\exp(at)\,\gamma(\alpha,at)}{\Gamma(\alpha)}.$$

It can be seen from these relations that the formula for the differentiation of a power function remains the same, but the integer n is replaced by the fractional α, the α-order fractional derivative of a constant C is nonzero.

The book by Oldham and Spanier (1974) is devoted to theoretical and applied aspects of differentiation and integration of an arbitrary order. They analyze numerical fractional differentiation algorithms. In particular, the following approximate formula is deemed possible:

$$D_{at}^{\alpha}(\varphi(t))\approx\frac{t^{-\alpha}n^{\alpha}}{\Gamma(-\alpha)}\sum_{k=0}^{n-1}\frac{\Gamma(k-\alpha)}{k!}\varphi_k(t),\quad t>0, \tag{5.136}$$

where

$$\varphi_n(t)=\varphi(0),\quad \varphi_{n-1}(t)=\varphi(t/n),$$

$$\varphi_k(t)=\varphi_n(t-kt/n),\ldots\ldots,\ \varphi_0(t)=\varphi(t).$$

If $0<\alpha<1$, then the following approximate equality can be used:

$$\begin{aligned}D_{at}^{\alpha}(\varphi(t))\approx\frac{t^{-\alpha}n^{\alpha}}{\Gamma(2-\alpha)}&\left\{\frac{(1-\alpha)\varphi_n(t)}{n^{\alpha}}\right.\\&\left.+\sum_{k=0}^{n-1}\left[\varphi_k(t)-\varphi_{k+1}(t)\right]\left[(k+1)^{1-\alpha}-k^{1-\alpha}\right]\right\}.\end{aligned} \tag{5.137}$$

Thus, a fractional-order derivative is a nonlocal characteristic of the function φ (t), because it depends not only on its values near the point under consideration (as in the case of an integer derivative) but also on its values in the entire interval $(a, 0)$. The papers by Shkhanukov (1996) and Shkhanukov-Lafishev and Nakhusheva (1998) are devoted to the algorithms for approximating fractional derivatives and to the difference methods for solving various transfer equations. The various aspects associated with the application of fractional differential equations can be found elsewhere (see, e.g., Samko et al. 1987; Nakhushev 2003).

5.3.3 FFPK Equation to Describe Evolutionary Processes in Fractal Time

Let us now turn to the spatially homogeneous continuity equation (5.13) (in the space of integral coordinates q)

$$\frac{\partial P_2(q,t)}{\partial t}+\frac{\partial J(q,t)}{\partial q}=0 \tag{5.138}$$

in which (5.125) for the probability flux $J(q,t)$ can be written using (5.38) and (5.40) as

$$\begin{aligned} J &= -A_\xi D_{0t}^{-D}\left(L(q)\cdot\frac{\partial\mu_{\text{turb}}(q,t)}{\partial q}\right) \\ &= -A_\xi D_{0t}^{-D}\left(\frac{\varepsilon^2}{2}Q(q)\frac{\partial P_2(q,t)}{\partial q}-K(q)P_2(q,t)\right). \end{aligned} \tag{5.139}$$

where $A_\xi\equiv\left[\sqrt{2}(1-\xi)\right]^{-D}$. The presence of a fractional time integral on the right-hand side of this equation reflects the influence of the process memory on the probability density $P_2(q,t)$: the probability flux $J(q,t)$ at a given time t is determined not by the local density gradient at the same time t but by its evolution during the entire preceding diffusion period $0<t'$. The case of an empty Cantor set ($\xi\to 0$, $D\to 0$) corresponds to a linear combination of two half-intensity delta functions localized at the ends of the chosen interval $[0,t]$, i.e., a process without any memory (a Markovian process) takes place. As the similarity parameter ξ increases, the index in (1.2) grows and the Laplace transform for the memory function becomes an increasingly rapidly changing function. The limiting value for the similarity parameter is $\xi=1/j$; in this case, it follows from (1.2) that the Hausdorff dimension of the Cantor set is $D=1$. In view of (5.126), the quantity $J(q,t)$ is related to $\partial\mu_{\text{turb}}(q,t)/\partial q$ via the complete integral, which corresponds to the case of ideal memory. The fractional-order integro-differential equation (5.138) is very difficult to analyze.

To pass to the fractional differential equation, let us apply the differential operator D_{0t}^{D} with an index $D\in(0,1]$ to both sides of (5.138); as a result, we obtain the following fractional-order partial differential equation:

$$\frac{\partial^\omega P_2(q,t)}{\partial t^\omega}+A_\xi\frac{\partial}{\partial q}\left(K(q)P_2(q,t)-\frac{\varepsilon^2}{2}Q(q)\frac{\partial P_2(q,t)}{\partial q}\right)=0, \tag{5.140}$$

where the parameter ω is related to the fractal dimension of the active time, $\omega\equiv 1-D$. This equation may be called an anomalous diffusion equation in the space of internal coordinates for the subsystem of turbulent chaos. Equation (5.140) takes the ordinary differential form (5.39) at $\omega=1$ ($D=0$). Introducing the fractional derivative $\partial^{\omega}/\partial t^{\omega}$ into the kinetic equation (5.137) allows the temporal element of the "strange" kinetic processes in a chaotic fractal vortex subsystem (the fractal-time random walks of the subsystem (FTRWs)) that manifest themselves as memory effects in a turbulent fluid to be taken into account. Modeling the evolution of turbulent chaos for $0 \leq D < 1$ based on (5.140) reveals yet another property of turbulence—the presence of turbulent bursts (bursts of pulsations) associated with temporal intermittency (see Boiko et al. 2006). In whatever scale we observe the distribution of points on the time axis, it appears discrete. Regions of crowding alternate (interleave) with voids. The mean number of points in the interval $(0, t)$ increases proportionally to t^D, i.e., here we are dealing with a stochastic fractal (see Cronover, 2000) with a fractal dimension D.

Chapter 6
Self-Organization of Developed Turbulence and Formation Mechanisms of Coherent Structures

The ultimate goal of the phenomenological modeling of hydrodynamic turbulence is to construct the most realistic semiempirical model that corresponds to various dynamical states of natural media. Many difficulties are yet to overcome in this way, because developing a universal turbulence model seems problematic at present. In this regard it is pertinent to recall what we earlier quoted from A. Buckingham review our book (Marov and Kolesnichenko 2002) who said that when the problem of turbulence is investigated, it is important to put emphasis not only on the elucidation of its nature but also on the construction of models that allow the influence of turbulence on the evolution of accompanying physical processes to be studied, a problem that not only has enormous academic interest but is also "fraught with potential disappointments."

In this monograph, we took into account such considerations when constructing a synergetic model of developed turbulence in a fluid. A model of turbulence as a self-organization process in open systems can be constructed by using the concept of a two-level description of structured turbulence (see Chap 5). In this case, the subsystem of turbulent chaos (a vortex continuum with an internal structure) is designed both to model the stochastic small-scale fluctuating motion of a turbulized fluid and to describe the ensemble of mesoscale ordered vortex structures in the process of its evolution. The formation of granules in the solar photosphere can serve as a vivid example of the existence of an extensive family of coherent structures in a turbulized natural medium appearing against the background of a small-scale fluctuating motion.

As we showed in the preceding chapter, introducing a set of stochastic internal parameters $\boldsymbol{q}$ characterizing to a certain extent the microstructure of a turbulized fluid into the thermodynamic description of turbulent chaos makes it possible to obtain the Fokker–Planck–Kolmogorov equations (5.69) for the probability density functions $P_2 \equiv P_2(\boldsymbol{q}_0|\boldsymbol{q}, t)$ of various statistical characteristics of vortex structures in the space of internal coordinates by stochastic-thermodynamic methods. These kinetic equations describe the evolution of the probability density function for homogeneous (in time) random processes $\boldsymbol{q}(t)$. In particular, they can be used to analyze the various mechanisms of the transition (as a result of the loss of flow stability when varying

M.Y. Marov and A.V. Kolesnichenko, *Turbulence and Self-Organization: Modeling Astrophysical Objects*, Astrophysics and Space Science Library 389,
DOI 10.1007/978-1-4614-5155-6_6, © Springer Science+Business Media New York 2013

control parameters) from one stationary-nonequilibrium state of turbulent chaos to another. In this case, various scenarios for the evolution of the stochastic coordinates $\boldsymbol{q}(t)$ of turbulent chaos are possible in the neighborhood of bifurcation points: depending on the intensity of random fluctuations at each instant of time, the corresponding dynamical system can visit some attractors (corresponding to certain vortex structures in the coordinate space) and bypass others.

In Sect. 6.1 of this chapter, we show that the coherent behavior of turbulent chaos giving rise to mesoscale vortex structures can also be modeled in the space of internal coordinates $\boldsymbol{q}$ by Markovian diffusion processes that are the solutions of the FPK equations. Thus, we demonstrate that self-organization (the formation of ordered mesoscale structures with a lower symmetry than that of the initial state) is possible in principle in a thermodynamically open quasi-equilibrium vortex subsystem, when the generation of new temporal coherent structures (CSs) is probable in the course of its evolution. In particular, such CS generation corresponds to the process of multiplicative-noise-induced nonequilibrium "phase transitions" emerging in a structured turbulent flow (see Horsthemke and Lefever 1984). The multiplicative noise resulting in this case from the "interaction" between the subsystems of averaged motion and turbulent chaos depends, by definition, on the random process $\boldsymbol{q}(t)$ itself. Below, when analyzing the evolution of vortex chaos, we proceed from the fact that it is admissible to replace the real noise of a turbulent flow (whose pulsations are attributable to the cumulative action of numerous factors determining the state of a turbulized medium and cannot be assigned to any one quite specific factor) by Gaussian white noise. We use such noise as the simplest model of the total noise effect on $\boldsymbol{q}(t)$ resulting both from the subsystem of averaged motion and from the internal noise of the subsystem of turbulent chaos produced by its "thermal" structure. An argument for this choice of the random force in SDE (5.86) is that only for the white noise model will the resulting stochastic process $\boldsymbol{q}(t)$ be Markovian and only this choice leads to complete equivalence of the SDE method to the method of the FPK equation for the transition probability. Note also that white noise, which is a convenient mathematical idealization of real noise, is a generalized random process—it is not differentiable and has an infinite variance.

Under these assumptions, we show that if the multiplicative noise of turbulent chaos is intense enough, then the extrema of the probability density function describing the behavior of the stochastic vortex subsystem differ significantly in both number and position from the stationary states corresponding to deterministic chaos. Moreover, multiplicative noise can give rise to new stationary states, thereby changing the properties (in particular, the bifurcation diagrams) of the local stability of deterministic chaos themselves—the transition points can move in the space $\boldsymbol{q}$ under the influence of intense noise in a turbulent fluid. This qualitative change is reflected most clearly in the extrema of the stationary probability density P_2^{st}. According to Horsthemke and Lefever (1984), the latter are the most suitable indicators of the transition: the extrema of the stationary probability density correspond to the system's macroscopic "phases" and serve as an order parameter of the transition. Thus, multiplicative noise not only has a disorganizing effect on the system of defining internal parameters $\boldsymbol{q}$ of turbulent chaos but can also both stabilize unstable macroscopic states and cause new nonequilibrium phase transitions unpredictable by the theory of deterministic chaos.

In Sect. 6.2, we consider one of the specific mechanisms for the formation and evolution of large concentrated vortex structures that is interpreted from the viewpoint of the theory of dynamical systems. Here, we formulate the general concept of synergetic generation of CSs (mesoscale clusters) from turbulent chaos associated with the phenomenon of phase-frequency synchronization of a number of small-scale vortex structures by the main incoherent component of the subsystem of turbulent chaos and consider scenarios for the dynamical action of the fine-grained fluctuating field of chaos on the evolution of such CSs. We show the interrelationship between nonequilibrium phase transitions and the self-organization of vortex clusters with a lower symmetry than that of the initial state. We explain the emergence of some effective force field of chaos as a self-consistent process whereby a nonzero initial field traps individual small-scale clusters of synchronized oscillators (modeling the small-scale vortices of the incoherent component of turbulent chaos). As a result, the actions of these trapped objects will be summed coherently and this coherent group will generate a nonzero contribution to the effective force field acting on the system. Finally, we reach the important conclusion that whereas the growth of the sizes of solid particles during collisions in chaotic turbulence is hampered, they coalesce and enlarge within coherent vortex structures. In particular, the formation of mesoscale vortex clusters facilitates the solution of the problem of solid particle enlargement in a turbulized gas suspension through the mechanism of collisions even at comparatively low relative velocities. As a rule, this encounters difficulties in attempting to reproduce similar processes in laboratory experiments.

6.1 Role of Nonequilibrium Phase Transitions in Structuring Hydrodynamic Turbulence

Below, we distinguish the large-scale fluctuations of turbulent chaos proper from the statistical molecular fluctuations attributable to the atomic structure of the medium. In contrast to the molecular fluctuations (negligible for any macroscopic systems), the stochasticity of the subsystem of vortex chaos is not microscopic in origin but is a manifestation of its turbulent nature and reflects the cumulative action of a large number of interrelated fluid turbulization factors. In particular, the amplitudes of the molecular fluctuations of some passive flow admixture (if they are taken as additional variables) near some stable macroscopic state behave as V^{-1}, where V is the system's volume. These fluctuations are known to grow at the critical point as quantities $\sim V^{-1/2}$ and, consequently, they become negligible in the thermodynamic limit $V \to \infty$ (Leontovich 1954). At the same time, the turbulent fluctuations of chaos behave not as the inverse powers of the system's characteristic size V^{-1} but can reach $\sim V^0$ and, hence, do not disappear at the macroscopic level of the system's description. Thus, the scales of the molecular and turbulent fluctuations are very different and do not overlap. That is why it is necessary to refine the stochastic-thermodynamic description of the subsystem of turbulent chaos so as to be able to include the effects of its stochasticity in the model of structured turbulence (Kolesnichenko 2002). The variants of the dynamical behavior of various nonnegative small-scale turbulence characteristics with increasing supercriticality are associated with the influence of the random properties of vortex chaos (attributable, in particular, to its natural noise). The next important step in modeling structured turbulence consists in explicitly relating the probabilistic characteristics of the fluctuations of its small-scale characteristics to the "physical properties" of the quasi-stationary subsystem of vortex chaos.

In our phenomenological description of chaos, we use the formalism of generalized statistical thermodynamics that requires a probabilistic approach (Stratonovich 1985) and suggests investigating a statistical ensemble of macroscopically identical systems (i.e., systems with the same generalized extensive state parameters like the internal energy $E_{\text{turb}}(r,t)$, entropy $S_{\text{turb}}(r,t)$, specific volume $1/\bar{\rho}(r,t)$, and a number of internal variables). The turbulent fluctuations (relative to the means) of internal coordinates $q_k(r,t) \quad (k=\overline{1,\ n})$ for the state of chaos, which actually serve as a measure of the differences in any set of such thermodynamically identical systems, are responsible for stochasticity. We assign the continuously changing local random parameters adequately characterizing a turbulized fluid within a physically infinitesimal volume $\mathrm{d}r$ to the internal coordinates describing the macroscopic state of chaos. However, we assume that some of the internal coordinates $q_k(r,t)$ can bear a relation to the incoherent component of the subsystem of turbulent chaos, while others can characterize individual mesoscale CSs.

In accordance with the approach being developed here, we assume that for a complete statistical description of the total (vector) random process

$$\boldsymbol{q}(r,t) = \{q_k(r,t),\ \ k=\overline{1,\ n}\}$$

(a set of n state parameters q_k) in a vortex continuum, it suffices to know the one-point probability density $W_1(\boldsymbol{q},t)$ and the joint two-point probability density $W_2(\boldsymbol{q}_0,t_0;\boldsymbol{q},t)$. The stochastic processes that are completely described only by these two functions are known to be Markovian. In addition, we use the two-point conditional probability density $P_2(\boldsymbol{q}_0,t_0|\boldsymbol{q},t) = W_2(\boldsymbol{q}_0,t_0;\boldsymbol{q},t)/W_1(\boldsymbol{q}_0,t_0)$ that establishes the probable value of the stochastic characteristics for vortex structures $\boldsymbol{q}$ in the space of internal coordinates at time t if $\boldsymbol{q}=\boldsymbol{q}_0$ at time t_0 with a probability equal to unity. In accordance with the methods of statistical nonequilibrium thermodynamics, the coherent behavior of the subsystem of turbulent chaos, which gives rise to dissipative structures, can then be modeled in the space of internal coordinates $\boldsymbol{q}$ by Markovian diffusion processes that are the solutions of the Fokker–Planck–Kolmogorov equations (5.39) defining the evolution of the transition probability density $P_2 \equiv P_2(\boldsymbol{q}_0|\boldsymbol{q},t)$ for stationary random processes (i.e., processes for which all statistical characteristics do not change with time)

$$P_2(\boldsymbol{q}_0,t_0|\boldsymbol{q},t') \equiv P_2(\boldsymbol{q}_0 0|t'-t_0) \equiv P_2(\boldsymbol{q}_0|\boldsymbol{q},t).$$

In particular, these equations allow one to analyze the processes of the transition from one stationary-nonequilibrium state of chaos $\boldsymbol{q}^{\mathrm{st}}$ to another resulting from a successive loss of stability when varying control parameters (e.g., the Reynolds number ***Re***) and to calculate the probability density near bifurcation points. The transition to an ordered state is associated with the instability of a turbulized system and the breaking of its symmetry: self-organization arises in the cases where the stationary state becomes unstable and gives way to a new state with a lower symmetry than that of the initial one. Accordingly, the system can be said to enter the domain of synergy beyond the thermodynamic threshold of self-organization (Horsthemke and Lefever 1984).

We showed in Chap. 5 that the self-organizing subsystem of turbulent chaos could also have a much more complex behavior than the multiple stationary-nonequilibrium states $\boldsymbol{q}_k^{\mathrm{st}}$. In particular, at least one relaxation mode that either is constant or oscillates with a frequency equal to the imaginary part of the eigenvalue of this mode exists at the so-called critical point. In the other potentially possible case where a pair of complex-conjugate eigenvalues crosses the imaginary axis, the normal Andronov–Hopf bifurcation (cycle generation bifurcation) whereby the corresponding deterministic equation has an isolated stable periodic solution $\boldsymbol{q}^{\mathrm{c}}(t) = \boldsymbol{q}^{\mathrm{c}}(t) + \tau^{\mathrm{c}}$ with a period τ^{c} is one of the two typical situations at the critical point. This implies that the phase equilibrium point jumps from the stable focus (stationary state) at the instant of stability loss to the asymptotically stable limit cycle (attractor) to which all trajectories from its neighborhood tend when $t \to \infty$. Thus, the dynamics of chaos at critical points is determined by the nonlinearities existing in the system (see Kolesnichenko 2004). In this case, the enormous number of macroscopic degrees of freedom for a real turbulent flow is sharply reduced, billions or more excitations of the system turn out to obey a small number of modes, i.e., turbulence is determined by a finite number of collective excitations (associated, for example, with the limit cycle), while the remaining excitations simply adjust to them, acting as more intense dissipation. Various bifurcations of these periodic states like the period-doubling bifurcation, the formation of an invariant torus, the bifurcation with the loss of symmetry, etc. are possible during the subsequent temporal evolution of the subsystem of turbulent chaos. Consequently, when the subsystem of fluctuating chaos emerges from some attractor under the action of fairly strong turbulent pulsations, it can evolve to one of the possible attractors (whose zone of attraction covers the point of the system's initial state), including strange attractors.

Thus, when varying control parameters, various evolution scenarios (associated with the formation of various attractors referring to certain vortex structures) in the space of internal coordinates describing the macroscopic states of fluctuating chaos can unfold in the system. In this case, the transitions from one dissipative structure to another are similar in mathematical properties to equilibrium phase transitions if the concept of a phase transition is extended to the new class of nonequilibrium phenomena (see, e.g., Kadomtsev and Ryazanov 1983). Moreover, as has already been said above, turbulent noise can give rise to new states of the system, thereby changing the properties of the local stability of turbulent chaos themselves—the critical transition points can shift under the influence of multiplicative noise. Such transitions, which are usually called noise-induced nonequilibrium phase transitions, bear the characteristic features of its turbulent origin.

In this section, we analyze the relationships of the instabilities caused by variations in control parameters to the self-organization processes (phase transitions) in a developed turbulized flow whose reflection in the $\boldsymbol{q}$-space is the formation of coherent structures. The properties of phase transitions are investigated for a random medium with Gaussian white noise used as a model for the random force action of the subsystem of averaged motion and vortex chaos itself on the evolution of the fluctuating small-scale turbulence characteristics $\boldsymbol{q}(t)$. Below, we also assume that the internal coordinates $\boldsymbol{q}$ of turbulent chaos characterizing to a certain extent the structure of a turbulized fluid are distributed lognormally in a stationary state (see Monin and Yaglom 1996).

6.1.1 Basic Mathematical Apparatus

Let us now briefly formulate the most significant features of the mathematical apparatus on which the approach used below is based. To avoid unnecessary complications in describing the noise-induced modifications of macroscopic properties for the subsystem of fluctuating turbulent chaos, here we restrict our analysis only to the temporal evolution of chaos, i.e., we assume the vortex motions at all points of space to be completely identical. In addition, we assume that the stochastically changing properties of vortex chaos can be satisfactorily described only by one continuously changing intensive variable, for example, the dissipation rate of turbulent energy in vortices $q = \varepsilon_{\mathrm{T}}$ or any other related quantity (see Monin and Yaglom 1996). In particular, this assumption stems from the fact that it is undesirable to use approximate methods in investigating noise-induced transitions, because even small errors of the numerical method can lead to an imaginary qualitative rearrangement of the system's behavior. On the other hand, as it well known, exact analytic solutions of the FPK equations can be obtained mainly for one-dimensional systems (see, e.g., Gardiner 1983).

In our view, the patterns established when investigating Markovian fluctuations of the energy dissipation and related positive characteristics are also important for understanding the evolution of arbitrary random velocity and temperature fields in a real turbulent fluid motion. Moreover, they may turn out to be very useful in constructing the so-called subgrid turbulence models with intermittency. Below, we also distinguish the internal fluctuations of a random variable q generated by the "thermal" structure of the subsystem of vortex chaos from the external (with respect to the variable q) fluctuations produced by the subsystem of averaged motion and simulating the numerous excited degree of freedom for the turbulized fluid as a whole that are not taken into account explicitly (Kolesnichenko 2002). The qualitative changes (i.e., the noise-induced "phase transitions") in the dynamical behavior of various nonnegative small-scale turbulence characteristics are associated precisely with the influence of intense external fluctuations on the stochastic subsystem of vortex chaos.

Under the assumptions made above, we can use the stochastic differential equation (SDE) for a homogeneous (in time) one-dimensional Markovian process $q(t)$

$$\frac{\mathrm{d}q(t)}{\mathrm{d}t} = K_\alpha(q) + \varepsilon\sqrt{Q(q)}\cdot\xi(t), \quad q>0 \tag{6.1}$$

or the equivalent direct Fokker–Planck–Kolmogorov (FPK) equation

$$\frac{\partial P_2(q_0|q,t)}{\partial t} + \frac{\partial J(q_0|q,t)}{\partial q} = 0, \tag{6.2}$$

defining the evolution of the transition probability density $P_2 \equiv P_2(q_0|q,t)$ for a stationary random process $q(t)$ as a basis for the dynamical description of the stochastic properties of the subsystem of turbulent chaos fluctuating near some stationary-nonequilibrium state q^{st}. Recall that the form of the drift coefficient $K(q,\varepsilon)$ in the FPK equation (6.2) and the form of the regular force $K_\alpha(q)$ in the Langevin equation (6.1) [see Chap. 5] depend on the sense in which the corresponding stochastic integral relations corresponding to the SDE are understood (Ito's or Stratonovich's sense). Here,

$$J(q_0|q,t) \equiv [K(q,\varepsilon)P_2(q_0|q,t)] - \frac{1}{2}\varepsilon^2\frac{\partial}{\partial q}[Q(q)P_2(q_0|q,t)], \tag{6.3}$$

is the conditional probability flux for the process $q(t)$ whose value at the initial time $t_0 = 0$ is $q(0) = q_0$, $\varepsilon^2 Q(q)$, and

$$K(q,\varepsilon) \equiv \begin{cases} K_\alpha(q) & - \quad \text{Ito} \\ K_\alpha(q) + \frac{1}{4}\varepsilon^2\partial Q(q)/\partial q & - \text{ Stratonovich}, \end{cases} \tag{6.4}$$

are, respectively, the time-independent diffusion and drift coefficients in the space of configurations; $\varepsilon^2 \equiv \gamma\, k_B T_{turb}$ is a parameter characterizing the intensity of the noise effect from the subsystem of turbulent chaos generated by its "thermal" structure on the random process $q(t)$; T_{turb} is the thermodynamic temperature of chaos related to the Favre-averaged fluctuating kinetic energy of a turbulized fluid flow $\langle b\rangle \equiv \overline{\rho\,(u'')^2}/2\bar{\rho}$ by the relation $\frac{3}{2}\gamma\; k_B T_{turb} = \langle b\rangle$ (the turbulence energy $\langle b\rangle$ is assumed to coincide with the thermodynamic internal energy of chaos, $E_{turb} \equiv \langle b\rangle$)

(see Kolesnichenko and Marov 1999); $K_\alpha(q) \equiv K(q) + \alpha\, g(q)$ is the regular force in the space of stochastic variable q, with $K_\alpha(q)$ in (6.4) depending not only on the variable q but also on some control parameter α describing the cumulative action of an averaged turbulent flow on the stochastic process $q(t)$ in a vortex continuum (for example, the Reynolds number corresponding to large-scale motions in a turbulized flow or a number related to Re, for example, the supercriticality index, can act as the order parameter α);

$$K_\alpha(q) = \frac{-\partial V(q)}{\partial q}$$

is the regular force (the "friction force" counteracting the fluctuations of chaos) generated by the so-called local synergetic (deterministic) potential $V(q) = -\left[\int^q K_\alpha(x)\,\mathrm{d}x\right]$ of the subsystem of vortex chaos in the limit $\varepsilon^2 \to 0$ of weak noise; $\xi(t)$ is the stochastic cofactor in the expression for the random force $\tilde{f}(q) \equiv \varepsilon\sqrt{Q(q)} \cdot \xi(t)$ (allowing both for the internal additive ($\sqrt{Q(q)} = \text{const}$) noise in the subsystem of chaos attributable to small-scale fluctuations of the parameter $q(t)$ relative to the quasi-stationary state of a vortex continuum and for the source of external multiplicative ($\sqrt{Q(q)} \neq \text{const}$) noise simulating all possible excited degrees of freedom of a turbulized hydrodynamic system that are not taken into account explicitly and associated with the action of the subsystem of averaged motion on the process $q(t)$). Below, we approximate the latter by Gaussian white noise with a zero mean and a delta-like (in time) correlation function:

$$\langle \xi(t) \rangle = 0, \quad \langle \xi(t)\xi(t_1) \rangle = \delta(t - t_1).$$

(Here, the average is taken over the stochastic process $\xi(t)$). The drift coefficient $K_\alpha(q)$ in relation (6.4) corresponds to a degenerate (in the deterministic limit $\varepsilon^2 \to 0$) random process $q(t)$ described by the Landau–Khalatnikov deterministic dynamical regression equation (an autonomous ordinary differential equations of the first order in time)

$$\frac{\mathrm{d}q(t)}{\mathrm{d}t} = K_\alpha(q) \tag{6.5}$$

with the initial condition $q(0) = q_0$, while the term $\frac{1}{4}\varepsilon^2 \partial Q(q)/\partial q$ allows for the drift induced by multiplicative ($Q(q) \neq \text{const}$) noise when the influence of the random force $\tilde{f}(q)$ depends on the state of the process $q(t)$. In what follows, we assume that the deterministic equation (6.5) is stable, meaning the boundedness of the solution $q(t)$ by the stability.

The FPK equation (6.2) is designed to analyze the temporal evolution of Markovian diffusion processes related to initial conditions of the type $P_2(q_0|q,0) = \delta(q - q_0)$, which correspond to a delta-like probability density concentrated at point $q = q_0$. Furthermore, particular boundary conditions in q that are determined, as a rule, by the essence of the physical problem should be imposed on the function $P_2(q_0|q,t)$. Below, we assume that the interval $[q_1,q_2]$ serves as the space of states for the diffusion process $q(t)$; the upper boundary can become infinite, $q_2=\infty$. If not the initial state $q(0)=q_0$ but the initial distribution $W(q)$ is specified at the initial time $t=0$, then it is convenient to seek a solution of the FPK equation for the one-point probability density $W_1(q,t)$. Using the relation

$$\int W_1(q_0,t_0)P_2(q_0,t_0|q,t)\mathrm{d}q_0 = W_1(q,t),$$

it is easy to verify that the one-point probability density $W_1(q,t)$ of a Markovian diffusion process also satisfies the FPK equation in form (6.2) and (6.3).

In this section, we restrict our analysis to the processes $q(t)$ that never reach the boundaries of their state space and, consequently, the boundary conditions may not be imposed on the function $P_2(q_0|q,t)$. The existence and uniqueness of a solution to SDE (6.1) on $[q_1,q_2]$ at $\forall t$ is known to be guaranteed in this case (see, e.g., Gardiner 1983). The unattainable boundaries q_i include the natural boundaries (attainable only for $t\to\infty$ with a zero probability, $\lim\limits_{t\to\infty,\, q\to q_i} P_2(q,t)=0$) and the attractive boundaries when $q(t)\to q_i$ for $t\to\infty$. According to the Gikhman and Skorokhodov (1968) classification, for example, the lower boundary q_1 will be natural if $L_1(q_1)=+\infty$ and attractive if $L_1(q_1)<+\infty$ and $L_2(q_1)=+\infty$. Here, the constants L_1 and L_2 are defined by the relations

$$L_1(q_1)=\int_{q_1}^{\beta}\phi(x)\mathrm{d}x,\quad L_2(q_1)=\frac{1}{\varepsilon^2}\int_{q_1}^{\beta}\frac{1}{Q(y)\phi(y)}\left\{\int_{q_1}^{y}\phi(x)\mathrm{d}x\right\}\mathrm{d}y, \tag{6.6}$$

in which β is a point near the boundary q_1 and

$$\phi(x)=\exp\left\{-\int_{\beta}^{x}\frac{2K_\alpha(z)}{\varepsilon^2 Q(z)}\mathrm{d}z\right\},\qquad \beta\in(q_1,q_2) \tag{6.7}$$

is a nonintegrable function near the boundary q_1.

Recall once again that the form of the drift coefficient in the FPK equation depends on the sense (Ito's or Stratonovich's sense) in which the corresponding stochastic integral relations written out for the SDE are understood (see, e.g., Gardiner 1983). Additional physical considerations should be invoked to determine the most suitable (in the context of the specific problem being considered) drift and diffusion coefficients for the random process $q(t)$. In what follow, we use SDE (6.1) written in Stratonovich's (1961) symmetrized form, when the definition of the corresponding stochastic integrals is consistent with the habitual rules of mathematical analysis. It is well known that one can always pass from Stratonovich's interpretation of SDE to Ito's interpretation of SDE by adding $\frac{1}{4}\varepsilon^2\partial Q(q)/\partial q$ to $K_\alpha(q)$ [see Chap. 5]. Note also that no difference exists between Ito's and Stratonovich's integrals in the case of additive noise (i.e., at $Q(q) = \text{const}$) (see, e.g., Tikhonov and Mironov 1977).

Regarding the convenience of using Stratonovich's interpretation of SDE in our model of structured turbulence, we note the following: in the general case of strongly nonlinear fluctuating fields of the velocity or other thermohydrodynamic parameters in a real turbulized fluid, the statistics of the subsystem of turbulent chaos is neither Gaussian nor Markovian, especially on large scales (see Monin and Yaglom 1996). Nevertheless, in the approach adopted below, we proceed from the fact that Gaussian white noise used as a model for the intensity of the random force action $\tilde{f}(q)$ from the noise of turbulent chaos still provides an acceptable approximation when modeling the correlations between the random processes $q(t)$ and $\xi(t)$. In other words, we assume that real noise with short memory can be replaced by white noise with zero memory. This replacement is known (see, e.g., Stratonovich 1961) to be possible if the correlation time τ_{corr} (noise memory time) of the random process $\xi(t)$ is much shorter than the time scale of the macroscopic evolution of chaos τ_{macro}, which is usually identified with the relaxation time of the subsystem of vortex chaos to a reference stationary state, $\tau_{\text{corr}} \ll \tau_{\text{macro}}$. Thus, for the adopted idealization of white noise (for which $\tau_{\text{corr}} \to 0$), the stochastic differential equations (6.1) understood in Stratonovich's sense can be interpreted as the limits of certain virtual equations written for non-Markovian (but nearly Markovian) random processes $q(t)$ in a real turbulized flow. In conclusion, note that Stratonovich's interpretation of SDE should be applied every time when delta-correlated Gaussian white noise in SDE is an idealization of a real random process $\xi(t)$ with a very short but finite correlation time (Tikhonov and Mironov 1977). This allows one to use the powerful apparatus of the theory of Markovian diffusion processes to obtain exact analytical results describing the behavior of systems under the action of noise.

6.1.1.1 Stationary Solution of the FPK Equation

For the one-dimensional processes under consideration, it is very easy to find the stationary probability density $P^{\text{st}}(q) = \lim_{t\to\infty} P_2(q,t)$ that characterizes the stationary behavior of a system under the action of white noise after a fairly long time interval. By definition, the one-dimensional stationary probability density $P^{\text{st}}(q)$, if it exists, does not depend on time t and initial conditions $q(0) = q_0$. Therefore, the FPK equation (6.2) in a stationary state degenerates into the equation $\partial P^{\text{st}}(q)/\partial t = 0$. It follows from the latter that the stationary probability flux $J^{\text{st}}(q)$ is constant for any states $q \in [q_1, q_2]$, i.e., $J^{\text{st}}(q) \equiv J^{\text{st}} = \text{const}$. Thus, the FPK equation in the stationary case transforms into a first-order equation for $P^{\text{st}}(q)$:

$$\varepsilon^2 \frac{\partial}{\partial q}[Q(q)P^{\rm st}(q)] - 2[K(q,\varepsilon)P^{\rm st}(q)] = -2J^{\rm st}, \tag{6.8}$$

with the probability flux inside the space of state being equal to that through the boundaries: $J^{\rm st} = J(q_1) = J(q_2)$. Below, we consider a situation where no probability flux from the space of states exists and, hence, $J^{\rm st} = 0$. Recall that when deriving the FPK equation in Chap. 5 to describe the evolution of the distribution functions for nonnegative small-scale turbulence characteristics, we assumed zero boundary conditions: $J(q_1) = J(q_2) = 0$. The stationary probability density, the solution of (6.8), is then

$$P^{\rm st}(q) = \frac{N}{Q(q)} \exp\left[\frac{2}{\varepsilon^2}\int_{q_1}^{q} {\rm d}x \frac{K(x,\varepsilon)}{Q(x)}\right] = \frac{N}{\sqrt{Q(q)}} \exp\left[\frac{2}{\varepsilon^2}\int_{q_1}^{q} {\rm d}x \frac{K_\alpha(x)}{Q(x)}\right], \tag{6.9}$$

where the constant of integration N can be determined from the normalization condition:

$$N^{-1} = \int_{q_1}^{q_2} \frac{1}{Q(q)} \exp\left[\frac{2}{\varepsilon^2}\int_{q_1}^{q} {\rm d}x \frac{K(x,\varepsilon)}{Q(x)}\right] {\rm d}q < \infty. \tag{6.10}$$

The expression for $P^{\rm st}(q)$ can be transformed using (6.9) to the form

$$\begin{aligned} P^{\rm st}(q) &= N \exp\left[\frac{2}{\varepsilon^2}\int_{q_1}^{q} {\rm d}x \left(\frac{K_\alpha(x)}{Q(x)}\right) - \ln\sqrt{Q(q)}\right] \\ &= N\exp\left(-2\Phi(q)/\varepsilon^2\right) = N\exp\left(-2\Phi(q)/\gamma\, k_B T_{\rm turb}\right), \end{aligned} \tag{6.11}$$

where

$$\Phi(q) = -\left(\int_{q_1}^{q} \frac{K_\alpha(x)}{Q(x)} {\rm d}x - \frac{\varepsilon^2}{2} \ln\sqrt{Q(q)}\right) \tag{6.12}$$

is the so-called probabilistic potential.

It is important to note that distribution (6.11) extends the standard Boltzmann–Planck formula (for the equilibrium state of the thermodynamic parameters of a closed system when $\Phi = -S$) to the stationary states of the ensemble corresponding to an open subsystem of turbulent chaos. In this case, the inverted formula

$$2\Phi(q) = -\varepsilon^2 \ln\left(\frac{P^{\mathrm{st}}(q)}{N}\right) \tag{6.13}$$

allows, in particular, (6.1) and (6.2) for the chosen stochastic small-scale turbulence characteristics to be concretized if some hypotheses of the distribution in a stationary state are adopted for the latter. Here and below, we assume the time-independent solutions q^{st} of the deterministic equation (6.5) to be stationary states. At the same time, the words "stationary state" can also be used with regard to the distribution in possible states. For example, distribution (6.9) may be called a stationary distribution of states. In the case of additive noise of chaos, no difference exists between the deterministic stationary state q^{st} and the extrema of the stationary distribution P^{st}, although it should be kept in mind that the system in a stationary state actually fluctuates near some average state. This is because the distribution of states (6.9) in this case has such a sharp maximum that, in essence, only the stationary state q^{st} makes a major contribution to this distribution. In the case of multiplicative noise of chaos, the random variable defining the system's stationary state qualitatively differs greatly from the degenerate random variable corresponding to deterministic chaos.

6.1.2 H-*Theorem for Stationary States*

When the probabilistic aspects of the macroscopic stationary states for the vortex system under consideration are studied, an initial physical ensemble corresponding to the subsystem of turbulent chaos is required. If quite a specific stationary state q^{st} is investigated, then it is necessary that the initial conditions q_0 represented by the ensemble be concentrated precisely in the domain of its attraction. The following question arises: Is this stationary state q^{st} actually stable according to Lyapunov under the initial condition q_0 (or orbitally stable according to Poincare), much like the thermodynamic equilibrium states in classical thermodynamics? Otherwise, the phase trajectories $q(t)$ that start near q^{st} will not reach the needed stationary state. In turn, this implies with a high probability that the unstable stationary states of a vortex continuum without any domain of attraction cannot be described by a stationary statistical ensemble, because it cannot even be properly organized.

To answer the posed question, let us consider the following functional of $P_2(q,t)$:

$$H(t) = \gamma\, k_{\mathrm{B}} T_{\mathrm{turb}} \int\limits_{q_1}^{q_2} \mathrm{d}q P_2(q,\,t) \ln\left[\frac{P_2(q,\,t)}{P^{\mathrm{st}}(q)}\right]. \tag{6.14}$$

Such quantities were first introduced by Kullback (1959). Therefore, $H(t)$ is often called the Kullback entropy. However, in contrast to the normal entropy, the latter is not a measure of the uncertainty in the parameters q (characterizing their statistical scatter) but is a measure of the closeness of the distributions P_2 and P^{st}. Indeed, it is easy to show that the functional H is the Lyapunov one, because it has the following two properties: $H(t) \geq 0$ and $\partial H(t)/\partial t \leq 0$ at $\forall t$, with the equality sign taking place only for $P_2(q,t) \equiv P^{st}(q)$.

Indeed, due to the inequality $\ln(1/x) \geq 1 - x$ valid for $x > 0$ and the normalization condition for the functions $P_2(q,t)$ and $P^{st}(q)$, we have

$$H(t) = \gamma k_B T_{turb} \int_{q_1}^{q_2} dq P_2(q,t) \left\{ \ln\left[\frac{P_2(q,t)}{P^{st}(q)} + \frac{P^{st}(q)}{P_2(q,t)} - 1\right] \right\} \geq 0, \quad (6.15)$$

with inequality (6.15) transforming into the equality only for $P_2(q,t) \equiv P^{st}(q)$.

Let us now show that $H(t)$ decreases monotonically. For this purpose, let us differentiate relation (6.14) with respect to time and use the FPK equation (6.2) and relation (6.3) for the probability flux, which is assumed to be zero at the boundary. Integration by parts then yields

$$\begin{aligned}\frac{\partial H(t)}{\partial t} &= \varepsilon^2 \int_{q_1}^{q_2} \ln\left(\frac{P_2(q,t)}{P^{st}(q)}\right) \frac{\partial P_2(q,t)}{\partial t} dq = -\varepsilon^2 \int_{q_1}^{q_2} \ln\left(\frac{P_2(q,t)}{P^{st}(q)}\right) \frac{\partial J(q,t)}{\partial q} dq \\ &= \varepsilon^2 \int_{q_1}^{q_2} J(q,t) \frac{\partial}{\partial q} \ln\left(\frac{P_2(q,t)}{P^{st}(q)}\right) dq = \varepsilon^2 \int_{q_1}^{q_2} J(q,t) \frac{P^{st}(q)}{P_2(q,t)} \frac{\partial}{\partial q}\left(\frac{P_2(q,t)}{P^{st}(q)}\right) dq \\ &= \varepsilon^2 \int_{q_1}^{q_2} J^{st}(q) \frac{\partial}{\partial q}\left(\frac{P_2(q,t)}{P^{st}(q)}\right) dq - \frac{\varepsilon^4}{2} \int_{q_1}^{q_2} Q(q) \frac{[P^{st}(q)]^2}{P_2(q,t)} \left[\frac{\partial}{\partial q}\left(\frac{P_2(q,t)}{P^{st}(q)}\right)\right]^2 dq \leq 0,\end{aligned} \quad (6.16)$$

because the first integral is equal to zero. This can be easily verified using the condition $J^{st} = 0$ and performing integration by parts:

$$\begin{aligned}\int_{q_1}^{q_2} J^{st}(q) \frac{\partial}{\partial q}\left(\frac{P_2(q,t)}{P^{st}(q)}\right) dq &= -\int_{q_1}^{q_2} dq \left(\frac{P_2(q,t)}{P^{st}(q)}\right) \frac{\partial J^{st}(q)}{\partial q} = \int_{q_1}^{q_2} dq \left(\frac{P_2(q,t)}{P^{st}(q)}\right) \frac{\partial P^{st}(q)}{\partial t} \\ &= 0.\end{aligned}$$

Consequently, if the fluctuations inside the space of states nowhere become zero, i.e., the inequality $Q(q) > 0$ holds for $\forall q \in [q_1, q_2]$, then $\partial H(t)/\partial t \leq 0$.

Thus, the functional $H(t)$ is strictly positive and decreases with time. Hence it follows that $H(t) \to 0$ when $t \to 0$ and since $H(t) = 0$ if and only if $P_2(q,t) = P^{st}(q)$, we arrive at the equality $P^{st}(q) = \lim_{t\to\infty} P_2(q,t)$. In other words, the global asymptotic stability of the stationary state q^{st} follows from the existence of the Lyapunov functional. In addition, it can be shown (see, e.g., Stratonovich 1961) that if the stationary probability density $P^{st}(q)$ exists and if the diffusion process at $t = 0$ begins from it, then $q(t)$ is an ergodic process. This implies that the mathematical expectation $E\{\phi(q)\}$ of a stationary diffusion process $q(t)$ can be determined from the observation of only one arbitrary trajectory of the process—by taking a time average,

$$E\{\phi(q)\} \equiv \int \phi(q)P^{st}(q)\mathrm{d}q = \lim_{T\to\infty} \frac{1}{T} \int \phi\{q(t)\}\mathrm{d}t.$$

In this case, the product $P^{st}(q)\mathrm{d}q$ is equal to the fraction of the time that an arbitrary trajectory of the diffusion process spends in an infinitesimal neighborhood of the state q (Horsthemke and Lefever 1984).

6.1.2.1 Additive Noise of Turbulent Chaos

If the noise of chaos is additive, i.e., $Q(q) = \text{const}$, then the probabilistic potential $\Phi(q)$ and the deterministic potential $V(q)$ coincide to within an insignificant constant. In this case, it follows from (6.8)

$$\varepsilon^2\text{const} \cdot \frac{\partial P^{st}(q)}{\partial q} - 2K_\alpha(q)P^{st}(q) = 0$$

that the extrema q_{ext} of the stationary probability density determined from the condition

$$\frac{\partial P^{st}(q)}{\partial q} = 0 \tag{6.17}$$

coincide with the stationary solutions

$$\frac{\partial q^{st}}{\partial t} = K_\alpha(q^{st}) = 0 \tag{6.18}$$

of the deterministic equation (6.5) corresponding to the degenerate case of $\varepsilon^2 \to 0$ where the stochastic process $q(t)$ is not affected by the natural noise of turbulent chaos. Thus, the deterministic stationary states q^{st} (recall that a singular point of a certain type—a center, a stable focus, a saddle, an unstable node, etc., serves as the image of a stationary state in the space of configurations) are determined either by the zeros of the function $K_\alpha(q^{st})$ [accordingly by the extrema of the deterministic potential $V(q)$] or by the extrema q_{ext} of the stationary probability density $P^{st}(q)$.

In the deterministic case (i.e., for degenerate turbulence, $\varepsilon^2 \to 0$), a nonequilibrium phase transition occurs at the point at which the potential $V(q)$ changes qualitatively, for example, the number of local extrema changes. The simplest method of investigating this possibility is to check the stability of state q_k^{st} against small perturbations δq_k. When an isolated stationary state q_k^{st} is analyzed in the first approximation for asymptotic stability $\lim_{t\to\infty} q(q_0, t) = q_k^{\mathrm{st}}$, the temporal evolution of the perturbation δq_k is specified by the solution of the equation

$$\frac{\partial\, \delta q_k}{\partial t} = -\left.\frac{\partial^2 V(q)}{\partial q \partial q}\right|_{q=q_k^{\mathrm{st}}} \delta q_k$$

obtained by discarding the nonlinear terms in the Taylor expansion of the right-hand part of (6.5). The solutions of this equation $\delta q_k(t) = A_k \exp(\omega_k t)$ (where $\omega_k = -\partial^2 V(q)/\partial q \partial q\big|_{q=q_k^{\mathrm{st}}}$ is a normal mode) can have a lower symmetry than the initial stationary state q_k^{st} ; the stable stationary states for which $\partial^2 V(q)/\partial q \partial q\big|_{q=q_k^{\mathrm{st}}}>0$ correspond to the maxima of $P^{\mathrm{st}}(q)$, while the unstable stationary states ($\partial^2 V(q)/\partial q \partial q\big|_{q=q_k^{\mathrm{st}}}<0$) correspond to the minima of $P^{\mathrm{st}}(q)$. Consequently, a qualitative change of the stationary state has an unequivocal effect on the extrema of the probability density. As has already been noted above, in the deterministic case, the stationary probability density $P^{\mathrm{st}}(q)$ consists of delta-like peaks concentrated in stationary states. In the stochastic case, the additive noise of chaos, by analogy with molecular fluctuations, also leads to some (depending on the intensity ε^2) smearing of the probability density function $P^{\mathrm{st}}(q)$ near the point $q_{\mathrm{min}}^{\mathrm{st}}$ corresponding to the minimum of the potential $V(q)$. Nevertheless, one may expect the transition probability density to reach its maximum precisely at this point, $P^{\mathrm{st}}(q_{\mathrm{min}}^{\mathrm{st}})>P^{\mathrm{st}}(q)$. If more than one minimum of the potential $V(q)$ exist, then there is a polymodal probability density $P^{\mathrm{st}}(q)$ with the vertices corresponding to various local minima of the deterministic potential; the largest maximum of the function and the position of the deepest well of the potential $V(q)$ are not shifted relative to each other. When $t \to \infty$ (this passage to the limit is needed to find the stationary probability density function), very intense fluctuations (the higher their intensity, the less probable they are) will necessarily take place, as a result of which the system will visit (infinitely many times) all minima of the potential $V(q)$. Thus, the number and positions of extrema of the probability density $P^{\mathrm{st}}(q)$ in the stochastic case and of the potential $V(q)$ in the deterministic case are the most characteristic distinctive features of the system's stationary behavior, i.e., the extrema corresponding to the system's macroscopic phases are to a certain extent indicators of the transition—order parameters.

To conclude this subsection, we emphasize once again that additive noise does not change qualitatively the system's stationary behavior. The extrema of the probability density always coincide with the deterministic stationary states irrespective of the intensity of white noise. Therefore, in the case of additive noise, in order to ascertain, for example, the locally stable state for which the inequality $P^{\mathrm{st}}(q_{\mathrm{min}}^{\mathrm{st}})>P^{\mathrm{st}}(q)$ holds at $\forall q \neq q_{\mathrm{min}}^{\mathrm{st}}$, an analysis based only on the deterministic description of chaos is quite sufficient. The influence of the natural noise of turbulent chaos on the evolution of a multidimensional Markovian diffusion process $\boldsymbol{q}(t)$ was analyzed by Kolesnichenko (2004).

6.1.2.2 Multiplicative Noise

Some modification of the correspondence between the macroscopic stationary states q^{st} and the extrema q_{ext} of the stationary probability density $P^{st}(q)$ is required for the less idealized (more realistic) multiplicative noise of turbulent chaos, whose fluctuations depend on the evolution of the random process $q(t)$ (i.e., $Q(q) \neq \text{const}$). The extrema of the function $P^{st}(q)$ can be found in this case from the condition [see (6.11)]

$$h(q_{\text{ext}}, \varepsilon) = K(q_{\text{ext}}) - \frac{\varepsilon^2}{2}\sqrt{Q(q_{\text{ext}})}\frac{\partial\sqrt{Q(q)}}{\partial q}\bigg|_{q=q_{\text{ext}}} = 0. \tag{6.19}$$

Equation (6.19) is the main one for analyzing the influence of the multiplicative noise of chaos on the stationary behavior of the stochastic system (6.1) under consideration. The first term in this equation, if it is set equal to zero, corresponds to (6.18) for deterministic stationary states and the second term describes the influence of multiplicative noise. Thus, the intensity ε^2 of the natural noise of turbulent chaos turns out to be an important parameter of the system.

The stationary probability density has a maximum at $\partial h(q)/\partial q|_{q=q_{\text{ext}}}<0$ and a minimum at $\partial h(q)/\partial q|_{q=q_{\text{ext}}}>0$. The set of points in the space of configurations for which $\partial h(q)/\partial q|_{q=q_{\text{ext}}} = 0$ corresponds to structurally unstable situations (called P-bifurcations). Because of P-bifurcations, the correspondence between the attractors of the deterministic system and the maxima of the stationary probability density breaks down. If ε^2 is small, then the roots of (6.19) do not differ in number and position from the deterministic stationary states. However, if the noise intensity ε^2 is high enough, then the extrema of the probability density $P^{st}(q)$ can differ significantly in both number and position from the deterministic stationary state. In turn, this implies that (in contrast to additive noise) multiplicative noise not only has a disorganizing effect on the system but also can stabilize new macroscopic states and cause new nonequilibrium phase transitions unpredictable within the framework of deterministic chaos.

6.1.3 *Phenomenology of Small-Scale Turbulence*

Let us now analyze the peculiar features of the transitions induced by the multiplicative noise of turbulent chaos for a lognormal distribution of fluctuating parameters q in the stationary state of chaos.

The basis ideas of Kolmogorov's (1941) cascade theory of small-scale vortex turbulence can be used as a basis for the macroscopic model of turbulent chaos proper associated with the stochastic small-scale fluctuating motion of a turbulized fluid. For this reason, it deserves here a brief overview. According to this theory, in the case of a turbulized fluid considered as a continuous medium, the small-scale structure of turbulence is determined by the cascade pattern of energy transfer over the spectrum of vortices with various spatiotemporal scales through the inertial range from the energy range to the viscous one, where it dissipates into heat. Small vortices gain energy through successive fragmentation of large vortices as the control parameter, the Reynolds number $Re = Lu_0/\nu$ corresponding to large-scale motions in the flow (L is the integral turbulence scale, u_0 is the characteristic flow velocity, ν is the kinematic molecular viscosity), increases. The specific kinetic energy of the carrier flow ($\sim u_0^3/L$) is continuously redistributed from large-scale vortices to smaller ones down to the smallest vortices with a characteristic size of the order of the internal turbulence scale $\eta \sim (\nu^3/\bar{\varepsilon})^{1/4}$, which characterizes the influence of viscous effects on the structure of small-scale turbulence. It is important to clearly realize that in the limit of large Reynolds numbers Re, despite the anisotropy, inhomogeneity, and nonstationarity of the averaged flow, the random fragmentation of vortices and the chaoticity of their energy transfer over the cascade downward cause the stochastic regime of turbulent fluctuations within the spatiotemporal region over which the instantaneous hydrodynamic equations are averaged to be almost locally isotropic—homogeneous, isotropic, and quasi-stationary, i.e., changing only with the number Re eventually determining the number of cascades itself in the hierarchy of vortices with various orders. Naturally, there can be no complete local isotropy due to the presence of mesoscale coherent structures.

6.1.3.1 Some Results of Kolmogorov's Theory

Kolmogorov's theory of 1941 (*K41*) is concerned with the statistical properties of homogeneous and isotropic turbulence at large Reynolds numbers. According to the first similarity hypothesis, the statistical regime of small-scale locally isotropic turbulence on scales $r \ll L$ is uniquely determined by three dimensional parameters: the energy dissipation rate averaged over an ensemble of possible fluid flow realizations $\bar{\varepsilon}$ (the key characteristic of locally isotropic turbulence), the viscosity ν, and the scale r itself. The above estimate for the dissipation length scale η is a corollary of precisely this hypothesis. At the same time, there is no noticeable production and dissipation of kinetic energy in the inertial range ($\eta \ll L$); a quasi-stationary regime of turbulence that is universal and depends only on $\bar{\varepsilon}$ and r is established here (Kolmogorov's second similarity hypothesis).

The quantitative description of small-scale locally isotropic turbulence in the inertial range is based on the use of structure functions

$$S_p(r) = \overline{[u''_r(x) - u''_r(x+r)]^p} \tag{6.20}$$

and their spectra (in view of flow isotropy, the structure functions do not depend on the direction of the segment r). One of the most important laws of small-scale turbulent motions (the two thirds law) follows from the second similarity hypothesis and the assumption that the turbulence parameters change only slightly at distances of the order of $r = |r|$ if $r<\Lambda \ll L$: in any turbulent flow with a sufficiently large Reynolds number Re, the mean square of the velocity difference at two points separated by a distance r at r not too small but also not too large (comparable to the hydrodynamic length scale Λ of the corresponding averaged flow) must be proportional to $r^{2/3}$: $S_2(r) \sim (\bar{\varepsilon}\, r)^{2/3}$. This law has now been well confirmed experimentally for a wide variety of turbulent flows (see Monin and Yaglom 1996).

As regards the pth-order structure functions, Kolmogorov's (1941) similarity theory leads to the relation

$$S_p(r) \sim (\bar{\varepsilon}\, r)^{p/3}, \tag{6.21}$$

which, in general, is not confirmed experimentally, particularly for $p \gg 1$, and at best is their weak estimate. The shortcoming of the similarity hypotheses for locally isotropic turbulence in their original form is related to the significant assumption that the energy influx to small-scale perturbations in the inertial range is constant, when the Kolmogorov parameter $\bar{\varepsilon}$, whose exact definition is $\bar{\varepsilon} = \bar{\varepsilon}(r,t) = \frac{\nu}{2}\overline{\left(\partial u''_i/\partial r_j + \partial u''_j/\partial r_i\right)^2}$, was assumed to be a universal constant for a given flow and to characterize the energy flux pumped along the entire inertial range to dissipative scales. In reality, however, $\bar{\varepsilon}$ is not a constant but is a random variable characterized by its own distribution function. Thus, experimental measurements of structure functions of relatively low orders confirmed the validity of Landau's "Kazan" remark—local variations in energy dissipation rate violate Kolmogorov's scenario of homogeneous turbulence (see, e.g., Landau and Lifshitz 1988; Frisch 1995). In particular, this violation of homogeneous turbulence is related to the so-called flow intermittency, when turbulized regions coexist with quasi-laminar ones at arbitrarily large Reynolds numbers.

6.1.3.2 Lognormal Model

In view of the difficulties noted above, Kolmogorov's (1941) views of a random vortex cascade were refined by Obukhov (1962). He suggested abandoning the condition $\bar{\varepsilon}(r,t) = \text{const}$ in the region dr (with the center at point r and a characteristic scale $\Lambda \ll L$) and, to take into account the structure of the energy dissipation field, proceeding from the fact that the statistical characteristics of small-scale motions (e.g., the structure functions) are not determined by the probability-theoretic mean $\bar{\varepsilon}\,(r,t)$ of the random variable ε but depend on the local dissipation $\varepsilon_r(r,t)$ averaged over some volume V_r with a characteristic size r small compared to the typical inhomogeneity scale of the averaged flow, $r \ll \Lambda$. Turbulence corresponding to this approach was called locally homogeneous one. If a sphere of radius r is chosen as the averaging region lying within dr (the results obtained depend weakly on the shape of the averaging region), then

$$\varepsilon_r(r,t) = \frac{3}{4\pi r^3} \int\limits_{|r^*|\le r} dr^* \frac{1}{2} \nu \sum_{i,j} \left(\frac{\partial u''_i(r+r^*)}{\partial r_j} + \frac{\partial u''_j(r+r^*)}{\partial r_i} \right)^2. \tag{6.22}$$

Using the parameter ε_r, Kolmogorov (1962) reformulated the first and second similarity hypotheses (establishing the so-called refined hypotheses in which the constant $\bar{\varepsilon}$ was replaced by the random variable ε_r) and, in addition, supplemented them with the third hypothesis postulating (for a large scale ratio $L : r$) a lognormal probability density function for ε_r and a linear dependence of the variance $\sigma^2{}_{\ln \varepsilon_r}$ on $\ln(L/r)$:

$$W_1(\varepsilon_r) = \frac{1}{\sqrt{2\pi}\ \varepsilon_r \sigma_{\ln \varepsilon_r}} \exp\left(-\frac{\ln^2(\varepsilon_r / m_{\ln \varepsilon_r})}{2\sigma^2_{\ln \varepsilon_r}} \right), \tag{6.23}$$

$$\ln m_{\ln \varepsilon_r} = \frac{-\sigma^2_{ln\varepsilon}}{2} + \ln \overline{\varepsilon_r}, \quad \sigma^2_{\ln \varepsilon_r} = \mu \ln\left(\frac{L}{r}\right) + B(r,t), \tag{6.24}$$

where $\sigma^2_{\ln \varepsilon_r}$ is the variance of the random variable $\ln \varepsilon_r$; $m_{\ln \varepsilon_r} = \exp(\overline{\ln \varepsilon_r})$ is the median of the distribution for the random variable $\ln \varepsilon_r$; $\overline{\varepsilon_r} = \bar{\varepsilon}$; μ is a universal (intermittency) parameter; $B(r,t)$ is a term dependent on the characteristics of the averaged (large-scale) fluid motion. The coefficient μ is the lognormal model is a fitting parameter whose initial estimate was $\mu \approx 0.5$; this estimate then decreased to $\mu \approx 0.2$ (see Anselmet et al. 1984; Monin and Yaglom 1996). Such a stationary probability density function $W_1^{st}(\varepsilon_r)$ in Kolmogorov's refined theory (*K62*) can be additionally justified by the fact that the model for the cascade successive fragmentations of turbulent vortices is akin to the fragmentation of dust particles to which a lognormal size distribution corresponds asymptotically. In addition, there are numerous experimental confirmations for the lognormal probability distribution of the energy dissipation or related nonnegative small-scale turbulence characteristics in a wide range of moderate values of the argument. The validity of the lognormal distribution for nonnegative small-scale turbulence characteristics was questioned in a number of publications, because it suggests, in particular, the appearance of supersonic speeds at very large Reynolds numbers (Frisch 1995). In recent years, attempts were made to describe random turbulent fields by other distribution functions, for example, log-Poissonian, log-Levy, etc. (see, e.g., She and Leveque 1994; Dubrulle 1994). However, as yet no ultimate answer to the question about the true probability distribution laws for ε_r-type quantities has been gotten. Therefore, it seems natural to use in this approach Kolmogorov's assumption that the distribution for ε_r is lognormal as the initial step. A comprehensive overview of the corresponding works is given, for example, in the monograph by Monin and Yaglom (1996), to which we refer the interested reader.

It is commonly assumed in the literature that (6.23) also gives the probability density of the averaged dissipation if $r \approx \eta$ (Frenkiel and Klebanoff 1975). Using the well-known formula $\eta = LRe^{-3/4}$ expressing the dissipative turbulence scale in terms of the turbulence macroparameters (see Monin and Yaglom 1996), this assumption allows the variance $\sigma^2_{\ln \varepsilon_r}$ to be related to the logarithm of the Reynolds number:

$$\sigma^2_{\ln \varepsilon_r} \cong 0.75\mu \ln Re + B(r,t). \tag{6.24*}$$

Distribution (6.23) allows one to calculate important statistical characteristics of stationary small-scale turbulence, in particular, the moments of the dissipation rate $\overline{\varepsilon_r{}^p}$:

$$\overline{\varepsilon_r{}^p} = \int\limits_0^\infty \varepsilon_r{}^n W_1(\varepsilon_r)\mathrm{d}\varepsilon_r = (\overline{\varepsilon_r})^p \exp\left[\tfrac{1}{2}p(p-1)\sigma^2_{\ln \varepsilon_r}\right]$$
$$= F_p(r,t)(\overline{\varepsilon})^p (r/L)^{-\mu p(p-1)/2}. \tag{6.25}$$

(the coefficients $F_p(r,t)$ depend on the turbulent flow macrostructure). Hence it follows that the intermittency coefficient μ has a simple physical meaning—to within the sign, this is the exponent r for the second-order moment of the energy dissipation field. Using (6.25), we can obtain a refined expression for the pth-order structure function $S_p(r) \sim \overline{\varepsilon^{p/3}} \cdot r^{p/3}$:

$$S_p(r) \sim F_p(r,t)(\overline{\varepsilon})^{p/3}(r/L)^{\mu p(3-p)/18}(r)^{p/3}. \tag{6.26}$$

For example, at $p = 2$ it turns out to be close to the accuracy limits of the available experimental data, while for higher-order structure functions it differs greatly from the dependence predicted by Kolmogorov's original theory for locally isotropic turbulence (but, at the same time, agrees satisfactorily with the available experimental data). The formula $S_p(r) \sim \overline{\varepsilon^{p/3}} \cdot r^{p/3}$ generalizes (6.21) in the sense that not the constant quantity ε to the $p/3$ power but the $p/3$-order statistical moment characterizing the structure of the random energy dissipation field on the corresponding scales r appears on the right-hand side. The theory for the local structure of the fluctuating temperature, density, and concentration fields for a chemically active admixture mixed by turbulence was corrected in a similar way (see, e.g., Monin and Yaglom 1996).

There is one more important reason why it is convenient to use (as an internal parameter) the fluctuating locally averaged dissipation ε_r when phenomenologically describing turbulent chaos in terms of the two-level macroscopic model of structured turbulence with self-organization processes. In particular, it is related to the fact that the original cascade theory (*K41*) implying uniform filling of the physical space with vortices of each scale is unsuitable for this purpose, because it does not consider explicitly any coherent structures (e.g., vortex spirals, vortex filaments, streaks, burstings, etc.) in which the most intense viscous energy dissipation takes place (the Kolmogorov parameter $\overline{\varepsilon}$ in this theory is a constant). Meanwhile, as the most recent experiments showed (see, e.g., Crow and

Champagne 1971; Brown and Roshko 1974; Khlopkov et al. 2002), any adequate theory of turbulence should take into account the existence of vortex dissipative CSs and the nonuniformity of their spatiotemporal distribution in a fluctuating flow. In addition, the phenomenon of hydrodynamic intermittency inherent in a turbulent flow, whereby the regions occupied by the so-called turbulent spots (where intense fluctuations in velocity gradients are observed, $\bar{\varepsilon}_r>0$) are closely intertwined with the regions with a weakly turbulized or completely vortex-free fluid (where there are virtually no such fluctuations, $\bar{\varepsilon}_r \cong 0$), is associated with the chaotic nature of the kinetic energy transfer over a cascade of vortices. In other words, a multiscale system (a fractal set) of interacting active and passive regions with different scaling laws is formed in developed turbulent fields. As is well known, multifractality can be determined and measured in terms of local dissipation fluctuations (see Frisch 1995). Thus, the energy dissipation ε_r or related quantities (quadratic in velocity gradients) can serve both as an important turbulent flow structuring characteristic and as an intermittency indicator.

6.1.4 Model Stochastic Differential Equations and the Fokker–Planck–Kolmogorov Equation for the Turbulent Energy Dissipation Rate

In view of the aforesaid, it seems reasonable to use the locally averaged turbulent energy dissipation rate, $q \equiv \varepsilon_r>0$, as a single fluctuating coordinate of turbulent chaos, which is assumed below. In this case, using (6.23) for the stationary distribution of ε_r and designating $\alpha \equiv N\sqrt{\pi}\sigma_{\ln q}$, we can concretize the expressions for the probabilistic potential (6.12) and the diffusion and drift coefficients in (6.2). As a result, we have

$$\begin{aligned}\Phi(q) &= \frac{\varepsilon^2}{2}\ln\sqrt{Q(q)} - \int^{q}\frac{K_\alpha(x)}{Q(x)}\mathrm{d}x = -\frac{\varepsilon^2}{2}\ln\left(\frac{P^{\mathrm{st}}(q)}{N}\right)\\ &= \frac{\varepsilon^2}{2}\ln(\alpha q) + \frac{1}{2\alpha}\ln^2\left(\frac{q}{m_{\ln q}}\right),\end{aligned} \tag{6.27}$$

$$\begin{gathered}Q(q) = \alpha^2 q^2, \quad K(q) = -\alpha q \ln\left(\frac{q}{m_{\ln q}}\right),\\ K(q,\varepsilon) = -\alpha q\left[\ln\left(\frac{q}{m_{\ln q}}\right) - \sigma^2_{\ln q}\right],\end{gathered} \tag{6.28}$$

where

$$\sigma^2_{\ln q} = \frac{1}{2}\alpha\varepsilon^2. \tag{6.29}$$

Consequently, as a basis for the evolutionary description of the Markovian diffusion process $q(t) \equiv \varepsilon_r(t)$ modeling the stochastic behavior of the subsystem of turbulent chaos with multiplicative noise ($Q(q) = \alpha^2 q^2$), we can use the SDE

$$\frac{\partial q(t)}{\partial t} = -\alpha\, q(t) \ln\left(\frac{q(t)}{m_{\ln q}}\right) + \varepsilon\alpha\, q(t) \cdot \xi(t), \tag{6.30}$$

which we interpret in Stratonovich's sense, and the corresponding FPK equation for the transition probability density $P_2 \equiv P_2(q_0|q,t)$

$$\frac{\partial}{\partial t} P_2(q,t) = -\alpha \frac{\partial}{\partial q}\left\{\left[\ln\left(\frac{q}{m_{\ln q}}\right) - \frac{\alpha}{2}\varepsilon^2\right] q P_2(q,t)\right\} + \frac{\varepsilon^2}{2}\alpha^2 \frac{\partial^2}{\partial q_2}\left[q^2 P_2(q,t)\right]. \tag{6.31}$$

6.1.4.1 Description in Unified Terms of the K41 and K62 Models

Let us now return to our main problem—investigating the mechanism for the synergetic generation of CSs in the space of internal coordinates (as a result of a significant change in Reynolds number) associated with the phenomenon of non-equilibrium phase transitions induced by the noise of turbulent chaos. As has already been noted above, the incoherent (fine-grained) component of chaos produced by a successive cascade of a large number of spatial and temporal bifurcations ultimately turns out to be organized in such a complex way that the concept of the K41 theory about homogeneous and isotropic turbulence at a constant Kolmogorov parameter $\bar{\varepsilon}$ can be used to characterize its statistical properties (in the inertial and dissipative ranges). At the same time, to investigate the possible self-organization processes in the system (i.e., the formation of meso-scale vortex CSs in a real turbulized flow), it is necessary, as was shown, to depart from such a high symmetry and to introduce a random dissipation rate ε_r whose fluctuations violate Kolmogorov's scenario of homogeneous turbulence.

To be able to describe the locally isotropic state (background deterministic state) of turbulent chaos and its structuring mechanisms in unified terms, we assume below that the state of chaos in the deterministic case is described by a degenerate random variable $\bar{\varepsilon}_r = \text{const}$ characterized by its probability distribution. One would think what "distribution" we are talking about if there is only one admissible value, i.e., the variable being investigated is a constant. Nevertheless, such a distribution can be interpreted as the limit of a continuous distribution when the domain of existence of the density function $P(\varepsilon_r)$ shrinks to only one point $\bar{\varepsilon}_r$, i.e.,

$$P(\varepsilon_r) = \delta(\varepsilon_r - \bar{\varepsilon}_r) = \begin{cases} \infty & \text{at } \varepsilon_r = \bar{\varepsilon}_r \\ 0 & \text{at } \varepsilon_r \neq \bar{\varepsilon}_r, \end{cases} \tag{6.32}$$

and by the random variable we mean a quantity that is almost always equal to $\bar{\varepsilon}_r$. The relations $M(\varepsilon_r) = \bar{\varepsilon}_r$ and $D(\varepsilon_r) = 0$ hold for the degenerate distribution (6.32) (see, e.g., Bostandzhian 2000). The reverse is also true: if the variance of some distribution with a finite mean is equal to zero, then it can be only degenerate. Thus, in the deterministic case, the stationary "probability density" consists of delta-like peaks concentrated at the stationary state $\bar{\varepsilon}_r$. Note that the lognormal distribution (6.23) tends to the degenerate distribution (6.32) with the variance $D(\varepsilon_r) = 0$ and mean $M(\varepsilon_r) = \exp(\ln m_{\ln \varepsilon_r}) = \bar{\varepsilon}_r$ when $\sigma^2_{\ln \varepsilon_r} \to 0$.

Let us now determine the coefficient α in (6.30) and (6.31). Using (6.24*) and (6.29), we obtain the relation

$$\sigma^2_{\ln \varepsilon_r} = \tfrac{1}{2}\alpha\varepsilon^2 = \tfrac{3}{4}\mu \ln Re + B.$$

In the special case of degenerate turbulence ($\sigma^2_{\ln \varepsilon_r} \to 0, \quad Re = Re_{\text{det}}$), it allows the universal term B (dependent on the characteristics of large-scale motion) to be written as $B = -\frac{3}{4}\mu \ln Re_{\text{det}}$. Here, Re_{det} is some limiting value of the Reynolds number at which non-structured turbulent chaos (our background state) is possible. For the external parameter α, we then have

$$\alpha = \frac{3}{2\varepsilon^2}\mu \ln(1+\gamma) \tag{6.33}$$

where $\gamma = (Re - Re_{\text{det}})/Re_{\text{det}}$ is the supercriticality index. Thus, the coefficient α is a control parameter for (6.30) and (6.31). The specificity of these equations is such that for some regime of averaged motion ($\alpha \geq \alpha_{\text{crit}}$), the large-scale fluctuations of turbulent chaos can reach critical values at which the phase trajectories of the random process $q(t)$ jump to a different attractor. As a result, a certain rearrangement of the state of the subsystem of turbulent chaos associated, in particular, with the formation of new relatively stable collective motions (mesoscale vortex structures) occur, which, in turn, changes the macroscopic behavior of the turbulized fluid as a whole.

6.1.5 *Phase Transitions Induced by the Multiplicative Noise of Turbulent Chaos*

6.1.5.1 Degenerate Turbulence

Let us first analyze the degenerate case ($\sigma^2_{\ln \varepsilon_r} \to 0$) where the intensity of the white noise of turbulent chaos is extremely low, i.e., $\varepsilon^2 \to 0$. In this case, SDE (6.30) is reduced to the deterministic equation

$$\frac{\partial q(t)}{\partial t} = -\alpha\, q(t)\left[\ln\left(\frac{q(t)}{\bar{q}}\right) + \frac{\sigma^2_{\ln \varepsilon_r}}{2}\right] \cong -\alpha\, q(t) \cdot \ln\left(\frac{q(t)}{\bar{q}}\right), \quad q(0) = \bar{q}, \quad (6.34)$$

As has been noted above, the changes occurring in the regime of averaged fluid motion affect the "random" process $q(t)$ via the external (control) parameter α. The time-dependent solution of the deterministic equation (6.34) can be found in explicit form: $q = \bar{q}\,\exp[\exp(-\alpha t)]$. Hence it follows that the macroscopic time scale of the system's evolution to the reference state (the lifetime of possible perturbations) is a quantity of order $\tau_{\text{macro}} = 1/\alpha$. For $\alpha<0$ (i.e., for $Re<Re_{\text{det}}$), (6.34) admits the unique asymptotically stable stationary solution $q_1^{\text{st}} = 0$ [for the stability (instability) conditions for stationary states, see the text below (6.18)] that corresponds to the laminar flow regime with the highest symmetry. At point $\alpha = 0$ ($Re = Re_{\text{det}}$), the lifetime of the fluctuations in the first approximation tends to infinity, the solution $q_1^{\text{st}} = 0$ becomes unstable and undergoes a bifurcation, i.e., the phase transition occurs, because the stationary point $q_1^{\text{st}} = 0$ for $\alpha>0$ ($Re>Re_{\text{det}}$) becomes unstable, but a new stable stationary state $q_2^{\text{st}} = \bar{q}$ appears. Consequently, $\alpha = 0$ is the bifurcation value of the parameter: the quantity q jumps from the lower branch of stationary states $q_1^{\text{st}} = 0$ to the upper branch of stationary states $q_2^{\text{st}} = \bar{q}$. Thus, the deterministic system (6.34) undergoes an equilibrium first-order phase transition. This corresponds to spontaneous generation (at $Re = Re_{\text{det}}$) of incoherent background (homogeneous and isotropic) turbulence—a flow in which there is a tendency for all symmetries to be restored in the statistical sense with increasing Reynolds number far from the boundaries.

6.1.5.2 Noise-Induced Critical Points

Let us now investigate the stationary behavior of a stochastic system under noise of chaos with an arbitrary intensity ε^2. We assume below that the fluctuations of real turbulent chaos affecting the system are fast compared to $\tau_{\text{macro}} = 1/\alpha$. This allows us to pass to the limit of white noise and to model the truly random process $q(t)$ using SDE (6.30) and the corresponding FPK equation (6.31). Note that the stationary solution (6.23) of (6.31) exists if and only if $\alpha>0$. The space of states to which the process $q(t)$ should be limited is a nonnegative real half-line. From a physical viewpoint, (6.30) and (6.31) are unlikely to be valid at arbitrarily large q. Therefore, it would be more appropriate to bound the space of states by the interval $[0, q_{\max}]$, where $q_{\max}$ is the largest local dissipation that is physically possible in the system. However, such a procedure is not needed, because the stationary probability density decreases exponentially when $q \to \infty$, i.e., the transition to large q is completely safe from a physical viewpoint.

It can be seen from (6.6) that the boundaries of the process ($q_1 = 0$ and $q_2 = \infty$) will be natural if for $\beta>0$ (β is a point near the boundary, lower or upper)

$$L_1(0) = \int_0^\beta \phi(x)\mathrm{d}x = C\int_0^\beta x^{\left(\frac{1}{2}-\frac{2\ln\bar{\varepsilon}_r}{\varepsilon^2\alpha}\right)} \exp\left[\frac{1}{\varepsilon^2\alpha}\ln^2 x\right]\mathrm{d}x = \infty, \quad \beta \in (0,\infty), \qquad (6.35)$$

and

$$L_1(\infty) = \int_0^\infty \phi(x)\mathrm{d}x = C\int_0^\infty x^{\left(\frac{1}{2}-\frac{2\ln\bar{\varepsilon}_r}{\varepsilon^2\alpha}\right)} \exp\left[\frac{1}{\varepsilon^2\alpha}\ln^2 x\right]\mathrm{d}x = \infty. \qquad (6.36)$$

Condition (6.36) is always met, i.e., infinity is a natural boundary at all values of α and ε. It can be seen from relation (6.35) that zero is a natural boundary only if $\alpha>0$, which coincides with the existence condition for the stationary probability density (6.23) (for $\alpha>0$ and $\sigma^2_{\ln q}>0$). On the other hand, for $\alpha<0$ it follows from (6.7) and (6.35) that $L_1(0)<\infty$ and $L_2(0) = \infty$, i.e., zero is an attractive boundary. In addition, it should be kept in mind that zero is not only the boundary but also the stationary point of the random process $q(t)$: at $q = 0$ the drift and diffusion vanish simultaneously. Since zero is an attractive point, the entire stationary "mass" of the probability must be concentrated at zero. In terms of the probability density, this implies that $P^{\mathrm{st}}(q) = \delta(q)$ for $\alpha<0$ [see (6.32)].

As has been noted above, the number and positions of extrema for the stationary probability density $P^{\mathrm{st}}(q)$ in the stochastic case are the most characteristic distinctive features of the system's stationary behavior, i.e., they can be the order parameters for nonequilibrium phase transitions. In our model, the extrema of the probability density $P^{\mathrm{st}}(q)$ coincide with the zeros of (6.19)

$$h(q,\varepsilon) \equiv -\alpha\, q(t)\ln\left(\frac{q(t)}{\bar{q}}\right) - \frac{3}{4}\varepsilon^2\alpha^2 q(t) = 0, \qquad (6.37)$$

i.e., with

$$(q_{\mathrm{ext}})_1 = 0 \qquad (6.38)$$

and

$$(q_{\mathrm{ext}})_2 = \bar{q}\exp\left[-\frac{3}{4}\varepsilon^2\alpha\right] = \bar{q}\exp\left[-\frac{9}{8}\mu\ln(1+\gamma)\right] \qquad (6.39)$$

(the second root exists if $\alpha>0$). The points in the space of configurations for which

$$\left.\frac{\partial h(q,\varepsilon)}{\partial q}\right|_{q=q_{\mathrm{ext}}} = -\alpha\,\ln\left(\frac{q_{\mathrm{ext}}}{\bar{q}}\right) - \alpha - \left(\frac{3}{4}\right)\varepsilon^2\alpha^2 = 0$$

correspond to structurally unstable situations where phase transitions are possible. Since $\partial h(q,\varepsilon)/\partial q|_{q=(q_{\text{ext}})_2} \equiv -\alpha$ is always less than zero, the stationary probability density at point $(q_{\text{ext}})_2$ reaches its maximum,

$$P^{\text{st}}(q) = \frac{1}{\sqrt{2\pi}\bar{q}\sigma_{\ln q}} \exp\left(\frac{\alpha\varepsilon^2}{2}\right);$$

the derivative

$$\left(\frac{\partial h(q,\phi)}{\partial q}\right)\Bigg|_{q=(q_{\text{ext}})_1} \equiv -\alpha \ln\left[\frac{(q_{\text{ext}})_1}{\bar{q}}\right] - \alpha - \left(\frac{3}{4}\right)\varepsilon^2\alpha^2$$

and, hence, the root $(q_{\text{ext}})_1$ will be the maximum only for $\alpha<0$ (when $\partial h(q)/\partial q|_{q=(q_{\text{ext}})_1}<0$). Hence it follows that when the parameter α passes through zero from left to right, the stable focus $(q_{\text{ext}})_1$ becomes unstable and the system passes to a new stationary regime $(q_{\text{ext}})_2$, i.e., in contrast to the deterministic case, there are two transition points in fluctuating turbulent chaos. The transition at $\alpha = 0$ ($Re = Re_{\text{det}}$) is determined by the nature of the boundary at zero.

Summarizing the aforesaid, it can be asserted that the probability density is characterized by the following features:

- If $\alpha<0$ ($Re<Re_{\text{det}}$), then the stationary point $(q_{\text{ext}})_1 = 0$ (corresponding to a laminar flow) is stable.
- At $\alpha = 0$ ($Re = Re_{\text{det}}$), the stationary point $(q_{\text{ext}})_1 = 0$ becomes unstable; the phase transition determined by the nature of the boundary at zero occurs.
- At $\alpha = 0$, a new stable stationary solution $q_2^{\text{st}} = \bar{q}$ corresponding to incoherent (background) turbulence emerges; the parameter q jumps from the lower branch of stationary states $q_1^{\text{st}} = 0$ to the upper branch of stationary states $q_2^{\text{st}} = \bar{q}$. In this case, the lognormal distribution (6.23) turns into the degenerate distribution (6.32) with variance $D(q) = 0$ and mean $M(q) = \bar{q}$. Although the stationary state $q_2^{\text{st}} = \bar{q}$ is not stable in the presence of noise, it nevertheless remains the most probable value. In a sense, the δ-function (6.32) begins to spread rightward when α passes this transition point.
- If the parameter α is greater than zero (i.e., for $Re>Re_{\text{det}}$), then the pattern of distribution (6.32) changes sharply; a new branch of stationary states

$$(q_{\text{ext}})_2 = \bar{q}\exp\left(-\frac{3}{4}\varepsilon^2\alpha\right) = \bar{q}\exp\left(-\frac{9}{8}\mu\ln\left(\frac{Re}{Re_{\text{det}}}\right)\right),$$

shifted relative to the deterministic case of $(q_{\text{ext}})_2 = \bar{q}$ appears.

Thus, inducing a nonequilibrium phase transition, the multiplicative noise of chaos causes a shift in the deterministic transition point to turbulence. This transition corresponds to situations where the system no longer adapts its macroscopic behavior to the average properties of the medium (as in the deterministic case) but responds to a change in parameter Re in a more definite and active way. The random variable (6.39) defining the system's stationary state in the stochastic case differs qualitatively from the degenerate random variable $(q_{\text{ext}})_2 = \bar{q}$ corresponding to background turbulence. Thus, here we have a nontrivial truly noise effect: noise (when its intensity increases) lowers the threshold for the transition of a stochastic system to a truly stationary state.

6.1.6 *Analysis of Verhulst's Mathematical Model for the Dissipation of Turbulent Energy*

To consider the noise-induced transitions near the average state $\bar{q}$ in more detail, let us now turn to Verhulst's well-known model whose deterministic (degenerate) version is

$$\frac{\partial q(t)}{\partial t} = \lambda\, q(t) - q^2(t) \tag{6.40}$$

(here, $\forall q \in [0, \infty)$; λ is the control parameter) (see Horsthemke and Malek Mansour 1976). This model was initially designed to describe the growth of a biological population, but, in the course of time, found applications in a great variety of sciences. For $\lambda<0$, (6.40) admits the unique asymptotical stable solution $q^{\text{st}} = 0$ that corresponds to the regime with the highest (compatible with the constraints imposed on the system) symmetry. At $\lambda = 0$, the solution $q^{\text{st}} = 0$ becomes unstable and undergoes a bifurcation: a new branch of stationary states $q^{\text{st}} = \lambda$ separates from it in a continuous (but non-differentiable) way. Thus, the deterministic system undergoes an equilibrium second-order phase transition, which corresponds to the formation of an equilibrium coherent structure with a lower symmetry.

Let us show how this model can be used in our case where the fluctuations in energy dissipation rate $\delta\varepsilon_r = \varepsilon_r - \bar{\varepsilon}_r$ are small compared to its mean value, $\delta\varepsilon_r/\bar{\varepsilon}_r \ll 1$. Applying (6.23) and (6.29), we transform (6.28) for the drift coefficients as follows:

$$K(q) = -\alpha q\left(\ln\frac{q}{\bar{q}} + \frac{\alpha\varepsilon^2}{4}\right) \approx \frac{\alpha}{\bar{q}}\left[\bar{q}\left(1 - \frac{\alpha\varepsilon^2}{4}\right)q - q^2\right] = \frac{\alpha}{\bar{q}}(\lambda q - q^2) \tag{6.41}$$

and

$$K(q,\varepsilon) = K(q) + \frac{1}{2}\alpha^2\varepsilon^2\, q \approx \frac{\alpha}{\bar{q}}\left(\lambda q - q^2 + \frac{1}{2}\alpha\varepsilon^2\bar{q}\, q\right). \tag{6.42}$$

Here, we use the expansion of the logarithmic function $\ln(q/\bar{q})$ into a series near the mean $\bar{q}$, $\ln(q/\bar{q}) \approx \Delta q/\bar{q} = -1 + q/\bar{q}$, and introduce the (external) parameter

$$\lambda \equiv \bar{q}\left(\frac{1-\alpha\varepsilon^2}{4}\right) = \bar{q}\left[1 - \frac{3}{8}\mu \ln\left(\frac{Re}{Re_{\text{det}}}\right)\right] \tag{6.43}$$

describing the action of the subsystem of averaged motion on the random process $q(t)$. Using the similarity transformation in time $t' = t\alpha/\bar{q}$ and the designations $\varepsilon' = \sqrt{\alpha\bar{q}}\varepsilon$ and $\xi'(t) = \sqrt{\bar{q}/\alpha}\xi(t)$, (6.30) and (6.31) are then brought to the following standard from of the stochastic Verhulst system:

$$\frac{\partial q(t)}{\partial t} = \lambda q(t) - q^2(t) + \varepsilon' q(t) \cdot \xi'(t), \tag{6.44}$$

$$\frac{\partial}{\partial t}P_2(q,t) = -\frac{\partial}{\partial q}\left\{\left[\lambda q - q^2 + \frac{\varepsilon'^2}{2}q\right]P_2(q,t)\right\} + \frac{\varepsilon'^2}{2}\frac{\partial^2}{\partial q^2}\left[q^2 P_2(q,t)\right]. \tag{6.45}$$

The stationary solution $P^{\text{st}}(q)$ of (6.45) is defined by the expression

$$P^{\text{st}}(q) = \begin{cases} Nq^{2(\lambda/\varepsilon'^2)-1}\exp(-2q/\varepsilon'^2), & \lambda>0 \\ \delta(q), & \lambda<0 \end{cases}, \tag{6.46}$$

where

$$N^{-1} = \left(\frac{2}{\varepsilon'^2}\right)^{-2(\lambda/\varepsilon^2)} \Gamma\left(\frac{2\lambda}{\varepsilon'^2}\right).$$

$\Gamma(x)$ is the gamma function. The stationary solution is thus seen to exist [i.e., the quantity $P^{\text{st}}(q)$ is integrable in the interval $[0, \infty)$] if and only if $2\lambda/\varepsilon'^2$, i.e., if $\lambda>0$. In addition, it should be kept in mind that zero ($q = 0$) is not only the boundary point but also the stationary one. Therefore, the entire stationary "mass" of the probability must be concentrated at zero.

In Verhulst's model, the extremal of the probability density $P^{\text{st}}(q)$ coincide with the zeros of the equation

$$h(q, \varepsilon') \equiv \lambda q - q^2 - \frac{\varepsilon'^2}{2}q = 0, \tag{6.47}$$

i.e., with

$$(q_{\text{ext}})_1 = 0 \text{ and } (q_{\text{ext}})_2 = \lambda - \varepsilon'^2/2 \tag{6.48}$$

(the second root exists if $\lambda > \varepsilon'^2/2$). The points in the space of configurations for which

$$\left.\frac{\partial h(q)}{\partial q}\right|_{q=q_{\text{ext}}} = \lambda - 2q_{\text{ext}} - \varepsilon'^2/2 = 0$$

correspond to structurally unstable situations where phase transitions are possible. Since $\partial h(q)/\partial q|_{q=(q_{\text{ext}})_2} \equiv -\lambda + \varepsilon'^2/2$ is always less than zero, the root $(q_{\text{ext}})_2$ is the maximum of the stationary probability density $P^{\text{st}}(q)$; the derivative $\partial h(q)/\partial q|_{q=(q_{\text{ext}})_1} \equiv \lambda - \varepsilon'^2/2$ and, hence, the root $(q_{\text{ext}})_1$ will be the maximum only for $0 < \lambda < \varepsilon'^2/2$ (when $\partial h(q)/\partial q|_{q=(q_{\text{ext}})_1} < 0$). Hence it follows that, in contrast to the deterministic case, there are two transition points in fluctuating turbulent chaos (within the framework of Verhulst's model). Thus, we can assert the following:

1. If $\lambda < 0$ (with $\alpha > 4/\varepsilon^2$), then the zero stationary point is stable.
2. At $\lambda = 0$ (when $\alpha = 4/\varepsilon^2$) the transition occurs, because for $\lambda > 0$ the stationary point $(q_{\text{ext}})_1 = 0$ becomes unstable and a new truly stationary probability density emerges.
3. The stationary probability density becomes infinite at zero if $0 < \lambda < \varepsilon'^2/2$ (or $4/\varepsilon^2 > \alpha > 4/3\varepsilon^2$), i.e., it retains some of the properties of the δ-function.
4. If λ is greater than $\varepsilon'^2/2$ ($\alpha < 4/3\varepsilon^2$), then the pattern of the density distribution again changes sharply, i.e., $\lambda = \varepsilon'^2/2$ ($\alpha = 4/3\varepsilon^2$)—the second transition point.

Thus, the transition in the system can be caused by maintaining the average state of fluctuating chaos but increasing or reducing the noise intensity in it.

In conclusion, let us briefly summarize the main results of this section. The introduction of a set of stochastic internal parameters adequately characterizing the internal structure and temporal evolution of a turbulized fluid (including the ensemble of mesoscale CSs) within a physically infinitesimal volume into the thermodynamic description of the subsystem of turbulent chaos made it possible to obtain the FPK equations for the probability density functions of the statistical characteristics of vortex structures in the space of internal coordinates by the methods of stochastic nonequilibrium thermodynamics. These equations are designed to analyze the Markovian diffusion processes of the transition from one stationary-nonequilibrium state in the space of internal coordinates to another through a successive loss of stability as the supercriticality parameter increases and to ascertain how the probability density actually appears in the neighborhood of critical points.

At the same time, an alternative approach to investigating such bifurcation transitions based on stochastic Langevin equations statistically equivalent to the kinetic FPK equations is possible. Based on this approach, we demonstrated that flow self-organization is possible in principle when the generation of coherent structures related to the effect of second-order "phase transitions" induced by the natural noise of fluctuating turbulent chaos is probable during the temporal evolution of a quasi-equilibrium vortex subsystem. We showed that if the noise of chaos is multiplicative and if its intensity is large enough, then the extrema of the probability density characterizing the stationary behavior of a stochastic system differ significantly in both number and position from the stationary states corresponding to deterministic chaos. This implies that the multiplicative noise of turbulent chaos not only has a disorganizing effect on the system of defining internal parameters but also can both stabilize the unstable macroscopic states and cause new nonequilibrium phase transitions unpredictable in the deterministic approach.

6.2 Generation of Structured Turbulence Through Phase Synchronization

Let us now analyze the existing tendency for the vortex structure of a real turbulized flow to become more complicated and try to describe some of its underlying mechanisms. In Sect. 5.1, we thermodynamically established that self-organization, along with the formation and complication of spatiotemporal vortex structures in an initially symmetric dissipative medium, is also inherent in a nonlinear open subsystem of turbulent chaos. In particular, stable coherent structures in the subsystem of chaos can result from a continuous redistribution of the turbulent energy coming into it from outside (i.e., from the subsystem of averaged motion) both through the dissipation and turbulent diffusion flux of the averaged kinetic energy of turbulent fluctuations and through the generation of small-scale turbulence energy under the action of buoyancy forces and nonlinear interaction of various small-scale vortex structures inside this subsystem itself. At the same time, unfortunately, the thermodynamic approach does not allow any physical mechanisms for the enlargement of dissipative CSs in the vortex continuum to be found and described quantitatively. Invoking additional considerations, using, for example, the so-called basic mathematical models of the self-organization theory (see Klimontovich 1990), is required to find such mechanisms. The simplest model of such a kind can be obtained by considering turbulent chaos as an active (excitable) medium modeled, in turn, by a distributed system of interacting generalized Van der Pol oscillators in which an enormous number of oscillatory modes can be immediately excited following the forced emergence from rest after some cycle of transformations, which is typical of developed turbulence.

As has been mentioned in Chap. 1, four "scenarios" for the turbulization of an incompressible fluid flow have been developed at present (see, e.g., Landau and Lifshitz 1988). Three of them refer entirely to the initial stage of turbulence generation when the number of excited macroscopic (collective) degrees of freedom is still small. In the preceding chapter, we adhered mainly to the non-obsolete proposition of the Landau–Hopf theory (developed more than five decades ago) that the generation of turbulence is determined by the formation of a chain of consecutive instabilities each of which is accompanied by the appearance of a new flow velocity oscillation frequency. The formation of developed turbulence in this approach can be interpreted as the excitation of a very large number of collective degrees of freedom, with the set of fundamental oscillation frequencies being replenished by a newly formed incommensurate frequency after each normal Andronov–Hopf bifurcation. Accordingly, the solution of the Navier–Stokes equations passes (in the phase space) to a new torus whose topological dimension coincides with the number of independent frequencies; the fluid flow becomes even more irregular and, when an enormous set of modes is excited, turbulent (which corresponds to the motion of the phase point over a multidimensional torus—a quasi-periodic attractor). Thus, according to the Landau–Hopf theory, the transition to fully developed turbulence is executed through an extremely large cascade of bifurcations associated with the successive excitation of irrationally coupled frequencies of oscillations in thermohydrodynamic parameters, i.e., the turbulent motion in this scenario is quasi-periodic and, in general, is characterized by a discrete spectrum of oscillation frequencies. For all that, it is important to clearly realize that the Landau–Hopf theory for the formation of hydrodynamic turbulence turns out to be valid only in part, because, in reality, developed turbulence has a continuous oscillation spectrum, with the transition from the discrete spectrum to the continuous one being abrupt. In addition, such multifrequency oscillations, even in the presence of fluctuations, are nevertheless not chaotic, because the Landau–Hopf scenario does not suggest the development of instability and mixing in the system. Actually, because of this inconsistency with the true state of affairs, it was shown in a number of publications (collected, e.g., in Strange Attractors 1981) that an attractor in the form of an open winding on the multidimensional torus (which is the image of turbulence in the Landau–Hopf theory) for dissipative dynamical systems of any nature is structurally unstable (i.e., it is destroyed when the system's parameters are slightly changed): as a rule, the chain of normal Andronov–Hopf bifurcations breaks already after the first three steps. In this case, either the open winding on the three-dimensional torus is closed, turning into a limit cycle (mutual synchronization of modes), or the torus is destroyed to abruptly produce a strange attractor—a manifold with a complex topology on which true dynamical chaos with a continuum of frequencies, a sensitive dependence on initial conditions, mixing, etc. is realized (see Ruelle and Takkens 1971).

At the same time, the theory of strange attractors (Rabinovich 1978; Landa 1983; Rabinovich and Trubetskov 1984) need not be invoked to explain the abrupt generation of structured turbulence. The point is that the interaction of excited oscillatory modes in the case of turbulence is similar to the interaction of oscillators with hard excitation having initially different eigenfrequencies of oscillations (or rotations). It is well known from the theory of oscillations (see, e.g., Bogolyubov and Mitropolsky 1963; Pikovsky et al. 2001) that an excited oscillator acting on the remaining underexcited ones can excite self-oscillations in them at the eigenfrequencies (analogous to the phenomenon of "asynchronous excitation"). It is in this way that many oscillatory modes can be immediately excited in a stochastic ensemble of mutually coupled Van der Pol oscillators, which is typical of developed turbulence (a "gas" of self-oscillatory modes).

In this section, we use such considerations for the synergetic interpretation of possible turbulence structuring mechanisms. Within the framework of the model being developed, this implies the existence of a certain mechanism for the enlargement of small-scale vortex structures during the temporal evolution (here, for simplicity, we also restrict our analysis only to temporal chaos, i.e., assume the vortex motions to be completely identical at all points of space) of a quasi-equilibrium subsystem of turbulent chaos. We mean the mechanism associated with the mutual phase synchronization (coherence) of some set of small-scale oscillatory modes with close frequencies (stochastic self-oscillations of internal coordinates), i.e., with the tendency for frequency clusters (or, more precisely, intermediate cluster structures) to be formed, for which the distribution of individual oscillation frequencies can have several maxima. The formation of frequency clusters is characteristic of active media consisting of oscillators not only with a periodic behavior but also with a chaotic one, as suggested by the results from Kuramoto (1984), Pikovsky et al. (1996), and Osipov et al. (1997). The ensemble of intermediate cluster structures emerged in such a way makes a nonzero contribution to the so-called effective (amplitude-phase) mean field of chaos oscillating at some characteristic frequency. The mean field arises from the interaction in an ensemble of globally coupled oscillators (see Kuramoto 1984). In turn, this effective field acting on each vortex cluster in the ensemble as a certain external force (which, depending on the model parameters, can generally lock its frequency) causes a growth of the individual synchronous cluster and a further growth of the mean field component at a certain frequency. As a result of this scenario (known as Kuramoto's transition), the increasing component of the driving force induces subsequent (additional) synchronization in the ensemble and, thus, contributes to the formation of several mesoscale vortex clusters (corresponding to the coherent component of turbulent chaos) that can eventually both merge together and coexist together. It is important to realize that the specificity of such synchronization of self-oscillations also lies in the fact that it occurs in the presence of natural fluctuating noise (associated, in particular, with the "thermal" structure of the incoherent component of turbulent chaos). The latter tends to uniformly distribute the oscillation phases in the ensemble, thereby reducing the efficiency of the amplitude-phase mean field of chaos. It is quite admissible that the clash of these opposite tendencies for a real turbulized flow can give rise to mesoscale coherent vortex structures under certain conditions (dependent, naturally, on the characteristics of the medium and the critical values of control parameters). Thus, the mutual phase synchronization of some set of chaotic self-oscillations may be considered as a possible mechanism for the self-organization of interacting vortex structures in an open subsystem of turbulent chaos.

In Sect. 6.1, we focused our attention on analyzing the transition mechanisms from one stable stationary-nonequilibrium state to another associated with the successive loss of stability by the subsystem of turbulent chaos far from complete chaos of the thermodynamic equilibrium. Such transitions were described there as nonequilibrium second-order phase transitions in a vortex continuum; as a result, the internal coordinates at bifurcation points change abruptly. At the same time, as the supercriticality grows, stationary states can generally give way not only to stationary ones but also to periodic, quasi-periodic, and nonperiodic (in time) motions with which the great variety of spatiotemporal structures in a fluid flow is ultimately associated. In this section, we discuss the general concept of synergetic generation of mesoscale CSs from turbulent chaos associated with the phase-frequency synchronization of the self-oscillations of those internal coordinates that refer to the coherent component of chaos. We also consider some of the scenarios for the dynamical influence of the incoherent component (fine-grained fluctuating field) of turbulent chaos on the evolution of mesoscale vortex structures.

Negative viscosity

It is quite obvious that the generation of mesoscale CSs from turbulent chaos is eventually related to the negativity of the turbulent viscosity (see Starr 1968) characterizing not the physical properties of natural media but the statistical properties of their turbulent motions. Whereas a positive turbulent viscosity is usually accompanied by the conversion of the kinetic energy of a large-scale motion into the energy of small-scale turbulent fluctuations, the direction of the energy transport is reversed in the presence of a negative viscosity: in this case, larger vortex structures in a turbulent flow (e.g., an ensemble of mesoscale CSs) gain kinetic energy from small-scale irregular perturbations. The decaying small-scale turbulence (unless it is related to small-scale vortices of the incoming flow) must be maintained by some internal processes, for example, by the systematic conversion of heat into kinetic energy within individual small-scale vortices (much as is the case, for example, for cyclones in the terrestrial atmosphere gaining energy through local heat influxes). In addition, the inverse cascade of energy transport can be affected by the anisotropy and helicity of small-scale turbulence and by weak compressibility of the medium (see, e.g., Rabinovich and Sushchik 1990). In the approach to modeling structured turbulence in thermodynamically active media we develop, the realization of quasi-stationary flows with a negative viscosity was shown to be possible in Kolesnichenko (2002; see also Chap. 8). For example, the differential rotations of the disks in spiral galaxies and circumstellar disks (in particular, the rotation of the solar protoplanetary cloud), the differential rotation of the solar photosphere and the differential rotation of a planetary atmosphere (which, e.g., for Jupiter was discovered by the motion of its spots—vortex structures in local motions of the Jovian atmosphere), etc. are real natural media with a negative viscosity. This phenomenon is very important for quantitatively explaining some features in the general circulation of the Earth's atmosphere (e.g., the maintenance of global jet streams) and for explaining the stream nature of oceanic currents like the Gulf Stream and Kuroshio.

6.2.1 *Stable Limit Cycles and the Related Synchronization of Periodic Self-Oscillations (Phase Dynamics Approximation)*

Recall now some of the characteristic properties of the synchronization phenomenon that is useful to us below. Assuming that some active medium consisting of a large number of globally coupled identical Van der Pol oscillators models the state of the coherent component of vortex chaos, let us consider the forced and mutual synchronization of these oscillators, including the influence of noise on the synchronization effect. The Van der Pol synchronizer, as a rule, serves as a basic model for investigating the synchronization effects observed both under the action of an external periodic force (forced synchronization) and during the interaction of two coupled oscillators (mutual synchronization).

Let us first consider the simplest case of the forced synchronization of stable periodic self-oscillations $\boldsymbol{q}^{c}(t) = \boldsymbol{q}^{c}(t+\tau)$ of an oscillator by a periodic force $\alpha K_1(\boldsymbol{q},t) = \alpha K_1(\boldsymbol{q},t+\tau^{B})$, as a result of which the autonomous oscillator frequency ω^{c} can adjust to the frequency $\omega^{B} = 2\pi/\tau^{B}$ of the external force (oscillation phase locking). In our description of structured turbulence, this case can bear a relation, for example, to the generation of a vortex CS under the action of an external periodic force (e.g., an external sine wave). As has already been repeatedly emphasized, the mathematical images of such vortex structures in the space of configurations are orbitally asymptotically stable limit cycles—isolated trajectories that have a finite length and when moving along which the system returns to the same point in a finite time.

Below, for simplicity, we restrict our analysis to a weak perturbing force [see (5.71)]

$$\alpha K_1(\boldsymbol{q},t) \equiv -L(\boldsymbol{q}) \cdot \partial U(\boldsymbol{q},t)/\partial \boldsymbol{q}$$

proportional to a small parameters α that below is interpreted as the force amplitude. A general approach to studying the action of external periodic forces on nonlinear dynamical systems with slowly changing parameters is presented, for example, in the monograph by Bogolyubov and Mitropolsky (1963). Under the influence of a small force αK_1, the trajectory $\boldsymbol{q}(t)$ is expelled from the limit cycle $\boldsymbol{q}^{c}(t)$, but this trajectory nevertheless remains in its small neighborhood because of the cycle stability. In other words, the perturbations in the direction transverse to the cycle (oscillation amplitude perturbations) remain small, while the phase perturbations can be large. That is why the description in this case can be reduced only to phase dynamics.

Thus, let us first assume that the evolution of the internal coordinate $\boldsymbol{q}(t)$ of turbulent chaos is described by the deterministic equation (5.71), which we rewrite here as

$$\frac{\partial q_m}{\partial t} = K_m(\boldsymbol{q},0) + \alpha K_{m1}(\boldsymbol{q},t), \quad (m = 1,2,\ldots) \tag{6.49}$$

Let the unperturbed ($\alpha = 0$) equation (6.49) have a periodic trajectory $\boldsymbol{q}^c(t)$ with a period τ^c. It is well known (see, e.g., Bogolyubov and Mitropolsky 1963; Pikovsky et al. 2001) that the standard definition $\partial\phi(t)/\partial t = \omega^c$ of the phase on the limit cycle $\boldsymbol{q}^c$ can also be generalized to trajectories close to this limit cycle (here, $\boldsymbol{q} = \boldsymbol{q}^c(\phi)$ is the parametrically specified orbit of the limit cycle; ϕ is the coordinate along the cycle increasing by 2π after each turn). For this purpose, one usually chooses a thin tube Σ in its neighborhood and defines such a scalar field $\phi(\boldsymbol{q})$ in it that

$$\frac{\partial\phi(\boldsymbol{q})}{\partial t} = \sum_m \frac{\partial\phi}{\partial q_m}\frac{\partial \boldsymbol{q}_m}{\partial t} \cong \sum_m \frac{\partial\phi}{\partial q_m} K_m(\boldsymbol{q},0) = \omega^c. \tag{6.50}$$

Let us now consider the perturbed ($\alpha \neq 0$) system (6.49) and, using the unperturbed definition of the phase, substitute (6.49) into (6.50). As a result, we obtain the following basic equation to describe the phase dynamics of self-oscillations in the presence of a small periodic external force, to within terms proportional to α^2:

$$\frac{\partial\phi}{\partial t} = \sum_m \frac{\partial\phi}{\partial q_m}\frac{\partial q_m}{\partial t} = \sum_m \frac{\partial\phi}{\partial q_m}(K_m(\boldsymbol{q},0) + \alpha K_{1m}(\boldsymbol{q},t)) \cong \omega^c + \alpha f(\phi,t), \tag{6.51}$$

where the function

$$f(\phi,t) \equiv \sum_m \frac{\partial\phi(\boldsymbol{q}^c(\phi))}{\partial q_m} K_{1m}(\boldsymbol{q}^c(\phi),t)$$

calculated on the limit cycle is uniquely related to the phase ϕ.

6.2.1.1 Slow Phase Dynamics

Since $f(\phi,t)$ is a 2π-periodic function of ϕ and a periodic function with respect to t with a period τ^B, it can be expanded into a double Fourier series, $f(\phi,t) = \sum_{l,k} p_{l,k} \exp(ik\phi + il\omega^B t)$. In the absence of any perturbation (at $\alpha = 0$), the solution of (6.51) is $\phi(t) = \omega^c t + \phi^c$, i.e., the phase rotates uniformly with an autonomous unperturbed frequency ω^c. The expansion of the function $f(\phi,t)$ into a series in the first approximation in α

$$f(\phi,t) = \sum_{l,k} p_{l,k} \exp(ik\phi^c) \exp i(k\omega^c + l\omega^B)t \tag{6.52}$$

then contains rapidly oscillating (compared to the time scale $1/\alpha$) terms, except the terms satisfying the resonance condition $k\omega^c + l\omega^B \approx 0$. The rapidly oscillating terms lead to phase deviations of order $O(\alpha)$, while the resonance terms in series

(6.52) can lead to large (though slow due to the smallness of α) phase changes and, for this reason, turn out to have the most significant influence on the ultimate pattern of oscillations.

It can be shown (see Bogolyubov and Mitropolsky 1963) that only the resonances for which the frequencies are related by the relation $\omega^{\mathrm{B}} \approx (m/n)\omega^{\mathrm{c}}$, where m and n are (usually small) coprime integers, manifest themselves in the first approximation. The following resonance cases are distinguished: (1) $m = n = 1$, i.e., $\omega^{\mathrm{c}} \approx \omega^{\mathrm{B}}$; this case corresponds to an ordinary (main) resonance; (2) $n = 1$, i.e., $\omega^{\mathrm{c}} \approx \omega^{\mathrm{B}}/m$; this case corresponds to a resonance at the eigenfrequency overtone (parametric resonance); (3) $m = 1$, i.e., $\omega^{\mathrm{c}} \approx \omega^{\mathrm{B}} n$; in this case, there is a resonance at the external frequency overtone. Thus, all terms in sum (6.52) with the indices $k = mj$ and $l = -nj$ are the resonance ones and, when this sum is averaged over a sufficiently long time interval, they will make a noticeable contribution to the averaged force. As a result, we can obtain the following averaged equation for the phase:

$$\frac{\partial \phi}{\partial t} = \omega^{\mathrm{c}} - \alpha\,\hat{q}(m\phi - n\omega^{\mathrm{B}} t), \tag{6.53}$$

where

$$\begin{aligned} f(\phi, t) &= \sum_{l=-nj,\,k=mj} p_{l,k} \exp(ik\phi + il\omega^{\mathrm{B}}\, t) = \sum_j p_{-nj,mj} \exp\{ij(m\phi - n\omega^{\mathrm{B}}\, t)\} \\ &\equiv -\hat{q}(m\phi - n\omega^{\mathrm{B}}\, t) \end{aligned}$$

is the averaged force, which is a 2π-periodic function of the argument. This function takes into account all features of the limit cycle in the autonomous system (5.71) and all features of the external force.

Finally, introducing the phase difference as $\Psi(t) = m\phi(t) - n\omega^{\mathrm{B}} t$ (the parameter Ψ can be treated as a slow phase in a rotating frame of reference), from (6.53) we obtain the so-called reduced equation

$$\frac{\partial \Psi}{\partial t} = \Delta - \alpha\, q(\Psi), \tag{6.54}$$

where the force $q(\Psi) \equiv m\,\hat{q}(\Psi)$ is a 2π-periodic function of Ψ;

$$\Delta \equiv m\omega^{\mathrm{c}} - n\omega^{\mathrm{B}}$$

is the so-called detuning between the autonomous oscillator frequency and the external signal frequency. When deriving (6.53), we assumed the detuning to be small—of the order of α. Below, $\sin\Psi$, the simplest 2π-periodic function, is considered as the function $q(\Psi)$ (again for simplicity); in this case, (6.54) acquires the simplest form of an averaged phase equation coincident with the well-known Adler equation:

$$\frac{\partial \Psi}{\partial t} = \Delta - \alpha \sin \Psi .$$

In particular, the case of $q(\Psi) = \sin \Psi$ is valid for the coupled Van der Pol equations.

Equation (6.54) can be rewritten in the following potential form:

$$\frac{\partial \Psi}{\partial t} = \frac{-\partial \Phi(\Psi)}{\partial \Psi},$$

where

$$\Phi(\Psi) = -\Delta \Psi + \alpha \int^{\Psi} q(x) \mathrm{d}x$$

is the potential function. Thus, the dynamics of the phase difference Ψ may be considered as the motion of an overdamped particle in a given potential. The detuning Δ defines the slope of the potential, while the parameter α specifies the height of the potential barriers. Let $q_{\min}$ and $q_{\max}$ be, respectively, the minimum and maximum of the function $q(\Psi)$ in the interval $[0, 2\pi)$. Then, there exists at least one pair of fixed points of (6.54) for $-\gamma\, q_{\min} < \Delta < \gamma\, q_{\max}$ (the synchronization region), i.e., a pair of stationary states of the phase Ψ takes place. It is easy to see that one of these points is (asymptotically) stable, while the other one is unstable (in the general case where the function $q(\Psi)$ in the interval $[0, 2\pi)$ has more than two extrema, there can be several pairs of stable and unstable points). Therefore, the system evolves with time to the stable fixed point and remains at it, so that the rotating phase is constant, $\Psi = \Psi^{\mathrm{c}}$. For the initial phase ϕ, this implies rotation with the frequency of the external force $\phi = \omega^{\mathrm{B}} t + \Psi^{\mathrm{c}}$ and this is precisely the synchronization regime (the self-oscillation phase ϕ follows the external force $\omega^{\mathrm{B}}\, t$). In particular, for the main resonance when $m = n = 1$ in the above formulas and a zero detuning Δ (when $\omega^{\mathrm{c}} \approx \omega^{\mathrm{B}}$), the stable value of the phase difference is zero, $\phi = \omega^{\mathrm{B}} t$. The phases in some finite domain of parameters for the system can be said to attract to each other (oscillation frequency and phase locking). In a more general case, frequency locking implies a rational ratio of two initially independent frequencies, $\omega_1 : \omega_2 = m : n$, everywhere in the synchronization region. In the configuration space of turbulent chaos, the induced CS $\boldsymbol{q}(t)$ with the period of the external action $2\pi/\omega^{\mathrm{B}}$ corresponds to this.

It should be emphasized that the averaged equation for the phase difference $\Psi = m \phi_1 - n\phi_2$ describing the mutual synchronization of two interacting self-oscillating clusters with different eigenfrequencies in the case of a small coupling parameter α between them has exactly the same form as (6.54). For our problem, the coalescence of two localized CSs (interacting, for example, according to the Biot-Savart law) in a fully developed turbulent shear flow into a single CS with a larger energy content due to the mutual synchronization of oscillations can be described in a similar way.

6.2.1.2 Phase Dynamics Approximation in the Presence of Noise

The presence of noise leads to diffusion of the phase difference Ψ in the potential $\Phi(\Psi)$, thereby destroying the periodicity: the phase $\Psi(t)$ fluctuates near the minima of the potential $\Psi_{\min,k}$ and makes random transitions from one potential well to another, changing abruptly by 2π. The larger the detuning, the smaller the amplitude of the synchronizing action (the shallower the wells), and the higher the noise intensity, the shorter the time during which the oscillator and action phases remain locked. Nevertheless, one may expect the tendency for synchronization to triumph at some critical amplitude of the external force α_{crit}, i.e., phase locking will also take place. It is important to note that the phase dynamics approximation can also be used in this case, because no force regularity was assumed above when deriving (6.51).

To quantitatively describe the synchronization in the presence of fluctuations on limit cycles, it is necessary to proceed from the stochastic Langevin equation (6.1) in which the random force is $\tilde{f}_k(\boldsymbol{q},t) \equiv \sqrt{k_{\text{B}} T_{\text{turb}} Q}\,\xi_k(t), \quad (k = \overline{1,\ n})$ — normal white noises with zero means and δ-like correlation functions. For simplicity, we assume the components of the random force to be spatially homogeneous ($Q_{km}(\boldsymbol{q}) = Q\delta_{km}$) and independent for different oscillators; then, $\overline{\xi_k}^0 = 0$ and $\overline{\xi_k(t)\xi_m(t_1)}^0 = \delta(t - t_1)\delta_{km}$. Repeating the procedure for deriving (6.51) and (6.54), we obtain the following phase equation in the form of a Langevin equation from (6.1):

$$\begin{aligned}\frac{\partial \phi}{\partial t} &= \sum_m \frac{\partial \phi}{\partial q_m}\left(K_m(\boldsymbol{q},0) + \alpha K_{1m}(\boldsymbol{q},t) + \tilde{f}_m(\boldsymbol{q},t)\right)\\ &\cong \omega^{\text{c}} + \alpha f(\phi,t) + \sqrt{k_{\text{B}} T_{\text{turb}} Q}\,\xi(\phi,t),\end{aligned} \tag{6.55}$$

where $\xi(\phi,t) \equiv \sum_m \frac{\partial \phi(\boldsymbol{q}^{\text{c}}(\phi))}{\partial q_m}\xi_k(t)$ is a random function of time. In the simplest case, the stochastic term in (6.55) does not depend on the phase at all, so that $\xi(t)$ is a stationary random process. The main properties of the phase dynamics in the presence of noise can then be described either by the stochastic equation for the phase,

$$\frac{\partial \Psi}{\partial t} = \Delta - \alpha\, q(\Psi) + \sqrt{k_{\text{B}} T_{\text{turb}} Q}\,\xi(t) \tag{6.56}$$

or by the corresponding FPK equation for the phase probability density $P_1(\Psi,t)$:

$$\frac{\partial P_1(\Psi,t)}{\partial t} = -\frac{\partial}{\partial \Psi}[(\Delta - \alpha q(\Psi))P_1(\Psi,t)] + k_B T_{\text{turb}} \frac{Q}{2} \frac{\partial^2 P_1(\Psi,t)}{\partial \Psi^2}. \quad (6.57)$$

It should be noted that the synchronization of various self-oscillating systems is among the well-studied phenomena. The synchronization of self-oscillatory modes by a periodic signal is a classical problem of the theory of nonlinear oscillations. Weakly nonlinear systems (in the sense that the self-oscillations are close to linear oscillations) are described in detail in the literature (see, e.g., Bogolyubov and Mitropolsky 1963; Rabinovich and Trubetskov 1984). The phase dynamics approximation considered above is based on the assumption that the amplitude of the external force is small. For a small oscillation amplitude, the synchronization properties can also be investigated analytically. The existence of a small parameter (the ratio of the oscillation period to the characteristic time of change in the amplitude) allows the phenomenon to be described in this case using the reduced system equations that, apart from (6.53), also includes the oscillation-period-averaged amplitude equations. The synchronization mechanism both at small and large amplitudes of the external force, which is more complex, was well studied in terms of this more general approach (see, e.g., Bogolyubov and Mitropolsky 1963; Pikovsky et al. 2001). In recent years, interest in studying the influence of various noises on the synchronization processes, i.e., in the statistical dynamics of synchronization, has increased (see, e.g., Pikovsky et al. 1996; Anishchenko et al. 2003).

6.2.2 Mechanism for the Formation of Mesoscale Coherent Structures (Clusters) in the Subsystem of Turbulent Chaos

Let us now examine whether an ensemble of mesoscale CSs can be formed in principle through the mutual synchronization of some set of intermediate cluster structures in the subsystem of turbulent chaos and investigate the influence of the internal noise of chaos on the regime of phase synchronization. According to the Landau–Hopf theory, an enormous number of oscillatory modes N_{turb} (macroscopic degrees of freedom) are excited in a developed turbulent flow, with their number depending on the Reynolds number Re as $N_{\text{turb}} \sim Re^{9/4}$(see Landau and Lifshitz 1988). In reality, however, the chaotic dynamics of a flow is determined only by a small number of independent small-scale excitations; the remaining ones are coupled with them rigidly, so that small-scale vortex clusters $\boldsymbol{q}^{\text{c}}(t)$ (intermediate cluster structures of synchronized modes) consisting of groups of elements with strictly identical or close mean frequencies $\omega^{\text{c}} = 2\pi/\tau^{\text{c}}$ are formed as a result of the interaction between oscillatory modes. In other words, the dimension of turbulent chaos correlates with the number of interacting small-scale cluster structures. Such vortex structures during their interaction in a large ensemble can produce a host of complex nonlinear effects—such as, for example, the formation of mesoscale CSs (for the two-level model under consideration, these vortex structures will have characteristic sizes l of the order of or smaller than the averaging scale, $l \leq \Lambda$, $dr \sim \Lambda^3$), the quenching of oscillations (the so-called oscillation death), the emergence of regimes of frozen states, etc. (see, e.g., Anishchenko et al. 2003). All these effects are directly related to the manifestation of phase synchronization properties for an ensemble of oscillating autostructures.

To construct a model theory of such a regime in a vortex continuum, we proceed from the experimental fact that the intensities of the small-scale fluctuations in thermohydrodynamic parameters for a turbulized flow have maxima within some spectral energy range (Monin and Yaglom 1996). The relatively large autostructures should then have the highest energy among the small-scale dissipative vortex structures appearing under the action of hydrodynamic nonlinearity. Such autostructures should be fed most intensely through the beats between small-scale fluctuations. We associate the initial state of the coherent component of the subsystem of turbulent spatiotemporal chaos precisely with these vortex structures having the largest energy content in the $\boldsymbol{q}$-space.

Thus, we assume below that as the supercriticality grows, the system of dynamical equations (6.49) has M stable limit cycles $\boldsymbol{q}_k^{\mathrm{c}}(t)$ ($k = \overline{1, M}$) with which M small-scale CSs (intermediate cluster structures) in the volume $\mathrm{d}\boldsymbol{r}$ of the subsystem of turbulent chaos are associated. Obviously, the individual autostructures in such an ensemble of small-scale clusters that was produced by a successive cascade of a large number of spatial and temporal bifurcations and that is under the constant action of quasi-homogeneous background noise of the incoherent component of turbulized chaos can initially have completely different partial frequencies of natural oscillations (rotations) ω_k^{c}.

We analyze the solutions of the deterministic problem (6.49) in a small neighborhood of a periodic trajectory $\boldsymbol{q}_k^{\mathrm{c}}(t)$ in the so-called normal coordinates (see, e.g., Shilnikov et al. 2004), when system (6.49) is reduced to a system of reduced equations for the amplitudes and phases. Although the bifurcation scenario generally depends on the behavior of the instantaneous oscillation amplitude, the phase synchronization effect in the presence of fluctuations in many cases is described satisfactorily in terms of phase dynamics, to which we restrict ourselves below. This approach allows the frequency-phase locking phenomena to be described qualitatively and, in a number of cases, makes an analytical solution of the problem possible (see Bogolyubov and Mitropolsky 1963; Tikhonov 1986; Anishchenko et al. 2003; Pikovsky et al. 2001).

6.2.2.1 Medium of Globally Coupled Oscillators

Consider the next basic mathematical model for the generation of mesoscale CSs in a physically small volume $\mathrm{d}\boldsymbol{r}$ of a vortex continuum. Suppose that some active media consisting of a large number of interacting oscillatory elements, for example, M globally coupled identical Van der Pol oscillators, models the state of the coherent component of chaos in the volume $\mathrm{d}\boldsymbol{r}$. The most complete statistical description of this medium can be given based on the kinetic equation for the M-particle distribution function $f_M(X_1,\ldots X_M,t)$, where X_k is the set of variables characterizing the internal motion of the k-th oscillator and its motion as a whole (Klimontovich 1995). To simplify our analysis, we assume that the interaction of oscillators (with a common power supply and a common feedback) is homogeneous, i.e., all pairs of oscillators are coupled identically. In the case where the interaction is completely absent, a multifrequency quasi-periodic regime is observed in the ensemble of M oscillators. In the other extreme case where the coupling between oscillators is strong, the difference between the autonomous frequencies may be neglected, because all oscillators are synchronized. The appearance of partially synchronized regimes is potentially admissible between these extreme (idealized) cases, because the couplings tend to synchronize the nearest neighbors (N oscillators). As a result, mesoscale clusters of synchronized oscillators to which the mesoscale CSs in the subsystem of turbulent chaos correspond are formed. The transition from independent small-scale oscillations to a partially synchronized regime depends significantly on the initial distribution of frequencies ω_k. Below, we assume that the oscillator eigenfrequencies ω_k (for $M \to \infty$) are specified by a distribution function $p(\omega)$ that is symmetric with respect to the unique maximum at frequency ω_0 in the absence of any interaction, e.g., by the Lorentz distribution:

$$p(\omega) = \left[\frac{d}{\pi[(\omega-\omega_0)^2+d^2]}\right].$$

The force interaction between such mesoscale clusters of synchronized oscillators, in turn, can lead to subsequent phase synchronization and, thus, to the formation of mesoscale phase clusters with an even larger energy content (corresponding to the relatively large mesoscale CSs in the subsystem of turbulent chaos) and, along with them, to an enhancement of some mean field (see below) with amplitude A and phase Θ. Such a field acts on each oscillating cluster as a certain periodic external force that, depending on the ensemble parameters (in particular, on the coupling α between autostructures), will or will not be able to lock its frequency. On the other hand, the distribution of phases in the ensemble of small-scale clusters under the influence of noise (corresponding to the internal noise of the incoherent component of turbulent chaos) tends to a uniform one, which can lead to complete disappearance of the mean field. Clearly, the tendency for the absence of coherence will win in the case of strong noise. However, one may expect the force interaction and noise to balance each other at some critical coupling α_{crit} in such a way that the synchronization of small-scale clusters will still arise—the phenomenon with which we associate the mechanism for the generation of mesoscale CSs (the coherent component of turbulent chaos).

Let us now confirm this qualitative picture analytically for the simplest case of a model based on the phase approximation. It should be noted that the phase description of weakly coupled oscillators is well known (see, e.g., Kuramoto 1984) and is used to construct a wide variety of models. Our objective is to use this approach to explain the formation of mesoscale coherent structures in developed turbulized fluid flows.

6.2.3 Phase Dynamics Equations

Let us first analyze the phase synchronization of intermediate small-scale clusters (synchronized vortices) in the absence of any fluctuation effects of internal noise (i.e., for some idealization of "real processes") whereby the internal coordinates $\boldsymbol{q}(t)$ (related, e.g., to coherent vorticity) satisfy the system of autonomous first-order (in time) ordinary differential equations (6.49). We assume that a certain ensemble of N $(N \ll M)$ coherent small-scale vortex structures $\boldsymbol{q}_k^{\mathrm{c}}(t)$ $(k = \overline{1, N})$ with N different averaged partial frequencies ω_k^{c} is contained in a physically infinitesimal volume d$\boldsymbol{r}$ of a vortex continuum at some Reynolds number whose synchronization gives rise to mesoscale CSs. An ensemble of N oscillators capable of being synchronized corresponds to this ensemble of N similar (in properties) weakly coupled vortex clusters in the model approach. Assuming that all pairs of oscillators interact identically, we write the dynamical equations in the phase approximation using the analogy with (6.53). As a result, we obtain the system of reduced equations

$$\frac{\partial \phi_k}{\partial t} = \omega_k^{\mathrm{c}} + \frac{\alpha}{N} \sum_{j=1}^{N} \hat{q}(\phi_j - \phi_k), \quad (k = 1, 2, \ldots, N), \tag{6.58}$$

where ϕ_k and ω_k^{c} are, respectively, the phase and mean frequency of the oscillations of oscillator k. The parameter α characterizes the coupling force $\hat{q}(\phi_j - \phi_k)$ between each pair of oscillators that is chosen here to be proportional to N^{-1}. Only in this case do we obtain the results independent of N in the thermodynamic limit $N \to \infty$ used below for mathematical simplicity of the ultimate formulas.

The coupling function $\hat{q}(\phi_j - \phi_k)$, which generally depends on the CS topology, can be represented by a Fourier series,

$$\hat{q}(\phi_j - \phi_k) = \sum\nolimits_l \hat{q}_l \exp\{i2\pi l(\phi_j - \phi_k)\}.$$

Assuming that the phases of all synchronous oscillators will eventually rotate with frequency ω_0, let us introduce the so-called order parameter

$$Z_l = \frac{1}{N} \sum_{j=1}^{N} \exp\{i2\pi l(\phi_j - \omega_0 t)\}$$

and rewrite (6.58) as

$$\frac{\partial \phi_k}{\partial t} = \omega_k^c - \alpha H(\phi_k - \omega_0 t), \quad (k = 1, 2, \ldots, N), \tag{6.59}$$

where

$$H(\phi_k - \omega_0 t) = -\sum_l \hat{q}_l Z_l \exp\{-i2\pi l(\phi_k - \omega_0 t)\}$$

is the so-called mean field (force) acting on each oscillator (order function). A nonzero mean field points to the possibility of synchronization in the ensemble of oscillators.

Equations (6.59) can be concretized if we specify the interaction force between oscillators (corresponding to the interaction of excitations on neighboring vortex structures). Below, we consider only the simplest force $\hat{q}$ proportional to the sine of the phase difference, $\hat{q}(\phi_j - \phi_k) = \sin(\phi_j - \phi_k)$. Although this assumption is to some extent artificial, it accurately conveys typical features of the real dependences that arise between the properties of active systems with multiple limit cycles (in particular, for coupled Van der Pol oscillators). In this case, the equations of motion (6.59) can be rewritten as a system of equations for the phases of oscillators under the action of some complex mean field

$$Z = A \exp(i\Theta) = \frac{1}{N} \sum_{k=1}^{N} \exp(i\phi_k)$$

(with amplitude A and phase $\Theta = \omega_0 t$) in the following form (the model by Kuramoto 1984)

$$\frac{\partial \phi_k}{\partial t} = \omega_k^c - \alpha A \sin(\phi_k - \Theta), \quad (k = 1, 2, \ldots, N), \tag{6.60}$$

Since the initial distribution of cluster eigenfrequencies $p(\omega)$ is symmetric, the mean field oscillates at the central frequency ω_0; below, we make sure that this frequency is actually the solution.

It follows from (6.60) that the mean field determines the dynamics of autostructures by acting on the k-th oscillator as some external force that will or will not be able to lock its frequency, depending on the system's parameters. When all self-oscillations are independent ($\alpha = 0$), the state space of system (6.60) breaks up into N planes on each of which there is only one stable limit cycle with a period $2\pi/\omega_k^c$. In the original N-dimensional configuration space, the attraction of phase trajectories to an N-dimensional torus, the product of independent cycles, corresponds to such independent self-oscillations of oscillators. If all ω_k^c are incommensurate, then the phase trajectories on the torus represent a dense, nowhere closing winding—a quasi-periodic motion. When, alternatively, the coupling appears between autostructures ($\alpha \neq 0$), such a simple quasi-periodic motion, in general, can be destroyed: in particular, multimode generation can appear. In particular, the effect of synchronization whereby the oscillators mutually shift the frequencies in such a way that they either will become commensurate or will coincide (given the nonlinear corrections) leads to such ordering of the oscillation spectrum. The mutual synchronization of modes is also possible in wave numbers. The generation of complex spatially ordered structures in non-one-dimensional self-oscillating systems that we do not consider here is associated with the synchronization of modes in wave numbers. Limit cycles appear on the torus instead of a quasi-periodic winding.

When the oscillators are synchronized at some frequency, their oscillations are added coherently and a nonzero mean field emerges. If it is periodic ($A \cong$ const, $\Theta = \omega_0 t$), then each of (6.60) is equivalent to the phase equation for an oscillator under the action of a periodic force. Substituting $\Theta = \omega_0 t$, $A =$ const, and $\Psi_k = \phi_k - \omega_0 t$ into (6.60) yields

$$\frac{\partial \Psi_k}{\partial t} = \Delta_{k0} - \Delta \cdot \sin \Psi_k, \quad (k = 1, 2, \ldots, N), \tag{6.61}$$

where $\Delta_{k0} = \omega_k^c - \omega_0$ and $\Psi_k(t)$ are, respectively, the initial frequency detuning and the phase difference between the oscillations of the oscillator being synchronized and the periodic "external" force Z; $\Delta \equiv \alpha A$ is the synchronization (locking) band.

Equation (6.61) coincides with the Adler equation and can have both synchronous and asynchronous solutions. In particular, from (6.61) in a stationary state we have $\Delta \cdot \sin \Psi_k = \Delta_{k0}$. Consequently, the synchronous solution exists provided that the eigenfrequency of the kth oscillator is close to $\omega_0 : |\Delta_{k0}| \leq \Delta$; the corresponding oscillators can then be trapped by the mean field. If $|\Delta_{k0}| > \Delta$, then there is an asynchronous solution of (6.61) for which the oscillator phases are distributed nonuniformly. Since the sine is a periodic function, the characteristic feature of system (6.61) is that at $|\Delta_{k0}| \leq \Delta$ it has a denumerable number of equilibrium states out of which

$$(\Psi_k)^*_{\pm j} = \arcsin\left(\frac{\Delta_{k0}}{\Delta}\right) \pm 2j\pi \quad (j = 0, 1, 2, \ldots) \tag{6.62}$$

and

$$(\Psi_k)^{**}_{\pm j} = \pi - \arcsin\left(\frac{\Delta_{k0}}{\Delta}\right) \pm 2j\pi \tag{6.62*}$$

correspond to stable and unstable states, respectively.

The complex mean field $Z = A \exp(i\Theta) = N^{-1} \sum_{k=1}^{N} \exp(i\phi_k)$ in (6.60) can be calculated (in the limit $N \to \infty$) just from the probability density of the phase difference $W(\Psi)$ from the groups of synchronized and non-synchronized clusters

$$Z = A\exp(i\Theta) = \int_{-\pi}^{\pi} \exp(i\Psi + i\omega_0 t)W(\Psi)\mathrm{d}\Psi, \tag{6.63}$$

where $W(\Psi, t) = W_s(\Psi) + W_{as}(\Psi)$ is the sum of the synchronous, $W_s(\Psi)$, and asynchronous, $W_{as}(\Psi)$, components. Since the phase difference for synchronized clusters does not depend on time, the distribution $W_s(\Psi)$ can be derived from the initial probability density of the eigenfrequencies $p(\omega)$ using the formula

$$W_s(\Psi) = p(\omega) \cdot \left|\frac{\mathrm{d}\omega}{\mathrm{d}\Psi}\right| = \Delta \cos \Psi \cdot p(\omega_0 + \Delta \sin \Psi).$$

where $-\pi/2 \leq \Psi \leq \pi/2$ (see Hahn and Shapiro 1967). It can be shown (see, e.g., Pikovsky et al. 2001) that the distribution of the phase difference for the oscillators that are not trapped by the mean field $W_{as}(\Psi)$ (which follows directly from (6.61) for each ω^c) has a period π in Ψ and, hence, does not contribute to integral (6.63). Thus, two real equations [the real and imaginary parts of (6.63)] are obtained to determine the amplitude A and frequency ω_0 of the mean field:

$$A = \Delta \int_{-\pi/2}^{\pi/2} \cos^2 \Psi \cdot p(\omega_0 + \Delta \sin \Psi)\mathrm{d}\Psi, \tag{6.64}$$

$$0 = \Delta \int_{-\pi/2}^{\pi/2} \cos \Psi \cdot \sin \Psi \cdot p(\omega_0 + \Delta \sin \Psi)\mathrm{d}\Psi, \tag{6.65}$$

which can be solved analytically only for some specific forms of the distribution function $p(\omega)$. In particular, for an ensemble of coupled oscillators with a Lorentz frequency distribution, (6.65) is satisfied automatically, because this distribution is symmetric in frequency (see above), while (6.64) leads to the following expression for the mean field amplitude: $A = (1 - 2d/\alpha)^{1/2}$. Thus, a nontrivial mean field for an active medium of globally coupled oscillators exists if the coupling exceeds its critical value $\alpha_{\text{crit}} = 2d$, while the transition to synchronization is analogous to a second-order phase transition with a critical index of $1/2 : A \sim (\alpha - \alpha_{\text{crit}})^{1/2}$.

6.2.3.1 Synchronization in the Presence of Noise

Above, we have demonstrated that mesoscale clusters of coherent oscillators can be formed in some abstract system without any noise. However, this deterministic description of the dynamics of a stochastic vortex continuum is approximate, because the internal coordinates $\boldsymbol{q}(t)$ (along with the phases $\phi(t)$ on the cycles and in their neighborhoods) are generally random variables, i.e., they continuously undergo turbulent fluctuations. Internal and external noise is inherently present in the "real" subsystem of turbulent chaos (in a stationary-nonequilibrium state) characterized by the thermodynamic temperature $T_{\text{turb}}(r, t)$ (Kolesnichenko 2002).

Let us consider what will happen to the theory developed in this section when the system of oscillators is exposed to noise. In this case, the following first-order stochastic differential equation can be obtained for the phase different Ψ_k between the oscillations of a cluster of type k being synchronized and a periodic "external force" Z [see (6.56)]

$$\frac{\partial \Psi_k}{\partial t} = \Delta_{k0} - \Delta \cdot \sin \Psi_k + \sqrt{k_{\mathrm{B}} T_{\text{turb}} Q} \xi_k(t), \quad (k = 1, 2, \ldots, N). \tag{6.66}$$

It describes the self-oscillations in the presence of two forces—the periodic Z and stochastic $\tilde{f}_k(t) \equiv \sqrt{k_{\mathrm{B}} T_{\text{turb}} Q} \xi_k(t)$. The phase-locked loop for the frequency ω_k^c described by this equation is usually called a first-order loop.

The probability density for one oscillator now also depends on frequency $\omega : P_2(\Psi, \omega, t)$. Therefore, the mean field can be determined by averaging over the phase and frequency distributions; in the thermodynamic limit, we have

$$Z = A \exp(i\Theta) = \int_{-\pi}^{\pi} \mathrm{d}\Psi \int_{-\infty}^{\infty} \mathrm{d}\omega \exp i(\Psi + \omega_0 t) P_2(\Psi, \omega, t). \tag{6.67}$$

Here (and occasionally below), we omit the index k on the parameters characterizing one of the interacting oscillators. As applied to the stochastic equation (6.66), the FPK equation for the probability density of the total phase difference Ψ has the form of (6.57):

$$\frac{\partial P(\Psi, t)}{\partial t} = -\frac{\partial J}{\partial \Psi}, \quad J = (\Delta_0 - \Delta \cdot \sin \Psi) P(\Psi, t) + k_{\mathrm{B}} T_{\text{turb}} \frac{Q}{2} \frac{\partial P(\Psi, t)}{\partial \Psi}, \tag{6.68}$$

where J is the probability flux in the space Ψ. Since the phase variable Ψ is generally unbounded, the random process described by (6.68) is nonstationary. Equations (6.67) and (6.68) constitute a self-consistent ultimate system of equations for the unknown distribution function and the mean field.

6.2.4 Solution of the Stochastic Equations for the Phase Difference Between the Oscillations of a Synchronized Cluster in a Stationary State

Let us first analyze the qualitative behavior of system (6.66) in a stationary state ($\partial \Psi / \partial t = 0$). The acting noise affects the behavior of the ensemble of oscillators in two ways. First, because of weak noise effects, the values of the phase difference $\Psi(t)$ will be scattered in the course of time near the initial ($t = 0$) stable equilibrium $\Psi_0^* = \arcsin(\Delta_0/\Delta)$ [see (6.62)]. Second, intense noise effects can remove the system from the domain of attraction to the stable equilibrium state Ψ_0^*, causing the system to pass to the vicinities of neighboring equilibrium states $\Psi_{\pm1}^*$, $\Psi_{\pm2}^*$, etc. The probabilities of such transitions (which are arbitrarily called phase jumps) increase with time; in turn, the system will pass from the equilibrium state $\Psi_{\pm j}^*$ to the neighboring states $\Psi_{\pm(j+1)}^*$, $\Psi_{\pm(j-1)}^*$, etc. Such phase drifts (jumps) by a whole number of periods (2π) from the initial equilibrium position are possible in both directions.

Thus, the phase dynamics can be represented as a sequence of independent jumps. The number of such jumps in a finite time interval is always different in different directions. In systems with noise the stochastic synchronization condition is determined using the mean angular frequency (see, e.g., Anishchenko et al. 2003); therefore, the "mean" oscillator frequency $\langle\omega\rangle$ determined in a finite interval

$$\langle\omega\rangle = \int \mathrm{d}\phi\, P(\phi)\, \frac{\partial\phi(t)}{\partial t}$$

will not coincide with the frequency ω_0 of the synchronizing complex mean field Z; increasing the initial detuning and decreasing the signal-to-noise ratio lead to an increasingly clear manifestation of this effect. Since the preceding jump is quickly forgotten, the entire process is usually approximated by a Poissonian one and is characterized by only one parameter—the frequency of jumps (see Stratonovich 1961). Let N^+ be the mean number of jumps of the phase difference by 2π per unit time in the direction of its increase and N^- is the analogous number of jumps by 2π in the direction of its decrease. The difference between the mean frequency of the synchronized oscillator and the frequency of the acting mean field is then expressed in terms of the frequency of jumps as the mean phase drift rate

$$\langle\omega\rangle - \omega_0 = 2\pi(N^+ - N^-) = 2\pi J_1. \tag{6.69}$$

Let us now consider the quantitative description of the phase dynamics. According to the formula $\Psi(t) = \phi(t) - \omega_0 t$, the mean oscillation frequency of the oscillator being synchronized can be defined by the relation

$$\langle\omega\rangle \equiv \left\langle \frac{\partial\phi}{\partial t} \right\rangle = \omega_0 + \left\langle \frac{\partial\Psi}{\partial t} \right\rangle. \tag{6.70}$$

Until now, we have everywhere talked about the total phase difference Ψ. At the same time, since the coefficients of the FPK equation (6.68) in our problem are periodic in Ψ, we can deal with the reduced phase difference $\tilde{\Psi}(t)$ bounded in the interval $[-\pi, \pi]$ and seek the probability density for the reduced phase $\tilde{P}(\tilde{\Psi}, t)$, because this replacement will not affect the behavior of periodic functions with a period of 2π (see Stratonovich 1961). Indeed, if $P(\Psi, t)$ is a solution of (6.68) under the initial condition $\Psi = \Psi_0$, then the function $P(\Psi + 2\pi n, t)$ is also a solution under the initial condition $\Psi = \Psi_0 + 2\pi n$, where n is an arbitrary integer (below, the tilde over Ψ is omitted). Let us introduce the probability density

$$\tilde{P}(\Psi, t) = \sum_{n=-\infty}^{n=\infty} P(\Psi + 2\pi n, t) \tag{6.71}$$

of the phase difference Ψ reduced to the interval $[-\pi, \pi]$. $\tilde{P}(\Psi, t)$ is then a periodic function with a period of 2π; the FPK equation for it is the same in form as (6.68), but it can now be solved in the interval $[-\pi, \pi]$ under the initial condition $\tilde{P}(\Psi, 0) = \delta(\Psi - \Psi_0)$, where $-\pi \leq \Psi_0 \leq \pi$, the boundary condition $\tilde{P}(-\pi, t) = \tilde{P}(\pi, t)$, and the normalization condition $\int_{-\pi}^{\pi} \tilde{P}(\Psi, t)\, d\Psi$. This implies that the probability density of the reduced phase difference is obtained by "winding" the straight line Ψ around a circumference.

To find $\langle \partial\Psi/\partial t \rangle$, let us turn to the stationary solution of (6.68), which is possible, if at all, only in the limit $t \to \infty$. According to (6.68), the stationary probability density $W(\Psi) = \lim_{t \to \infty} \tilde{P}(\Psi, t)$ is defined by the following first-order ordinary differential equation:

$$\begin{aligned} J &= (\varDelta_0 - \varDelta \cdot \sin\Psi) W(\Psi) + k_B T_{\text{turb}} \frac{Q}{2} \frac{\partial W(\Psi)}{\partial \Psi} \\ &= \frac{\varDelta}{D} \left[(D_0 - D \cdot \sin\Psi) W(\Psi) + \frac{\partial W(\Psi)}{\partial \Psi} \right] = \text{const.} \end{aligned} \tag{6.72}$$

Here, $D_0/D = (\varDelta_0/\varDelta)$; the parameters D $(= 2\varDelta/k_{\mathrm{B}} T_{\text{turb}}\, Q)$ and D_0 characterize, respectively, the signal-to-noise ratio in the synchronization band and the relative initial detuning. The solution of (6.72) (under the boundary condition $W(-\pi) = W(\pi)$) can be written as (Stratonovich 1961)

$$W(\Psi) = \frac{\exp(D_0 \Psi + D \cos\Psi)}{4\pi^2 \exp(-\pi D_0) |I_{iD_0}(D)|^2} \int_{\Psi}^{\Psi + 2\pi} \exp(-D_0 \phi - D \cos\phi) \mathrm{d}\phi, \tag{6.73}$$

where $-\pi \leq \Psi \leq \pi$; $I_{\pm i\mu}(x)$ is the tabulated Bessel function of an imaginary argument and imaginary index (see Janke et al. 1977).

Given the stationary probability density $W(\Psi)$ of the phase difference, the mean oscillation frequency of a partial oscillator $\langle \omega \rangle = \omega_0 + \langle \partial\Psi/\partial t \rangle$ can be found. To find $\langle \partial\Psi/\partial t \rangle$ in a stationary state, we use the respective Langevin equation. Since the mean value of noise is zero,

$$\left\langle \frac{\partial \Psi}{\partial t} \right\rangle = \langle \Delta_0 - \Delta \cdot \sin \Psi \rangle = \int_{-\pi}^{\pi} (\Delta_0 - \Delta \cdot \sin \Psi) W(\Psi) \mathrm{d}\Psi. \tag{6.74}$$

We define the integrand in (6.74) via the constant probability flux J (in a stationary state, the probability flux does not depend on Ψ) equal to the quantity J_1 introduced above: $J = J_1 = N^+ - N^-$(Stratonovich 1961). According to (6.72), for the probability flux we have

$$J = (\Delta_0 - \Delta \cdot \sin \Psi) W(\Psi) + k_B T_{\text{turb}} \frac{Q}{2} \frac{\partial W(\Psi)}{\partial \Psi}, \tag{6.75}$$

whence follows

$$(\Delta_0 - \Delta \cdot \sin \Psi) W(\Psi) = J_1 - k_B T_{\text{turb}} \frac{Q}{2} \frac{\partial W(\Psi)}{\partial \Psi}$$

Substituting this expression into (6.74) and performing integration using the condition $W(-\pi) = W(\pi)$, we eventually obtain

$$\left\langle \frac{\partial \Psi}{\partial t} \right\rangle = 2\pi J_1. \tag{6.76}$$

The difference between the mean numbers of jumps in opposite directions J_1 (stationary probability flux) can be found by substituting (6.73) for the probability density into (6.72); as a result, we obtain

$$J_1 = J = N^+ - N^- = \frac{\Delta\, sh(\pi D_0)}{|I_{iD_0}(D)|^2 2\pi^2 D} = \frac{\Delta_0\, sh(\pi D_0)}{|I_{iD_0}(D)|^2 2\pi^2 D_0}. \tag{6.77}$$

Thus, for the mean oscillation frequency of an oscillator, we have

$$\langle \omega \rangle = \omega_0 + 2\pi J = \omega_0 + \frac{\Delta_0\, sh(\pi D_0) |I_{iD_0}(D)|^{-2}}{\pi D_0}. \tag{6.78}$$

This formula allows the most complete and substantive results on the evolution of the reduced-phase characteristics to be obtained. Let us consider here three limiting cases of this formula.

In the special case where there is no initial detuning ($\Delta_0 = D_0 = 0$), irrespective of D, we have $\langle \omega \rangle - \omega_0 = 0$, i.e., when the frequency of the synchronizing signal coincides with the oscillation frequency of the oscillator being synchronized, noise does not change the mean oscillation frequency, i.e., in this case, despite the presence of phase jumps, the mean numbers of forward and backward phase jumps are equal ($N^+ = N^-$) in view of complete symmetry and, hence, there is no systematic probability flux ($J_1 = 0$). In this case, a cluster of synchronized oscillators emerges. Under very intense noise: $D \to 0$ and $D_0 \to 0$ (recall that the parameter D characterizes the signal-to-noise ratio in the synchronization band). Using the equality $I_0(D) \approx 1$, we obtain $\langle \omega \rangle = \omega_0 + \Delta_0 = \omega^c$ from (6.78). Thus, in the case of very intense noise, there is no synchronization effect and the mean oscillation frequency of an oscillator remains equal to its initial value. If the asymptotic representations of Bessel functions of an imaginary index are used, then it can be shown that in the opposite extreme case, under very weak noise ($D \to \infty$), the following formula is valid (Stratonovich 1961):

$$\langle \omega \rangle - \omega_0 = \begin{cases} 0, & \text{for} \quad |\Delta_0| \leq |\Delta| \\ \sqrt{\Delta_0^2 - \Delta^2}, & \text{for} \quad |\Delta_0| > |\Delta|, \end{cases} \tag{6.79}$$

which describes the synchronization effect in the absence of noise. It can be seen from this formula that if the initial detuning is within the synchronization band, then the oscillators become coherent due to synchronization in the ensemble; when, alternatively, the initial detuning is outside the synchronization band, the oscillation frequency of an oscillator increases partially. In addition, note that when the signal-to-noise ratio is great ($D >> 1$) and, hence, the phase of the mean field is accurately tracked ($\Psi \ll 1$), the approximate relations $\cos \Psi \approx 1 - \Psi^2/2$ and $I_0(D) \approx \sqrt{2\pi D}\, \exp(D)$ are valid. In this case, the stationary probability density (6.74) takes a Gaussian form

$$W(\Psi) = \sqrt{2\pi\sigma_\Psi^2}\, \exp(-\Psi^2/2\sigma_\Psi^2)$$

with the center at $\Psi_0 = -\pi/2$ (here, $\sigma_\Psi^2 = 1/D$ is the variance of the phase difference). In the limit $D \to \infty$, this probability density turns into a δ-function, $\lim\limits_{D \to \infty} W(\Psi) = \delta(\Psi)$. The distinct maximum of the Gaussian probability distribution for the phase difference suggests once again that the oscillator phase is locked.

When the evolution of the model system of weakly interacting oscillators is nonstationary, the characteristics of the establishment of the phase difference can be determined by taking into account the numerical solution of the FPK equation (6.68). Several papers are devoted to a numerical analysis of the nonstationary equation (6.68) (see, e.g., Akopyan 1966; La Frieda and Lindsey 1973).

In conclusion, note the following. Although the problem of developed turbulence formally refers to the theory of dissipative dynamical systems, the potentialities of this theory (developed in the world literature mostly for systems with a small number of degrees of freedom) have remained virtually untapped for the purposes of modeling structured turbulence until recently. However, as has now become clear, developed turbulence is determined by a moderate number of collective excitations (macroscopic degrees of freedom), because the remaining mass of excitations simply adjust to them, acting as more intense dissipation. These collective degrees of freedom can be characterized by some stochastic parameters $\boldsymbol{q}$

that act as the internal coordinates of the subsystem of turbulent chaos. The parameters $\boldsymbol{q}$ describe both small-scale vortex structures and mesoscale CSs. Certain stability of the ensemble of individual CSs and, in some cases, autonomy of their dynamics allow one to pass (when describing the evolution of vortex structures) from the hydrodynamic partial differential equations to the stochastic ordinary differential equations for the parameters of structures. Thus, the combined stochastic-thermodynamic approach to modeling structured fluid turbulence proposed here, where the CS motion is considered not in the coordinate space but in the functional $\boldsymbol{q}$-space of the corresponding dynamical system (corresponding to the internal structure of the subsystem of turbulent chaos) allows one to a certain extent reduce the problem of describing random hydrodynamic fields in the turbulence theory to the stochastic dynamics of localized structures.

This approach correlates well with the approach by Migdal (1987), who used the analogy with gluon gauge fields in the theory of quark motion to derive the stochastic equations of motion for vortex filaments (considered as elementary excitations of a turbulent flow) and the corresponding Fokker–Planck equations in the loop space (in the space of vortex filaments) for the probability distribution $W(C_1, \ldots, C_N)$ of an ensemble of fluctuating vortex filaments $(C_1, \ldots, C_N)$ with intensities $\Gamma_k = \int_{C_k} \omega_k(\rho)\mathrm{d}\rho$ ($k = 1, \ldots, N$; $\omega_k(n')$ is the vorticity distribution). In this way, he showed how the formalism based on the probability distribution functions for the shape of all possible closed loops could be developed for hydrodynamic turbulence. Unfortunately, Migdal's abstract theory has not yet gained deserved recognition in the literature and has not been developed further.

At the same time, it should be realized that the mechanism for the generation of mesoscale CSs was analyzed in this section under fairly limited assumptions. It is hoped that a further study of this problem will allow one, in particular, to ascertain the situations in which small-scale turbulence is crucial for the formation and maintenance of mesoscale CSs and those in which it has no effect on the qualitative picture of the phenomena. Note also that it will be difficult to construct an efficient hydrodynamic model of structured turbulence (suitable for engineering calculations) until the kinematic, thermohydrodynamic, and topological characteristics of mesoscale CSs are concretized and described most adequately in terms of a set of internal coordinates. In addition, allowance for the CS interaction dynamics in the coordinate space, of course, is needed, which is not reflected in the developed approach. All of this is the subject matter of future studies.

Chapter 7
Foundations of Mechanics of Heterogeneous Media for Accretion Disks

In Chaps. 3 and 4 we dealt with some approaches to semi-empirical modeling of the developed turbulence in multicomponent turbulent media. Here an attempt is undertaken of the model of turbulent heterogeneous medium development with application to reconstruction of the protoplanetary gas–dust nebula and its evolution. The earlier developed approaches to the mathematical description of multicomponent turbulence and to the construction of the new classes of models of natural environment are utilized with the account for influence of inertial properties of polydisperse dust particles mixture, heat and mass transfer, chemical reaction and radiation on the turbulence development. This allowed us to extend possibilities of numerical simulation of physical phenomena in such complicated space media like gas dust accretion discs formed around differentially rotated stars of some classes and to study in detail their structure, physical–chemical and hydrodynamic properties, and evolution.

Special attention is given to in-depth treatment of the mechanism of turbulent gas and dust interaction in the rotating accretion disc that surrounded the proto-Sun after the system setup. We formulate a complete system of equations of two-phase multicomponent mechanics including the relative motion of the phases, coagulation processes, phase transitions, chemical reactions, and radiation in terms of the problem of reconstructing the evolution of the protoplanetary gas–dust cloud that surrounded the proto-Sun at an early stage of its existence. These equations are intended for schematized formulations and numerical solutions of special model problems on mutually consistent modeling of the structure, dynamics, thermal regime, and chemical composition of the circumsolar disk at various stages of its evolution, in particular, the developed turbulent motions of a coagulating gas suspension that lead to the formation of a dust subdisk, its gravitational instability, and the subsequent formation and growth of planetesimals.

To phenomenologically describe the turbulent flows of disk material, we perform a Favre probability-theoretical averaging of the stochastic equations of heterogeneous mechanics and derive defining relations for the turbulent flows of interphase diffusion and heat as well as for the "relative" and Reynolds stress tensors needed to close the equations of mean motion. The inertial effects of dust

M.Y. Marov and A.V. Kolesnichenko, *Turbulence and Self-Organization: Modeling Astrophysical Objects*, Astrophysics and Space Science Library 389, DOI 10.1007/978-1-4614-5155-6_7, © Springer Science+Business Media New York 2013

particles on the properties of turbulence in the disk, in particular, on the additional generation of turbulent energy by large particles near the equatorial plane of the proto-Sun are specially addressed. We develop a semiempirical method of modeling the coefficient of turbulent viscosity in a two-phase disk medium by taking into account the inverse effects of the transfer of a dispersed phase (or heat) on the growth of turbulence in the stratified by temperature or concentration subdisk and its atmosphere. For steady motion when solid particles settle to the midplane of the disk under gravity, the parametric method of moments for solving the Smoluchowski integro-differential coagulation equation for the particle size distribution function is studied assuming that distribution function a priori belongs to a certain parametric class of distributions. We also analyze the possible "regime of limiting saturation" of the subdisk atmosphere by fine dust particles (when the boundary condition for particle concentration on the disc central plane is not required) that is responsible for the intensification of various coagulation mechanisms in a turbulent medium. Such an approach underlies the development of refined models of stellar-planetary cosmogony thus getting insight in solving the challenging problem of planetary systems origin and evolution.

7.1 Theoretical Prerequisites to Modeling the Evolution of Turbulized Accretion Disks

7.1.1 Basic Concepts of Planetary Cosmogony

The material of a protoplanetary gas–dust cloud is a complex multiphase medium with regions of various densities, temperatures, and degrees of ionization. This material, generally a dust plasma, is magnetized and is in a state of strong turbulization. Understanding the evolution of a protoplanetary cloud is a necessary prerequisite for solving the question about the formation of the Earth and planets, a question intimately related to the fundamental problem of cosmogony whose solution is presently the appealing problem of the fundamental science.

The first attempts to understand the solar system origin go back to Ancient and Middle Ages, the latter being intimately connected with outstanding philosopher Iogannes Kant and mathematician Pier-Simon Laplas. Basically, their hypothesis about primary gas–dust protoplanetary nebula and forming rotating disk then breaking down into separate clusters giving rise eventually to planets formation kept its actuality until present time. In Russia, this concept was thoroughly developed by O. Shmidt and his followers (Schmidt 1957; Safronov 1969, 1982) who put forward quite solid arguments in favor of planet embryos origin of protosolar disk-shape cold matter and placed some important quantitative constraints on such a model.

In these and similar models developed by western scientists (see, e.g., Cameron 1973, 1978, 1985; Lynden-Bell and Pringle 1974; Lin and Papaloizou 1980) different approximation were involved when considering numerous physical processes accompanying accretion disc evolution. In particular, this is the case when addressing hydrodynamic turbulence which mainly determines disc structure and dynamics. Obviously, the situation becomes even more complicated when dealing with an actual nonuniform disperse medium of the turbulent disc composed of multicomponent gas and polydisperse dust particles. In particular, dusts significantly contributes to gas turbulization and drift motion in both radial and orthogonal directions relative to equatorial solar plane, as well as influence phase transitions (evaporation, condensation) depending on temperature stratification in the disk. Turbulence is apparently responsible for the uniform disk structure as it is evident from limited data of observations shown as examples in Fig. 7.1

The planets are believed to set up the differentially rotating turbulent protoplanetary disc formed in an orbit around a solar-type star in due course of continuing accretion of original nebula matter on proto-Sun, followed by the dust subdisk origin produced when dust settles to the midplane of the disk perpendicular to the rotation axis. Subdisk continues to flatten and ultimately loses its gravitational stability and decomposes to dusty clumps. In other words, it was the subdisk material from which the Solar planetary system was formed through the formation of discrete compaction centers and their subsequent growth (see, e.g., Toomre 1964; Safronov 1969, 1982, 1987; Goldrich and Ward 1973; Nakagawa et al. 1986; Makalkin 1994; Youdin and Shu 2002). Let us notice that the flattening of a rotating protoplanetary cloud and then disc–subdisk results from the competition between two main dynamical forces, the gravitational and centrifugal ones. When there is equilibrium between these forces, weaker factors, such as the thermal and viscous processes, the disk self-gravity, and the electromagnetic phenomena, become important for the evolution of the cloud/disc. Due to the viscous forces of friction (arising from the relative shear of gas-suspension elements during their orbital motion), the disk material drifts toward the proto-Sun along a flat spiral trajectory as its angular momentum is transferred outward, from the inner disk regions to the outer ones.

As we already noted in Chap. 1, one of the key viewpoints in astrophysics regarding the origin and structure of circumstellar gas–dust accretion disk of any type is their turbulent nature (Zel'dovich 1981; Fridman 1989; Dubrulle 1993; Balbus and Hawley 1998; Richard and Zahn 1999; Bisnovaty-Kogan and Lovelace 2001). Basically, protoplanetary accretion discs possess substantial viscosity that jointly with differential rotation of disc matter results in permanent source of intrinsic internal thermal energy origin. Viscosity of the differentially rotating disc appears to cause by the shear turbulence (Gor'kavyi and Fridman 1994; Fridman et al. 2003), as well as chaotic magnetic fields (see Armitage et al. 2001), energy of the latter may be comparable with hydrodynamic turbulence energy. Therefore, adequate numerical simulations of the evolution of the protoplanetary cloud that surrounded the young Sun generally require taking into account the dynamical processes of interaction between turbulized gas and dust. These are, in particular, the effects of modification of the carrier-phase turbulence energy by solid particles (i.e., the inverse effect of the dust component on the turbulent and thermal regimes of the disk gas component) and the influence of

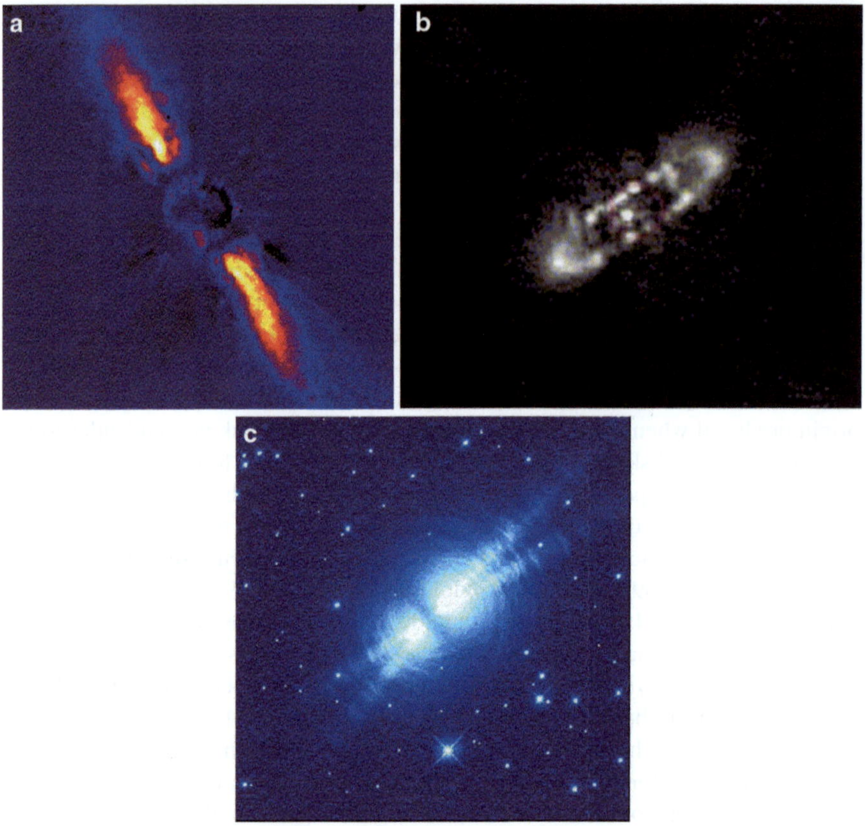

Fig. 7.1 Inhomogeneous structure of circumstellar gas–dust disks. (**a**) Disk around the star β Pictoris. Its extent in every direction from the star is 25 AU; the clearly distinguishable inhomogeneous structure is attributable to turbulent processes in the gas–dust medium on which gravitational perturbations from the planets forming inside the disk may be superimposed. The image was obtained at the 3.6-m European Observatory telescope with adaptive optics. (**b**) The dust disk around the star HR 4796A. The image was obtained with the Hubble Space Telescope in the near infrared at $\lambda = 1.1$ μm. The brightness peak lies at a distance of ~ 70 AU from the central star (Schneider et al. 1999). (c) The Egg Protoplanetary Nebula with an F-type supergiant inside. The outflow of gas and dust from the star produces radial columns in the form of two symmetric gas–dust disks with an inhomogeneous structure. Courtesy of NASA

turbulence on the rates of phase transitions (evaporation and condensation in the cooling disk) and on the jumplike disperse particle accumulation processes (coagulation and fragmentation when particles mutually collide with one another in a flow), and, finally, on the settling of solid particles through the gas to the midplane of the disk, where they form a flattened dust layer—subdisk.

The model encompassing continuing accretion of protoplanetary nebula matter on proto-Sun and surrounding disc, forming subdisk where solid particles consisting of refractory compounds (CAIs) are partially running outward, and the formation bipolar plasma flows at poles stipulated by magnetic field, are shown in

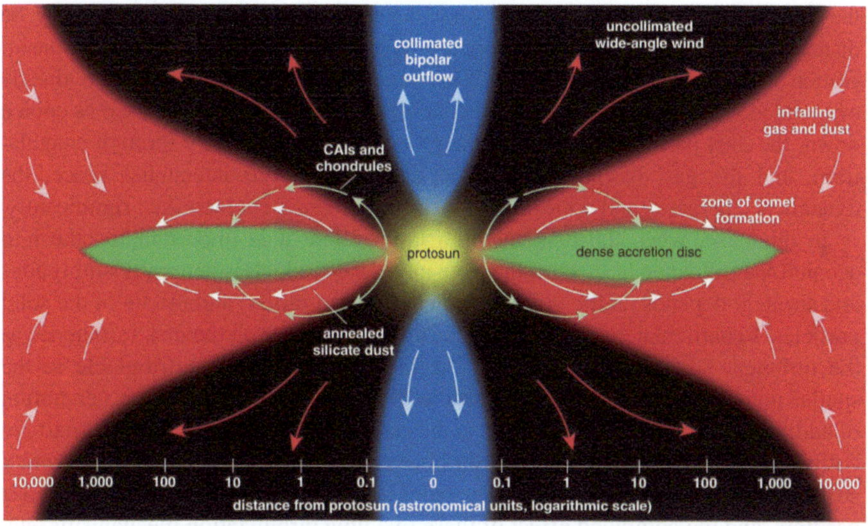

Fig. 7.2 Scheme for the formation of a gas–dust accretion disk and a subdisk. The proto-Sun onto which matter from the protoplanetary nebula (the *red color*) continues to accrete is at the center. The *green color* indicates the forming dust subdisk near which the outflow of gas and dust, including the formed high-temperature condensates in the inner zone, such as refractory CAIs, takes place. The *blue color* indicates the bipolar flows of matter attributable to the solar magnetic field NASA Courtesy

Fig. 7.2. As mass of protostar and density of disc matter increase, tidal interactions begin to play important role, which exert, in particular, concentric gas–dust layers formation (Bryden et al. 1999).

There is extensive literature on modeling the evolution of the circumsolar protoplanetary disk without dust [see, e.g., the vast bibliography to the review paper by Bisnovaty-Kogan and Lovelace (2001)]. At the same time, the few publications on dusty disk systems cover a comparatively narrow range of problems pertaining to the problem under consideration and the results obtained in them are limited, since the turbulence models for two-phase "gas–solid particles" media discussed in them cannot be recognized to be quite satisfactory (see, e.g., Weidenschilling 1977, 1980; Sekiya and Nakagawa 1988; Cuzzi et al. 1993; Dubrulle 1993; Dubrulle et al. 1995; Stepinski and Valageas 1996, 1997; Goodmann and Pindor 2000; Takeuchi and Lin 2002, 2003; Youdin and Goodman 2004). In particular, this is because the currently existing theory of turbulence of heterogeneous flows is imperfect due to both the incompleteness of the "classical" theory of hydrodynamic turbulence and the various additional regimes of two-phase turbulent flows realized in the disk when varying the volume content and sizes of solid particles in the gas-suspension flow.

In general, fine solid particles (a relatively low-inertia gas suspension component) have a laminarizing effect on a two-phase turbulent flow (via the growth of additional dissipation), while coarse particles enhance the generation of pulsational energy via the formation of a vortex wake. It should be noted that the dust phase may be disregarded only at the initial evolutionary stage of the cosmic system under consideration when almost all of the primordial (interstellar) solid particles have already evaporated. At later evolutionary stages of the protoplanetary cloud, as the disk cooled down, the dust particles condensed, their sizes increased (mainly through coagulation), and the gas dissipated from the disk system into interstellar space, the dynamical, energetic, and optical roles of the dust component increased significantly.

In this case, when the disk medium is modeled, it is important to take into account the influence of dust on the flow turbulence, which is generally ambiguous and strongly depends on the volume content (concentration) and inertia of the solid particles. In particular, such effects of the dust component on the disk turbulence as the turbulent "diffusion" transfer of the disperse component attributable to the spatial nonuniformity of the dust particle distribution in the disk, the generation of additional turbulent disturbances via the collective effects related to interparticle collisions between solid particles (Shraiber et al. 1980), the formation of vortex structures behind the streamlined large particles during the separation of the carrier gas flow, and the combined effect of these two flow turbulization mechanisms, etc. become important at certain evolutionary stages of a heterogeneous mixture. Besides, the very presence of a polydisperse admixture in a turbulent flow complicates significantly the disk hydrodynamics, contributing to the realization of additional cosmic flow regimes. In particular, an increase in the concentration of solid particles in a heterogeneous flow related to the dust settling to the midplane of the disk under the vertical gravity of the proto-Sun leads to an additional local enhancement of the generation of turbulent flow energy attributable to the growth of the transverse relative phase velocity gradient near the midplane, i.e., to flow returbulization (cf. Goldrich and Ward 1973).

In addition, the efficiency of the accretion mechanisms in a protoplanetary cloud (particularly at the subdisk formation stage) also depends significantly on the intensity of its turbulization; turbulence can have a completely unexpected effect on the particle coagulation in various situations, but it probably always contributes to the coagulation (Voloshchuk and Sedunov 1975). Thus, for example, if the internal Kolmogorov turbulence scale length λ_K is smaller than or comparable to the disperse particle size, then there is a turbulent motion of particles (similar to Brownian motion) that leads to their mutual collisions, i.e., to turbulent coagulation (which complements the effective gravitational coagulation in a quiet gas). For particles whose sizes are much smaller than λ_K, turbulence affects the evolution of fine dust through different channels. In this case, the enhancement of various coagulation processes (produced by factors other than turbulence) will result from intense turbulent particle mixing at distances larger than the Kolmogorov scale length, when the number of mutual collisions between solid particles per unit time increases significantly compared to a laminar flow due to chaotic turbulent pulsations. In turn, turbulent pulsations can contribute to drawing fine particles into the hydrodynamic wake or into the zone of action of the induction forces in the case of likely charged particles and can also promote the electrostatic coagulation via the destruction of the screening (Voloshchuk and Sedunov 1975).

Based on the results of laboratory study and modeling available (see, e.g., Dominik et al. 2007) one, as the first approximation, may admit that collisions of relatively small (roughly about 1 mm in size) particles with the velocities <1 m/s lead to their association/coalescence. It can be promoted by the so-called fast coagulation (when van der Waals repulsion forces are not taking into account although adhesion mechanism should be investigated in more detail) or by fluffy structure of collided particles. In such a case integration assumes to domain over destruction of fluffy particles up to velocities <10 m/s. Another effective mechanism of solid particles accumulation, apart from the gravitational accretion, could be nongravitational accretion due to Brownian, electrical, and turbulent-Brownian coagulation of charged and neutral particles, etc. These plausible mechanisms are discussed in detail elsewhere (see Marov and Kolesnichenko 2002). Obviously, as inertia of growing particles increases they will be less affected by turbulent (vortex) pulsations in the gas-suspension motion to a progressively lesser extent and hence in the long run experience efficient sedimentation towards equatorial plane of protostar and ultimately, agglomeration. Therefore, contrary to widely accepted views that turbulence in the gas–dust medium prevents particles to grow (see, e.g., Makalkin 2003), an opposite situation may occur that is in accord with recent results of observations (Natta et al. 2007).

Basically, this concept parallels the authors ideas developed throughout the monograph that high probability of self-organization processes in the chaotic gas–dust medium is pertinent to the turbulent accretion disc and dusty subdisk. These processes are assumed to be especially important in creation of dust clusters within ordered (eddy) structures caused by strong interaction of relatively slow moving particles, which results in their integration. In other words, we argue for self-organization regions in the dynamical nonequilibrium system ("islands of order") setup rather than for less-probable mechanism of direct particle associations in the turbulent flow. Nonetheless, it still remains unclear how embryos of planetesimals of about 1 m in size begin to grow up in the turbulent disc medium and what time frame would be required for this process (Dullemond and Dominik 2005; Cuzzi 2004; Cuzzi et al. 2003, 2005; Cuzzi and Weidenschilling 2006; Russell et al. 2006; Wadhwa et al. 2007).

Finally, there are no doubts that plasma effects play significant role in the formation of circumsolar protoplanetary disk and its early evolution. In general, the cosmic plasma is a dust plasma, i.e., contains very fine dust particles (see Fortov et al. 2006). Since any accretion disk contains solid particles of various sizes, there probably exists a boundary linear scale (usually of the order of 10^{-5}–10^{-7} m dependent on the electromagnetic and gravitational fields as well as on the particle charge and density) that separates the small particles comprising the dust plasma and large particles whose motion is stipulated by nonelectromagnetic forces. Photoelectron emission and collisions with plasma electrons and positive ions are known to be the main physical processes that determine the grain surface charge (see, e.g., Alfven and Arrhenius 1976). At the same time, a solid particle in cosmic plasma is more frequently charged negatively to potential of the order of several volts due to collisions with electrons. In plasma region with a large amount of super-thermal electrons particle can acquire negative potential of several thousand volts and then becomes captured by plasma.

In the processes of disc formation and evolution, electrization of settled dust particles in due course of their interaction with the magnetic field of mother star plays very important role. Respective electrodynamics effects in differentially rotating medium setup which results in more reliable estimate of the turbulent exchange coefficients compared to simplified values deduced from dimensions ratio in the widely used gradient models of turbulence. Obviously, when an electrically conducting two-phase medium moves in an electromagnetic field, the Lorentz ponderomotive force acts on the charged particles, which gives rise to a number of additional effects, specifically during turbulization of the flow (see, e.g., Vereshchagin et al. 1974; Busroyd 1975). In particular, such unique properties as high dissipativity, the capability for the self-organization and formation of ordered structures are characteristic of a turbulent dust plasma. The problems intrinsically related to the combined magnetohydrodynamic effects and turbulence in the heterogeneous medium of differentially rotating disc involving those of heat and mass transfer, the inertial properties of a polydisperse admixture of solid particles, coagulation, and radiation are specially addressed in Chap. 9 where an attempt to develop the continuum model of disc early evolution is undertaken.

Let us note that complexity of turbulence modeling in the conducting multiphase medium is not confined by the necessity of derivation of the reliable turbulent exchange coefficients. Not less important is the study of large-scale turbulent structures responding the ideas of appearance of coherent orderings at the background of chaotic turbulent motion. In contrast to the chaotic turbulent flow where particles grow in collisions is hampered, eddy coherent structures promote particles association and growth. Hence as we already mentioned, eddy clusters setup facilitate solution of the problem how dust particles can grow in collisions under moderate velocities because mechanism of direct integration is supported neither by theoretical treatment nor by the laboratory experiments.

Obviously, the synenergetic processes of self-organization in the thermodynamically open system of a protoplanetary cloud against the background of a large-scale shear flow of cosmic material related to its differential rotation is also a very important mechanism that might shape the properties of the cloud at various stages of its evolution, including the formation of a viscous accretion disk around the young Sun that was passing through the T Tauri stage, the formation of a dust–gas subdisk, the destruction of the latter due to gravitational instability, and the production of discrete compaction centers followed by the formation of and growth of planetesimals. Let us emphasize again that we are focusing on the gravity instability setup just in the dust subdisk rather than in the disk itself because small mass of the latter at the T Tauri stage is insufficient to provide gravity instability. However it could experience instability of Balbus–Hawly type mentioned in Chap. 1.

We come therefore to the conclusion that in the developed disk turbulence of various mesoscales, relatively stable gas–dust coherent structures can form that probably provide the most favorable conditions for the mechanical and

Fig. 7.3 Gas–dust disk around central star with solid particles and larger fragments, planetesimals, forming inside it. An artist's concept, NASA courtesy

physical–chemical interaction between dust particles (see Barge and Sommeria 1995; Tanga et al. 1996; Chavanis et al. 1999; Kolesnichenko 2004). As a result, spontaneous formation and growth intensification of phase transitions and heat and mass transfer at various thermohydrodynamic parameters of the carrier and disperse phases, as well as significant modification of the oscillation spectrum in a heavily dusted medium, etc., occurs. One of the plausible scenarios of the formation and growth of dust particles in the partially ionized protoplanetary disc could be as follows: formation of primary dust clusters followed by heterogeneous condensation, then coagulation and agglomeration, and finally, ions surface recombination giving rise to matter sedimentation, especially on the surface of isolated multicharged particles. Collective self-organization processes are assumed to contribute to all these stages. Such a scenario involving larger size bodies formation is shown schematically in Fig. 7.3.

In this chapter we shall not touch upon plasma effects which are specially addressed in Chap. 9. We focus mainly on the following four aspects of the problem of modeling a gas dust disk medium:

- Formulating a basic system of mass, momentum, and energy conservation equations for the instantaneous (actual) parameters of the flow of a gas–dust mixture and radiation that are intended for numerical simulations of the circumsolar protoplanetary disk at various stages of its evolution (in particular, the laminar subdisk formation stage) and in spatial zones located at various distances from a protostar.

- Weighted (Favre) averaging of the stochastic equations of motion of two-phase mechanics to phenomenologically describe the averaged gas-suspension flow and the processes of turbulent heat and mass transfer in a gas–dust disk.
- Deriving the defining relations for the correlation parameters of a turbulent two-phase flow needed to close the hydrodynamic equations of mean motion.
- Modeling the turbulent transfer coefficients in a gas–dust disk by taking into account the inverse effect of the polydisperse component on the turbulence intensity of the carrier gas.

7.1.2 Basic Assumptions of the Model

The motion of a gas suspension in a gas–dust accretion disk can probably be modeled most adequately in terms of the mechanics of heterogeneous turbulized media by taking into account the peculiarities of the physical–chemical properties of the phases, heat and mass transfer and radiation, chemical reactions, phase transitions, coagulation, fragmentation, etc. The evolution of such media is studied by invoking new thermohydrodynamic parameters and by solving more complex equations than those that we deal with in "ordinary" hydrodynamics (Kolesnichenko and Marov 2006). In this case, the intraphase and interphase interactions in heterogeneous media are occasionally very difficult to describe in detail, and rational schematizations leading to manageable and solvable equations are particularly needed here to obtain reliable results and to understand them.

In general, the continuity and energy equations and the equations of motion for each individual phase are used in the above and other known works on the continuum modeling of turbulized accretion disks with dust (Nakagawa et al. 1981, 1986; Weidenschilling 1984; Hayashi et al. 1985; Schmitt et al. 1997; Balbus and Hawley 1998). The authors have to heuristically specify the laws of interphase interactions, in particular, the rates of momentum and energy transfer between the phases. This approach is an analog of Grad's 13-moment method (Grad 1949), which gained wide acceptance, for example, in the kinetic theory of a multicomponent plasma. The subsequent averaging of the mutually coupled hydrodynamic equations for the individual phases (to describe the turbulized motions) leads not only to cumbersome equations of mean motion, which is related to the need for retaining a large number of correlation moments of the pulsating parameters (such as, $\overline{\rho'_{g}\boldsymbol{u}'_{g}}$, $\overline{\rho'_{d}\boldsymbol{u}'_{d}\boldsymbol{u}'_{d}}$, $\overline{T'\boldsymbol{u}'_{d}}$, $\overline{\rho'_{g}T'}$, and the like) in their structure, but also to difficulties in physically interpreting each individual term in the averaged equations. To overcome these difficulties and to simplify the problem, some of the authors often unjustifiably discard a number of correlation terms, which, of course, narrows the range of application of such an approach.

At the same time, the evolution of a turbulized gas–dust cloud can be modeled in the single-velocity approximation of heterogeneous mechanics, which is similar to the Chapman–Enskog moment method of solving the system of Boltzmann kinetic equations for multicomponent gas mixtures (see, e.g., Chapman and Cowling 1952; Marov and Kolesnichenko 1987). In the case under consideration, this approach is peculiar in that, despite the hydrodynamic velocity difference between the individual phases and the related necessity of allowing for the dynamical and inertial effects of their relative motion, the continuum description of a disk medium can be made using the laws of conservation of mass, momentum, and energy for the system as a whole supplemented with the defining (closing) relations for a number of thermohydrodynamic flows, both intraphase and interphase ones. In particular, the generalized Stefan–Maxwell relations that were derived by Kolesnichenko and Maksimov (2001) for heterogeneous media with sufficient completeness and logical coherence by the methods of nonequilibrium thermodynamics can be used for the interphase diffusion flows (or the relative velocities of the phases). It is important to emphasize that using the total continuum alone to model the gas–dust cosmic material allows us, when using the Favre weighted averaging (Favre 1969), to average the hydrodynamic equations for the disk medium as a whole fairly accurately (Marov and Kolesnichenko 2002).

The huge variety, mutual influence, and complexity of the multiphase effects in the solar protoplanetary cloud (phase transitions, chemical reactions, heat transfer, gravitational interaction, pulsational and random motions, rotation, radiation, coagulation, etc.) necessarily require appropriately schematizing the description of the motion of a gas–dust medium. Therefore, in this paper, we will assume that the motion of a disperse mixture in a protoplanetary disk can be adequately described under the following assumptions:

- The dust particle (by which we mean below a solid-phase condensate) are solid, nondeformable, spherical in shape, and polydisperse.
- The dust particle material is assumed to be incompressible, $\rho_{\mathrm{d}} = \mathrm{const}$.
- The true dust density is much higher than the true gas density of the system, $\rho_{\mathrm{d}} \gg \rho_{\mathrm{g}}$.
- The volume concentration of the disperse phase is moderately high ($s^2 \ll 1$), so the terms of order s^2 may be neglected.
- The carrier phase is a compressible multicomponent perfect gas.
- The diffusion transfer of molecules of all chemical types relative to one another may be ignored, $\boldsymbol{u}_{\alpha(k)} \equiv \boldsymbol{u}_{\alpha}$.
- The viscosity and thermal conductivity of the disperse phase may be disregarded, $\Pi_{\mathrm{d}} = 0$ and $\boldsymbol{q}_{\mathrm{d}} = 0$.
- The gas and disperse phases are assumed to be in thermal equilibrium, $T_{\mathrm{g}} = T_{\mathrm{d}} = T$.
- The total heterogeneous continuum is considered in the single-pressure approximation, $p_{\mathrm{g}} = p_{\mathrm{d}} = p(\rho_{\mathrm{g}}, T)$.
- The heterogeneous reactions on the surfaces of solid particles may be disregarded; the contribution from the interphase boundaries (the near-surface layer of solid particles) to the energetics of the disk system as a whole may be neglected; the presence in heterogeneous systems of interphase boundaries modeled by mathematical surfaces on which the fields of various thermodynamic

parameters become discontinuous severely complicates the continuum theory of multiphase multicomponent systems (see Nigmatulin 1978).

- It is assumed that the rotation of the solid particles may be ignored when describing the dynamical interaction between the phases.
- The heat transfer between the disperse particles and the carrier gas may be disregarded.

Thus, we are going to model a heterogeneous continuum composed of two contacting phases, a carrier gas phase of solar composition and a disperse phase of solid condensed particles of complex chemical composition (see, e.g., Dorofeeva and Makalkin 2004), at absolute temperature T and pressure p and well mixed inside each macrovolume element $\delta\mathcal{V}$ of the disk medium. Note that this is the condition only for thermal and mechanical equilibrium of the phases, but not for total phase equilibrium, which additionally requires that the chemical potentials of the phases (which are the key concept of the theory of phase equilibrium) be equal. In addition, in chemical equilibrium, i.e., for an equilibrium distribution of chemical components between the two phases, their chemical potentials must have a constant value in both phases. For continuum approximation to be applicable, we assume that linear dimension of macrovolume element $\delta\mathcal{V}$ of disc matter is much larger than that of disperse inclusions but much smaller than characteristic hydrodynamic scale length L_{hydr}. Each phase is assumed to be a homogeneous n-component mixture (with each geochemically significant material of the protoplanetary cloud being present in each phase α). Below, we use the Greek subscripts $\alpha, \beta, \ldots$ to denote the phases and refer the Latin subscripts in parentheses of type (k) or (i) for any quantity to the molecular component of the phase. The gas phase ($\alpha = \mathrm{g}$) is the carrier medium described by the model of a viscous fluid. The disperse phase ($\alpha = \mathrm{d}$) present in the form of solid inclusions is inviscid and non-heat-conductive while we will not ignore the collisions between them. Occasionally, we will use numbers instead of letters to denote the gas and condensed phases, referring the subscript $\alpha = 1$ to the gas phase and $\alpha = 2$ to the disperse-phase parameters.

7.2 Equations of Mechanics of Heterogeneous Media to Describe a Protoplanetary Gas–Dust Cloud

Assuming a local thermodynamic equilibrium within each phase and a local thermal equilibrium between radiation and matter, we will use a phenomenological theory of multifluid interpenetrating continua to describe the hydrodynamic motions in a gas–dust medium (with the corresponding physical–chemical properties). In particular, this theory includes the dynamical effects due to the inequality of the hydrodynamic velocities $\boldsymbol{u}_\alpha$ of the phases in the system (see, e.g., Nigmatulin 1978; Kolesnichenko and Maksimov 2001). For each of the two phases at each space–time point $(\boldsymbol{r}, t)$, we define the mass density, hydrodynamic velocity, internal energy, and other thermohydrodynamic parameters pertaining to its own continuum and its own chemical component of the mixture. As the phase parameters, we will use quantities averaged both over the total macrovolume element $\delta\mathcal{V} = \sum_\alpha \delta\mathcal{V}_\alpha$ pertaining to the heterogeneous system as a whole and over the part $\delta\mathcal{V}_\alpha$ of the volume element occupied by the individual phase α. In particular, apart from the distributed (spread over the total volume $\delta\mathcal{V}$) mass density $\tilde{\rho}_\alpha$ of phase α, we will use below the true (physical) density ρ_α (equal to the ratio of the mass of the phase-α particles in the macrovolume element $\delta\mathcal{V}$ to the part of this volume $\delta\mathcal{V}_\alpha$ occupied by the phase). The latter is defined by the expression

$$\rho_\alpha = \frac{\tilde{\rho}_\alpha}{s_\alpha}, \quad s_\alpha \equiv \frac{\delta\mathcal{V}_\alpha}{\delta\mathcal{V}}, \quad \sum_\alpha s_\alpha = 1, \tag{7.1}$$

where s_α is the so-called volume content (or volume concentration) of phase α. The true (ρ_α) rather than distributed ($\tilde{\rho}_\alpha$) phase density, together with other parameters of the state, such as the temperature T_α, internal energy e_α, and entropy S_α, determines the thermodynamic properties of an elementary phase-α macroparticle in its various states. In addition, the values of s_α also directly affect the hydrodynamic motion of the phases, since it appears in the corresponding equations of motion. Concurrently, we will assume that r-independent chemical reactions, including the interphase reactions and the cases where the chemical transformations are reduced to the displacement of component k from one phase to the other, are possible between the individual chemical components k of the disk system.

7.2.1 Mass Conservation: A Monodisperse Gas–Dust Medium

Let us first consider the case where all condensed particles of the protoplanetary gas–dust disk in each macrovolume element δ' have the same instantaneous hydrodynamic velocity $\boldsymbol{u}_\mathrm{d}(\boldsymbol{r}, t)$, irrespective of their sizes. The mass density $\rho(\boldsymbol{r}, t)$ and the weighted mean hydrodynamic velocity $\boldsymbol{u}(\boldsymbol{r}, t)$ (the instantaneous center-of-mass velocity of the gas-suspension macrovolume element centered at point $\boldsymbol{r}$) of the gas–dust mixture as a whole are defined by

$$\rho = \sum_\alpha \rho_\alpha s_\alpha = \rho_\mathrm{g}(1 - s) + \rho_\mathrm{d}\, s, \tag{7.2}$$

$$\boldsymbol{u} = \rho^{-1} \sum_\alpha \rho_\alpha s_\alpha \boldsymbol{u}_\alpha = \frac{\rho_\mathrm{g}(1 - s)}{\rho} \boldsymbol{u}_\mathrm{g} + \frac{\rho_\mathrm{d} s}{\rho} \boldsymbol{u}_\mathrm{d}, \tag{7.3}$$

where $\rho_\alpha(\boldsymbol{r}, t)$ and $\boldsymbol{u}_\alpha(\boldsymbol{r}, t)$ are the true mass density and hydrodynamic velocity of phase α, respectively; $s_\mathrm{d}(\boldsymbol{r}, t)$ is the instantaneous volume concentration of the disperse phase ($s_\mathrm{g} + s_\mathrm{d} = 1$); below, we omit the subscript "d" of the parameter s_d, $s_\mathrm{d} \equiv s$. To model the chemical composition of a protoplanetary cloud, particularly at early stages of its evolution (see, e.g., Willacy et al. 1998; Dorofeeva and Makalkin 2004), we must generally invoke the mass balance equations for each chemical component of the phase, which can be represented as the equations for the conservation of particles of type k in phase α. Under the above assumptions, these equations take the form

$$\frac{\partial}{\partial t}\left(s_\alpha n_{\alpha(k)}\right) + \operatorname{div}\left(s_\alpha n_{\alpha(k)}\,\boldsymbol{u}_\alpha\right) = \sigma_{\alpha(k)} \equiv \sum_{\rho=1}^{r} \nu_{\alpha(k),\rho}\xi_\rho + \delta_{2\alpha}\,\overset{\circ}{n}_{\alpha(k)}$$

$$(\alpha = 1, 2; \quad k = 1, 2, \ldots, n) \tag{7.4}$$

or

$$\rho\frac{\mathrm{d}}{\mathrm{d}t}\left(\frac{s_\alpha n_{\alpha(k)}}{\rho}\right) + \operatorname{div}\left(s_\alpha n_{\alpha(k)}\boldsymbol{w}_\alpha\right) = \sigma_{\alpha(k)}, \quad \boldsymbol{w}_\alpha \equiv (\boldsymbol{u}_\alpha - \boldsymbol{u}). \tag{7.5}$$

Here, $d(\ldots)/\mathrm{d}t \equiv \partial(\ldots)/\partial t + \boldsymbol{u}\cdot\partial(\ldots)/\partial\boldsymbol{r}$ is the substantial derivative related to the motion of the macrovolume element of the gas–dust medium as a whole; $\partial(\ldots)/\partial\boldsymbol{r} = \sum_l \boldsymbol{i}_l\partial(\ldots)/\partial x_l$ is a vector differential operator; $\boldsymbol{i}_l$ $(l = 1, 2, 3)$ are the Cartesian unit vectors along the corresponding coordinate axes; the quantity $(\partial/\partial\boldsymbol{r})\cdot\boldsymbol{b}$ is the divergence of $\boldsymbol{b}$; $n_{\alpha(k)}(\boldsymbol{r}, t)$ is the number of particles of chemical substance k in a unit volume occupied by phase α(countable concentration); $\boldsymbol{w}_\alpha(\boldsymbol{r}, t)$ is the instantaneous diffusion velocity of phase α that, by the definition of the weighted mean velocity $\boldsymbol{u}$, satisfies the relation

$$\sum_\alpha \rho_\alpha s_\alpha \boldsymbol{w}_\alpha = 0, \quad \boldsymbol{w}_\alpha \equiv (\boldsymbol{u}_\alpha - \boldsymbol{u}) \tag{7.6}$$

or

$$\sum_\alpha \boldsymbol{J}_\alpha = 0, \quad \boldsymbol{J}_\alpha \equiv \rho_\alpha s_\alpha \boldsymbol{w}_\alpha = \rho_\alpha s_\alpha(\boldsymbol{u}_\alpha - \boldsymbol{u}), \tag{7.7}$$

where $\rho_\alpha = \sum_k m_{(k)} n_{\alpha(k)}$; $\boldsymbol{J}_\alpha(\boldsymbol{r}, t)$ is the mass diffusion flux of phase-α particle; $\sigma_{\alpha(k)}$ is the production rate of particles of component k per unit volume via chemical reactions and phase transitions (evaporation and condensation) as well as the fragmentation and coagulation of the disperse component; $\xi_\rho(\boldsymbol{r}, t)$ is the rate of chemical reaction ρ (including the interphase reactions and phase transitions), $\rho = 1, 2, \ldots, r$; $\nu_{\alpha(k),\rho}$ is the stoichiometric coefficient of component k in phase α with respect to chemical reaction ρ whose stoichiometric equation can be symbolically written as (see, e.g., Prigogine and Defay 1954)

$$\sum_\alpha\sum_k \nu_{\alpha(k),\rho} m_{(k)} = 0, \quad (\rho = 1, 2, \ldots, r) \tag{7.8}$$

the principle of conservation of total mass in chemical reaction ρ; $m_{(k)}$ is the molecular mass of component k; $\overset{\circ}{n}_{\alpha(k)}$ is a quantity that describes the change in the number density of chemical component k in the dust phase related to the fragmentation or adhesion of condensed particles in the gas–dust cloud. Stoichiometric coefficients of the components derivative related to the motion of the macrovolume produced during the reaction (from left to right) are taken positive, while the coefficients of the expendable components are negative.

If the mass of all chemical components in the dust phase is conserved during the transformation of solid particles ($\sum_k m_{(k)}\, \overset{\circ}{n}_{\alpha(k)} = 0$), then the following differential conservation equation follows from (7.4):

$$\rho \frac{\mathrm{d}}{\mathrm{d}t}\left(\frac{\rho_\alpha s_\alpha}{\rho}\right) + \operatorname{div}(\rho_\alpha s_\alpha\, \boldsymbol{w}_\alpha) = \sigma_{\alpha\beta} = \sum_{\rho=1}^{r} \nu_{\alpha,\rho}\xi_\rho, \quad (\alpha, \beta = 1, 2) \tag{7.9}$$

for the distributed mass density

$$\tilde{\rho}_\alpha \equiv s_\alpha \sum_k m_{(k)} n_{\alpha(k)} = \rho_\alpha s_\alpha \tag{7.10}$$

of phase α. Here $\nu_{\alpha,\rho} \equiv \sum_k m_{(k)} \nu_{\alpha(k),\rho}$; the quantity $\sigma_{\alpha\beta}(\boldsymbol{r}, t)$ characterizes the rate of mass transformation from phase α to phase β (or vice versa, then $\sigma_{\alpha\beta} < 0$) via chemical reactions and the evaporation or condensation of material in the protoplanetary cloud; in this case, $\sigma_{\alpha\beta} = -\sigma_{\beta\alpha}$. For our subsequent purposes, it is appropriate to introduce the mass phase concentrations $C_\alpha(\boldsymbol{r}, t)$,

$$C_\alpha \equiv \frac{\tilde{\rho}_\alpha}{\rho} = \frac{\rho_\alpha s_\alpha}{\rho}, \quad \sum_\alpha C_\alpha = 1 \tag{7.11}$$

and the relative velocity of the dust and gas, $\boldsymbol{w} \equiv \boldsymbol{w}_{\mathrm{dg}} = (\boldsymbol{u}_{\mathrm{d}} - \boldsymbol{u}_{\mathrm{g}})$; (7.9) can then be written in a more compact form,

$$\rho \frac{\mathrm{d}C_\alpha}{\mathrm{d}t} = -\operatorname{div}\boldsymbol{J}_\alpha + \sigma_{\alpha\beta} \quad (\alpha,\, \beta = 1,\, 2), \quad \boldsymbol{J}_{1,2} = \begin{cases} \rho C_1 \boldsymbol{w}_1 = -\rho C_1 C_2 \boldsymbol{w}, \\ \rho C_2 \boldsymbol{w}_2 = \rho C_1 C_2 \boldsymbol{w}. \end{cases} \tag{7.12}$$

Given (7.7) and (7.8), the law of conservation of total mass obtained by summing (7.4) over the indices k and α takes a standard form,

$$\rho \frac{\mathrm{d}}{\mathrm{d}t}\left(\frac{1}{\rho}\right) - \operatorname{div}\boldsymbol{u} = 0 \quad \text{or} \quad \frac{\partial \rho}{\partial t} + \operatorname{div}(\rho\, \boldsymbol{u}) = 0 \tag{7.13}$$

as in a single-phase continuum. Note that radiation does not modify the continuity equation (7.13), since it "does not possess" any mass.

Next, we will assume that the material of the solid inclusions remains incompressible during the evolution of the gas–dust cloud, i.e., the true (physical) dust density is $\rho_{\mathrm{d}}=$ const. The instantaneous equation (7.12) for dust is then be reduced to the equation

$$\rho \frac{\mathrm{d}}{\mathrm{d}t}\left(\frac{s}{\rho}\right) = -\mathrm{div}(s\,\boldsymbol{w}_{\mathrm{d}}) + \rho_{\mathrm{d}}^{-1}\sigma_{\mathrm{dg}}, \quad \boldsymbol{w}_{\mathrm{d}} = C_{\mathrm{g}}\boldsymbol{w} = (1-s)\frac{\rho_{\mathrm{g}}}{\rho}\boldsymbol{w} \tag{7.14}$$

which allows the volume content $s(\boldsymbol{r},t)$ of the dust component in a two-phase flow to be found at a given relative velocity of the phases $\boldsymbol{w}$. Thus, (7.13) and (7.14) can be used instead of the two equations (7.9) to calculate the parameters ρ and s (and, hence, the gas density ρ_{g}) (see (7.2)).

The intensity of the force interaction between the phases and the radiation parameters in a gas–dust cloud strongly depend on the characteristic size of the solid inclusions (e.g., the characteristic volume of one dust particle $\tilde{U}_{\mathrm{d}}(\boldsymbol{r},t)$) and their total number $N_{\mathrm{d}} \equiv s\sum_k n_{\mathrm{d}(k)}$ in a unit total gas-suspension volume. If all solid-phase condensates are spherical or nearly spherical with the characteristic Feret diameter $\tilde{d}(\boldsymbol{r},t)$, then $\tilde{U}_{\mathrm{d}} = (\pi/6)\tilde{d}^3$. The balance equation for the total number of disperse particles $N_{\mathrm{d}}(\boldsymbol{r},t)$ can be derived from (7.4):

$$\rho \frac{\mathrm{d}}{\mathrm{d}t}\left(\frac{N_d}{\rho}\right) + \mathrm{div}(N_d W_d) = \sigma_{N_d} \equiv \sum_k \sum_{\rho=1}^{r} \nu_{d(k),\rho}\xi_\rho + \overset{\circ}{N}_d \tag{7.15}$$

where the source term $\overset{\circ}{N}_{\mathrm{d}} \equiv \sum_k \overset{\circ}{n}_{\alpha(k)}$, which characterizes the change in the total number density of different-scale solid particles via coagulation and fragmentation, is generally defined by the Smoluchowski kinetic equation (see (7.29)). An important parameter of the coagulating mixture in a two-phase flow, the characteristic volume $\tilde{U}_{\mathrm{d}}(\boldsymbol{r},t)$ (or the mean linear diameter $\tilde{d}_{\mathrm{d}}$) of the solid inclusions, can be determined from the known parameters N_{d} and s:

$$\tilde{U}_{\mathrm{d}} = \frac{s}{N_{\mathrm{d}}}, \quad \tilde{d}_{\mathrm{d}} = \sqrt[3]{\left(\frac{6}{\pi}\right)\left(\frac{s}{N_{\mathrm{d}}}\right)}. \tag{7.16}$$

If we disregard the fragmentation and coagulation of solid particles ($\overset{\circ}{N}_{\mathrm{d}} = 0$) and the evaporation and condensation processes ($\sigma_{\mathrm{dg}} = 0$) in our numerical simulations of the evolution of a gas–dust cloud and assume that the material of the inclusions is incompressible ($\rho_{\mathrm{d}}=$ const), then we will have $\tilde{U}_{\mathrm{d}} =$ const . Under these assumptions, $\tilde{\rho}_{\mathrm{d}} = s\rho_{\mathrm{d}} = \rho_{\mathrm{d}}\tilde{U}_{\mathrm{d}}N_{\mathrm{d}}$, and (7.15) (with a zero right-hand side) is just a corollary of the mass conservation equation (7.14) for the second phase, which in this case takes a simple form,

$$\rho \frac{\mathrm{d}}{\mathrm{d}t}\left(\frac{s}{\rho}\right) + \mathrm{div}(s\,\boldsymbol{w}_{\mathrm{d}}) = 0, \quad \boldsymbol{w}_{\mathrm{d}} = C_{\mathrm{g}}\boldsymbol{w}. \tag{7.14*}$$

7.2.2 *Interphase Diffusion: The Aerodynamic Drag Coefficient for Disk Dust Particles*

As we already said above, the generalized Stefan–Maxwell relations can serve as the basic equations in analyzing the phase diffusion in a gas–dust cloud. Using the kinetic theory, Hirschfelder et al. (1954) considered in detail the diffusion in a multicomponent mixture of gases in their monograph. Their main result is that the relative diffusion velocity $(\boldsymbol{u}_{(j)} - \boldsymbol{u}_{(k)})$ of two gas components can be caused by factors that do not directly affect these components, for example, by the thermodynamic forces acting on the molecules of the other mixture components. The velocities $(\boldsymbol{u}_{(j)} - \boldsymbol{u}_{(k)})$ can be found (to the first approximation of the Chapman–Enskog method of solving the *Boltzmann* equations) from the system of *Stefan–Maxwell* equations

$$\sum_j \frac{n_{(j)}n_{(k)}}{n^2\mathcal{D}_{(jk)}}(\boldsymbol{u}_{(j)} - \boldsymbol{u}_{(k)}) - k_{T(k)}\frac{\partial \ln T}{\partial \boldsymbol{r}} = \boldsymbol{d}_{(k)}$$

$$\equiv \frac{1}{p}\left\{-\rho_{(k)}\boldsymbol{F}_{(k)} + \frac{\partial p_{(k)}}{\partial \boldsymbol{r}} - C_{(k)}\sum_j\left(-\rho_{(j)}\boldsymbol{F}_{(j)} + \frac{\partial p_{(j)}}{\partial \boldsymbol{r}}\right)\right\}, \quad (k = 1, 2, \ldots)$$

which can be written in the form of equations of motion for the individual mixture components [see, e.g., Note I in the monograph by Chapman and Cowling (1952)],

$$\rho_{(k)}\frac{\mathrm{d}\boldsymbol{u}}{\mathrm{d}t} = -\frac{\partial p_{(k)}}{\partial \boldsymbol{r}} + \rho_{(k)}\boldsymbol{F}_{(k)} + \sum_j k_\mathrm{B}T\frac{n_{(j)}n_{(k)}}{n\mathcal{D}_{(jk)}}(\boldsymbol{u}_{(j)} - \boldsymbol{u}_{(k)}) - k_\mathrm{B}k_{T(k)}n\frac{\partial T}{\partial \boldsymbol{r}}.$$

Here, $\rho_{(k)} = m_{(k)}n_{(k)}$, $C_{(k)} = \rho_{(k)}/\rho$, and $p_{(k)}$ are the mass density, mass concentration, and partial pressure of the particles of type k, respectively; $p = \sum_k p_{(k)}$ is the total pressure of the mixture (the *Dalton* law), $p_{(k)} = n_{(k)}k_\mathrm{B}T$; k_B is the *Boltzmann* constant; $n = \sum_k n_{(k)}$, and $\rho = \sum_k \rho_{(k)}$ are the total number and mass densities of the multicomponent mixture, respectively; $\mathcal{D}_{(jk)}$ are the binary diffusion coefficients; $\boldsymbol{F}_{(k)}$ is the external bulk force (per unit mass of component k); and $k_{T(k)}$ is the thermodiffusion ratio.

These Stefan–Maxwell relations for multicomponent mixtures were derived by the methods of nonequilibrium thermodynamics and then generalized them to heterogeneous mixtures (Kolesnichenko 1998; Kolesnichenko and Maksimov 2001). They can also be written in the form of equations of motion for the individual phases of the system. In the single-pressure approximation ($p_\mathrm{g} = p_\mathrm{d}$) considered here, these relations for the interphase diffusion take the form

$$\sum_\beta R_{\alpha\beta}(\boldsymbol{u}_\beta - \boldsymbol{u}_\alpha) - pk_{T\alpha}\frac{\partial \ln T}{\partial \boldsymbol{r}} = \boldsymbol{d}_\alpha \equiv \tilde{\rho}_\alpha\boldsymbol{K}_\alpha + s_\alpha\frac{\partial p_\alpha}{\partial \boldsymbol{r}} - C_\alpha\sum_\beta\left(-\tilde{\rho}_\alpha\boldsymbol{K}_\beta + s_\beta\frac{\partial p_\beta}{\partial \boldsymbol{r}}\right)$$

$$\equiv \rho_\alpha s_\alpha\frac{d_\alpha\boldsymbol{u}_\alpha}{dt} + s_\alpha\frac{\partial p}{\partial \boldsymbol{r}} - \rho_\alpha s_\alpha\boldsymbol{F}_\alpha - \left(\frac{\partial}{\partial \boldsymbol{r}}\cdot\Pi_\alpha\right) + \frac{1}{2}\sum_\beta\sigma_{\alpha\beta}(\boldsymbol{u}_\alpha - \boldsymbol{u}_\beta), \quad (\alpha = 1, 2, \ldots), \tag{7.17}$$

where $d_\alpha(\ldots)/dt = \partial(\ldots)/\partial t + \boldsymbol{u}_\alpha \cdot \partial(\ldots)/\partial \boldsymbol{r}$ is the substantial derivative along the trajectory of the center of mass of phase α contained inside the macrovolume element δV of a multiphase medium; Π_α is the partial viscous stress tensor; $R_{\alpha\beta}$ is the coefficient of interphase friction for phases α and β (since the coefficient $R_{\alpha\beta}$ reflects the interaction between two phase continua, it is often convenient to write in a symmetric form, $R_{\alpha\beta} = \tilde{\rho}_\alpha \tilde{\rho}_\beta \theta_{\alpha\beta}$, where the parameters $\theta_{\alpha\beta}$ do not depend, at least roughly, on the mixture proportions), $R_{\alpha\beta} = R_{\beta\alpha}$; $\boldsymbol{F}_\alpha$ is the external bulk force per unit mass of phase α;

$$\tilde{\rho}_\alpha \boldsymbol{K}_\alpha \equiv -\tilde{\rho}_\alpha \frac{d_\alpha \boldsymbol{u}_\alpha}{dt} + \tilde{\rho}_\alpha \boldsymbol{F}_\alpha + \left(\frac{\partial}{\partial \boldsymbol{r}} \cdot \Pi_\alpha\right) - \frac{1}{2}\sum_\beta \sigma_{\alpha\beta} \boldsymbol{w}_{\alpha\beta} \tag{7.18}$$

is the generalized thermodynamic force coupled with the diffusion flux $\boldsymbol{J}_\alpha$; the quantities $\boldsymbol{d}_\alpha$ also have the meaning of generalized thermodynamic forces that cause the relative motion of the phases $\left(\sum_\alpha \boldsymbol{d}_\alpha = 0\right)$; $k_{T\alpha}$ is the thermophoretic ratio, with $\sum_\alpha k_{T\alpha} = 0$. A more detailed analysis of the interphase interaction shows that large gradients in macroscopic parameters, the rotation of solid particles, nonstationary establishment of the velocity profile near the particles, the deformation of disperse particles, and some other effects can give rise to additional forces on the left-hand side of (7.17). Apart from the *Stokes* force of friction, the *Safman* force (related to the nonuniformity of the carrier-gas velocity profile), the "hereditary" *Basset* force (which takes into account the prehistory of the motion on the behavior of disperse particles), and the *Magnus* or *Zhukowski* force (the force of additional action on rotating disperse particles due to the gradients in the mean velocity field of the carrier phase), etc., can be such forces.

A number of theoretical formulas were suggested (see, e.g., Soo et al. 1960; Mednikov 1981) to determine the thermophoretic force. It should be noted, however, that the thermophoresis-related force may be disregarded for disperse particles whose sizes are generally much smaller than the temperature nonuniformity scale length in the disk. The last term on the right-hand side of (7.17), $(1/2)\sum_\beta \boldsymbol{w}_{\alpha\beta} \sum_{\rho=1}^{r} \nu_{\alpha,\rho}\xi_\rho$, describes the change in the momentum of phase α via phase transitions (recall that $\boldsymbol{w}_{\alpha\beta} \equiv \boldsymbol{u}_\alpha - \boldsymbol{u}_\beta$). In the case of a two-phase gas–dust disk system under consideration, this term take into account the loss of momentum by the dust-phase particles through the mass transformation of some of them to the gas phase during evaporation or the acquisition of additional momentum by the disperse phase through the formation (from the gas) of new solid particles during condensation. Nevertheless, below we will also disregard this term in most cases, since it is almost always much smaller than the *Stokes* force of friction, $\boldsymbol{F}_{\text{fric},\alpha} \equiv \sum_\beta R_{\alpha\beta} \boldsymbol{w}_{\alpha\beta}$ arising from phase viscosity effects (see, e.g., Nigmatulin 1978) and is exactly equal to zero in the absence of phase transitions. It is important to emphasize once again that, in contrast to the classical inertia-free Stefan–Maxwell relations for the relative velocities of the components $\boldsymbol{w}_{(jk)} = \boldsymbol{u}_{(j)} - \boldsymbol{u}_{(k)}$, the interphase diffusion laws described by the generalized equations (7.17) incorporate the inertia of the relative phase motion.

For a two-phase gas–dust disk medium, (7.17) take the form of equations of motion for gas and dust

$$\begin{cases} R_{\rm gd}\boldsymbol{w} \equiv (1-s)s\rho_{\rm d}\theta_{\rm dg}\boldsymbol{w} = \rho_{\rm g}(1-s)\frac{d_{\rm g}\boldsymbol{u}_{\rm g}}{{\rm d}t} + (1-s)\frac{\partial p}{\partial \boldsymbol{r}} - \rho_{\rm g}(1-s)\boldsymbol{g} - \left(\frac{\partial}{\partial \boldsymbol{r}}\cdot \Pi_{\rm g}\right) \\ -R_{\rm dg}\boldsymbol{w} = \rho_{\rm d}s\frac{d_{\rm d}\boldsymbol{u}_{\rm d}}{{\rm d}t} + s\frac{\partial p}{\partial \boldsymbol{r}} - \rho_{\rm d}s\boldsymbol{g}, \end{cases} \qquad (7.17^*)$$

where $\boldsymbol{F}_{\rm d} = \boldsymbol{F}_{\rm g} \equiv \boldsymbol{g}(\boldsymbol{r},t)$ is the bulk force per unit mass, which is generally related to both the gravitational attraction by the star and the gravitational attraction by the gas–dust cloud itself. Note that the general forms of the equations of motion and the continuity equations in our paper and those of various authors who studied the gas suspensions in accretion disks are identical. The only difference is the presence of the term $-s(\partial p/\partial \boldsymbol{r})$ in the equation of motion (7.17*) for gas and the absence of the term $s(\partial p/\partial \boldsymbol{r})$ in the equation of motion (7.17*) for dust. Their appearance in continuum heterogeneous mechanics is eventually related to allowance for the effect of associated masses (due to the accelerated motion of the solid particles relative to the carrier gas, when disturbances of the order of the particle size emerge in the latter) and the buoyant force, which are often much smaller than the other terms in the equations of motion at large $\rho_{\rm d}/\rho_{\rm g}$—typical of gas flows with solid particles (Nigmatulin 1978). At the same time, since these forces are proportional not only to the gas density, but also to the local acceleration of the medium or the difference between the local accelerations of the gas medium and the solid particles, the situations where these additional terms of (7.17*) will be comparable to the aerodynamic Stokes force are quite possible.

The sought-for diffusion relation for the relative velocity vector of the dust and the gas $\boldsymbol{w} \equiv (\boldsymbol{u}_{\rm d} - \boldsymbol{u}_{\rm g})$ can be derived from the terms of (7.18) that describe the action of friction if we divide each of them by the corresponding quantity $s_\alpha\rho_\alpha$, subtract one from the other, and separate out the term with $\boldsymbol{w}$. Writing the true velocities of the phases as $\boldsymbol{u}_{\rm d} = (C_{\rm g}\boldsymbol{w} + \boldsymbol{u})$ and $\boldsymbol{u}_{\rm g} = (-C_{\rm d}\boldsymbol{w} + \boldsymbol{u})$ and assuming that $\rho_{\rm d} \gg \rho_{\rm g}$, we obtain a defining relation for $\boldsymbol{w}$ (an analog of the Darcy law for filtration),

$$\frac{\rho}{\tilde{\rho}_{\rm d}\tilde{\rho}_{\rm g}}R_{\rm dg}\boldsymbol{w} \equiv \rho\theta_{\rm gd}\boldsymbol{w} = \frac{d_{\rm g}\boldsymbol{u}_{\rm g}}{dt} - \frac{d_{\rm d}\boldsymbol{u}_{\rm d}}{dt} + \frac{\rho_{\rm d}-\rho_{\rm g}}{\rho_{\rm g}\rho_{\rm d}}\frac{\partial p}{\partial \boldsymbol{r}} - \frac{1}{\tilde{\rho}_{\rm g}}\left(\frac{\partial}{\partial \boldsymbol{r}}\cdot \Pi_{\rm g}\right)$$
$$\cong -\frac{{\rm d}\boldsymbol{w}}{dt} + \frac{1}{\rho_{\rm g}}\frac{\partial p}{\partial \boldsymbol{r}}, \qquad (7.19)$$

which is considered below as the main equation in modeling the phase diffusion in the disk. Note that the gravitational forces after this subtraction canceled out, but their action on the motion of the gas–dust medium manifests itself through a pressure gradient. In writing (7.19), we made no distinction between the substantial derivatives for the individual phases and the system as a whole; i.e., we assumed that $d_{\rm d}/{\rm d}t \cong d_{\rm g}/{\rm d}t \cong d/{\rm d}t$, which is valid in the mechanics of mixtures only in the so-called diffusion approximation. In general, i.e., when the acceleration of the diffusion flows relative to the center of mass is taken into account, the following exact transformation should be used:

$$\begin{aligned}\frac{d_{\mathrm{d}}\boldsymbol{u}_{\mathrm{d}}}{dt}-\frac{d_{\mathrm{g}}\boldsymbol{u}_{\mathrm{g}}}{dt}&=\frac{\mathrm{d}\boldsymbol{w}}{\mathrm{d}t}+\left(\boldsymbol{w}_{\mathrm{d}}\cdot\frac{\partial}{\partial\boldsymbol{r}}\right)\boldsymbol{u}_{\mathrm{d}}-\left(\boldsymbol{w}_{\mathrm{g}}\cdot\frac{\partial}{\partial\boldsymbol{r}}\right)\boldsymbol{u}_{\mathrm{g}}\\&=\frac{\mathrm{d}\boldsymbol{w}}{\mathrm{d}t}+\left(\boldsymbol{w}\cdot\frac{\partial}{\partial\boldsymbol{r}}\right)\boldsymbol{u}+C_{\mathrm{g}}\left(\boldsymbol{w}\cdot\frac{\partial}{\partial\boldsymbol{r}}\right)C_{\mathrm{g}}\boldsymbol{w}-C_{\mathrm{d}}\left(\boldsymbol{w}\cdot\frac{\partial}{\partial\boldsymbol{r}}\right)C_{\mathrm{d}}\boldsymbol{w}\\&=\frac{\mathrm{d}\boldsymbol{w}}{\mathrm{d}t}+\left(\boldsymbol{w}\cdot\frac{\partial}{\partial\boldsymbol{r}}\right)\boldsymbol{u}+\Im(\boldsymbol{w}^2)\approx\frac{\mathrm{d}\boldsymbol{w}}{\mathrm{d}t}+\left(\boldsymbol{w}\cdot\frac{\partial}{\partial\boldsymbol{r}}\right)\boldsymbol{u},\end{aligned}\tag{7.20}$$

where $\Im(\boldsymbol{w}^2)\equiv C_{\mathrm{g}}\left(\boldsymbol{w}\cdot\frac{\partial}{\partial\boldsymbol{r}}\right)C_{\mathrm{g}}\boldsymbol{w}-C_{\mathrm{d}}\left(\boldsymbol{w}\cdot\frac{\partial}{\partial\boldsymbol{r}}\right)C_{\mathrm{d}}\boldsymbol{w}$ is a quadratic (in $\boldsymbol{w}$) function; the latter may often be omitted (see Youdin and Goodman 2004), in particular, for a fine passive admixture, since $|\boldsymbol{w}|$ is small for them. The defining diffusion relation for the relative phase velocity vector then takes a more complex form:

$$\rho\theta_{\mathrm{gd}}\boldsymbol{w}\cong-\frac{\mathrm{d}\boldsymbol{w}}{\mathrm{d}t}-\left(\boldsymbol{w}\cdot\frac{\partial}{\partial\boldsymbol{r}}\right)\boldsymbol{u}-C_{\mathrm{g}}\left(\boldsymbol{w}\cdot\frac{\partial}{\partial\boldsymbol{r}}\right)C_{\mathrm{g}}\boldsymbol{w}+C_{\mathrm{d}}\left(\boldsymbol{w}\cdot\frac{\partial}{\partial\boldsymbol{r}}\right)C_{\mathrm{d}}\boldsymbol{w}+\frac{1}{\rho_{\mathrm{g}}}\frac{\partial p}{\partial\boldsymbol{r}}\tag{7.19*}$$

7.2.2.1 Aerodynamic Drag Coefficient

The coefficient of friction R_{dg} between gas and dust continua is used to define by various formulas, depending on the characteristic diameter $\tilde{d}_{\mathrm{d}}$ of the disperse-phase particles (Sternin et al. 1980; Shraiber et al. 1987). If the characteristic diameter of spherical solid particles $\tilde{d}_{\mathrm{d}}\ll\lambda_{\mathrm{g}}$, where λ_{g} is the mean free path of molecules in the gas phase, then R_{dg} is given by the Epstein formula (Epstein 1924). For course spherical condensates with diameters larger than the mean free path of gas molecules, the coefficient of resistance is defined by the Stokes law (Stokes 1851). Thus, for the coefficients of resistance R_{dg} (or θ_{dg}) of a smooth spherical particle, we have (see, e.g., Weidenschilling 1977; Garaud et al. 2005),

$$R_{\mathrm{d\,g}}=\begin{cases}\dfrac{2\tilde{\rho}_{\mathrm{d}}\tilde{\rho}_{\mathrm{g}}c_{\mathrm{sg}}}{\tilde{d}_{\mathrm{d}}\,\rho_{\mathrm{d}}}, & \text{when}\quad\tilde{d}_{\mathrm{d}}\ll\lambda_{\mathrm{g}}(\text{regime of Epstein}),\\ \dfrac{2\tilde{\rho}_{\mathrm{d}}\tilde{\rho}_{\mathrm{g}}}{\tilde{d}_{\mathrm{d}}\rho_{\mathrm{d}}}C_{\mathrm{D}}(\boldsymbol{Re}_{\mathrm{d}})|\boldsymbol{w}|, & \text{when}\quad\tilde{d}_{\mathrm{d}}\gg\lambda_{\mathrm{g}}(\text{regime of Stokes}),\end{cases}\tag{7.21}$$

or

$$\theta_{\mathrm{d\,g}}=\begin{cases}\dfrac{2c_{\mathrm{sg}}}{\tilde{d}_{\mathrm{d}}\rho_{\mathrm{d}}}, & \text{when}\quad\tilde{d}_{\mathrm{d}}\ll\lambda_{\mathrm{g}},\\ \dfrac{2}{\tilde{d}_{\mathrm{d}}\rho_{\mathrm{d}}}C_{\mathrm{D}}(\boldsymbol{Re}_{\mathrm{d}})|\boldsymbol{w}|, & \text{when}\quad\tilde{d}_{\mathrm{d}}\gg\lambda_{\mathrm{g}}.\end{cases}\tag{7.21*}$$

Here, c_{sg} is the speed of sound in the gas (see (7.57) below); $\boldsymbol{Re}_{\mathrm{d}} = \tilde{d}_{\mathrm{d}}|\boldsymbol{w}|/\nu_{\mathrm{g}}$ is the Reynolds number for the dust; ν_{g} is the coefficient of molecular kinematic viscosity for the gas component of the mixture $\nu_{\mathrm{g}} = \lambda_{\mathrm{g}} c_{\mathrm{sg}}/2$; $C_{\mathrm{D}}(\boldsymbol{Re}_{\mathrm{d}})$ is the coefficient of aerodynamic resistance (the so-called standard resistance curve), which has a fairly complex form. A considerable number of formulas that fit this curve are known (Schlichting 1964; Sternin et al. 1980; Mednikov 1981). In particular, the following expression gained wide acceptance in astrophysics (Whipple 1972):

$$C_{\mathrm{D}}(\boldsymbol{Re}_{\mathrm{d}}) = \begin{cases} 9\,\boldsymbol{Re}_{\mathrm{d}}^{-1}, & \boldsymbol{Re}_{\mathrm{d}} \le 1, \\ 9\,\boldsymbol{Re}_{\mathrm{d}}^{-0.6}, & 1 \le \boldsymbol{Re}_{\mathrm{d}} \le 800, \\ 0.165, & \boldsymbol{Re}_{\mathrm{d}} \ge 800. \end{cases} \tag{7.22}$$

In our view, the following trinomial formula is no less convenient:

$$C_{\mathrm{D}}(\boldsymbol{Re}_{\mathrm{d}}) = 9\boldsymbol{Re}_{\mathrm{d}}^{-1}(1 + 0.179\boldsymbol{Re}_{\mathrm{d}}^{1/2} + 0.013\boldsymbol{Re}_{\mathrm{d}}), \quad (0.1 < \boldsymbol{Re}_{\mathrm{d}} < 10^3) \tag{7.22*}$$

its advantage is that it is applicable over a wide $\boldsymbol{Re}_{\mathrm{d}}$ range.

It should be noted that, in general, the conditions for the flow around particles in actual multiphase flows differ significantly from the idealized conditions in which the standard curve is applicable. The particles in a gas–dust cloud generally have an irregular shape and a rough surface and move nonuniformly in a turbulized flow of rarefied and compressible gas. Of course, each of these factors changes (sometimes significantly) the conditions for the flow around a particle in a disk and the force of aerodynamic resistance. Let us briefly consider their effect that, as a rule, is disregarded in astrophysical literature.

1. It is customary to characterize the degree of deviation of the particle shape from a sphere in heterogeneous mechanics by the shape factor β ($\beta \ge 1$), the ratio of the surface area of an actual particle to the surface area of a sphere of the same volume. Gorbis (1970) suggested formulas for calculating the coefficients of aerodynamic resistance $C_{\mathrm{D}}(\boldsymbol{Re}_{\mathrm{d}}, \beta)$, which have higher values than those of the standard curve for essentially nonisometric dust particle shapes.
2. Sternin et al. (1980) established that the coefficient of resistance $C_{\mathrm{D}}(\boldsymbol{Re}_{\mathrm{d}})$ for particles with appreciable roughness increases (compared to the standard curve) if the latter is comparable to the thickness of the boundary layer.
3. Flow turbulization also affects significantly $C_{\mathrm{D}}(\boldsymbol{Re}_{\mathrm{d}})$. As was pointed out by Sternin and Shraiber (1994), according to the data of various authors, for example, for $20 < \boldsymbol{Re}_{\mathrm{d}} < 100$, C_{D} varies within the range (0.01–3) C_{D}^* (in what follows, C_{D}^* corresponds to the standard curve). It is important to note that the effect of turbulence decreases with decreasing Reynolds number $\boldsymbol{Re}_{\mathrm{d}}$. For comparatively small $\boldsymbol{Re}_{\mathrm{d}}$, the Lopez–Dackler formulas (see, e.g., Sternin et al. 1980) can be used:

$$C_D(\boldsymbol{Re}_d) = \begin{cases} 60.75\,\varepsilon^{1/3}\boldsymbol{Re}_d^{-1}, & \boldsymbol{Re}_d < 50; \quad 0.05 < \varepsilon < 0.5; \\ 0.0498(1 + 150/\boldsymbol{Re}_d)^{1.565} + 1.5\varepsilon, & 50 < \boldsymbol{Re}_d < \boldsymbol{Re}_d^*; \quad 0.07 < \varepsilon < 0.5, \end{cases}$$

where ε is the relative turbulence level, i.e., the ratio of the root-mean-square pulsation velocity to the averaged sliding velocity; $\boldsymbol{Re}_d^* = \min\{0.9\boldsymbol{Re}_{crit}, 700\}$; $\boldsymbol{Re}_{crit}$ is the critical Reynolds number $\ln \boldsymbol{Re}_{crit} = 5.477 - 15.8\varepsilon$ $(\varepsilon \leq 0.15)$; $\ln \boldsymbol{Re}_{crit} = 3.371 - 1.75\varepsilon$ $(\varepsilon > 0.15)$.

4. The onflow compressibility and rarefaction affect significantly the aerodynamic resistance of the particles. The role of these factors is determined primarily by the Mach, $\boldsymbol{Ma} = |\boldsymbol{u}_g|/c_{sg}$, and Knudsen, $\boldsymbol{Kn}$, numbers. The compressibility of the carrier gas plays a significant role in a high-velocity flow of gas suspension in a disk. From the numerous generalized relations available in the literature, we give the following formula (Sternin et al. 1980):

$$C_D(\boldsymbol{Re}_d) = 9\boldsymbol{Re}_d^{-1}(1 + 0.179\boldsymbol{Re}_d^{1/2} + 0.013\boldsymbol{Re}_d) \times \frac{[1 + \exp(-0.427\boldsymbol{Ma}^{-4.63} - 3\boldsymbol{Re}_d^{-0.88})]}{1 + \boldsymbol{Re}_d^{-1}\boldsymbol{Ma}[3.82 + 1.28\exp(-1.25\boldsymbol{Ma}^{-1}\boldsymbol{Re}_d)]}, \qquad (7.22^{**})$$

where $\boldsymbol{Re}_d < 100$ and $\boldsymbol{Ma} < 2$. Here, the first two factors in the numerator correspond to the standard curve, the third factor takes into account the compressibility effect, and the denominator take into account the rarefaction effect. Hunter et al. (1981) gave the following relation for $\boldsymbol{Re}_d < 10^3$:

$$C_D(\boldsymbol{Re}_d) = (C_D^* - 2)\exp\left[-3.07\gamma^{1/2}\frac{\boldsymbol{Ma}}{\boldsymbol{Re}_d}\cdot\frac{1 + \boldsymbol{Re}_d(12/28 + 0.584\boldsymbol{Re}_d)}{1 + 11.28\boldsymbol{Re}_d}\right] + \frac{1}{\gamma^{1/2}\boldsymbol{Ma}}\left[\frac{5.6}{\boldsymbol{Ma} + 1} + 1.7\left(\frac{T_d}{T}\right)^{1/2}\right]\exp\left(-\frac{\boldsymbol{Re}_d}{2\boldsymbol{Ma}}\right) + 2,$$

where T_d is the dust particle temperature.

5. At a reduced gas pressure in the disk (for example, according to Wasson (1985) the following pressure estimates are pertinent to the midplane of the circumsolar disk: 2×10^{-5}–10^{-1} g/cm^3 at $r = 1$ AU and 5×10^{-7}–2×10^{-6}g/cm^3 at $r = 3$ AU), the gas component of the medium can slide over the surface of a solid particle, which also causes the coefficient of aerodynamic resistance to decrease. The rarefaction of the medium is characterized by the Knudsen number, $\boldsymbol{Kn} = \lambda_g/\tilde{d}_d$. Four $\boldsymbol{Kn}$ ranges are usually distinguished: Knudsen flow ($\boldsymbol{Kn} > 10$), transition regime ($10 > \boldsymbol{Kn} > 0.25$), sliding flow ($0.25 > \boldsymbol{Kn} > 0.01$), and continuum flow (there is no rarefaction effect, $\boldsymbol{Kn} < 0.01$). For the first three ranges, the coefficient of aerodynamic resistance may be represented as $C_D = \varphi C_D^*$, where the coefficient φ is defined by the well-known Millikan formula (see, e.g., Fuks 1955)

$$\varphi = \{1 + \boldsymbol{Kn}[1.155 + 0.471\exp(-0.596/\boldsymbol{Kn})]\}^{-1}$$

As the rarefaction increases, the effect of compressibility on the coefficient of resistance degenerates. All the above improvements of (7.22) can be easily taken into account when numerically simulating the structure of a protoplanetary gas–dust disk, for example, at the subdisk formation stage.

Returning to (7.21*) for the coefficient $\theta_{\mathrm{d\,g}}(\boldsymbol{Re}_{\mathrm{d}})$, note that (7.21*) are convenient only for monodisperse dust with a given characteristic linear diameter of the inclusions $\tilde{d}_{\mathrm{d}}$, since in this case $\theta_{\mathrm{d\,g}}$ does not depend on the volume concentration of the disperse phase s and the total number density of the solid particles N_{d}. However, when the coagulation processes in the gas–dust protoplanetary cloud are taken into account, i.e., given that the dust is multifractional, it is appropriate to rewrite (7.21*) in a form that explicitly depends on the parameters s and N_{d}, which are defined by the Smoluchowski equation. When using (7.16), (7.21*) for $\theta_{\mathrm{d\,g}}(s, N_{\mathrm{d}}, \boldsymbol{Re}_{\mathrm{d}})$ transforms to

$$\theta_{\mathrm{d\,g}}(s, N_{\mathrm{d}}, \boldsymbol{Re}_{\mathrm{d}}) = \begin{cases} \left(\frac{4}{3}\pi\right)^{1/3} \rho_{\mathrm{d}}^{-1} s^{-1/3} c_{\mathrm{sg}} N_{\mathrm{d}}^{1/3}, & \text{when} \quad \tilde{d}_{\mathrm{d}} \ll \lambda_{\mathrm{g}}, \\ \left(\frac{4}{3}\pi\right)^{1/3} \rho_{\mathrm{d}}^{-1} s^{-1/3} N_{\mathrm{d}}^{1/3} C_{\mathrm{D}}(\boldsymbol{Re}_{\mathrm{d}})|\boldsymbol{w}|, & \text{when} \quad \tilde{d}_{\mathrm{d}} >> \lambda_{\mathrm{g}}. \end{cases} \tag{7.23}$$

7.2.3 *Allowance for the Multifractional Composition of Dust: The Kinetic Coagulation Equation*

Let us now consider in more detail the technique of calculating $N_{\mathrm{d}}(\boldsymbol{r}, t)$ when the dust multifractionality of the system is taken into account. The actual protoplanetary cloud is polydisperse; i.e., condensed particles of various sizes $d_{\mathrm{d},k}$ are present in the macrovolume element $\delta\mathcal{V}$. This factor can be taken into account by breaking down the dust component into a finite number of fractions each of which is gene rally characterized by its own thermohydrodynamic parameters; i.e., instead of one disperse phase, we must consider m phases (where m is the number of fractions) each of which has its own macroparameters,

$$d_{\mathrm{d},k}, \quad n_{\mathrm{d},k}, \quad s_{\mathrm{d},k} = n_{\mathrm{d},k}\left(\frac{\pi}{6}\right)d_{\mathrm{d},k}^3, \quad \rho_{\mathrm{d},k}, \quad \boldsymbol{u}_{\mathrm{d},k}\cdots \quad (k = 1, \ldots, m), \tag{7.24}$$

where $\boldsymbol{u}_{\mathrm{d},k}(\boldsymbol{r}, t)$ is the hydrodynamic velocity of the fraction-k solid particles.

Next, we assume that the material of different fractions is the same ($\rho_{\mathrm{d},1} = \rho_{\mathrm{d},2} = \ldots = \rho_{\mathrm{d},m} = \rho_{\mathrm{d}} = \mathrm{const}$) and that the solid-phase condensates of fraction 1 constitute the group of the smallest (primary) particles, those of fraction 2 constitute the group of double particles, etc., up to the maximum size. To simplify our analysis of the coagulation process in an $(m + 1)$-phase polydisperse flow, we also assume that all solid particles are spherical or nearly spherical with the Feret diameter $d_{\mathrm{d},k}$. Since the size of chemically identical solid particles after their adhesion increases as the cubic root of the number of its constituent primary condensates ($d_{\mathrm{d},k} = d_{\mathrm{d},1}\sqrt[3]{k}$), the volume concentration of the fraction-k disperse particles is defined by the relation

$$s_{\mathrm{d},k} = n_{\mathrm{d},k}(\pi/6)d_{\mathrm{d},k}^{3} = U_1\, \boldsymbol{k}\, \boldsymbol{n}_{\mathrm{d},k} \tag{7.25}$$

where $n_{\mathrm{d},k}(\boldsymbol{r}, t)$ is the number density of the fraction-k particles (their number in a unit total gas-suspension volume); $U_1 = (\pi/6)d^3$ and $d \equiv d_{\mathrm{d},1}$ are, respectively, the volume and diameter of one smallest particle. The volume content $s(\boldsymbol{r}, t)$, the distributed mass density $\tilde{\rho}_{\mathrm{d}}(\boldsymbol{r}, t)$, and the hydrodynamic velocity $\boldsymbol{u}_{\mathrm{d}}(\boldsymbol{r}, t)$ of the entire dust continuum can then be expressed as

$$s \equiv \sum_{k=1}^{m} s_{\mathrm{d},k} = U_1 \sum_{k=1}^{m} k\, n_{\mathrm{d},k}, \quad \tilde{\rho}_{\mathrm{d}} = \rho_{\mathrm{d}} \sum_{k=1}^{m} s_{\mathrm{d},k}, \quad s\, \boldsymbol{u}_{\mathrm{d}} = \sum_{k=1}^{m} s_{\mathrm{d},k} \boldsymbol{u}_{\mathrm{d},k}. \tag{7.26}$$

In a disperse mixture in which the macroscopic velocities of the fractions differ, i.e., fractions j and k move relative to one another with the velocity $\boldsymbol{u}_{\mathrm{d},j} - \boldsymbol{u}_{\mathrm{d},k}$ ($j, k = 1, \ldots, m$), there will be collisions between particles of different fractions, which will lead to the mass, momentum, and energy transfer between the fractions. Allowance for this fact, which is important at the final subdisk formation stage (when the fraction of particles "reflected" from the subdisk with a mean or macroscopic velocity different from that of the "incident" particles can appear in the flow) and at the planetesimal formation stage (after the subdisk disintegration), severely complicates the problem of modeling the evolution of a protoplanetary gas–dust disk (Kolesnichenko 2001). Here we assume, however, that the particles of the material ("pseudomolecules") belonging to different dust continua (fractions) move with the same hydrodynamic velocity, $\boldsymbol{u}_{\mathrm{d},k} \equiv \boldsymbol{u}_{\mathrm{d}}$ ($k = 1, \ldots, m$).

We have already mentioned above that the fragmentation and coagulation processes are the main size formation mechanisms of large solid particles as they accumulate in a gas–dust cloud. The fragmentation mechanism of colliding solid bodies was thoroughly studied (Sternin and Shraiber 1994) and can be taken into account, when necessary; therefore, below, we will not consider the particle disintegration in order not to overload the problem of modeling the disk evolution with details. In this case, the number density $n_{\mathrm{d},k}$ of fraction k can change only through a decrease in the number of particles of this fraction as they combine with other dust particles and through an increase in the number of particles of this fraction due to the adhesion of smaller condensates. The system of kinetic equations that describes the coagulation can then be written as (Smoluchowski 1936)

$$\overset{\circ}{n}_{\mathrm{d},k} = \frac{1}{2}\sum_{j=1}^{k-1} K_{j(k-j)} n_{\mathrm{d},j} n_{\mathrm{d},(k-j)} - \sum_{j=1}^{m} K_{kj} n_{\mathrm{d},k} n_{\mathrm{d},j} \quad (k = 1, 2, \ldots, m), \tag{7.27}$$

where $\overset{\circ}{n}_{\mathrm{d},k}(\boldsymbol{r},t)$ is the total rate of change in the number density $n_{\mathrm{d},k}(\boldsymbol{x},t)$ of fraction-k dust particles via the coagulation processes; $K_{kj}(d_k, d_j)$ is the coagulation coefficient (kernel) for particles of sizes k and j that characterizes the coagulation interaction efficiency; it is defined as the mean number of collisions between particles of sizes d_k and d_j in a unit volume per unit time for a unit number density of one and the other type. Since such an interaction between two different-size particles in the flow is complicated by the influence of the ambient medium, the pattern of interaction in a laminar or turbulent flow, and the force fields (gravitation, electromagnetic field, molecular interaction), determining the coagulation kernel is a challenging problem of its own (see, e.g., Voloshchuk 1984; Mazin 1971).

Kolesnichenko (2001) analyzed various coagulation mechanisms for a turbulized gas–dust cloud and provided the corresponding expressions for the coefficients K_{kj}. Given (7.27), the system of instantaneous equations for the conservation of the number of fraction-k dust particles takes the form

$$\frac{\partial}{\partial t} n_{\mathrm{d},k} + \operatorname{div}\left(n_{\mathrm{d},k}\,\boldsymbol{u}_{\mathrm{d}}\right) = \overset{\circ}{n}_{\mathrm{d},k} = \frac{1}{2}\sum_{j=1}^{k-1} K_{j(k-j)} n_{\mathrm{d},j}\, n_{\mathrm{d},(k-j)} - n_{\mathrm{d},k}\sum_{j=1}^{m} K_{kj}\, n_{\mathrm{d},j}$$
$$(k = 1, 2, \ldots, m). \tag{7.28}$$

The balance equation for the total number $N'_{\mathrm{d}} = \sum_k n_{\mathrm{d},k}$ of disperse particles in a unit total gas-suspension volume determined only by the coagulation processes (see (7.15)) follows from (7.28):

$$\rho\frac{\mathrm{d}}{\mathrm{d}t}\left(\frac{N'_{\mathrm{d}}}{\rho}\right) + \operatorname{div}\left(N'_{\mathrm{d}}\boldsymbol{w}_{\mathrm{d}}\right) = \sum_k \overset{\circ}{n}_{\mathrm{d},k} = -\frac{1}{2}\sum_{k=1}\sum_{j=1} K_{kj} n_{\mathrm{d},k} n_{\mathrm{d},j}. \tag{7.29}$$

Note that the right-hand side of (7.29) is equal to half the second term on the right-hand side of (7.28), since the total number of dust particles in a unit volume does not increase during the coagulation. In the spatially uniform case where all coagulation constants are approximately equal, $K_{kj} = K$, (7.29) $\partial N'_{\mathrm{d}} \partial\, t = -(K/2)\, N'^{2}_{\mathrm{d}}$ (with the initial condition $N'_{\mathrm{d}}(0) = N'_{\mathrm{d}0}$) has a simple solution, $N'_{\mathrm{d}}(t) = N'_{\mathrm{d}0}/(1+qt)$, where $q = KN'_{\mathrm{d}0}/2$, which allows the coagulation constant K to be determined experimentally (from the slope of the straight line).

Given that the total mass of the dust particles during the coagulation is conserved, $\sum_k m_{\mathrm{d},k}\, \overset{\circ}{n}_{\mathrm{d},k} = 0$, the summation of the left- and right-hand sides of the equations of system (7.28) over k that were first multiplied by the mass of an individual fraction-k particle, $m_{\mathrm{d},k} = \rho_{\mathrm{d}} U_1 k$, leads to (7.14*), which allows the total dust volume concentration s in a two-phase polydisperse flow to be calculated.

It is also important to keep in mind that the number of nonlinear differential equations (7.28) required to describe the space–time distribution of the entire set of dust particle sizes in the disk is generally infinite. At the same time, we have to use a finite (m) number of equations when numerically simulating the coagulation processes based on system (7.28). Of course, the "loss of material" is possible in this case, since a number of particles can coagulate to sizes exceeding the largest size $d_{\mathrm{d},m}$ taken into account in this approach. Therefore, for our purposes, a different (integral) form of the system of coagulation equations (7.28) is preferred.

To obtain this form, we assume that the number of particles located at time t in a volume element from U to $U + \mathrm{d}U$ in the vicinity of point $\boldsymbol{r}$ is $f(U, \boldsymbol{r}, t)\,\mathrm{d}U$. By definition, the function $f(U, \boldsymbol{r}, t)$, which characterizes the particle size spectrum, satisfies the normalization relation

$$N_{\mathrm{d}}(\boldsymbol{r}, t) = \int_0^\infty f(U, \boldsymbol{r}, t)\,\mathrm{d}U. \tag{7.30}$$

Clearly, the formula

$$s(\boldsymbol{r}, t) = \int_0^\infty U f(\boldsymbol{r}, t, U)\mathrm{d}U \tag{7.31}$$

defines the total volume concentration of the dust particles. Since the volume of the size-k particles is equal to kU_1, the number density $n_{\mathrm{d},k}$ of particles k can be expressed in terms of $f(\boldsymbol{r}, t, U)$ as

$$n_{\mathrm{d},k} = f(kU_1, \boldsymbol{r}, t)\,U_1. \tag{7.32}$$

Using this relation, we can derive the following kinetic coagulation equation from (7.28) after the operation $U_1 \to \mathrm{d}U$:

$$\frac{\partial f(U,\boldsymbol{r},t)}{\partial t} + \mathrm{div}[f(U,\boldsymbol{r},t)\,\boldsymbol{u}_{\mathrm{d}}] \equiv \rho \frac{\mathrm{d}}{\mathrm{d}t}\left(\frac{f(U,\boldsymbol{r},t)}{\rho}\right) + \mathrm{div}[f(U,\boldsymbol{r},t)\,\boldsymbol{w}_{\mathrm{d}}]$$
$$= \frac{1}{2}\int_0^U f(W,\boldsymbol{r},t) f(U-W,\boldsymbol{r},t)\,K(W,U-W)\,\mathrm{d}W - f(U,\boldsymbol{r},t)\int_0^\infty f(W,\boldsymbol{r},t)\,K(W,U)\,\mathrm{d}W, \tag{7.33}$$

which is a generalization of the well-known Muller equation for describing a coagulating disperse medium (see, e.g., Voloshchuk 1984) to the spatially nonuniform motions of a gas suspension. Here, $K(W, U)$ is the symmetric (in arguments) coagulation kernel that determines the behavior of a dispersed medium in time. To solve this equation, we must require the satisfaction of the conditions $f(U, \boldsymbol{r}, t) \to 0$ for $U \to 0$ and $U \to \infty$ and specify the initial condition $f(U, \boldsymbol{r}, 0) = f_0(U, \boldsymbol{r})$ and the boundary conditions.

The kinetic equation (7.33) is a nonlinear integro-differential equation whose solution can generally be obtained only by numerical methods, since, unfortunately, the terms that describe the convection of dust particles severely complicate the standard coagulation equation (see, e.g., Lissauer and Stewart 1993). Several exact analytical solutions of the nonstationary spatially uniform analog of (7.33) for some of the structurally simple coagulation kernels (linear in each individual argument) based on the Laplace integral transform are known (see, e.g., Safronov 1969; Voloshchuk 1984). Therefore, the following should be noted. The analyses of the coagulation processes for kernels $K(W,U) = \Lambda_0$ that do not depend on the coagulating-particle volumes are currently most advanced theoretically. The solution of the coagulation equation with the kernel $K(W,U) = \Lambda_1 WU$ can hardly be considered physically feasible, since it is not continuous in time (starting from a certain time, the number of particles in the system becomes negative; Voloshchuk 1984). An analytical solution of the kinetic equation with a kernel proportional to the sum of the coagulating-particle volumes $K(W,U) = \Lambda_2(W+U)$, was obtained by Safronov (1969) when studying the evolution of a protoplanetary gas–dust cloud. However, as yet no disperse system for which the coagulation microphysics would exactly lead to kernels of such a type has been found.

At the same time, when hydrodynamically modeling a gas–dust disk, the full knowledge of the particle size distribution function is often not required, and only information about the behavior of its first several moments of type $N_{\mathrm{d}}(\boldsymbol{r},t)$, $s(\boldsymbol{r},t)$, and the like in time and space will suffice. In this case, one of the possible approximate methods for solving the kinetic coagulation equation, in particular, the method of moments, can be used. In Appendix A, we illustrate the potentialities of this method by solving the kinetic coagulation equation (7.33) for the case where the particle size distribution depends on one space coordinate z, which corresponds to steady dust motion when solid particles settle under gravity to the subdisk.

7.2.4 Momentum Conservation Equation for Gas–Dust Matter and Radiation

In modeling a protoplanetary cloud, we have to solve the equations of radiation hydrodynamics for large space–time scales of motion that define the averaged thermohydrodynamic and radiation parameters of the gas–dust disk medium. When the linear size of the total volume element $\delta\mathcal{V}$ is much larger than the radiation mean free path λ_{rad}, the radiation energy and pressure cannot be disregarded. It is quite clear that in the case of local equilibrium between radiation and matter, where the radiation energy density is $E_{\mathrm{rad}} = aT^4/\rho$ (per unit mass) and the radiation pressure is

$$p_{\mathrm{rad}} = \frac{\rho E_{\mathrm{rad}}}{3} = \frac{1}{3} a T^4 \tag{7.34}$$

we should everywhere add the radiation energy and pressure to the internal energy $E(\boldsymbol{r}, t)$ and the thermal pressure $p(\boldsymbol{r}, t)$ in the equations of heterogeneous mechanics and consider the process of radiative heat conduction. Here, $a = 4\sigma/c$, σ, and c are the radiation density constant, the Stefan–Boltzmann constant, and the speed of light, respectively.

The instantaneous equation for the conservation of total momentum of the gas–dust material can be derived, for example, by adding the equations of motion for the individual phases (7.17*). As a result, the differential equation for the momentum conservation of the disk medium as a whole (including the radiation field), which, in contrast to the continuity equation (7.13), depends on the relative motion of the phases, can be written as

$$\rho\frac{\mathrm{d}\boldsymbol{u}}{\mathrm{d}t} \equiv \frac{\partial(\rho\boldsymbol{u})}{\partial t} + \left(\frac{\partial}{\partial\boldsymbol{r}}\cdot(\rho\,\boldsymbol{u}\boldsymbol{u})\right) = -\frac{\partial p_{\mathrm{sum}}}{\partial\boldsymbol{r}} + \left(\frac{\partial}{\partial\boldsymbol{r}}\cdot\Pi^*_{\mathrm{sum}}\right) + \rho\boldsymbol{g}, \qquad (7.35)$$

where $(\partial(\boldsymbol{ab})/\partial\boldsymbol{r}) \equiv \sum_k\sum_l \boldsymbol{i}_l\partial(a_k b_l)/\partial x_k$ is the divergence of the dyad $\boldsymbol{ab} \equiv \sum_k \sum_l \boldsymbol{i}_l\boldsymbol{i}_k(a_k b_l)$ (see Appendix); $p_{\mathrm{sum}}(\boldsymbol{r}, t)$ is the total pressure equal to the sum of the thermal pressure of the gas–dust mixture and the radiation pressure, $p_{\mathrm{sum}} = p + p_{\mathrm{rad}}$;

$$\Pi^*_{\mathrm{sum}} \equiv \Pi_{\mathrm{sum}} + \Pi_{\mathrm{rel}} = \Pi_{\mathrm{g}} + \Pi_{\mathrm{rad}} - (1-s)\rho_{\mathrm{g}}\boldsymbol{w}_{\mathrm{g}}\boldsymbol{w}_{\mathrm{g}} - s\rho_{\mathrm{d}}\boldsymbol{w}_{\mathrm{d}}\boldsymbol{w}_{\mathrm{d}}. \qquad (7.36)$$

$\Pi_{\mathrm{sum}}(\boldsymbol{r}, t)$ is the total viscous stress tensor equal to the sum of the viscous stress tensors for a heterogeneous mixture, $\Pi = \sum_\alpha \Pi_\alpha \cong \Pi_{\mathrm{g}}$ (since we assumed that $\Pi_{\mathrm{d}} \cong 0$), and the radiative shear stress tensor Π_{rad}; Π_α is the viscous stress tensor (a tensor of the second rank or a dyad) for phase α, which depends on the deformation rate tensor determined by the velocity field of the corresponding phase;

$$\Pi_{\mathrm{rel}} \equiv -\sum_\alpha \tilde{\rho}_\alpha\boldsymbol{w}_\alpha\boldsymbol{w}_\alpha = -(1-s)\rho_{\mathrm{g}}\boldsymbol{w}_{\mathrm{g}}\boldsymbol{w}_{\mathrm{g}} - s\rho_{\mathrm{d}}\boldsymbol{w}_{\mathrm{d}}\boldsymbol{w}_{\mathrm{d}} \qquad (7.37)$$

is the "relative" stress tensor that arises from the dynamical effects of the relative motion of solid particles and gas; $\boldsymbol{g}(\boldsymbol{r}, t) = -\partial\Psi/\partial\boldsymbol{r}$ is the vector of acceleration by the external bulk force (gravity); $\Psi(\boldsymbol{r}, t)$ is the Newtonian gravitational potential. Let us note that in heterogeneous media, the laws that describe the relative motion of the phases become more complicated, because this motion is determined not by the diffusion mechanism (collisions between molecules during their random motion), but by the interaction between the phases as macroscopic systems (Kolesnichenko and Maksimov 2001). These processes can be described using forces and a more consistent allowance for the phase inertia. The relative stress tensor in the total equation of motion for the mixture leads to a cardinal difference between heterogeneous mechanics and multicomponent mechanics, for which the terms containing quantities of the second order relative to the diffusion velocities $\boldsymbol{w}_\alpha$ may be disregarded (the so-called diffusion approximation in the mechanics of mixtures).

When the mass of the gas–dust cloud accounts for a few percent of the mass of the central body or, more precisely, when $\mathcal{M}_{\text{disk}}/\mathcal{M}_{\odot} \leq h_{\text{disk}}/R$, where h_{disk} and R are the disk half-thickness and radius, (see, e.g., Hersant et al. 2004), the dust particle self-gravity may be ignored; in this case, we have

$$\Psi = \frac{G\mathcal{M}_{\odot}}{|\tilde{\boldsymbol{r}}|}, \quad \boldsymbol{g} = -\frac{\partial}{\partial \boldsymbol{r}}\Psi = \frac{G\mathcal{M}_{\odot}}{|\tilde{\boldsymbol{r}}|^3}\tilde{\boldsymbol{r}} \tag{7.38}$$

where $\mathcal{M}_{\odot}$ is the mass of the central body (star); G is the gravitational constant; $|\tilde{\boldsymbol{r}}| = \sqrt{r_1^2 + r_2^2 + r_3^2}$ is the central radius vector $\tilde{\boldsymbol{r}} = \sum_k \boldsymbol{i}_k r_k$; in what follows, the center of mass of the protostar is taken as the coordinate origin. In the cases where the self-gravity effects are important,

$$\Psi = G\mathcal{M}_{\odot}/|\tilde{\boldsymbol{r}}| + \Psi_{\text{cr}} \tag{7.38*}$$

and the self-gravity potential Ψ_{cr} satisfies the Poisson equation $\nabla^2\Psi_{\text{cr}} = 4\pi G\rho$, where ∇^2 is the Laplace operator.

The relative stress tensor Π_{rel} for a gas–dust disk can be written in several equivalent forms convenient for writing the model equations of motion in various coordinate systems. Using (7.6) and (7.12), we have

$$\begin{aligned}\Pi_{\text{rel}} &\equiv -(1-s)\rho_{\text{g}}\boldsymbol{w}_{\text{g}}\boldsymbol{w}_{\text{g}} - s\rho_{\text{d}}\boldsymbol{w}_{\text{d}}\boldsymbol{w}_{\text{d}} = -(1-s)\rho_{\text{g}}\boldsymbol{u}_{\text{g}}\boldsymbol{u}_{\text{g}} - s\rho_{\text{d}}\boldsymbol{u}_{\text{d}}\boldsymbol{u}_{\text{d}} + \rho\boldsymbol{u}\boldsymbol{u}\\ &= -s\rho_{\text{d}}C_{\text{g}}\boldsymbol{w}\boldsymbol{w} = -s\rho_{\text{d}}C_{\text{g}}(\boldsymbol{i}_1\boldsymbol{i}_1 w_1 w_1 + \boldsymbol{i}_1\boldsymbol{i}_2 w_1 w_2 + \boldsymbol{i}_1\boldsymbol{i}_3 w_1 w_3 + \boldsymbol{i}_2\boldsymbol{i}_1 w_2 w_1\\ &\quad + \boldsymbol{i}_2\boldsymbol{i}_2 w_2 w_2 + \boldsymbol{i}_2\boldsymbol{i}_3 w_2 w_3 + \boldsymbol{i}_3\boldsymbol{i}_1 w_3 w_1 + \boldsymbol{i}_3\boldsymbol{i}_2 w_3 w_2 + \boldsymbol{i}_3\boldsymbol{i}_3 w_3 w_3).\end{aligned} \tag{7.39}$$

It is important to note that when the dynamics of a protoplanetary cloud is modeled, these additional stresses must be taken into account when fractions of relatively large solid particles ($\geq$1 mm) are present in it, since in this case there is a significant velocity difference between the phases, i.e., the relative velocity of the phases $\boldsymbol{w}$ can be equal in order of magnitude to the hydrodynamic velocity of the total continuum $\boldsymbol{u}$. At the same time, for very small particles ($\ll$1 mm at a *Stokes* number ***Stk*** $\ll 1$) when the particles have time to respond to a change in the parameters of the carrier medium, the approximation of a passive admixture can be used—a two-phase gas–dust flow is approximated by a flow of a single-phase (generally multicomponent) medium with certain effective thermophysical properties (density, gas constant, specific heat, etc.) (see Kolesnichenko 2000). In the other extreme case (at ***Stk*** $\gg 1$), where the large solid particles in the disk system do not change their state as the gas parameters vary, we can also consider a single-phase flow, but already of pure gas; the inverse effect of the large bodies can be taken into account by introducing distributed sources of resistance. Finally, when $C_{\text{d}} \ll 1$, the presence of rare particles of the gas–dust mixture does not affect the gas flow parameters and therefore, the approximation of a single particle can be used; here, first, the equations of motion for the gas are solved, and then the particle trajectories and the change in their state along the trajectories are determined from known gas parameters (Garaud et al. 2005). Below, we will use mainly the representation of the tensor Π_{rel} via the velocity vector of the relative phase motion, $\boldsymbol{w} \equiv \boldsymbol{u}_{\mathrm{d}} - \boldsymbol{u}_{\mathrm{g}}$.

It is known (see Tassoul 1979) that the radiative shear stress tensor Π_{rad} is structurally similar to the viscous stress tensor for material Π. Hence if the matter–radiation interaction up to the terms of the lowest order in $|\boldsymbol{u}|/c$ is taken into account, then the following terms also enter into the radiative shear stress tensor components $(\boldsymbol{\Pi}_{\mathrm{rad}})_{ik}$: $-c^{-2}(u_i(\boldsymbol{q}_{\mathrm{rad}})_k + u_k(\boldsymbol{q}_{\mathrm{rad}})_i + \delta_{ik}u_s(\boldsymbol{q}_{\mathrm{rad}})_s)$, where $\boldsymbol{q}_{\mathrm{rad}}$ is the radiative heat flux vector defined by (7.48) (see, e.g., Hazlehurst and Sargent 1959). We therefore, may write

$$\Pi_{\mathrm{sum}} \equiv (\Pi + \Pi_{\mathrm{rad}}) \cong 2(\mu_{\mathrm{g}} + \mu_{\mathrm{rad}})\overset{\circ}{\boldsymbol{D}} + \left(\xi_{\mathrm{g}} + \frac{5}{3}\mu_{\mathrm{rad}}\right)(\mathrm{div}\boldsymbol{u})\boldsymbol{U}$$

$$\overset{\circ}{\boldsymbol{D}} \equiv \boldsymbol{D} - \frac{1}{3}\boldsymbol{U}\mathrm{div}\boldsymbol{u} \tag{7.40}$$

where $\overset{\circ}{\boldsymbol{D}}$ is the deformation rate tensor; $\boldsymbol{D} \equiv \frac{1}{2}\left((\partial \boldsymbol{u}/\partial \boldsymbol{r}) + (\partial \boldsymbol{u}/\partial \boldsymbol{r})^{\mathrm{transp}}\right)$ is the deformation tensor; $\boldsymbol{I}$ is a unit vector (or a unit dyad, $\boldsymbol{U} = \boldsymbol{i}_1\boldsymbol{i}_1 + \boldsymbol{i}_2\boldsymbol{i}_2 + \boldsymbol{i}_3\boldsymbol{i}_3$); $\mu_{\mathrm{g}}(\rho, T)$ and $\xi_{\mathrm{g}}(\rho, T)$ are the molecular coefficients of dynamic and bulk viscosities of the gas, respectively; $\mu_{\mathrm{rad}} = 4aT^4/15\, c\tilde{\kappa}\rho$ is the coefficient of radiative viscosity; $\tilde{\kappa}$ is the total Rosseland mean opacity of the medium, which, in turn, also depends on ρ, s, N_{d}, T, and the chemical composition of the gas (see (7.72) and (7.73)).

7.2.5 *Heat Influx Equation for a Heterogeneous Gas–Dust Medium and Radiation in a Disk*

The instantaneous heat influx equation (the equation of internal energy) for a heterogeneous gas–dust medium as a whole under the above assumptions can be written as (Kolesnichenko and Maksimov 2001)

$$\rho \frac{\mathrm{d}}{\mathrm{d}t}(E_{\mathrm{sum}}) = -\mathrm{div}\boldsymbol{q}_{\mathrm{sum}} - p_{\mathrm{sum}}\mathrm{div}\boldsymbol{u} + \Phi_u + \sum_{\alpha}(\boldsymbol{J}_\alpha \cdot \boldsymbol{K}_\alpha). \tag{7.41}$$

Here, $\boldsymbol{K}_\alpha \cong -(d_\alpha \boldsymbol{u}_\alpha/\mathrm{d}t) + \boldsymbol{F}_\alpha - (1/(2\tilde{\rho}_\alpha))\sigma_{\alpha\beta}\boldsymbol{w}_{\alpha\beta}$ is the generalized thermodynamic diffusion force that includes the "inertial term" and the term due to phase transitions (see (7.18)); $E_{\mathrm{sum}} \equiv E + E_{\mathrm{rad}}$ is the total internal energy of the disk system (matter plus radiation) per unit mass; $E(\boldsymbol{r}, t) \equiv \sum_\alpha C_\alpha e_\alpha$ is the internal energy of the material (note that the internal energy of the gas–dust mixture as a whole that we introduced is its true internal energy, since it does not contain the contribution from the kinetic energy of the interphase diffusion); $e_\alpha(\boldsymbol{r}, t)$, $h_\alpha(\boldsymbol{r}, t)$ $(= e_\alpha + p/\rho_\alpha)$ are, respectively, the partial internal energy and enthalpy (per unit mass) of the phase-α material; E_{rad} is the radiation energy density (per unit mass) defined by the Stefan–Boltzmann law, $E_{\text{rad}} = aT^4/\rho$; $\boldsymbol{q}_{\text{sum}} \equiv \boldsymbol{q} + \boldsymbol{q}_{\text{rad}}$ is the total energy flux density in the system; $\boldsymbol{q}_{\text{rad}}$ is the specific radiatively transferred energy flux (which should be always be taken into account, since it is large even at a low radiation energy density (due to the high photon velocity)); $\boldsymbol{q}$ is the specific energy flux related to the thermal motion of the particles of the phase material (i.e., determined by the heat conductivity) and to the transfer of particle enthalpies by phase diffusion flows: $\Phi_u \cong (\Pi_{\text{sum}} : \nabla u)$ is the dissipative function, the rate at which heat is generated by the viscous friction of the gas in a unit volume per unit time. In writing (7.41), we assumed that the thermodynamic functions (internal energy, enthalpy, etc.) are additive in the masses of the phases in the heterogeneous system, which is admissible only when the contribution from the near-surface (*Knudsen*) layer of solid particles is disregarded.

Using (7.17) and (7.18), the last term in (7.41) can be rewritten as

$$\sum_\alpha \boldsymbol{J}_\alpha \cdot \boldsymbol{K}_\alpha = \sum_\alpha \boldsymbol{w}_\alpha \cdot (-\boldsymbol{d}_\alpha + s_\alpha \partial p/\partial \boldsymbol{r}),$$

where $\boldsymbol{d}_{\text{g}} = -\boldsymbol{d}_{\text{d}} = R_{\text{gd}}\boldsymbol{w}$ (without thermophoresis). For the additional source of heat associated with the dissipation of kinetic diffusion energy, we will then have (an analog of the *Joule* heating for plasma)

$$\begin{aligned}\sum_\alpha \boldsymbol{J}_\alpha \cdot \boldsymbol{K}_\alpha &= -C_{\text{d}}\boldsymbol{w} \cdot \left(-\boldsymbol{d}_{\text{g}} + s_{\text{g}}\frac{\partial p}{\partial \boldsymbol{r}}\right) + C_{\text{g}}\boldsymbol{w} \cdot \left(-\boldsymbol{d}_{\text{d}} + s_{\text{d}}\frac{\partial p}{\partial \boldsymbol{r}}\right) \\ &= R_{\text{gd}}|\boldsymbol{w}|^2 - s\sigma\left(\boldsymbol{w} \cdot \frac{\partial p}{\partial \boldsymbol{r}}\right),\end{aligned}$$

since $s^2 \ll 1$. Here, $\sigma \equiv (\rho_{\text{d}} - \rho_{\text{g}})/\rho$ is the relative excess of the dust particle density above the gas density; for small solid particles, $s\sigma \ll 1$ and the last term in this relation may be neglected.

It is important to note that the heat influx equation (7.41) contains the true internal energy of the gas–dust medium E per unit mass, which was determined by subtracting the potential and kinetic energies of all phases from the total energy U_{tot} of the material of the disk system (Kolesnichenko and Maksimov 2001):

$$\begin{aligned}E = U_{\text{tot}} - \sum_\alpha C_\alpha \psi_\alpha - \sum_\alpha \tfrac{1}{2} C_\alpha |\boldsymbol{u}_\alpha|^2 &= U_{\text{tot}} - \Psi - \tfrac{1}{2}|\boldsymbol{u}|^2 - \sum_\alpha C_\alpha \tfrac{1}{2}|\boldsymbol{w}_\alpha|^2 \\ &= U_{\text{tot}} - \Psi - \tfrac{1}{2}|\boldsymbol{u}|^2 - C_{\text{d}}C_{\text{g}}|\boldsymbol{w}|^2/2.\end{aligned} \tag{7.42}$$

At the same time, if the internal energy of the gas–dust system is defined by the relation $E^* = U_{\text{tot}} - \Psi - \frac{1}{2}|\boldsymbol{u}|^2$, then it will also include the macroscopic kinetic energy of the phases in the center-of-mass system, i.e., $E^* = E + C_{\text{d}}C_{\text{g}}|\boldsymbol{w}|^2/2$. If we now write (7.41) via the internal energy E^* redefined in this way, then it will take a usual form, i.e., it will not contain the terms $R_{gd}|\boldsymbol{w}|^2 - s\sigma\boldsymbol{w}\cdot(\partial p/\partial\boldsymbol{r})$. Indeed, using (7.19*) and the vector transformation $\boldsymbol{ab}{:}(\partial/\partial\boldsymbol{r})\boldsymbol{c} = (\boldsymbol{a}\cdot(\boldsymbol{b}\cdot(\partial/\partial\boldsymbol{r}))\boldsymbol{c}$, we can write the balance equation for the kinetic energy of the interphase diffusion as

$$\rho\frac{\mathrm{d}}{\mathrm{d}t}(C_\mathrm{d}C_\mathrm{g}|w|^2/2) \approx \rho C_\mathrm{d}C_\mathrm{g}\frac{\mathrm{d}}{\mathrm{d}t}(|w|^2/2)$$
$$\cong -R_{gd}|w|^2 + \left(\Pi_\mathrm{rel} : \frac{\partial}{\partial\boldsymbol{r}}\boldsymbol{u}\right) + s\sigma\left(\boldsymbol{w}\frac{\partial p}{\partial\boldsymbol{r}}\right). \qquad (7.43)$$

since, to the terms of the second-order relative $\boldsymbol{w}$, we have

$$\rho\frac{\mathrm{d}}{\mathrm{d}t}\left(C_\mathrm{d}C_\mathrm{g}|\boldsymbol{w}|^2/2\right) = \rho C_\mathrm{d}C_\mathrm{g}\frac{\mathrm{d}}{\mathrm{d}t}\left(|\boldsymbol{w}|^2/2\right) + (|\boldsymbol{w}|^2/2)\left\{C_\mathrm{d}\rho\frac{\mathrm{d}C_\mathrm{g}}{\mathrm{d}t} + C_\mathrm{g}\rho\frac{\mathrm{d}C_\mathrm{d}}{\mathrm{d}t}\right\}$$
$$= \rho C_\mathrm{d}C_\mathrm{g}\frac{d}{\mathrm{d}t}\left(|\boldsymbol{w}|^2/2\right) + (|\boldsymbol{w}|^2/2)(C_\mathrm{d} - C_\mathrm{g})\mathrm{div}(\rho\, C_\mathrm{g}C_\mathrm{d}\boldsymbol{w}) \approx \rho C_\mathrm{d}C_\mathrm{g}\frac{\mathrm{d}}{\mathrm{d}t}\left(|\boldsymbol{w}|^2/2\right).$$

Still, the quantity E in the heat influx equation (7.41) probably deserves the name "internal energy" more than E^*, since the internal energy must contain only the contribution from the thermal motion and the short-range molecular interactions and no macroscopic terms (see de Groot and Mazur 1962).

7.2.5.1 Other Forms of the Energy Equation for Gas Suspension

Below, we will need the energy equations written in several other forms. Let us introduce the total enthalpy $H_\mathrm{sum} \equiv H + H_\mathrm{rad}$ of the matter and radiation in the disk, where

$$\begin{cases} H \equiv \sum_\alpha C_\alpha h_\alpha = \sum_\alpha C_\alpha(e_\alpha + p/\rho_\alpha) = E + p/\rho, \\ H_\mathrm{rad} = E_\mathrm{rad} + p_\mathrm{rad}/\rho = \tfrac{4}{3}aT^4/\rho, \qquad H_\mathrm{sum} = E_\mathrm{sum} + p_\mathrm{sum}/\rho. \end{cases} \qquad (7.44)$$

Using (7.42) and the transformation $\rho\mathrm{d}E_\mathrm{sum}/\mathrm{d}t + p_\mathrm{sum}\nabla\cdot\boldsymbol{u} = \rho\mathrm{d}H_\mathrm{sum}/\mathrm{d}t - \mathrm{d}\, p_\mathrm{sum}/\mathrm{d}t$, which is a corollary of definitions (7.44) and the mixture continuity equation (7.13), we will then obtain

$$\rho\frac{\mathrm{d}H_\mathrm{sum}}{\mathrm{d}t} = \frac{\mathrm{d}p_\mathrm{sum}}{\mathrm{d}t} - \mathrm{div}\boldsymbol{q}_\mathrm{sum} + \Phi_u + R_{gd}|\boldsymbol{w}|^2 - s\sigma\left(\boldsymbol{w}\cdot\frac{\partial p}{\partial\boldsymbol{r}}\right). \qquad (7.45)$$

This equation corresponds to the first law of thermodynamics (i.e., the law of conservation of thermal energy).

Let us now rewrite (7.45) in the variables $T(\boldsymbol{r},t)$ and $p(\boldsymbol{r},t)$. For most of the purposes pertaining to our problem of modeling the evolution of an accretion disk, it will suffice to approximate the partial enthalpies of the gas and dust (per unit mass) using the expressions $h_{\mathrm{g}} = c_{\mathrm{Pg}}T + h_{\mathrm{g}}^0$, and $h_{\mathrm{d}} = c_{\mathrm{Pd}}T + h_{\mathrm{g}}^0$, where h_α^0 is the enthalpy of phase α at zero temperature (the so-called heat of formation) and $c_{\mathrm{P}\alpha}$ is the specific heat (at constant pressure) of phase α. Below, the thermophysical quantities $c_{\mathrm{P}\alpha}$ and h_α^0 are assumed to be the constants that approximate the actual disk-specific heats $c_{\mathrm{P}\alpha}(T)$ and partial heats of formation $h_\alpha^0(T)$ in a limited temperature range. We can then write

$$H = c_{\mathrm{P}}T + \sum_\alpha C_\alpha h_\alpha^0, \quad H_{\mathrm{rad}} = E_{\mathrm{rad}} + p_{\mathrm{rad}}/\rho = \frac{4}{3}aT^4/\rho, \tag{7.46}$$

where

$$c_{\mathrm{P}} = \sum_\alpha c_{\mathrm{P}\alpha} C_\alpha = \rho^{-1}\left(\rho_{\mathrm{g}}(1-s)c_{\mathrm{Pg}} + s\rho_{\mathrm{d}}c_{\mathrm{d}}\right)$$

is the total specific heat of the "gas–solid particles" system at constant pressure. Using now (7.46), along with (7.9), (7.12), and (7.13), we obtain

$$\begin{aligned} \rho\frac{\mathrm{d}H}{\mathrm{d}t} &\equiv \rho\sum_\alpha C_\alpha\frac{\mathrm{d}h_\alpha}{\mathrm{d}t} + \rho\sum_\alpha h_\alpha\frac{\mathrm{d}C_\alpha}{\mathrm{d}t} = \rho c_{\mathrm{P}}\frac{\mathrm{d}T}{\mathrm{d}t} + \sum_\alpha h_\alpha(-\mathrm{div}\boldsymbol{J}_\alpha + \sigma_{\alpha\beta}) \\ &= \rho c_{\mathrm{P}}\frac{\mathrm{d}T}{\mathrm{d}t} - \mathrm{div}\left[\sum_\alpha h_\alpha\boldsymbol{J}_\alpha\right] + \sum_{\rho=1}^{r} q_\rho\xi_\rho + \frac{\partial T}{\partial \boldsymbol{r}}\cdot\left[\sum_\alpha c_{\mathrm{P}\alpha}\boldsymbol{J}_\alpha\right], \end{aligned} \tag{7.47}$$

where the relation $q_\rho \equiv \sum_\alpha h_\alpha \nu_{\alpha,\rho} = q_\rho^0 + \sum_\alpha c_{\mathrm{P}\alpha}\nu_{\alpha,\rho}$ introduces the so-called heat of reaction ρ, which is equal to the difference between the products of the partial enthalpies of the reaction products by the corresponding stoichiometric coefficients and the analogous sum for the reactants ($\nu_{\alpha,\rho} \equiv \sum_{k=1}^N m_{(k)}\nu_{\alpha(k),\rho}$); note that $q_\rho^0 = \sum_\alpha h_\alpha^0\nu_{\alpha,\rho}$ can be interpreted as the heat of phase transition ρ at zero temperature. The last term on the right-hand side of (7.47) represents the effect of the so-called diffusing specific heats, which is negligible and, hence, is commonly ignored.

According to Kolesnichenko and Maksimov (2001), the total energy flux $\boldsymbol{J}_q \equiv \boldsymbol{q} - \sum_\alpha h_\alpha\boldsymbol{J}_\alpha$ related to the thermal motion of particles in a heterogeneous continuum can be written in the standard form

$$\boldsymbol{J}_q \equiv \boldsymbol{q} - \sum_\alpha h_\alpha\boldsymbol{J}_\alpha \equiv \boldsymbol{q} - \rho C_{\mathrm{g}}C_{\mathrm{d}}(h_{\mathrm{d}} - h_{\mathrm{g}})\boldsymbol{w} = -\chi_{\mathrm{g}}\frac{\partial T}{\partial \boldsymbol{r}}$$

or

$$\boldsymbol{q} = \sum_\alpha h_\alpha\boldsymbol{J}_\alpha - \chi_{\mathrm{g}}\frac{\partial T}{\partial \boldsymbol{r}} \tag{7.48}$$

which generalizes the analogous relation for multicomponent homogeneous mixtures (see Hirschfelder et al. 1954) to heterogeneous media. Note that in (7.48) we disregarded the thermophoretic effect, $k_{P\alpha} = 0$. Let us recall that the heat flux for diffusing mixtures can be defined in various ways, with the specific form of the expression for the rate of entropy production $\sigma_{(S)}$ corresponding to each definition of the heat flux; the choice in each specific case depends on the convenience of considering the problem. In a similar way, by excluding the disk regions close to the surface of the protostar, we can write the law of heat conduction for the radiative heat flux vector,

$$\boldsymbol{q}_{\rm rad} = -\chi_{\rm rad} \frac{\partial T}{\partial \boldsymbol{r}}. \tag{7.48*}$$

Here, $\chi_{\rm g}$ is the molecular coefficient of heat conductivity for the gas, and $\chi_{\rm rad} = 4acT^3/(3\tilde{\kappa}\rho)$ is the coefficient of radiative (nonlinear) heat conductivity, which strongly depends on the temperature and density of the material (see (7.71)).

Substituting (7.47), (7.48), and (7.48*) in (7.44) ultimately yields

$$\rho c_{\rm P,sum} \frac{\mathrm{d}T}{\mathrm{d}t} = \frac{\mathrm{d}p_{\rm g}}{\mathrm{d}t} + \mathrm{div}\left(\chi_{\rm sum} \frac{\partial T}{\partial \boldsymbol{r}}\right) - 4p_{\rm rad}\, \mathrm{div}\boldsymbol{u}$$
$$+ \Phi_u + R_{\rm gd}\boldsymbol{w}^2 - s\sigma\, \boldsymbol{w} \cdot \frac{\partial p_{\rm g}}{\partial \boldsymbol{r}} - \sum_{\rho=1}^{r} q_\rho \xi_\rho, \tag{7.49}$$

where we use the notation $\chi_{\rm sum} = \chi_{\rm g} + \chi_{\rm rad}$ and $c_{\rm P,sum} = c_{\rm P} + 16aT^3/3\rho$.

Finally, let us derive the balance equation for the specific entropy $S = \sum_\alpha C_\alpha S_\alpha$ of the total continuum modeling the gas–dust disk medium as a whole, which is commonly called the general heat transfer equation (here, S_α is the entropy per unit mass of phase α). For this purpose, we will use the fundamental Gibbs relation (see, e.g., Prigogine and Defay 1954) for a single-temperature heterogeneous multicomponent radiative continuum in the single-pressure approximation. Being written along the center-of-mass trajectory of the volume element $\delta\mathcal{V}$, it takes the form (Kolesnichenko and Maksimov 2001)

$$T\frac{\mathrm{d}S_{\rm sum}}{\mathrm{d}t} = \frac{\mathrm{d}E_{\rm sum}}{\mathrm{d}t} + p_{\rm sum}\frac{d}{d}\left(\frac{1}{\rho}\right) - \sum_\alpha \sum_{k=1}^{n} \mu_{\alpha(k)} \frac{\mathrm{d}}{\mathrm{d}t}\left(\frac{s_\alpha n_{\alpha(k)}}{\rho}\right), \tag{7.50}$$

where $\mu_{\alpha(k)}$ is the chemical potential of component k in phase α. Using (7.9), (7.13), and (7.41), we can write the Gibbs relation (7.50) in the form of a balance equation,

$$\rho\frac{\mathrm{d}S_{\rm sum}}{\mathrm{d}t} + \mathrm{div}\left\{\frac{1}{T}\left(\boldsymbol{q}_{\rm sum} - \sum_\alpha G_\alpha \boldsymbol{J}_\alpha\right)\right\} = \sigma_{(S)} \tag{7.51}$$

where

$$0 \leq T\sigma_{(S)} = -(\boldsymbol{J}_q + \boldsymbol{q}_{\mathrm{rad}}) \cdot \frac{\partial \ln T}{\partial \boldsymbol{r}} - \sum_{\alpha} \boldsymbol{w}_\alpha \cdot \boldsymbol{d}_\alpha + \left(\Pi_{\mathrm{sum}} : \frac{\partial}{\partial \boldsymbol{r}} \boldsymbol{u} \right) + \sum_{\alpha} \left(\Pi_\alpha : \frac{\partial}{\partial \boldsymbol{r}} \boldsymbol{w}_\alpha \right) + \sum_{\rho=1}^{r} A_\rho \xi_\rho \tag{7.52}$$

is the energy dissipation in irreversible processes, which is a local measure of nonequilibrium of the system; $G_\alpha = \rho_\alpha^{-1} \sum_{k=1}^{n} \mu_{\alpha(k)} n_{\alpha(k)} = e_\alpha + p/\rho_\alpha - T S_\alpha$ is the Gibbs free energy of an elementary macroparticle of phase α;

$$A_\rho \equiv - \sum_{\alpha=1} \sum_{k=1}^{n} \mu_{\alpha(k)} \nu_{\alpha(k),\rho} \tag{7.53}$$

is the chemical affinity of reaction ρ that generally proceeds between components in different phases.

Note that the specific representation of the rate of entropy production ($T\sigma_{(S)}$) as a bilinear form is used in nonequilibrium thermodynamics to establish the defining relations that linearly relate the thermodynamic fluxes and the conjugate thermodynamic forces in the irreversible process under consideration by the Onsager method. In particular, the generalized Stefan–Maxwell relations (7.17) for heterogeneous media were derived by Kolesnichenko and Maksimov (2001) precisely in this way.

7.2.6 *Thermodynamic Equation of State for Disk Matter*

In what follows, we will use a baroclinic equation of state for a mixture of perfect gases as the thermal state of the multicomponent gas phase of the disk (the equation for pressure):

$$p_{\mathrm{g}} = \sum_{(k)} p_{(k)} = k_{\mathrm{B}} T \sum_{k} n_{\mathrm{g}(k)} = \rho_{\mathrm{g}} \Re_{\mathrm{g}} T, \tag{7.54}$$

where $\Re_g = k_{\mathrm{B}} \sum_k n_{g(k)} / \rho_g = k_{\mathrm{B}} / \mathcal{M}_{\mathrm{g}}$; $\mathcal{M}_{\mathrm{g}}$ is the mean molecular mass of the gas particles, which below is assumed to be constant.

Assuming that the partial pressures in the phases are equal, $p_{\mathrm{g}} = p_{\mathrm{d}} = p$, let us write the equation of state for the disk material as

$$p = \rho \Re T, \quad \Re(C_{\mathrm{g}}, s) = \Re_{\mathrm{g}} \rho_{\mathrm{g}} / \rho = \Re_{\mathrm{g}} C_{\mathrm{g}} / (1 - s) \cong \Re_{\mathrm{g}} C_{\mathrm{g}} \tag{7.55}$$

(in the case considered here, $\Re$ is not a constant). The approximate equality in (7.55) holds for a gas suspension with a low volume concentration of the condensed phase (i.e., when $s \ll 1$, which is assumed in this paper; nevertheless, the dynamical effect of solid particles on the gas flow may turn out to be significant even in this case due to the enormous influence of the gravitational force). Thus, the gas–dust disk medium as a whole may be treated as a perfect gas with the adiabatic index γ and the speed of sound c_s defined by the relations

$$\gamma \equiv \frac{c_{\mathrm{P}}}{c_{\mathrm{P}} - \Re} \cong \frac{\rho_{\mathrm{g}} c_{\mathrm{Pg}} + s\,\rho_{\mathrm{d}} c_{\mathrm{Pd}}}{\rho_{\mathrm{g}}(c_{\mathrm{Pg}} - \Re_{\mathrm{g}}) + s\rho_{\mathrm{d}} c_{\mathrm{Pd}}},$$
$$1 \le \gamma \le \gamma_{\mathrm{g}} \equiv \frac{c_{\mathrm{Pg}}}{c_{\mathrm{Pg}} - \Re_{\mathrm{g}}}, \tag{7.56}$$

$$c_{\mathrm{s}}^2 \equiv \left(\frac{\partial p}{\partial \rho}\right)_S = \gamma \frac{p}{\rho} = \gamma \Re T \cong c_{\mathrm{s\,g}}^2 \frac{\gamma \rho_{\mathrm{g}}}{\gamma_{\mathrm{g}} \rho}, \quad c_{\mathrm{s}} < c_{\mathrm{s\,g}} \equiv \left(\frac{\partial p}{\partial \rho_{\mathrm{g}}}\right)_{S_{\mathrm{g}}} = (\gamma_{\mathrm{g}} \Re_{\mathrm{g}} T)^{1/2}, \tag{7.57}$$

where γ_{g} and $c_{\mathrm{s\,g}}$ are the adiabatic index and the isothermal speed of sound in a pure gas. For a gas of solar composition, with hydrogen and helium comprising 98% of it, the adiabatic index is $\gamma_{\mathrm{g}} = 1.45$ and the mean molecular mass is $\mathcal{M}_{\mathrm{g}} = 2.39$.

7.2.7 *Radiative Transfer Equation for a Gas–Dust Disk: Optical Properties of Dust Grains*

Radiative heat transfer has a decisive effect on the state and motion of a turbulized high-temperature protoplanetary cloud. Meanwhile, the interaction between radiative heat transfer and turbulence has been studied inadequately; as a result, it has been disregarded in the literature and in modeling the evolution of a gas–dust disk until recently. Since this interaction can actually be significant (see, e.g., Ievlev 1975), below we attempt to take it into account (at least in part) using the approach being developed here. Therefore, let us consider in more detail some of the basic concepts of the theory of radiative transfer that we will need for the above purposes. The scattering is known to have no direct effect on the thermal regime of the medium. That is why, in general, the scattering of radiation is disregarded in the problems of radiation hydrodynamics (which we will do below), and only the true attenuation coefficient κ_ν and the true source function of the radiation B_ν (without scattering) are considered. The emission and absorption of photons are described by a radiative transfer equation that for a gas–dust medium in local equilibrium (at any point in space and at any instant in time) takes the form

$$\frac{1}{c}\frac{\partial I_\nu}{\partial t} + \left(\Omega \cdot \frac{\partial I_\nu}{\partial \boldsymbol{r}}\right) = \rho\kappa_\nu(B_\nu - I_\nu) \tag{7.58}$$

Here, $I_\nu \equiv I_\nu(\boldsymbol{r}, \Omega, t)$ is the spectral intensity of the radiation defined in such a way that $I_\nu \mathrm{d}\nu \mathrm{d}\Omega$ describes the energy of the photons in the frequency range from ν to $\nu + \mathrm{d}\nu$ that cross a unit surface element with normal Ω within the solid angle $\mathrm{d}\Omega$ oriented along Ω per unit time; $B_\nu(T) \equiv (2h\nu^3/c^2)(\exp(h\nu/k_\mathrm{B}T) - 1)^{-1}$ is the *Planck* function; h is the Planck constant; $\kappa_\nu(\boldsymbol{r}, t)$ is the total spectral attenuation coefficient (opacity) expressed in terms of the cross-sections for the elementary physical processes in the gas–dust mixture as

$$\rho\kappa_\nu = (1 - s)\sum_k n_{\mathrm{g}(k)}\sigma_{(k)}(\nu) + N_\mathrm{d}\frac{\pi\tilde{d}_\mathrm{d}^2}{4}Q_\mathrm{d}(m(\nu), \tilde{d}_\mathrm{d});$$
$$\sigma_{(k)}(\nu) = \sigma_{a\,(k)}(\nu)\,[1 - \exp\,(h\,\nu/\,k_B T)] + \sigma_{s(k)}^{\mathrm{eff}} \tag{7.59}$$

is the cross-section for the attenuation of radiation at frequency ν per one gas molecule of type k, which is equal to the photon absorption cross-section (corrected for the induced emission of radiation) plus the effective scattering cross-section; $Q_\mathrm{d} = Q_\mathrm{ds} + Q_\mathrm{da}$; Q_ds and Q_da are, respectively, the efficiency factors for the scattering and absorption of light by dust particles (the dimensionless quantities calculated using the *Mie* theory); $m(\nu)$ is the complex refractive index of the grain material. The following moments used here are related to the function I_ν:

$$E_{\mathrm{rad},\nu}(\boldsymbol{r}, t) \equiv \frac{1}{c\rho}\int_{4\pi} I_\nu(\boldsymbol{r}, \Omega, t)\mathrm{d}\Omega \tag{7.60}$$

the spectral energy density (per unit mass); and

$$\boldsymbol{q}_{\mathrm{rad},\nu}(\boldsymbol{r}, t) \equiv \int_{4\pi} I_\nu(\boldsymbol{r}, \Omega, t)\Omega\mathrm{d}\Omega \tag{7.61}$$

the spectral energy density along Ω. The total energy density and flux can be obtained by integrating the corresponding monochromatic quantities over the frequency,

$$E_\mathrm{rad}(\boldsymbol{r}, t) \equiv \int_{\nu=0}^{\infty} E_{\mathrm{rad},\nu}(\boldsymbol{r}, t)\mathrm{d}\nu, \quad \boldsymbol{q}_\mathrm{rad}(\boldsymbol{r}, t) \equiv \int_{\nu=0}^{\infty} \boldsymbol{q}_{\mathrm{rad},\nu}(\boldsymbol{r}, t)\mathrm{d}\nu \tag{7.62}$$

Equation (7.58) written along the direction of propagation of the radiation takes a simpler form,

$$\frac{\mathrm{d}I_\nu}{\mathrm{d}l} = \rho\kappa_\nu(B_\nu - I_\nu). \tag{7.63}$$

Here, l is the coordinate along the ray; below, we will omit the term $c^{-1}\partial I_\nu/\partial t$ in (7.58), since the characteristic time scales for the motion of a gas–dust medium are much longer than l^*/c, where l^* is the ray length. If the optical depth of the layer of the gas–dust medium (with the ray length l) along the direction of propagation of the radiation is defined by

$$\tau_\nu = \int_0^l \rho\kappa_\nu \mathrm{d}l, \tag{7.64}$$

then (7.63) can be easily integrated to give the following expression for the intensity of the radiation from a region with a total optical depth τ_ν:

$$I_\nu(\tau_\nu) = I_\nu(0)\exp(-\tau_\nu) + \int_{\tau_{\nu1}=0}^{\tau_\nu} B_\nu(\tau_{\nu1})\exp[-(\tau_\nu - \tau_{\nu1})]\mathrm{d}\tau_{\nu1} \tag{7.65}$$

where $I_\nu(0)$ is the integration constant; the latter has the meaning of the radiation intensity at some point on the ray at which we set the coordinate l equal to zero, $l = 0$. When removing the point $l = 0$ to a great distance, it follows from (7.65) that

$$I_\nu \approx \int_{-\infty}^{l} B_\nu(l_1)\exp\left(-\int_{l_2=l_1}^{l}\rho\kappa_\nu(l_2)\,\mathrm{d}l_2\right)\cdot\rho\kappa_\nu(l_1)\,\mathrm{d}l_1. \tag{7.66}$$

Ideally, (7.65) and (7.66) allow the intensity $I_\nu(\boldsymbol{r},\Omega,t)$ at various points and in various directions to be found for known optical properties of the medium [i.e., the distribution $\kappa_\nu(\boldsymbol{r},t)$ and boundary conditions in the midplane of the disk; the radiative heat flux distribution $\boldsymbol{q}_{\mathrm{rad}}(\boldsymbol{r},t)$ can then be calculated using (7.61) and (7.62)].

At the same time, the heat influx equations (7.41) and (7.45) include the divergence of the radiation flux $\mathrm{div}\boldsymbol{q}_{\mathrm{rad}}$. For a known I_ν distribution, this quantity can often be found without calculating $\boldsymbol{q}_{\mathrm{rad}}$ from (7.62). Integrating the steady-state equation (7.58) over the entire frequency spectrum and over the solid angle Ω yields the following general expression for the contribution of the radiation to the thermal balance of the gas–dust disk medium:

$$\mathrm{div}\boldsymbol{q}_{\mathrm{rad}} = -\int_{\nu=0}^{\infty}\int_{4\pi}\left(\Omega\cdot\frac{\partial I_\nu}{\partial \boldsymbol{r}}\right)\mathrm{d}\Omega\,\mathrm{d}\nu = \int_{\nu=0}^{\infty}\left\{4\pi\rho\kappa_\nu B_\nu(T) - \int_{4\pi}\rho\kappa_\nu I_\nu \mathrm{d}\Omega\right\}\mathrm{d}\nu, \tag{7.67}$$

where the first and second terms correspond, respectively, to the spontaneously emitted and absorbed radiation energies in a unit volume per unit time. The contribution of the radiation to the heat influx equation (7.41) is generally difficult to calculate using (7.67). However, these calculations are simplified significantly in the following two cases that are important for modeling the various evolutionary stages of a protoplanetary gas–dust cloud:

1. At small optical depths of the gas–dust disk. In this case, the term with I_ν in (7.67) may be ignored, i.e., we can assume in the heat influx equation that

$$\mathrm{div}\boldsymbol{q}_{\mathrm{rad}} \cong 4\pi \int_{\nu=0}^{\infty} \rho\kappa_\nu B_\nu(T)\,\mathrm{d}\nu. \tag{7.68}$$

   At high temperatures, this term can be significant even at a small optical depth of the gas (Ievlev 1975), for example, in the near-surface layer of the disk.

2. At large optical depths of the gas–dust disk for the radiation of all energetically significant frequencies ν. In this case, the diffusion approximation is applicable for the radiative heat transfer (the approximation of radiative heat conduction), where the radiation field I_ν is anisotropic only slightly. Multiplying the radiative transfer equation (7.58) by Ω and integrating it by all angles yields (given that the isotropic term with $\rho\kappa_\nu B_\nu$ does not depend on the direction and, hence, does not contribute to the integral)

$$\int_{4\pi} \Omega\left(\Omega\cdot\frac{\partial I_\nu}{\partial \boldsymbol{r}}\right)\mathrm{d}\Omega = -\rho\kappa_\nu \boldsymbol{q}_{\mathrm{rad},\nu}; \tag{7.69}$$

   whence we obtain for the total heat flux

$$\boldsymbol{q}_{\mathrm{rad}} = -\int_{\nu=0}^{\infty}\frac{1}{\rho\kappa_\nu}\int_{4\pi}\Omega\left(\Omega\cdot\frac{\partial I_\nu}{\partial \boldsymbol{r}}\right)\mathrm{d}\Omega. \tag{7.70}$$

   If we retain only the most significant isotropic part for the slightly anisotropic radiation field on the left-hand side of (7.69), then

$$\begin{aligned}\boldsymbol{q}_{\mathrm{rad}} &= -\int_{\nu=0}^{\infty}\frac{1}{\rho\kappa_\nu}\int_{4\pi}\Omega\left(\Omega\cdot\frac{\partial I_\nu}{\partial \boldsymbol{r}}\right)\mathrm{d}\Omega \cong -\frac{c}{4\pi}\int_{\nu=0}^{\infty}\frac{1}{\rho\kappa_\nu}\left(\frac{\partial B_\nu}{\partial \boldsymbol{r}}\int_{4\pi}\cdot\Omega\right)\Omega\mathrm{d}\Omega\\ &= -\frac{c}{3\rho}\int_{\nu=0}^{\infty}\frac{1}{\kappa_\nu}\frac{\partial B_\nu}{\partial \boldsymbol{r}}\mathrm{d}\nu = -\frac{c}{3\rho}\int_{\nu=0}^{\infty}\frac{1}{\kappa_\nu}\frac{\mathrm{d}B_\nu}{\mathrm{d}T}\frac{\partial T}{\partial \boldsymbol{r}}\mathrm{d}\nu = -\frac{4caT^3}{3\rho\tilde{\kappa}}\frac{\partial T}{\partial \boldsymbol{r}} = -\chi_{\mathrm{rad}}\frac{\partial T}{\partial \boldsymbol{r}},\end{aligned} \tag{7.71}$$

where we introduced the so-called total opacity of the medium $\tilde{\kappa}(\rho, s, T, N_{\mathrm{d}})$, which is defined as the Rosseland mean (for the inverse quantities $1/\kappa_\nu$) spectral opacity (see Pollack et al. 1985),

$$\frac{1}{\tilde{\kappa}} \equiv \frac{1}{4aT^3} \int_{\nu=0}^{\infty} (1/\kappa_\nu)(\mathrm{d}B_\nu/\mathrm{d}T)\mathrm{d}\nu = \frac{\int_{\nu=0}^{\infty} (1/\kappa_\nu)(\mathrm{d}B_\nu/\mathrm{d}T)\mathrm{d}\nu}{\int_{\nu=0}^{\infty} (\mathrm{d}B_\nu/\mathrm{d}T)\mathrm{d}\nu} \tag{7.72}$$

(since $\int_{\nu=0}^{\infty} \mathrm{d}B_\nu/\mathrm{d}T\,\mathrm{d}\nu = 4aT^3$). The diffusion approximation is valid if the radiation field is isotropic at distances comparable to or smaller than the photon mean free path, $\lambda_\nu = 1/\kappa_\nu$. Note also that (7.71) expresses the radiative flux vector $\boldsymbol{q}_{\mathrm{rad}}$ in the inner regions of the gas–dust disk with an excellent accuracy. However, in the near-surface layers of the disk, the optical depth is of the order of or smaller than unity, and the flux is no longer defined by this local expression. Therefore, we must use the nonlocal solution (7.68) of the transfer equation, which is commonly used to study the stellar atmospheres.

7.2.7.1 Optical Properties of Dust Grains

It is convenient to rewrite the spectral opacity of the medium related to the dust component that is defined by (7.59) as

$$\begin{aligned} \kappa_\nu(\rho, s, N_{\mathrm{d}}) &= \tfrac{\pi \tilde{d}_{\mathrm{d}}^2}{4\rho} N_{\mathrm{d}} Q_{\mathrm{d}}(m(\nu), d_{\mathrm{d}}) \\ &= \tfrac{\pi^{1/3} 6^{2/3}}{4\rho} s^{2/3} N_{\mathrm{d}}^{1/3} [Q_{\mathrm{d}a}(m(\nu), d_{\mathrm{d}}) + Q_{\mathrm{ds}}(m(\nu), d_{\mathrm{d}})], \end{aligned} \tag{7.73}$$

and this form explicitly depends on the first moments s and N_{d} (see (7.30) and (7.31)) of the dust particle size distribution function $f(U, \boldsymbol{r}, t)$. We are going to calculate the scattering and absorption of light by spherical solid bodies with a complex refractive index m using the Mie theory. Note that, for example, $m = \infty$ corresponds to an infinite permittivity, $m = 1.33$ corresponds to ice particles (for visual wavelengths), $m = 1.33$–$0.09i$ corresponds to dirty ice (ice with absorbing admixtures), and $m = 1.27$–$1.37i$ corresponds to iron grains.

The size of a spherical solid particle is commonly expressed in terms of the dimensionless parameter $x(\nu) = \pi\, d_{\mathrm{d}}/\lambda$, where $\lambda = c/\nu$ is the wavelength of light. For small x, the efficiency factor for the scattering of light by dust particles $Q_{\mathrm{d\,s}}$ is very small; at $|mx| \ll 1$, we have the standard formula for Rayleigh scattering:

$$Q_{\mathrm{d\,s}} = \frac{8}{3} x^4 \left| \frac{m^2 - 1}{m^2 + 2} \right|^2, \tag{7.74}$$

and the absorption efficiency factor in this case is given by

$$Q_{\mathrm{d}\,a} = -4x \mathrm{Im} \left[\frac{m^2 - 1}{m^2 + 2} \right], \tag{7.75}$$

where Im means that the imaginary part should be taken.

7.2.8 *Basic System of Laminar Equations of Motion for a Gas–Dust Disk*

Medium Let us now summarize (for the convenience of referencing) the above equations of motion for a two-phase polydisperse gas–dust medium. These equations (the reference basis of the model), which include the relative motion of the phases, coagulation, phase transitions, and various physical–chemical and radiative processes, are intended, in particular, for a continuum description of the space–time evolution of the composition, dynamics, and thermal regime of a gas–dust cloud at the final laminar evolutionary stage of the protoplanetary disk (after the decay of turbulent motions though turbulence may not decay completely in the disk regions close to the proto-Sun due to the disturbing effect of magnetic fields, corpuscular fluxes, etc. on the medium) in subdisk zones located at various distances from the proto-Sun (see Nakagawa and Sekiya 1986). It is also important to notice that these equations, which describe the instantaneous state of a turbulized protoplanetary cloud at any stage of its evolution, may be considered as the basic ones in studying the mean motion of the disk system when the probability-theoretical averaging of the stochastic equations of motion has to be performed to phenomenologically describe the hydrodynamic and physical–chemical processes. Thus, we have

$$\begin{cases} \frac{\mathrm{d}\rho}{\mathrm{d}t} + \rho \mathrm{div}\boldsymbol{u} = 0, \quad \left(\frac{d}{\mathrm{d}t} \equiv \frac{\partial}{\partial t} + \boldsymbol{u} \cdot \frac{\partial}{\partial r}\right), \\ \rho \frac{\mathrm{d}}{\mathrm{d}t}\left(\frac{s\rho_\mathrm{d}}{\rho}\right) = -\mathrm{div}\boldsymbol{J}_\mathrm{d} + \sum\limits_{\rho=1}^{r} \nu_{\mathrm{d},\rho}\zeta_\rho, \quad \left(\boldsymbol{J}_\mathrm{d} = \rho C_\mathrm{d} C_\mathrm{g} \boldsymbol{w}, \quad C_\mathrm{g} = 1 - \frac{s\rho_\mathrm{d}}{\rho}, \rho_\mathrm{d} = \mathrm{const}\right), \\ \rho \frac{\mathrm{d}}{\mathrm{d}t}\left(\frac{N_\mathrm{d}}{\rho}\right) = -\mathrm{div}(N_\mathrm{d} C_\mathrm{g} \boldsymbol{w}) - \frac{1}{2}\int_0^\infty \int_0^\infty K(W,U) f(W) f(U) \mathrm{d}W \mathrm{d}U + \sum\limits_k \sum\limits_{\rho=1}^{r} \nu_{\mathrm{d}(k),\rho}\zeta_\rho, \\ \rho \frac{\mathrm{d}\boldsymbol{u}}{\mathrm{d}t} = -\frac{\partial(p + p_\mathrm{rad})}{\partial r} + \left(\frac{\partial}{\partial \boldsymbol{r}} \cdot (\Pi_\mathrm{sum} - \boldsymbol{J}_\mathrm{d}\boldsymbol{w})\right) + \rho \frac{G\mathcal{M}_\odot}{|\tilde{r}|^3}\tilde{\boldsymbol{r}}, \\ (\rho c v_\mathrm{g} + 4aT^3/3)\frac{\mathrm{d}T}{\mathrm{d}t} = -\mathrm{div}(\boldsymbol{J}_q + \boldsymbol{q}_\mathrm{rad}) - (p + 4p_\mathrm{rad})\mathrm{div}\boldsymbol{u} + \Phi_u \\ -s\rho_\mathrm{d} C_\mathrm{g} \frac{\mathrm{d}}{\mathrm{d}t}\left(\frac{w^2}{2}\right) - \sum\limits_{\rho=1}^{r} q_\rho \zeta_\rho, \quad (\rho_\mathrm{g} \cong \rho - s\rho_\mathrm{d}), \\ p = p_\mathrm{g} = \rho_\mathrm{g} \Re_\mathrm{g} T, \quad p_\mathrm{rad} = aT^4/3. \end{cases} \tag{7.76}$$

The hydrodynamic equations of motion (7.76) must be complemented by the corresponding expressions for the phase transition rates ξ_ρ and the defining relations for the thermodynamic fluxes,

$$\begin{cases} \Pi_\mathrm{sum} = (\mu_\mathrm{g} + \mu_\mathrm{rad})\left[\frac{\partial \boldsymbol{u}}{\partial \boldsymbol{r}} + \left(\frac{\partial \boldsymbol{u}}{\partial \boldsymbol{r}}\right)^\mathrm{transp}\right] + (\zeta_\mathrm{g} - \frac{2}{3}\mu_\mathrm{g} + \mu_\mathrm{rad})(\mathrm{div}\boldsymbol{u})U, \\ \boldsymbol{w} \cong \frac{1}{\rho\theta_\mathrm{dg}}\left(-\frac{\mathrm{d}\boldsymbol{w}}{\mathrm{d}t} - \left(\boldsymbol{w} \cdot \frac{\partial}{\partial t}\right)\boldsymbol{u} + \frac{1}{\rho_\mathrm{g}}\frac{\partial p_\mathrm{g}}{\partial \boldsymbol{r}}\right), \\ \boldsymbol{J}_q = -\chi_\mathrm{g}\frac{\partial T}{\partial \boldsymbol{r}}, \quad \boldsymbol{q}_\mathrm{rad} = -\chi_\mathrm{rad}\frac{\partial T}{\partial \boldsymbol{r}}, \end{cases} \tag{7.77}$$

as well as by the expressions for the coagulation coefficients $K(W,U)$ (see Kolesnichenko 2001) and the coefficients of molecular transfer $\mu_g(s,T)$, $\xi_g(s,T)$, $\chi_g(s,T)$, $\theta_{dg}(s,N_d,\boldsymbol{Re})$ and radiative heat conductivity $\chi_{rad}(s,N_d,T)$. For the above system of equations of two-phase mechanics (7.76)–(7.77), we must specify the initial and boundary conditions whose choice requires a special analysis in each specific case, since, in general, not the disk system as a whole with, say, such natural boundaries as the midplane of the disk or its outer boundary, but its separate regions are modeled.

It is important to emphasize that the system of equations (7.76)–(7.77), which also describes all particular features of the instantaneous state of the stochastic thermohydrodynamic fields of the turbulized flow of a gas–dust disk medium and their variations under given initial and boundary conditions, often cannot be solved using currently available computers. This is because using numerical methods entails the approximation of the colossal space–time field of turbulized flow parameters by a finite number of mesh points that should be used to solve the finite-difference approximations of the differential equations. At present, there is only one economically justified way out: to solve the hydrodynamic equations (7.76)–(7.77) for large space–time scales of motion that determine the averaged structure parameters of such a stochastic medium and to model all of the smaller scales phenomenologically. In this case, stochasticity means the existence of an ensemble of possible realizations of pulsating gas-suspension flow fields for which the concept of statistical mean (mathematical expectation) is defined for all thermohydrodynamic parameters.

7.3 Averaged Equations of Two-Phase Mechanics to Describe a Turbulized Gas–Dust Disk

Before developing the basics of the phenomenological theory for the turbulence of multiphase media, as applied to the problem of modeling the evolution of the circumsolar protoplanetary disk, note once again that the currently available approaches to describing multiphase turbulent flows are incomplete (see, e.g., Shraiber et al. 1987; Varaksin 2003). This is attributable both to incompleteness of the "classical" theory of turbulence in "ordinary" hydromechanics (see. Chap. 4) and to a dramatic complication of the pattern of turbulent gas flow in the presence of disperse admixture. It should be kept in mind that, being a fundamental problem of heterogeneous mechanics, the problem of the inverse effect of solid particles on the flow parameters has not yet been solved completely. In particular, this concerns the methods of allowance for the collective effects related to interparticle interactions whose role increases with particle concentration and size. For example, the mechanism of intense randomization of the motion of large particles (the so-called pseudoturbulence), which are weakly entrained by the turbulent pulsations of the carrier medium (Shraiber et al. 1980), is related to interparticle collisions. Thus, in view of the above peculiarities of turbulent flows in heterogeneous media, any theoretical approaches to their description and mathematical models based on them will always be limited, since they essentially pertain to a strictly definite range of concentrations and inertias of the disperse phase. This also applies to all of the currently existing models for the evolution of a gas–dust disk that cover a relatively narrow range of problems pertaining to the problem in question. Let us now turn to deriving the basic balance equations of matter, momentum, and energy for a disk gas–dust turbulized medium intended to formulate and numerically solve the specific problems of consistently modeling the thermohydrodynamic parameters of a protoplanetary cloud at various stages of its evolution and analyze the physical meaning of the individual terms in these equations. For the functions in these equations to be smooth and continuous with continuous first derivatives, (7.76) must be averaged over time or an ensemble of possible realizations. The progress made in recent years in developing and using semiempirical models for turbulence of the first order of closure (the so-called gradient models) for a homogeneous compressible fluid (see, e.g., Townsend 1956; van Mieghem 1973; Kolesnichenko and Marov 1999) allows us to generalize some of these models to the shear flows of a two-phase gas–dust mixture that we describe in terms of a single-fluid continuum. We will derive the closing (defining) relations for the turbulent flows of phase diffusion, heat, and the Reynolds turbulent stress tensor by a standard method based on the concept of mixing length.

As it was already said, various methods of averaging the fields of physical quantities are known to be used in the theories of fluid and gas turbulence. These include, for example, the time averaging (3.1) where the averaging interval Δt of the pulsating parameter $\mathcal{A}(\boldsymbol{r}, t)$ is assumed to be large compared to the characteristic pulsation period and essentially small compared to the variation period of the averaged field $\bar{\mathcal{A}}(\boldsymbol{r}, t)$; the space averaging through integration over the space volume; the space–time averaging; the probability-theoretical averaging over the ensemble of possible realizations (see (3.3)); etc. (see, e.g., Monin and Yaglom 1992). The latter approach using the concept of an ensemble, i.e., an infinite set of stochastic hydrodynamic systems of the same nature that differ by the state of the field of velocities and/or other thermohydrodynamic parameters of motion at a given instant in time, is most fundamental and it will be used by us in the modeling of heterogeneous turbulence. Let us recall again that, according to the well-known ergodicity hypothesis, the time and ensemble averages for a stationary stochastic process are identical. Here, without further discussing the advantages and shortcomings of various averaging methods, we only note that the practice of constructing phenomenological models to study the turbulent motions shows that, in general, the methods of introducing the averaged parameters of motion are unimportant for setting up a complete system of averaged hydrodynamic equations if we require that the Reynolds postulates be satisfied during any averaging.

In the classical theories of the turbulence of homogeneous incompressible fluids that have been developed to date fairly completely (Monin and Yaglom 1992), the averagings for all thermohydrodynamic parameters without exception are usually introduced in an identical way and, as a rule, without weighting coefficients. For example, in the time averaging (3.1) or the probability-theoretical averaging over the ensemble of possible realizations (3.3) the actual value of the parameter $\mathcal{A}$ is represented as the sum of the averaged, $\bar{\mathcal{A}}$, and pulsational, $\mathcal{A}'$, components: $\mathcal{A}= \bar{\mathcal{A}} + \mathcal{A}'$ (with $\bar{\mathcal{A}}' = 0$). In this case, the separation of the actual stochastic motion into a slowly changing continuous mean motion and a rapidly oscillating turbulent (irregular, pulsating about the means) motion depends entirely on the choice of the space–time region for which the means are defined. This is the size of the region that fixes the scale of mean motion. Let us note that various procedures of deriving the equations describing large-scale turbulence could be suggested. For example "smoothed" values of thermohydrodynamic parameters may be introduced using filter-function with the formula

$$\bar{\mathcal{A}}(\boldsymbol{r},t) = \int G(\boldsymbol{r}-\boldsymbol{r}')\mathcal{A}(\boldsymbol{r}',t)\mathrm{d}\boldsymbol{r}'$$

(Leonard 1974), or in numerical simulation composing balance equations (mass, momentum, and energy) for every grid of the calculation net (Ievlev 1970). In this approach all vortices of larger sizes contribute to the averaged motion determined by the averaged thermohydrodynamic parameters, whereas all vortices of smaller sizes filtered out in the averaging process contribute to the small-scale turbulent motion determined by the corresponding pulsations of the same structure parameters.

At the same time, this averaging method (the same for all variables) for a two-phase continuum with a pulsating total density ρ leads not only to cumbersome hydrodynamic equations of mean motion, which is related to the need for retaining correlations of the type $\overline{\rho' \boldsymbol{u}'}$, $\overline{\rho' \boldsymbol{u}' \boldsymbol{u}'}$, $\overline{\rho' C'_\alpha \boldsymbol{u}'}$, etc. (which emerge, because the convective terms of the basic equations for instantaneous motion are nonlinear) in their structure but also to difficulties in physically interpreting the individual terms of the averaged equations. Therefore, when developing our models for a gas–dust disk medium, apart from the "ordinary" means for some of the pulsating parameters, we will use below the Favre (1969) weighted mean averaging for several other parameters specified by the relation

$$\langle \mathcal{A} \rangle \equiv \overline{\rho \mathcal{A}}/\overline{\rho}, \tag{7.78}$$

in this case, $\mathcal{A}= \langle \mathcal{A} \rangle + \mathcal{A}''$, $\bar{\mathcal{A}}'' \neq 0$, $\mathcal{A}''$ being the corresponding turbulent pulsation. Thus, likewise in the case of modeling turbulence of multicomponent mixture (see Chap. 3), in the averaged equations of heterogeneous mechanics we will use two symbols to denote the mean quantities: the overbar denotes the ensemble (time and/or space) averaging, while the angular brackets denote the weighted mean averaging. The double prime is used below to denote the pulsations about the Favre averaged quantity. Note that the Favre averaging of several pulsating thermohydrodynamic parameters for a heterogeneous continuum simplifies significantly the form and analysis of the averaged hydrodynamic equations. This is because for the ordinary averaging, correlations of the type $\overline{\rho'\mathcal{A}'}$, $\overline{\rho'\mathcal{A}'\boldsymbol{u}'}$, etc., appear in the averaged equations of motion in explicit form, while for the Favre averaging, these correlations are hidden in the corresponding terms of the equations that have a simpler (traditional) form. In Chap. 3 some of the properties of the weighted mean averaging (see (3.7)) that we will widely use when compiling heterogeneous turbulence models.

7.3.1 *Averaged Mass Balance Equations for Gas–Dust Matter: The Turbulent Transport Coefficient*

We will consider the turbulized two-phase disk medium as a continuum whose instantaneous motions can be described by the system of hydrodynamic equations (7.76) for a random sample of initial and boundary conditions. This is possible for the space–time scales enclosed between the scales of molecular motions and the minimum turbulence scales (the linear size and lifetime of the smallest vortices), which are generally several (at least three) orders of magnitude larger than the scales of molecular motions, i.e., the separation between the molecules, let alone the sizes of the molecules (Van Migem 1977).

The macroscopic equations of turbulent motion for the gas–dust medium can then be obtained (in a form convenient for the subsequent analysis) by the theoretical-probability ensemble averaging of stochastic (7.76) using the weighted means for such flow parameters as the velocity $\langle \boldsymbol{u} \rangle$, temperature $\langle T \rangle$, mass concentrations $\langle C_\alpha \rangle$, etc. However, it is convenient to average the pressure p and density ρ of the medium as well as all "molecular" thermodynamic fluxes $\boldsymbol{J}_\alpha, \boldsymbol{q}, \Pi$ and ξ_ρ in the "ordinary" way, i.e., without using any weighting coefficients. It is worthwhile to notice that the Favre averaging of the equations of motion for a two-phase flow of a gas suspension described in terms of a single-fluid continuum was probably performed by us for the first time.

7.3.1.1 The Averaged Continuity Equation

Obviously, the averaged density $\bar{\rho}$ and the weighted mean hydrodynamic velocity $\langle \boldsymbol{u} \rangle = \overline{\rho \boldsymbol{u}}/\bar{\rho}$ satisfy the continuity equation for mean motion

$$\frac{\partial \bar{\rho}}{\partial t} + \operatorname{div}(\bar{\rho}\langle \boldsymbol{u} \rangle) = 0 \quad \text{or} \quad \bar{\rho}\frac{D}{Dt}\left(\frac{1}{\bar{\rho}}\right) - \operatorname{div}\,\langle \boldsymbol{u} \rangle = 0 \tag{7.79}$$

Regarding this equation, it is important to emphasize the following: with known difficulties in modeling the binary correlations $\overline{\rho'\boldsymbol{u}'}$ that appear after the "ordinary" averaging of (7.13) for the true density and hydrodynamic velocity of a two-phase system without any weight, the preservation of the standard form (7.79) for the averaged continuity equation is a convincing argument for using the weighted mean $\langle \boldsymbol{u} \rangle$ for the total hydrodynamic flow velocity (Kolesnichenko and Marov 1999).

In particular, (7.80) widely used below can be derived only using (7.79) by averaging the operator relation $\rho \mathrm{d}\mathcal{A}/\mathrm{d}t = \partial(\rho\mathcal{A})/\partial t + (\partial/\partial \boldsymbol{r}) \cdot (\rho \mathcal{A} \boldsymbol{u})$; as a result, we have

$$\overline{\rho \frac{\mathrm{d}\mathcal{A}}{\mathrm{d}t}} = \frac{\partial(\overline{\rho \mathcal{A}})}{\partial t} + \left(\frac{\partial}{\partial \boldsymbol{r}} \cdot (\overline{\rho \mathcal{A} \boldsymbol{u}}) \right)$$
$$= \frac{\partial(\overline{\rho \mathcal{A}})}{\partial t} + \left(\frac{\partial}{\partial \boldsymbol{r}} \cdot \left[\bar{\rho} \langle \mathcal{A} \rangle \langle \boldsymbol{u} \rangle + \overline{\rho \mathcal{A}'' \boldsymbol{u}''} \right] \right) = \bar{\rho} \frac{D}{Dt} \langle \mathcal{A} \rangle + \left(\frac{\partial}{\partial \boldsymbol{r}} \cdot \boldsymbol{J}_A^{\mathrm{turb}} \right), \tag{7.80}$$

where we denote

$$\boldsymbol{J}_A^{\mathrm{turb}} \equiv \overline{\rho \mathcal{A}'' \boldsymbol{u}''} = \bar{\rho} \langle \mathcal{A}'' \boldsymbol{u}'' \rangle \tag{7.81}$$

for the second single-point moments of the flow velocity pulsations and some transferred substance $\mathcal{A}$. Thus, (7.81) defines the so-called turbulent flux related to the transfer of substance $\mathcal{A}$ by the turbulent pulsations of the system's hydrodynamic velocity.

Let us introduce here formula for the turbulent flux $\boldsymbol{J}_v^{\mathrm{turb}}$ of the specific volume $\mathcal{V}(\boldsymbol{r}, t) (\equiv 1/\rho)$. The flux $\boldsymbol{J}_v^{\mathrm{turb}}$ plays an important role in our approach and appears in many averaged equations of motion, for example, in the averaged energy equation (see (7.122*)). Using the relation $v'' = -\rho'/\rho\bar{\rho}$, which directly follows from the definition of pulsations v'' ($v'' = v - \langle v \rangle = 1/\rho - 1/\bar{\rho} = -\rho'/\rho\bar{\rho}$), we obtain from (7.81)

$$\boldsymbol{J}_v^{\mathrm{turb}} \equiv \overline{\rho v'' \boldsymbol{u}''} = -\overline{\rho' \boldsymbol{u}''}/\bar{\rho} = \overline{\boldsymbol{u}''} \tag{7.82}$$

Below, we everywhere assume that only the dust volume content s and the true gas density ρ_{g} fluctuate in the gas–dust flow (this is a cardinal assumption of the approach developed here); it then follows from (7.2) that

$$\frac{\rho'}{\bar{\rho}} \cong (1 - \bar{s}) \frac{\rho'_{\mathrm{g}}}{\bar{\rho}} + \langle \sigma \rangle s' = \langle C_{\mathrm{g}} \rangle \frac{\rho'_{\mathrm{g}}}{\bar{\rho}_{\mathrm{g}}} + \langle \sigma \rangle s', \tag{7.83}$$

where we denoted

$$\langle \sigma \rangle \equiv \frac{\rho_{\mathrm{d}} - \bar{\rho}_{\mathrm{g}}}{\bar{\rho}} \cong \rho_{\mathrm{d}}/\bar{\rho} \tag{7.84}$$

for the averaged excess of the dust particle density above the gas-suspension density and use the expression

$$\langle C_{\rm g}\rangle \equiv \frac{\overline{(1-s)\rho_{\rm g}}}{\bar{\rho}} \cong \frac{(1-\bar{s})\bar{\rho}_{\rm g}}{\bar{\rho}} \cong \frac{\bar{\rho}_{\rm g}}{\bar{\rho}} \tag{7.85}$$

for the averaged mass concentration of the gas phase; moreover

$$\langle C_{\rm d}\rangle \equiv \rho_{\rm d}\bar{s}/\bar{\rho} \cong \bar{s}\langle\sigma\rangle, \quad (\langle C_{\rm g}\rangle + \langle C_{\rm d}\rangle = 1)$$

Then the following expression for the turbulent flux of the specific volume $\boldsymbol{J}_v^{\rm turb}$ in the gas–dust medium follows from (7.83) and (7.82):

$$\boldsymbol{J}_{\rm v}^{\rm turb} \equiv -\overline{\rho'\boldsymbol{u}''}/\bar{\rho} = -\langle\sigma\rangle\overline{s'\boldsymbol{u}''} - \left(1 - \frac{\rho_{\rm d}\bar{s}}{\bar{\rho}}\right)\frac{\overline{\rho_{\rm g}'\boldsymbol{u}''}}{\bar{\rho}_{\rm g}} \tag{7.86}$$

It should be noted that the above expressions for ρ' and $\langle C_{\rm g}\rangle$ (as many similar quantities that will appear below) are valid only when the inequalities $\overline{\mathcal{A}'\mathcal{B}'}/\bar{\mathcal{A}}\bar{\mathcal{B}} \ll 1$ and $\langle\mathcal{A}''\mathcal{B}''\rangle/\langle\mathcal{A}\rangle\langle\mathcal{B}\rangle \ll 1$ hold for any pulsating thermodynamic parameters $\mathcal{A}$ and $\mathcal{B}$ not equal to the gas–dust flow velocity $\boldsymbol{u}$. In what follows, we everywhere assume that the ratios of this kind are small without any special stipulations.

7.3.1.2 The Averaged Diffusion Equation for the Disperse Component of Disk System

Applying the averaging operator (7.80) to the diffusion equation (7.12) for disperse particles, we obtain the balance equation for the dust concentration

$$\bar{\rho}\frac{D\langle C_{\rm d}\rangle}{Dt} + {\rm div}\left(\overline{\boldsymbol{J}_{\rm d}} + \boldsymbol{J}_{\rm d}^{\rm turb}\right) = \overline{\sigma_{\rm dg}}, \quad \overline{\sigma_{\rm dg}} \equiv \sum_{\rho=1}^{r} \nu_{{\rm d},\rho}\bar{\xi}_{\rho}. \tag{7.87}$$

Here, $\langle C_{\rm d}\rangle = \rho_{\rm d}\bar{s}/\bar{\rho}$; $\overline{\boldsymbol{J}_{\rm d}}$ is the averaged "molecular" diffusion flux of the dust defined by the relation (see (7.12)):

$$\overline{\boldsymbol{J}_{\rm d}} \equiv \overline{\rho\, C_{\rm d}C_{\rm g}\boldsymbol{w}} \cong \bar{\rho}\langle C_{\rm d}\rangle\langle C_{\rm g}\rangle\bar{\boldsymbol{w}} \cong \rho_{\rm d}\frac{\bar{\rho}_{\rm g}}{\bar{\rho}}\bar{s}\,\bar{\boldsymbol{w}}. \tag{7.88}$$

whereas

$$\boldsymbol{J}_{\rm d}^{\rm turb} \equiv \overline{\rho\, C_{\rm d}''\boldsymbol{u}''} = \rho_{\rm d}\overline{s\,\boldsymbol{u}''} \tag{7.89}$$

is the so-called turbulent diffusion flux of the disperse phase. Note that for the gas diffusion flux, we may write $\boldsymbol{J}_{\rm g}^{\rm turb} \equiv \overline{\rho\, C_{\rm g}''\boldsymbol{u}''} = -\overline{\rho\, C_{\rm d}''\boldsymbol{u}''} = -\boldsymbol{J}_{\rm d}^{\rm turb}$.

If we write the turbulent dust flux as $\boldsymbol{J}_{\mathrm{d}}^{\mathrm{turb}} = \rho_{\mathrm{d}}\bar{s}\boldsymbol{J}_{\mathrm{v}}^{\mathrm{turb}} + \rho_{\mathrm{d}}\overline{s'\boldsymbol{u}''}$, then using (7.86) we can represent the turbulent flux of the specific volume in slightly different form:

$$\boldsymbol{J}_{\mathrm{v}}^{\mathrm{turb}} = -\langle\sigma\rangle\frac{\bar{\rho}}{\rho_{\mathrm{d}}\bar{\rho}_{\mathrm{g}}}\boldsymbol{J}_{\mathrm{d}}^{\mathrm{turb}} - (1-\bar{s})\frac{\overline{\rho_{\mathrm{g}}'\boldsymbol{u}''}}{\bar{\rho}_{\mathrm{g}}} \tag{7.90}$$

We will need this expression for $\boldsymbol{J}_{\mathrm{v}}^{\mathrm{turb}}$ below.

To close the averaged equation (7.87), we must have a defining relation for the turbulent diffusion flux of dust, $\boldsymbol{J}_{\mathrm{d}}^{\mathrm{turb}} \equiv \bar{\rho}\langle\, C_{\mathrm{d}}''\boldsymbol{u}''\rangle$. There are several approaches to modeling second-order correlation moments of this type that differ in complexity (see Marov and Kolesnichenko 2002 and Chap. 3). Here, we restrict our analysis to the simplest gradient relation that we will derive in a traditional way, by introducing the concept of mixing length. It is worth to note, however, that in the past decade, deeper (in physical sense) differential turbulence models have come to be used to model turbulent single-phase flows in thin accretion disks analogous to those discussed in Chap. 4. Apart from the equations for averaged quantities, these models include the additional differential transfer equations for the most important parameters of the turbulent structure (Dubrulle 1992). For this purpose, we will assume that the transfer of particular flow field characteristics $\mathcal{A}$ by the turbulent pulsations of the medium takes place as a diffusion process and that the existence of an effective mixing length ξ_A of substance $\mathcal{A}$, the distance to which the turbulent moles (vortices) in the flow move before they are destroyed through the interaction with other disturbances, may be admitted. If we denote the Lagrangian turbulent pulsation of the transferred substance $\mathcal{A}$ corresponding to the Eulerian pulsation $\mathcal{A}''$ by $\mathcal{A}''_{\mathrm{L}}$ and the effective mixing length by ξ_A, then we may write $\mathcal{A}''_{\mathrm{L}} = \mathcal{A}'' + \left(\xi_A\cdot\frac{\partial}{\partial\boldsymbol{r}}\langle\mathcal{A}\rangle\right)$ (see Chap. 3). Assuming also that dust substance is indestructible and/or changed in an elementary volume when moving in the carrying gas, we will further admit that vortices being displaced by the distance ξ_A preserve in the Lagrangian volume the same specific dust density as at the original level. Then $(C_{\mathrm{d}})''_{\mathrm{L}} \cong 0$, or

$$C_{\mathrm{d}}'' = -\left(\xi_{\mathrm{d}}\cdot\frac{\partial}{\partial\boldsymbol{r}}\langle C_{\mathrm{d}}\rangle\right).$$

Hence, the diffusion flux $\boldsymbol{J}_{\mathrm{d}}^{\mathrm{turb}}$ of the system's dust component in terms of the gradient representations is

$$\boldsymbol{J}_{\mathrm{d}}^{\mathrm{turb}} \equiv \overline{\rho\, C_{\mathrm{d}}''\boldsymbol{u}''} = -\bar{\rho}\langle\boldsymbol{u}''\xi_{\mathrm{d}}\rangle\cdot\frac{\partial}{\partial\boldsymbol{r}}\langle C_{\mathrm{d}}\rangle$$
$$-\bar{\rho}\left(\boldsymbol{D}_{\mathrm{d}}^{\mathrm{turb}}\cdot\frac{\partial\langle C_{\mathrm{d}}\rangle}{\partial\boldsymbol{r}}\right) = -\bar{\rho}\rho_{\mathrm{d}}\left(\boldsymbol{D}_{\mathrm{d}}^{\mathrm{turb}}\cdot\frac{\partial}{\partial\boldsymbol{r}}\left(\frac{\bar{s}}{\bar{\rho}}\right)\right) \tag{7.91}$$

where the dyad $\boldsymbol{D}_{\mathrm{d}}^{\mathrm{turb}} \equiv \langle \boldsymbol{u}''\xi_{\mathrm{d}} \rangle$ defines the nonsymmetric turbulent diffusivity tensor of the dust that, in the general anisotropic case, allows for the differences in the intensities of the turbulent pulsations of the solid particle velocity and concentration along different coordinate axes. Relation (7.91) is equivalent to the assertion that the turbulent flux of the dust phase is proportional to the gradient in mean concentration $\langle C_{\mathrm{d}} \rangle$ and has the direction opposite to it. It is important to keep in mind (see Chap. 4 for more detail) that using the gradient hypothesis does not eliminate all of the difficulties associated with the closure problem, since the corresponding turbulent transfer coefficients must also be determined (experimentally or based on a qualitative physical analysis).

For an isotropic turbulent field, we may assume that the tensor $\boldsymbol{D}_{\mathrm{d}}^{\mathrm{turb}}$ is spherical, $\boldsymbol{D}_{\mathrm{d}}^{\mathrm{turb}} = D_{\mathrm{d}}^{\mathrm{turb}}\boldsymbol{U}$, i.e., is defined by the turbulent diffusivity of the dust $D_{\mathrm{d}}^{\mathrm{turb}}(\boldsymbol{r})$ alone, which is statistical parameter of turbulence. Then

$$\boldsymbol{J}_{\mathrm{d}}^{\mathrm{turb}} = -\bar{\rho} D_{\mathrm{d}}^{\mathrm{turb}} \frac{\partial}{\partial \boldsymbol{r}} \langle C_{\mathrm{d}} \rangle \cong -D_{\mathrm{d}}^{\mathrm{turb}} \frac{\rho_{\mathrm{d}} \bar{\rho}_{\mathrm{g}}}{\bar{\rho}} \left(\frac{\partial \bar{s}}{\partial \boldsymbol{r}} - \bar{s} \frac{\partial \ln \bar{\rho}_{\mathrm{g}}}{\partial \boldsymbol{r}} \right) \tag{7.92}$$

and the averaged diffusion equation (7.87) takes the form

$$\bar{\rho} \frac{D\langle C_{\mathrm{d}} \rangle}{Dt} + \operatorname{div}\left\{ \bar{\rho}_{\mathrm{g}} \langle C_{\mathrm{d}} \rangle \bar{\boldsymbol{w}} - \frac{\bar{\rho} \nu^{\mathrm{turb}}}{\boldsymbol{Sc}^{\mathrm{turb}}} \frac{\partial}{\partial \boldsymbol{r}} \langle C_{\mathrm{d}} \rangle \right\} = \bar{\sigma}_{\mathrm{dg}}, \tag{7.87*}$$

where ν^{turb} is a turbulent analog of the coefficient of kinematic viscosity for a gas–dust mixture (see below); $\boldsymbol{Sc}^{\mathrm{turb}} \equiv \nu^{\mathrm{turb}}/D_{\mathrm{d}}^{\mathrm{turb}}$ is the turbulent Schmidt number for the disperse phase (a dimensionless factor of the order of unity that depends on the nature of the dust substance and that is a function of the dimensionless flow parameters). In the gradient theory, the Schmidt number is calculated using the formula $\boldsymbol{Sc}^{\mathrm{turb}} \equiv \xi_u/\xi_{\mathrm{d}}$, where ξ_u is the velocity mixing length. The dependence of the Schmidt number on the dust particle concentration was first obtained by Abramovich and Girshovich (1973).

7.3.1.3 The Turbulent Transfer Coefficient: The Stocks Number

First, note that the coefficients of turbulent transfer in any turbulized medium, in contrast to the corresponding molecular transfer coefficients, describe not just its thermophysical properties, but also the state of the turbulent field and, hence, depend on the averaging scale of the pulsating thermohydrodynamic parameters. For this reason, the way of introducing the averaged characteristics of turbulent motion is crucial in developing the methods for experimentally determining this kind of transfer coefficients. The assumption that the particles are completely entrained by the turbulent pulsations of the scale that plays a major role in particle encounter mechanics (the approximation of a passive admixture) underlies the most advanced approach to modeling the turbulent diffusivity. If the solid particles are very small and, hence, their motion does not differ in any way from the motion of the gas carrier moles, then the turbulent diffusivity of the dust particles $D_{\mathrm{d}}^{\mathrm{turb}}$ and the coefficient of turbulent viscosity ν^{turb} of the gas are equal for them. In this case, $D_{\mathrm{d}}^{\mathrm{turb}}$ depends only on the turbulent pulsation scale length of the carrier gas and can be estimated, for example, as

$$D_{\mathrm{d}}^{\mathrm{turb}} \sim \nu^{\mathrm{turb}} = \sqrt{b}l \sim (\varepsilon l)^{1/3} l = \varepsilon^{1/3} l^{4/3} \quad \text{when} \quad l < \lambda_k. \tag{7.93}$$

The following notation is used in this expression: b is the turbulent energy of the gas–dust medium as a whole (see (7.108)); $\varepsilon \cong b^{3/2}/l \cong v_{\mathrm{g}}^3/\lambda_K^4$ is the dissipation rate of the turbulent gas energy (see (7.121*)); $\lambda_K \cong (v_{\mathrm{g}}^3/\varepsilon)^{1/4}$ is the Kolmogorov (internal) turbulence scale length; and $l(\boldsymbol{r})$ is the Prandtl mixing length (a numerical factor that can be included in l). Below, we call l the turbulence scale length at a given point in the flow.

It should be noted, however, that the numerous experimental data (e.g., Mednikov 1981) confirm he equality $D_{\mathrm{d}}^{\mathrm{turb}} \cong \nu^{\mathrm{turb}}$ only for very small particles, when the dimensionless Stokes number in the large-scale pulsational motion $\boldsymbol{Stk} \ll 1$. In general, several Stokes numbers $\boldsymbol{Stk}$, which are equal to the ratio of the dynamic relaxation time of the dust particles to particular flow time scales (e.g., to the Kolmogorov turbulence time scale $\tau_K \cong (v_{\mathrm{g}}/\varepsilon)^{1/2}$ or to the time scale of the large-scale pulsational motion of the medium $\tau_{\mathrm{L}} \propto b/\varepsilon$) that characterize the particle inertia relative to the chosen flow scale in the turbulent flow, can be introduced for a turbulized heterogeneous flow. In the case of Keplerian differential rotation of the solid particles in the disk, where there is a radial gradient in averaged velocity, it is important to take into account the particle inertia when analyzing the relaxation of the averaged phase velocities. For this purpose, it is convenient to introduce the Stokes number in the averaged motion, which we will write as $\boldsymbol{Stk} = \omega_{\mathrm{turb}}\tau_{\mathrm{relax}}$, where τ_{relax} is the dynamic relaxation (dynamic inertia) time of the particles; ω_{turb} is the lower frequency limit for the turbulent pulsations of the carrier gas in the disk belonging to the largest vortices with the (macroscopic turbulence) scale length L; the frequency ω_{turb} then determines the slow macroscopic variations in flow parameters (which are generally not related to turbulence) and, according to Safronov (1969), is specified in the form $\omega_{\mathrm{turb}} = \Omega_{K,\mathrm{mid}}$, where $\Omega_{K,\mathrm{mid}} \equiv \sqrt{G\mathcal{M}_{\odot}/\tilde{r}^3}$ is the orbital frequency (the Keplerian angular velocity near the midplane of the disk) (Cuzzi et al. 1993) slightly refined this estimate: $\omega_{\mathrm{turb}} \approx \zeta\, \Omega_{K,\mathrm{mid}}$, where $\zeta \approx 0.0126$.

For small spherical particles (e.g., with diameters $\ll$ 1 cm at 1 AU or ~600 cm at 10 AU), the dynamic relaxation time scale is defined by the Epstein law (see (7.21))

$$\tau_{\mathrm{relax}}^{Ep} = \frac{\tilde{\rho}_{\mathrm{d}}\tilde{\rho}_{\mathrm{g}}}{\rho R_{\mathrm{dg}}} = \frac{\rho_{\mathrm{d}}\tilde{d}_{\mathrm{d}}}{2\rho c_{\mathrm{sg}}}, \quad \tilde{d}_{\mathrm{d}} < \lambda_{\mathrm{g}} \tag{7.94}$$

(the mean free path of gas molecules at 1 AU is $\lambda_g \sim 1$ cm). However, for coarse spherical particles, this formula is slightly modified. The simplest expression for τ_{relax} can be obtained when the Reynolds number for dust, $\boldsymbol{Re}_d = \tilde{d}_d|\boldsymbol{w}|/\nu_g = 2\tilde{d}_d|\boldsymbol{w}|/\lambda_g c_{sg}$, is fairly small, $\boldsymbol{Re}_d < 1$ (which is the case for the so-called Stokes particles). This inequality is true, for example, for particles with diameters from 1 to 10 cm at 1 AU and with diameters from 600 to 1,000 cm at 10 AU (Dubrulle et al. 1995). In this case, according to (7.22), the coefficient of aerodynamic drag is $C_D(\boldsymbol{Re}_d) = 9\,\boldsymbol{Re}_d{}^{-1}$ and the dynamic relaxation time scale τ_{relax} will be defined by the Stokes law ($\tilde{d}_d > \lambda_g$)

$$\tau_{relax}^{St} = \frac{\tilde{\rho}_d \tilde{\rho}_g}{\bar{\rho} R_{dg}} = \frac{\tilde{d}_d\,\rho_d}{2\bar{\rho} C_D(\boldsymbol{Re}_d)|\boldsymbol{w}|} \cong \frac{\tilde{d}_d \rho_d \boldsymbol{Re}_d}{18\bar{\rho}|\boldsymbol{w}|} = \frac{\rho_d \tilde{d}_d^2}{9\bar{\rho} c_{sg}\lambda_g} = \frac{\rho_d \tilde{d}_d^2}{18\bar{\rho}\nu_g} \tag{7.95}$$

Thus, the inertia of a Stokes particle depends on the parameters of the medium in which it moves. In addition, if the particles are not too small (and, hence, are not entirely entrained by the gas carrier moles), then their relative velocities acquired through turbulent pulsations depend significantly on their masses. For the motion of a non-Stokes particle, its inertia depends on the Reynolds number for dust $\boldsymbol{Re}_d$ and may be written as

$$\tau_{relax} = \tau_{relax}^{St}/C(\boldsymbol{Re}_d) \tag{7.95*}$$

where

$$C(\boldsymbol{Re}_d) = \begin{cases} 1 + 0.179\boldsymbol{Re}_d^{1/2} + 0.013\boldsymbol{Re}_d, & \boldsymbol{Re}_d \le 10^3; \\ 0.0183\boldsymbol{Re}_d, \quad \boldsymbol{Re}_d > 10^3 \end{cases}$$

is a correction function that allows for the influence of inertial forces on the relaxation time scale of the non-Stokes particle (the coefficient of aerodynamic resistance for the particle is $C_D(\boldsymbol{Re}_d) = 9\boldsymbol{Re}_d^{-1}C(\boldsymbol{Re}_d)$). The difference between the pulsation velocities of particles of different sizes determines their encounter and increases the collision probability. Note that the inertial coagulation of particles in a turbulized flow is also related to this fact. Thus, the following formula (cf., e.g., Cuzzi et al. 1993) is valid for a polydisperse disk medium:

$$\boldsymbol{Sc}^{turb} = \frac{\nu^{turb}}{D_d^{turb}} \cong (1 + \boldsymbol{Stk})\sqrt{1 + 3|\bar{\boldsymbol{w}}|^2/2b} \tag{7.96}$$

where

$$\boldsymbol{Stk} = \zeta \Omega_{K,mid} \frac{\rho_d \tilde{d}_d^2}{18\rho \nu_g C(\boldsymbol{Re}_d)}.$$

7.3.1.4 Defining Equation for the Averaged Relative Velocity of Phases

Averaging the defining equation (7.19*) for the actual values of vector $\boldsymbol{w}$ yields

$$\bar{\rho}\theta_{\mathrm{gd}}\bar{\boldsymbol{w}} \cong -\frac{D\bar{\boldsymbol{w}}}{Dt} - \overline{\left(\boldsymbol{w}\cdot\frac{\partial}{\partial \boldsymbol{r}}\right)\boldsymbol{u}} + \frac{1}{\bar{\rho}_{\mathrm{g}}}\frac{\partial \bar{p}}{\partial \boldsymbol{r}}$$
$$= -\frac{D\bar{\boldsymbol{w}}}{Dt} - \left(\boldsymbol{J}_{\mathrm{v}}^{\mathrm{turb}}\cdot\frac{\partial}{\partial \boldsymbol{r}}\bar{\boldsymbol{w}}\right) - \overline{\left(\boldsymbol{u}''\cdot\frac{\partial}{\partial \boldsymbol{r}}\boldsymbol{w}'\right)} - \left(\bar{\boldsymbol{w}}\cdot\frac{\partial}{\partial \boldsymbol{r}}\right)\langle\boldsymbol{u}\rangle$$
$$- \overline{\left(\boldsymbol{w}'\cdot\frac{\partial}{\partial \boldsymbol{r}}\right)\boldsymbol{u}''} + \frac{1}{\bar{\rho}_{\mathrm{g}}}\frac{\partial \bar{p}}{\partial \boldsymbol{r}} - \left(\bar{\boldsymbol{w}}\cdot\frac{\partial}{\partial \boldsymbol{r}}\right)\boldsymbol{J}_{\mathrm{v}}^{\mathrm{turb}} = -\frac{D\bar{\boldsymbol{w}}}{Dt} - \left(\bar{\boldsymbol{w}}\cdot\frac{\partial}{\partial \boldsymbol{r}}\right)\langle\boldsymbol{u}\rangle + \frac{1}{\bar{\rho}_{\mathrm{g}}}\frac{\partial \bar{p}}{\partial \boldsymbol{r}}. \tag{7.97}$$

In writing this relation, we disregarded the pulsations $\boldsymbol{w}'$ of the relative velocity (which is valid only when the velocity of the averaged relative phase motion $\bar{\boldsymbol{w}}$ is much higher than the pulsation velocity $\boldsymbol{w}'$, i.e., for fairly large particles) and the products of the averaged thermodynamic fluxes of various natures as terms of the second order of smallness. In addition, we used the identical transformation

$$\frac{\overline{\mathrm{d}\mathcal{A}}}{\mathrm{d}t} \equiv \frac{D\bar{\mathcal{A}}}{Dt} + \left(\boldsymbol{J}_{\mathrm{v}}^{\mathrm{turb}}\cdot\frac{\partial}{\partial \boldsymbol{r}}\bar{\mathcal{A}}\right) + \overline{\left(\boldsymbol{u}''\cdot\frac{\partial \mathcal{A}'}{\partial \boldsymbol{r}}\right)} \tag{7.98}$$

which by averaging the substantial derivative can be tentatively written as

$$\frac{\mathrm{d}\mathcal{A}}{\mathrm{d}t} = \frac{D\mathcal{A}}{Dt} + \boldsymbol{u}''\cdot\left(\frac{\partial \mathcal{A}}{\partial \boldsymbol{r}}\right) = \frac{D\mathcal{A}}{Dt} + \left(\boldsymbol{u}''\cdot\left(\frac{\partial \bar{\mathcal{A}}}{\partial \boldsymbol{r}}\right)\right) + \left(\boldsymbol{u}''\cdot\left(\frac{\partial \mathcal{A}''}{\partial \boldsymbol{r}}\right)\right).$$

7.3.2 *Averaged Smoluchowski Coagulation Equation*

Turbulence leads to two types of phenomena that affect the coagulation in a disperse system. First, the particles acquire an additional relative velocity under the effect of turbulent pulsations, which, in turn, changes the coagulation kernel $K(W,U)$ that characterizes the particle collision probability in the system (Voloshchuk 1984). Here, so far we can talk about two coagulation-accelerating effects with certainty. The first effect is related to an increase in the trapping coefficient through turbulent mixing; as a result, the number of collisions between solid particles increases significantly compared to a laminar flow. The second effect is related to the presence of a shear in the turbulent flow velocity field that causes a change in the trapping conditions at $K(W,U)$ close to zero and increases the coagulation probability of small particles (Woods et al. 1972).

The phenomena of the second type are related to the collective behavior of particles in a turbulized system. Turbulence increases the local nonuniformities in the distribution of coagulating particles to scales comparable to the mean separation between the particles, thereby giving rise to fluctuations in the particle size distribution function $f(U, \boldsymbol{r}, t)$ at macroscopic distances. From a physical point of view, since the coagulation equation (7.33) is nonlinear, this variation in concentration of particles with volume U results in an acceleration of the coagulation in a region with an enhanced particle concentration and in its deceleration in a region with a reduced particle concentration, so, on average, this leads to a different coagulation rate than in the uniform case ($U = \text{const}$) and contributes to the faster appearance of large particles.

This process can be described by the formal averaging of the coagulation equation (7.33)

$$\begin{aligned}
&\bar{\rho}\frac{D}{Dt}\left(\frac{\overline{f(U)}}{\bar{\rho}}\right) + \operatorname{div}\left[\boldsymbol{J}_f^{\text{turb}}(U) + \overline{f(U)}\langle C_{\text{g}}\rangle\bar{\boldsymbol{w}}\right] \\
&\quad = \frac{1}{2}\int_0^U \overline{f(W)}\,\overline{f(U-W)}\,K(W,U-W)\,\mathrm{d}W - \overline{f(U)}\int_0^\infty \overline{f(W)}\,K(W,U)\,\mathrm{d}W \\
&\quad = \frac{1}{2}\int_0^U \overline{f'(W)f'(U-W)}\,K(W,U-W)\,\mathrm{d}W - \int_0^\infty \overline{f'(U)f'(W)}\,K(W,U)\,\mathrm{d}W,
\end{aligned} \tag{7.99}$$

where

$$\boldsymbol{J}_f^{\text{turb}}(U) \equiv \overline{\rho(f/\rho)''\boldsymbol{u}''} = -\bar{\rho}D_U^{\text{turb}}\frac{\partial}{\partial \boldsymbol{r}}\left(\frac{\overline{f(U)}}{\bar{\rho}}\right) \tag{7.100}$$

is the turbulent flux of the dust particles of volume U; $\langle C_{\text{g}}\rangle = (1-\bar{s})\bar{\rho}_{\text{g}}/\bar{\rho}$; D_U^{turb} is the turbulent diffusivity coefficient for the fraction-U particles, the expression for which was derived, for example, by Schmitt et al. (1997). Equation (7.99) is not closed, since the function $\gamma(U, W, \boldsymbol{r}, t) \equiv \overline{f'(U,\,\boldsymbol{r},\,t)f'(W,\,\boldsymbol{r},\,t)}$ is undefined. The equation for $\gamma(U, W, \boldsymbol{r}, t)$ can be derived by multiplying the basic equation (7.33) by f and by its subsequent stochastic averaging over the ensemble of possible realizations to give an equation that contains the mean of the product of three functions f'. This operation leads to an infinite chain of equations. The problem of closing the latter was solved only by introducing a particular hypothesis.

If we integrate (7.99) over U, then the equation for the averaged total number of disperse particles $\bar{N}_{\text{d}}(\boldsymbol{r}, t)$ will take the form

$$\bar{\rho}\frac{\mathrm{d}}{\mathrm{d}t}\left(\frac{\bar{N}_{\mathrm{d}}}{\bar{\rho}}\right) = -\mathrm{div}\left(\boldsymbol{J}_{N_{\mathrm{d}}}^{\mathrm{turb}} + \frac{\bar{N}_{\mathrm{d}}(1-\bar{s})\bar{\rho}_{\mathrm{g}}}{\bar{\rho}}\bar{\boldsymbol{w}}\right)$$
$$-\frac{1}{2}\int_0^\infty\int_0^\infty K(W,U)\bar{f}(W,\boldsymbol{r},t)\bar{f}(U,\boldsymbol{r},t)\,\mathrm{d}W\mathrm{d}U$$
$$-\frac{1}{2}\int_0^\infty\int_0^\infty K(W,U)\,\gamma(U,W,\boldsymbol{r},t)\,\mathrm{d}W\mathrm{d}U + \sum_k\sum_{\rho=1}^{r}\nu_{\mathrm{d}(k),\rho}\bar{\xi}_\rho, \quad (7.101)$$

where

$$\boldsymbol{J}_{N_{\mathrm{d}}}^{\mathrm{turb}} = \int_U \boldsymbol{J}_f^{\mathrm{turb}}(U)\mathrm{d}U \quad (7.102)$$

is the turbulent flux of the number of dust particles for which the representation $\boldsymbol{J}_{N_{\mathrm{d}}}^{\mathrm{turb}} = \overline{N_{\mathrm{d}}\boldsymbol{u}''} = \boldsymbol{J}_{\mathrm{d}}^{\mathrm{turb}}/\rho_{\mathrm{d}}\tilde{U}_{\mathrm{d}}$ is valid. Since the function $\gamma(U,W,\boldsymbol{r},t)$ must be positively defined in view of its symmetry in U and W, the coagulation in a turbulized medium with nonuniformly distributed particles will be faster. In conclusion, note that, despite its importance, the question of how the fluctuations affect the coagulation rate has not yet been fully developed and requires the further study.

7.3.3 Averaged Equation of Motion for a Gas–Dust Disk Medium

Given (7.35), the Favre averaging of the instantaneous equation of motion for the gas–dust mixture considered as a single entity and accounting for (7.80) yields

$$\bar{\rho}\frac{D\langle\boldsymbol{u}\rangle}{Dt} \equiv \frac{\partial}{\partial t}(\bar{\rho}\langle\boldsymbol{u}\rangle) + \left(\frac{\partial}{\partial r}\cdot(\bar{\rho}\langle\boldsymbol{u}\rangle\langle\boldsymbol{u}\rangle)\right)$$
$$= -\frac{\partial\bar{p}_{\mathrm{sum}}}{\partial\boldsymbol{r}} + \left(\frac{\partial}{\partial\boldsymbol{r}}\cdot(\boldsymbol{R}+\bar{\Pi}_{\mathrm{sum}}+\bar{\Pi}_{\mathrm{rel}})\right) + \bar{\rho}\frac{G\mathcal{M}_\odot}{|\tilde{\boldsymbol{r}}|^3}\tilde{\boldsymbol{r}}, \quad (7.103)$$

where

$$\boldsymbol{R} \equiv -\overline{\rho\boldsymbol{u}''\boldsymbol{u}''} = -\bar{\rho}\langle\boldsymbol{u}''\boldsymbol{u}''\rangle \quad (7.104)$$

is the turbulent (*Reynolds*) stress tensor, which takes the following form in Cartesian coordinates:

$$R_{ij} \equiv -\overline{\rho u_i'' u_j''} = \begin{pmatrix} -\overline{\rho u_1''2} & -\overline{\rho u_1'' u_2''} & -\overline{\rho u_1'' u_3''} \\ -\overline{\rho u_2'' u_1''} & -\overline{\rho u_2''2} & -\overline{\rho u_2'' u_3''} \\ -\overline{\rho u_3'' u_1''} & -\overline{\rho u_3'' u_2''} & -\overline{\rho u_3''2} \end{pmatrix}. \quad (7.105)$$

The Reynolds tensor is a symmetric tensor of the second rank and describes the turbulent stresses produced by the pulsations of the turbulent velocity field in the gas–dust continuum as a whole. It is well known (see, e.g., Monin and Yaglom 1992) that in a developed turbulent single-phase flow, i.e., at large values of the global Reynolds number $\boldsymbol{Re}_{\mathrm{glob}} = Lu_0/\nu$ corresponding to large-scale motions (here, u_0 are typical changes in the velocity of the gas–dust mixture at distances of the order of the macroscopic turbulence scale length L; and ν is the effective coefficient of kinematic viscosity for the gas suspension), we may disregard the averaged viscous stress tensor of the medium $\bar{\Pi}$ compared to the Reynolds stress tensor $\boldsymbol{R}$, except the thin regions of the so-called viscous sublayer that border the solid substrate (in our case, the thin layer of dust adjacent to the midplane of the disk is this substrate). This is also valid for a differentially rotating Keplerian protoplanetary disk with a typical Reynolds number $\boldsymbol{Re}_{\mathrm{glob}} \geq 10^{10}$, since the turbulent viscosity of its material is larger than the molecular one by eight orders of magnitude or more, as follows from the observed distribution of angular momentum and mass in the Solar system and in numerous systems of young stars with disks (Richard and Zahn 1999). However, it should be kept in mind that the aforesaid does not apply to "the averaged relative stress tensor of the phases $\bar{\Pi}_{\mathrm{rel}}$," whose effect on the two-phase flow of the disk medium can be comparable in order of magnitude to the Reynolds tensor $\boldsymbol{R}$. In particular, near the subdisk, where the concentration of dust particles of fairly large sizes (and, hence, $\bar{\boldsymbol{w}} \gg 0$) is significant, these shear stresses act particularly effectively, leading to additional flow turbulization, but in a volume comparable to the volume of the dust layer, i.e., small compared to the total disk volume (Goldrich and Ward 1973).

Using the fact that the Lagrangian pulsations of the weighted mean gas–dust flow velocity are conservative, $(\boldsymbol{u}'')_{\mathrm{L}} \cong 0$, it can be shown that the Reynolds tensor $\boldsymbol{R}$ (for an isotropic turbulent field) is related to the gradients of Favre averaged flow velocity $\partial\langle\boldsymbol{u}\rangle/\partial\boldsymbol{r}$ by the following defining relation (Kolesnichenko and Marov 1999)

$$\boldsymbol{R} = -\frac{2}{3}\bar{\rho}\, b\, \boldsymbol{U} + 2\bar{\rho}\, \nu^{\mathrm{turb}} \overset{\circ}{\boldsymbol{D}}, \quad \overset{\circ}{\boldsymbol{D}} \equiv \boldsymbol{D} - \frac{1}{3}\boldsymbol{U}\,\mathrm{div}\langle\boldsymbol{u}\rangle) \tag{7.106}$$

or

$$\boldsymbol{R} = -\frac{2}{3}\bar{\rho}\, b\, \boldsymbol{U} + \bar{\rho}\, \nu^{\mathrm{turb}} \left(\frac{\partial\langle\boldsymbol{u}\rangle}{\partial\boldsymbol{r}} + \left(\frac{\partial\langle\boldsymbol{u}\rangle}{\partial\boldsymbol{r}}\right)^{\mathrm{transp}} \right) - \frac{2}{3}\bar{\rho}\, \nu^{\mathrm{turb}} \boldsymbol{U}\,\mathrm{div}\langle\boldsymbol{u}\rangle, \tag{7.107}$$

where $\boldsymbol{D} \equiv \frac{1}{2}\left(\frac{\partial\langle\boldsymbol{u}\rangle}{\partial\boldsymbol{r}} + \left(\frac{\partial\langle\boldsymbol{u}\rangle}{\partial\boldsymbol{r}}\right)^{\mathrm{transp}} \right)$ is the averaged deformation tensor; $\overset{\circ}{\boldsymbol{D}}$ is the averaged deformation rate tensor; ν^{turb} is the kinematic coefficient of turbulent viscosity for the gas–dust mixture. Let us notice that the defining relations for the Reynolds stress tensor $\boldsymbol{R}$ (7.106) and the averaged viscous stress tensor $\bar{\Pi}$ were earlier derived by Marov and Kolesnichenko (2002) by the methods of nonequilibrium thermodynamics from the averaged Gibbs relation. The possible anisotropy in the coefficients of turbulent viscosity ν^{turb} in a differentially rotating gas–dust cloud (Safronov 1969) was analyzed in detail by Kolesnichenko (2000). Here and in what follows, we retain the designations $\boldsymbol{D}$ and $\overset{\circ}{\boldsymbol{D}}$ for the averaged deformation and deformation rate tensor to avoid ambiguity (cf. (7.40)).

Relation (7.106) includes a key parameter in the theory of turbulence,

$$b \equiv \overline{\rho \left|\boldsymbol{u}''\right|^2} / 2\bar{\rho} \tag{7.108}$$

the averaged kinetic energy of the turbulent pulsations in the weighted mean velocity of the gas–dust continuum (turbulent energy); in general, the corresponding balance equation (see (7.144)) is needed to determine it. Since a continuous distribution of velocity pulsations $\boldsymbol{u}''$ at various frequencies f (from the minimum ones determined by the viscous forces to the maximum ones determined by the boundary conditions for the flow) is produced in a developed turbulent flow, it is often convenient to represent dispersion (7.108) as the sum of the corresponding quantities pertaining to different frequencies,

$$b = \int_0^\infty b(f)\mathrm{d}f, \tag{7.109}$$

where $b(f)$ is the fraction of the turbulent energy of the gas–dust mixture that corresponds to the frequency band df (the energy spectrum for b).

Under the above assumptions, the averaged relative stress tensor $\bar{\Pi}_{\mathrm{rel}}$ in (7.103) (recall that the tensor $\bar{\Pi}_{\mathrm{rel}}$ arises from the inertial effects of the relative motion of the coarse dust particle fraction and gas (see (7.37))) can be transformed to

$$\begin{aligned} \overline{\Pi_{\mathrm{rel}}} &= -\overline{\rho C_{\mathrm{d}} C_{\mathrm{g}} \boldsymbol{w}\boldsymbol{w}} = -\overline{\rho C_{\mathrm{d}} C_{\mathrm{g}}}(\bar{\boldsymbol{w}}\bar{\boldsymbol{w}} + \overline{\boldsymbol{w}'\boldsymbol{w}'}) - \overline{\boldsymbol{w}\boldsymbol{w}(\rho C_{\mathrm{d}} C_{\mathrm{g}})'} \\ &= -(\bar{\boldsymbol{w}}\bar{\boldsymbol{w}} + \overline{\boldsymbol{w}'\boldsymbol{w}'})(\bar{\rho}\langle C_{\mathrm{d}}\rangle\langle C_{\mathrm{g}}\rangle + \overline{\rho C''_{\mathrm{d}} C''_{\mathrm{g}}}) - 2\bar{\boldsymbol{w}}\overline{\boldsymbol{w}'(\rho C_{\mathrm{d}} C_{\mathrm{g}})'} - \overline{\boldsymbol{w}'\boldsymbol{w}'(\rho C_{\mathrm{d}} C_{\mathrm{g}})'} \\ &\cong -\bar{s}\rho_{\mathrm{d}}\langle C_{\mathrm{g}}\rangle(\bar{\boldsymbol{w}}\bar{\boldsymbol{w}} + \overline{\boldsymbol{w}'\boldsymbol{w}'}) - 2\bar{\boldsymbol{w}}(\rho_{\mathrm{d}}\overline{\boldsymbol{w}'s'}\langle C_{\mathrm{g}}\rangle + \overline{\rho C''_{\mathrm{g}}\boldsymbol{w}'}\langle C_{\mathrm{d}}\rangle + \overline{\rho C''_{\mathrm{d}} C''_{\mathrm{g}}\boldsymbol{w}'}). \end{aligned} \tag{7.110}$$

For a two-phase turbulent flow, all correlations in relations (7.110) are usually disregarded and only the first term is retained (see Zuev and Lepeshinskii 1981; Kartushinskii 1984):

$$\overline{\Pi_{\mathrm{rel}}} \cong -\bar{s}\rho_{\mathrm{d}}\langle C_{\mathrm{g}}\rangle\bar{\boldsymbol{w}}\bar{\boldsymbol{w}} \tag{7.111}$$

which is certainly valid only if the averaged relative velocity of the phases $\bar{\boldsymbol{w}}$ is much higher than the pulsation velocity $\boldsymbol{w}'$, i.e., for fairly large solid particles. For lower-inertia fine and medium particles, the first two terms in (7.110) should generally be retained; then,

$$\overline{\Pi_{\rm rel}} = -\bar{\rho}\langle C_{\rm d}\rangle \bar{\boldsymbol{w}}\bar{\boldsymbol{w}} + \boldsymbol{R}_{\rm rel} \tag{7.112}$$

where $\boldsymbol{R}_{\rm rel} \equiv -\bar{\rho}_{\rm g}\langle C_{\rm d}\rangle\overline{\boldsymbol{w}'\boldsymbol{w}'}$ is the additional Reynolds stress tensor attributable to the pulsations of the relative velocity field of the phases. In gradient models, two methods are used to determine the pair correlations of this type. According to the first method, the correlation moments of $\boldsymbol{R}_{\rm rel}$ for relatively small particles are expressed directly in terms of the Reynolds stresses $\boldsymbol{R}$ for the gas–dust continuum as a whole, i.e. $\boldsymbol{R}_{\rm rel} = \beta\boldsymbol{R}$, where β is the coefficient of entrainment of disperse particles into the pulsational gas motion (Gavin et al. 1984). The second method of determining the additional turbulent stresses $\boldsymbol{R}_{\rm rel}$ uses gradient relations of type (7.107) with $\bar{\boldsymbol{w}}$ substituted for the averaged velocities $\langle\boldsymbol{u}\rangle$ and with the determination of the corresponding coefficient of turbulent viscosity (see, e.g., Melville and Bray 1979).

7.3.4 *Balance Equation for the Averaged Internal Energy of a Gas Suspension*

Now we will obtain the averaged energy equation for the gas–dust disk system as a whole by averaging the energy equation (7.45) for instantaneous motion over the ensemble of possible realizations. As a result, we have

$$\bar{\rho}\frac{D<H_{\rm sum}>}{Dt} = \overline{\frac{{\rm d}p_{\rm sum}}{{\rm d}t}} - {\rm div}\left(\boldsymbol{q}_{\rm sum}^{\rm turb} + \bar{\boldsymbol{q}}_{\rm sum}\right) + \overline{\Phi_u} + \overline{R_{\rm gd}|\boldsymbol{w}|^2 - s\sigma\left(\boldsymbol{w}\cdot\frac{\partial p}{\partial \boldsymbol{r}}\right)}, \tag{7.113}$$

where

$$\boldsymbol{q}_{\rm sum}^{\rm turb} = \boldsymbol{q}^{\rm turb} + \boldsymbol{q}_{\rm rad}^{\rm turb};\quad \boldsymbol{q}^{\rm turb} \equiv \overline{\rho H''\boldsymbol{u}''} \cong \langle c_{\rm P}\rangle\overline{\rho T''\boldsymbol{u}''} + \sum_{\alpha}\langle h_\alpha\rangle\boldsymbol{J}_\alpha^{\rm turb} \tag{7.114}$$

is the turbulent heat flux arising from the correlation between the enthalpy H'' and hydrodynamic velocity $\boldsymbol{u}''$ pulsations;

$$\boldsymbol{q}_{\rm rad}^{\rm turb} \equiv \overline{\rho H''_{\rm rad}\boldsymbol{u}''} \cong \langle c_{\rm P,rad}\rangle\overline{\rho T''\boldsymbol{u}''},\quad \langle c_{\rm P,rad}\rangle \equiv 16a\,\langle T\rangle^3/3\bar{\rho} \tag{7.115}$$

is the turbulent radiative heat flux. The approximate relations for the thermal energy fluxes (7.114) and (7.115) are written up to the terms containing triple correlations.

Formula (7.114) can be easily derived using the algebraic property (3.7) of the Favre averaging and the expression

$$H'' = \sum_\alpha \left[h''_\alpha \langle C_\alpha \rangle + \langle h_\alpha \rangle C''_\alpha + (C''_\alpha h''_\alpha)'' \right] = \langle c_\mathrm{p} \rangle T'' + \sum_\alpha C''_\alpha \langle h_\alpha \rangle + (c''_\mathrm{p} T'')'' \tag{7.116}$$

for the specific enthalpy pulsations of the disk material. Here, $h''_\alpha = c_{\mathrm{p}\alpha} T''$ is the partial enthalpy pulsation for phase α ($c_{\mathrm{p}\alpha} \cong \mathrm{const}$);

$$\langle c_\mathrm{p} \rangle = \sum_\alpha c_{\mathrm{p}\alpha} \langle C_\alpha \rangle, \quad c''_\mathrm{p} = \sum_\alpha c_{\mathrm{p}\alpha} C''_\alpha \tag{7.117}$$

are, respectively, the averaged and pulsation components of the specific heat of the mixture at constant pressure. The averaged values for the enthalpies of radiation and matter appearing in (7.113) are defined by the relations

$$\langle H_\mathrm{rad} \rangle \cong \frac{4}{3} a \langle T \rangle^4 / \bar{\rho}, \quad \langle H \rangle \cong \langle c_\mathrm{p} \rangle \langle T \rangle + \sum_\alpha h^0_\alpha \langle C_\alpha \rangle, \tag{7.118}$$

that follow from (7.43) and (7.44).

It is convenient to represent the substantial derivative of the total pressure in (7.113) using (7.98) as

$$\begin{aligned} \overline{\frac{\mathrm{d}p_{sum}}{\mathrm{d}t}} &\equiv \frac{D\bar{p}_\mathrm{sum}}{Dt} + \left(\boldsymbol{J}^\mathrm{turb}_v \cdot \frac{\partial \bar{p}_\mathrm{sum}}{\partial \boldsymbol{r}} \right) + \overline{\left(\boldsymbol{u}'' \cdot \frac{\partial p'_\mathrm{sum}}{\partial \boldsymbol{r}} \right)} \\ &= \frac{D\bar{p}_\mathrm{sum}}{Dt} + \left(\boldsymbol{J}^\mathrm{turb}_v \cdot \frac{\partial \bar{p}_\mathrm{sum}}{\partial \boldsymbol{r}} \right) + \operatorname{div} \left(\overline{p'_\mathrm{sum} \boldsymbol{u}''} \right) - \overline{p'_\mathrm{sum} \operatorname{div} \boldsymbol{u}''} \end{aligned} \tag{7.119}$$

In addition, $\overline{\Phi_u}$ can be transformed as

$$\overline{\Phi_u} \equiv \left(\bar{\Pi}_\mathrm{sum} : \frac{\partial \langle \boldsymbol{u} \rangle}{\partial \boldsymbol{r}} \right) + \overline{\left(\Pi_\mathrm{sum} : \frac{\partial \boldsymbol{u}''}{\partial \boldsymbol{r}} \right)} = (\bar{\Pi}_\mathrm{sum} : \boldsymbol{D}) + \bar{\rho} \langle \varepsilon_\mathrm{e} \rangle, \tag{7.120}$$

where

$$\bar{\rho} \langle \varepsilon_\mathrm{e} \rangle \equiv \overline{\left(\Pi_\mathrm{sum} : \frac{\partial \boldsymbol{u}''}{\partial \boldsymbol{r}} \right)} \tag{7.121}$$

is the dissipation rate of the turbulent kinetic energy of the gas–dust mixture into heat under the effect of "molecular" viscosity (a second key parameter in the theory of turbulence). It can be shown (see Marov and Kolesnichenko 2002 and Chap. 3) that the dissipative term (7.121) for developed turbulence is slightly simplified,

$$\bar{\rho}\langle\varepsilon_{\rm e}\rangle \cong \left(\bar{\varPi}_{\rm sum} : \frac{\partial}{\partial \boldsymbol{r}} \boldsymbol{J}_{\rm v}^{\rm turb}\right) + \overline{\left(\varPi' : \frac{\partial}{\partial \boldsymbol{r}} \boldsymbol{u}'\right)} \approx \overline{\left(\varPi' : \frac{\partial}{\partial \boldsymbol{r}} \boldsymbol{u}'\right)} \equiv \bar{\rho}\varepsilon \geq 0 \quad (7.121^*)$$

note that ε (the "true" dissipation of turbulent energy) is always positive. Substituting (7.119) and (7.121*) in (7.113) yields an averaged energy equation for the gas–dust mixture in the form

$$\bar{\rho}\frac{D\langle H_{\rm sum}\rangle}{Dt} = \frac{D\bar{p}_{\rm sum}}{Dt} - {\rm div}(\boldsymbol{q}_{\rm sum}^{\rm turb} - p'_{\rm sum}\boldsymbol{u}'' + \bar{\boldsymbol{q}}_{\rm sum}) + (\bar{\varPi}_{\rm sum} : \boldsymbol{D})$$
$$+ \overline{R_{\rm gd}|\boldsymbol{w}|^2 - s\sigma\left(\boldsymbol{w}\cdot\frac{\partial p}{\partial r}\right)} + \left(\boldsymbol{J}_{\rm v}^{\rm turb}\cdot\frac{\partial \bar{p}_{\rm sum}}{\partial \boldsymbol{r}}\right) - \overline{p'_{\rm sum}{\rm div}\boldsymbol{u}''} + \bar{\rho}\varepsilon. \quad (7.122)$$

Defining relations for the turbulent heat fluxes are required to close (7.122); these relations derived by Marov and Kolesnichenko (2002) are

$$\boldsymbol{q}^{\rm turb} = \overline{p'\boldsymbol{u}''} - \chi^{\rm turb}\left\{\frac{\partial\langle T\rangle}{\partial \boldsymbol{r}} - \frac{1}{\bar{\rho}\langle c_{\rm P}\rangle}\frac{\partial \bar{p}}{\partial \boldsymbol{r}}\right\} + \sum_{\alpha}\langle h_\alpha\rangle\boldsymbol{J}_\alpha^{\rm turb}, \quad (7.123)$$

$$\boldsymbol{q}_{\rm rad}^{\rm turb} = \overline{p'_{\rm rad}\boldsymbol{u}''} - \chi_{\rm rad}\left\{\frac{\partial\langle T\rangle}{\partial \boldsymbol{r}} - \frac{1}{\bar{\rho}\langle c_{\rm P,rad}\rangle}\frac{\partial \bar{p}_{\rm rad}}{\partial \boldsymbol{r}}\right\} \quad (7.124)$$

where

$$\chi^{\rm turb} = \bar{\rho}\langle c_{\rm P}\rangle\frac{\nu^{\rm turb}}{\boldsymbol{Sc}^{\rm turb}}, \quad \chi_{\rm rad} = \frac{4ac\langle T\rangle^3}{3\kappa\bar{\rho}} \quad (7.125)$$

are, respectively, the coefficient of turbulent heat conductivity for the gas–dust medium and the coefficient of turbulent radiative heat conductivity;

$$\langle c_{\rm P}\rangle = \frac{\bar{\rho}_{\rm g}(1-\bar{s}) + \rho_{\rm d}\bar{s}c_{\rm Pd}}{\bar{\rho}}$$

is the averaged specific heat (at constant pressure) for the total continuum. Below, we will assume that $\boldsymbol{Sc}^{\rm turb} = \boldsymbol{Pr}^{\rm turb}$ in the disk, since the turbulent diffusivity and the coefficient of turbulent thermal diffusivity in a turbulized mixture are commonly assumed to be equal $\chi^{\rm turb}/\bar{\rho}\langle c_{\rm P}\rangle = D^{\rm turb}$ which is equivalent to the equality of the mixing lengths for matter and heat. According to (7.123), there are two mechanisms of thermal energy transfer through the gas suspension:

1. under the effect of the averaged temperature (more precisely, potential temperature) gradient

$$\theta \equiv {\rm const}\cdot\langle T\rangle/\bar{p}^{<\Re>/\langle c_{\rm P}\rangle} \quad (7.126)$$

since

$$\frac{1}{\theta}\frac{\partial\theta}{\partial \boldsymbol{r}}=\frac{1}{\langle T\rangle}\left(\frac{\partial\langle T\rangle}{\partial \boldsymbol{r}}-\frac{1}{\bar{\rho}\langle c_{\mathrm{P}}\rangle}\frac{\partial\bar{p}}{\partial \boldsymbol{r}}\right)\approx\frac{1}{\langle T\rangle}\left(\frac{\partial\langle T\rangle}{\partial z}+G_{\mathrm{a}}\right),$$

where $G_{\mathrm{a}}\equiv g_z/\langle c_{\mathrm{P}}\rangle$ is the adiabatic temperature gradient in the gas–dust disk.

2. by the turbulent diffusion fluxes

$$\boldsymbol{J}_{\alpha}^{\mathrm{turb}}=-\bar{\rho}D^{\mathrm{turb}}\partial\langle C_{\alpha}\rangle/\partial\boldsymbol{r}$$

(see (7.92)), with each particle of phase α transferring, on average, the thermal energy $\langle h_{\alpha}\rangle$ (since $\sum_{\alpha}\boldsymbol{J}_{\alpha}^{\mathrm{turb}}=0$, $D_{\mathrm{d}}^{\mathrm{turb}}=D_{\mathrm{g}}^{\mathrm{turb}}=D^{\mathrm{turb}}$).

It should also be noted that the first terms in (7.123) and (7.124) do not act as the energy flux, because $\overline{p'_{\mathrm{sum}}\boldsymbol{u}''}$ drops out of the complete energy equation (7.122), and were retained in (7.123) and (7.124) only for convenience.

Let us now write (7.123) in a form useful for modeling a turbulized gas suspension. Using (7.92) to transform (7.123) yields

$$\begin{aligned}\boldsymbol{q}^{\mathrm{turb}}&=\overline{p'\boldsymbol{u}''}-\chi^{\mathrm{turb}}\left(\frac{\partial\langle T\rangle}{\partial\boldsymbol{r}}-\frac{1}{\bar{\rho}\langle c_{\mathrm{p}}\rangle}\frac{\partial\bar{p}}{\partial\boldsymbol{r}}\right)-D^{\mathrm{turb}}\bar{\rho}\sum_{\alpha}\langle h_{\alpha}\rangle\frac{\partial\langle C_{\alpha}\rangle}{\partial\boldsymbol{r}}\\&=\overline{p'\boldsymbol{u}''}-\frac{\chi^{\mathrm{turb}}}{\langle c_{\mathrm{p}}\rangle}\left(\frac{\partial\langle H\rangle}{\partial\boldsymbol{r}}-\frac{1}{\bar{\rho}}\frac{\partial\bar{p}}{\partial\boldsymbol{r}}\right),\end{aligned}\qquad(7.123^*)$$

where we made the usual (for the theory of turbulence) assumption that the turbulent Lewis number is equal to unity, $\boldsymbol{Le}^{\mathrm{turb}}\equiv\chi^{\mathrm{turb}}/\bar{\rho}\langle c_{\mathrm{P}}\rangle D^{\mathrm{turb}}=1$ (Monin and Yaglom 1992).

It is occasionally convenient to write (7.122) via the averaged total energy $\langle E_{\mathrm{sum}}\rangle$ of the matter and radiation. Using for this purpose the transformation

$$\bar{\rho}D\langle E_{\mathrm{sum}}\rangle/Dt+\bar{p}_{\mathrm{sum}}\mathrm{div}\langle\boldsymbol{u}\rangle=\bar{\rho}\,D\langle H_{\mathrm{sum}}\rangle/Dt-D\,\bar{p}_{\mathrm{sum}}/Dt\qquad(7.127)$$

(which is a corollary of the relation $\langle H_{\mathrm{sum}}\rangle=\langle E_{\mathrm{sum}}\rangle+\bar{p}_{\mathrm{sum}}/\bar{\rho}$ and the averaged continuity equation (7.79)), we obtain for a developed turbulent flow in disc

$$\begin{aligned}&\bar{\rho}\frac{D\langle E_{\mathrm{sum}}\rangle}{Dt}+\mathrm{div}(\boldsymbol{q}_{\mathrm{sum}}^{\mathrm{turb}}-\overline{p'_{\mathrm{sum}}\boldsymbol{u}''}+\bar{\boldsymbol{q}}_{\mathrm{sum}})=-\bar{p}_{\mathrm{sum}}\mathrm{div}\langle\boldsymbol{u}\rangle+(\bar{\Pi}_{\mathrm{sum}}{:}\boldsymbol{D})\\&\overline{+R_{\mathrm{gd}}|\boldsymbol{w}|^2-s\sigma\left(\boldsymbol{w}\cdot\frac{\partial p}{\partial\boldsymbol{r}}\right)}+\left(\boldsymbol{J}_{\mathrm{v}}^{\mathrm{turb}}\cdot\frac{\partial\bar{p}_{\mathrm{sum}}}{\partial\boldsymbol{r}}\right)-\overline{p'_{\mathrm{sum}}\mathrm{div}\boldsymbol{u}''}+\bar{\rho}\varepsilon.\end{aligned}\qquad(7.122^*)$$

7.3.4.1 The Turbulent Flux of the Specific Volume in Gas–Dust Medium

Let us now derive the final relation for the turbulent flux of the specific volume $\boldsymbol{J}_{\mathrm{v}}^{\mathrm{turb}}$ (see (7.86)). For its derivation, let us first find the expression for the turbulent density pulsations ρ_{g}' in the gas component of the mixture; as the equation of state for the latter, we will take, as previously, the equation of state for a perfect multicomponent gas

$$p = p_{\mathrm{g}} = k_{\mathrm{B}} T \sum_k n_{\mathrm{g}(k)} = \rho_{\mathrm{g}} \Re_{\mathrm{g}} T \tag{7.128}$$

where

$$\Re_{\mathrm{g}} = k_{\mathrm{B}} \sum_k n_{\mathrm{g}(k)} / \rho_{\mathrm{g}} = k_B \sum_k Z_k \tag{7.129}$$

is the so-called gas constant for the mixture of gases; $Z_k = n_{\mathrm{g}(k)}/\rho_{\mathrm{g}}$ is the specific (per unit mass of the gas continuum) number density of component k. Representing the actual values of $\Re_{\mathrm{g}}$ and T as the sums of the averaged and pulsational values ($\Re_{\mathrm{g}} = \langle \Re_{\mathrm{g}} \rangle + \Re_{\mathrm{g}}''$, $T = \langle T \rangle + T''$), let us rewrite (7.128) as

$$\begin{aligned} p &= \langle \Re_{\mathrm{g}} \rangle \rho_{\mathrm{g}} \langle T \rangle + \Re_{\mathrm{g}}'' \rho_{\mathrm{g}} \langle T \rangle + \langle \Re_{\mathrm{g}} \rangle \rho_{\mathrm{g}} T'' + \Re_{\mathrm{g}}'' \rho_{\mathrm{g}} T'' = \\ &\cong \langle \Re_{\mathrm{g}} \rangle \rho_{\mathrm{g}} \langle T \rangle + k_{\mathrm{B}} \rho_{\mathrm{g}} \langle T \rangle \sum_{k=1}^{n} Z_k'' + \langle \Re_{\mathrm{g}} \rangle \rho_{\mathrm{g}} T'' + k_{\mathrm{B}} \rho_{\mathrm{g}} \sum_{k=1}^{n} (Z_k'' T''). \end{aligned} \tag{7.130}$$

If we now apply the statistical averaging operator to (7.130), then we will obtain the averaged equation of state for the gas-suspension pressure

$$\bar{p} = \langle \Re_{\mathrm{g}} \rangle \bar{\rho}_{\mathrm{g}} \langle T \rangle + k_{\mathrm{B}} \bar{\rho}_{\mathrm{g}} \sum_{k=1}^{n} \langle Z_k'' T'' \rangle \cong \langle \Re_{\mathrm{g}} \rangle \bar{\rho}_{\mathrm{g}} \langle T \rangle. \tag{7.131}$$

Note that the pulsational term in the averaged equation of state (7.131) is usually discarded in the theory of turbulence. We will use this expression to eliminate the product $\langle \Re_{\mathrm{g}} \rangle \langle T \rangle$ from (7.130); as a result, we will have for the pulsations ρ_{g}' (Marov and Kolesnichenko 1987)

$$\frac{\rho_{\mathrm{g}}'}{\bar{\rho}_{\mathrm{g}}} = \frac{p'}{\bar{p}} - \frac{\rho_{\mathrm{g}} T''}{\bar{\rho}_{\mathrm{g}} \langle T \rangle} - \frac{k_B \langle T \rangle \rho_{\mathrm{g}}}{\bar{p}} \sum_{k=1}^{n} Z_k'' \tag{7.132}$$

It is well known that the relative pressure pulsations for gas flows with small Mach numbers ($\boldsymbol{Ma}$) may be disregarded almost always compared to the relative temperature pulsations. This hypothesis (Morkovin 1961), which has been tested up to $\boldsymbol{Ma} = 5$, is probably also valid for turbulent motions in thin accretion disks: the motion along the r and z directions is subsonic, while the rotation velocity u_φ exceeds the speed of sound c_{gs} (the thin-disk condition $h_{disk} \ll r$, together with the expression $h_{disk} \approx c_{gs}/\Omega_{K,\mathrm{mid}}$ for the disk thickness, requires that $h_{disk}/r \approx c_{gs}/u_\varphi \ll 1$). Below, we will also assume that the mean mass of the gas component of the gas suspension does not fluctuate; therefore, $\sum_{k=1}^{n} Z'' = (n_g/\rho_g)'' = 0$. Using (7.132), the correlation term (containing the true gas density pulsations in (7.90)) can then be rewritten as

$$-\frac{\overline{\rho'_g \boldsymbol{u}''}}{\bar{\rho}_g} = \frac{\overline{\rho\, C_g T'' \boldsymbol{u}''}}{\bar{\rho}_g \langle T \rangle} \cong \langle C_g \rangle \frac{\overline{\rho T'' \boldsymbol{u}''}}{\bar{\rho}_g \langle T \rangle} = \frac{\langle T'' \boldsymbol{u}'' \rangle}{\langle T \rangle} \tag{7.133}$$

(here, as everywhere below, the terms with triple correlations were discarded). For the final transformation of this expression, we will use the defining relation (7.123) for the turbulent heat flux (7.114):

$$-\frac{\overline{\rho'_g \boldsymbol{u}''}}{\bar{\rho}_g} \cong \frac{1}{\langle T \rangle} \overline{\rho T'' \boldsymbol{u}''} \cong \frac{1}{\langle T \rangle \langle c_P \rangle} \left(\boldsymbol{q}^{turb} - \sum_\alpha \langle h_\alpha \rangle \boldsymbol{J}_\alpha^{turb} \right) \tag{7.134}$$

Finally, substituting (7.92) and (7.134) in (7.90) and using (7.125), we will obtain the following defining relation for the turbulent flux of the specific volume for a heterogeneous mixture:

$$\begin{aligned} \boldsymbol{J}_v^{\mathrm{turb}} &= -\langle \sigma \rangle \frac{\bar{\rho}}{\rho_d \bar{\rho}_g} \boldsymbol{J}_d^{\mathrm{turb}} + \frac{1}{\langle T \rangle \langle c_P \rangle} \left(\boldsymbol{q}^{\mathrm{turb}} - \sum_\alpha \langle h_\alpha \rangle \boldsymbol{J}_\alpha^{\mathrm{turb}} \right) = \\ &\cong \bar{\rho} \frac{\nu^{\mathrm{turb}}}{\boldsymbol{Sc}^{\mathrm{turb}}} \left[\frac{\rho_d}{\bar{\rho}_g} \frac{\partial}{\partial \boldsymbol{r}} \left(\frac{\bar{s}}{\bar{\rho}} \right) - \frac{1}{\langle T \rangle} \left(\frac{\partial \langle T \rangle}{\partial \boldsymbol{r}} - \frac{1}{\bar{\rho} \langle c_P \rangle} \frac{\partial \bar{p}}{\partial \boldsymbol{r}} \right) \right]. \end{aligned} \tag{7.135}$$

7.3.5 *Energy Balance Equations for Disk Matter*

In a turbulized flow of disk material, compared to its laminar analog, there is a great variety of possible transfer mechanisms (conversion rates) between the various forms of the energies of motion of solid particles and gas that contribute to the conserved averaged total energy. To interpret the individual terms of the energy balance most accurately, let us consider the complete system of energy equations for the averaged fields of pulsating thermohydrodynamic parameters for a gas–dust cloud, including the balance equation for the kinetic energy of turbulent pulsations.

7.3.5.1 The Balance Equation for the Averaged Kinetic Energy of Gas–Dust Flow

Given (7.103) for the gravitational force, a scalar multiplication of the equation of motion (7.38) by the velocity $\langle \boldsymbol{u} \rangle$ after the necessary transformations yields the following substantial form of the equation of live forces for the averaged motion of disk material (the momentum theorem): yields the following transfer equation for the averaged mechanical energy of a turbulized gas–dust flow:

$$\bar{\rho}\frac{D}{Dt}\left(\frac{|\langle \boldsymbol{u}\rangle|^2}{2}\right) + \operatorname{div}\left(\left[\boldsymbol{U}\bar{p}_{\text{sum}} - \boldsymbol{R} - \overline{\Pi^*_{\text{sum}}}\right] \langle \boldsymbol{u}\rangle\right)$$
$$= \bar{p}_{\text{sum}}\operatorname{div}\langle \boldsymbol{u}\rangle - \left(\left[\boldsymbol{R} + \overline{\Pi^*_{\text{sum}}}\right]:\frac{\partial\langle \boldsymbol{u}\rangle}{\partial \boldsymbol{r}}\right) - \bar{\rho}\left(\langle \boldsymbol{u}\rangle \cdot \frac{\partial\langle \Psi\rangle}{\partial \boldsymbol{r}}\right). \qquad (7.136)$$

Here, $-\,\partial\langle\Psi\rangle/\partial\boldsymbol{r} = \boldsymbol{g} = G\mathcal{M}_\odot\tilde{\boldsymbol{r}}/|\tilde{\boldsymbol{r}}|^3$; $|\langle \boldsymbol{u}\rangle|^2/2$ is the specific kinetic energy of the averaged motion of disk material. Although (7.136) is of an energy nature, it is not the law of conservation of energy in a turbulized continuum: (7.136) describes the law of conversion of the kinetic energy of the averaged gas-suspension motion into the work of external bulk and surfaces forces and the work of internal forces (and conversely) without allowance for the irreversible conversion of the mechanical disk energy into thermal and other forms of energy.

Let us explain now the physical meaning of the individual terms in (7.136): the quantity $\operatorname{div}(\bar{p}_{\text{sum}}\langle \boldsymbol{u}\rangle)$ is related to the outflow of mechanical energy from a unit volume of the disk medium per unit time; the divergence $(\partial/\partial\boldsymbol{r})\cdot[(\boldsymbol{R} + \overline{\Pi^*_{\text{sum}}})\langle \boldsymbol{u}\rangle]$ is the rate with which the total surface stress $(\boldsymbol{R} + \overline{\Pi^*_{\text{sum}}})$ in the averaged moving "gas suspension plus radiation" system does the work in a unit volume; the quantity $\bar{p}_{\text{sum}}\operatorname{div}\langle \boldsymbol{u}\rangle$ (>0 or <0) is related to the rate of inverse adiabatic transformation of the averaged internal energy (heat) $\langle E_{\text{sum}}\rangle$ into the system's mechanical energy (see (7.122*)) and is the work done per unit time in a unit volume against the averaged total pressure $\bar{p}_{\text{sum}}$ by the flow of a moving gas suspension; the sign of $\bar{p}_{\text{sum}}\operatorname{div}\langle \boldsymbol{u}\rangle$ depends on whether the mixture flow will expand ($\operatorname{div}\langle \boldsymbol{u}\rangle > 0$) or contract ($\operatorname{div}\langle \boldsymbol{u}\rangle < 0$); the quantity $(\boldsymbol{R} + \overline{\Pi^*_{\text{sum}}} : \partial\langle \boldsymbol{u}\rangle/\partial\boldsymbol{r})$ is the total rate of irreversible transformation of the kinetic energy of mean motion into other forms of energy (see (7.122*), (7.139), and (7.144)), with the energy dissipating under the effects of both "molecular" viscosity with the rates $(\bar{\Pi}_{\text{sum}} : \partial\langle \boldsymbol{u}\rangle/\partial\boldsymbol{r})$ and $(\bar{\Pi}_{\text{rel}} : \partial\langle \boldsymbol{u}\rangle/\partial\boldsymbol{r})$ and turbulent viscosity with the rate $(\boldsymbol{R}:\partial\langle \boldsymbol{u}\rangle/\partial\boldsymbol{r})$.

Adding (7.136) and the averaged balance equation for the potential energy of disk material

$$\bar{\rho}\frac{D\langle\Psi\rangle}{Dt} \equiv \frac{\partial\bar{\rho}\langle\Psi\rangle}{\partial t} + \operatorname{div}(\bar{\rho}\langle \boldsymbol{u}\rangle\langle\Psi\rangle) = \bar{\rho}\left(\langle \boldsymbol{u}\rangle \cdot \frac{\partial\langle\Psi\rangle}{\partial\boldsymbol{r}}\right), \qquad (7.137)$$

yields the equation of the averaged mechanical energy of turbulized gas–dust flux transfer

$$\bar{\rho}\frac{D}{Dt}\left(\frac{|\langle \boldsymbol{u}\rangle|^2}{2}+\langle\Psi\rangle\right)+\operatorname{div}([U\bar{p}_{\text{sum}}-R-\bar{\Pi}_{\text{sum}}-\bar{\Pi}_{\text{rel}}]\langle\boldsymbol{u}\rangle)$$

$$=\bar{p}_{\text{sum}}\operatorname{div}\langle\boldsymbol{u}\rangle-([\boldsymbol{R}+\bar{\Pi}_{\text{sum}}-\bar{\Pi}_{\text{rel}}]:\partial\boldsymbol{D}/\partial\boldsymbol{r}). \tag{7.138}$$

7.3.5.2 The Balance Equation for the Averaged Kinetic Energy of the Relative Phase Motion

Averaging (7.43) and disregarding the third-order correlation terms (and, thus, the turbulent kinetic energy of the interphase diffusion and the products of the thermo-hydrodynamic fluxes (e.g., the terms $(\bar{\Pi}_{\text{rel}}:\partial\boldsymbol{J}_{\text{v}}^{\text{turb}}/\partial\boldsymbol{r})$)) in the averaged gas–dust continuum as quantities of the second order of smallness, we obtain

$$\bar{\rho}\frac{D}{Dt}(\langle C_{\text{d}}\rangle\,\langle C_{\text{g}}\rangle|\boldsymbol{w}|^2/2)\approx\bar{\rho}\langle C_{\text{d}}\rangle\,\langle C_{\text{g}}\rangle\frac{D}{Dt}(|\boldsymbol{w}|^2/2)$$

$$\cong -R_{\text{gd}}|\boldsymbol{w}|^2-s\sigma\left(\boldsymbol{w}\cdot\frac{\partial p}{\partial\boldsymbol{r}}\right)+(\bar{\Pi}_{\text{rel}}:\boldsymbol{D})-\bar{\rho}\sigma_{\text{rel}}, \tag{7.139}$$

where

$$\bar{\rho}\sigma_{\text{rel}}\equiv-\left(\Pi'_{\text{rel}}:\frac{\partial\boldsymbol{u}'}{\partial\boldsymbol{r}}\right). \tag{7.140}$$

Let us note that when neglecting turbulent kinetic energy of the interphase diffusion as a third-order quantity, we thus do not take into account an additional dissipative term caused by fine particles (Danon et al. 1977) in the equation of turbulent energy of gas–dust mixture (7.144). This is even more justified when dealing with the gas flow involving large-size particles because they ensure much smaller additional turbulent energy dissipation compared to other terms in (7.144) (Varaksin 2003).

In (7.139) the heat dissipation $(\bar{\Pi}_{\text{relrel}}:\boldsymbol{D})$ (representing the mean work done by the relative stress tensor on the averaged velocity gradient $(\partial\langle\boldsymbol{u}\rangle/\partial\boldsymbol{r})\neq 0$ due to the relative phase velocity shear in the orbital motion of the disk material) is related to the conversion rate of the averaged kinetic diffusion energy into the kinetic energy of the averaged motion of the gas mixture as a whole (cf. (7.136)); σ_{rel} is the additional generation source of turbulent energy b related to the presence of medium and large inertial particles in the flow (see (7.144*)). It should be kept in mind that, according to Gore and Crowe (1989), the turbulent vortex wakes that destabilize the gas flow and transform the energy of the averaged relative motion into the high-frequency components of the turbulence energy spectrum are formed only behind large particles (at Reynolds numbers of the flow around the particles $\boldsymbol{Re}_{\rm d} > 400$). In contrast, small particles ($\boldsymbol{Re}_{\rm d} < 110$) predominantly suppress the turbulence energy, spending it on their own acceleration (i.e., their entrainment into the pulsational motion of a polydisperse flow); the laminarizing effect of the disperse phase on the flow increases with decreasing particle inertia. As regards medium-sized particles ($110 < \boldsymbol{Re}_{\rm d} < 400$), they have a mixed effect on the disk turbulence.

7.3.5.3 The Turbulent Energy Balance

Let us now consider the transfer equation for the turbulent energy $b \equiv \overline{|\boldsymbol{u}''|^2}/2$ of the gas–dust material of an accretion disk. This fundamental (in the theory of turbulence) equation or some of its modifications underlie many present-day semiempirical theories of turbulence (see Monin and Yaglom 1971; Kolesnichenko and Marov 1999). Using the equation for b in the case of a heterogeneous medium, we can, in particular, analyze the dynamical effect of the disperse phase on the intensity of turbulence in a gas–dust disk medium and develop a phenomenological method of modeling the coefficient of turbulent viscosity for the medium by taking into account the influence of the inverse effects of the dust transfer and "potential" temperature on the decay (maintenance) of shear turbulence in a protoplanetary cloud. The balance equation for b can be derived by various methods (Marov and Kolesnichenko 2002), one of which we will use in the case of a two-phase medium considered here.

Let $\mathcal{A}(\boldsymbol{r}, t)$ be the actual value of a scalar quantity (in particular, this can be the vector components) whose substantial balance is $\rho \mathrm{d}\mathcal{A}/\mathrm{d}t = -(\partial/\partial \boldsymbol{r}) \cdot \boldsymbol{J}_A + \sigma_A$, where $\boldsymbol{J}_{\rm A}$ and $\sigma_{\rm A}$ are, respectively, the substantial flux density vector and the volume density of source $\mathcal{A}$. For example, for the equation of motion (7.112)

$$\mathcal{A} \equiv \boldsymbol{u}, \quad \boldsymbol{J}_A \equiv -\Pi^*_{\rm sum}, \quad \sigma_A \equiv -p_{\rm sum}\boldsymbol{U} + \rho \boldsymbol{g} \qquad (7.141)$$

It is easy to show (this requires multiplying the identity $\mathrm{d}\mathcal{A}''/\mathrm{d}t = \mathrm{d}\mathcal{A}/\mathrm{d}t - D\langle \mathcal{A}\rangle/Dt - \boldsymbol{u}'' \cdot (\partial/\partial \boldsymbol{r})\langle \mathcal{A}\rangle$ by $\rho\mathcal{A}''$ and the Reynolds averaging of the result) that the balance equation for the root-mean-square pulsation $\rho\mathcal{A}''$ has the following general form (Kolesnichenko 1995):

$$\bar{\rho}\frac{D\langle \mathcal{A}''^2/2\rangle}{Dt} + \frac{\partial}{\partial \boldsymbol{r}} \cdot \left(\overline{\rho \mathcal{A}''^2 \boldsymbol{u}''/2} + \overline{\mathcal{A}'' \boldsymbol{J}_{(A)j}} \right)$$
$$= -\boldsymbol{J}_A^{\rm turb} \cdot \frac{\partial}{\partial \boldsymbol{r}} \langle \mathcal{A}\rangle + \overline{\mathcal{A}''\sigma_A} - \bar{\rho}\langle \varepsilon_A\rangle, \qquad (7.142)$$

where

$$\bar{\rho}\langle \varepsilon_A\rangle \equiv -\overline{\left(\boldsymbol{J}_A \cdot \frac{\partial}{\partial \boldsymbol{r}} \mathcal{A}'' \right)} \qquad (7.143)$$

is the rate of scalar dissipation of the dispersion $\langle \mathcal{A}''^2 \rangle$. The generalized transfer equation (7.142) contains the terms that reflect the influence of the following processes on the space–time distribution of the dispersion $\langle \mathcal{A}''^2 \rangle$: convective transfer, diffusion, the dispersion formation through the energy transfer between the averaged and pulsational motions, redistribution (between the pulsational motions in various directions), and the dissipation of the turbulent characteristic $\langle \mathcal{A}''^2 \rangle$ through the "molecular" transfer processes.

Let us now substitute (7.141) in (7.142) and (7.143); as a result, we obtain the following transfer equation for the turbulent energy of a gas–dust mixture:

$$\bar{\rho}\frac{Db}{Dt} + \mathrm{div}\boldsymbol{J}_b^{\mathrm{turb}} = (\boldsymbol{R} : \boldsymbol{D}) - \left(\boldsymbol{J}_{\mathrm{v}}^{\mathrm{turb}} \cdot \frac{\partial \bar{p}_{\mathrm{sum}}}{\partial \boldsymbol{r}}\right) + \overline{p'_{\mathrm{sum}}\mathrm{div}\boldsymbol{u}''} - \bar{\rho}\langle \varepsilon_b \rangle, \qquad (7.144)$$

where

$$\begin{cases} \boldsymbol{J}_b^{\mathrm{turb}} \equiv \overline{\rho(|\boldsymbol{u}''|^2/2 + p'_{\mathrm{sum}}/\rho)\,\boldsymbol{u}''} - \overline{(\Pi_{\mathrm{sum}} + \Pi_{\mathrm{rel}}) \cdot \boldsymbol{u}''} \\ \bar{\rho}\langle \varepsilon_b \rangle \equiv \overline{(\Pi_{\mathrm{sum}} + \Pi_{\mathrm{rel}}):\partial \boldsymbol{u}''/\partial \boldsymbol{r}} \end{cases} \qquad (7.145)$$

The estimates of the individual terms in (7.144) obtained for developed turbulence in our monograph (Marov and Kolesnichenko 2002) allow it to be slightly simplified:

$$\bar{\rho}\frac{\overline{D|\boldsymbol{u}''|^2/2}}{Dt} + \mathrm{div}\left\{\overline{\rho(|\boldsymbol{u}''|^2/2 + p'_{\mathrm{sum}}/\rho)\boldsymbol{u}''} - \overline{(\Pi^*_{\mathrm{sum}})' \cdot \boldsymbol{u}''}\right\}$$
$$= (\boldsymbol{R} : \boldsymbol{D}) + \overline{p'_{\mathrm{sum}}\mathrm{div}\,\boldsymbol{u}'} - \left(\boldsymbol{J}_{\mathrm{v}}^{\mathrm{turb}} \cdot \frac{\partial \bar{p}_{\mathrm{sum}}}{\partial \boldsymbol{r}}\right) - \bar{\rho}\varepsilon + \bar{\rho}\sigma_{\mathrm{rel}}, \qquad (7.144^*)$$

where

$$\bar{\rho}\langle \varepsilon_b \rangle = (\bar{\Pi}_{\mathrm{sum}} + \bar{\Pi}_{\mathrm{rel}}):\partial \boldsymbol{J}_{\mathrm{v}}^{\mathrm{turb}}/\partial \boldsymbol{r} + \overline{\Pi'_{\mathrm{sum}} : \partial \boldsymbol{u}'/\partial \boldsymbol{r}} + \overline{\Pi'_{\mathrm{rel}}:\partial \boldsymbol{u}'/\partial \boldsymbol{r}} \cong \bar{\rho}\varepsilon - \bar{\rho}\sigma_{\mathrm{rel}}.$$

The first term on the left-hand side of (7.144*) describes the change with time (and the convective transfer by the averaged motion) in the kinetic energy of disk turbulence b, and the second term reflects the turbulent pulsation energy transfer through the turbulent "diffusion" processes; the quantity (energy dissipation)

$$\boldsymbol{R}:\boldsymbol{D} \equiv -\frac{2}{3}\bar{\rho}\, b\, \mathrm{div}\langle \boldsymbol{u} \rangle + 2\bar{\rho}\, \nu^{\mathrm{turb}}\left(\overset{\circ}{\boldsymbol{D}}: \overset{\circ}{\boldsymbol{D}}\right) \qquad (7.146)$$

on the right-hand sides of (7.136) and (7.144*) appear with different signs and, therefore, it may be interpreted as the conversion rate of the kinetic energy of the averaged motion into the turbulence energy of the gas–dust disk medium considered as a whole (this hydrodynamic generation mechanism of turbulence in a differentially rotating Keplerian disk is considered here as the main mechanism). Let us emphasize that such energy transfer is regarded as entirely kinematic process dependent only on the temporal–spacious scale of averaging. In the case of small-scale turbulence, the quantity $(\boldsymbol{R}:\boldsymbol{D})$ is always positive, and hence, the small-scale turbulence transforms kinetic energy of the mean motion into kinetic energy of turbulent pulsations (this is the so-called dissipative effect of the small-scale turbulence; at the same time, large-scale turbulent vortices can transfer the turbulent kinetic energy to the energy of mean motion (Van Migem 1977). The quantity $\overline{p'_{\text{sum}}\text{div}\boldsymbol{u}'}$ is related to the transformation rate of the internal energy of the gas suspension into the kinetic energy of turbulent vortices and is the work done per unit time in a unit volume of the pulsating medium on the vortices as a result of the existence of pulsations in the total pressure p'_{sum} of the disk medium and the expansion or contraction of turbulent vortices ($\text{div}\boldsymbol{u}' > 0$ or $\text{div}\boldsymbol{u}' < 0$); the quantity $(\boldsymbol{J}_{\text{v}}^{\text{turb}} \cdot \partial\bar{p}_{\text{sum}}/\partial\boldsymbol{r})$ is the conversion rate (in a unit volume of the medium) between the turbulent and averaged internal energies of the disk; small-scale vortices transform the turbulence energy into heat, since $\boldsymbol{J}_{\text{v}}^{\text{turb}} \cdot \partial\bar{p}_{\text{sum}}/\partial\boldsymbol{r} > 0$ for them, while large vortex structures related to thermal convection [for which $\boldsymbol{J}_{\text{v}}^{\text{turb}} \cdot \partial\bar{p}_{\text{sum}}/\partial\boldsymbol{r} < 0$ (Kolesnichenko and Marov 1999)] transform the thermal energy of the gas–dust flow into the averaged kinetic energy of the velocity pulsations. It should be noted that such a generation mechanism of turbulence in the disk suggested by Lin and Papaloizou (1980) as the main mechanism cannot be considered in this capacity, since it is not universal and temporary in nature (Ruden and Pollack 1991; Nomura 2002). The pair correlation $\bar{\rho}\varepsilon \equiv \overline{\mathit{\Pi}'_{\text{sum}} : \partial\boldsymbol{u}'/\partial\boldsymbol{r}} > 0$ in a developed turbulent flow (the mean work done by the pulsations of the viscous stress tensor on the turbulent vortices with a pulsation velocity gradient, $\text{div}\boldsymbol{u}' \neq 0$) is the dissipation rate of the turbulent kinetic energy into heat under the effect of molecular viscosity (see (7.122*)); let us note that pulsations quantity $\mathit{\Pi}'$ of the viscosity stress tensor is defined by the effective coefficient of kinematic viscosity of the gas–dust medium taking into account an existence of disperse matter and considering as a whole. Finally, the quantity $\bar{\rho}\sigma_{\text{rel}} \equiv -\overline{\mathit{\Pi}'_{\text{rel}} : \partial\boldsymbol{u}'/\partial\boldsymbol{r}} > 0$, the work in the turbulent flow (per unit time in a unit volume) done by the pulsations of the relative stress tensor on the turbulent vortices (see 7.3.63)), may be interpreted as the additional generation of turbulence in the gas–dust disk arising from the inertial effects of the relative motion of the disperse and gas phases and related to the formation of a vortex wake behind large particles with sizes >1 cm (see, e.g., Zaichik and Varaksin 1999). This mechanism could be referred to as an additional source of turbulization of the gas-suspension flow near the thin dust layer generated by large-scale particles of centimeter sizes or larger produced through coagulation and settling near the midplane of the protoplanetary cloud (Goldrich and Ward 1973). It appears to prevent further settling of small (micron-size) dust particles to the subdisk and thus, can be associated with delay of the onset of direct gravitational instability of this layer (see, e.g., Safronov 1969; Weidenschilling 1984; Goodmann and Pindor 2000). Indeed, a high rate of thinning and relaxation of the dusty subdisk is necessary to accomplish critical density.

The quantity $\bar{\rho}\sigma_{\text{rel}}$, up to the triple correlation terms, can be transformed to

$$\begin{aligned}\bar{\rho}\sigma_{\text{rel}} &= \overline{(s\rho_{\text{d}}\,C_{\text{g}}\boldsymbol{w}\boldsymbol{w})' : \frac{\partial \boldsymbol{u}'}{\partial \boldsymbol{r}}} \\ &= \rho_{\text{d}}\bar{\boldsymbol{w}}\,\bar{\boldsymbol{w}}\,\overline{s\,C_{\text{g}} : \frac{\partial \boldsymbol{u}'}{\partial \boldsymbol{r}}} + 2\rho_{\text{d}}\,\bar{\boldsymbol{w}}\,\overline{s\,C_{\text{g}}\boldsymbol{w}' : \frac{\partial \boldsymbol{u}'}{\partial \boldsymbol{r}}} + \rho_{\text{d}}\overline{s\,C_{\text{g}}\boldsymbol{w}'\boldsymbol{w}' : \frac{\partial \boldsymbol{u}'}{\partial \boldsymbol{r}}} \\ &\cong \rho_{\text{d}}\bar{\boldsymbol{w}}\,\bar{\boldsymbol{w}} : \left(\langle C_{\text{g}}\rangle \overline{s' \frac{\partial \boldsymbol{u}'}{\partial \boldsymbol{r}}} + \bar{s}\overline{C_{\text{g}}' \frac{\partial \boldsymbol{u}'}{\partial \boldsymbol{r}}} \right) + 2\rho_{\text{d}}\bar{s}\langle C_{\text{g}}\rangle \bar{\boldsymbol{w}}\,\overline{\boldsymbol{w}' : \frac{\partial}{\partial \boldsymbol{r}}\boldsymbol{u}'},\end{aligned} \tag{7.147}$$

We see from this relation that the additional generation of turbulence in the dusty disk (in particular, in the subdisk where relatively large solid particles are present) can arise from the averaged dynamical sliding of the phases, the correlation of the pulsations of the dust particle volume content and the gas concentration of the mixture with the flow pulsation velocity, and the pulsational interphase sliding. As we have repeatedly emphasized above, this additional source of turbulization is small for fine particles, for which the inertia effects may be ignored ($\Pi_{\text{rel}} \cong 0$).

To conclude this section, note that, in general, the energy equation (7.122*) for a gas–dust disk system is used to use in astrophysics by assuming a steady-nonequilibrium state of the turbulent field, where an internal equilibrium exists in the turbulence structure at which the production of turbulent entropy S^{turb} of the gas–dust matter is approximately equal to its dissipation. Unfortunately, in such a case energetic equation for laminar flow is utilized substituting coefficient of turbulent viscosity instead of that of molecular viscosity neglecting the very delicate approach of deducing energy equation for the real turbulent gas suspension. Meantime, if we take the above condition for the balance of S^{turb}, then we will have (see (7.162))

$$2\bar{\rho}\,\nu^{\text{turb}}\left(\overset{\circ}{\boldsymbol{D}} : \overset{\circ}{\boldsymbol{D}}\right) + \overline{p'_{\text{sum}}\operatorname{div}\boldsymbol{u}'} - \left(\boldsymbol{J}_{\text{V}}^{\text{turb}} \cdot \frac{\partial \bar{p}_{\text{sum}}}{\partial \boldsymbol{r}}\right) - \bar{\rho}\varepsilon + \bar{\rho}\sigma_{\text{rel}} \approx 0, \tag{7.148}$$

and (7.122*) for a developed turbulent flow can be transformed to

$$\begin{aligned}&\bar{\rho}\frac{D\langle E_{\text{sum}}\rangle}{Dt} + \operatorname{div}\left(\boldsymbol{q}_{\text{sum}}^{\text{turb}} - \overline{p'_{\text{sum}}\boldsymbol{u}''}\right) \cong -\bar{p}_{\text{sum}}\operatorname{div}\langle\boldsymbol{u}\rangle + 2(\bar{\rho}\,\nu^{\text{turb}} + \mu_{\text{rad}})\left(\overset{\circ}{\boldsymbol{D}} : \overset{\circ}{\boldsymbol{D}}\right) \\ &\quad + \bar{\rho}\,\sigma_{\text{rel}} + R_{\text{gd}}|\boldsymbol{w}|^2 - \overline{s\sigma\left(\boldsymbol{w} \cdot \frac{\partial p}{\partial \boldsymbol{r}}\right)},\end{aligned} \tag{7.149}$$

or, using (7.139),

$$\begin{aligned}&\bar{\rho}\frac{D}{Dt}\left(\langle E_{\text{sum}}\rangle + \langle C_{\text{d}}\rangle\,\langle C_{\text{g}}\rangle|\bar{\boldsymbol{w}}|^2/2\right) \cong -\operatorname{div}\left(\boldsymbol{q}^{\text{turb}} - \overline{p'_{\text{sum}}\boldsymbol{u}''} + \boldsymbol{q}_{\text{rad}}^{\text{turb}}\right) \\ &\quad - \bar{p}_{\text{sum}}\operatorname{div}\langle\boldsymbol{u}\rangle + 2(\bar{\rho}\,\nu^{\text{turb}} + \mu_{\text{rad}})(\overset{\circ}{\boldsymbol{D}} : \overset{\circ}{\boldsymbol{D}}) + (\bar{\Pi}_{\text{rel}}:\boldsymbol{D}).\end{aligned} \tag{7.150}$$

Thus, the equation for the internal energy of the averaged turbulized gas–dust continuum written via the absolute temperature takes the form:

$$\bar{\rho}\langle c_{\mathrm{P}}\rangle\frac{D\langle T\rangle}{Dt}-\operatorname{div}\left\{\chi^{\mathrm{turb}}\left(\frac{\partial\langle T\rangle}{\partial\boldsymbol{r}}-\frac{1}{\bar{\rho}\langle c_{\mathrm{P}}\rangle}\frac{\partial\bar{p}}{\partial\boldsymbol{r}}\right)+\chi_{\mathrm{rad}}\frac{\partial\langle T\rangle}{\partial\boldsymbol{r}}\right\}$$
$$=\frac{D\bar{p}_{\mathrm{sum}}}{Dt}+2(\bar{\rho}\,\nu^{\mathrm{turb}}+\mu_{\mathrm{rad}})\left(\overset{\circ}{\boldsymbol{D}}:\overset{\circ}{\boldsymbol{D}}\right)+(\bar{\Pi}_{\mathrm{rel}}:\boldsymbol{D})-\sum_{s=1}^{r}\langle q_s\rangle\bar{\xi}_s+\bar{\rho}\,\sigma_{\mathrm{rel}}+R_{\mathrm{gd}}|\bar{\boldsymbol{w}}|^2 \tag{7.149*}$$

Here, we disregarded the energy of the dissipation $\bar{\Phi}_u$ (in a unit volume per unit time) through the "molecular" viscosity of the gas–dust mixture compared to the "frictional heat" $2\bar{\rho}\,\nu^{\mathrm{turb}}\left(\overset{\circ}{\boldsymbol{D}}:\overset{\circ}{\boldsymbol{D}}\right)$ through the viscous Reynolds stresses arising from the relative shear of the gas-suspension elements in the orbital motion of the disk material and the kinetic diffusion energy compared to the internal energy of the gas suspension. It should be noted that the presence of an internal heating source of the protoplanetary disk related to turbulence viscosity is in satisfactory agreement with the currently available astrophysical data and with all cosmochemical constraints. In addition, under the assumptions made, an additional heating source of the gas–dust medium related to the energy dissipation under the effect of "molecular" diffusion $(\bar{\Pi}_{\mathrm{rel}}{:}\boldsymbol{D})$ appears in (7.149*); it plays an important role in the subdisk, where the relative velocities of the phases can be significant.

7.3.5.4 The Law of Conservation of Total Energy for a Turbulized Mixture

Adding the balance equations for the internal energy (7.122*), mechanical energy (7.138), kinetic energy of the interphase diffusion (7.139), and turbulent energy of the disk system (7.144) yields the law of conservation of total averaged energy for a two-phase gas–dust mixture and radiation in the disk in a substantial form (note that here we retained the designation U_{tot} for the total energy of the "matter plus radiation" system (cf. (7.42))):

$$\bar{\rho}\frac{D}{Dt}\langle U_{\mathrm{tot}}\rangle=\operatorname{div}\left(\bar{\boldsymbol{J}}_U+\boldsymbol{J}_U^{\mathrm{turb}}\right), \tag{7.151}$$

where

$$\langle U_{\mathrm{tot}}\rangle=\langle E_{\mathrm{sum}}\rangle+\langle\Psi\rangle+\frac{1}{2}|\langle\boldsymbol{u}\rangle|^2+\langle C_{\mathrm{d}}\rangle\,\langle C_{\mathrm{g}}\rangle|\bar{\boldsymbol{w}}|^2/2+\overline{|\boldsymbol{u}''|^2/2} \tag{7.152}$$

is the averaged total energy of the gas suspension and radiation (see (7.42));

$$\bar{J}_U \equiv \overline{q_{\mathrm{sum}} + (U\, p_{\mathrm{sum}} - \Pi_{\mathrm{sum}} - \Pi_{\mathrm{rel}}) \cdot u}$$
$$\cong \bar{q}_{\mathrm{sum}} + (U\, \bar{p}_{\mathrm{sum}} - \bar{\Pi}_{\mathrm{sum}} - \bar{\Pi}_{\mathrm{rel}}) \cdot \langle u \rangle + \overline{p_{\mathrm{sum}} u''} - \overline{\Pi'_{\mathrm{sum}} \cdot u''} - \overline{\Pi'_{\mathrm{rel}} \cdot u''} \tag{7.153}$$

is the averaged actual total energy flux in the two-phase gas suspension;

$$J_U^{\mathrm{turb}} \equiv \bar{\rho} \langle U''_{\mathrm{tot}} u'' \rangle = \overline{\rho \left(H_{\mathrm{sum}} - p_{\mathrm{sum}}/\rho + \Psi + \frac{1}{2}|u|^2 + C_{\mathrm{d}}\, C_{\mathrm{g}} |w|^2/2 \right) u''}$$
$$= q_{\mathrm{sum}}^{\mathrm{turb}} - \overline{p_{\mathrm{sum}} u''} + \overline{\rho |u''|^2 u''/2} - (R \cdot \langle u \rangle) + q_{\mathrm{sum}}^{\mathrm{turb}} \tag{7.154}$$

is the turbulent total energy flux of the mixture.

Combining (7.153) and (7.154), we obtain the following expression for the total energy flux of the turbulized flow of a gas–dust mixture:

$$\bar{J}_U + J_U^{\mathrm{turb}} = \overline{\rho |u''|^2 u''/2} - \overline{(\Pi'_{\mathrm{sum}} + \Pi'_{\mathrm{rel}}) \cdot u''} + q_{\mathrm{sum}}^{\mathrm{turb}} + \bar{q}_{\mathrm{sum}}$$
$$+ (U\, \bar{p}_{\mathrm{sum}} - \bar{\Pi}_{\mathrm{sum}} - \bar{\Pi}_{\mathrm{rel}} - R) \cdot \langle u \rangle \tag{7.155}$$

Here, $\bar{q}_{\mathrm{sum}} + q_{\mathrm{sum}}^{\mathrm{turb}}$ is the total heat flux attributable to both the averaged molecular transfer and the turbulent transfer; $\bar{p}_{\mathrm{sum}} \langle u \rangle$ is the "mechanical" energy flux; $(\bar{\Pi}_{\mathrm{sum}} + \bar{\Pi}_{\mathrm{rel}} + R) \cdot \langle u \rangle$ is the total energy flux attributable to the work of viscous, relative, and turbulent stresses; $\overline{\rho |u''|^2 u''/2} - \overline{(\Pi'_{\mathrm{sum}} + \Pi'_{\mathrm{rel}}) \cdot u''}$ is the "diffusion" flux of vortex turbulent energy. Note that the term $\overline{p_{\mathrm{sum}} u''}$ in (7.154) and (7.155) does not act as the energy flux, because it drops out of the total energy flux (7.155).

7.3.6 Modeling the Turbulent Viscosity Coefficient in a Dust Subdisk

Let us now consider a semiempirical method for determining the coefficient of kinematic turbulent viscosity ν^{turb} in a two-phase disk medium that includes the influence of the inertial effects of medium-size and coarse particles on the additional generation of turbulence in a gas–dust cloud. Below, the velocity shear of the cosmic material (when the kinetic energy of turbulence is extracted from the kinetic energy of averaged motion) related to its differential pattern of its rotation around the proto-Sun (see Dubrulle 1993; Gor'kavyi and Fridman 1994) is assumed to be the main source of turbulence in the disk. Each layer of material with radius r of a differentially rotating thin disk ($h_{\mathrm{disk}}(r) \ll r$) that lies near the $r\varphi$ plane (located at $z = 0$ in cylindrical coordinates) moves almost exactly according to Kepler's third law, i.e., it rotates increasingly fast as the central body (with mass $\mathcal{M}_{\odot}$) is approached: the Keplerian orbital velocity is $u_{\varphi}(r) = r\Omega_{K,\mathrm{mid}}\,(r) = \sqrt{G\mathcal{M}_{\odot}/r}$

and the angular velocity of orbital rotation $\Omega_{K,\mathrm{mid}}(r)$ increases as $r^{-3/2}$. Note that generally speaking, the disc thickness does not remain constant but increase with the distance from proto-Sun. Such motion is a typical case of shear flow, which can be analyzed in terms of the invariant modeling of developed turbulent flows in inhomogeneous media (Kolesnichenko and Marov 1999).

7.3.6.1 α—Parameterization of the Viscosity of Protoplanetary Disk

The coefficient of turbulent viscosity in an astrophysical gas-phase disk was at the first time modeled by Shakura and Sunyaev (1973). They used Kolmogorov's concept of dynamic turbulent viscosity $\mu_{\mathrm{g}}^{\mathrm{turb}} = \bar{\rho}_{\mathrm{g}}\, u_{\mathrm{g}}^{\mathrm{turb}} l_{\mathrm{g}}^{\mathrm{turb}}$, where $u_{\mathrm{g}}^{\mathrm{turb}} \leq c_{\mathrm{sg}}|_{z=0} \cong \sqrt{\bar{p}_{\mathrm{g}}/\bar{\rho}_{\mathrm{g}}}|_{z=0}$ is the root-mean-square velocity of turbulent pulsations limited by the speed of sound in the gas calculated in the midplane of the disk; $l_{\mathrm{g}}^{\mathrm{turb}}$ is the so-called Prandtl mixing length limited by the disk half-thickness h_{disk}, $l_{\mathrm{g}}^{\mathrm{turb}} \leq h_{\mathrm{disk}} \approx c_{\mathrm{sg}}|_{z=0}/\Omega_{K,\mathrm{mid}}$, $\rho_{\mathrm{g}}, p_{\mathrm{g}}$ are, respectively, the mass density and pressure in the gas-phase disk. The following relation between the r, φ component of the Reynolds turbulent stress tensor $R_{r\varphi}$ and the thermal gas pressure p_{g} was obtained ($\alpha \leq 1$):

$$R_{r\varphi} = \bar{\rho}_{\mathrm{g}} v_{\mathrm{g}}^{\mathrm{turb}}(r)\, r\partial_r(\boldsymbol{u}_{\varphi}/r) = -\frac{3}{2}\bar{\rho}_{\mathrm{g}}\, u_{\mathrm{g}}^{\mathrm{turb}}\, l_{\mathrm{g}}^{\mathrm{turb}}\, \Omega_{K,\mathrm{mid}}(r) = -\alpha \bar{p}_{\mathrm{g}}|_{z=0}. \quad (7.156)$$

The Shakura–Sunyaev disk parameter (a dimensionless free parameter) α, which characterizes the degree of excitation of turbulent motions, can be calibrated empirically using time-dependent spectra obtained, in particular, during observations of outbursts in binary systems with mass transfer containing dwarf novae. For this case, it was found that $0.01 \leq \alpha \leq 1$ (Eardley et al. 1978). The models of turbulized accretion disks constructed using relation (7.156) pertain to the so-called viscous α -disks. Determining the parameter a under various assumptions about the nature of the physical processes in the disk was the subject of many studies [see, e.g., extensive bibliography to the paper by Makalkin (2004)]. In particular, a number of authors (Dubrulle 1993; Cabot et al. 1987) who used the α -model when considering such physical mechanisms of turbulence in a protoplanetary disk as differential rotation, thermal convection, etc., obtained $\alpha \sim 10^{-3}$, the value that satisfies best the astrophysical constraints.

The main advantage of such an heuristic approach to describing the disk turbulence is that it is relatively simple: it will suffice to substitute the coefficient of turbulent viscosity $v^{\mathrm{turb}}(r)$ for the coefficient of molecular viscosity v in the equations of stellar hydrodynamics to somehow take into account the turbulization of the medium in an accretion disk (in fact, this is what most astrophysicists do by essentially ignoring almost all correlation terms in the averaged equations of motion). At the same time, it is important to keep in mind that the Shakura–Sunyaev approach, which was specially developed by the authors to model thin (vertically uniform, i.e., structureless) astrophysical disks and which disregards the height dependence of the coefficient of turbulent viscosity, is appropriate to use only for global (one-dimensional in r) modeling of the evolution of the solar protoplanetary disk with parameters averaged over its thickness. In recent years, however, this approach has come to be used often uncritically and in two-dimensional (r, z) models that are associated with modeling of particular features of the vertical disk structure, in particular, with calculations of the height distribution of thermohydrodynamic parameters in the dusty subdisk, which is hardly justified.

In addition, it should be borne in mind that (7.156), which was derived for gas-phase disks, naturally disregards the inverse effect of dust and heat transfer on the development of turbulence in the disk, which should be done when modeling many phenomena important for cosmogony. This effect is that additional buoyant forces facilitating or preventing the growth of turbulence in the disk arise due to the difference between the concentrations of dust material mixed with the medium (at turbulent diffusion) or the temperature difference (at heat transfer) at separate points of the disk medium. Thus when accurately modeling the evolution of the protoplanetary cloud as a viscous gas–dust disk it is important to take into account the dynamical processes of gas–dust interaction and, in particular, the inverse effect of the inertial properties of dust particles on the intensity of turbulence and the thermal regime of the subdisk. The following is an argument for such a general approach: the dust particles, which account for only ~2% of the mass of the circumsolar protoplanetary cloud, may be disregarded only at the initial evolutionary stage of the cosmic system under consideration, when almost all of the primordial (interstellar) dust particles evaporated. At later stages of its evolution, as the protoplanetary cloud cooled down, the solid particles condensed, their sizes increased significantly through coagulation, the disperse phase settled to the midplane of the disk, and the gas dissipated from the disk system into interstellar space, the dynamical, energetic, and optical roles of the dust component of the gas suspension increased significantly (Cuzzi et al. 1993).

At first glance, turbulent mixing hinders the diffusive separation of the dust and gas components in the gravitational field of the proto-Sun, preventing the settling of fine solid particles to its equatorial plane (where they form a flattened dust layer), and thus, inhibits the formation of a critical mass of the subdisk at which it becomes gravitationally unstable (Safronov 1969). However, on the other hand, as we have repeatedly noted above, nongravitational accretion related, in particular, to the growth of particles through various turbulent coagulation mechanisms becomes an efficient accumulation mechanism of medium- and large-scale solid particles for turbulent flow. In addition, turbulence promotes the formation of mesoscale, relatively stable gas–dust coherent structures that provide the most favorable conditions for the adhesion of dust particles. In such vortex structures, the number of collisions (per unit time) increases significantly, while the relative collision velocities decrease significantly compared to laminar conditions (through the combined coherent mesoscale motion of particles and small-scale turbulent pulsations of their relative velocities inside vortex structures), which also contributes to the growth of a condensed subdisk component (see Barge and Sommeria 1995; Tanga et al. 1996; Chavanis et al. 1999; Kolesnichenko 2004). As the inertia of solid particles increases, they are drawn into the pulsational motion of the gas carrier flow to a progressively lesser extent. Thus, turbulence eventually contributes to an efficient settling of dust particles to the midplane of the disk and, thus, to the formation of a critical mass of the disk whose gravitational instability and disintegration lead to the formation of planetesimals.

Basically, the pattern of the disperse-phase effect on the dynamics of turbulent gas-suspension flow is not unique, but depends significantly on the inertia and volume content (concentration) of dust particles, since they can both laminarize and turbulize the flow (Shraiber et al. 1987). Kolesnichenko (2000) investigated the rotation-generated flows of disk material where the solid particles of the gas suspension begin to have an inverse effect on the parameters of the latter. Formula (7.156) for the coefficient of turbulent viscosity was generalized to the case of allowance for the low-inertia dust component where the approximation of a passive admixture (in which the two-phase gas–dust flow is approximated by the flow of a single-phase "multicomponent" medium with known effective thermophysical properties) could be used. It was recommended to apply the derived corrections to the coefficient of turbulent viscosity that incorporate the inverse effect of the transfer of a fine admixture and heat on the growth of turbulence when modeling the formation of a flattened dust layer in the disk.

At the same time, the problem of influence of medium-size and coarse particles on the processes of turbulent heat and mass transfer in a protoplanetary gas–dust disk and their contribution to the correction to the coefficient of turbulent viscosity of a gas suspension remains unresolved. Determining this kind of correction is challenging indeed and the problem requires an in-depth study of the gas-suspension turbulence structure. In the next section, we attempt to theoretically determine the coefficient of turbulent viscosity ν^{turb} in a gas-suspension flow with large inertial dust particles.

7.3.6.2 Modeling the Coefficient of Turbulent Viscosity in a Dusty Subdisk

Before determining the above correction to ν^{turb}, we will recall that the relationship between the coefficient of turbulent viscosity and the gas-suspension turbulence energy is defined by Kolmogorov (1942) (see (7.93))

$$\nu^{\text{turb}} = \gamma^* \, l \sqrt{b}, \tag{7.157}$$

where $l = l(\boldsymbol{r})$ is the turbulence scale length at a given point of the flow (the numerical factor γ^* may be included in l). For a turbulized shear flow around an infinite plane (in our case, the equatorial plane of the disk, $z = 0$), the local turbulence scale length $l(\boldsymbol{r})$ may be assumed to be proportional to the thickness of the thin layer under consideration,

$$l(z) = \gamma^* \kappa z \tag{7.158}$$

or

$$l(z) = \gamma^* \kappa z \Phi(\boldsymbol{Re}_{\text{glob}}, \boldsymbol{Ri}, \boldsymbol{K}), \tag{7.159}$$

where Φ is a dimensionless function; κ is the Karman constant, which may be set equal to ~0.4.

It should be kept in mind that deriving an adequate differential equation for the scale length $l(\boldsymbol{r})$ is one of the most complex problems in the semiempirical theory of shear turbulence. The point is that, in general, $l(\boldsymbol{r})$ cannot be defined only via the one-point moments of the pulsating velocity. Being a measure of the distance between two points, $\boldsymbol{r}_1$ and $\boldsymbol{r}_2$, in a turbulized flow at which the two-point correlators $\langle \boldsymbol{u}''(\boldsymbol{r}_1)\boldsymbol{u}''(\boldsymbol{r}_2)\rangle$ are still differ markedly from zero, the scale length $l(\boldsymbol{r})$ can be found from the complex differential equations for these moments by their integration over the distance between $\boldsymbol{r}_1$ and $\boldsymbol{r}_2$ (Ievlev 1975). At the same time, even in the case of a given differential equation for $l(\boldsymbol{r})$, there is a complex problem of boundary conditions at the free boundary of the turbulent flow region, where the scale $l(\boldsymbol{r})$ does not tend to zero (Laikhtman 1970). For these reasons, to ensure the efficiency of practical calculations, the local turbulence scale length $l(\boldsymbol{r})$ is often specified in the form of purely empirically found functions or calculated using an algebraic formula of type (7.159) that includes only the flow geometry, the distance to the wall, etc., and that does not depend on the peculiarities of the fluid flow. In some cases the formula of Prandtle–Nikuradze for the quantity $l(z)$ can be apparently used which in application to the plane case under consideration take the form

$$l(z) = \gamma 0.4z\left[1 - 1.1(z/h_{\text{disk}}) + 0.6(z/h_{\text{disk}})^2 - 0.15(z/h_{\text{disk}})^3\right].$$

7.3.6.3 Internal Equilibrium in the Disk Turbulence Structure

Let us now derive the sought-for correction to $\nu^{\text{turb}}(\boldsymbol{r})$. We will restrict our analysis to a simplified statistical scheme of turbulence in a two-phase medium based on the turbulent energy transfer equation (7.144*) alone (a one-parameter turbulence model). Let us note that two-parametric model of turbulence *b*–*c* for two-phase flows was firstly developed by Elghobashi and Abou-Arab (1982, 1983). If we disregard the small terms in (7.144*) [all necessary estimates can be found in the monograph after Marov and Kolesnichenko (2002)], then, given (7.106) for the

Reynolds tensor and (7.135) for a turbulent flow of specific volume $\boldsymbol{J}_{\mathrm{v}}^{\mathrm{turb}}$, it can be rewritten as

$$\bar{\rho}\left(\frac{\partial b}{\partial t}+\langle\boldsymbol{u}\rangle\cdot\nabla b\right)+\operatorname{div}\boldsymbol{J}_{b}^{\mathrm{turb}}\cong-\frac{2}{3}b\bar{\rho}\operatorname{div}\langle\boldsymbol{u}\rangle+2\bar{\rho}\nu^{\mathrm{turb}}\left(\overset{\circ}{\boldsymbol{D}}:\overset{\circ}{\boldsymbol{D}}\right)$$
$$-\langle\sigma\rangle\frac{\bar{\rho}}{\rho_{\mathrm{d}}\bar{\rho}_{\mathrm{g}}}\boldsymbol{J}_{\mathrm{dz}}^{\mathrm{turb}}g_{z}+\frac{g_{z}}{\langle T\rangle\langle c_{\mathrm{P}}\rangle}\left(\boldsymbol{q}^{\mathrm{turb}}-\sum_{\alpha}\langle h_{\alpha}\rangle\boldsymbol{J}_{\alpha}^{\mathrm{turb}}\right)_{z}-\bar{\rho}\varepsilon+\bar{\rho}\sigma_{\mathrm{rel}},\quad(7.160)$$

where g_z is the vertical gravity component of the proto-Sun. As we already emphasized above, the contribution of additional dissipation to ε is significant only for relatively fine dust particles; as a result, it plays no prominent role in the balance equation (7.160) for coarse dust (this is partially accounted for the term $2\bar{\rho}\,\nu^{\mathrm{turb}}\left(\overset{\circ}{\boldsymbol{D}}:\overset{\circ}{\boldsymbol{D}}\right)$).

To write (7.160) in the form required for our subsequent purposes, we will use the basic concept of a two-level macroscopic description of a turbulized medium in the form of two interconnected continua (open subsystems) that simultaneously fill the same volume of coordinate space of the disk continuously, the subsystem of averaged motion and the subsystem of turbulent chaos. Using the concept of two-level macroscopic description of a turbulized fluid was the starting point that allowed us to phenomenologically develop a hydrodynamic model of structured turbulence as a self-organization in open nonequilibrium systems associated with fluctuating media (Kolesnichenko 1998, 2004; Marov and Kolesnichenko 2006). For the disk material, the continuum of averaged motion obtained through the probability-theoretical averaging of the instantaneous equations of motion for a heterogeneous medium (7.76) is intended to study the evolution of the averaged fields of thermohydrodynamic gas-suspension parameters (including also the possible large vortex structures). The subsystem of turbulent chaos for the disk (a vortex continuum with an internal structure) generally consists of two components: the turbulent chaos proper (the so-called incoherent turbulence) associated with the stochastic small-scale pulsational motion of an eddy fluid and the coherent component embedded in this almost uniform pulsating field, an ensemble of mesoscale ordered vortex structures ("multimolecular" formations). According to the currently available experimental data, the coherent turbulent structure can be defined as a connected, fluid mass with vorticity correlated in phase (i.e., coherent) in the entire region of coordinate space occupied by the structure (see Khlopkov et al. 2002). The formation of granules in the solar photosphere is a clear example of the existence of a vast family of coherent structures in a turbulent flow that appears against the background of small-scale turbulent motion.

For the subsystem of turbulent chaos, we postulate the Gibbs fundamental identity (Kolesnichenko 1998)

$$\delta E^{\mathrm{turb}}=T^{\mathrm{turb}}\delta S^{\mathrm{turb}}-p^{\mathrm{turb}}\delta(1/\bar{\rho}).\quad(7.161)$$

Using this identity, we can determine in a well-known way (de Groot and Mazur 1962) the thermodynamic structure of the vortex continuum, i.e., introduce the specific internal energy $E^{\mathrm{turb}}(\boldsymbol{r},t)$, specific entropy $S^{\mathrm{turb}}(\boldsymbol{r},t)$, pressure $p^{\mathrm{turb}}(\boldsymbol{r},t) \equiv \frac{2}{3} \bar{\rho} E^{\mathrm{turb}}$, and temperature $T^{\mathrm{turb}}(\boldsymbol{r},t)$ of turbulization. The various relations between the parameters E^{turb}, S^{turb}, T^{turb}, and p^{turb} that can be derived in a standard way may then be interpreted as the "equations of state" for the subsystem under consideration. Below, we assume that the internal energy of the subsystem of turbulent chaos $E^{\mathrm{turb}}(\boldsymbol{r},t)$ is identical to the turbulence energy, $E^{\mathrm{turb}}(\boldsymbol{r},t) \equiv \overline{\rho\,|\boldsymbol{u}''|^2}/2\bar{\rho} \equiv b(\boldsymbol{r},t)$, and that the subsystem of turbulent chaos is an ideal gas in the thermodynamic sense, $\bar{\rho} b = \frac{3}{2} p^{\mathrm{turb}} = \frac{3}{2} \Re_{\mathrm{gd}} \bar{\rho}\, T^{\mathrm{turb}}$, where $\Re_{\mathrm{gd}} = k_{\mathrm{B}}/\mathcal{M}_{\mathrm{gd}}$, $\mathcal{M}_{\mathrm{gd}}$ is the mean molecular mass of the gas-suspension particles (the cardinal assumptions of the model). Using the Gibbs identity (7.161), (7.160) takes the form

$$\bar{\rho}\, T^{\mathrm{turb}} \frac{DS^{\mathrm{turb}}}{Dt} + \mathrm{div} \boldsymbol{J}_b^{\mathrm{turb}} + \frac{g_z}{\langle T\rangle \langle c_{\mathrm{P}}\rangle} \left(\boldsymbol{q}^{\mathrm{turb}} - \sum_\alpha \langle h_\alpha \rangle \boldsymbol{J}_\alpha^{\mathrm{turb}} \right)_z - \bar{\rho}\varepsilon + \bar{\rho}\sigma_{\mathrm{rel}}, \tag{7.162}$$

which, by analogy with (7.51), can be written in the form of a balance equation,

$$\frac{\partial}{\partial t}\left(\bar{\rho} S^{\mathrm{turb}}\right) + \mathrm{div}\left(\bar{\rho} S^{\mathrm{turb}} \langle \boldsymbol{u} \rangle + \frac{\boldsymbol{J}_b^{\mathrm{turb}}}{T^{\mathrm{turb}}} \right) = \sigma_{S^{\mathrm{turb}}} \equiv \sigma^e_{S^{\mathrm{turb}}} + \sigma^i_{S^{\mathrm{turb}}}$$

Let us now use (7.162) to analyze the steady-non-equilibrium regime of developed ($\boldsymbol{Re}_{\mathrm{glob}} \gg 1$) turbulence of the gas–dust mixture in a disk system. Obviously, its realization requires the existence of a continuously acting turbulization mechanism (e.g., the large-scale velocity shear of the flow of material related to its differential rotation around the proto-Sun or the thermal-convective large-scale instability) that transfers the kinetic energy of the averaged flow to the vertex motion on large scales and that does not allow the subsystem of turbulent chaos to reach a complete thermodynamic equilibrium for a long time. This energy source must have a power that would compensate, in particular, for the expenditure of the turbulent energy dissipated into heat through "molecular" viscosity. For this steady-nonequilibrium regime, almost all of the expendable turbulence energy is known to be transferred without noticeable losses through the inertial interval to the dissipative interval (Landau and Lifshitz 1988). Then, an internal equilibrium is established in the structure of the subsystem of turbulent chaos at which $DS^{\mathrm{turb}}/D t \cong 0$ (the entropy of chaos does not change along the path, $\boldsymbol{J}_{(S^{\mathrm{turb}})} \equiv \boldsymbol{J}_b^{\mathrm{turb}}/T^{\mathrm{turb}} \cong \mathrm{const}$; Kolesnichenko 2003). In fact, all of the existing semiempirical theories of turbulence assume (explicitly or implicitly) the existence of an internal equilibrium in the turbulence structure, when the production of turbulence energy is equal to its dissipation at each point. This implies that the production $\sigma^i_{S^{\mathrm{turb}}}$ of turbulization entropy (due to internal dissipative processes) is offset by its outflow $\sigma^e_{S\text{turb}}$; i.e., there is no total production of entropy S^{turb}, $\sigma_{S\text{turb}} = \sigma^e_{S\text{turb}} + \sigma^i_{S\text{turb}} \cong 0$. Thus, the subsystem of turbulent chaos exports its entropy into the "external medium," i.e., gives it up to the subsystem of averaged motion. It is important to keep in mind that the conditions of this kind are sufficient for the formation of dissipative coherent structures in an "open" eddy continuum (Prigogine and Stengers 1994).

7.3.6.4 Deriving the Correction Function to the Coefficient ν^{turb}

Thus, for the locally steady state of a developed turbulized flow in the disk, (7.162) can be written as

$$\nu^{\text{turb}}(1 - \boldsymbol{R_f} - \boldsymbol{K_f})\left(\overset{\circ}{\boldsymbol{D}}:\overset{\circ}{\boldsymbol{D}}\right) + \sigma_{\text{rel}} - \varepsilon \approx 0. \tag{7.163}$$

Here, by analogy with the dimensionless dynamic Richardson number (see (3.184))

$$\boldsymbol{R_f} \equiv -\frac{g_z\left(\boldsymbol{q}^{\text{turb}} - \sum_\alpha \langle h_\alpha \rangle \boldsymbol{J}^{\text{turb}}_\alpha\right)_z}{\langle c_P \rangle \langle T \rangle 2\,\bar{\rho}\nu^{\text{turb}}\left(\overset{\circ}{\boldsymbol{D}}:\overset{\circ}{\boldsymbol{D}}\right)} = \frac{1}{\boldsymbol{Sc}^{\text{turb}}}\,\frac{\frac{g_z}{\langle T \rangle}\left(\frac{\partial \langle T \rangle}{\partial z} + \frac{g_z}{\langle c_P \rangle}\right)}{2\left(\overset{\circ}{\boldsymbol{D}}:\overset{\circ}{\boldsymbol{D}}\right)} = \frac{\boldsymbol{Ri}}{\boldsymbol{Sc}^{\text{turb}}}, \tag{7.164}$$

that allows for the effect of thermal convection on the generation of turbulence in the disk compared to the dynamical factors ($\boldsymbol{Ri}$ is the gradient Richardson number), we introduced the dimensionless dynamic Kolmogorov number (cf. (3.185))

$$\begin{aligned} \boldsymbol{K_f} &\equiv \frac{(g_z/\bar{\rho}_g)(\boldsymbol{J}^{\text{turb}}_{\text{d}})_z}{2\,\bar{\rho}\nu^{turb}\left(\overset{\circ}{\boldsymbol{D}}:\overset{\circ}{\boldsymbol{D}}\right)} = -\frac{1}{\boldsymbol{Sc}^{\text{turb}}}\,\frac{(g_z/\bar{\rho}_g)((\partial \langle C_d \rangle)/\partial z)}{2\left(\overset{\circ}{\boldsymbol{D}}:\overset{\circ}{\boldsymbol{D}}\right)} \\ &= -\frac{1}{\boldsymbol{Sc}^{\text{turb}}}\,\frac{\langle \sigma \rangle g_z\left[((\partial \bar{s})/(\partial z)) - \bar{s}((\partial \ln \bar{\rho}_g)/(\partial z))\right]}{2\,\overset{\circ}{\boldsymbol{D}}:\overset{\circ}{\boldsymbol{D}}} = \frac{\boldsymbol{K}}{\boldsymbol{Sc}^{\text{turb}}}, \end{aligned} \tag{7.165}$$

It is a criterion for the dynamical activity of dust particles in a turbulized shear flow of disk material (Kolesnichenko 2000). Basically, only the first term in (7.165) is the Kolmogorov number (at $\bar{\rho}_g = \text{const}$, the number $\boldsymbol{K}$ expresses the relative expenditure of turbulent energy on the suspension of particles by the carrier flow); the second term, which describes the influence of gas inhomogeneity on turbulence, is the so-called Prandtl criterion.

The introduced dimensionless parameters are criteria for the dynamical activity of baroclinic disk material. It follows from (7.164) that $\boldsymbol{Ri} < 0$ for $-\partial\langle T \rangle/\partial z > g_z/\langle c_P \rangle$ (i.e., for unstable thermal stratification of the disk material) and $\boldsymbol{Ri} < 0$ for $-\partial\langle T \rangle/\partial z < g_z/\langle c_P \rangle$ (for stable stratification); for indifferent stratification, $\boldsymbol{Ri} = 0$. However, the presence of suspended fine particles in the flow always causes a decrease in turbulent energy, since the gradient Kolmogorov number is always positive, $\boldsymbol{K}>0$ (see Barenblatt and Golitsyn 1974). Thus, the dimensionless Kolmogorov number $\boldsymbol{K}$ allows for the inverse effect of the stratification (in disk thickness) of the volume concentration of small dust particles on the growth of turbulence in the disk.

Assuming, according to the Kolmogorov (1942) hypotheses, that the kinematic coefficient of turbulent viscosity ν^{turb} and the dissipation rate of turbulent energy into heat ε depend only on two flow parameters, the turbulence energy b and the local turbulence scale length $l(\boldsymbol{r})$, we obtain (see (7.93))

$$\nu^{\text{turb}} = l\sqrt{b}, \quad \varepsilon = \frac{1}{\alpha^2}\frac{b^{\frac{3}{2}}}{l}. \tag{7.166}$$

Here, since the scale length $l(\boldsymbol{r})$ is uncertain, the constant in the expression for ν^{turb} is taken to be unity and the numerical factor $1/\alpha^2$ is assumed, to a first approximation, to be constant. Let us represent the term (see (7.147))

$$\bar{\rho}\sigma_{\text{rel}} \equiv -\overline{\left(\Pi'_{\text{rel}}:\frac{\partial \boldsymbol{u}'}{\partial \boldsymbol{r}}\right)} = \overline{s\rho_{\text{d}}\, C_{\text{g}}(\boldsymbol{w}\boldsymbol{w})':\frac{\partial \boldsymbol{u}'}{\partial \boldsymbol{r}}} \cong \rho_{\text{d}}\bar{\boldsymbol{w}}\,\bar{\boldsymbol{w}}:\left(\bar{s}\overline{C'_{\text{g}}\frac{\partial \boldsymbol{u}'}{\partial \boldsymbol{r}}} + \langle C_{\text{g}}\rangle \overline{s'\frac{\partial}{\partial \boldsymbol{r}}\boldsymbol{u}'}\right),$$

which is responsible for the additional generation of turbulence energy at large Reynolds numbers (in the wakes behind the moving large particles) as

$$\sigma_{\text{rel}} = \beta\bar{s}\frac{|\bar{\boldsymbol{w}}|^2\sqrt{b}}{l}, \tag{7.167}$$

where β is an empirical constant. It should be noted that (7.167) is similar in form to the expression that was derived by Varaksin (2003) using a self-similar solution for the far axisymmetric turbulent wake (Yarin and Hetsroni 1994) and by Derevich (1994) using a semiempirical approach. The latter formulas valid, however, only at a very low volume concentration $\bar{s}$ of the disperse phase, when there is no interference of the wakes behind individual particles.

Substituting (7.166) and (7.167) in (7.163) yields

$$2l\sqrt{b}(1-\boldsymbol{R}_f-\boldsymbol{K}_f)\left(\overset{\circ}{\boldsymbol{D}}:\overset{\circ}{\boldsymbol{D}}\right) + \beta\bar{s}\frac{|\bar{\boldsymbol{w}}|^2\sqrt{b}}{l} - \frac{1}{\alpha^2}\frac{b^{3/2}}{l} = 0. \tag{7.168}$$

Equation (7.168) breaks up into two equations: $b=0$ that corresponds to a laminar flow in the disk system, and

$$b = 2\alpha^2 l^2\left(1-\alpha^2\beta\bar{s}\frac{|\bar{\boldsymbol{w}}|^2}{b}\right)^{-1}\left(1-\frac{\boldsymbol{Ri}+\boldsymbol{K}}{\boldsymbol{Sc}^{\text{turb}}}\right)\left(\overset{\circ}{\boldsymbol{D}}:\overset{\circ}{\boldsymbol{D}}\right) \tag{7.169}$$

that describes the steady turbulent flow of a gas suspension. Equation (7.169) has real solutions only at $\boldsymbol{Ri}+\boldsymbol{K}<(\boldsymbol{Ri}+\boldsymbol{K})_{\mathrm{cr}}\equiv \boldsymbol{Sc}^{\mathrm{turb}}$ (while at $\boldsymbol{Ri}+\boldsymbol{K}\geq \boldsymbol{Sc}^{\mathrm{turb}}$ only one real solution, $b=0$, pertaining to a laminar flow exists). Let the regime be turbulent; then,

$$\sqrt{b}=\alpha l\,\frac{\sqrt{1-(\boldsymbol{Ri}+\boldsymbol{K})/\boldsymbol{Sc}^{\mathrm{turb}}}}{\sqrt{1-\alpha^2\beta^*(d_{\mathrm{d}}^2/l^2)\bar{s}}}\sqrt{2\left(\overset{\circ}{\boldsymbol{D}}:\overset{\circ}{\boldsymbol{D}}\right)} \tag{7.170}$$

We see from (7.169) that the presence of large solid particles in the flow always causes an increase in turbulent energy, since the averaged kinetic energy of the relative phase motion ($\propto |\bar{\boldsymbol{w}}|^2$) for them is comparable to the turbulence energy b.

Thus, we obtain for the local coefficient of turbulent viscosity in a gas–dust disk

$$\nu^{\mathrm{turb}}(\boldsymbol{r})=\alpha l^2(\boldsymbol{r})\,\varphi\varphi_1\sqrt{2\left(\overset{\circ}{\boldsymbol{D}}:\overset{\circ}{\boldsymbol{D}}\right)}, \tag{7.171}$$

where

$$\varphi=\varphi(\boldsymbol{Ri},\boldsymbol{K})\equiv\sqrt{1-(\boldsymbol{Ri}+\boldsymbol{K})/\boldsymbol{Sc}^{\mathrm{turb}}}, \tag{7.172}$$

$$\varphi_1=\varphi_1(\bar{s},C_D,\delta/\xi_{\mathrm{d}},d_{\mathrm{d}}/l,)\equiv\sqrt{1-\alpha^2 a(C_D\delta/\xi_{\mathrm{d}})^{4/3}(d_{\mathrm{d}}/l)^2\bar{s}} \tag{7.173}$$

are the correction dimensionless functions that allow, respectively, for the inverse effect of the dust and heat transfer on the growth of turbulence in the protoplanetary disk (φ) and for the influence of medium and coarse particles on the turbulent heat and mass transfer in the flow and their contribution to the coefficient of turbulent viscosity for the gas suspension (φ_1).

7.4 Steady Motions in a Turbulized Gas–Dust Subdisk

As an illustration of the approach developed here, let us now use the general relations derived above to model the protoplanetary gas–dust cloud that rotated around the proto-Sun at an early stage of its evolution, until it lost the gas component, based on a schematized description of a steady axisymmetric turbulized flow of disk material, which, however, leads to manageable and numerically solvable equations. Since we are eventually interested in the spatial distribution of thermo-hydrodynamic parameters inside the dust layer (in the subdisk) formed when solid particles settled to the equatorial plane of the proto-Sun in the presence of developed turbulence in the disk, for completeness, it will also be important to consider simple mechanical properties of the rotating gas–dust cloud as a whole. We will analyze the disk system under the following assumptions:

(a) We investigate a slowly evolving gas–dust cloud that rotates around a fixed (in space) z axis with an angular velocity $\Omega(r, z)$.
(b) The rotation is assumed to be so slow that the meridional circulation of the protoplanetary cloud material may be ignored (in essence, the velocity has only the φ component for Keplerian accretion disks, i.e., $\boldsymbol{u}_z \ll \boldsymbol{u}_r \ll \boldsymbol{u}_\varphi$).
(c) The magnetic fields play no significant role (the figure of the cloud is known to become flattened in the absence of a macroscopic magnetic field).
(d) The disk configuration is assumed to be stationary in an inertial frame of reference with the origin at the center of mass of the proto-Sun.
(e) ThZexistence of a midplane of symmetry that coincides with the equatorial plane of the proto-Sun defined by the condition $z = 0$ is postulated for a baroclinic disk (the equation of state (7.54) is valid for its material).
(f) The ratio of the disk half-thickness $h_{\text{disk}}(r)$ to its radius is assumed to be much smaller than unity, $h_{\text{disk}}(r)/r \ll 1$ (the thin-disk condition).
(g) We neglect the self-gravity of the disk material compared to the gravitational field of the proto-Sun.
(h) The radiation pressure in the disk is assumed to be much lower than the gas pressure, $p_R \ll p_g$.
(i) The gas–dust disk is optically thick to radiation at all frequencies.
(j) We disregard the chemical reactions and phase transitions and assume the composition of the disk gas phase to be homogeneous.
(k) The large-scale velocity shear of the material related to its differential rotation around the proto-Sun is assumed to be responsible for the turbulization of a Keplerian protoplanetary disk.

7.4.1 Axisymmetric Motion in a Gas–Dust Disk

When rotating around the proto-Sun almost exactly according to Kepler's law, each gas-suspension element in the disk slowly moves radially inward, since the deceleration related to the forces of viscous friction between the adjacent cylindrical layers rotating with different angular velocities leads to a redistribution of specific angular momentum and the appearance of a radial mass flow. Thus, the main mass of the disk material slowly (compared to the orbital motion) drifts toward the center of mass along a flat spiral trajectory as the angular momentum, together with the smaller mass, is transferred outward (in view of the conservation law), from the inner disk regions to the outer ones. Concurrently, the turbulent stresses arising from the relative shear of the separate layers of disk material during their orbital motion lead to viscous heat dissipation. The thin-disk condition is known to imply that the disk temperature is relatively low and the pressure gradient is much smaller than the two main mechanical forces, the gravitational and centrifugal ones (see, e.g., Shapiro and Teukolsky 1983). However, low temperatures are maintained only if the viscous heat dissipated in the turbulized system is effectively radiated outward and is not accumulated in the disk. In the steady state, the bulk of this heat must be radiated by the upper and lower disk surfaces (because the disk is thin and the radiation is directed mainly vertically rather than radially). Thus, a thin accretion disk must be highly nonadiabatic.

Below, we will use an inertial cylindrical (r, φ, z) coordinate system with the coordinate origin at the center of mass; the $z = 0$ plane is assumed to coincide with the midplane of symmetry of the disk. We will also assume that the averaged motion of the cosmic fluid has only an azimuthal component,

$$\langle u_r \rangle = 0, \quad \langle u_\varphi \rangle = \Omega(r, z)\, r, \quad \langle u_z \rangle = 0, \tag{7.174}$$

and that the true flow velocity of the gas–dust mixture randomly pulsates about its mean value, varying irregularly in the meridional and azimuthal directions. It can be shown that, if the disk material is in a state of quasi-steady rotation in an inertial frame of reference, then it necessarily possesses axial symmetry ($\partial/\partial\varphi = 0$): $\bar{s} = \bar{s}(r, z)$, $\bar{\rho} = \bar{\rho}(r, z)$, $\bar{p} = \bar{p}(r, z)$, $\langle T \rangle = \langle T \rangle(r, z)$, $\Omega = \Omega(r, z)$, etc. (Tassoul 1979). Note that the mass conservation law (7.79) in the steady-state case under consideration always holds, since the motions are axisymmetric and there are no meridional flows, $\mathrm{div}\langle \boldsymbol{u} \rangle = 0$.

7.4.1.1 Momentum Conservation Equations

Since T Tauri stars probably have a high dust concentration throughout the disk thickness (see, e.g., Beckwith et al. 2000) the additional stresses $\bar{\Pi}_{\mathrm{rel}}$ related to the relative motion of the gas and the coarse dust effectively act only in the region immediately adjacent to the equatorial plane of the disk (small compared to the entire disk) and, therefore, can be omitted in the equation of motion.

If we take into account the matter–radiation interaction inside the disk up to the terms of the lowest order in $|\langle \boldsymbol{u} \rangle|/c$ (see Sect. 7.2.3), then the three components of the averaged equation of motion (7.103) can be written as

$$\frac{1}{\bar{\rho}} \frac{\partial}{\partial r} \left(\bar{p} + p^{\mathrm{turb}} \right) = r\Omega^2(r, z) - \frac{G\mathcal{M}_{\odot} r}{(r^2 + z^2)^{3/2}} \equiv g_r \cong 0 \tag{7.175}$$

$$\frac{1}{\bar{\rho}} \frac{\partial}{\partial z} \left(\bar{p} + p^{\mathrm{turb}} \right) = -\frac{G\mathcal{M}_{\odot} z}{(r^2 + z^2)^{3/2}} \equiv -g_z, \tag{7.176}$$

$$\bar{\rho}\frac{\partial}{\partial t}[r\Omega(r,z,t)] \cong \frac{1}{r}\frac{\partial}{\partial r}\left\{r\left[(\boldsymbol{R}+\bar{\boldsymbol{\Pi}}_{\text{rad}})_{r\varphi} - \frac{1}{c^2}(\boldsymbol{q}_{\text{rad}})_r r\,\Omega(r,z)\right]\right\}$$
$$+\frac{\partial}{\partial z}\left[(\boldsymbol{R}+\bar{\boldsymbol{\Pi}}_{\text{rad}})_{z\varphi} - \frac{1}{c^2}(\boldsymbol{q}_{\text{rad}})_z r\Omega(r,z)\right]$$
$$+\frac{1}{r}\left[(\boldsymbol{R}+\bar{\boldsymbol{\Pi}}_{\text{rad}})_{\varphi r} - \frac{1}{c^2}(\boldsymbol{q}_{\text{rad}})_r r\Omega(r,z)\right] \cong 0, \qquad (7.177)$$

where

$$g_z = \frac{G\mathcal{M}_{\odot}z}{r^3}\left(1+\frac{z^2}{r^2}\right)^{-3/2} \cong \Omega^2_{K,\text{mid}}(r)\,z. \qquad (7.178)$$

$\boldsymbol{g} = \{g_r,\,0,\,-g_z\}$ is the effective gravity (per unit mass) corrected for the centrifugal acceleration; $\Omega_K(r,z) \equiv \sqrt{G\mathcal{M}_{\odot}/(r^2+z^2)^{3/2}}$ is the Keplerian angular velocity; $\Omega_{K,\text{mid}}(r) \equiv \Omega_K(r,0) = \sqrt{G\mathcal{M}_{\odot}/r^3}$ is the Keplerian angular velocity in the midplane of the disk; $\bar{p}(r,z)$ is the thermal pressure of the disk midplane related to the density $\bar{\rho}(r,z)$ and temperature $\langle T(r,z)\rangle$ by the averaged equation of state (7.55), $\bar{p} = \bar{\rho}\langle\Re\rangle\langle T\rangle$; $p^{\text{turb}}(r,z) \equiv \frac{2}{3}\bar{\rho}b = \frac{1}{3}\overline{\rho\,|\boldsymbol{u}''|^2}$ is the turbulization pressure (see (7.106)); c is the speed of light in a vacuum;

$$\boldsymbol{R}+\bar{\boldsymbol{\Pi}}_{\text{rad}} \equiv -p^{\text{turb}}\boldsymbol{U} + 2(\bar{\rho}\,\nu^{\text{turb}}+\mu_{\text{rad}})\overset{\circ}{\boldsymbol{D}} = -\boldsymbol{i}_r\boldsymbol{i}_r p^{\text{turb}} - \boldsymbol{i}_\varphi\boldsymbol{i}_\varphi p^{\text{turb}} - \boldsymbol{i}_z\boldsymbol{i}_z p^{\text{turb}}$$
$$+(\bar{\rho}\,\nu^{\text{turb}}+\mu_{\text{rad}})\left\{\boldsymbol{i}_r\boldsymbol{i}_\varphi r\frac{\partial\Omega(r,z)}{\partial r} + \boldsymbol{i}_\varphi\boldsymbol{i}_r r\frac{\partial\Omega(r,z)}{\partial r} + \boldsymbol{i}_\varphi\boldsymbol{i}_z r\frac{\partial\Omega(r,z)}{\partial z} + \boldsymbol{i}_z\boldsymbol{i}_\varphi r\frac{\partial\Omega(r,z)}{\partial z}\right\}, \qquad (7.179)$$

(see (7.106) and Appendix). In writing (7.175), we took into account the fact that the gravitational force is balanced by the centrifugal force in the radial direction (perpendicular to the rotation axis); i.e., the total gas-suspension pressure gradient $\partial(\bar{p}+p^{\text{turb}})/\partial r$ is very small and the rotation is almost Keplerian (however, this gradient eventually serves as the driving force of the radial drift of dust particles toward the disk center). On the other hand, since there is no net motion of the gas suspension in the vertical direction (perpendicular to the midplane of the disk), the momentum conservation along the $\boldsymbol{i}_z$ axis is reduced to the hydrostatic equilibrium condition under which the equilibrium in the z direction is maintained by the pressure gradient. Let us notice that the following expression for the half-thickness of a turbulized disk can be derived by substituting the differentials in (7.173) with finite differences (i.e., by substituting $\Delta\,(\bar{p}+p^{\text{turb}})$ for $\approx \bar{p}+p^{\text{turb}}$, where $(\bar{p}+p^{\text{turb}})$ is the total pressure calculated at $z=0$, and substituting $\Delta\, z \approx h_{\text{disk}}$):

$$h_{\text{disk}} = \sqrt{((\bar{p}+p^{\text{turb}})/\bar{\rho})_{z=0}}/\Omega_{K,\text{mid}} \cong c_s\sqrt{1+\tfrac{2}{3}bc_s^{-2}}/\Omega_{K,\text{mid}} \qquad (7.180)$$

(cf. (7.156)). In this relation, $\Omega_{K,\mathrm{mid}} = \sqrt{G\mathcal{M}_\odot/r^3}$ is the Keplerian angular velocity; c_s is the speed of sound of the gas suspension (see (7.57)) in the midplane of the disk. This is the gradient $\partial(\bar{p} + p^{\mathrm{turb}})/\partial z$ that determines the main direction of the buoyant force in the gravitational field of the central mass which contributes, in particular, to the additional generation of turbulent energy through convective instability in the vertical direction (between the midplane and the surfaces of the disk). Thus, it follows from (7.175) and (7.176) that viscous dissipation does not affect the r and z components of the equation of motion for the entire disk, which break up into separate equations for the averaged radial and vertical motions.

In contrast, using the following transformations (in what follows, the total coefficient of shear viscosity is denoted by $\mu(r,z) \equiv \bar{\rho}\,\nu^{\mathrm{turb}}(r,z) + \mu_{\mathrm{rad}}(r,z)$ for short)

$$\left[\frac{\partial}{\partial \boldsymbol{r}} \cdot (\boldsymbol{R} + \bar{\Pi}_{\mathrm{rad}})\right]_\varphi \equiv \left\{\frac{1}{r}\frac{\partial}{\partial r}\left[r(\boldsymbol{R} + \bar{\Pi}_{\mathrm{rad}})_{r\varphi}\right] + \frac{\partial}{\partial z}\left[(\boldsymbol{R} + \bar{\Pi}_{\mathrm{rad}})_{z\varphi}\right] + \frac{(\boldsymbol{R} + \bar{\Pi}_{\mathrm{rad}})_{\varphi r}}{r}\right\}$$
$$= \frac{1}{r}\mathrm{div}\left(\mu(r,z)\, r^2 \frac{\partial}{\partial r}\Omega(r,z,t)\right), \tag{7.181}$$

$$-\frac{1}{c^2}\left\{\frac{\partial}{\partial \boldsymbol{r}} \cdot \left[r\Omega(r,z,t)\boldsymbol{q}_{\mathrm{rad}}\right]\right\}_\varphi = -\frac{1}{c^2 r}\frac{\partial}{\partial \boldsymbol{r}} \cdot \left[r^2\Omega(r,z,t)\boldsymbol{q}_{\mathrm{rad}}\right], \tag{7.182}$$

the φ component of the equation of motion (7.177) (in essence, the equation for determining the angular velocity $\Omega(r,z)$ for given boundary conditions) can be reduced (see Appendix) to

$$\bar{\rho}\frac{\partial J(r,s,t)}{\partial t} = \mathrm{div}\left(\mu(r,z)r^2\frac{\partial}{\partial \boldsymbol{r}}\Omega(r,z,t)\right) - c^{-2}\mathrm{div}(r^2\Omega(r,z,t)\boldsymbol{q}_{\mathrm{rad}}) \cong 0 \tag{7.183}$$

Here $\mathrm{div}\boldsymbol{A} \equiv r^{-1}\partial(rA_r)/\partial r + \partial A_z/\partial z$ is the divergence in cylindrical coordinates, describes the irreversible change in specific angular momentum $J(r,z,t) \equiv [r^2\Omega(r,z,t)]$ through viscous friction and, in addition, allows for the transfer of angular momentum by the total radiative flux $\boldsymbol{q}_{\mathrm{rad}}$ (the second term on the right-hand side of (7.179) allows for the loss of angular momentum via the radiation emitted by rotating disk regions). The deceleration by radiation is known (see, e.g., Tassoul 1978) to be stronger than the angular velocity diffusion through viscosity if $\langle T\rangle/|\partial\langle T\rangle/\partial \boldsymbol{r}|$ is small compared to $\langle\Omega\rangle/|\partial\langle\Omega\rangle/\partial \boldsymbol{r}|$, which is probably the case only in the disk regions adjacent to its emitting surface.

Below, however, we will restrict our analysis to steady or quasi-steady motions inside the disk, where, to terms of the order of $|\langle \boldsymbol{u}\rangle|/c$, (7.183) can be reduced to

$$\mathrm{div}\left[\mu(r,z)r^2\frac{\partial}{\partial \boldsymbol{r}}\Omega(r,z)\right] = 0. \tag{7.184}$$

Since the shear stress vector must become zero at the outer disk boundary $\sum$, the following condition must be satisfied

$$\boldsymbol{n}\cdot\frac{\partial}{\partial \boldsymbol{r}}\Omega(r,z)=0 \quad \text{on} \quad \Sigma, \tag{7.185}$$

where $\boldsymbol{n}$ is the outer normal to the disk surface.

In connection with (7.184) and (7.185) let us formulate now one curious result pertaining to the problem of a slowly evolving viscous disk. For this purpose, we will write the total dissipative function of the disk in the steady state as

$$D(\Omega)\equiv\int_{\wp}\Phi_{\langle\boldsymbol{u}\rangle}\,\mathrm{d}\boldsymbol{r}\equiv\int_{\wp}2\mu\left(\overset{\circ}{\boldsymbol{D}}:\overset{\circ}{\boldsymbol{D}}\right)\mathrm{d}\boldsymbol{r}$$

$$=\int_{\wp}4\mu(r,z)(D^2_{r\varphi}+D^2_{z\varphi})\,\mathrm{d}\boldsymbol{r}=\int_{\wp}\mu(r,z)r^2\left\{\left[\frac{\partial}{\partial r}\Omega(r,z)\right]^2+\left[\frac{\partial}{\partial z}\Omega(r,z)\right]^2\right\}\mathrm{d}\boldsymbol{r}, \tag{7.186}$$

where $\wp$ is the disk volume (see Appendix). If we now consider an arbitrary disk rotation law, $Q(r, z) + \delta Q(r, z, t)$ on which we impose only the constraint that the configuration surface Σ and volume $\wp$ are conserved, then the classical results (which, in essence, was first obtained by Helmholtz) is that each solution of (7.184) and (7.185) has the property that the total power (7.186) dissipated by friction is an absolute minimum compared to the power for any other motion that agrees with the boundary Σ and the volume $\wp$.

7.4.1.2 The Conservation of Energy

To model the internal thermal structure of the protoplanetary disk around the young Sun at the T Tauri stage, we must invoke the energy equation (7.150), in which the dissipation of turbulent energy is the main internal source of heating. If we disregard the chemical reactions and the evaporation and condensation of disk material, then this equation for quasi-steady axisymmetric motion takes the form

$$\frac{1}{r}\frac{\partial}{\partial r}\left[r\left(q^{\mathrm{turb}}_r+q_{\mathrm{rad},r}\right)\right]+\frac{\partial}{\partial z}\left(q^{\mathrm{turb}}_z+q_{\mathrm{rad},z}\right)$$
$$=\mu(r,z)r^2\left\{\left[\frac{\partial}{\partial r}\Omega(r,z)\right]^2+\left[\frac{\partial}{\partial z}\Omega(r,z)\right]^2\right\}+Q_{\odot},$$

or

$$\frac{\partial}{\partial z}\left(q_z^{\text{turb}} + q_{\text{rad},z}\right) = \mu(r,z)r^2\left\{\left[\frac{\partial}{\partial r}\Omega(r,z)\right]^2 + \left[\frac{\partial}{\partial z}\Omega(r,z)\right]^2\right\} + Q_\odot, \quad (7.187)$$

since a thin disc radiation is mostly directed in vertical rather than in radial direction. Note that possible contribution of an additional source of heating connected with the term $\bar{\Pi}_{\text{rel}}:\boldsymbol{D}$ is efficient only in the dusty subdisk (i.e. in the volume small compared to the whole disc) and therefore can be omitted in (7.187).

In (7.185) and (7.187) the term $Q_\odot$ designates possible local heat sources in the gas–dust cloud related, in particular, to the absorption of solar electromagnetic and corpuscular radiation by the gas–dust disk components and its subsequent transformation through various radiative processes, re-radiation, scattering, photochemical and chemical reactions, etc., are denoted by $Q_\odot$. The complexity and multiplicity of chemical and photochemical reactions in the protoplanetary disk medium is generally attributable to the presence of basic chemical elements of the Solar system that constituted the initial gas-mixture components and to the existence of ionization (dissociation) agents in the form of energetic photons and photoelectrons (photolysis products) (see, e.g., Willacy et al. 1998). Their absorption leads to the dissociation, ionization, and/or excitation of rotational and vibrational levels of the gas-mixture components; each of these reactions can proceed in both forward and inverse directions. In practical calculations, by no means all of the elementary processes responsible for the thermal balance of a disk medium can be adequately taken into account in the corresponding models. For this reason, when physically self-consistent problems of modeling the evolution of the chemical composition and hydrodynamics of a disk are formulated, one of the most important problems is to accurately take into account the contributions from the matter–radiation interaction in the structure of the energy equation to determine the so-called heating function of the material that allows for the fraction of the absorbed solar radiation transformed into heat (see, e.g., Marov and Kolesnichenko 1987). Estimating this function involve well-known difficulties and requires concretizing the chemical stage of the disk evolution.

The turbulent heat flux q_z^{turb} and the radiative energy flux $q_{\text{rad},z}$ emitted by the disk are defined by the (7.123) and (7.124)

$$\begin{cases} q_z^{\text{turb}}(r,z) = -\bar{\rho}\langle c_P\rangle \dfrac{\nu^{\text{turb}}(r,z)}{\boldsymbol{Sc}^{\text{turb}}}\left(\dfrac{\partial}{\partial z}\langle T\rangle + \dfrac{\Omega_{K,\text{mid}}^2(r)}{\langle c_P\rangle}z\right), \\ q_{\text{rad},z}(r,z) = -\chi_{\text{rad}}(r,z)\dfrac{\partial}{\partial z}\langle T\rangle, \end{cases} \quad (7.188)$$

where $\chi_{\text{rad}}(r,z) = 4ca\langle T\rangle^3/3\tilde{\kappa}\,\bar{\rho}$ is the coefficient of radiative heat conductivity for the disk medium; $\tilde{\kappa}$ is the total Rosseland mean opacity of the gas suspension (see (7.72)), which depends significantly on the presence and height distribution of dust particles in the pro-toplanetary cloud (see, e.g., Pollack et al. 1985); $\mu_{\text{rad}}(r,z) = 4$

$a\langle T\rangle^4/15\,c\bar{\kappa}\bar{\rho}$ is the coefficient of radiative viscosity. Equation (7.188) should be complemented by the following boundary conditions in the midplane (in view of the disk symmetry) and on the upper surface of the disk:

$$q_z^{\rm turb}\big|_{z=0} = \frac{\partial}{\partial z}\langle T\rangle = 0, \qquad (*)$$

$$\boldsymbol{n}\cdot\left(\boldsymbol{q}^{\rm turb}+\boldsymbol{q}_{\rm rad}\right) = \sigma\langle T\rangle^4 - f_{\rm o}\frac{L_\odot}{4\pi r^2}, \quad \text{on} \quad \Sigma \quad (**) \qquad (7.189)$$

(where σ is the Stefan–Boltzmann constant; $L_\odot$ is the solar luminosity), which allow for the heat balance at the boundaries. The first term on the right-hand side of (**) allows for the blackbody radiation of the disk surface, while the second term describes the attenuated radiative flux from the protostar incident on the disk surface; the attenuation factor (Kusaka et al. 1970)

$$f_{\rm o} = \left[\frac{r}{2}\frac{\partial}{\partial r}\left(\frac{h_{\rm disk}(r,z)}{r}+\frac{2R_\odot^{\rm proto}}{3\pi r}\right)\right]\frac{L_\odot^{\rm proto}}{L_\odot} \qquad (7.190)$$

depends on the disk geometry and the radius $R_\odot^{\rm proto}$ and luminosity $L_\odot^{\rm proto}$ of the (r) proto-Sun (in particular, for $L_\odot^{\rm proto} = 7L_\odot$ and $R_\odot^{\rm proto} = 5R_\odot$ (see Watanabe et al. 1990), $f_{\rm o} = 0.1$ at $r = 1$ AU).

7.4.1.3 The Dust Transfer Equation

The averaged transfer equation (7.91) for the dust concentration $\langle C_{\rm d}\rangle = \rho_{\rm d}\bar{s}/\bar{\rho}$ of the disk material should be used to model the evolution of a turbulized gas–dust cloud, particularly during the formation of a dust subdisk (of thickness $2h_{\rm subdisk}$, where $h_{\rm subdisk}$ is the upper boundary of the dust subdisk, $h_{\rm disk} > h_{\rm subdisk}$) near its midplane. If we disregard the evaporation and condensation ($\bar{\sigma}_{\rm dg} = 0$) of solid particles, then there is a balance between the dust settling $\bar{\boldsymbol{J}}_{\rm d} \equiv \overline{\rho_{\rm d}\boldsymbol{w}_{\rm d}} \cong \bar{\rho}_{\rm g}\langle C_{\rm d}\rangle\bar{\boldsymbol{w}}$ and the turbulent mixing $\boldsymbol{J}_{\rm d}^{\rm turb} = -\bar{\rho}D_{\rm d}^{\rm turb}\partial\langle C_{\rm d}\rangle/\partial\boldsymbol{r}$ in the steady state (see (7.92)); (7.87*) then takes the form

$$\operatorname{div}\left(\bar{\rho}_{\rm g}\langle C_{\rm d}\rangle\bar{\boldsymbol{w}} - \frac{\bar{\rho}\nu^{\rm turb}}{\boldsymbol{Sc}^{\rm turb}}\frac{\partial}{\partial\boldsymbol{r}}\langle C_{\rm d}\rangle\right) = 0, \qquad (7.191)$$

or

$$\frac{1}{r}\frac{\partial}{\partial r}\left[r\left(\bar{s}\langle C_{\rm g}\rangle\bar{w}_r - \frac{\bar{\rho}\nu^{\rm turb}}{\boldsymbol{Sc}^{\rm turb}}\frac{\partial}{\partial r}\left(\frac{\bar{s}}{\bar{\rho}}\right)\right)\right] + \frac{\partial}{\partial z}\left(\bar{s}\langle C_{\rm g}\rangle\bar{w}_z - \frac{\bar{\rho}\nu^{\rm turb}}{\boldsymbol{Sc}^{\rm turb}}\frac{\partial}{\partial z}\left(\frac{\bar{s}}{\bar{\rho}}\right)\right) = 0, \qquad (7.192)$$

where $\bar{\boldsymbol{w}} \equiv (\bar{\boldsymbol{u}}_{\mathrm{d}} - \bar{\boldsymbol{u}}_{\mathrm{g}})$ is the averaged relative gas–dust velocity defined by (7.97) which in cylindrical coordinates takes the form:

$$\begin{aligned}\bar{\rho}\theta_{\mathrm{gd}}\bar{w}_r(r,z) &\cong \bar{w}_\varphi(r,z)\Omega(r,z) + \frac{1}{\bar{\rho}_{\mathrm{g}}}\frac{\partial\bar{p}}{\partial r}\\ &\cong \bar{w}_\varphi(r,z)\Omega_{K,\mathrm{mid}}(r) - \frac{\bar{\rho}}{\bar{\rho}_{\mathrm{g}}} r\eta\Omega^2_{K,\mathrm{mid}}(r),\end{aligned} \tag{7.193}$$

$$\bar{\rho}\theta_{\mathrm{gd}}\bar{w}_\varphi(r,z) = -\frac{1}{2}\bar{w}_r(r,z)\Omega(r,z) \cong -\frac{1}{2}\bar{w}_r(r,z)\Omega_{K,\mathrm{mid}}(r), \tag{7.194}$$

$$\bar{\rho}\theta_{\mathrm{gd}}\bar{w}_z(r,z) = \frac{1}{\bar{\rho}_{\mathrm{g}}}\frac{\partial\bar{p}}{\partial z} \cong -\frac{\bar{\rho}}{\bar{\rho}_{\mathrm{g}}}\Omega^2_{K,\mathrm{mid}}(r)\, z. \tag{7.195}$$

Then the r, φ, z components of the averaged relative velocity $\bar{\boldsymbol{w}}$ derived using the equations of motion (7.172) and (7.173) appear as

$$\begin{cases} \bar{w}_r(r,z) \cong -\dfrac{\bar{\rho}}{\bar{\rho}_{\mathrm{g}}}\eta\dfrac{\zeta}{1+\zeta^2} r\Omega_{K,\mathrm{mid}}(r), \\ \bar{w}_\varphi(r,z) \cong \dfrac{\bar{\rho}}{2\bar{\rho}_{\mathrm{g}}}\eta\dfrac{\zeta^2}{1+\zeta^2} r\Omega_{K,\mathrm{mid}}, \\ \bar{\mathrm{w}}_{\mathrm{z}}(r,z) \cong -\zeta\, z\Omega_{K,\mathrm{mid}}(r), \end{cases} \tag{7.196}$$

where the parameter $\zeta = \Omega_{K,\mathrm{mid}}/\bar{\rho}\theta_{\mathrm{gd}} \ll 1$, because the time it takes for quasi-equilibrium motion of the dust and gas to be established $(1/\bar{\rho}\theta_{\mathrm{gd}})$ in the disk is much shorter than the Keplerian period $(2\pi/\Omega_{K,\mathrm{mid}})$, which determines the slow variation time scales of the macroscopic flow parameters. Here, we used the following approximate equality that follows from (7.175):

$$\Omega(r,z) = \Omega_{K,\mathrm{mid}}(r)[1-\eta]^{\,1/2} \cong \Omega_{K,\mathrm{mid}}(r), \tag{7.197}$$

In this equality, the small parameter is defined as

$$\begin{aligned}\eta &\equiv -\left(r\Omega^2_{K,\mathrm{mid}}\bar{\rho}\right)^{-1}\partial\bar{p}/\partial r\\ &= -\bar{\gamma}\left(\frac{\mathcal{H}}{r}\right)^2\left(f + q + \frac{q+3}{2}\frac{z^2}{\mathcal{H}^2}\right) = 3.62\times 10^{-3} r^{1/2}_{\mathrm{a.e.}}\end{aligned} \tag{7.198}$$

where the second estimation representation was obtained using (7.210) (cf. Nakagawa et al. 1986; Takeuchi and Lin 2002).

The diffusion equation (7.191) can be simplified depending on whether the gas or dust component dominates in the disk region under consideration.

7.4.2 Turbulent Viscosity Coefficient in a Gas–Dust Disk

The coefficient of turbulent viscosity in (7.179), (7.183), (7.188), and (7.188) is defined by relation (7.171). In the axisymmetric case under consideration, the latter takes the form

$$\nu^{\mathrm{turb}}(r,z) = \alpha l^{*2}\, r \sqrt{\left\{ \left[\frac{\partial}{\partial r}\Omega(r,z)\right]^2 + \left[\frac{\partial}{\partial z}\Omega(r,z)\right]^2 \right\}} \qquad (7.199)$$

where

$$l^*(z) \equiv l(z)\left[1 - (\boldsymbol{Ri} + \boldsymbol{K})/\boldsymbol{Sc}^{\mathrm{turb}}\right]^{0.25}, \qquad (7.200)$$

$$\boldsymbol{Ri} \equiv \frac{\Omega^2_{K,\mathrm{mid}}\, z}{r^2}\,\frac{1}{\langle T\rangle}\,\frac{\frac{\partial\langle T\rangle}{\partial z} + G_a}{\left[(\partial/\partial r)\Omega(r,z)\right]^2 + \left[(\partial/\partial z)\Omega(r,z)\right]^2} \qquad (7.201)$$

$$\boldsymbol{K} \equiv -\langle\sigma\rangle\frac{\Omega^2_{K,\mathrm{mid}} z}{r^2}\,\frac{\frac{\partial\bar{s}}{\partial z} - \bar{s}\frac{\partial}{\partial z}\ln\bar{\rho}_{\mathrm{g}}}{\left[\frac{\partial}{\partial r}\Omega(r,z)\right]^2 + \left[\frac{\partial}{\partial z}\Omega(r,z)\right]^2} \qquad (7.202)$$

$$G_{\mathrm{a}} \equiv \frac{g_z}{\langle c_{\mathrm{P}}\rangle} = -\frac{1}{\langle c_{\mathrm{P}}\rangle}\frac{G\mathcal{M}_{\odot} z}{r^3}\left(1 + \frac{z^2}{r^2}\right)^{-3/2} \cong \frac{1-\bar{\gamma}}{\bar{\gamma}}\,\frac{1}{\langle\Re\rangle}\,\Omega^2_{K,\mathrm{mid}} z \qquad (7.203)$$

is the adiabatic temperature gradient in the protoplanetary gas–dust disk; $\bar{\gamma} = \langle c_{\mathrm{P}}\rangle / (\langle c_{\mathrm{P}}\rangle - \langle\Re\rangle)$ and $\langle\Re\rangle = \Re_{\mathrm{g}}\bar{\rho}_{\mathrm{g}}/\bar{\rho}$ are, respectively, the adiabatic index and the "gas constant" for the averaged two-phase continuum. We see from (7.199)–(7.201) that for an adiabatic temperature distribution in height, the *Richardson* number is $\boldsymbol{Ri} = 0$ and the temperature gradient in the disk has no effect on the turbulent transfer coefficients. In the case of temperature-unstable stratification ($\boldsymbol{Ri} < 0$) of the gas–dust disk, where the temperature gradients are superadiabatic,

$$-\frac{\partial\langle T\rangle}{\partial z} = \left(-\frac{\partial\langle T\rangle}{\partial z}\right)_{\mathrm{ad}} = \frac{\Omega^2_{K,\mathrm{mid}}(r)}{\langle c_{\mathrm{P}}\rangle} z \qquad (7.204)$$

$$\frac{\partial\langle T\rangle}{\partial z} + \frac{\Omega^2_{K,\mathrm{mid}}(r)}{\langle c_{\mathrm{P}}\rangle} z = f\,\frac{\partial\langle T\rangle}{\partial z} \qquad (7.205)$$

(the factor f, which characterizes the excess of the vertical temperature gradient in the disk above the adiabatic one, can reach $f = 0.2$ at $r = 10$ a.u. (see Makalkin and Dorofeeva 1995, 1996)). The turbulence energy increases through the energy of instability in the direction perpendicular to the midplane of the disk (a convective source of turbulence) and the coefficient of turbulent viscosity increases simultaneously. At the same time, gas-suspension inhomogeneity always causes turbulent energy to decrease, since the Kolmogorov number is greater than zero, $\boldsymbol{K} > 0$. The reciprocal Schmidt number $1/\boldsymbol{Sc}^{\text{turb}}$ in (7.200) may be set equal to unity when the shear stresses in the case of differential Keplerian disk rotation are the main turbulence mechanism; however, it can be a factor of 2 or 3 larger when thermal convection in the vertical direction is responsible for the turbulence (see, e.g., Shakura et al. 1978).

Let us now show that the vertical gradient in angular velocity $\partial\Omega(r,z)/\partial z$ in the dissipative function $\Phi_{\langle u\rangle} = 2\bar{\rho}\nu^{\text{turb}}\overset{\circ}{\boldsymbol{D}}:\overset{\circ}{\boldsymbol{D}}$ of an axisymmetric disk may occasionally be disregarded compared to its radial gradient $\partial\Omega(r,z)/\partial r$ for the following reasons. If we assume (for our estimation) that the disk is isothermal in the vertical direction and neglect the terms of order $(z/r)^2$ or higher, then the following formula (known for a laminar flow) can be derived for the vertical gas-suspension density distribution in a turbulized disk from (7.173):

where

$$\bar{\rho}(r,z) = \bar{\rho}(r,0)\exp\left\{-\frac{\Omega^2_{K,\text{mid}}(r)}{2(\langle\Re\rangle(\langle T\rangle + T^{\text{turb}})}z^2\right\}$$

$$= \bar{\rho}(r,0)\exp\left\{-\frac{z^2}{2\mathcal{H}^2}\right\} \tag{7.206}$$

$$\mathcal{H} = \frac{\sqrt{\langle\Re\rangle(\langle T\rangle + T^{\text{turb}})}}{\Omega_{K,\text{mid}}(r)} = \frac{\tilde{c}_s}{\bar{\gamma}^{1/2}\Omega_{K,\text{mid}}(r)} \tag{7.207}$$

is the local scale height for the disk,

$$\tilde{c}_s \equiv \sqrt{(\partial(\bar{p} + p^{\text{turb}})/\partial\bar{\rho})_{\langle S\rangle}} = \sqrt{\bar{\gamma}\langle\Re\rangle(\langle T\rangle + T^{\text{turb}})} \tag{7.208}$$

is the isothermal speed of sound in the turbulized medium. The density, pressure, temperature, opacity, etc., in any accretion disk have different spatial distributions, depending on its nature and the distance from protostar and/or from the midplane of the disk. We will assume that the radial distributions of such structure parameters follow a power law (this is a common assumption in astrophysical literature; see, e.g., Takeuchi and Lin 2002); then,

$$\begin{cases} \tilde{c}^2_s(r) = \tilde{c}^2_{s,\text{a.u.}}r^q_{\text{a.u.}}, \quad q = -0.5 \\ \bar{\rho}(r,z) = \bar{\rho}_{\text{a.u.}}r^f_{\text{a.u.}}\exp\{-z^2/2\mathcal{H}^2\}, \quad \bar{\rho}_{\text{a.u.}} = 2.83\times10^{-10}\text{g/cm}, \; f = -2.25 \\ \mathcal{H}(r) = \mathcal{H}_{\text{a.u.}}r^{(q+3)/2}_{\text{a.u.}}, \quad \mathcal{H}_{\text{a.u.}} = 3.33\times10^{-2}\text{a.u.} \end{cases} \tag{7.209}$$

where $r_{\mathrm{a.u.}}$ is the radius measured in AU. Using (7.206), (7.175) can be rewritten as

$$\begin{aligned}\Omega^2(r,z) &= \Omega^2_{K,\mathrm{mid}}\left[1+\frac{1}{\bar{\rho} r\Omega^2_{K,\mathrm{mid}}}\frac{\partial}{\partial r}\left(\bar{p}+p^{\mathrm{turb}}\right)\right]\\ &= \Omega^2_{K,\mathrm{mid}}\left[1+\frac{\tilde{c}_s^2}{\bar{\rho} r\Omega^2_{K,\mathrm{mid}}}\frac{\partial\bar{\rho}}{\partial r}\right] \cong \Omega^2_{K,\mathrm{mid}}\left[1+\frac{\bar{\gamma}\mathcal{H}^2}{r^2}\left(f+q+\frac{q+3}{2}\frac{z^2}{\mathcal{H}^2}\right)\right],\end{aligned}$$

whence

$$\Omega(r,z) = \Omega_{K,\mathrm{mid}}\left[1+\frac{\bar{\gamma}}{2}\left(\frac{\mathcal{H}}{r}\right)^2\left(f+q+\frac{q+3}{2}\frac{z^2}{\mathcal{H}^2}\right)\right]. \tag{7.210}$$

It follows from (7.210) that

$$\frac{\partial}{\partial z}\Omega(r,z) \sim \frac{\mathcal{H}}{r}\frac{\partial}{\partial r}\Omega(r,z), \quad (\mathcal{H}/r \ll 1) \tag{7.211}$$

which allows the vertical angular velocity gradient in the dissipative function $\Phi_{\langle \boldsymbol{u}\rangle}$ to be disregarded.

Thus, the following approximate relation for the coefficient of turbulent viscosity is valid for the bulk of the disk (except the regions close to the proto-Sun):

$$\nu^{\mathrm{turb}}(r,z) = \alpha l^{*2}\, r\left|\frac{\partial}{\partial r}\Omega(r,z)\right|, \quad l^*(z) \equiv l(z)\left[1-(\boldsymbol{Ri}+\boldsymbol{K})/\boldsymbol{Sc}^{\mathrm{turb}}\right]^{0.25}. \tag{7.212}$$

For (7.212) to be formally identical to the Shakura–Sunyaev formula (7.156), which is valid for gas-phase accretion disks, one must set $\boldsymbol{K}=0$ and $\boldsymbol{Ri}=0$ in (7.212). If we now substitute the angular velocity of Keplerian rotation $\Omega_{K,\mathrm{mid}}(r) = (G\mathcal{M}_\odot)^{\frac{1}{2}} r^{-\frac{3}{2}}$ in (7.212) (note that $r\left|\partial\Omega_{K,\mathrm{mid}}(r)/\partial r\right| = -\frac{3}{2}\Omega_{K,\mathrm{mid}}$) and use $l = h_{\mathrm{disk}} = \sqrt{(\bar{p}/\bar{\rho})|_{z=0}}/\Omega_{K,\mathrm{mid}}$ as the turbulence scale length, then we will obtain

$$\nu^{\mathrm{turb}} = \frac{3}{2}(\bar{p}/\bar{\rho})\Big|_{z=0}\Big/\Omega_{K,\mathrm{mid}}.$$

Hence,

$$R_{r\varphi} = \bar{\rho}\nu^T r(\partial\Omega_{K,\mathrm{mid}}(r)/\partial r) = -\frac{9}{4}\alpha\bar{p}|_{z=0}, \tag{7.213}$$

which is identical to (7.156) (since the dimensionless parameter α cannot be accurately calculated and remains a free parameter in the disk structure equations, the factor 9/4 in (7.213) is of no fundamental importance).

So, (7.55), (7.172), (7.173), (7.184), (7.187), and (7.191) form a system of six equations with six unknown functions, $\bar{p}(r,z)$, $\bar{\rho}(r,z)$, $\langle T\rangle(r,z)$, $\bar{s}$, $\bar{\rho}_g$, and $\Omega(r,z)$. Thus, in principle, the structure of a gas–dust disk with the dust component is completely defined by these equations together with the boundary conditions and relations (7.199) and (7.21*) for the turbulent transfer coefficients ν^{turb} and the coefficient of resistance θ_{dg} for a smooth spherical particle. A complete solution of the formulated problem requires using numerical evaluation.

7.4.3 Limiting Saturation of a Rotating Gas–Dust Disk with Fine Dust Particles

As a simple example that illustrates the potentialities of the approach developed here, let us qualitatively consider the model problem of the height distribution of suspended fine dust particles in a steady gas–dust flow (for temperature-neutral disk stratification, $\boldsymbol{Ri}=0$) in a thin layer of "cosmic fluid" located near a dust-gas subdisk (a layer of enhanced dust concentration, but lower than the critical value at which gravitational instability arises). We will assume that the concentration of solid particles in the subdisk atmosphere and in the subdisk itself is so high that the inverse effect of the disperse phase on the turbulent flow dynamics should be taken into account to describe the gas-suspension motion. Below, for simplicity, we assume the disk material in the subdisk atmosphere to be isothermal and the gas phase to be incompressible. In addition, we will keep in mind that the sedimentation of solid particles takes place without their large radial migration; therefore, the dust diffusion flux in the vertical direction $\partial\overline{\bar{\rho}_{\text{d}}\text{w}_{\text{d}z}}/\partial z \gg \partial\overline{\bar{\rho}_{\text{d}}\text{w}_{\text{d}r}}/\partial r$. For steady motion of the dust component in the atmosphere, the regular gravitational settling of particles to the subdisk (the value $a\boldsymbol{Sc}^{\text{turb}}$, where $a=-w_z$ is the gravitational settling velocity, is assumed to be constant below) must then be balanced by their upward turbulent transfer, i.e., $\bar{J}_{\text{d}z}+J_{\text{d}z}^{\text{turb}}=\text{const}=0$; hence, using (7.92) and (7.191), we obtain for the relative velocity in the vertical direction $\bar{s}$

$$-a=\frac{\nu^{\text{turb}}}{\boldsymbol{Sc}^{\text{turb}}}\frac{\partial \ln(\bar{s}/\bar{\rho}_g)}{\partial z}\cong\frac{\nu^{\text{turb}}}{\boldsymbol{Sc}^{\text{turb}}}\frac{\partial \ln\bar{s}}{\partial z}. \tag{7.214}$$

Since the additional stresses $\bar{\Pi}_{\text{rel}}$ associated with the relative motion of the gas and coarse dust effectively act in the subdisk, the meridional φ component of the Reynolds equation (7.178) for the inner subdisk regions is reduced to the equation

$$\frac{1}{r}\frac{\partial}{\partial r}\left\{r\left[\left(\boldsymbol{R}+\bar{\Pi}_{\text{rel}}\right)_{r\varphi}\right]\right\}+\frac{\partial}{\partial z}\left[\left(\boldsymbol{R}+\bar{\Pi}_{\text{rel}}\right)_{z\varphi}\right]+\frac{1}{r}\left[\left(\boldsymbol{R}+\bar{\Pi}_{\text{rel}}\right)_{\varphi r}\right]\cong 0, \tag{7.215}$$

where, in view of relations (7.196), the corresponding components of the relative stress tensor $\bar{\Pi}_{\text{rel}}\cong-\bar{s}\rho_{\text{d}}\langle C_g\rangle\bar{\boldsymbol{w}}\,\bar{\boldsymbol{w}}$ take the form

$$(\Pi_{\text{rel}})_{z\varphi} = \bar{s}\rho_{\text{d}} \langle C_{\text{g}} \rangle a\bar{w}_{\varphi} = \bar{s}\rho_{\text{d}} \frac{\zeta^3 \eta}{2(1+\zeta^2)} zr\Omega^2_{K,\text{mid}}(r) \tag{7.216}$$

$$(\Pi_{\text{rel}})_{\varphi r} = -\bar{s}\rho_{\text{d}} \langle C_{\text{g}} \rangle \bar{w}_r \bar{w}_{\varphi} = \frac{\bar{s}\rho_{\text{d}}\bar{\rho}}{2\bar{\rho}_{\text{g}}} \frac{\zeta^3\eta^2}{(1+\zeta^2)^2} r^2\Omega^2_{K,\text{mid}}(r). \tag{7.217}$$

Since the main direction in the entire disk is radial (see Safronov and Guseinov 1990), we may set for the subdisk

$$\frac{\partial}{\partial z}\left[(\boldsymbol{R} + \bar{\Pi}_{\text{rel}})_{z\varphi}\right] \cong 0 \tag{7.218}$$

This equation shows that the flux density of the φ momentum component along the vertical axis will be the same at all distances from the $z = 0$ plane to the $z = h_{\text{subdisk}}$ "surface" of the subdisk: $(\boldsymbol{R} + \bar{\Pi}_{\text{rel}})_{z\varphi} = (\boldsymbol{R} + \bar{\Pi}_{\text{rel}})_{z\varphi}\big|_{z=0} = \text{const}$. Since near the midplane and at moderately large z, in view of (7.158) and (7.212), the Reynolds equation (7.215) for the subdisk atmosphere can be taken in the form

$$R_{z\varphi}\big|_{z\approx 0} \equiv \bar{\rho}(r,0)\nu^{\text{turb}}(r,0)r\left\{\frac{\partial\Omega(r,z)}{\partial z}\right\}\bigg|_{z\approx 0} = 0, \tag{7.219}$$

$$\nu^{\text{turb}}\frac{\partial}{\partial z}[r\partial\Omega(r,z)/\partial z] \equiv V_*^2 \tag{7.220}$$

where

$$V_*^2 = (\bar{\Pi}_{\text{rel}})_{z\varphi}\big|_{z=h_{\text{subdisk}}} \cong \frac{\rho_d\zeta^3\eta}{2\bar{\rho}(r,0)(1+\zeta^2)} r\Omega^2_{K,\text{mid}}(r)[z\bar{s}(r,z)]_{z=h_{\text{subdisk}}}, \tag{7.221}$$

V_* is the so-called dynamic velocity (a natural velocity scale for the flow near the subdisk "surface"). In (7.220), we disregarded the contribution from the relative stresses compared to the Reynolds turbulent stresses.

Eliminating the meridional velocity gradient $\partial[r\Omega(r,z)]/\partial z$ from the expression $b = \alpha^2 l^2(1-K_f)[r\partial\Omega(r,z)/\partial z]^2$ (see (7.169)) and using (7.220) and the relation $\nu^{\text{turb}} = l\sqrt{b}$, we obtain

$$b = \alpha V_*^2\sqrt{1-\boldsymbol{K}_f} \tag{7.222}$$

where

$$\boldsymbol{K}_f \equiv -\frac{\langle\sigma\rangle}{\boldsymbol{Sc}^{\text{turb}}}\frac{G\mathcal{M}_{\odot}z}{r^3}\frac{\partial\bar{s}/\partial z}{[r\partial\Omega(r,z)/\partial z]^2} = \frac{G\mathcal{M}_{\odot}z}{r^4}\frac{\langle\sigma\rangle\bar{s}\,a}{V_*^2\partial\Omega(r,z)/\partial z}. \tag{7.223}$$

is the dynamic Kolmogorov number. Thus, the z distribution of the disk structure parameters can be calculated using (7.214), (7.220), (7.222), and (7.223), which we will write as

$$\begin{cases} l\sqrt{b}\,\frac{\partial}{\partial z}\left[r\partial\Omega(r,z)/\partial z\right] = V_*^2, \quad \frac{1}{\boldsymbol{Sc}^{\mathrm{turb}}}\, l\sqrt{b}\,\frac{\partial \bar{s}}{\partial z} + a\bar{s} = 0, \quad l(z) = \frac{\gamma^*\kappa}{\alpha^{1/2}}\, z\, \Phi(K_f) \\ b = \alpha V_*^2\sqrt{1-K_f}, \quad K_f = \langle\sigma\rangle \frac{G\mathcal{M}_\odot z}{r^4}\, \frac{\vec{s}a}{V_*^2\partial\Omega(r,z)/\partial z}. \end{cases} \tag{7.224}$$

Here, the local turbulence scale length $l(z)$ is expressed (under the assumption of complete self-similarity in local, $\boldsymbol{Re} = u_0 z/\nu$, and global, $\boldsymbol{Re}_{\mathrm{glob}} = u_0 L/\nu$, Reynolds numbers) in terms of a universal function $\Phi(\boldsymbol{K}_f)$ of the Kolmogorov number (clearly, $\Phi(0) = 1$). Since the turbulence scale length under the effect of the gas-suspension dust component decreases (see (7.199)), the function $\Phi(\boldsymbol{K}_f)$ must also decrease with increasing argument.

Let us now determine the z distribution of the dust volume concentration and the angular velocity of a turbulized flow carrying fine suspended particles for a steady flow in the subdisk atmosphere. To integrate the system of equations (7.224) generally requires knowledge of the boundary conditions on the subdisk surface. For example, if we eliminate ν^{turb} from (7.214) and (7.220), then we will obtain

$$\frac{\partial \ln \bar{s}}{\partial z} = -\omega \frac{\gamma^*\kappa}{V_*}\frac{\partial}{\partial z}\left[r\partial\Omega(r,z)\right], \tag{7.225}$$

where

$$\omega \equiv \frac{\boldsymbol{Sc}^{\mathrm{turb}} a}{\gamma^*\kappa V_*} \tag{7.226}$$

is a dimensionless parameter. Integrating (7.225) yields

$$\bar{s}(r,z) = \bar{s}(r,z)|_{z=h_{\mathrm{subdisk}}} \exp\left\{-\frac{\omega\,\gamma^*\kappa}{V_*} r\left[\Omega(r,z) - \Omega_{K,\mathrm{mid}}(r,z)|_{z=h_{\mathrm{subdisk}}}\right]\right\}. \tag{7.227}$$

At the same time, system (7.224) has properties peculiar to this type of gas-suspension flow that allow us to find a self-similar solution independent of the boundary conditions. Note that patterns of turbulent flow with solid admixture was at the first time studied by Kolmogorov (1954) with application to evaluation of dust deposits formation and by Barenblatt and Golitsyn (1973) with application to atmospheric problems (in particular, dust storms formation in the Martian atmosphere). This is because it contains only the angular velocity gradient rather than the velocity itself. This, in turn, implies that for an unlimited store of particles near the dust subdisk, in view of the inverse effect of particles on the flow dynamics, one might expect the existence of a gas-suspension flow regime at which the turbulent flow absorbs the maximum possible amount of dust at a given dynamic velocity and other flow parameters (Barenblatt and Golitsyn 1974). This regime, which was called the "regime of limiting saturation," must be described by a special solution of system (7.224) that is defined by the parameters appearing only in the differential equations (7.224) themselves.

A group analysis of system (7.224) shows that it has the solution

$$\frac{\partial \Omega(r,z)}{\partial z} = C_1(r)/r\,z, \quad \bar{s} = \frac{C_2(r)}{z^2}, \tag{7.228}$$

where C_1 and C_2 are the z-independent functions of r to be determined. Substituting these expressions in (7.224) yields whence follows the functional equation

$$C_1 = \frac{V_*}{\gamma^* \kappa \Phi(\boldsymbol{K}_f)(1-\boldsymbol{K}_f)^{1/4}} = \frac{2V_*}{\gamma^* \kappa \omega} \tag{7.229}$$

$$\Phi(\boldsymbol{K}_f)(1-\boldsymbol{K}_f)^{1/4} = \frac{\omega}{2}, \tag{7.230}$$

intended to determine the specific (for the regime of motion under study) Kolmogorov number $\boldsymbol{K}_f$ (calculated at a given distance r from the protostar).

Since Φ is a nonincreasing function of its argument, $\Phi(0) = 1$, and since the number $\boldsymbol{K}_f$ by its physical meaning, lies between zero and unity, the functional equation at $\omega > 2$ (which corresponds to a low flow velocity or large particles) has no root; however, at $\omega < 2$ (the condition for the existence of the limiting saturation regime), one root exists, $\boldsymbol{K}_f = \boldsymbol{K}_f^*$. Given (7.229), it follows from the first relation (7.228) (at $\omega < 2$) that

$$\Omega(r,z) = \Omega_{K,\mathrm{mid}}(r) + \frac{V_*(r)}{\gamma^* \kappa \omega\, r} \ln z. \tag{7.231}$$

Using this relation and (7.226), we can easily obtain the following limiting steady-state height distribution of fine dust particles in the thin equatorial layer of the subdisk:

$$\bar{s} = \frac{C_2(r)}{z^2} = \frac{V_*^4 K^*}{\langle\sigma\rangle a^2 \Omega_{K,\mathrm{mid}}^2} \cdot \frac{1}{z^2}, \tag{7.232}$$

which the flow approaches for an unlimited store of dust on the underlying surface.

Formula (7.231) shows that the z velocity distribution in the "near-surface" subdisk atmosphere in a turbulent flow heavily loaded with particles is logarithmic (as must be the case in a turbulized fluid near a "wall"), but the presence of dust seemingly results in a decrease in the Karman coefficient κ. This can be interpreted in such a way that the gas-suspension flow under the effect of dust particles for the same external conditions (the same dynamic velocity V_*) is accelerated compared to a flow of "pure" gas. In other words, the velocity gradients near the subdisk surface increase, which contributes to the saltation effect, the separation and rise of a large number of fine dust particles into its atmosphere In turn, the presence of such a dust cloud with an enhanced concentration of suspended fine particles contributes to the intensification of all the possible processes of turbulent coagulation, which lead to an increase in the inertia of solid particles and to their effecting settling to the subdisk. Thus, the possible regime of limiting saturation of a rotating gas–dust cloud near the subdisk by small dust particles is an additional mechanism that speeds up the formation of the subdisk itself by relatively large solid particles on which turbulent pulsations have a weaker effect.

7.4.4 *Analytical Solution of the Coagulation Kinetics Equation by the Method of Moments*

Let us consider now the method of moments for solving the kinetic coagulation equation (7.33) for the case where the particle size distribution depends on one space coordinate z. For the problem considered here, this corresponds to steady motion in a dust layer as particles settle to the midplane of the disk under gravity.

The sense of the method of moments implies reducing the kinetic coagulation equation (7.33) to a system of ordinary differential equations for the moments

$$M_p = \int_0^\infty U^p f(U,z)\,\mathrm{d}U, \quad (p = 0,\,1,\,2,\,\ldots) \tag{7.233}$$

of the distribution function $f(U,z)$. To derive such a system, let us multiply both sides of the simplified equation (7.33)

$$w_z \frac{\partial f(U,z)}{\partial z} = \frac{1}{2}\int_0^U f(W,z) f(U-W,z) K(W,U-W)\,\mathrm{d}W - f(U,z)$$
$$\int_0^\infty f(W,z) K(W,U)\,\mathrm{d}W,$$

where w_z is a constant velocity of dust settling from height h_{disk} to the midplane of the disk, by U^{p} and integrate the result over U from 0 to ∞; as a result, we obtain

$$w_z \frac{\partial M_0}{\partial z} = w_z \frac{\partial N_{\mathrm{d}}}{\partial z} = -\frac{1}{2}\int_0^\infty\int_0^\infty K(U,W) f(U,z) f(W,z)\mathrm{d}W\mathrm{d}U \tag{7.234}$$

$$w_z \frac{\partial M_1}{\partial z} = w_z \frac{\partial s}{\partial z} = \frac{1}{2} \int_0^\infty \int_0^\infty (W - U) K(U, W) f(U, z) f(W, z) \mathrm{d}W \mathrm{d}U = 0, \qquad (7.235)$$

$$w_z \frac{\partial M_p}{\partial z} = \int_0^\infty \int_0^\infty K(U, W) \left[\frac{1}{2} (U + W)^p - U^p \right] f(U, z) f(W, z) \mathrm{d}W \mathrm{d}U$$
$$(p = 2, 3, \ldots) \qquad (7.236)$$

To express the right-hand sides of the equations of this system in terms of the moments, we must specify the coagulation kernel and make the assumption regarding the form of the distribution function. Let us consider kernels of the type

$$K(U, W) = \Lambda \sum_{j=0}^{K} \beta_j (U^{\alpha - \alpha_j} W^{\alpha_j} + U^{\alpha_j} W^{\alpha - \alpha_j}) \qquad (7.237)$$

where Λ is the factor determined by the external conditions under which coagulation takes place. The kernels for which the kinetic equation admits of exact solutions belong to this class of kernels. In addition, formulas of type (7.237) can approximate the kernels that correspond to many coagulation mechanisms in a gas–dust disk, in particular, those analyzed by Kolesnichenko (2001). Indeed, all of the kernels considered by this author are uniform functions for which we may write $K(U, W) = U^\alpha K(1, x)$, where α is the degree of uniformity, $x \equiv W/U$. It is well known that any such function can be approximated by a polynomial:

$$K(U, W) = \Lambda U^\alpha \sum_{j=0}^{K} \beta_j (x^{\alpha_j} + x^{\alpha - \alpha_j}); \qquad (7.238)$$

to determine the unknown coefficients β_j, we should choose s (interpolation) points x_i whose number is equal to the number of unknown coefficients β_j and assume that

$$K(1, x_i) = \Lambda \sum_{j=0}^{K} \beta_j (x_i^{\alpha_j} + x_i^{\alpha - \alpha_j}), \quad (i = 1, 2, \ldots, s) \qquad (7.239)$$

In order that no moments with an order higher than the degree of uniformity of the kernel appear after the subsequent integration of expansion (7.238), the quantity α_j should be defined by the equality $\alpha_j = \alpha j / (K + 1)$. Different interpolation polynomials can be obtained at different values of K.

Substituting (7.237) in (7.234)–(7.236) yields the following infinite system of equations for the moments:

$$w_z \frac{\partial M_0}{\partial z} = w_z \frac{\partial N_\mathrm{d}}{\partial z} = -\Lambda \sum_{j=0}^{K} \beta_j M_{\alpha - \alpha_j} M_{\alpha_j}, \qquad (7.240)$$

$$w_z \frac{\partial M_1}{\partial z} = w_z \frac{\partial s}{\partial z} = 0, \tag{7.241}$$

$$w_z \frac{\partial M_p}{\partial z} = \frac{\Lambda}{2} \sum_{j=0}^{K} \beta_j \sum_{k=1}^{p-1} C_{\mathrm{p}}^k (M_{\alpha-\alpha_j+p-k} M_{\alpha_j+k} + M_{\alpha-\alpha_j+k} M_{\alpha_j+p-k}),$$
$$(p = 2, 3, \ldots). \tag{7.242}$$

To solve this system, whose right-hand sides generally include fractional moments, we must complement it by the coupling equations between the fractional and integer moments. Here, different approaches are also possible: an approach related to the approximation (by Lagrangian polynomials) of fractional moments via integer moments (see, e.g., Loginov 1979) and a parametric method.

Below, we restrict our analysis to the parametric method, which is based on the fact that the sought-for distribution function $f(U, z)$ belongs to a certain parametric class of distributions. Let us assume, for simplicity, that the distribution $f(U, z)$ remains in the class of distributions to which the initial distribution belongs after coagulation and that only its statistical parameters, the mean, variance, etc., change (with height) as the particles settle to the midplane of the disk. By analogy with the atmospheric aerosol, we choose a two-parameter lognormal distribution as the initial dust particle size (diameter d) distribution.

The probability density of the lognormal law depends on the mean $\langle \ln d \rangle$ and variance of the logarithm of the diameter $\sigma_{\mathrm{L}}^2 \equiv \langle (\ln d - \langle \ln d \rangle)^2 \rangle$:

$$p(d; \mu^*, \sigma_L) = \frac{N_{\mathrm{d}}}{\sigma_{\mathrm{L}} d\sqrt{2\pi}} \exp\left\{ -\frac{(\ln d - \langle \ln d \rangle)^2}{2\sigma_{\mathrm{L}}^2} \right\}$$
$$= \frac{N_{\mathrm{d}}}{\sigma_{\mathrm{L}} d\sqrt{2\pi}} \exp\left\{ -\frac{\ln^2(d/\mu^*)}{2\sigma_{\mathrm{L}}^2} \right\}. \tag{7.243}$$

The median of the distribution is known to be defined by the relation $\mu^* = \exp(\langle \ln d \rangle)$, while the mean value of the diameter itself and its variance are, respectively,

$$\langle d \rangle = \exp\left(\frac{1}{2} \sigma_{\mathrm{L}}^2 + \ln \mu^* \right), \tag{7.244}$$

$$\sigma^2 \equiv \langle (d - \langle d \rangle)^2 \rangle = \langle d \rangle^2 \left[\exp \sigma_{\mathrm{L}}^2 - 1 \right]. \tag{7.245}$$

Using the above relations, we can derived formulas for the statistical parameters (σ_{L}^2 and μ^*) of the lognormal distribution (7.243) only via the mean particle diameter $\langle d \rangle$ and its relative variance $\beta^2 \equiv \langle (d - \langle d \rangle)^2 \rangle / \langle d \rangle^2$:

$$\sigma_{\mathrm{L}}^2 = \ln(1 + \beta^2), \quad \mu^* = \frac{\langle d \rangle}{\sqrt{1 + \beta^2}}. \tag{7.246}$$

To determine the density of the initial distribution of the dust particle volume $U = (\pi/6)d^3$, we use the formula

$$f(U) = p[d(U)]\ |\mathrm{d}d/\mathrm{d}U| \tag{7.247}$$

(which is valid for a strictly increasing function $U = U(d)$ of the random variable d (Hahn and Shapiro 1967)) and distribution (7.243); as a result, we obtain

$$f(U;\sigma_{\mathrm{L}},\mu) = \frac{N_{\mathrm{d}}}{3\sqrt{2\pi}\sigma_{\mathrm{L}}U}\exp\left[-\frac{\ln^2(U/\mu)}{18\sigma_{\mathrm{L}}^2}\right], \tag{7.248}$$

where $\mu = (\pi/6)\mu^{*3}$.

Let the coagulation in the disk do not change this distribution and only the parameters $\mu(z)$ and $\sigma_{\mathrm{L}}^2(z)$ change with height. We introduce the moments of the lognormal distribution

$$M_p(z) = \frac{N_{\mathrm{d}}}{3\sqrt{2\pi}\sigma_{\mathrm{L}}(z)}\int\limits_0^\infty U^{p-1}\exp\left[-\frac{\ln^2[U/\mu(z)]}{18\,\sigma_{\mathrm{L}}^2(z)}\right]\mathrm{d}U. \tag{7.249}$$

According to Lee (1983), the following representation is valid for any moment of order p:

$$M_p = M_1\,\mu^{p-1}\exp\left[\frac{3}{2}(p^2-1)\sigma_{\mathrm{L}}^2\right], \tag{7.250}$$

$$M_1 = s = \mathrm{const}, \tag{7.251}$$

which allows the fractional moments in (7.240) – (7.242) to be expressed in terms of $M_1,\ \mu,\ \sigma_{\mathrm{L}}^2$. As a result, we obtain the following parametric system of two ordinary differential equations (the number of equations must be equal to the number of unknown coefficients) to determine the parameters $\mu(z),\ \sigma_{\mathrm{L}}^2(z)$ from given boundary values of $\mu(h_{\mathrm{disk}}),\ \sigma_{\mathrm{L}}^2(h_{\mathrm{disk}})$:

$$\begin{aligned} w_z\frac{\partial M_0}{\partial z} &= -\Lambda\sum_{j=0}^{K}\beta_j M_{\alpha-\alpha_j}M_{\alpha_j} \\ &= -\Lambda M_1^2\mu^{\alpha-2}\sum_{j=0}^{K}\beta_j\exp\left[\frac{3}{2}[\alpha_j^2+(\alpha-\alpha_j)^2-2]\sigma_{\mathrm{L}}^2\right] \\ &= -\Lambda\mu^{\alpha}M_0^2\sum_{j=0}^{K}\beta_j\exp\left[\frac{3}{2}[\alpha_j^2+(\alpha-\alpha_j)^2]\sigma_{\mathrm{L}}^2\right], \end{aligned} \tag{7.252}$$

$$
\begin{aligned}
w_z \frac{\partial M_2(z)}{\partial z} &= 2\Lambda \sum_{j=0}^{K} \beta_j M_{\alpha-\alpha_j+1} M_{\alpha_j+1} \\
&= 2\Lambda M_1^2 \mu^\alpha \sum_{j=0}^{K} \beta_j \exp\left[\frac{3}{2}[(\alpha_j+1)^2 + (\alpha-\alpha_j+1)^2 - 2]\sigma_{\mathrm{L}}^2\right] \\
&= 2\Lambda M_1^2 \mu^\alpha \sum_{j=0}^{K} \beta_j \exp\left[\frac{3}{2}[(\alpha_j+1)^2 + (\alpha-\alpha_j+1)^2 - 2]\sigma_{\mathrm{L}}^2\right] \\
&= 2\Lambda \mu^{\alpha+2} \sum_{j=0}^{K} \beta_j \exp\left[\frac{3}{2}[\alpha_j^2 + (\alpha-\alpha_j)^2]\sigma_{\mathrm{L}}^2\right].
\end{aligned}
\tag{7.253}
$$

$$
N_{\mathrm{d}} \equiv M_0 = s\mu^{-1} \exp\left(-\frac{3}{2}\sigma_{\mathrm{L}}^2\right); \quad M_2 = s\mu \exp\left(\frac{9}{2}\sigma_{\mathrm{L}}^2\right);
\tag{7.254}
$$

This parametric system of nonlinear equations can be solved numerically, which allows us to model dust settling to the midplane of the disk. Let us note that the change in the mean number of particles with height $N_{\mathrm{d}}(z)$ can be also estimated by assuming that the variance σ_{L}^2 remains constant. In this case, restricting ourselves to the first two moments, we find from (7.253)

$$
w_z \frac{\partial N_{\mathrm{d}}}{\partial z} = -\Lambda \mu^\alpha N_{\mathrm{d}}^2 \sum_{j=0}^{K} \beta_j \exp\left[\frac{3}{2}[\alpha_j^2 + (\alpha-\alpha_j)^2]\sigma_{\mathrm{L}}^2\right].
\tag{7.255}
$$

The solution of this equation obtained using the boundary condition $N_{\mathrm{d}}(h_{\mathrm{disk}}) \equiv N_{\mathrm{d},h_{\mathrm{disk}}} = s/\tilde{U}(h_{\mathrm{disk}})$ is

$$
N_{\mathrm{d}}(z) = \frac{s}{\tilde{U}(h_{\mathrm{disk}})} \frac{1}{1 + q[(h_{\mathrm{disk}} - z)/w_z]}
\tag{7.256}
$$

where

$$
q = \Lambda \mu^\alpha \frac{s}{\tilde{U}(h_{\mathrm{disk}})} \sum_{j=0}^{K} \beta_j \exp\left[\frac{3}{2}[\alpha_j^2 + (\alpha-\alpha_j)^2]\sigma_L^2\right].
\tag{7.257}
$$

Here, $\tilde{U}(h_{\mathrm{disk}}) = (\pi/6) < d >^3 = \mu(h_{\mathrm{disk}}) \exp(\frac{3}{2}\sigma_{\mathrm{L}}^2)$ is the upper limit of the mean volume (this formula follows from (7.246)). Thus, using the relation $\tilde{U}(z) = s/N_{\mathrm{d}}$, we can determine the change in the mean particle volume with height. For relatively low z (i.e., when the particles are already near the midplane of the disk), $q[(h_{\mathrm{disk}} - z)/w_z] \gg 1$, it follows from (7.257)

$$N_{\mathrm{d}}(z) = \left\{ \Lambda \mu^{\alpha}[(h_{\mathrm{disk}} - z)/w_z] \sum_{j=0}^{K} \beta_j \exp\left[\frac{3}{2}[\alpha_j^2 + (\alpha - \alpha_j)^2]\sigma_{\mathrm{L}}^2\right] \right\}^{-1} \tag{7.258}$$

We see from this expression that the mean number of particles in the system for a fairly long coagulation time ceases to depend on their initial distribution, i.e., as it were, "forgets its past," and can be described by a universal function whose form is determined only by the coagulation kernel. A similar analysis can also be performed with other possible dust particle volume distributions in a coagulating turbulent flow, for example, with the gamma distribution.

Summarizing the results of this section let us underline again that adequate models of the solar and other planetary systems origin and evolution require the comprehensive study of the dynamical and thermal evolution of the heterogeneous gas–dust matter of a differentially rotating protoplanetary disk with the inclusion of magnetohydrodynamic, turbulent, and radiative effects as well as with the involvement of phase transitions, chemical reactions, and coagulation processes. The aggregate state of the main components of the protoplanetary material, the location of their condensation-sublimation fronts, and, hence, the chemical composition of the planets, their satellites, asteroids, and comets depend on the space–time distribution of thermohydrodynamic parameters for the disk medium. The cosmochemical data to be obtained by directly studying the extraterrestrial material should serve as an important constraints in reconstruction of the key processes we address when dealing with the modeling.

Refinement of mathematical approach to developing numerical models for such a dynamical system in which the evolution of an initial protoplanetary cloud sequentially leads to the formation of an accretion gas–dust disk around a young solar type star following by the compacted dust–gas subdisk, as well as progressively growing bodies to form eventually planetesimals and ultimately planets, are of the primary importance in the study. Certain aspects of this extremely complicated problem were embodied in this section involving the efficient methods of invariant modeling of turbulent flows in multicomponent reactive gas mixtures. Such an approach earlier developed by the authors (Kolesnichenko and Marov 1999; Marov and Kolesnichenko 2002) was further generalized to heterogeneous media. This study offers prospects for considerably more complete and more realistic modeling of the various processes of the evolution of a differentially rotating protoplanetary gas–dust turbulized disk and protoplanetary system setup.

Chapter 8
Influence of Hydrodynamic Helicity on the Evolution of Turbulence in an Accretion Disk

In this chapter we analyze the influence of hydrodynamic spirality that develops in a rotating disk on the synergetic structurization of disc matter. It is shown that comparatively slow damping of turbulence in the disk is caused by the lack of reflective symmetry of the anisotropic field of turbulent velocities relative its equatorial plane. The general concept of the development of energy-intensive coherent mesoscale eddy structures in the thermodynamically open system of turbulent chaos associated with the realization of inverse cascade of kinetic energy in mirror-nonsymmetrical disk turbulence is formulated. Because of energy release, the inverse cascade produces a hierarchical system of mass concentrations with a fractal density distribution, which ultimately initiate the mechanisms of triggered cluster formation. In turn, eddy coherent formations are responsible for intensification of mechanical and physical-chemical interactions of particles in the flow, which results in spontaneous setup and growth of dust clusters. Condensation, phase transition, and the processes of mass and heat transfer between different regions of the heterogeneous medium as well as significant modification of oscillation patterns (density wave spectrum) are encouraged. It is also shown that negative viscosity in a rotating disk system appears to be a manifestation of cascade processes in spiral turbulence where inverse energy transfer from small to larger vortices occurs.

8.1 Theoretical Prerequisites to Modeling Hydrodynamic Helicity

Here we have to emphasize again that turbulence is commonly regarded as the key feature of circumstellar discs origin and evolution (Zel'dovich 1981; Dubrulle 1993; Balbus and Hawley 1998; Richard and Zahn 1999). In particular, for the solar

M.Y. Marov and A.V. Kolesnichenko, *Turbulence and Self-Organization; Modeling Astrophysical Objects*, Astrophysics and Space Science Library 389,
DOI 10.1007/978-1-4614-5155-6_8, © Springer Science+Business Media New York 2013

protoplanetary disc of radius R rotating with angular velocity Ω Reynolds number is $\boldsymbol{Re}_{\text{glob}} \equiv \Omega R^2/\nu > 10^{10}$ (hereafter kinematic viscosity ν is assumed to be constant).

According to modern concepts, the most likely causes of turbulence generation in astrophysical disks are a large-scale shear flow of differentially rotating cosmic matter (see, e.g., Gor'kavyi and Fridman 1994; Fridman et al. 2003) and chaotic magnetic fields (Armitage et al. 2001), whose energy is often comparable to the energy of hydrodynamic turbulence. Indeed, accretion disks have substantial turbulent viscosity, which, combined with differential rotation of matter, provides a permanent internal source of thermal energy. As far as chaotic magnetic fields is concerned, they are dragged along with the accreting plasma, mixed by the differential rotation of the disk, and undergo reconnection at the interfaces between chaotic cells and must also appreciably contribute to the viscosity in the inner region of the disk and in the outer layers of its atmosphere, where the matter reaches sufficiently high degree of ionization. Large-scale magnetic fields (Eardley and Lightman 1975) may also play an important part in the physics of accretion.

Basically, synergetic processes of self-organization (which ultimately result in the disk structurization) in a thermodynamically open subsystem of turbulent chaos against the background of averaged large-scale shear flow of cosmic matter associated with its differential rotation form the properties of the protoplanetary disc at various stages of its evolution including viscous accretion disc around the young Sun at the T Tauri stage and more dense dust subdisk setup, followed by its fragmentation due to gravity instability. These processes are also related to the formation of mesoscale coherent eddy structures which in turn, would ensure favorable conditions for both mechanical and physical–chemical interactions between disc gas and dust particles (see, e.g., Barge and Sommeria 1995; Tanga et al. 1996; Chavanis 1999; Kolesnichenko 2005a, b). Intrinsically, intensification of phase transitions and heat-and-mass transfer between different parts of disc matter occurs resulting in spontaneous origin and growth of condensed gas–dust clusters, wave spectrum modification, etc. One may assume the following scenario of the progressive growth of dust particles in the disc plasma: first, formation of primary clusters followed by heterogenic condensation after some critical size is accomplished; then, processes of coagulation and agglomeration domain; and finally, surface recombination of ions leading to permanent sedimentation of matter on the surface of isolated multi-charged ions. As the disc continues to cool down, condensation of solid particles and their size growth (mainly due to coagulation), as well as gas dissipation from the disc system into interstellar space, dynamic, energetic, and optical contribution of dust particles increases. In parallel, because of growing inertia, particles become less important in eddy motion of gas–dust mixture patterns, the latter increasing an efficiency of their sedimentation to the disc equatorial plane, i.e., subdisk formation.

We may see that the problem of turbulence (unordered chaotic motions) maintenance is really important in terms of the disc formation and evolution. Moreover, evolution of the structurized large-scale turbulence is associated with redistribution of the initial angular momentum and the cloud material (outer parts outward and inner parts toward the Sun) along the disk radius. In other words, because of the forces of viscous friction the matter of the inner disk regions drifts toward the proto-Sun along a very low-angle spiral trajectory, whereas its angular momentum is transferred outward. To produce the present-day distribution of angular momentum, a starward mass flow must exist in the entire disk or in its substantial part over the entire *T* Tauri phase. In a differentially rotating Keplerian disk (which can be used as a first approximation model of the solar protoplanetary disk), the angular velocity of average rotation $\Omega(\boldsymbol{r})$ increases as $|\boldsymbol{r}|^{-3/2}$, i.e., the rotation velocity of mass layers increases toward the central body. It thus follows that the turbulent flow of momentum (mass) directed toward the inner disk layers is, generally speaking, a manifestation of negative turbulent viscosity because the flow transfers averaged momentum from slower rotating outer parts of the disk to its faster rotating inner parts (see Starr 1968). Negative viscosity is a property of a statistic ensemble of chaotic eddy motions of rotating gaseous matter to transfer statistically averaged momentum from the spatial domains of higher to lower density. The averaged flow takes kinetic energy from irregular eddy motions, these motions slowly weakening or maintaining by some internal processes, for example, by the continuous heat transformation in kinetic energy at the boundaries of single perturbations. Since averaged flow has a longer scale length of motion than the pulsation flow; therefore, negative viscosity involves energy transfer along the spectrum from longer to shorter scale lengths. Although various sources for generation of turbulence during different stages of the evolution of the circumsolar protoplanetary cloud (or its individual parts) and various mechanisms for the transfer of mass and angular momentum in disks have been suggested (see, e.g., a thorough review by Makalkin (2003) and extensive list of references therein), these results still require verification and further development.

It is worth to notice that numerous studies of disk turbulence carried out by various authors are based mostly on the classical concept of its statistical homogeneity and local isotropy (Kolmogorov 1941, 1962). Local isotropy of small-scale turbulence implies the invariance of the pulsation velocity field $\boldsymbol{u}$ both with respect to rotations of the reference frame and a mirror reflection in arbitrary plane. At the same time, according to modern concepts, the turbulence pattern in free shear layers of differentially rotating disk matter has to a certain extent the form of a "double" anisotropic system consisting of an ensemble of moving and interacting macro- and mesoscale spiral vortical formations superimposed on a background of small-scale pulsational velocities (turbulent chaos); note that small-scale vortical motions may be partially well organized themselves. Taunsend (1969) called coherent eddy formations and small-scale turbulence a manifestation of the "double" structure of turbulence, thereby emphasizing the effect of organized vortical motions on the processes of turbulent transport in shear layers. Such a phenomenon is due to the insufficiently studied tendency of turbulent flows to self-organize into various coherent structures in the case of high Reynolds numbers (see, e.g., Khlopkov et al. 2002).

Moreover, in a turbulized protoplanetary cloud, like in any rotating gaseous object with internal sources of heat, the so-called density of spirality $\boldsymbol{u} \cdot \mathrm{rot}\boldsymbol{u}$ (a pseudoscalar, which reverses its sign in the case of mirror reflection) appears to develop, which also leads to anisotropy of small-scale turbulence, which in this case has a gyrotropic nature (Vainshtein et al. 1980; Krause and Radler 1980). The latter means that, in the case of a small-scale eddy motion, left-rotating motions in the ensemble may more likely occur than right-rotating, and vice versa. It thus follows that many important hydrodynamic parameters of cosmic matter depend on the magnitude and direction of the vector of the angular velocity of the rotating protoplanetary cloud $\Omega(\boldsymbol{r})$. These parameters include statistical characteristics of the field of pulsational velocity such as the mean spirality H (this quantity has no reflection symmetry). Conservation of the mean spirality in nonviscous liquid flows was discovered not very long ago by Moreau (1961). We show below that the effect of the mean spirality on cascade energy processes in three-dimensional gyrotropic turbulence may explain the possible effect of negative viscosity in the disk. Basically, this phenomenon is the hydrodynamic analogue of the alpha effect in magnetic hydrodynamics (Steenbeck et al. 1966), which explains the growth of the large-scale magnetic field (dynamo effect) in the case of turbulent motion of a conducting continuum with the break down of invariance with respect to the change of parity.

Note that this phenomenon is usually explained in terms of the theory of two-dimensional turbulence, which provides for inverse transport of energy from small to larger vortices (Monin and Yaglom 1996). Negative viscosity in the solar system is known to show up, in particular, in global mass circulations on the Sun, Jupiter, Saturn, Venus (and, probably, on Neptune), and in circulations of the terrestrial atmosphere and ocean (see Monin et al. 1989). Large-scale motions on the spherical surfaces of these cosmic bodies can be analyzed in terms of two-dimensional hydrodynamics, because the horizontal sizes of flows far exceed the local density scale height (Starr 1968; Sivashinsky and Frenkel 1992; Vergassola et al. 1993; Gama et al. 1994). Such an approach can apparently also be partially implemented in the case of the disk turbulence under consideration, because shear mass motions in thin astrophysical disks with the thickness-to-radius ratio much smaller than unity ($h_{\mathrm{disk}}/R << 1$) must also possess certain properties of two-dimensional turbulence (Bodenheimer 1995; Klahr and Bodenheimer 2003). It would be more correct, however, to speak about quasi-two-dimensional turbulence, where motions are approximately two-dimensional; i.e., they can be described by two-dimensional hydrodynamic equations, but with additional special terms.

Unfortunately, the application of the theory of two-dimensional turbulence to the models of the evolution of a thin astrophysical disk faces a purely formal problem: one has to expect condensation of energy on a certain maximally accessible scale length, which lies between the pumping scale and the scale length of the system. In geophysical hydrodynamics, such problems can be overcome because of skin friction, orography, radiation energy cooling, etc., which result in energy sink and determine, in the long run, the quasi-equilibrium state of the turbulent field. The manifestations of such factors include, in particular, the existence of certain external scale lengths (e.g., the Rossby–Obukhov radius), which limit the propagation of energy toward increasingly larger-scale motions and thereby determine the typical sizes of observed coherent vortical formations. However, it is not so easy to avoid such problems when using two-dimensional turbulence in an evolutionary accretion-disk model (a system without clearly defined boundaries), because this would require us to introduce a virtual long-wavelength dissipation (and such an approach would put us onto a purely speculative path) or agree with the presence of rapidly decaying turbulence, which is inconsistent with the relatively long-term (up to assumed time of planetesimals formation) sustenance of the chaotic velocity field in the disk. Hence, the theory of two-dimensional turbulence is of limited interest as far as disks are concerned.

At the same time, we have already mentioned above the real turbulence in an astrophysical disk has a gyrotropic nature, because, in the case of fast rotation of the disk matter and inhomogeneous distribution of the intensity of turbulent pulsations, the field of pulsational velocities $\boldsymbol{u}$, in the general case, is reflection-symmetric with respect to the transformation $z \to -z$. Moffat (1969) was the first to point out on importance of the spirality of localized vortex perturbations for three-dimensional hydrodynamics of turbulized liquid. He was also the first to discover the integral invariant $H \equiv \langle(\boldsymbol{u} \cdot \text{rot}\boldsymbol{u})/2\rangle$ associated with this spirality, which characterizes the degree of connectedness of eddy formations in the flow (see, e.g., Saffman 1992; Alekseenko et al. 2005) and remains constant along the trajectory of motion of any liquid particle of a nonviscous medium. Let us recall that, according to Kelvin's theorem, vortex lines are frozen into liquid if $\nu = 0$, $p = p(\rho)$ and external forces per unit mass are conservative, implying that knots and linkages of vortex lines are inevitably conserved (Alekseenko et al. 2005).

The existence of this additional nonviscous invariant for three-dimensional turbulence immediately implies a certain degree of freedom for the energy cascade process, because the two quantities (the average energy $E \equiv \langle|\boldsymbol{u}|^2/2\rangle$ and the average spirality H), which are conserved in nonlinear interactions in the inertial interval of the energy spectrum, participate simultaneously in the cascade turbulent process. By analogy with two-dimensional flows of an incompressible nonviscous fluid (which also involves cascade transfer of two velocity-quadratic integrals—the integrals of energy, E, and enstrophy $\Omega \equiv \langle(\text{rot}\boldsymbol{u})^2/2\rangle$); in principle, a mode of turbulent motion is possible that involves a cascade of these conserved quantities moving toward opposite ends of the spectrum where the direct cascade of spirality toward shorter scales should be accompanied by a reverse cascade of energy toward longer scales.

It follows from the above discussion that an adequate mathematical model of the evolution of the real anisotropic field of velocity pulsations in a protoplanetary cloud is apparently difficult to develop without the allowance for the symmetry laws of rotating turbulence. The effect of fluctuations of background spirality on the development of negative viscosity in the disk and on the structurization of differentially rotating disk matter has not yet been discussed in detail, and this gap was at the first attempted to fill up by the authors (Kolesnichenko and Marov 2006).

8.2 Energy Cascade in Isotropic Turbulence with Reflection Symmetry

Hereafter, we proceed from the concept of the two level macroscopic description of the turbulized medium of a protoplanetary cloud in terms of two interacting continua (mutually open subsystems), which simultaneously and continuously fill the same volume of coordinate space—the subsystems of averaged motion and of turbulent chaos. This approach is essentially an application of multiscale methods to turbulence (Dubrulle and Frisch 1991; Fannjiang and Papanicolaou 1994). This very description of developed hydrodynamic turbulence served as the starting point that allowed us to begin the development of models of structurized mesoscale turbulence as a process of self-organization in open non-equilibrium fluctuating media (see Chap. 5 and Kolesnichenko 2002, 2003, 2005a, b; Marov and Kolesnichenko 2006). The continuum of averaged motion, obtained by theoretic-probabilistic averaging of instantaneous hydrodynamic equations, serves to study the evolution of averaged hydrodynamic fields, including large eddy formations in the disk. In Chap. 6 we argued that the subsystem of turbulent chaos (in the general case, the eddy anisotropic continuum with an internal structure is actually the turbulent velocity field $\boldsymbol{u}(\boldsymbol{r},t)$) associated with stochastic small-scale pulsational motion of whirling liquid (for which $\omega \equiv \mathrm{rot}\boldsymbol{u} \neq 0$). Note that vorticity plays the crucial role in the mechanics of turbulence, because it makes possible the cascade process of generation of smaller vortices by larger ones.

As it was shown in Chap. 6, in the process of temporal evolution of the quasi-equilibrium subsystem of turbulent chaos, mesoscale coherent structures can be generated because of the mutual phase synchronization (coherence) of a certain ensemble of small-scale oscillatory modes with close frequencies. Such a division of the real turbulent flow into the imaginary averaged and pulsational flows depends, generally speaking, on the choice of the spatial–temporal averaging domain (provided that ergodicity conditions are satisfied), for which the mean values of local hydrodynamic variables (which are continuous functions of coordinates $\boldsymbol{r} = (x, y, z)$ and time t) are known; hence, this division is arbitrary to a certain degree.

To compose the full set of equations of motion for turbulized liquid, which is characterized by two linear scale lengths of motion—L (external) and l_0 (internal)—it is convenient to introduce two coordinate systems—the microscale coordinate system x'_j ($\delta x'_j \sim l$) and the macroscale coordinate system x_j ($\mathrm{d}x_j \approx \Lambda >> l$). These coordinate systems subdivide space into elementary volumes $\delta\,\boldsymbol{r}'$ and $\mathrm{d}\,\boldsymbol{r}'$, respectively. Hereafter, we assume that $L >> \Lambda \geqslant l_0$ and $l_0 >> l >> l_v$. The quantity L is the integral turbulence scale (the scale length of motion of the system); l_0 (the size of a turbulent "mole") is equal to the scale length of the internal motion or the state of the medium, and l_v is the molecular microscale, which is almost equal to zero. It is important to bear in mind that here we do not speak about absolute dimensions. For example, in a hydrodynamic model of a disk medium, a physically infinitely small volume may be much greater than the volume of an entire planet. One can formulate the problem of deriving hydrodynamic equations of motion on macroscale x_j from the known Navier–Stokes equations on microscale x'_j. The associated problem of averaging is one of the central problems in continuum mechanics, and, in the case of such a complex system as a turbulized liquid, the method of averaging often determines the very construction of the macroscopic model. The most clear description of modern methods of spatial averaging for turbulized liquid, which correspond to the transition from equations of motion of small elements of the continuum to the description of the same motions on macroscales, can be found in the paper by Nikolaevskiy (2003).

Hereafter, we assume that the hydrodynamic scale length of the averaged motion Λ (e.g., the mesh size of the difference grid) is such that the subsystem of turbulent chaos in the averaging domain $d\boldsymbol{r} \sim \Lambda^3$ contains the entire ensemble of coherent mesoscale structures whose sizes are smaller than the size of the averaging domain. According to available estimates, for the averaged flow to contain the bulk (80 or 90 %) of the total energy of the turbulent motion, the averaging scale length Λ must be ten to twenty times shorter than the integral scale length L. In this case, the effect of turbulent chaos on the averaged motion manifests itself as an additional turbulent transfer of momentum and energy by small-scale vortical formations, which requires the construction of semiempiric closure models (defining relations) for the averaged hydrodynamic equations and modeling of the effective coefficients of turbulent exchange, which allow, in particular, for the anisotropy of chaos associated with the presence of coherent mesoscale structures (Kolesnichenko 2002).

8.2.1 *Equations of Turbulent Chaos in the Presence of a Mean Flow*

Let the random component $\boldsymbol{u}(\boldsymbol{r}, t)$ of the velocity field $\boldsymbol{U}(\boldsymbol{r}, t)$ have a scale length l_0 (in the case of small-scale turbulence, the scale length l_0 may coincide with the size of energy-containing vortices) which is small compared to the scale of hydrodynamic averaging Λ. Hereafter, we assume that the averaged velocity $\boldsymbol{U}(\boldsymbol{r}, t)$ only slightly varies on any intermediate scale length l satisfying the inequality $l_\nu << l << l_0 << \Lambda$, and hence the methods of the theory of homogeneous turbulence (see Monin and Yaglom 1996) can be applied on such intermediate scales. In this paper, we also assume that the disk medium is not subject to electromagnetic forces, the disk rotates about the z-axis with the Keplerian angular velocity $\Omega(r)$, and the origin of the coordinate system coincides with the center of mass of the system. We further assume, for the sake of simplicity, that the large-scale (averaged) flow reduces to differential rotation exclusively. The components of the vector of the averaged velocity $\langle \boldsymbol{U} \rangle$ then have the following form in cylindrical coordinates $\langle \boldsymbol{U} \rangle_r = 0$, $\langle \boldsymbol{U} \rangle_\varphi = \Omega(r) r$, $\langle \boldsymbol{U} \rangle_z = 0$.

Hereafter, for the sake of simplicity, we restrict our analysis to incompressible fluids $<\rho> = \rho = \text{const}$. Analysis of turbulence in a compressible medium would require more mathematical efforts. We thus proceed in our analysis of three-dimensional disk turbulence from the following set of instantaneous equations of motion for incompressible homogeneous liquid, which includes the Navier–Stokes equation, the continuity equation, and the equation of state:

$$\frac{\partial}{\partial t}\boldsymbol{U} + (\boldsymbol{U}\cdot\frac{\partial}{\partial \boldsymbol{r}})\boldsymbol{U} = -\rho^{-1}\frac{\partial P}{\partial \boldsymbol{r}} + \nu\nabla^2\boldsymbol{U} + \boldsymbol{g}, \quad \mathrm{div}\boldsymbol{U} = 0, \quad p = p(\rho). \tag{8.1}$$

Here, $P(\boldsymbol{r},t)$ is the true (instantaneous) pressure of the disk matter; $\boldsymbol{g}(\boldsymbol{r},t) = -\partial\Psi/\partial\boldsymbol{r}$ is the vector of acceleration due to the external mass force (gravity), and $\Psi(\boldsymbol{r},t)$ is the Newtonian gravitational potential. If the mass of the protoplanetary cloud is equal to several percent of the mass of the central body (or, more precisely, if $\mathcal{M}_{\mathrm{disk}}/\mathcal{M}_{\odot} \le h_{\mathrm{disk}}/R$, where h_{disk} and R are the disk half thickness and radius, respectively), one can neglect the self-gravity of disk particles; in this case, we have

$$\Psi = G\mathcal{M}_{\odot}/|\boldsymbol{r}|, \quad \boldsymbol{g} = -\partial\Psi/\partial\boldsymbol{r} = G\mathcal{M}_{\odot}\boldsymbol{r}/|\boldsymbol{r}|^3, \tag{8.2}$$

(where $\mathcal{M}_{\odot}$ is the mass of the star, G is the gravitational constant, and $|\boldsymbol{r}|$ is the central radius-vector). In the cases where self-gravity effects are important, $\Psi = G\mathcal{M}_{\odot}/|\boldsymbol{r}| + \Psi_{\mathrm{cr}}$, and the potential of self-gravity Ψ_{cr} satisfies the Poisson equation.

Angular braces in this chapter mean averaging

$$\langle\mathcal{A}(\boldsymbol{r},t)\rangle \equiv \frac{1}{l^3}\int\limits_{|\xi|<l} \mathcal{A}(\boldsymbol{r}+\xi,t)d^3\,\xi; \tag{8.3}$$

of an l-periodic parameter $d(\boldsymbol{r},\,t)$ over some cubic spatial domain with the edge l (such that periodic boundary conditions are satisfied on the faces of this cube) under the assumption that this averaging does not depend on the exact value of the scale; from a purely mathematical viewpoint, this definition can be considered to be identical to averaging over an ensemble of identical hydrodynamical systems in the asymptotic limit $l/\Lambda \to 0$. Hereafter, we use the following identity relations for periodic functions that can be proved by integration by parts:

$$\langle\partial_i\mathcal{A}\rangle = 0; \quad \langle(\partial_i\mathcal{A})\mathcal{B}\rangle = -\langle\mathcal{B}\partial_i\mathcal{A}\rangle; \quad \langle(\nabla^2\mathcal{A})\mathcal{B}\rangle = -\langle(\partial_i\mathcal{B})(\partial_i\mathcal{A})\rangle;$$

and then, if $\mathrm{div}\nu = 0$,

$$\langle\boldsymbol{u}\cdot(\mathrm{rot}\nu)\rangle = \langle(\mathrm{rot}\boldsymbol{u})\cdot\nu\rangle; \quad \langle\boldsymbol{u}\cdot\nabla^2\nu)\rangle = -\langle(\mathrm{rot}\boldsymbol{u})\cdot(\mathrm{rot}\nu)\rangle, \tag{8.4}$$

We now average (8.1) over an ensemble of identical hydrodynamic systems to derive the Reynolds equation:

$$\frac{\partial}{\partial t}\langle\boldsymbol{U}\rangle + \left(\langle\boldsymbol{U}\rangle\cdot\frac{\partial}{\partial\boldsymbol{r}}\right)\langle\boldsymbol{U}\rangle = -\rho^{-1}\frac{\partial}{\partial\boldsymbol{r}}\cdot(\boldsymbol{I}\langle P\rangle - \boldsymbol{R}) + \nu\nabla^2\langle\boldsymbol{U}\rangle - \frac{\partial}{\partial\boldsymbol{r}}\langle\Psi\rangle,$$
$$\mathrm{div}\langle\boldsymbol{U}\rangle = 0, \tag{8.5}$$

where $\boldsymbol{R}(\boldsymbol{r},t) \equiv -\rho\langle\boldsymbol{uu}\rangle$ is the Reynolds stress tensor and $\boldsymbol{I}$ is the unit tensor.

The equations for pulsational velocity $\boldsymbol{u}$ can be derived by subtracting the corresponding (8.5) from (8.1). If we further restrict our analysis to the so-called second-order correlation approximation (see Krause and Radler 1980), when the terms that are quadratic in velocity fluctuations can be neglected, and use the above assumption that the averaged velocity $\langle \boldsymbol{U}(\boldsymbol{r},t)\rangle$ remains unchanged on any intermediate microscale l inside the inertial interval $l_\nu << l << l_0$, then we have to consider the following equations:

$$\frac{\partial}{\partial t}\boldsymbol{u} + \left((\langle \boldsymbol{U}\rangle + \boldsymbol{u})\cdot\frac{\partial}{\partial \boldsymbol{r}}\right)\boldsymbol{u} = -\rho^{-1}\frac{\partial p}{\partial \boldsymbol{r}} + \nu\nabla^2\boldsymbol{u} - \frac{\partial \psi}{\partial \boldsymbol{r}}, \quad \mathrm{div}\boldsymbol{u} = 0, \tag{8.6}$$

which, in the case of the elementary volume $\mathrm{d}\boldsymbol{r}'$ moving at the average velocity $\langle \boldsymbol{U}(\boldsymbol{r},t)\rangle$ of the turbulized flow, acquire (in the case of the appropriate redetermination of the velocity $\boldsymbol{u}$ in this coordinate system) the following form:

$$\frac{\mathrm{d}}{\mathrm{d}t}\boldsymbol{u} \equiv \frac{\partial}{\partial t}\boldsymbol{u} + \left(\boldsymbol{u}\cdot\frac{\partial}{\partial \boldsymbol{r}}\right)\boldsymbol{u} = -\frac{\partial}{\partial \boldsymbol{r}}\left(\frac{p}{\rho} + \psi\right) + \nu\nabla^2\boldsymbol{u}, \quad \mathrm{div}\boldsymbol{u} = 0. \tag{8.7}$$

This means that the subsystem of turbulent chaos has zero hydrodynamic velocity relative to the subsystem of averaged motion. Here, $p \equiv P - \langle P\rangle$ is the pressure pulsation; $\psi \equiv \Psi_{\mathrm{cr}} - \langle \Psi_{\mathrm{cr}}\rangle$ (hereafter, we assume that ψ excites $\boldsymbol{u}$ in some way that we do not specify here). It goes without saying that (8.7) must be supplemented by periodic boundary conditions in spatial variable: $\boldsymbol{u}(x+nl, y+ml, z+ql) = \boldsymbol{u}(x, y, z)$ for all x, y, z and all integer n, m, q.

8.2.2 Conservation Laws in Locally Isotropic Turbulence

Consider now integrated conservation laws associated with the homogeneity, isotropy, and mirror symmetry of the turbulent field $\boldsymbol{u}$. Using (8.3), (8.4) we may obtain from the (8.7), by integrating it over the spatial periodicity cell (a spatial cubic domain with the edge l such that periodic boundary conditions are satisfied on its faces), the following conservation laws for the averaged kinetic energy $E = \langle |\boldsymbol{u}|^2/2\rangle$, enstrophy $\Omega = \langle |\omega|^2/2\rangle$ and the total spirality $H = \langle \boldsymbol{u}\cdot\omega/2\rangle$ (see, e.g., Frisch (1995)):

$$\frac{\mathrm{d}E}{\mathrm{d}t} = -\varepsilon = -2\nu\Omega, \qquad \frac{\mathrm{d}\Omega}{\mathrm{d}t} = -\varepsilon_\Omega, \qquad \frac{\mathrm{d}H}{\mathrm{d}t} = -\varepsilon_H, \tag{8.8}$$

where

$$\varepsilon \equiv \frac{\nu}{2}\langle \sum_{i,j}\left(\partial_i u_j + \partial_j u_i\right)^2\rangle, \quad \varepsilon_\Omega \equiv \nu\langle |\mathrm{rot}\omega|^2\rangle, \quad \varepsilon_H \equiv \nu\langle \omega\cdot\mathrm{rot}\omega\rangle. \tag{8.9}$$

Note that the balance (8.8) for enstrophy is valid only in the two-dimensional case. Here, the quantities ε, ε_Ω, and ε_H denote the dissipation rates of averaged kinetic energy, enstrophy, and spirality per unit mass, respectively. It is evident from (8.8) that the quantities E, Ω, and H remain constant in the nonviscous limit $\nu \to 0$ (and, in particular, throughout the entire inertial interval) if no dissipation and motion pumping are present.

8.2.3 *Vorticity Dynamics and Energy Cascade*

Before analyzing the possible effect of spirality on the dynamics of disk turbulence, we must recall some of the concepts and quantitative spectral characteristics of small-scale turbulence (see Batchelor 1953; Monin and Yaglom 1996).

As we have already pointed out above, most of the studies of turbulent motion within a disk, carried out within the framework of the problem considered here, were based mostly on the concept of Kolmogorov (1941, 1962). According to this concept, in the limit of large Reynolds numbers (which correspond to large-scale motions in the flow of cosmic matter), the random nature of eddies fragmentation and the chaotic nature of the transfer of their energy downscale the cascade result in almost locally isotropic stochastic regime of turbulent fluctuations within the boundaries of the spatiotemporal domain of averaging of instantaneous hydrodynamic equations despite the anisotropy, inhomogeneity, and nonstationarity of the averaged flow. Recall that in locally isotropic turbulence, any average quantity that characterizes it is invariant with respect to any parallel translations, rotations, and mirror reflections.

In this case, the energy structure of the three-dimensional small-scale field of velocity pulsations is statistically similar for large Reynolds numbers $\boldsymbol{Re} \equiv u_0\, l_0/\nu$ (where $u_0 = \sqrt{\langle|\boldsymbol{u}|^2\rangle}$ is the characteristic velocity of pulsational velocity field and l_0 is the characteristic scale length of energy-bearing vortices), and the inertial interval of wavenumbers $k_0 << k << k_\nu$ that separates the domains of dissipation and generation of turbulent energy $E = u_0^2/2$ in the space of wave numbers k is the wider the higher is the Reynolds number $\boldsymbol{Re}$. The dissipation rate of the averaged kinetic energy per unit mass ε, which is given by (8.9) in the initial Kolmogorov theory (*K41*), is considered to be a universal constant for the turbulent motion studied. The quantity ε also characterizes the flow of kinetic energy, which is transferred in a cascade mode without any loss along successively increasing wavenumbers $k_n >> k_{n-1}$ $(n = 1, 2, \ldots)$ (decreasing scale lengths, $l_n = 1/k_n$) within the inertial interval until the flow reaches the dissipation scale length $\mathrm{l}_\nu = 1/k_\nu \sim (\nu^3/\varepsilon)^{1/4}$, for which the energy dissipation rate due to kinematic viscosity is equal to ε.

Under the assumption of homogeneity and stationarity of the field of pulsation velocities $\boldsymbol{u}(\boldsymbol{r}, t)$, the most important turbulence characteristics are the (two-point and two-time) correlation tensor of the velocity field $\tilde{R}_{ij}(\boldsymbol{r}, \boldsymbol{x}, t, \tau) \equiv \langle u_i(\boldsymbol{r}, t) u_j(\boldsymbol{r} + \boldsymbol{x}, t + \tau)) \rangle$ and the spectral energy tensor (Batchelor 1953):

$$\Phi_{ij}(\boldsymbol{k}, \omega) = \frac{1}{(2\pi)^4} \int\int \tilde{R}_{ij}(\boldsymbol{x}, \tau) \exp[-i(\boldsymbol{k} \cdot \boldsymbol{x} - \omega\tau)] \, \mathrm{d}\boldsymbol{x} \mathrm{d}\tau. \tag{8.10}$$

which is in fact the Fourier image of the correlation tensor $\tilde{R}_{ij}$. Note that components of the turbulent Reynolds stress tensor can be written in the form $R_{ij}(\boldsymbol{r}, t) = -\rho \tilde{R}_{ij}$ $(\boldsymbol{r},\ 0, t,\ 0)$, and the total spirality, in the form $H = \varepsilon_{ijk}(\partial \tilde{R}_{ik}/\partial x_j)|_{x=0,\tau=0}$. In turn, the complex tensor $\Phi_{ij}(\boldsymbol{k}, \omega)$ for incompressible fluid ($\mathrm{div}\boldsymbol{u} = 0$) has the following properties (which we use below):

$$k_j \Phi_{ij}(\boldsymbol{k}, \omega) = k_i \Phi_{ij}(\boldsymbol{k}, \omega) = 0. \tag{8.11}$$

Hereafter, summation is to be performed over repeated indices if not otherwise stated.

The energy spectral function (density) $E(k, \omega)$, which is the most important quantity in the problem of homogeneous turbulence, is determined by the following integral:

$$E(k, \omega) = \frac{1}{2} \int_{S_k} \Phi_{ii}(\boldsymbol{k}, \omega) \mathrm{d}S, \tag{8.12}$$

where integration is performed in the $\boldsymbol{k}$-space over the sphere S_k of radius $k \equiv |\boldsymbol{k}|$. The total kinetic energy per unit mass transferred along the cascade can then be written in the following form

$$E \equiv \frac{1}{2} \langle |\boldsymbol{u}|^2 \rangle = \frac{1}{2} \tilde{R}_{ii}(0, 0) = \frac{1}{2} \int\int \Phi_{ii}(\boldsymbol{k}, \omega) \mathrm{d}\boldsymbol{k} \mathrm{d}\omega = \int\int E(k, \omega) \mathrm{d}k \mathrm{d}\omega, \tag{8.13}$$

and $\Phi_{ii} \geqslant 0$ for almost all $\boldsymbol{k}$ and ω, because these diagonal elements represent the density of the kinetic energy components in the wave space. Hence, the quantity $E(k, \omega) \mathrm{d}k \mathrm{d}\omega$ can be interpreted as the kinetic energy (from all wave numbers with a fixed absolute value) contained in the wave number interval $(k, k + \mathrm{d}k)$ and in the frequency interval $(\omega,\ \ \omega + \mathrm{d}\omega)$.

In the wave number interval $k_0 << k << k_\nu$, the spectral energy tensor $\Phi_{ij}(\boldsymbol{k}, \omega)$, which is a second-rank isotropic tensor, is statistically unrelated to the source of energy, which is restricted by the wave number k_0, and therefore we can assume that it must be determined by the energy dissipation rate ε exclusively:

$$\Phi_{ij}(\boldsymbol{k}, \omega) = \frac{E(k, \omega)}{4\pi k^4} (k^2 \delta_{ij} - k_i k_j), \tag{8.14}$$

As follows from dimensional considerations, the spectral energy density must be described by the Kolmogorov formula

$$E(k) = C\varepsilon^{2/3}k^{-5/3} \quad (k_0 \ll k \ll k_\nu), \tag{8.15}$$

where C is a dimensionless constant on the order of unity.

8.2.4 Two-Dimensional Turbulence

In the case of developed two-dimensional turbulence in a noncompressible fluid (which, as already said, strictly is only a mathematical idealization and is never realized in nature), Kolmogorov-type theories have two positive definite quantities that are conserved in the nonviscous limit. They are the averaged kinetic energy $E = \langle |\boldsymbol{u}|^2/2\rangle$ and the enstrophy $\Omega \equiv \langle |\mathrm{rot}\boldsymbol{u}|^2/2\rangle$, which, if generated in a uniform stream on some intermediate scale lengths of energy pumping k_I which are far from the dissipation scale k_ν, both become involved in the cascade process. As is evident from (8.8), in the case of finite viscosity in a two-dimensional flow, the enstrophy can only monotonically increase with time along with the quantity $\varepsilon = 2\nu\Omega$. This is due to the fact that the mechanism of stretching of vortex tubes, which ensures the growth of enstrophy in three-dimensional flows, is blocked in the case of a two-dimensional flow.

In the general three-dimensional case, the Fourier transform of the vortex field $\omega \equiv \mathrm{rot}\boldsymbol{u}$, whose components have the form $\omega_i = \varepsilon_{ijk}\,\partial_j u_k$ is evidently equal to $\omega \equiv i\boldsymbol{k} \times \boldsymbol{u}$ and its spectral tensor has the form $\Omega_{ij}(\boldsymbol{k}, \omega) = \varepsilon_{imn}\varepsilon_{jpq}k_m k_p \Phi_{nq}(\boldsymbol{k}, \omega)$, where ε_{jpq} is the asymmetric Levi-Civita tensor. In particular, we derive from this, in view of (8.11):

$$\Omega_{ij}(\boldsymbol{k}, \omega) = k^2\Phi_{ij}(\boldsymbol{k}, \omega) = \frac{E(k, \omega)}{4\pi k^2}(k^2\delta_{ij} - k_i k_j). \tag{8.16}$$

Note that ε_{ijk} is the tensor, which is equal to $\varepsilon_{ijk} = 0$ if i, and are not all different, and $\varepsilon_{ijk} = 1$ or -1 if i, and are all different and arranged in a cyclic or anticyclic order, respectively. Basically, the formula $\Omega = \varepsilon/2\nu$ implies that the spectrum of the root-mean-square velocity vortex Ω coincides with the spectrum of viscous dissipation of kinetic energy ε.

The following formula is an immediate implication of (8.16):

$$\Omega \equiv \langle |\omega|^2/2\rangle = \int\!\!\int \Omega(k)\mathrm{d}k = \int\!\!\int k^2 E(k, \omega)\mathrm{d}k\mathrm{d}\omega, \tag{8.17}$$

where the corresponding spectral density for the root-mean-square velocity vortex has the form

$$\Omega(k) = k^2E(k) = C\varepsilon^{2/3}k^{1/3}. \tag{8.18}$$

It is shown in the theory of two-dimensional turbulence (see Monin and Yaglom 1996; Chap. 26) that relationship (8.18) between the spectral densities of energy $E(k)$ and enstrophy $\Omega(k)$ forbids the simultaneous downscale transfer of these quantities. Kraichnan (1967) was the first to propose the hypothesis of inverse cascade of energy in two-dimensional turbulence. The same author (Kraichnan 1976b) interpreted inverse cascade in terms of negative turbulent viscosity. According to some estimates, inverse cascade, which has been confirmed many times by numerical simulations, is one of the most important results in the theory of developed turbulence since Kolmogorov's works (Gama et al. 1994; Frisch 1995).

Energy in the two-dimensional case is transferred to large and not to small (as in the three-dimensional case) scale lengths, whereas the flow of enstrophy is directed toward short scales. Note that there are two inertial intervals in the case of developed two-dimensional turbulence. For small wave numbers $k_0<k<k_I$, the cascade process is determined by the energy dissipation rate and the dimension analysis yields classical formula (8.15) for the spectral density with the only essential difference from the three-dimensional case that the energy flow in the inertial interval with spectrum (8.15) is directed from shorter to longer scales. For large wave numbers ($k_\nu>k>k_I$), the enstrophy dissipation rate $\varepsilon_\Omega \equiv \nu|\mathrm{rot}\omega|^2$ becomes an additional determining quantity, and dimension analysis yields a different spectral distribution of kinetic energy by the absolute values of the wave number k:

$$E(k) = C_\Omega \varepsilon_\Omega^{2/3} k^{-3}, \tag{8.19}$$

which describes the inertial interval of enstrophy transfer. Note that, in the two-dimensional case, the enstrophy cascade is a direct one; i.e., enstrophy is transferred from longer to shorter scale lengths. Formula $k_\nu \sim (\varepsilon_\Omega/\nu^3)^{1/6}$ determines the boundary of the interval of the enstrophy transfer.

Thus, in the case of two-dimensional turbulence, we have a reverse cascade of kinetic energy E, where the energy of small-scale chaotic motion is used for the energy pumping of mesoscale vortex structures. This effect, which came to be known as negative viscosity (see Starr 1968; Vergassola et al. 1993; Gama et al. 1994), is typical, as we pointed out above, for many quasi-two-dimensional cosmic objects. We now proceed to the analysis of real three-dimensional hydrodynamic turbulence in natural media.

8.3 On the Energy and Helicity Cascades in Disk Reflection-Invariant Turbulence

8.3.1 Mirror Symmetry Breaking in a Protoplanetary Disk

Hence, the analysis of turbulence fields, which statistically satisfy certain symmetry conditions (as in the case of locally isotropic turbulence), can be seen to result in a number of important mathematical simplifications. Although mirror symmetry is a fundamental property of hydrodynamic equations, the definition of some of the important hydrodynamic parameters includes the notions of right-handedness (left-handedness). Such quantities include, in particular, the vorticity of the field of velocity pulsations $\omega \equiv \mathrm{rot}\boldsymbol{u}$. As we pointed out above, such a statistical characteristic of small-scale turbulence devoid of mirror symmetry as the total spirality (a pseudoscalar of the velocity field) determined by the mean

$$H \equiv \frac{1}{2}\langle \boldsymbol{u} \cdot \boldsymbol{\omega} \rangle = \frac{1}{2l^3} \int\limits_{|\xi|<l} \boldsymbol{u}(\boldsymbol{r}+\boldsymbol{\xi}, t) \cdot \boldsymbol{\omega}(\boldsymbol{r}+\boldsymbol{\xi}, t) d^3 \boldsymbol{\xi} \tag{8.20}$$

may be of considerable interest for differentially rotating matter. Recall that vectors $\boldsymbol{A}$, which behave as $\boldsymbol{A}^{\mathrm{ref}}(\boldsymbol{r}, t) = -\boldsymbol{A}(-\boldsymbol{r}, t)$, are referred to as polar vectors, whereas vectors satisfying the relation $\boldsymbol{A}^{\mathrm{ref}}(\boldsymbol{r}, t) = A(-\boldsymbol{r}, t)$ are referred to as axial vectors or pseudovectors (the superscript "ref" hereafter means the operator of reflection in an arbitrary plane or in an arbitrary point). The scalar $V^{\mathrm{ref}} \equiv (\boldsymbol{A}^{\mathrm{ref}} \times \boldsymbol{B}^{\mathrm{ref}}) \cdot \boldsymbol{C}^{\mathrm{ref}} = -(\boldsymbol{A} \times \boldsymbol{B}) \cdot \boldsymbol{C} = -V$, which depends on the use of right-handedness, is a pseudoscalar; the last statement means that this scalar reverses its sign if we change over from a right-handed coordinate system to a left-handed one.

The field of pulsation velocities $\boldsymbol{u}$ with non-zero average spirality has the form of anisotropic continuum formed by the ensemble of arbitrarily oriented small-scale spiral vortices where, e.g., right-handed motions (eddy structures) are more likely than left-handed structures. Let us notice that, in his studies of small-scale properties of flows (the random-cascade model), Kolmogorov (1941, 1962) did not take into account any spatial structures that could possibly exist in the turbulent flow. However, as we pointed out above, according to modern concepts, vortical structures (threads) are almost always present in a turbulent flow on small scales and these structures may affect the flow properties in the inertial interval as well. Thus, the quantity H, related to the topological structure of a complex vorticity field, is a fundamental measure of the "lack of reflection symmetry" in a turbulent flow.

According to Moffat (1981), the spirality of turbulent field, which is a quadratic invariant of some localized motion of a fluid (under the conditions determined above), has a status comparable to the status of the perturbation kinetic energy. The question so far remains open about other possible integral invariants (Edwards 1967, 1968) besides the classical invariants of energy, momentum, and angular momentum, which characterize to a certain degree the preserved topological configuration of eddy threads. Note that only the analysis of the background anisotropic turbulence with a magnetic field and devoid of mirror symmetry (whose simplest measure is provided, in particular, by the hydrodynamic spirality H which is distinguished from magnetic spirality) allowed a breakthrough to happen in the understanding of turbulent magnetic dynamo in astrophysics (the so-called α-effect) (see Vainshtein et al. 1980; Parker 1979; Krause and Radler 1980). At the same time, as we show below, the quantity H can be also considered as a statistical characteristic of an anisotropic field of pulsation velocities, which is capable of developing the effect of negative viscosity in a rotating medium (in particular, in a protoplanetary disk).

It is known that the spirality of the velocity field can be efficiently generated in a mirror asymmetric (i.e., not invariant with respect to parity reversal) field of small-scale random velocities $\boldsymbol{u}$, e.g., in turbulence that rapidly rotates about a fixed axis (Steenbeck et al. 1966). In particular, spirality may naturally develop in a proto-planetary cloud because of its rotation and nonuninform distribution of density (density stratification) or intensity of turbulent pulsations in zones with developed convection. Here we give some preliminary arguments in favor of the hypothesis that, at some heliocentric distances, vertical thermal turbulent convection in the disk (during certain stages of the disk evolution) is very likely to result in left-handed spiral motions in the regions between its equatorial plane and the "upper" surface. Indeed, the ascending matter should expand and rotate under the action of the Coriolis forces and this process results in left handed spiral motion. It is evident that right-handed spiral motions should dominate in the "lower" part of the disk. The balance of left- and right-handed spiral motions may be established in the vicinity of the equatorial plane of the disk only in the absence of gradients of turbulence intensity, i.e., at the very late stages of the evolution of a differentially rotating protoplanetary cloud.

8.3.2 *Influence of Helicity on the Energy Cascade*

Let us imagine a situation where the source of kinetic energy (e.g., as a result of the interaction of Archimedes and Coriolis forces in the differentially rotating disk matter) at wave numbers k_0 generates a nonzero total spirality H. We have shown above that the total spirality of the pulsational velocity field in the inertial interval is a conserved quantity (like the total energy), $\mathrm{d}\langle \boldsymbol{u} \cdot \omega\rangle/\mathrm{d}t = 0$. It is important to keep in mind that generation of spirality is associated with the development of large-scale engagements of eddy flow lines (Alekseenko et al. 2005), which are conserved in a cascade (not nonviscous) process in the inertial domain, but are destroyed by viscosity on scale lengths $l_\nu = k_\nu^{-1}$. As we already pointed out above, spirality characterizes the degree of connectedness of vortex lines in a flow. The number of turns n of a thread around another thread is characterized by the engagement $H = \pm 2n\Gamma_1\Gamma_2$, where Γ_k is the intensity of the thread and the "sign" refers to the right or left engagement, respectively. If a single vortex tube twines itself before closing, it has got a knot on it; the conservation of spirality also means that the knot structure of the vortex field is conserved. It has been also shown that the spirality invariant H is associated with a more general topological property—the so-called Hopf invariant (see Moffat 1984).

By analogy with the definition of quantity $E(k)$ (8.12), we can define the spectral spirality density

$$F(k,\tau) \equiv i \int\limits_{S_k} \varepsilon_{jkl} \boldsymbol{k}_k \Phi_{jl}(k,\tau)\mathrm{d}S, \tag{8.21}$$

where integration is performed in the $\boldsymbol{k}$ space over a sphere S_k of radius $k \equiv |\boldsymbol{k}|$. We thus have for the total spirality H

$$H = i\varepsilon_{jkl} \int\int k_k \Phi_{jl}(\boldsymbol{k}, \tau) \mathrm{d}\boldsymbol{k}\mathrm{d}\tau = \int\int F(k, \tau) \mathrm{d}k\mathrm{d}\tau. \tag{8.22}$$

The function $F(k, \tau)$ is, in view of relation (8.21),real and it is a pseudoscalar; it is equal to zero if the field of pulsational velocities $\boldsymbol{u}$ is statistically invariant with respect to the reflection transform (e.g., a parity transform of the form $x' = x$, $y' = y$, $z' = -z$, which describes a mirror reflection in the $z = 0$ plane).

In the case of isotropic and reflection non-symmetric turbulence, the functions $E(k, \tau)$ and $F(k, \tau)$ are sufficient for the complete determination of the spectral tensor $\Phi_{il}(\boldsymbol{k}, \tau)$. In this case, the most general formula for the uniform field (which is stationary in t) of small scale turbulence that satisfies equalities (8.11), (8.12), and (8.21) has the following form (Moffat 1978)

$$\Phi_{il}(\boldsymbol{k}, \tau) = \frac{E(k, \tau)}{4\pi k^4}(k^2\delta_{ij} - k_i k_j) + \frac{iF(k, \tau)}{8\pi k^4}\varepsilon_{ijk}k_k. \tag{8.23}$$

Unlike the spectral density of kinetic energy $E(k, \tau)$, the spectral spirality density $F(k, \tau)$ may be both positive and negative. And the ambiguous role of spirality in three-dimensional cascade processes is due to this very feature, because simple arguments that imply the existence of two inertial intervals (as in the case of two-dimensional turbulence) do not work in this case. In the two-dimensional case, the spectral densities of energy and enstrophy are related by (8.18), whereas only an upper limit (the so-called "realizability" condition) can be obtained for the function $F(k)$:

$$|F(k)| \leq 2kE(k), \tag{8.24}$$

This constraint follows, e.g., from the Cauchy–Bunyakovsky–Schwartz inequality written in the form

$$\left|\int_{S_k} \langle \hat{\mathbf{u}} \cdot \hat{\omega}^* + \hat{\mathbf{u}}^* \cdot \hat{\omega} \rangle \mathrm{d}S\right|^2 \leq 4 \int_{S_k} \left\langle |\hat{\mathbf{u}}|^2 \right\rangle \mathrm{d}S \int_{S_k} \left\langle |\hat{\omega}|^2 \right\rangle dS$$

and from equalities (8.13), (8.16), and (8.22). Here asterisk denotes complex conjugation and the "$\wedge$" sign above a symbol denotes Fourier transform.

Generally speaking, inequality (8.24) allows the realization of two scenarios of the behavior of spirality in a turbulent flow (Brissaund et al. 1973). First, in some cases, by analogy with two-dimensional turbulence, there is a cascade of conserved quantities toward the opposite ends of the inertial interval of the spectrum, and the direct downscale cascade of spirality H is accompanied by the synchronous inverse cascade of energy E toward larger scales. Second, there is the possibility of the simultaneous direct cascade of both quantities toward small scales. Which of the two processes occurs at the given time instant, depends on the integral properties of the system considered and also on the boundary and initial conditions.

In the first case, we use the hypothesis that the energy spectrum $E(k)$ may depend only on the wave-number k and the spirality dissipation rate ε_H ; dimension considerations yield a spectral law of the form

$$E(k) \sim \varepsilon_H^{2/3} k^{-7/3}. \tag{8.25}$$

The spectral spirality function $F(k)$ is determined by the process of spirality transfer from the source acting at wave numbers k_0 to the viscous sink on wave numbers k_ν and further. During the spirality generation, large-scale engagements of vortex lines of the flow under consideration develop, which survive in the cascade process in the inertial domain, but are destroyed by viscosity on scale lengths $l_\nu = k_\nu^{-1}$.

The second scenario assumes passive behavior of spirality in the turbulent flow. This means the realization of common Kolmogorov energy cascade $E(k)$ toward small scales with law (8.15). Let the spirality generation rate at wave numbers $ik_0 \sim k_0$ be equal to ε_H (see (8.9)). Spirality is generated simultaneously with energy, and therefore it is evidently constrained by an inequality of the form $|\varepsilon_H| \leq k_0\varepsilon$ (Brissaund et al. 1973). If spirality is injected at the maximum rate, then

$$|\varepsilon_H| \approx ik_0\varepsilon \approx iu_0^3/l_0^2. \tag{8.26}$$

The spirality spectrum $F(k)$ must be proportional to ε_H (because of the pseudoscalar nature of both quantities), and the only possible additional parameters that may determine $F(k)$ in the inertial domain $k_0 << k << k_\nu$ are ε and k. It thus follows from dimension considerations that

$$F(k) = C_H \varepsilon_H \varepsilon^{-1/3} k^{-5/3} \quad (k_0 << k << k_\nu). \tag{8.27}$$

where C_H is the universal constant similar to the Kolmogorov constant C. Note that equalities (8.26), (8.15), and (8.27) also imply inequality $|F(k)| \leq 2kE(k)$, in full agreement with (8.24).

It follows from the above that the presence of spirality in this case has a weak effect on the cascade transfer of energy, because, independently of the spirality injected into the flow at wave numbers ik_0, the relative value of spirality determined by the dimensionless ratio $F(k)/2kE(k)$ decreases monotonically with increasing k. It is safe to assume that, in the case of sufficiently large k/k_0, the spirality should have only a minor effect on the dynamics and should be transferred and diffuse in the same way as a dynamically passive scalar admixture (Monin and Yaglom 1996). At the same time, Kraichnan (1973) and Andre and Lesieur (1977) showed that, if the mode of generation of almost maximum spirality is realized for each wave number in the liquid flow under consideration, then the total transfer of kinetic energy toward higher wave numbers should be weakened, and the process of the decay of turbulence should be extended in time. This leads us to an important conclusion that the relatively long existence of turbulence in a rotating protoplanetary cloud may be partially due to the lack of reflection symmetry with respect to the equatorial plane of the field of vortex velocities in the disk.

Hence, gyrotropic turbulence behaves in a way that is qualitatively different from the behavior of non-spiral turbulence. This allows us, in the case of possible realization of an inverse cascade of kinetic energy, not only to explain the development of the effect of negative viscosity in the differentially rotating (three-dimensional) protoplanetary cloud, but also to forecast the development of relatively stable and energy intensive coherent mesoscale eddy structures, which initiate the mechanisms of trigger cluster formation in the disk.

8.3.3 Generation of Hydrodynamic Helicity in a Rotating Disk

The lack of symmetry with respect to the $z = 0$ plane perpendicular to the vector of angular velocity Ω necessarily implies the breakdown of the mirror symmetry of random motions, which plays the crucial part in the generation of spirality (see Steenbeck et al. 1966). Let us show that such a symmetry is also absent in the case of a rotating protoplanetary disk stratified in z, where Archimedean forces $\rho' \boldsymbol{g}$ act on liquid elements whose density differs from the local ambient density ρ_0 by ρ' ($\boldsymbol{g} \cdot \Omega \neq 0$). In other words, let us analyze the role of the interaction of Archimedean and Coriolis forces in the generation of the mean spirality in the disk. Note in this connection that the lack of invariance with respect to the parity transform is a more general property than the presence of spirality, although it follows from the latter (Gilbert et al. 1988).

Let us adopt the viewpoint that small-scale vorticity in the disk is produced by convection. To find the pseudoscalar function $\langle \boldsymbol{u} \cdot \mathrm{rot}\boldsymbol{u} \rangle$ one has, in the general case, to solve the hydrodynamic problem in the Boussinesq approximation for the pulsating velocity field $\boldsymbol{u}$ in the presence of unstable density stratification and with a preferential direction of vortex twisting Ω (see, e.g., Eltayeb 1972). We describe the motion of matter within the framework of this approach in a reference frame rotating at the average angular velocity Ω of the disk rotation. Equation of motion (8.1) then acquires the following form:

$$\frac{\partial}{\partial t}\boldsymbol{U} + \left(\boldsymbol{U} \cdot \frac{\partial}{\partial \boldsymbol{r}}\right)\boldsymbol{U} + 2\Omega \times \boldsymbol{U} = -\rho^{-1}\frac{\partial}{\partial \boldsymbol{r}}P + \boldsymbol{g} + \nu\nabla^2\boldsymbol{U}, \tag{8.28}$$

where $2\Omega \times \boldsymbol{U}$ and $\frac{1}{2}|\Omega \times \boldsymbol{r}|^2$ are the Coriolis acceleration and the potential of the centrifugal force, respectively, and $\boldsymbol{g} = -\boldsymbol{U} \cdot \frac{\partial}{\partial \boldsymbol{r}}\left(\psi - \frac{1}{2}|\Omega \times \boldsymbol{r}|^2\right)$. The problem is characterized by two dimensionless numbers: the Rossby number $\boldsymbol{Ro} = U_0/\Omega L$ and the Ekman number $\boldsymbol{Ek} = \nu/\Omega L^2$. Here $L = O(R)$ is the scale length of the variation of the characteristic velocity U_0. The Rossby and the Ekman numbers for the disk are much less than unity; and therefore, for the sake of simplicity, we restrict our analysis to the geostrophic motion, when we can neglect the translational acceleration and the viscous term in (8.28). In this case, the equation for the turbulent component of the velocity field $\boldsymbol{U}$ in the Boussinesq approximation takes the following form:

$$2\rho_0\Omega \times \boldsymbol{u} = -\boldsymbol{U} \cdot \frac{\partial}{\partial \boldsymbol{r}} p' + \rho' \boldsymbol{g}.$$

Recall that in the Boussinesq approximation we can assume, up to the first order of small quantities and in view of relation $\partial P_0/\partial \boldsymbol{r} = \rho_0 \boldsymbol{g}$ that $\rho^{-1}\partial P/\partial \boldsymbol{r} + \boldsymbol{g} \approx \rho_0^{-1}$ $(\partial p'/\partial \boldsymbol{r} + \rho' \boldsymbol{g})$ with the simultaneous replacement of the continuity equation by the nondivergence condition, $\mathrm{div}\boldsymbol{U} = 0$. Here, p' and ρ' are the pressure and density deviations from the main state P_0 and ρ_0 determined by the presence of winds and flows.

We now apply the curl operation to this equation to obtain, in view of

$$\mathrm{rot}(\rho' \boldsymbol{g}) = \boldsymbol{U} \cdot \frac{\partial}{\partial \boldsymbol{r}} \rho' \times \boldsymbol{g},$$

$$2\rho_0 \mathrm{rot}(\Omega \times \boldsymbol{u}) = -(\boldsymbol{g} \times \boldsymbol{U}) \cdot \frac{\partial}{\partial \boldsymbol{r}} \rho' \tag{8.29}$$

and, consequently,

$$\begin{aligned} 2\rho_0(\Omega \times \boldsymbol{u}) \cdot \mathrm{rot}(\Omega \times \boldsymbol{u}) &= -(\boldsymbol{\Omega} \times \boldsymbol{u}) \cdot \left(\boldsymbol{g} \times \boldsymbol{U} \cdot \tfrac{\partial}{\partial \boldsymbol{r}} \rho'\right) = -\boldsymbol{\Omega} \cdot \left[\boldsymbol{u} \times \left(\boldsymbol{g} \times \boldsymbol{U} \cdot \tfrac{\partial}{\partial \boldsymbol{r}} \rho'\right)\right] \\ &= -(\boldsymbol{\Omega} \cdot \boldsymbol{g}) \left(\boldsymbol{u} \cdot \boldsymbol{U} \cdot \tfrac{\partial}{\partial \boldsymbol{r}} \rho'\right) + \left(\boldsymbol{\Omega} \cdot \boldsymbol{U} \cdot \tfrac{\partial}{\partial \boldsymbol{r}} \rho'\right)(\boldsymbol{u} \cdot \boldsymbol{g}). \end{aligned} \tag{8.30}$$

We now set $\boldsymbol{u} = \boldsymbol{u}_\perp + \boldsymbol{u}_\parallel$ (where $\boldsymbol{u}_\parallel = u_z \boldsymbol{i}_z$ is the turbulent velocity component $\boldsymbol{u}$ that is parallel to the angular velocity vector $\boldsymbol{\Omega}$; $i_z = \boldsymbol{\Omega}/|\boldsymbol{\Omega}|$; $u_z = \boldsymbol{u} \cdot \boldsymbol{i}_z$; $\boldsymbol{u}_\perp = u_x$ $\boldsymbol{i}_x + u_y \boldsymbol{i}_y$ is the projection of velocity $\boldsymbol{u}$ onto the equatorial disk plane, which we hereafter assume to be uniform with respect to x and y; and $u_x = u_y = u_\perp$) and average formula (8.30) over horizontal planes $z = \text{const}$ to obtain

$$2\rho_0 |\boldsymbol{\Omega}|^2 \langle \boldsymbol{u}_\perp \cdot \mathrm{rot}\boldsymbol{u}_\perp \rangle = -(\boldsymbol{\Omega} \cdot \boldsymbol{g}) \langle \boldsymbol{u} \cdot \boldsymbol{U} \cdot \frac{\partial}{\partial \boldsymbol{r}} \rho' \rangle + \boldsymbol{g} \cdot \langle \boldsymbol{u}\boldsymbol{U} \cdot \frac{\partial}{\partial \boldsymbol{r}} \rho' \rangle \cdot \boldsymbol{\Omega}. \tag{8.31}$$

In the special case, where the vector $\boldsymbol{g}$ of gravity acceleration is parallel to the vector $\boldsymbol{\Omega}$ of angular rotation and ρ_0 is uniform in the horizontal planes, it follows from (8.31) that, if the continuity equation

$$\boldsymbol{U} \cdot \frac{\partial}{\partial \boldsymbol{r}} \cdot (\rho_0 \boldsymbol{u}) = \rho_0 \boldsymbol{U} \cdot \frac{\partial}{\partial \boldsymbol{r}_\perp} \boldsymbol{u}_\perp + \partial(\rho_0 u_z)/\partial z = 0,$$

is applied, the following formula (Hide 1976) is true:

$$\langle \boldsymbol{u}_\perp \cdot \mathrm{rot}\boldsymbol{u}_\perp/2\rangle \cong -\frac{\langle \rho' \partial u_z/\partial z\rangle}{4\rho_0|\boldsymbol{\Omega}|^2}(\boldsymbol{\Omega}\cdot\boldsymbol{g}), \tag{8.32}$$

which establishes the direct relation between the part $\langle \boldsymbol{u}_\perp \cdot \mathrm{rot}\boldsymbol{u}_\perp/2\rangle$ of the mean spirality H and the pseudoscalar $\Omega \cdot \boldsymbol{g}$. If we now use the estimate (Moffat 1978)

$$\begin{aligned}\langle \boldsymbol{u}\cdot\mathrm{rot}\boldsymbol{u}\rangle &= \langle \boldsymbol{u}_\perp\cdot\mathrm{rot}\boldsymbol{u}_\perp\rangle + \langle \boldsymbol{u}_\perp\cdot\mathrm{rot}\boldsymbol{u}_\parallel\rangle + \langle \boldsymbol{u}_\parallel\cdot\mathrm{rot}\boldsymbol{u}_\perp\rangle \\ &= \langle \boldsymbol{u}_\perp\cdot\mathrm{rot}\boldsymbol{u}_\perp\rangle\left(1+\frac{\langle \boldsymbol{u}_\parallel\cdot\mathrm{rot}\boldsymbol{u}_\perp\rangle}{\langle \boldsymbol{u}_\perp\cdot\mathrm{rot}\boldsymbol{u}_\perp\rangle}\right) = \langle \boldsymbol{u}_\perp\cdot\mathrm{rot}\boldsymbol{u}_\perp\rangle\left(1+O\left(\frac{U_\parallel/L_\perp}{U_\perp/L_\parallel}\right)\right) \\ &= \langle \boldsymbol{u}_\perp\cdot\mathrm{rot}\boldsymbol{u}_\perp\rangle\left(1+O\left(L_\parallel^2/L_\perp^2\right)\right) \cong \langle \boldsymbol{u}_\perp\cdot\mathrm{rot}\boldsymbol{u}_\perp\rangle,\end{aligned} \tag{8.33}$$

we that the spirality for the protoplanetary disk is approximately equal to $H \cong \langle \boldsymbol{u}_\perp \cdot \mathrm{rot}\boldsymbol{u}_\perp/2\rangle$, because the $L_\parallel/L_\perp$ ratio in this disk is sufficiently small. Here $U_\parallel, U_\perp, L_\parallel$ and $L_\perp$ are the characteristic velocities and spatial scale lengths parallel and perpendicular to the angular velocity vector Ω, which (by virtue of the continuity equation) are related by the formula $U_\parallel/L_\parallel = 0(U_\perp/L_\perp)$.

Note that the correlation $(g/\rho_0)<\rho' u_z>$ also appears in the definition of the coefficients of turbulent transfer in the vertical direction and, in particular, in the definition of the coefficient of turbulent heat exchange in the vertical direction (Van Mieghem 1973; Marov and Kolesnichenko 2002), where it characterizes the rate of transformation of turbulent energy into the internal energy of the medium (or vice versa depending on the stability or instability of the distribution of density and temperature in the system) and thereby the degree of decay or generation of turbulence. Thus, relation (8.32) proves once again that nonzero spirality in the disk, which is generated by the interaction of Archimedean and Coriolis forces, indeed affects the time of turbulence sustenance in the disk.

Hence, the correlation between the density and the vertical velocity $\langle \rho' \partial u_z/\partial z\rangle$ plays an important role in the determination of the magnitude and sign of the mean spirality in the disk. The explicit form of this correlation and the form of the function H can be found only in terms of an adequate hydrodynamic model of the turbulent flow in the disk, where the spatial distributions of all hydrodynamic parameters are known. We do not do this here, but consider the qualitative pattern of the development of spirality in some convection zone of the disk. Let a vortex with the velocity u_z shifts by the distance ζ in the convection zone of the upper part of the disk ($0<z<h_{\mathrm{disk}}$), whose mean density ρ_0 decreases from the equatorial plane toward the disk surface $[(-\partial\rho_0/\partial z)>0]$. In this case, positive fluctuations of density, $\rho' \cong -\zeta\, \partial\rho_0/\partial z>0$, must appear. Because of the gradient of the mean density, it must, in accordance with the continuity equation $\nabla_\perp \boldsymbol{u}_\perp \to -\rho_0^{-1}\partial(u_z\rho_0)/\partial z$ (in the Boussinesq approximation), expand, i.e., acquire horizontal velocity components.

The resulting momenta of Coriolis forces lead to the left-hand spiral rotation. The descending matter in the upper part of the disk should compress and rotate in the opposite direction under the action of the Coriolis forces; i.e., it must again undergo left-hand spiral motion. Note that in the above reasoning, we neglected the intensity of turbulent pulsations in the convective part of the disk, which, in the general case, may not be everywhere the same. The anisotropic intensity of small-scale turbulence may, in principle, reverse the direction of rotation of an individual vortex in the convection zone. However, the density gradient is the main contributor to the total spirality H.

Thus, convection in the upper half of the disk is more likely to lead to left-hand than to right-hand spiral motions. It is evident that right-hand spiral motions should dominate in the lower half of the disk ($-h_{\text{disk}}<z<0$). Hence, the lack of mirror symmetry with respect to the transform $z \rightarrow -z$ leads to the enhancement of vortex structures of a certain sign. The balance of left- and right-hand helical motions may be established in the region of the equatorial plane of the disk only in the absence of the gradient of turbulence intensity in this plane (because uniform turbulence is conductive to the establishment of a uniform distribution of differently oriented vortices), because one of the types of spiral motions comes from below and the other type, from above, i.e., only at the latest stages of the evolution of a differentially rotating protoplanetary cloud.

8.4 Negative Viscosity in Rotating Disk Turbulence as a Manifestation of the Helicity Cascade

Let us now proceed to the interpretation of the inverse cascade of the energy of pulsational velocity $\boldsymbol{u}$, which is possible in the case of three-dimensional spiral turbulence, in terms of negative viscosity.

8.4.1 *Difficulties of the Momentum Transfer Theory*

Let us begin with the fact that the authors of the vast majority of astrophysical papers use common hydrodynamic equations to model evolution of a rotating turbulent cloud using turbulent viscosity instead of molecular viscosity. In such a case, essentially the linear relation

$$R_{ij} - \frac{1}{3}R_{kl}\delta_{kl}\delta_{ij} = \rho K_{ijmn}\left(\partial_m\langle \boldsymbol{U}\rangle_n + \partial_n\langle \boldsymbol{U}\rangle_m\right) = \rho K_{ijmn}e_{mn} \tag{8.34}$$

between the symmetric Reynolds stress tensor R_{ij} and the symmetric tensor $e_{mn} \equiv \partial_m\langle \boldsymbol{U}\rangle_n + \partial_n\langle \boldsymbol{U}\rangle_m$ of deformation rates (i.e., the general Prandtl theorem of the momentum transfer) is used. The quantities K_{ijmn} (the components of some fourth-rank tensor that is symmetric in i, j and m, n) of this linear function have the meaning of the coefficients of turbulent viscosity and are determined by the statistical characteristics of small-scale turbulence. Basically, by the very definition of isotropic large-scale turbulence, all the mean quantities associated with it remain unchanged in the case of arbitrary rotations (but not necessarily in the case of reflections); the tensors that possess this property are isotropic. Recall that the Kronecker (δ_{ij}) and Levi-Civita (ε_{ijk}) tensors are examples of isotropic tensors of the second and third rank, respectively, and the tensor product $\delta_{ij}\delta_{mn}$ is a fourth-rank isotropic tensor. These tensors have the same components in all coordinate systems and hence have the invariable components in the case of arbitrary rotation. Let us further assume that the tensor of turbulent viscosity, K_{ijmn}, is isotropic (but not mirror-symmetric). In this case, the expansion $K_{ijmn} = a\delta_{ij}\delta_{mn} + b\delta_{im}\delta_{jn} + c\delta_{in}$ δ_{jm} is valid (Korenev 1996) and we obtain, by substituting it into (8.34),

$$R_{ij} = -\frac{2}{3}\rho E\delta_{ij} + \rho\nu^{\text{turb}} e_{mn} \quad (\nu^{\text{turb}} \equiv b + c). \tag{8.35}$$

Note that the coefficient of turbulent viscosity ν^{turb} in the (8.35), which is determined by the small-scale field of pulsational velocity $\boldsymbol{u}$, is usually assumed to be positive. However, we cannot rule out the exotic possibility $\nu^{\text{turb}}<0$, which, according to Kraichnan (1976a), may be realized owing to the fluctuations of the spirality of the turbulent field $\boldsymbol{u}$. Let us emphasize that the positiveness of the coefficients of turbulent exchange (and, in particular, of the coefficient ν^{turb}) has not been-proven in the general case for a turbulent liquid flow (unlike the case of molecular transfer coefficients, whose positiveness is deeply rooted in the thermodynamics of irreversible processes); turbulent viscosity can be negative in the case of a two-dimensional flow (see Vergassola et al. 1993; Gama et al. 1994). We now use general formula (8.35) to obtain for the particular model of the averaged disk motion that the tangential stress does not depend on the gradient of angular velocity

$$R_{r\varphi} = \rho\nu^{\text{turb}} r\frac{\partial}{\partial r}\left(\frac{\langle U\rangle_\varphi}{r}\right) = \rho\nu^{\text{turb}} r\frac{\partial\Omega(r)}{\partial r}. \tag{8.36}$$

Because the angular velocity in the Keplerian disk decreases with heliocentric distance, the direction of transfer of angular momentum (of matter) should also be directed away from the Sun. Thus, if one applies the Prandtl theory of the angular momentum transfer (or the theory of turbulent stress) to the entire cloud (without the allowance for the effect of the strong gravitational field of the Sun on the inner parts of the protoplanetary cloud) it leads us to a conclusion (an evidently wrong one) about the general outward transfer of matter throughout the entire rotating turbulent cloud. Since such problems faced by the theory of momentum transfer in the general case of curvilinear flows, the author of semi-empiric theory of turbulence Taylor (1915) and then Von Karman (1936) suggested a logical generalization of (8.36) where tangential stresses are assumed to depend on the gradient of momentum

$$R_{r\varphi} = \rho v_s^{\text{turb}} \frac{1}{r} \frac{\partial}{\partial r} \left[r^2 \Omega(r) \right]. \tag{8.37}$$

The difference between (8.36) and (8.37) proved to be very important for the model of the evolution of the protoplanetary cloud, because angular velocity in the disk decreases with heliocentric distance, whereas the angular momentum increases, and hence the two quantities should be transferred in the opposite directions according to the above two viewpoints. That is why (8.36) and (8.37), if used separately, cannot explain all the specific features of turbulent rotational motion of matter in all parts of the disk in the cases where outer parts of the cloud are effectively transferred outward and inner parts, toward the Sun (see Safronov 1969). In view of this, Wasiutynski (1946) suggested a more general formula for tangential stresses $R_{r\varphi}$ in a rotating medium

$$R_{r\varphi} = \rho K_r^r \frac{1}{r} \frac{\partial}{\partial r} \left[r^2 \Omega(r) \right] - 2\rho K_\varphi^\varphi \Omega(r), \tag{8.38}$$

which includes both cases considered above and is associated with the effect of anisotropic viscosity. In the case of a purely radial flow ($K_\varphi^\varphi = 0$), this formula transforms into (8.37), whereas in the case of an isotropic medium ($K_\varphi^\varphi = K_r^r$) it coincides with the common hydrodynamic formula (8.36). Note, however, that the form of the Wasiutynski stress tensor (of which (8.38) is basically not a component of any tensor) can be used only in a particular coordinate system. Its particular case is (8.36) which is widely used in the astrophysical applications (see, e.g., Tassoul 1979) to explain the differential rotation of various cosmic objects in terms of "anisotropic viscosity," so far lacks physical argumentation; i.e., it is still unclear whether it is only a formal generalization or whether it characterizes turbulent flow more precisely. Below, we give a possible substantiation of (8.38) in terms of asymmetric hydrodynamics of turbulized media. However, we shall first show that negative viscosity may appear in three-dimensional disk turbulence even within the framework of the classical Prandtl theory of turbulent stress (i.e., (8.36)).

8.4.2 Negative Viscosity (Thermodynamic Approach)

Turbulent chaos is much away from the full chaos of thermodynamic equilibrium because it possesses a certain orderliness: even in a developed locally isotropic turbulence, the Kolmogorov spectrum of the distribution of kinetic energy (of pulsation motion) in the space of wave numbers k in the inertial scale interval $E(k) \sim k^{-5/3}$ is far from uniform ($E(k) = \text{const}$). Nonetheless, in the case of phenomenological description of a quasi-equilibrium subsystem of structurized turbulent chaos, we proceed from the formalism of generalized statistical thermodynamics, which assumes the analysis of the ensemble of microscopically identical systems of turbulent chaos with identical generalized thermodynamical state parameters (such as the internal energy of chaos $U_{\text{turb}}(\boldsymbol{r},t)$, the generalized "turbulization temperature" $T_{\text{turb}}(\boldsymbol{r},t)$ (which is not reduced in the general case to the absolute temperature), the specific volume $1/\rho(\boldsymbol{r},t)$, etc.) and requires the use of a probabilistic approach (Kolesnichenko 2002; Marov and Kolesnichenko 2006). The latter is due to large-scale turbulent fluctuations of some additional parameters of the state of chaos $q_k(\boldsymbol{r},t)$ $(k=\overline{1,n})$ (the so-called internal coordinates) which should be distinguished from statistical molecular fluctuations, arising due to the atomic structure of the system and serve as a difference measure in any set of such thermodynamically identical systems. The internal coordinates, which describe the thermodynamic state of chaos, may include fluctuating positive-determined parameters, which adequately characterize a whirling liquid (including coherent mesoscale formations) inside a physically infinitesimal elementary volume $d\boldsymbol{r}$. In particular, one can choose the following stochastic quantities q_k the turbulent energy dissipation rate ε, the generalized angular velocities (which characterize coherent mesoscale vortex formations), the enstrophy Ω (in the case of a flat flow), etc. Note that some of the internal coordinates q_k may belong to the incoherent component of the subsystem of turbulent chaos, whereas other internal coordinates may characterize individual coherent mesoscale structures.

In Chap. 5 we used methods of non-equilibrium thermodynamics to represent our two-level macroscopic description of the turbulized medium of the protoplanetary cloud in terms of two interacting continua. We showed that, in the case of in the vortex continuum, which corresponds to small-scale components of pulsating thermohydrodynamic parameters, such a quasi-stationary regime is established between the takeoff of energy from the "external source" (which, in particular, is associated with the averaged differential rotation of the turbulized matter of the disk) and the energy dissipation due to internal dissipative processes in the subsystem of chaos. In the latter, the production of entropy of turbulization $S_{\text{turb}}(\boldsymbol{r},t)$ is compensated by its outflow to the subsystem of averaged motion so that the total production of entropy from chaos is minimal. Note that, according to Kolmogorov (1941), the energy flow η along the hierarchy of turbulent vortices down to the molecular level is a characteristic parameter of the subsystem of small-scale turbulence. In the stationary case, this flow coincides with the energy dissipation rate ε. It follows from this that the subsystem of turbulent chaos exports entropy into the "external medium", i.e., gives it to the subsystem of averaged motion. In other words, the inflow of negative entropy (negentropy) from the "external medium" (the subsystem of averaged motion) is required to maintain the stationary state inside the subsystem of turbulent chaos; this negentropy coming to the subsystem of chaos is used for the maintenance and improvement of its internal structure. Such a condition is known (Prigogine and Stengers 1984) to be sufficient for the development of coherent dissipative (mesoscale) structures in the vortex continuum.

It was showed (Kolesnichenko 2002) that in this quasi-stationary case the total production of turbulent entropy $\sigma_{\text{turb}}(\boldsymbol{r},t)$ (energy scattering) has the structure of a bilinear form $T_{\mathrm{turb}}\sigma_{\mathrm{turb}}(\boldsymbol{r},t) = \sum_{\alpha} \Im_{\alpha}(\boldsymbol{r},t)X_{\alpha}(\boldsymbol{r},t)$, whose explicit form is determined by the particular model of turbulized medium, i.e., by the set of hydrodynamic processes included into the model. According to the main postulate of non-equilibrium thermodynamics (see, e.g., de Groot and Mazur 1962), this form can be used to derive the defining (closing) relationships between the thermodynamic flows $\Im_{\alpha}$ and the forces X_{α} in the form of linear relations $\Im_{\alpha i} = \sum_{\beta} L^{ij}_{\alpha\beta} X_{\beta j}$ $(\alpha, \beta = 1, 2, \ldots)$. The peculiarity of the turbulized continuum is that the matrix of Onsager coefficients $L_{\alpha\beta}$ depends not only on averaged thermodynamical parameters of state of the medium (as in the laminar case), but also on the statistical characteristics of the subsystem of turbulent chaos and, in particular, on the energy flow ε along the cascade of turbulent vortices (which is thus one of the thermodynamic flows in the system) or on the flow of hydrodynamic spirality, which is effectively generated in the case of gyrotropic small-scale turbulence. Such a situation, which is typical of any self-organizing (synergetic) system, results, in the general case, in nonpositive-determined individual terms $\Im_{\alpha}(\boldsymbol{r},t)$ $X_{\alpha}(\boldsymbol{r},t)$ in the sum $T_{\mathrm{turb}}\sigma_{\mathrm{turb}}(\boldsymbol{r},t)$ which, however, is positive-determined, $\sigma_{\mathrm{turb}} \geq 0$. It is well known (see, e.g., Haken 1983) that in this case the superposition of various thermodynamic flows may in principle result in negative values of individual diagonal elements of matrix $L_{\alpha\beta}$ and, thereby, in a situation where some coefficients of turbulent exchange are negative.

Thus, in the case of the evolution of a turbulized protoplanetary cloud, we cannot exclude the possibility of the development of the situation when, in some parts of the cloud, modes of turbulent mass motion occur at which the coefficients of turbulent exchange may acquire negative values (e.g., the viscosity coefficient ν^{turb} in (8.36)) (Sivashinsky and Frenkel 1992; Vergassola et al. 1993; Gama et al. 1994). It follows from the above analysis of gyrotropic disk turbulence that the hydrodynamic spirality H may serve as a statistical characteristic of the turbulent field, which could provide the inverse cascade of kinetic energy and thereby the appearance of the effect of negative viscosity in a three-dimensional disk.

8.4.3 Rotational Viscosity

Let us now return to the difficulty of the Prandtl theory of momentum transfer in turbulized media that is faced in astrophysics when attempting to explain differential rotations of gaseous objects. The standard approach to the derivation of the averaged hydrodynamic equations (based on the Reynolds postulates), which, in particular, are intended for modeling the protoplanetary cloud, cannot apparently be considered quite adequate, because, as we mentioned above, the actual picture of turbulent transfer in the disk significantly differs from the classical one (Safronov 1969). Although the authors of published works, beginning from Reynolds, the founder of the phenomenological theory of turbulence, and then Mattioli earlier discussed the approaches involving the asymmetry of turbulent stress tensor ($R_{ij} \neq R_{ji}$) and such additional internal characteristics of the state of the turbulent field as the vortex, the moment of inertia, and the moment of internal forces, this line of research has been, unfortunately, neither deservedly appreciated nor further developed.

At the same time, recently interest has been renewed again in asymmetric hydrodynamics of turbulized media (hydrodynamics of moments). In this regard it worth to note that the asymmetric hydromechanics of Kosser (see, e.g., de Groot and Mazur 1962) has since long gained wide recognition, e.g., in the theory of liquid crystals and liquid helium. The new interest has been stipulated by certain achievements in the theory of spatial averaging of various equations of motion in the continuum mechanics including, e.g., liquid flows in porous media, suspension-carrying flows, deformation of composite materials, etc. It was shown, in particular, (Ferrari 1972; Nikolaevskiy 2003), that more accurate spatial averaging (without the traditional Reynolds postulate of the commutativity of averaging and differentiating operations) of hydrodynamic equations for small elements in order to describe the same motions on a macroscale, yield the generalized Reynolds equations. These averaged equations include, in particular, a term with rotational viscosity, which is associated with an asymmetric part of the turbulent stress tensor of a turbulized medium.

Nikolaevskiy (2003) used methods of hydromechanics of moments to derive for an asymmetric turbulent flow the following formula for energy dissipation associated with viscous processes:

$$T_{\mathrm{turb}}\sigma'_{\mathrm{turb}} = \left(R^{s} + \frac{2}{3}\rho E\boldsymbol{I}\right) : \boldsymbol{e} - \boldsymbol{R}^{a} \cdot \; + \boldsymbol{T} : \frac{\partial}{\partial \boldsymbol{r}}(\mathrm{rot}\langle \boldsymbol{U}\rangle + \;) \geq 0, \tag{8.39}$$

where $\boldsymbol{I}$ is the unit tensor, $\boldsymbol{e} = \frac{1}{2}\left(\frac{\partial\langle \boldsymbol{U}\rangle}{\partial \boldsymbol{r}} + \left(\frac{\partial\langle \boldsymbol{U}\rangle}{\partial \boldsymbol{r}}\right)^{\mathrm{transp}}\right)$ is the tensor of averaged deformations; $\boldsymbol{R}^{s}$ and $\boldsymbol{R}^{a}$ are the symmetric and asymmetric parts of the Reynolds stress tensor, respectively; $(\; \equiv \langle \mathrm{rot}\boldsymbol{u}\rangle)$ is the so-called vector of internal angular velocity, which is due to the intrinsic vorticity of the field of pulsation velocities $\boldsymbol{u}$ and characterizes the vortex "anisotropy" of the flow on the microscale l (recall that, in the case of a disk, an elementary volume of the scale length l may contain a large number of rotating vortex for motions (clusters)); $\boldsymbol{T}$ is the turbulent momentum stress tensor associated with pulsational transfer of fluctuations of the moment of momenta of small-scale vortices. Let us note that in asymmetric turbulent hydromechanics, the tensor of averaged deformations appears in the additional evolutionary equation of the internal angular momentum balance (in the equation for) (de Groot and Mazur 1962; Nikolaevskiy 2003). This, by the way, is a serious argument for the use of turbulent hydromechanics of moments and of the concept of two-level macroscopic description of turbulized medium in the models of a protoplanetary cloud.

In the general case of an anisotropic liquid, flows and thermodynamic forces appearing in (8.39) are related by the following simple set of defining relationships:

$$R_{ij}^{s} \equiv \tfrac{1}{2}(R_{ij} + R_{ji}) = -\tfrac{2}{3}\rho E\delta_{ij} + \rho K_{ijmn}e_{mn}, \tag{8.40}$$

$$R_{ij}^{a} \equiv \tfrac{1}{2}(R_{ij} - R_{ji}) = -\rho K_{ijmn}^{*}\varepsilon_{mnkk}, \tag{8.41}$$

$$T_{ij} = \rho K_{ijmn}^{**}\partial_n(\varepsilon_{mlk}\partial_l\langle U_k\rangle + {}_m), \tag{8.42}$$

typical of Kosser's asymmetric hydrodynamics. Here, the phenomenological turbulent coefficients K_{ijmn}, K_{ijmn}^{*}, and K_{ijmn}^{**} are strongly varying functions of averaged parameters of state of the medium and depend on the statistical characteristics of the turbulent velocity field $\boldsymbol{u}$.

Hence, here we addressed only the simplest case that follows from the isotropic (but not mirror symmetric) structure of turbulent transfer coefficients. In the case of an anisotropic small-scale turbulent field, the situation becomes much more complicated and a number of additional terms appear in (8.43), which are associated with the vector field (e.g., the field of the gradient of turbulence intensity, the field of density gradient, or the angular velocity of the system) that caused the anisotropy (Krause and Radler 1980). Note that in most of the liquid, after a short relaxation time the curl of the averaged velocity, $\mathrm{rot}\langle\boldsymbol{U}\rangle$, becomes equal to the angular velocity that determines the internal rotation of the mass elements of the continuum, $-\ \varepsilon_{mlk}\ \partial_l\langle U_k\rangle = {}_m$. In this case, the thermodynamic force in linear constitutive relations (8.43) also vanishes and so do the corresponding flux densities of the moments of momentum of small-scale vortices, i.e., the interaction between the vortices of the macroscopic velocity field and intrinsic rotational motion of the particles disappears (see, e.g., de Groot and Mazur 1962). However, the law of the parity of tangential stresses $R_{ij} = R_{ji}$ on a macrolevel breaks down in the case of the turbulized continuum (unlike the behavior of the laminar flow) and relation (8.35) should be replaced by the following formula:

$$R_{ik} = R_{ik}^{s} + R_{ik}^{a} = -\tfrac{2}{3}\rho E\delta_{ik} + \rho\nu^{\mathrm{turb}}e_{ik} + \rho\nu_r^{\mathrm{turb}}\varepsilon_{ikp}(\mathrm{rot}\langle\boldsymbol{U}\rangle)_p. \tag{8.43}$$

The coefficients ν^{turb} and ν_r^{turb} are determined by the field of turbulent velocity $\boldsymbol{u}$, and the coefficient ν^{turb} is a scalar, whereas the coefficient ν_r^{turb} is a pseudoscalar, because the tensor $\varepsilon_{ikp}(\mathrm{rot}\langle\boldsymbol{U}\rangle)_p$ is a second-rank pseudotensor.

Let us now analyze turbulence with mirror symmetry. In this case, on the one hand, the coefficients ν^{turb} and ν_r^{turb} should not change during reflection transform; however, on the other hand, the coefficient ν_r^{turb} must reverse its sign, because it is a pseudoscalar. Therefore, the coefficient $\nu_r^{\mathrm{turb}} = 0$ in the case of an isotropic and mirror symmetric small-scale turbulence. Hence, the rotational viscosity ν_r^{turb} may differ from zero only if the field of turbulent velocities is not statistically invariant with respect to the parity transform, in particular, if spirality $H \neq 0$.

In the disk turbulence model under consideration, relation (8.43) for shear stress acquires the following simple form:

$$\begin{aligned} R_{r\varphi} &= \rho \nu^{\mathrm{turb}} r \frac{\partial \Omega(r)}{\partial r} + \rho \nu_r^{\mathrm{turb}} \frac{1}{r} \frac{\partial r^2 \Omega}{\partial r} \\ &= \rho \nu^{\mathrm{turb}} (\frac{1}{r} \frac{\partial r^2 \Omega}{\partial r} - 2\Omega) + \rho \nu_r^{\mathrm{turb}} \frac{1}{r} \frac{\partial r^2 \Omega}{\partial r} \\ &= \rho (\nu^{\mathrm{turb}} + \nu_r^{\mathrm{turb}}) \frac{1}{r} \frac{\partial r^2 \Omega}{\partial r} - 2\rho \nu^{\mathrm{turb}} \Omega \end{aligned} \tag{8.44}$$

(compare it with (8.35)). Therefore, the formula for tangential stresses in a rotating medium suggested by Wasiutynski can be physically substantiated in terms of asymmetric mechanics of turbulized media with an asymmetric Reynolds stress tensor. One of the important implications of (8.44) is the conclusion about the mutual complementarity of the Prandtl theory of momentum transfer and the Taylor vortex transfer theory in a rotating medium. The appearance of the additional degree of freedom in asymmetric hydrodynamics makes both approaches necessary for solving particular problems. In particular, one of the competing mechanisms should dominate in a rotating protoplanetary cloud depending on the numerical values of the coefficients of shear and rotational viscosity in the corresponding disk regions.

Let us now briefly summarize the main results contained in this chapter. In framework of the problem of reconstruction of the evolution of the protoplanetary cloud we analyzed the possibility of the influence of hydrodynamical spirality that develops in a rotating disk on the synergetic structurization of matter and on the appearance of the effect of negative turbulent viscosity. It was showed that relatively long persistence of turbulence in a protoplanetary cloud may be partially due to the lack of mirror symmetry of the field of pulsating velocities with respect to the equatorial disk plane.

We advanced specific mechanism of the development of coherent mesoscale formations in the subsystem of turbulent chaos associated with the phenomenon of phase-frequency synchronization of auto-oscillations of stochastic internal coordinates (corresponding to the coherent component of chaos) that was earlier analyzed by us (Kolesnichenko 2004, 2005a, b; Marov and Kolesnichenko 2002, 2006). It was also shown the principal possibility of the self-organization of a flow, where the generation of coherent formations associated with the effect of "phase transitions" induced by the natural noise of the fine-grained fluctuating field of chaos is likely in the process of the temporal evolution of the quasi-equilibrium vortex subsystem. In this paper, we formulate the general concept of energy feeding of coherent mesoscale vortex structures in the thermodynamically open subsystem of turbulent chaos, which is associated with the realization of inverse cascade of kinetic energy in mirror asymmetric disk turbulence.

Because of the energy release, the inverse cascade generates the corresponding hierarchical system of gas condensations (with a fractal density distribution), which ultimately leads to the intensification of mechanical and physicochemical interactions between the particles of matter, resulting in spontaneous formation and growth of dust clusters, stimulation of condensation and phase transitions, of mass and heat exchange between different parts of the heterogeneous disk subsystem, and substantial modification of the oscillation spectrum, etc.

We used the methods of non-equilibrium thermodynamics and a two-level description of maximally developed disk turbulence to demonstrate the possible appearance of the effect of negative viscosity in the three-dimensional case. We suggested that hydrodynamic spirality, which develops owing to fast rotation of an unstably stratified disk medium, can be used as a statistical characteristic of isotropic, mirror-non-invariant small-scale turbulence. In this case, the appearance of negative viscosity in the disk can be due to cascade transfer of energy from small to larger vortices in spiral turbulence.

We derived, within the framework of asymmetric hydromechanics of turbulized media, a formula for the tensor of turbulent stress in the Wasiutynski form, which is widely used in astrophysical literature to explain the differential rotation of various cosmic objects in terms of "anisotropic viscosity." This phenomenological relation, which is used in astrophysics because of the well-known difficulties faced by the Prandtl theory of momentum transfer in a rotating turbulized medium, has until recently lacked physical substantiation. In this paper, we reveal the mutual complementarity of the Prandtl theory of momentum transfer and the Taylor theory of angular momentum transfer for a rotating protoplanetary cloud in the cases where one of these competing mechanisms dominates (depending on the numerical values of the shear and rotational viscosity coefficients) at a certain heliocentric distance.

Our interest in the hydrodynamic spirality as applied to a structured disk turbulence stems from the fact that the existence of such an additional nonviscous invariant implies, generally speaking, a certain modification of the classical energy cascade process in the inertial spectral domain, where inverse energy transfer is possible from small to larger vortices. This makes it possible not only to explain the phenomenon of negative viscosity in a differentially rotating protoplanetary cloud, but also to forecast the birth of energy active coherent vortex structures, which ultimately trigger the formation mechanisms of gas and dust clusters in the disk. Unfortunately, the effect of inverse cascade of energy for three-dimensional gyrotropic turbulence so far lacks a reliable confirmation in numerical simulations and hence, an unequivocal and conclusive answer to the key problem of approach to the criteria of realization of such a cascade in a differentially rotating disk is yet expected. Anyway, the development of a universal model of turbulence appears to be remained problematic and a great challenge.

Chapter 9
Thermodynamic Model of MHD Turbulence and Some of Its Applications to Accretion Disks

In this chapter we have derived a closed system of magnetohydrodynamic (MHD) equations for the scale of mean motion in the approximation of single-fluid magnetohydrodynamics. The system designed to model the shear and convective turbulent flows of electrically conducting media in the presence of a magnetic field and relevant to the more comprehensive study of accretion discs. In particular, these equations are applicable to model intense turbulent flows of cosmic plasma in accretion disks and associated coronas, in which the magnetic field noticeably affects the dynamics of astrophysical processes. In developing the model of a conducting turbulized medium, apart from the conventional probability-theoretical averaging of the MHD equations, we systematically use the weighted Favre averaging. The latter allows us to considerably simplify the writing of the averaged equations of motion for a compressible fluid and the analysis of the mechanisms of macroscopic field amplification by turbulent flows. To clearly interpret the individual components of the plasma and field-energy balance, we derive various energy equations that allow us to trace the possible energy conversions from one form into another, in particular, to understand the transfer mechanisms of the gravitational and kinetic energies of the mean motion into magnetic energy. Special emphasis is placed on the method for obtaining the closure relations for the total (with allowance made for the magnetic field) kinetic turbulent stress tensor in an electrically conducting medium and the turbulent electromotive force (or the so-called magnetic Reynolds tensor) in the framework of extended irreversible thermodynamics. This approach also makes it possible to analyze the constraints imposed on the turbulent transport coefficients by the entropy growth condition. We suggest a technique for modeling the turbulent transport coefficients, in particular, the coefficient of kinematic turbulent viscosity that allows us to take into account the magnetic field effect and the inverse effect of the heat transfer on the development of turbulence in a rotating electrically conducting disk.

M.Y. Marov and A.V. Kolesnichenko, *Turbulence and Self-Organization: Modeling Astrophysical Objects*, Astrophysics and Space Science Library 389,
DOI 10.1007/978-1-4614-5155-6_9, © Springer Science+Business Media New York 2013

9.1 Basic Equations of Magnetohydrodynamics for Modeling the Disk Structure and Corona

If the plasma is magnetohydrodynamically unstable, then the excitation of comparatively slow, large-scale motions that can be adequately described in the approximation of single-fluid magnetohydrodynamics is possible in it. These flows are similar to ordinary fluid and gas flows, but they are complicated by the action of the Lorentz body force. In this case, the motion of an electrically conducting medium in a magnetic field is accompanied by peculiar phenomena. These phenomena stem from the fact that induction currents are excited during the motion of conducting fluid masses; together with the magnetic field, these currents have an inverse effect on the fluid. If the dissipation mechanisms in the plasma are weak, then these flows become turbulent in nature. Since there are many papers that describe in detail the derivation of the MHD equations (see, e.g., Alfven 1950; Syrovatskii 1957; Kulikovskii and Lyubimov 1962; Biskamp 2003), we give here only the basic concepts for the reader convenience, thus addressing this section as preparatory for a more detailed consideration of the MHD turbulence in the main body of the text. These equations are given in the form that we use below in deriving the averaged equations of turbulent motion for an electrically conducting fluid in the presence of magnetic field.

9.1.1 *Magnetic Induction Equation*

The differential equations of non-relativistic magnetohydrodynamics are the conservation laws derived from the Maxwell electromagnetic equations and general hydrodynamic equations of fluid motion. Let us denote the magnetic-induction vector by $\boldsymbol{B}(\boldsymbol{r},t)$, the magnetic-field strength vector by $\boldsymbol{H}(\boldsymbol{r},t)$, the electric-field strength vector by $\boldsymbol{E}(\boldsymbol{r},t)$, the electric-current density by $\boldsymbol{j}(\boldsymbol{r},t)$, the charge density by $\rho_e(\boldsymbol{r},t)$, the magnetic permeability by $\mu(\boldsymbol{r},t)$, the permittivity by $\varepsilon(\boldsymbol{r},t)$ (the dimensionless coefficients ε and μ are related to the physical properties of the medium), the specific electrical conductivity by $\sigma_e(\boldsymbol{r},t)$, and the hydrodynamic velocity of the conducting fluid flow by $\boldsymbol{u}(\boldsymbol{r},t)$. Below, for simplicity, we assume the parameters μ, ε, and σ_e to be constant and the transfer properties of the medium to be isotropic, despite the presence of a magnetic field. The main assumption of magnetohydrodynamics is that the characteristic electromagnetic (plasma) velocity

$$U_0 \equiv \frac{l_0}{t_0} << c \quad \text{and} \quad \frac{E_0}{l_0} \approx \frac{B_0}{t_0}. \tag{9.1}$$

Here, E_0 and B_0 are typical values of $\boldsymbol{E}(\boldsymbol{r},t)$ and $\boldsymbol{B}(\boldsymbol{r},t)$; l_0 and t_0 are the electromagnetic length and time scales; and c is the speed of light in a vacuum.

In this case, the Maxwell equations (written in the absolute Gaussian system of units) to determine the fields $\boldsymbol{B}$, $\boldsymbol{H}$, $\boldsymbol{E}$, and $\boldsymbol{j}$ take the form

$$\frac{1}{c}\frac{\partial \boldsymbol{B}}{\partial t} = -\mathrm{rot}\,\boldsymbol{E}, \quad \mathrm{rot}\,\boldsymbol{H} = \frac{4\pi}{c}\boldsymbol{j}, \tag{9.2}$$

$$\mathrm{div}\,\boldsymbol{D} = 4\pi\rho_e, \quad \mathrm{div}\,\boldsymbol{B} = 0, \tag{9.3}$$

where the space-charge density ρ_e and the total current density $\boldsymbol{j}$ (measured by a stationary observer) induced by the motion of charged plasma particles are defined by the formulas

$$\rho_e \equiv \sum_k e_k n_k \cong 0, \quad \boldsymbol{j} \equiv \sum_k e_k n_k \boldsymbol{u}_k = \rho_e \boldsymbol{u} + \boldsymbol{j}^* \cong \boldsymbol{j}^*. \tag{9.4}$$

Here, $n_k(\boldsymbol{r},t)$ and $\boldsymbol{u}_k(\boldsymbol{r},t)$ are the number density (per unit volume of the medium) and mean hydrodynamic velocity of the plasma particles of type k, respectively, $\boldsymbol{j}^* \equiv \sum_k e_k n_k(\boldsymbol{u}_k - \boldsymbol{u})$ is the conduction current density (the current measured by an observer moving together with the gas). In (9.4), the summation is over the electrons (with charge $e_e = -e$) and over the charged particles of type k with charge $e_k = eZ_k$. A characteristic feature of the processes that take place, for example, in astrophysical disks, is that the plasma is electrically neutral ($\rho_e \cong 0$) with a high degree of approximation, implying that there is virtually no space charge outside the sphere of a Debye radius $R_D \equiv \sqrt{k_B T/(4\pi n_e e^2)}$, called a quasi-neutral region. In this case, the terms with the space charge ρ_e may be neglected in all Maxwell and hydrodynamic equations. In particular, the convective current density $\rho_e\boldsymbol{u}$ may be neglected by assuming that the electric-current density $\boldsymbol{j}$ is identically equal to the conduction current density $\boldsymbol{j}^*$. To close system of Maxwell equations (9.2) and (9.3), we need two relations describing the polarization and magnetization laws,

$$\boldsymbol{D} = \varepsilon\boldsymbol{E}, \quad \boldsymbol{B} = \mu\boldsymbol{H} \tag{9.5}$$

and the Ohm law giving the dependence of $\boldsymbol{j}^*$ on the various factors that cause the diffusion of the charged components of the electrically conducting mixture, including the fields $\boldsymbol{E}$ and $\boldsymbol{B}$, the hydrodynamic velocity $\boldsymbol{u}$, and the spatial gradients of certain parameters of the medium: $\boldsymbol{j}^* = \boldsymbol{j}^*[\boldsymbol{E}, \boldsymbol{B}, \boldsymbol{u}, \partial()/\partial\boldsymbol{r}]$. In the simplest case of magnetohydrodynamics to which we restrict our analysis, the following simplified form of the Ohm law is used:

$$\boldsymbol{j}^* = \sigma_e\left(\boldsymbol{E} + \frac{1}{c}\boldsymbol{u} \times \boldsymbol{B}\right). \tag{9.6}$$

Note that one of the corollaries of the main assumption of magnetohydrodynamics is that the term related to the displacement current, $c^{-1}\partial\boldsymbol{D}/\partial t$, could be omitted when writing the second equation in (9.2). Another corollary is that the continuity equation for the electric charge ρ_e takes the form div$\boldsymbol{j} = 0$ (the result of taking the divergence of the second equation in (9.2)); physically, this means that the local charge densities ρ_e are negligible in time and the electric currents flow along closed circuits. In this case, the Poisson equation (9.3) serves only to determine (from the electric field found from the remaining equations of system (9.2) and (9.3)) the small plasma deviation from quasi-neutrality. The latter should often be taken into account either near the system's boundary (in a layer on the order of the radius R_D in thickness) or in analyzing the high-frequency plasma oscillations.

It is convenient to eliminate the fields $\boldsymbol{E}$ and $\boldsymbol{j}$ from Maxwell equations (9.2) and Ohm law (9.6). When using the well-known formula of vector analysis curl rot rot $\boldsymbol{a} = (\partial/\partial\boldsymbol{r})\mathrm{div}\,\boldsymbol{a} - \nabla^2\boldsymbol{a}$, this leads to the so-called magnetic-induction equation

$$\frac{\partial \boldsymbol{B}}{\partial t} = \mathrm{rot}(\boldsymbol{u} \times \boldsymbol{B}) + \nu_M \nabla^2 \boldsymbol{B}, \tag{9.7}$$

which is one of the basic MHD equations. Here, ∇^2 is the Laplace operator and $\nu_M \equiv c^2/4\pi\mu\sigma_e$ is the so-called magnetic-diffusion coefficient (or magnetic-viscosity coefficient), which has the same dimensions as the kinematic-viscosity coefficient ν or the diffusion coefficient D, i.e., cm^2/s. Applying the divergence operator to induction equation (9.7), we find that the equation $(\partial/\partial t)\mathrm{div}\,\boldsymbol{B} = 0$ holds in magnetohydrodynamics. Therefore, in order that a particular solution satisfy the equation div $\boldsymbol{B} = 0$, it suffices to require that the initial conditions satisfy this equation.

Thus, the second equation in (9.3) holds in view of the induction equation and the initial conditions. For the subsequent analysis, it is convenient to write (9.7) in a different form. Using the formula $\mathrm{rot}(\boldsymbol{a} \times \boldsymbol{b}) = (\boldsymbol{b} \cdot (\partial/\partial\boldsymbol{r}))\boldsymbol{a} - (\boldsymbol{a} \cdot (\partial/\partial\boldsymbol{r}))\boldsymbol{b} + \boldsymbol{a}$ div $\boldsymbol{b} - \boldsymbol{b}$ div $\boldsymbol{a}$ and the continuity equation for the velocity of the medium (see (9.10)), we rewrite (9.7) as

$$\rho \frac{\mathrm{d}}{\mathrm{d}t}\left(\frac{\boldsymbol{B}}{\rho}\right) = \left(\boldsymbol{B} \cdot \frac{\partial}{\partial \boldsymbol{r}}\right)\boldsymbol{u} + \nu_M \nabla^2 \boldsymbol{B}, \tag{9.8}$$

where $\mathrm{d}/\mathrm{d}t \equiv \partial/\partial t + \boldsymbol{u} \cdot (\partial/\partial\boldsymbol{r})$ is the substantial derivative describing the rate of change of a particular parameter in a coordinate system moving at velocity $\boldsymbol{u}$.

We see from (9.8) that the evolution of the magnetic induction vector $\boldsymbol{B}$ depends on the motion of the electrically conducting medium, i.e., given the hydrodynamic velocity $\boldsymbol{u}$, (9.8) (div $\boldsymbol{B} = 0$) completely defines the vector $\boldsymbol{B}$ satisfying the condition div $\boldsymbol{B} = 0$. Thus, the induction vector $\boldsymbol{B}$ in magnetohydrodynamics is considered as the root cause of the generation of electromagnetic motions, while the generation of the current density and the electric field are considered as secondary phenomena: the corresponding values of $\boldsymbol{j}$ and $\boldsymbol{E}$ can be found from the Ampere and Ohm laws

$$\boldsymbol{j} = \frac{c}{4\pi\mu}\mathrm{rot}\,\boldsymbol{B}, \quad \boldsymbol{E} = c^{-1}(\nu_M \mathrm{rot}\,\boldsymbol{B} - \boldsymbol{u} \times \boldsymbol{B}). \tag{9.9}$$

In the cases of interest, the charge density ρ_e can be determined from the Poisson equation $\rho_e = (\varepsilon/4\pi)\mathrm{div}\,\boldsymbol{E}$.

It follows from magnetic-induction equation (9.8) that the similarity conditions for the flows of electrically conducting media are additionally determined by the magnetic Reynolds number $\boldsymbol{Re}_M \equiv V_0 L_0/\nu_M$, which characterizes the ratio of the convective term in (9.8) to the diffusive term. This parameter complements the similarity criteria for the flows of non-conducting media. Here, ρ_0, V_0, and L_0 are the certain density, velocity, and length, respectively, characteristic of the problem under consideration. If $\boldsymbol{Re}_M >> 1$, then the terms with the magnetic viscosity ν_M in (9.8) may be neglected. In turn, in the case of a perfectly conducting medium, Ohm law (9.6) is reduced to the form $\boldsymbol{E} = -c^{-1}\boldsymbol{u}\times\boldsymbol{B}$ with a very high degree of approximation. In magnetohydrodynamics, it was shown (see, e.g., Syrovatskii 1957) that, in this case, the quantity $(\boldsymbol{B}/\rho)$ at each point of the medium varies proportionally to the separation between two neighboring "fluid particles" located on a magnetic-field line; in other words, the magnetic-field lines are, as it were, frozen into the particles of the matter. If $\boldsymbol{Re}_M << 1$, then the induction equation takes the form $\partial\boldsymbol{B}/\partial t = \nu_M\nabla^2\boldsymbol{B}$, i.e., the field $\boldsymbol{B}$ is defined by the diffusion equation. In accretion disks, the "effective" magnetic Reynolds number is on the order of unity and the turbulent transfer is important. Let us now consider in more detail the MHD equations that describe satisfactorily the large-scale motions in plasma for which the difference in the motion of its constituent components is insignificant.

9.1.2 Mass and Momentum Conservation Equations

Let us address now in more detail magneto hydrodynamic equations applicable to describe large scale plasma flow when a difference in motions of the composed charged components is insignificant.

9.1.2.1 Mass Conservation Equation

The mass conservation equation can be written as

$$\frac{\partial\rho}{\partial t} + \mathrm{div}(\rho\boldsymbol{u}) = 0 \quad \text{or} \quad \rho\frac{\mathrm{d}}{\mathrm{d}t}\left(\frac{1}{\rho}\right) = \mathrm{div}\,\boldsymbol{u}. \tag{9.10}$$

Here, the mass density $\rho(\boldsymbol{r},t)$ and the total hydrodynamic velocity $\boldsymbol{u}(\boldsymbol{r},t)$ are defined by the relations

$$\rho \equiv \sum_k \rho_k = \sum_k n_k m_k, \quad \boldsymbol{u} \equiv \rho^{-1}\sum_k \rho_k \boldsymbol{u}_k, \tag{9.11}$$

where m_k is the molecular mass of the particle of type k.

9.1.2.2 Momentum Conservation Equation

The momentum balance equation for a conducting medium is

$$\rho \frac{\mathrm{d}\boldsymbol{u}}{\mathrm{d}t} \equiv \frac{\partial \rho \boldsymbol{u}}{\partial t} + \left(\frac{\partial}{\partial \boldsymbol{r}} \cdot (\rho \boldsymbol{u}\boldsymbol{u}) \right) = -\frac{\partial p}{\partial \boldsymbol{r}} + \left(\frac{\partial}{\partial \boldsymbol{r}} \cdot \Pi \right) + \sum_k n_k \boldsymbol{F}_k, \tag{9.12}$$

where $p(\boldsymbol{r}, t)$ the fluid; $\boldsymbol{u}\boldsymbol{u} = \{u_i u_j\}$ is a dyad (a tensor with nine Cartesian components); $\Pi(\boldsymbol{r}, t) (= \{\Pi_{ij}\})$ is the viscous stress tensor related to the molecular transfer of the momentum of all plasma components; $((\partial/\partial \boldsymbol{r}) \cdot \Pi)_j = \partial \Pi_{ij}/\partial x_i$, with the relation between the tensors Π and $\partial \boldsymbol{u}/\partial \boldsymbol{r}$ for an isotropic fluid having the standard form

$$\Pi_{jk} = \rho \nu \left(\frac{\partial u_k}{\partial x_j} + \frac{\partial u_j}{\partial x_k} - \frac{2}{3} \delta_{jk} \mathrm{div}\, \boldsymbol{u} \right) + \mu_\vartheta \delta_{jk} \mathrm{div}\, \boldsymbol{u},$$

where the molecular coefficients of kinematic viscosity $\nu(\boldsymbol{r}, t)$ and bulk viscosity μ_ϑ $(\boldsymbol{r}, t)$ depend only on the local thermodynamic properties of the medium; $\boldsymbol{F}_k(\boldsymbol{r}, t)$ $= \boldsymbol{F}_k^{\mathrm{pond}}(\boldsymbol{r}, t) + m_k \boldsymbol{g}(\boldsymbol{r})$ is the external force acting on one charged particle of type k; $\boldsymbol{F}_k^{\mathrm{pond}}$ is the Lorentz force (the so-called ponder motive force of the electromagnetic field), which in an inertial coordinate system in the non-relativistic approximation is

$$\boldsymbol{F}_k^{\mathrm{pond}} \equiv e_k \left(\boldsymbol{E} + \frac{1}{c} \boldsymbol{u}_k \times \boldsymbol{B} \right) \tag{9.13}$$

and it should be taken into account if the particles of type k carry the charge e_k; $\boldsymbol{g}(\boldsymbol{r}) = -\partial \Psi_G / \partial \boldsymbol{r}$ is the acceleration vector of the external mass force (gravity); and $\Psi_G(\boldsymbol{r})$ is the potential function of the gravitation field. For example, in the case of a gravitational disk where the self-gravity can be neglected (which takes place when $\mathcal{M}_{\mathrm{disk}}/\mathcal{M}_\odot \leq h_{\mathrm{disk}}/R$, where h_{disk} and R are, respectively, the disk half thickness and radius (see, e.g., Hersant et al. 2004)),

$$\Psi_G \equiv \frac{G\mathcal{M}_\odot}{|\tilde{\boldsymbol{r}}|}, \quad \boldsymbol{g} = -\frac{\partial}{\partial \boldsymbol{r}} \Psi_G = \frac{G\mathcal{M}_\odot}{|\tilde{\boldsymbol{r}}|^3} \tilde{\boldsymbol{r}}, \tag{9.14}$$

where $\mathcal{M}_\odot$ is the mass of the central body (star); G is the gravitational constant; $|\tilde{\boldsymbol{r}}| = \sqrt{r_1^2 + r_2^2 + r_3^2}$ is central radius vector; $\tilde{\boldsymbol{r}} = \sum_k \boldsymbol{i}_k \boldsymbol{r}_k$ (here and in what follows, the center of mass of the protostar is taken as the coordinate origin); and $\boldsymbol{i}_k$ are the unit coordinate vectors. When the self-gravity effects are important, $\Psi_G = G\mathcal{M}_\odot$ $/|\tilde{\boldsymbol{r}}| + \Psi_{cr}$, where the self-gravity potential Ψ_{cr} satisfies the Poisson equation ∇^2 $\Psi_{cr} = 4\pi G \rho$.

When (9.13) for the Lorentz force is used, the term in equation of motion (9.12) corresponding to the external electromagnetic and gravitational forces takes the following form for a quasi-neutral plasma:

$$\sum_k n_k \boldsymbol{F}_k = \rho \boldsymbol{g} + \rho_e \left(\boldsymbol{E} + \frac{1}{c} \boldsymbol{u} \times \boldsymbol{B} \right) + \frac{1}{c} \boldsymbol{j} \times \boldsymbol{B} \cong \rho \boldsymbol{g} + \frac{1}{c} \boldsymbol{j} \times \boldsymbol{B}.$$

Using the well-known relation of vector analysis $\boldsymbol{a} \times \operatorname{rot} \boldsymbol{a} = \frac{1}{2}(\partial/\partial \boldsymbol{r})|\boldsymbol{a}|^2 - ((\partial/\partial \boldsymbol{r}) \cdot (\boldsymbol{aa})) + \boldsymbol{a} \operatorname{div} \boldsymbol{a}$, force density $c^{-1}\boldsymbol{j} \times \boldsymbol{B}$ in this equation can be represented as

$$\begin{aligned} \frac{1}{c} \boldsymbol{j} \times \boldsymbol{B} &= -\frac{1}{4\pi\mu} \boldsymbol{B} \times \operatorname{rot} \boldsymbol{B} = -\frac{\partial}{\partial \boldsymbol{r}} \left(\frac{|\boldsymbol{B}|^2}{8\pi\mu} \right) + \left(\frac{\partial}{\partial \boldsymbol{r}} \cdot \left(\frac{\boldsymbol{BB}}{4\pi\mu} \right) \right) \\ &= -\frac{\partial p^M}{\partial \boldsymbol{r}} + \left(\frac{\partial}{\partial \boldsymbol{r}} \cdot \boldsymbol{T}^M \right). \end{aligned} \tag{9.15}$$

Here,

$$\boldsymbol{T}^M \equiv \frac{1}{4\pi\mu} \boldsymbol{BB}, \quad \boldsymbol{T}^M = \left\{ T^M_{ij} \right\} \equiv \left\{ \frac{B_i B_j}{4\pi\mu} \right\} \tag{9.16}$$

is the Maxwell magnetic tension tensor and

$$p^M \equiv \frac{|\boldsymbol{B}|^2}{8\pi\mu} \tag{9.17}$$

is the magnetic-field pressure. As a result, momentum conservation equation (9.12) can be rewritten as

$$\begin{aligned} \rho \frac{\mathrm{d}\boldsymbol{u}}{\mathrm{d}t} &= -\frac{\partial p}{\partial \boldsymbol{r}} + \left(\frac{\partial}{\partial \boldsymbol{r}} \cdot \Pi \right) + \frac{1}{c} \boldsymbol{j} \times \boldsymbol{B} + \rho \boldsymbol{g} \\ &= -\frac{\partial p^\Sigma}{\partial \boldsymbol{r}} + \left(\frac{\partial}{\partial \boldsymbol{r}} \cdot (\Pi + \boldsymbol{T}^M) \right) - \rho \frac{\partial}{\partial \boldsymbol{r}} \Psi_G, \end{aligned} \tag{9.18}$$

where $p^\Sigma \equiv p + p^M = p + |\boldsymbol{B}|^2/8\pi\mu$ is the total plasma pressure.

Let us now analyze this equation. From the dimensional considerations, we can easily derive the following expressions: $F_{\text{vis}} \propto \rho \nu V_0/L_0^2$ for the viscous force (per unit fluid volume), $F_{\text{iner}} \propto \rho V_0^2/L_0$ for the force of inertia, and $F_{\text{mag}} \propto \sigma_e B_0^2 V_0/c^2$ for the electromagnetic force (here, B_0 is the characteristic magnetic field induction that can be taken as the scale for the individual terms of the equations and σ_e is the electrical conductivity of the fluid). It thus follows that the quantity $\sigma_e B_0^2 L_0^2/\rho \nu c^2$ characterizes the ratio of the electromagnetic force to the viscous force in order of magnitude. The square root of this quantity, $\boldsymbol{Ha} \equiv B_0 L_0 \sqrt{\sigma_e/\rho\nu c^2}$ is called the Hartmann number. In turn, Stewart number $\boldsymbol{N} \equiv \sigma_e B_0^2 L_0/c^2 \rho V_0$ characterizes the ratio of the electromagnetic force to the force of inertia in order of magnitude. It is easy to see that $\boldsymbol{Ha}^2 = \boldsymbol{N} \cdot \boldsymbol{Re}$, where $\boldsymbol{Re} \equiv V_0 L_0/\nu$ is the Reynolds number. Combining the criteria $\boldsymbol{Re}$, $\boldsymbol{N}$, and $\boldsymbol{Re}_M$, we can obtain all of the other criteria characterizing the various phenomena observed during the motion of a conducting fluid in a magnetic field. In particular, by dividing the Stewart number by the magnetic Reynolds number, we obtain a criterion called the Alfven number, $\boldsymbol{Al} \equiv B_0^2/4\pi\mu\rho V_0^2$, which characterizes the magnetic-to-kinetic energy ratio. Note that when the questions of heat transfer are considered, the Nusselt, Peclet, and other numbers should be added to these criteria.

9.1.3 Various Forms of the Energy and Heat Influx Equations for an Electrically Conducting Medium

When constructing phenomenologically the single-fluid model of MHD turbulence, we have to average the various energy-balance equations. Here, we give their brief derivation.

9.1.3.1 The Equation for Mechanical Energy of the Conducting Matter

The equation for the mechanical energy (the kinetic energy of the center of mass plus the potential energy of the matter) of a plasma can be obtained through the scalar multiplication of (9.18) by the velocity $\boldsymbol{u}$. As a result, we obtain

$$\rho \frac{\mathrm{d}}{\mathrm{d}t}\left(\frac{|\boldsymbol{u}|^2}{2} + \Psi_G\right) + \mathrm{div}\{p^{\Sigma}\boldsymbol{u} - (\Pi + \boldsymbol{T}^M)\cdot\boldsymbol{u}\}$$
$$= p^{\Sigma}\mathrm{div}\,\boldsymbol{u} - \left((\Pi + \boldsymbol{T}^M) : \frac{\partial}{\partial \boldsymbol{r}}\boldsymbol{u}\right) \tag{9.19}$$

or (in the form of a balance equation)

$$\frac{\partial}{\partial t}\left\{\rho\left(\frac{|\boldsymbol{u}|^2}{2} + \Psi_G\right)\right\} + \mathrm{div}\,\boldsymbol{J}_{\mathrm{mech}} = p^{\Sigma}\mathrm{div}\,\boldsymbol{u} - \left((\Pi + \boldsymbol{T}^M) : \frac{\partial}{\partial \boldsymbol{r}}\boldsymbol{u}\right). \tag{9.19*}$$

Here,

$$\boldsymbol{J}_{\mathrm{mech}} \equiv \left\{\rho\left(\frac{|\boldsymbol{u}|^2}{2} + \Psi_G\right)\boldsymbol{u} + p^{\Sigma}\boldsymbol{u} - ((\Pi + \boldsymbol{T}^M)\cdot\boldsymbol{u})\right\},$$

is the total mechanical energy flux of the plasma, $(\Pi + \boldsymbol{T}^M) \cdot \boldsymbol{u}$ is a vector whose ith component is $(\Pi_{ik} + T_{ik}^M)\, u_k$, and $(\Pi + \boldsymbol{T}^M) : \partial \boldsymbol{u}/\partial \boldsymbol{r} \equiv (\Pi_{ik} + T_{ik}^M)\partial u_i/\partial x_k$ is a scalar. The energy on the right-hand side of (9.19*), which is the work spent on the deformation (velocity shear $\partial \boldsymbol{u}/\partial \boldsymbol{r} \neq 0$), expansion (div $\boldsymbol{u}$>0), or compression (div $\boldsymbol{u}$<0) of a unit volume of the medium, is converted into other forms of energy, for example, internal energy (see (9.20)) or magnetic energy (see (9.27*)).

9.1.3.2 The Equation for the Internal Energy of a Conducting Medium

In the MHD approximation, the differential equation (conservation law) for the internal (specific) energy $E(\boldsymbol{r}, t)$ of the matter can be written in the following substantial form (cf (2.22)):

$$\rho \frac{\mathrm{d}E}{\mathrm{d}t} + \operatorname{div}(\boldsymbol{q} + \boldsymbol{q}_R) = -p \operatorname{div} \boldsymbol{u} + \left(\Pi : \frac{\partial}{\partial \boldsymbol{r}} \boldsymbol{u}\right) + \frac{\nu_M}{4\pi\mu} |\operatorname{rot} \boldsymbol{B}|^2 \equiv Q_E, \qquad (9.20)$$

where $\boldsymbol{q}(\boldsymbol{r}, t)$ is the molecular heat-flux density and $\boldsymbol{q}_R(\boldsymbol{r}, t)$ is the radiation-transferred energy-flux density. The last term on the right-hand side of this equation is the Joule heat per unit time. Equation (9.20) is a modification of the energy equation (2.22) for a multicomponent mixture (Kolesnichenko and Marov 1999) to the case of a conducting medium that can be obtained using the transformation

$$\begin{aligned}\sum_{k=1} n_k(\boldsymbol{u}_k - \boldsymbol{u}) \cdot \boldsymbol{F}_k &= \sum_{k=1} n_k e_k(\boldsymbol{u}_k - \boldsymbol{u}) \cdot \left(\boldsymbol{E} + \frac{1}{c}\boldsymbol{u}_k \times \boldsymbol{B}\right) \\ &= \boldsymbol{j} \cdot \left(\boldsymbol{E} + \frac{1}{c}\boldsymbol{u} \times \boldsymbol{B}\right) = \frac{|\boldsymbol{j}|^2}{\sigma_e} = \frac{\nu_M}{4\pi\mu}|\operatorname{rot} \boldsymbol{B}|^2.\end{aligned}$$

For the subsequent analysis, we need the energy equation written in terms of the enthalpy $H(\boldsymbol{r}, t) (= E + p/\rho)$ of the material component of the plasma. This equation, which follows from (9.20) when the expression $\rho\, \mathrm{d}E/\mathrm{d}t + p \operatorname{div} \boldsymbol{u} = \rho\, \mathrm{d}H/\mathrm{d}t + \mathrm{d}p/\mathrm{d}t$ is taken into account, is

$$\rho \frac{\mathrm{d}H}{\mathrm{d}t} \equiv \rho c_p \frac{\mathrm{d}T}{\mathrm{d}t} = \frac{\mathrm{d}p}{\mathrm{d}t} - \operatorname{div}(\boldsymbol{q} + \boldsymbol{q}_R) + \left(\Pi : \frac{\partial}{\partial \boldsymbol{r}} \boldsymbol{u}\right) + \frac{\nu_M}{4\pi\mu}|\operatorname{rot} \boldsymbol{B}|^2, \qquad (9.20^*)$$

where c_p is the specific heat of the medium at a constant pressure.

Below, we also give the energy equation for conducting matter written in another useful form. Using the Gibbs identity $T\,\mathrm{d}S/\mathrm{d}t = \mathrm{d}E/\mathrm{d}t + p\mathrm{d}(1/\rho)/\mathrm{d}t$ for the entropy $S(\boldsymbol{r}, t)$ (per unit mass) of the system, which is the same function of the internal energy and specific volume as that for a non-conducting fluid (de Groot and Mazur 1962), let us write (9.20) in the form of the so-called general heat-transfer equation

$$\rho T \frac{\mathrm{d}S}{\mathrm{d}t} = -\mathrm{div}\,(\boldsymbol{q} + \boldsymbol{q}_R) + \left(\Pi : \frac{\partial}{\partial \boldsymbol{r}} \boldsymbol{u}\right) + \boldsymbol{j} \cdot \left(\boldsymbol{E} + \frac{1}{c} \boldsymbol{u} \times \boldsymbol{B}\right), \tag{9.21}$$

where, for an ideal fluid,

$$S = c_V \ln(p\rho^{-\gamma}) + \mathrm{const}, \tag{9.22}$$

where c_V is the specific heat at a constant volume, and $\gamma = c_P/c_V$ is the adiabatic index. Equation (9.21) can be rewritten in the divergence (balance) form

$$\frac{\partial(\rho S)}{\partial t} + \mathrm{div}(\rho S \boldsymbol{u} + \boldsymbol{J}_{(S)}) = \sigma_{(S)} \geq 0, \tag{9.23}$$

where

$$\boldsymbol{J}_{(S)} \equiv \frac{\boldsymbol{q} + \boldsymbol{q}_R}{T} \tag{9.24}$$

is the substantial entropy flux; and

$$0 \leq \sigma_{(S)} \equiv \frac{1}{T} \left\{ -\left((\boldsymbol{q} + \boldsymbol{q}_R) \cdot \frac{\partial \ln T}{\partial \boldsymbol{r}} \right) + \left(\Pi : \frac{\partial}{\partial \boldsymbol{r}} \boldsymbol{u} \right) + \boldsymbol{j} \cdot \left(\boldsymbol{E} + \frac{1}{c} \boldsymbol{u} \times \boldsymbol{B} \right) \right\} \tag{9.25}$$

is the entropy generation through the irreversible processes of heat conduction, internal friction, and electrical conduction of the medium. The last term in (9.25), which corresponds to the Joule heating, includes the contribution of electromagnetic phenomena to the production of the system's entropy. Note that, using the Onsager method of nonequilibrium thermodynamics, (9.25) allows us to obtain the closure (defining) relations for the irreversible processes in a conducting medium that linearly (to a first approximation) relate the thermodynamic fluxes $\boldsymbol{q}$, Π, and $\boldsymbol{j}$ to the thermodynamic forces $\partial \ln T/\partial \boldsymbol{r}$, $\partial \boldsymbol{u}/\partial \boldsymbol{r}$, and $\boldsymbol{E} + c^{-1}\boldsymbol{u} \times \boldsymbol{B}$.

9.1.3.3 The Equation for the Magnetic Energy of a Plasma

We derive the equation for the magnetic energy of a plasma $E^M \equiv |\boldsymbol{B}|^2/8\pi\mu\rho$ (per unit mass) through the scalar multiplication of induction equation (9.8) by $\boldsymbol{B}$. As a result, in view of the differential identity $\boldsymbol{ab} \cdot \partial \boldsymbol{c}/\partial \boldsymbol{r} = \boldsymbol{a} \cdot (\boldsymbol{b} \cdot \partial/\partial \boldsymbol{r})\boldsymbol{c}$, we obtain

$$\rho \frac{\mathrm{d}E^M}{\mathrm{d}t} = -p^M \mathrm{div}\, \boldsymbol{u} + \left(\boldsymbol{T}^M : \frac{\partial \boldsymbol{u}}{\partial \boldsymbol{r}} \right) + \frac{\nu_M}{4\pi\mu} \boldsymbol{B} \cdot \nabla^2 \boldsymbol{B} \tag{9.26}$$

or

$$\frac{\partial(\rho E^M)}{\partial t} + \operatorname{div}\left\{(\rho E^M + p^M)\boldsymbol{u} - \nu_M \frac{\partial p^M}{\partial \boldsymbol{r}} + \left(\frac{\partial}{\partial \boldsymbol{r}} \cdot \nu_M \boldsymbol{T}^M\right)\right\}$$
$$= \left(\boldsymbol{T}^M : \frac{\partial \boldsymbol{u}}{\partial \boldsymbol{r}}\right) - \frac{\nu_M}{4\pi\mu}|\mathrm{rot}\,\boldsymbol{B}|^2. \tag{9.27}$$

Balance form (9.27) of the magnetic-energy equation is written by taking into account the relation

$$\frac{\nu_M}{4\pi\mu}\boldsymbol{B}\cdot\nabla^2\boldsymbol{B} = -\operatorname{div}\left(\frac{\partial}{\partial \boldsymbol{r}}\cdot \nu_M(-p^M \boldsymbol{U} + \boldsymbol{T}^M)\right) - \frac{\nu_M}{4\pi\mu}|\mathrm{rot}\,\boldsymbol{B}|^2, \tag{9.28}$$

which can be obtained (using the formula of vector analysis div $(\boldsymbol{a}\times\boldsymbol{b}) = \boldsymbol{b}\,\mathrm{rot}\,\boldsymbol{a}$ $-\boldsymbol{a}\,\mathrm{rot}\,\boldsymbol{b}$) by taking the divergence of (9.15):

$$\operatorname{div}\left(\frac{\partial}{\partial \boldsymbol{r}}\cdot \nu_M(-p^M \boldsymbol{U} + \boldsymbol{T}^M)\right) = \frac{\nu_M}{c}\operatorname{div}(\boldsymbol{j}\times\boldsymbol{B}) = \frac{\nu_M}{4\pi\mu}\boldsymbol{B}\cdot\mathrm{rot}\,\mathrm{rot}\,\boldsymbol{B} - \frac{\nu_M}{c}\boldsymbol{j}\cdot\mathrm{rot}\,\boldsymbol{B}$$
$$= -\frac{\nu_M}{4\pi\mu}\boldsymbol{B}\cdot\nabla^2\boldsymbol{B} - \frac{|j|^2}{\sigma_e}.$$

Note that, in certain cases, it is convenient to write (9.27) using the so-called Umov–Poynting vector $\boldsymbol{q}_{\mathrm{Poynt}} \equiv (c/4\pi)\boldsymbol{E}\times\boldsymbol{H}$, which has the meaning of the electromagnetic field energy-flux density. In the MHD approximation, this vector can be transformed using the formulas (9.9) and (9.15) to

$$\boldsymbol{q}_{\mathrm{Poynt}} \equiv \frac{c}{4\pi}\boldsymbol{E}\times\boldsymbol{H} = \frac{1}{4\pi\mu}(\nu_M \mathrm{rot}\,\boldsymbol{B} - \boldsymbol{u}\times\boldsymbol{B})\times\boldsymbol{B}$$
$$= (\rho E^M + p^M)\boldsymbol{u} - (\boldsymbol{T}^M\cdot\boldsymbol{u}) - \nu_M\frac{\partial p^M}{\partial \boldsymbol{r}} + \left(\frac{\partial}{\partial \boldsymbol{r}}\cdot \nu_M \boldsymbol{T}^M\right). \tag{9.29}$$

In view of this expression, magnetic-energy balance equation (9.27) for a slowly moving medium ($|\boldsymbol{u}|/c << 1$) can be transformed to the following "classical" form:

$$\frac{\partial(\rho E^M)}{\partial t} = -\operatorname{div}\left(\frac{c}{4\pi}\boldsymbol{E}\times\boldsymbol{H} - p^M\boldsymbol{u} + \boldsymbol{T}^M\cdot\boldsymbol{u}\right) - p^M \operatorname{div}\boldsymbol{u} + \left(\boldsymbol{T}^M : \frac{\partial \boldsymbol{u}}{\partial \boldsymbol{r}}\right)$$
$$-\boldsymbol{j}\cdot\left(\boldsymbol{E} + \frac{1}{c}\boldsymbol{u}\times\boldsymbol{B}\right) = \operatorname{div}\boldsymbol{q}_{\mathrm{Poynt}} - \boldsymbol{j}\cdot\boldsymbol{E}. \tag{9.27*}$$

Finally, combining (9.19*) and (9.27*), we obtain the balance equation for the mechanical and magnetic energies of a conducting medium

$$\frac{\partial}{\partial t}\left\{\rho\left(\frac{|\boldsymbol{u}|^2}{2}+\Psi_G+E^M\right)\right\}+\operatorname{div}\left\{\rho\left(\frac{|\boldsymbol{u}|^2}{2}+\Psi_G\right)\boldsymbol{u}+p\boldsymbol{u}+\boldsymbol{q}_{\text{Poynt}}-\boldsymbol{\Pi}\cdot\boldsymbol{u}\right\}$$
$$=+p\operatorname{div}\boldsymbol{u}-\left(\boldsymbol{\Pi}:\frac{\partial}{\partial\boldsymbol{r}}\boldsymbol{u}\right)-\frac{\nu_M}{4\pi\mu}|\operatorname{rot}\boldsymbol{B}|^2\equiv -Q_E. \tag{9.30}$$

This equation expresses the fact that the sum of the plasma mechanical and magnetic energy densities is not conserved, more specifically, the amount of energy Q_E is converted into the internal energy of the system. Note that in the MHD approximation, the electric field produces no energy in the medium.

9.1.3.4 The Total Energy Conservation Law for an Electrically Conducting Medium

The equation of the first law of thermodynamics that expresses the conservation of the total energy (per unit mass) of a closed system (electrically conducting fluid plus magnetic field),

$$U_{(\text{tot})}(\boldsymbol{r},t)=\frac{1}{2}|\boldsymbol{u}|^2+\Psi_G+E+E^M \tag{9.31}$$

can be derived by adding the balance equations for mechanical (9.19), internal (9.20), and magnetic (9.27*) energies of the plasma. As a result, we obtain

$$\frac{\partial(\rho U_{(\text{tot})})}{\partial t}+\operatorname{div}\{\rho U_{(\text{tot})}\boldsymbol{u}+\boldsymbol{J}_{(U_{\text{tot}})}\}=0, \tag{9.32}$$

where

$$\boldsymbol{J}_{(U_{\text{tot}})}\equiv\boldsymbol{q}+\boldsymbol{q}_R+\frac{c}{4\pi}\boldsymbol{E}\times\boldsymbol{H}+p\boldsymbol{u}-\frac{|\boldsymbol{B}|^2}{8\pi\mu}\boldsymbol{u}-(\boldsymbol{\Pi}\cdot\boldsymbol{u})$$
$$=\boldsymbol{q}+\boldsymbol{q}_R+(p+p^M)\boldsymbol{u}-(\boldsymbol{\Pi}+\boldsymbol{T}^M)\cdot\boldsymbol{u}-\nu_M\frac{\partial p^M}{\partial\boldsymbol{r}}+\left(\frac{\partial}{\partial\boldsymbol{r}}\cdot\nu_M\boldsymbol{T}^M\right) \tag{9.33}$$

is the vector of the substantial total energy flux for a moving plasma (in the single-fluid MHD approximation, with the Ohm law in form (9.6)).

9.1.4 *Equations of State*

In addition to the electrodynamic and hydrodynamic equations presented above, the equation of state $p=p(\rho,T)$ and the caloric equation of state $E=E(\rho,T)$ should be invoked. We define the gas pressure by the equation of state for a perfect gas

$$p = \Re \rho T, \tag{9.34}$$

where $\Re = \mathcal{R}/\tilde{\mu}$; $\mathcal{R}$ is the gas constant, and $\tilde{\mu}$ is the mean atomic mass (the mean mass per particle in units of m_p). Here, the tilde over $\tilde{\mu}$ serves to distinguish the mean atomic mass from the magnetic permeability. We assume the internal energy of the gas to be proportional to the temperature,

$$E = c_V T = \frac{p}{(\gamma - 1)\rho}, \tag{9.35}$$

where $c_V = (\mathcal{R}/\tilde{\mu})/(\gamma - 1)$ is the specific heat of the gas at a constant volume, which is assumed to be a constant.

The system of MHD equations must be supplemented by the defining relations for the energy $\boldsymbol{q}$ (see (2.81)) and radiative-heat $\boldsymbol{q}_R$ (see (2.30)) fluxes for tensor of viscous stresses Π (see (2.64)) as well as by the expressions for all of the necessary thermodynamic and transfer parameters of the medium. Note that the initial and boundary conditions for the hydrodynamic parameters in magnetohydrodynamics do not differ from those for non-conducting media, but additional conditions for the magnetic induction $\boldsymbol{B}$ are required (see, e.g., Akhiezer et al. 1974).

9.2 Equations of Turbulent Motion for a Conducting Medium in the Presence of a Magnetic Field

Currently, there is no quite rigorous and widely accepted semi-empirical theory of turbulent electrical conductive media. Study and modeling of these media are mostly focused on the conductive fluids with permanent properties and isotropic tensor of conductivity, in other words, those pertinent to the case when there is no distinctive difference between plasma and liquid metals. It is known to be applicable, under some constraints, to the modeling of turbulent motion of high-temperature and very dense electrical conductive media or in the case of small magnetic field strength. Let us recall that high-temperature electrical conductive media (plasma) in the magnetic field are generally characterized by anisotropic coefficient of conductivity. In this chapter an attempt is undertaken to deduce thermodynamically the main set of MHD turbulence equations using systematically mean-weighted Favre averaging. Such an approach allows us to describe the all main specific peculiarities pertinent to turbulent motions of conductive isotropic media in magnetic field. This is important when addressing numerous astrophysical applications, in particular, magnetic accretion discs exhibiting turbulent flows as natural phenomena in objects of very large scale.

As we said above, the MHD flows with $\boldsymbol{Re}_M << 1$ and $\boldsymbol{Re}_M >> 1$ differ qualitatively. Naturally, this is also true for the properties of turbulence at $\boldsymbol{Re}_M << 1$ and $\boldsymbol{Re}_M >> 1$, when the pulsation velocity $w \equiv \sqrt{\overline{|\boldsymbol{u}''|^2}}$ and the turbulence scale length L are chosen as the characteristic flow velocity V_0 and scale length L_0, respectively.

If the magnetic Reynolds number $\boldsymbol{Re}_M \equiv Lw/\nu_M$ for turbulent pulsations is much larger than unity, then the idea of a "frozen-in" magnetic field is applicable both for considering the motions of large turbulent vortices and for studying the small-scale turbulence structure in a range of scale lengths in which $\boldsymbol{Re}_M$ for the corresponding vortices remains much larger than unity. This means that the magnetic-field lines are as tangled as the "material" lines formed by fluid particles. As a result, a pulsating magnetic field is generated. The turbulent electric currents produced in the medium by turbulent velocity pulsations in the presence of a magnetic field are physically responsible for its generation.

We restrict our analysis to the MHD turbulence of a low-temperature plasma under the following assumptions resulting from an analysis of the MHD equations (Ievlev 1975):

1. $\boldsymbol{Re}_M \leq 1$; in this case, the effect of only the external magnetic field $\boldsymbol{B}_0$ and the currents produced by it on the flow can be taken into account.
2. $\boldsymbol{N} \cdot \boldsymbol{Re}_M << 1$; in this case, the generation of MHD waves by turbulent pulsations is unessential.
3. $(c_A/c_s)(\boldsymbol{N} \cdot \boldsymbol{Re}_M)^{3/2} << 1$ and $w/c_s << 1$, where $c_A \equiv B_0/\sqrt{4\pi\mu\rho}$ is characteristic Alfven velocity and c_s is sound velocity; in such a case dynamics of the turbulence development can be evaluated with no account for generation of acoustic oscillations.

When any of these conditions is violated, the mechanism of turbulent motion becomes considerably more complicated. For example, when condition (2) is not valid, the turbulent pulsations in an electrically conducting medium should give rise to a turbulent field of both MHD and acoustic waves, as well as waves associated with the interaction between the MHD and acoustic oscillations; in this case, the energy of the pulsation motion of the medium can be less than or equal to the energy of the pulsation magnetic field and the energy of the acoustic oscillations. At the same time, even in the case of such restricted fields for which condition (2) is satisfied, the magnetic field effect on turbulence can be significant.

Basically, in a turbulized low-temperature plasma, the hydrodynamic velocity, temperature, density, and electromagnetic fields are fluctuating quantities. Below, similar to the previous chapters approach, we employ the Reynolds schematization using two symbols to designate the averaged physical quantities: a bar over a particular quantity $\mathcal{A}$ denotes its probability-theoretical averaging $\bar{\mathcal{A}} = \lim_{N\to\infty} (1/N) \sum_{p=1}^{N} \mathcal{A}^{(p)}$ over the ensemble of possible realizations of the stochastic system under consideration, i.e., the corresponding mean field $\langle \mathcal{A}''^2 \rangle$ is defined as the mathematically expected field $\mathcal{A}$ for an ensemble of identical MHD systems ($\mathcal{A} = \bar{\mathcal{A}} + \mathcal{A}'$, $\bar{\mathcal{A}}' = 0$), while the angular brackets denote the weighted-mean Favre averaging of the parameter $\mathcal{A}$: $\langle \mathcal{A} \rangle = \overline{\rho\mathcal{A}}/\bar{\rho}$ (in this case, $\mathcal{A} = \langle \mathcal{A} \rangle + \mathcal{A}''$, $\overline{\rho\mathcal{A}''} = 0$). It should be noted that introducing the weighted averaging turned out to also be convenient for the model description of the turbulent flow of an electrically conducting fluid with a variable density.

9.2.1 *Averaged Continuity Equation*

It can easily be verified that the averaged density $\bar{\rho}$ and the weighted-mean hydrodynamic velocity $\langle \boldsymbol{u} \rangle \equiv \overline{\rho \boldsymbol{u}}/\bar{\rho}$ satisfy the continuity equation for the mean motion

$$\frac{\partial \bar{\rho}}{\partial t} + \left(\frac{\partial}{\partial \boldsymbol{r}} \cdot \bar{\rho} \langle \boldsymbol{u} \rangle \right) = 0, \text{ or } \bar{\rho} \frac{D}{Dt} \left(\frac{1}{\bar{\rho}} \right) = \operatorname{div} \langle \boldsymbol{u} \rangle. \tag{9.36}$$

This equation can be derived by applying the Reynolds averaging operation to continuity equation (9.10), which is valid according to the assumption for the instantaneous (actual) density and hydrodynamic velocity.

Averaging the operator relation $\rho \, \mathrm{d}\mathcal{A}/\mathrm{d}t = \partial(\rho \mathcal{A})/\partial t + \partial(\rho \mathcal{A} \boldsymbol{u})/\partial \boldsymbol{r}$ between the substantial and local variations of the parameter d in the instantaneous plasma flow leads, given (9.36), to the identity

$$\begin{aligned} \overline{\rho \frac{\mathrm{d}\mathcal{A}}{\mathrm{d}t}} &= \frac{\partial}{\partial t} (\bar{\rho} \langle \mathcal{A} \rangle) + \left(\frac{\partial}{\partial \boldsymbol{r}} \cdot (\bar{\rho} \langle \mathcal{A} \rangle \langle \boldsymbol{u} \rangle) \right) + \left(\frac{\partial}{\partial \boldsymbol{r}} \cdot (\overline{\rho \mathcal{A}'' \boldsymbol{u}''}) \right) \\ &= \bar{\rho} \frac{D \langle \mathcal{A} \rangle}{Dt} + \left(\frac{\partial}{\partial \boldsymbol{r}} \cdot \boldsymbol{J}^{\text{turb}}_{(\mathcal{A})} \right). \end{aligned} \tag{9.37}$$

This specifies the relation between the substantial and local variations of the quantity $\mathcal{A}(\boldsymbol{r}, t)$ in the averaged flow. Here, $D(..)/Dt \equiv \partial(..)/\partial t + \langle \boldsymbol{u} \rangle \cdot \partial(..)/\partial \boldsymbol{r}$ is the substantial time derivative for the averaged continuum and

$$\boldsymbol{J}^{\text{turb}}_{(A)} \equiv \overline{\rho \mathcal{A}'' \boldsymbol{u}''} = \bar{\rho} \langle \mathcal{A}'' \boldsymbol{u}'' \rangle \tag{9.38}$$

is the turbulent flux of the quantity $\mathcal{A}(\boldsymbol{r}, t)$, which represents the transfer of the pulsating characteristic $\mathcal{A}''$ by the hydrodynamic velocity pulsations $\boldsymbol{u}''$.

9.2.2 *Magnetic Induction Equation for Mean Fields*

Averaging Maxwell equations (9.2) and (9.3) and Ohm law (9.6), which are assumed to be valid for the instantaneous values of the electrodynamic fields, over the ensemble of possible realizations (the stochastic electrically conducting fluid continuum), yields

$$\frac{1}{c} \frac{\partial \bar{\boldsymbol{B}}}{\partial t} = -\operatorname{rot} \bar{\boldsymbol{E}}, \quad \operatorname{rot} \bar{\boldsymbol{B}} = \frac{4\pi\mu}{c} \bar{\boldsymbol{j}}, \quad \operatorname{div} \bar{\boldsymbol{B}} = 0, \tag{9.39}$$

$$\bar{\boldsymbol{j}} = \sigma_e \left(\bar{\boldsymbol{E}} + \frac{1}{c} \overline{\boldsymbol{u} \times \boldsymbol{B}} \right) \equiv \sigma_e \left(\bar{\boldsymbol{E}} + \frac{1}{c} \langle \boldsymbol{u} \rangle \times \bar{\boldsymbol{B}} + \frac{1}{c} \overline{\boldsymbol{u}'' \times \boldsymbol{B}} \right). \tag{9.40}$$

We may see that the Maxwell equations also retain their previous form for the averaged fields due to their linearity. However, averaging nonlinear equation (9.6) gives rise to a new term $c^{-1}\overline{\boldsymbol{u}'' \times \boldsymbol{B}}$, which leads to an additional electromotive force in the Ohm law for the mean fields. Usually, it is referred to as the "turbulent electromotive force," which has a special designation,

$$\mathsf{G} \equiv \frac{1}{c}\overline{\boldsymbol{u}'' \times \boldsymbol{B}} \tag{9.41}$$

or, in tensor form,

$$\mathsf{G}_i \equiv -\frac{1}{2c}\varepsilon_{ijk}R^M_{jk}, \quad R^M_{jk} = -(\overline{u''_j B_k} - \overline{u''_k B_j}), \tag{9.41*}$$

where ε_{ijk} is a completely anti-symmetric third-rank tensor (an alternating Levi-Civitta tensor) and $\boldsymbol{R}^M$ is the magnetic Reynolds tensor. The quantity G has the dimensions of the electric field strength and it describes the mean result of interaction between the velocity field and magnetic-induction pulsations. Thus, the Ohm law for the mean fields takes the form

$$\bar{\boldsymbol{j}} = \sigma_e\left(\bar{\boldsymbol{E}} + \frac{1}{c}\langle\boldsymbol{u}\rangle \times \bar{\boldsymbol{B}} + \mathsf{G}\right) \equiv \sigma_e(\boldsymbol{E}^* + \mathsf{G}), \tag{9.42}$$

where $\boldsymbol{E}^* \equiv \bar{\boldsymbol{E}} + c^{-1}\langle\boldsymbol{u}\rangle \times \bar{\boldsymbol{B}}$ the electric field in a coordinate system moving at velocity $\langle\boldsymbol{u}\rangle$.

Now, in contrast to a regular flow, not only the averaged velocity $\langle\boldsymbol{u}\rangle$ but also the turbulent electromotive force G must be known to determine the averaged fields $\bar{\boldsymbol{B}}, \bar{\boldsymbol{E}}$, and $\bar{\boldsymbol{j}}$:

$$\bar{\boldsymbol{j}} = (c/4\pi\mu)\mathrm{rot}\,\bar{\boldsymbol{B}}, \tag{9.43}$$

$$\bar{\boldsymbol{E}} = \bar{\boldsymbol{j}}/\sigma_e - c^{-1}\langle\boldsymbol{u}\rangle \times \bar{\boldsymbol{B}} - \mathsf{G} = c^{-1}(\nu_M \mathrm{rot}\,\bar{\boldsymbol{B}} - \langle\boldsymbol{u}\rangle \times \bar{\boldsymbol{B}}) - \mathsf{G}. \tag{9.44}$$

Evolution of the averaged magnetic field is governed by the induction equation for $\bar{\boldsymbol{B}}$ that we obtain by Reynolds averaging (9.7). Using (9.37), it results in

$$\bar{\rho}\frac{D}{Dt}\left(\frac{\bar{\boldsymbol{B}}}{\bar{\rho}}\right) + \left(\frac{\partial}{\partial \boldsymbol{r}} \cdot \overline{\boldsymbol{B}\boldsymbol{u}''}\right) = \left(\bar{\boldsymbol{B}} \cdot \frac{\partial}{\partial \boldsymbol{r}}\right)\langle\boldsymbol{u}\rangle + \overline{\left(\boldsymbol{B} \cdot \frac{\partial}{\partial \boldsymbol{r}}\right)\boldsymbol{u}''} + \nu_M \nabla^2 \bar{\boldsymbol{B}}. \tag{9.45}$$

For the subsequent analysis, it is convenient to write this equation in a more compact from:

$$\bar{\rho}\frac{D}{Dt}\left(\frac{\bar{\boldsymbol{B}}}{\bar{\rho}}\right) = \left(\bar{\boldsymbol{B}} \cdot \frac{\partial}{\partial \boldsymbol{r}}\right)\langle\boldsymbol{u}\rangle + c\,\mathrm{rot}\,\mathsf{G} + \nu_M \nabla^2 \bar{\boldsymbol{B}}, \ (\mathrm{div}\,\boldsymbol{B} = 0) \tag{9.45*}$$

or, using the transformation

$$c\,\mathrm{rot}\,\mathsf{G} \equiv \mathrm{rot}(\overline{\boldsymbol{u}'' \times \boldsymbol{B}}) \equiv \overline{\left(\boldsymbol{B} \cdot \frac{\partial}{\partial \boldsymbol{r}}\right)\boldsymbol{u}''} - \overline{\left(\boldsymbol{u}'' \cdot \frac{\partial}{\partial \boldsymbol{r}}\right)\boldsymbol{B}} - \overline{\boldsymbol{B}\,\mathrm{div}\,\boldsymbol{u}''}$$
$$\equiv \frac{\partial}{\partial \boldsymbol{r}} \cdot \boldsymbol{R}^M, \tag{9.46}$$

in the form

$$\bar{\rho}\frac{D}{Dt}\left(\frac{\bar{\boldsymbol{B}}}{\bar{\rho}}\right) = \left(\bar{\boldsymbol{B}} \cdot \frac{\partial}{\partial \boldsymbol{r}}\right)\langle \boldsymbol{u}\rangle + \frac{\partial}{\partial \boldsymbol{r}} \cdot \boldsymbol{R}^M + \nu_M \nabla^2 \bar{\boldsymbol{B}}. \tag{9.45**}$$

Note that the term $c\,\mathrm{rot}\,\mathsf{G}$ which is defined by the product of the random fluctuations and plays the role of an additional source generating the field $\bar{\boldsymbol{B}}$, appears in averaged induction equation (9.45*). Clearly, a major goal of the semi-empirical theory of MHD turbulence is the construction of a special closure relation for the turbulent flux G as a function of the mean fields $\bar{\boldsymbol{B}}$ and $\langle \boldsymbol{u}\rangle$ in order that, given the field $\langle \boldsymbol{u}\rangle$, (9.45*) be solvable relative $\bar{\boldsymbol{B}}$.

9.2.3 *Averaged Equation of Motion*

We derive the averaged equation of motion for electrically conducting media by averaging momentum conservation equation (9.18) for regular flows over the ensemble of possible realizations (using the weighted-mean averaging for the flow velocity). As a result, we obtain

$$\bar{\rho}\frac{D\langle \boldsymbol{u}\rangle}{Dt} = -\frac{\partial}{\partial \boldsymbol{r}}(\bar{p} + \overline{p^M}) + \left(\frac{\partial}{\partial \boldsymbol{r}} \cdot (\bar{\Pi} + \boldsymbol{R} + \overline{\boldsymbol{T}^M})\right) - \bar{\rho}\frac{\partial}{\partial \boldsymbol{r}}\Psi_G. \tag{9.47}$$

Here, $\bar{\Pi}(\boldsymbol{r}, t)$ is the averaged viscous stress tensor that describes the momentum exchange between fluid particles due to the forces of molecular viscosity;

$$\boldsymbol{R}(\boldsymbol{r}, t) \equiv -\overline{\rho \boldsymbol{u}''\boldsymbol{u}''} = -\bar{\rho}\langle \boldsymbol{u}''\boldsymbol{u}''\rangle, \quad (R_{jk} = -\bar{\rho}\langle u_j'' u_k''\rangle) \tag{9.48}$$

is the standard Reynolds tensor for a fluid, which has the meaning of additional (turbulent) stresses;

$$\overline{\boldsymbol{T}^M} \equiv \frac{1}{4\pi\mu}\overline{\boldsymbol{BB}} = \boldsymbol{T}^M_{av} + \boldsymbol{T}^M_{\mathrm{turb}}, \quad \overline{p^M} \equiv p^M_{av} + p^M_{\mathrm{turb}} \tag{9.49}$$

are the averaged magnetic tension tensor and the averaged magnetic-field pressure, respectively;

$$\begin{cases} (\boldsymbol{T}^M_{av})_{jk} \equiv \frac{1}{4\pi\mu} \overline{B}_j \overline{B}_k, \\ (\boldsymbol{T}^M_{\text{turb}})_{jk} \equiv \frac{1}{4\pi\mu} \overline{B'_j B'_k} \end{cases} \tag{9.50}$$

are the components of the magnetic tension tensors for the averaged magnetic field and the pulsational magnetic-field component; and

$$\begin{cases} p^M_{av} \equiv |\bar{\boldsymbol{B}}|^2/8\pi\mu, \\ p^M_{\text{turb}} \equiv \overline{|\boldsymbol{B}'|^2}/8\pi\mu \end{cases} \tag{9.51}$$

is the pressure of the averaged magnetic field and the turbulent magnetic pressure of an electrically conducting fluid.

Then, using the transformation (cf. (9.15))

$$-\frac{\partial p^M_{av}}{\partial \boldsymbol{r}} + \frac{\partial}{\partial \boldsymbol{r}} \cdot \boldsymbol{T}^M_{av} = -\frac{\partial}{\partial \boldsymbol{r}}\left(\frac{|\bar{\boldsymbol{B}}|^2}{8\pi\mu}\right) + \left(\frac{\partial}{\partial \boldsymbol{r}} \cdot \left(\frac{\bar{\boldsymbol{B}}\bar{\boldsymbol{B}}}{4\pi\mu}\right)\right)$$
$$= -\frac{1}{4\pi\mu}\bar{\boldsymbol{B}} \times \operatorname{rot}\bar{\boldsymbol{B}} = \frac{1}{c}\bar{\boldsymbol{j}} \times \bar{\boldsymbol{B}},$$

we write averaged equation of plasma motion (9.47) as

$$\bar{\rho}\frac{D\langle \boldsymbol{u}\rangle}{Dt} = -\frac{\partial}{\partial \boldsymbol{r}}(\bar{p} + p^M_{\text{turb}}) + \left(\frac{\partial}{\partial \boldsymbol{r}} \cdot (\bar{\Pi} + \boldsymbol{R} + \boldsymbol{T}^M_{\text{turb}})\right) + \frac{1}{c}\bar{\boldsymbol{j}} \times \bar{\boldsymbol{B}} - \bar{\rho}\frac{\partial}{\partial \boldsymbol{r}}\Psi_G$$
$$\cong -\frac{\partial}{\partial \boldsymbol{r}}(\bar{p} + \overline{|\boldsymbol{B}'|^2}/8\pi\mu) + \left(\frac{\partial}{\partial \boldsymbol{r}} \cdot \boldsymbol{R}^K\right) + \frac{1}{4\pi\mu}\operatorname{rot}\bar{\boldsymbol{B}} \times \bar{\boldsymbol{B}} + \bar{\rho}\boldsymbol{g}, \tag{9.52}$$

where

$$\boldsymbol{R}^K \equiv \boldsymbol{R} + \boldsymbol{T}^M_{\text{turb}} \equiv \left(-\overline{\rho \boldsymbol{u}''\boldsymbol{u}''} + \frac{1}{4\pi\mu}\overline{\boldsymbol{B}'\boldsymbol{B}'}\right) \tag{9.53}$$

is the total turbulent stress tensor in plasma (the so-called kinetic Reynolds tensor for an electrically conducting fluid in a magnetic field). Approximate equation (9.53) is valid in the case of developed turbulence, where the averaged viscous (molecular) stress tensor $\bar{\Pi}$ can be neglected compared to the Reynolds stress tensor $\boldsymbol{R}$ (except for the viscous sublayer adjacent to the solid walls).

9.2.4 Energy Equations of Mean Motion for Electrically Conducting Matter

In the averaged turbulized flow of a conducting fluid, in contrast to its regular analog, there exist a large number of various mechanisms of exchange conversion rates between the various kinds of energy of moving-volume elements of matter that contribute to the conserved total energy of the matterfield plasma continuum. For a clear physical interpretation of the individual energy-balance components, let us first write out the various energy equations for the averaged motion of the material plasma component, including the balance equation for the turbulent kinetic energy of the matter.

9.2.4.1 The Heat Inflow Equation for the Mean Motion of Turbulent Plasma

Reynolds averaging of (9.20) for the internal energy of the conducting material and utilization of the easily derivable identities

$$\bar{\rho}\frac{D\langle E\rangle}{Dt} + \bar{p}\,\mathrm{div}\langle \boldsymbol{u}\rangle \equiv \bar{\rho}\frac{D\langle H\rangle}{Dt} - \frac{D\bar{p}}{Dt}$$

and

$$\overline{\frac{dp}{dt}} \equiv \frac{D\bar{p}}{Dt} + \left(\boldsymbol{J}^{\mathrm{turb}}_{(1/\rho)} \cdot \frac{\partial \bar{p}}{\partial \boldsymbol{r}}\right) + \mathrm{div}(\overline{p'\boldsymbol{u}''}) - \overline{p'\,\mathrm{div}\,\boldsymbol{u}''},$$

(where $\langle H\rangle \equiv \langle E\rangle + \bar{p}\,/\bar{\rho}$) allow us to obtain the following heat-inflow equation for the mean-fluid motion (cf. (3.1.54)):

$$\bar{\rho}\frac{D\langle E\rangle}{Dt} + \mathrm{div}(\bar{\boldsymbol{q}} + \bar{\boldsymbol{q}}_R + \boldsymbol{q}^{\mathrm{turb}}) = -\bar{p}\,\mathrm{div}\langle \boldsymbol{u}\rangle + \left(\bar{\Pi} : \frac{\partial \langle \boldsymbol{u}\rangle}{\partial \boldsymbol{r}}\right) + \frac{|\bar{\boldsymbol{j}}|^2}{\sigma_e}$$
$$+ \left(\boldsymbol{J}^{\mathrm{turb}}_{(1/\rho)} \cdot \frac{\partial \bar{p}}{\partial \boldsymbol{r}}\right) - \overline{p'\,\mathrm{div}\,\boldsymbol{u}''} + \frac{\overline{|\boldsymbol{j}'|^2}}{\sigma_e} + \bar{\rho}\langle \varepsilon_b\rangle \qquad (9.54)$$

or

$$\bar{\rho}\frac{D\langle H\rangle}{Dt} + \mathrm{div}\,\boldsymbol{J}_{\langle E\rangle} = \frac{D\bar{p}}{Dt} + \left(\bar{\Pi} : \frac{\partial \langle \boldsymbol{u}\rangle}{\partial \boldsymbol{r}}\right) + \frac{\nu_M}{4\pi\mu}|\mathrm{rot}\,\bar{\boldsymbol{B}}|^2 + \left(\boldsymbol{J}^{\mathrm{turb}}_{(1/\rho)} \cdot \frac{\partial \bar{p}}{\partial \boldsymbol{r}}\right)$$
$$- \overline{p'\,\mathrm{div}\,\boldsymbol{u}''} + \bar{\rho}(\langle \varepsilon_b\rangle + \langle \varepsilon_M\rangle). \qquad (9.54^*)$$

Here

$$\boldsymbol{J}_{\langle E\rangle} \equiv \bar{\boldsymbol{q}} + \bar{\boldsymbol{q}}_R + \tilde{\boldsymbol{q}}^{\mathrm{turb}}, \quad (\tilde{\boldsymbol{q}}^{\mathrm{turb}} \equiv \boldsymbol{q}^{\mathrm{turb}} - \overline{p'u''}) \qquad (9.55)$$

is the total internal-energy flux for the averaged plasma motion;

$$\boldsymbol{J}^{\mathrm{turb}}_{(1/\rho)}(\boldsymbol{r},t) \equiv \overline{\rho(1/\rho)''\boldsymbol{u}''} = \overline{\boldsymbol{u}''}, \quad \boldsymbol{q}^{\mathrm{turb}}(\boldsymbol{r},t) \equiv \overline{\rho H''\boldsymbol{u}''} \cong c_p\overline{\rho T''\boldsymbol{u}''} \qquad (9.56)$$

are the turbulent specific-volume and heat fluxes, respectively; and

$$\begin{cases} 0 \leq \bar{\rho}\langle\varepsilon_b\rangle \equiv \overline{\left(\Pi : \frac{\partial u''}{\partial r}\right)} = \overline{\Pi_{kj}(\partial u''_k/\partial x_j)} \\ 0 \leq \bar{\rho}\langle\varepsilon_M\rangle \equiv \frac{\overline{|j'|^2}}{\sigma_e} = \frac{\nu_M}{4\pi\mu}\overline{|\mathrm{rot}\,\boldsymbol{B}'|^2} \end{cases} \tag{9.57}$$

are, respectively, the specific rate of viscous dissipation of the turbulent kinetic energy into heat due to molecular kinematic viscosity and the specific rate of dissipation of the turbulence energy under the action of a pulsating magnetic field. The latter can be interpreted as the Joule heat release from the pulsational electric currents generated by turbulent magnetic filed pulsations in a conducting medium.

9.2.4.2 The Equation for the Averaged Mechanical Energy

The equation for the averaged mechanical energy of plasma can be derived by multiplying scalarly both sides of the averaged equation of motion (9.47) by the vector $\langle \boldsymbol{u}\rangle$. As a result, we obtain

$$\begin{aligned} &\bar{\rho}\frac{D}{Dt}\left(\frac{|\langle\boldsymbol{u}\rangle|^2}{2} + \Psi_G\right) + \mathrm{div}\{(\bar{p} + p^M_{\mathrm{turb}})\langle\boldsymbol{u}\rangle - (\bar{\Pi} + \boldsymbol{T}^M_{av} + \boldsymbol{R}^K)\cdot\langle\boldsymbol{u}\rangle\} \\ &= (\bar{p} + p^M_{\mathrm{turb}})\mathrm{div}\langle\boldsymbol{u}\rangle - \left((\bar{\Pi} + \boldsymbol{T}^M_{av} + \boldsymbol{R}^K) : \frac{\partial\langle\boldsymbol{u}\rangle}{\partial\boldsymbol{r}}\right). \end{aligned} \tag{9.58}$$

9.2.4.3 The Equation for the Turbulent Kinetic Energy of Matter

The equation for the turbulent kinetic energy of matter can be obtained using various methods (see Marov and Kolesnichenko 2002); we use one of them in the case of an electrically conducting medium considered here.

Let $\mathcal{A}(\boldsymbol{r}, t)$ be the actual value of a scalar quantity (in particular, this can be a vector component) whose substantial balance is $\rho\, \mathrm{d}\mathcal{A}/\mathrm{d}t = -(\partial/\partial\boldsymbol{r})\cdot J_{\mathcal{A}} + \sigma_{\mathcal{A}}$, where $\boldsymbol{J}_{\mathcal{A}}$ and $\sigma_{\mathcal{A}}$ are the substantial flux-density vector and the volume density of the source of quantity $\mathcal{A}$, respectively. For example, for the equation of motion (9.18) we have

$$\mathcal{A} \equiv \boldsymbol{u}, \quad \boldsymbol{J}_{\mathcal{A}} \equiv -\Pi, \quad \sigma_{\mathcal{A}} \equiv -\partial p/\partial\boldsymbol{r} + \frac{1}{c}\boldsymbol{j}\times\boldsymbol{B} + \rho\boldsymbol{g}. \tag{9.59}$$

It is easy to show (this requires multiplying the identity $\mathrm{d}\mathcal{A}''/\mathrm{d}t \equiv \mathrm{d}\mathcal{A}/\mathrm{d}t - D\langle\mathcal{A}\rangle/Dt - \boldsymbol{u}''\cdot(\partial/\partial\boldsymbol{r})\langle\mathcal{A}\rangle$ by $\rho\mathcal{A}''$ and Reynolds averaging of the result) that the transfer equation for the root-mean-square pulsation $\langle \mathcal{A}''^2 \rangle$ has the following general form (Kolesnichenko 1995):

$$\bar{\rho}\frac{D}{Dt}\left(\frac{\langle \mathcal{A}''^2 \rangle}{2}\right)+\frac{\partial}{\partial \boldsymbol{r}}\cdot\left(\overline{\frac{1}{2}\rho\mathcal{A}''^2\boldsymbol{u}''}+\overline{\mathcal{A}''\boldsymbol{J}_{(\mathcal{A})}}\right)$$
$$=-\left(\boldsymbol{J}_{\mathcal{A}}^{\text{turb}}\cdot\frac{\partial}{\partial \boldsymbol{r}}\langle \mathcal{A}\rangle\right)+\overline{\mathcal{A}''\sigma_{\mathcal{A}}}-\bar{\rho}\langle \varepsilon_{\mathcal{A}}\rangle, \tag{9.60}$$

where $\bar{\rho}\langle \varepsilon_{\mathcal{A}}\rangle \equiv -\overline{(\boldsymbol{J}_{\mathcal{A}}\cdot \partial \mathcal{A}''/\partial \boldsymbol{r})}$ is the so-called rate of scalar dissipation of the variance $\langle \mathcal{A}''^2 \rangle$. Transfer equation (9.60) contains the terms that implies an influence of the following processes on the spatial–temporal distribution of the variance $\langle \mathcal{A}''^2 \rangle$: convective transfer; turbulent diffusion—the second term on the left; formation through the energy exchange between the averaged and pulsation motions—the first term on the right; redistribution between pulsation motions in various directions and dissipation of the turbulent characteristic $\langle \mathcal{A}''^2 \rangle$ due to the "molecular" transfer processes.

Then the equation for the turbulent kinetic energy of matter for the averaged pulsational kinetic energy $b = |\boldsymbol{u}''|^2/2$ of a unit plasma mass

$$\langle b\rangle(\boldsymbol{r},t) \equiv \overline{\rho|\boldsymbol{u}''|^2}/2\bar{\rho} \tag{9.61}$$

can be derived from general equation for variance (9.60) by substituting (9.59) into (9.60); as a result, we obtain

$$\bar{\rho}\frac{D\langle b\rangle}{Dt}+\operatorname{div}\boldsymbol{J}_{\langle b\rangle}=\left(\boldsymbol{R}:\frac{\partial\langle \boldsymbol{u}\rangle}{\partial \boldsymbol{r}}\right)-\left(\boldsymbol{J}_{(1/\rho)}^{\text{turb}}\cdot\frac{\partial \bar{p}}{\partial \boldsymbol{r}}\right)+\overline{p'\operatorname{div}\boldsymbol{u}''}$$
$$-\mathrm{G}\cdot\bar{\boldsymbol{j}}-\frac{1}{c}\langle \boldsymbol{u}\rangle\cdot\overline{\boldsymbol{j}'\times\boldsymbol{B}'}+\overline{\boldsymbol{j}'\cdot\boldsymbol{E}'}-\bar{\rho}(\langle \varepsilon_b\rangle+\langle \varepsilon_M\rangle), \tag{9.62}$$

where

$$\boldsymbol{J}_{\langle b\rangle} \equiv \overline{\rho(|\boldsymbol{u}''|^2/2+p'/\rho)\boldsymbol{u}''}-\overline{\Pi\cdot\boldsymbol{u}''} \tag{9.63}$$

is the turbulent-diffusion flux of the turbulent kinetic energy of the material plasma component. In writing (9.62), we used the easily derivable transformation

$$c^{-1}\overline{\boldsymbol{u}''\cdot(\boldsymbol{j}\times\boldsymbol{B})} \equiv -c^{-1}\langle \boldsymbol{u}\rangle\cdot\overline{\boldsymbol{j}'\times\boldsymbol{B}'}+\overline{\boldsymbol{j}'\cdot\boldsymbol{E}'}-\bar{\rho}\langle \varepsilon_M\rangle-\mathrm{G}\cdot\bar{\boldsymbol{j}}. \tag{9.64}$$

The terms on the left-hand side of (9.62) express the total change in averaged pulsational kinetic energy per unit time. The terms on the right-hand side have the following physical meaning. The quantity $(\boldsymbol{R}:\partial\langle \boldsymbol{u}\rangle/\partial \boldsymbol{r})$ appears on the right-hand sides of (9.58) and (9.62) with opposite signs and, hence, it can be interpreted as the conversion rate of the kinetic energy of the averaged motion into the turbulence energy of the electrically conducting fluid. The quantity $\overline{p' \operatorname{div} \boldsymbol{u}''}$ is related to the conversion rate of the internal energy of the plasma into the kinetic energy of turbulent vortices and is the work done by the pulsating medium on the vortices per unit time in a unit volume due to the existence of hydrodynamic pressure p' pulsations in the system and the expansion ($\operatorname{div} \boldsymbol{u}''>0$) or compression ($\operatorname{div} \boldsymbol{u}''<0$) of the turbulent vortices. The quantity $(\partial\bar{p}/\partial\boldsymbol{r}) \cdot \boldsymbol{J}^{\text{turb}}_{(1/\rho)}$ is the conversion rate (per unit volume) between the turbulent and averaged internal energies of the system; the small-scale vortices convert the turbulence energy into heat, since this quantity is always positive for them. Note, however, that the large vortex structures related to the thermal convection (for which $\left(\boldsymbol{J}^{\text{turb}}_{(1/\rho)} \cdot \partial\bar{p}/\partial\boldsymbol{r}\right)<0$) can convert the thermal energy of the turbulized flow into the averaged kinetic energy of the velocity pulsations (see, e.g., Kolesnichenko and Marov 1999). In a developed turbulent flow, the pair correlation $\bar{\rho}\langle\varepsilon_b\rangle$ is the rate of viscous dissipation of the turbulent kinetic energy into heat under the effect of molecular viscosity (see (9.54)). Finally, the quantity $\bar{\rho}\langle\varepsilon_M\rangle$ which is the work done by the pulsations of the magnetic tension tensor on the turbulent vortices (per unit time and unit volume), can be interpreted as an additional scattering of the turbulent kinetic energy under the effect of magnetic viscosity in an electrically conducting, turbulized medium (Joule dissipation).

Thus from (9.62) follows that the effect of the magnetic induction $\boldsymbol{B}$ on the balance of the turbulent kinetic energy is determined by the influence of both direct and indirect factors. The direct action of the field $\boldsymbol{B}$ on the turbulence leads to the suppression of the turbulent pulsations and a change in the turbulence structure (the term $\bar{\rho}\langle\varepsilon_M\rangle$). The indirect action of the magnetic field is related to the influence of the changes in the mean velocity field and turbulence structure on the turbulence production and the energy dissipation and transfer. The various indirect effects of the magnetic field can generally decrease and increase the intensity of the turbulent pulsations.

9.2.4.4 The Equation for the Averaged Total Energy of an Electrically Conducting Matter

Combining (9.54), (9.58), and (9.62), we derive the balance equation for the averaged total specific energy of an electrically conducting matter

$$\langle U^{\text{sub}}_{\text{tot}}\rangle \equiv |\langle\boldsymbol{u}\rangle|^2/2 + \Psi_G + \langle E\rangle + \langle b\rangle, \tag{9.65}$$

in the form

$$\frac{\partial}{\partial t}\left(\bar{\rho}\langle U^{\text{sub}}_{\text{tot}}\rangle\right) + \operatorname{div}\left\{\bar{\rho}\langle U^{\text{sub}}_{\text{tot}}\rangle\langle \boldsymbol{u}\rangle + \boldsymbol{J}_{\langle U^{\text{sub}}_{\text{tot}}\rangle}\right\} = \overline{p^M}\operatorname{div}\langle \boldsymbol{u}\rangle - \left(\overline{\boldsymbol{T}^M} : \frac{\partial\langle \boldsymbol{u}\rangle}{\partial \boldsymbol{r}}\right)$$
$$+\frac{\nu_M}{4\pi\mu}\left|\operatorname{rot}\bar{\boldsymbol{B}}\right|^2 - G\cdot\bar{\boldsymbol{j}} - \frac{1}{c}\langle \boldsymbol{u}\rangle\cdot\overline{\boldsymbol{j}'\times\boldsymbol{B}'} + \overline{\boldsymbol{j}'\cdot\boldsymbol{E}'}, \tag{9.66}$$

where

$$\boldsymbol{J}_{\langle U^{\text{sub}}_{\text{tot}}\rangle} \equiv \left\{\bar{\boldsymbol{q}} + \bar{\boldsymbol{q}}_R + \boldsymbol{q}^{\text{turb}} + \overline{\rho(|\boldsymbol{u}''|^2/2)\boldsymbol{u}''} - \overline{\boldsymbol{\Pi}\cdot\boldsymbol{u}''} + \left(\bar{p} + p^M_{\text{turb}}\right)\langle \boldsymbol{u}\rangle\right.$$
$$\left. - \left(\bar{\boldsymbol{\Pi}} + \boldsymbol{T}^M_{av} + \boldsymbol{R}^K\right)\cdot\langle \boldsymbol{u}\rangle\right\} \tag{9.67}$$

is the substantial flux of the averaged total energy of the material turbulized-plasma component. It is evident from this equation that the averaged density of the total energy for an electrically conducting material is not conserved: indeed, the total energy of the matter $\langle U^{\text{sub}}_{\text{tot}}\rangle$ and the magnetic energy $\langle E^M\rangle$ can transform into one another.

9.2.5 Equations for the Magnetic Energy of a Turbulized Plasma

For a turbulized plasma, other kinds of energy related to magnetic field manifestations should also be taken into account. Let us first derive the equation for the averaged magnetic-energy density of the plasma $\langle E_M\rangle$, which is convenient to represent as the sum of two quantities: $\langle E_M\rangle \equiv \overline{|\boldsymbol{B}|^2/8\pi\mu}/\bar{\rho} = E^{av}_M + \langle b_M\rangle$, where $E^{av}_M \equiv |\bar{\boldsymbol{B}}|^2/8\pi\mu\bar{\rho}$ is the magnetic-energy density of the mean field, and $\langle b_M\rangle = \overline{|\boldsymbol{B}'|^2}/8\pi\mu\bar{\rho}$ —the turbulent magnetic-energy density. Performing the probability-theoretical averaging of (9.26) and taking into account (9.28) and (9.64), we have

$$\bar{\rho}\frac{D}{Dt}\langle E^M\rangle + \operatorname{div}\left\{\boldsymbol{J}^{\text{turb}}_{(E^M)} + \overline{p^M\boldsymbol{u}''} - \overline{\boldsymbol{T}^M\cdot\boldsymbol{u}''} - \nu_M\frac{\partial\overline{p^M}}{\partial \boldsymbol{r}} + \frac{\partial}{\partial \boldsymbol{r}}\cdot\nu_M\overline{\boldsymbol{T}^M}\right\}$$
$$= -\overline{p^M}\operatorname{div}\langle \boldsymbol{u}\rangle + \left(\overline{\boldsymbol{T}^M} : \frac{\partial\langle \boldsymbol{u}\rangle}{\partial \boldsymbol{r}}\right) - \frac{\nu_M}{4\pi\mu}\left|\operatorname{rot}\bar{\boldsymbol{B}}\right|^2 + G\cdot\bar{\boldsymbol{j}} + \frac{1}{c}\langle \boldsymbol{u}\rangle\cdot\overline{\boldsymbol{j}'\times\boldsymbol{B}'} - \overline{\boldsymbol{j}'\cdot\boldsymbol{E}'}, \tag{9.68}$$

where

$$\boldsymbol{J}^{\text{turb}}_{(E^M)} \equiv \overline{\rho E^M\boldsymbol{u}''} \tag{9.69}$$

is the turbulent magnetic-energy flux for the plasma.

Since the right-hand sides of (9.66) and (9.68) differ only in sign, the density of the total energy of the averaged continuum $\langle U_{(\text{tot})}\rangle$, which is the sum of the densities (per unit mass) of the averaged energy of the matter $\langle U^{\text{sub}}_{\text{tot}}\rangle$ and the averaged magnetic energy of the plasma $\langle E^M \rangle$, is conserved. The only process under the effect of which the total energy of the averaged continuum changes in a stationary volume is the inflow or outflow of energy through the surface of this volume, which the principle of conservation of total energy in a mechanically, thermally, and electromagnetically isolated system expresses.

Using the averaged Poynting vector (see (9.29))

$$\overline{\boldsymbol{q}_{\text{Poynt}}} \equiv \frac{c}{4\pi} \overline{\boldsymbol{E} \times \boldsymbol{H}} = \overline{\rho (E^M + p^M/\rho) \boldsymbol{u}} - \overline{\boldsymbol{T}^M \cdot \boldsymbol{u}} - \nu_M \frac{\partial \overline{p^M}}{\partial \boldsymbol{r}} + \frac{\partial}{\partial \boldsymbol{r}} \cdot \nu_M \overline{\boldsymbol{T}^M}, \quad (9.70)$$

we can write (9.68) for the averaged magnetic energy in the following balance form:

$$\frac{\partial}{\partial t} (\bar{\rho} \langle E^M \rangle) + \operatorname{div} \left\{ \frac{c}{4\pi} \overline{\boldsymbol{E} \times \boldsymbol{H}} - \overline{p^M} \langle \boldsymbol{u} \rangle + \overline{\boldsymbol{T}^M} \cdot \langle \boldsymbol{u} \rangle \right\}$$
$$= -\overline{p^M} \operatorname{div} \langle \boldsymbol{u} \rangle + \left(\overline{\boldsymbol{T}^M} : \frac{\partial \langle \boldsymbol{u} \rangle}{\partial \boldsymbol{r}} \right) - \frac{|\bar{\boldsymbol{j}}|^2}{\sigma_e} + \mathsf{G} \cdot \bar{\boldsymbol{j}} + \frac{1}{c} \langle \boldsymbol{u} \rangle \cdot \overline{\boldsymbol{j}' \times \boldsymbol{B}'} - \overline{\boldsymbol{j}' \cdot \boldsymbol{E}'}, \quad (9.71)$$

which, as can be easily verified, is equivalent to the averaged energy conservation law for an electromagnetic field

$$\frac{\partial}{\partial t} (\bar{\rho} \langle E^M \rangle) + \frac{c}{4\pi} \operatorname{div} \overline{\boldsymbol{E} \times \boldsymbol{H}} = -\overline{\boldsymbol{j} \cdot \boldsymbol{E}}. \quad (9.72)$$

9.2.5.1 The Equation for Magnetic Energy of the Mean Field

Let us now derive the equation for the magnetic energy of the mean field. For this purpose, we scalarly multiply averaged induction equation (9.45*) by $\bar{\boldsymbol{B}}/4\pi\mu$ and use the formula

$$\frac{\nu_M}{4\pi\mu} \bar{\boldsymbol{B}} \cdot \nabla^2 \bar{\boldsymbol{B}} = -\frac{|\bar{\boldsymbol{j}}|^2}{\sigma_e} - \operatorname{div} \left(-\nu_M \frac{\partial p^M_{av}}{\partial \boldsymbol{r}} + \frac{\partial}{\partial \boldsymbol{r}} \cdot \nu_M \boldsymbol{T}^M_{av} \right), \quad (9.73)$$

which, as can easily be verified, is also valid for the averaged fields (cf. (9.67)); as a result, we obtain

$$\bar{\rho} \frac{D E^M_{av}}{Dt} + \operatorname{div} \left\{ -\nu_M \frac{\partial p^M_{av}}{\partial \boldsymbol{r}} + \frac{\partial}{\partial \boldsymbol{r}} \cdot \nu_M \boldsymbol{T}^M_{av} \right\}$$
$$= -p^M_{av} \operatorname{div} \langle \boldsymbol{u} \rangle + \left(\boldsymbol{T}^M_{av} : \frac{\partial \langle \boldsymbol{u} \rangle}{\partial \boldsymbol{r}} \right) + \frac{c}{4\pi\mu} \bar{\boldsymbol{B}} \cdot \operatorname{rot} \mathsf{G} - \frac{|\bar{\boldsymbol{j}}|^2}{\sigma_e}. \quad (9.74)$$

Let us write this equation in a more convenient form for the subsequent analysis using the relation $|\bar{\boldsymbol{j}}|^2/\sigma_e = (\nu_M/4\pi\mu)|\operatorname{rot} \bar{\boldsymbol{B}}|^2$ and the formula of vector analysis $\operatorname{div} \boldsymbol{a} \times \boldsymbol{b} = \boldsymbol{b} \operatorname{rot} \boldsymbol{a} - \boldsymbol{a} \operatorname{rot} \boldsymbol{b}$; as a result, we obtain

$$\bar{\rho}\frac{DE^M_{av}}{Dt} + \operatorname{div}\left\{-\nu_M \frac{\partial p^M_{av}}{\partial \boldsymbol{r}}\frac{\partial}{\partial \boldsymbol{r}} \cdot \nu_M \boldsymbol{T}^M_{av} - \frac{c}{4\pi\mu}(\mathsf{G}\times\bar{\boldsymbol{B}})\right\}$$
$$= -p^M_{av}\operatorname{div}\langle\boldsymbol{u}\rangle + \left(\boldsymbol{T}^M_{av} : \frac{\partial\langle\boldsymbol{u}\rangle}{\partial\boldsymbol{r}}\right) + \bar{\boldsymbol{j}}\cdot\mathsf{G} - \frac{\nu_M}{4\pi\mu}\left|\operatorname{rot}\bar{\boldsymbol{B}}\right|^2. \tag{9.75}$$

We see from the right-hand side of this equation that the magnetic energy of the mean field E^M_{av} decreases due to dissipation (the last term) and increases due to the conversion of the kinetic energy of the mean motion (the term $\boldsymbol{T}^M_{av} : \partial\langle\boldsymbol{u}\rangle/\partial\boldsymbol{r}$) and the turbulent kinetic energy of the material (the term $\mathsf{G}\cdot\bar{\boldsymbol{j}}$) into the magnetic energy of the mean field (see (9.54*), (9.58), and (9.62)).

9.2.5.2 The Equation for the Turbulent Magnetic Energy

The equation for the turbulent magnetic energy $\langle b_M\rangle$ can then be derived from the difference of (9.68) and (9.75):

$$\bar{\rho}\frac{D\langle b_M\rangle}{Dt} + \operatorname{div}\boldsymbol{J}_{\langle b_M\rangle} = -p^M_{\text{turb}}\operatorname{div}\langle\boldsymbol{u}\rangle + \left(\boldsymbol{T}^M_{\text{turb}} : \frac{\partial\langle\boldsymbol{u}\rangle}{\partial\boldsymbol{r}}\right) + \frac{1}{c}\langle\boldsymbol{u}\rangle\cdot\overline{\boldsymbol{j}'\times\boldsymbol{B}'}$$
$$-\overline{\boldsymbol{j}'\cdot\boldsymbol{E}'}, \tag{9.76}$$

where

$$\boldsymbol{J}_{\langle b_M\rangle} \equiv \left\{\overline{\rho(E^M + p^M/\rho)\boldsymbol{u}''} - \overline{\boldsymbol{T}^M\boldsymbol{u}''} + \frac{c}{4\pi\mu}(\mathsf{G}\times\bar{\boldsymbol{B}})\right\} \tag{9.77}$$

is the turbulent-diffusion flux of the turbulent magnetic energy $\langle b_M\rangle$ of the plasma. From this equation follows that the turbulent electric currents (the term $c^{-1}\overline{\boldsymbol{j}'\times\boldsymbol{B}'} \equiv (\partial/\partial\boldsymbol{r})\cdot\boldsymbol{T}^M_{\text{turb}}$) produced in the medium by turbulent velocity pulsations in the presence of a magnetic field and by magnetic field pulsations are physically responsible for the generation and maintenance of turbulent magnetic energy (turbulent magnetic field). The last term in (9.76) describes the decrease in the turbulent magnetic energy due to its conversion into the kinetic turbulent energy of the medium (see (9.62)).

9.2.5.3 The Balance Equation for the Total Energy of the Plasma Turbulence

Adding now (9.62) and (9.76) yields a balance equation for the total energy of the turbulence $b_\Sigma \equiv \langle b\rangle + \langle b_M\rangle$ of an electrically conducting medium in the form

$$\frac{\partial(\bar{\rho} b_\Sigma)}{\partial t} + \operatorname{div}\{\bar{\rho}\, b_\Sigma \langle \boldsymbol{u} \rangle + \boldsymbol{J}_{b_\Sigma}\} = -p^M_{\text{turb}} \operatorname{div}\langle \boldsymbol{u} \rangle + \left(\boldsymbol{R}^K : \frac{\partial \langle \boldsymbol{u} \rangle}{\partial \boldsymbol{r}} \right)$$
$$+ \frac{1}{4\pi\mu} \left(\boldsymbol{R}^M : \frac{\partial \bar{\boldsymbol{B}}}{\partial \boldsymbol{r}} \right) - \left(\boldsymbol{J}^{\text{turb}}_{(1/\rho)} \cdot \frac{\partial \bar{p}}{\partial \boldsymbol{r}} \right) + \overline{p' \operatorname{div} \boldsymbol{u}''} - \bar{\rho} \varepsilon_\Sigma, \tag{9.78}$$

where

$$\boldsymbol{J}_{b_\Sigma} \equiv \boldsymbol{J}_{\langle b \rangle} + \boldsymbol{J}_{\langle b_M \rangle}$$
$$= \left\{ \overline{\rho(|\boldsymbol{u}''|^2/2 + p'/\rho)\boldsymbol{u}''} - \overline{\left(\Pi + \frac{1}{4\pi\mu} \boldsymbol{B}'\boldsymbol{B}' \right) \cdot \boldsymbol{u}''} - \frac{1}{4\pi\mu} \left[\overline{(\boldsymbol{B}' \cdot \boldsymbol{u}'')} \bar{\boldsymbol{B}} + \overline{\boldsymbol{u}''(\boldsymbol{B} \cdot \bar{\boldsymbol{B}})} \right] \right\} \tag{9.79}$$

is the turbulent-diffusion flux of the total (kinetic plus magnetic) turbulent energy of the electrically conducting medium;

$$\varepsilon_\Sigma \equiv \langle \varepsilon_M \rangle + \langle \varepsilon_b \rangle = \frac{1}{\bar{\rho}} \left\{ \overline{\left(\Pi : \frac{\partial \boldsymbol{u}''}{\partial \boldsymbol{r}} \right)} + \frac{\nu_M}{4\pi\mu} \overline{|\operatorname{rot} \boldsymbol{B}'|^2} \right\} \tag{9.80}$$

is the total specific rate of dissipation of the turbulent kinetic and turbulent magnetic energies into heat (under the action of molecular kinematic viscosity and magnetic field); and $\boldsymbol{R}^K \equiv \boldsymbol{R} + \boldsymbol{T}^{\text{turb}}$ is the total turbulent stress tensor for the plasma (the kinetic Reynolds tensor). In writing (9.78), we used the transformation

$$\mathsf{G} \cdot \bar{\boldsymbol{j}} \equiv -\frac{1}{4\pi\mu} \left(\boldsymbol{R}^M : \frac{\partial \bar{\boldsymbol{B}}}{\partial \boldsymbol{r}} \right), \tag{9.81}$$

which follows from (9.41*) and (9.43); indeed, we have

$$\mathsf{G}_i \bar{j}_i = -\frac{1}{2c} \varepsilon_{ijk} R^M_{jk} \bar{j}_i = -\frac{1}{8\pi\mu} \varepsilon_{ijk} R^M_{jk} (\operatorname{rot} \bar{\boldsymbol{B}})_i = -\frac{1}{8\pi\mu} \varepsilon_{kji} \varepsilon_{mli} R^M_{jk} \frac{\partial \bar{B}_m}{\partial x_l}$$
$$= -\frac{1}{8\pi\mu} R^M_{jk} \frac{\partial \bar{B}_m}{\partial x_l} (\delta_{km}\delta_{jl} - \delta_{jm}\delta_{kl}) = -\frac{1}{8\pi\mu} R^M_{jk} \left(\frac{\partial \bar{B}_k}{\partial x_j} - \frac{\partial \bar{B}_j}{\partial x_k} \right)$$
$$= -\frac{1}{4\pi\mu} R^M_{jk} \frac{\partial \bar{B}_k}{\partial x_j}.$$

Note that significant part (and, according to some estimates, the bulk) of the total turbulence energy of the system is contained in the pulsational magnetic field.

Equation (9.78) shows that Joule dissipation (the term $\bar{\rho}\langle \varepsilon_M \rangle$) leads to a faster decay of the plasma perturbation than in the case where only viscous dissipation is present, i.e., the direct interaction between the field and flow perturbations always increases the stability of the plasma flow. On the other hand, the magnetic field can also interact with the averaged fluid flow. In this case, the rate of kinematic exchange $(\boldsymbol{R}^K : \partial\langle \boldsymbol{u}\rangle/\partial \boldsymbol{r})$ between the kinetic energy of the mean-fluid motion (see (9.58)) and the kinetic energy of the eddy motion of the system depends both on the correlation between the pulsations of the velocity components $\boldsymbol{R}$ and the pulsations of the magnetic field components $\boldsymbol{R}^M$ and on the mean velocity shear $\partial\langle \boldsymbol{u}\rangle/\partial \boldsymbol{r}$, i.e., the stability is influenced through a change in the profile of the averaged flow velocity $\langle \boldsymbol{u}\rangle$. This influence can be much more significant than that of the direct interaction with perturbations, with the change in the averaged velocity profile leading to an increase in stability in the case of small-scale turbulence. As has been noted above, large-scale turbulence can convert the kinetic energy of turbulence into the energy of mean motion, as, for example, in the case of the action of a transverse magnetic field on the Couette flow (Kadomtsev 1976).

9.2.5.4 The Case of the Developed Turbulence

In solving specific problems of the turbulent hydrodynamics of electrically conducting media in the case of developed turbulence (where $\boldsymbol{q}^{\text{turb}} >> \bar{\boldsymbol{q}}$, and $\boldsymbol{R} >> \bar{\Pi}$), the total enthalpy by

$$\langle H_0\rangle \equiv \frac{|\langle \boldsymbol{u}\rangle|^2}{2} + \Psi_G + \langle H\rangle \tag{9.82}$$

and substituting the expression $\langle H\rangle \equiv \langle H_0\rangle - |\langle \boldsymbol{u}\rangle|^2/2 - \Psi_G$ in (9.54*), after simple transformations performed using the balance equation for averaged mechanical energy (9.58), we obtain

$$\begin{aligned}
&\bar{\rho}\frac{D\langle H_0\rangle}{Dt} + \operatorname{div}\left\{\bar{\boldsymbol{q}}_R + \tilde{\boldsymbol{q}}^{\text{turb}} + \overline{p^M}\langle \boldsymbol{u}\rangle - \left(\bar{\Pi} + \boldsymbol{T}^M_{av} + \boldsymbol{R}^K\right)\cdot\langle \boldsymbol{u}\rangle\right\}\\
&\quad = \frac{\partial\bar{p}}{\partial t} + \frac{\nu_M}{4\pi\mu}|\operatorname{rot}\bar{\boldsymbol{B}}|^2 - +\overline{p^M}\operatorname{div}\langle \boldsymbol{u}\rangle - \left(\left(\boldsymbol{T}^M_{av} + \boldsymbol{R}^K\right) : \frac{\partial\langle \boldsymbol{u}\rangle}{\partial \boldsymbol{r}}\right)\\
&\quad - \overline{p'\operatorname{div}\boldsymbol{u}''} + \left(\boldsymbol{J}^{\text{turb}}_{(1/\rho)}\cdot\frac{\partial\bar{p}}{\partial \boldsymbol{r}}\right) + \bar{\rho}\varepsilon_\Sigma.
\end{aligned} \tag{9.83}$$

Let us now make the important assumption (which often holds in practice) that, in the case of developed turbulence, where the hydrodynamic Reynolds number is great, $\boldsymbol{Re} >> 1$, a stationary non-equilibrium state is established in the structure of the pulsating fields $\boldsymbol{u}''$ and $\boldsymbol{B}'$ at which the total turbulent energy of the plasma b_Σ, which is the sum of the turbulent energy of the material plasma component $\langle b\rangle \equiv \overline{\rho|\boldsymbol{u}''|^2}/2\bar{\rho}$ and the turbulent energy of the pulsating magnetic field $\langle b_M\rangle \equiv \overline{|\boldsymbol{B}'|^2}/8\pi\mu\bar{\rho}$, is almost constant both in time and space. In this case, the following approximate relation (see (9.78)) holds:

$$-p_{\text{turb}}^{M}\text{div}\langle\boldsymbol{u}\rangle+\left(\boldsymbol{R}^{K}:\frac{\partial\langle\boldsymbol{u}\rangle}{\partial\boldsymbol{r}}\right)+\frac{1}{4\pi\mu}\left(\boldsymbol{R}^{M}:\frac{\partial\bar{\boldsymbol{B}}}{\partial\boldsymbol{r}}\right)$$
$$\cong\left(\boldsymbol{J}_{(1/\rho)}^{\text{turb}}\cdot\frac{\partial\bar{p}}{\partial\boldsymbol{r}}\right)-\overline{p'\,\text{div}\,\boldsymbol{u}''}+\bar{\rho}\varepsilon_{\Sigma}. \tag{9.84}$$

Then, taking the latter into account, the heat inflow equation for turbulized mixture (9.54*) acquires the following nearly "classical" form:

$$\bar{\rho}\frac{D\langle H\rangle}{Dt}+\text{div}\{\bar{\boldsymbol{q}}_{R}+\tilde{\boldsymbol{q}}^{\text{turb}}\}=\frac{D\bar{p}}{Dt}-p_{\text{turb}}^{M}\text{div}\langle\boldsymbol{u}\rangle+\left(\boldsymbol{R}^{K}:\frac{\partial\langle\boldsymbol{u}\rangle}{\partial\boldsymbol{r}}\right)$$
$$+\frac{1}{4\pi\mu}\left(\boldsymbol{R}^{M}:\frac{\partial\bar{\boldsymbol{B}}}{\partial\boldsymbol{r}}\right)+\frac{\nu_{M}}{4\pi\mu}\left|\text{rot}\,\bar{\boldsymbol{B}}\right|^{2}, \tag{9.85}$$

while (9.83) for the total enthalpy, given the identical transformation

$$\langle\boldsymbol{u}\rangle\cdot\frac{\partial p_{av}^{M}}{\partial\boldsymbol{r}}-\langle\boldsymbol{u}\rangle\cdot\frac{\partial\boldsymbol{T}_{av}^{M}}{\partial\boldsymbol{r}}=\frac{1}{c}\bar{\boldsymbol{j}}\cdot(\langle\boldsymbol{u}\rangle\times\boldsymbol{B})=\frac{\nu_{M}}{4\pi\mu}\left|\text{rot}\,\bar{\boldsymbol{B}}\right|^{2}+\frac{1}{4\pi\mu}\left(\boldsymbol{R}^{M}:\frac{\partial\bar{\boldsymbol{B}}}{\partial\boldsymbol{r}}\right)-\bar{\boldsymbol{j}}\cdot\bar{\boldsymbol{E}},$$

takes the form

$$\bar{\rho}\frac{D\langle H_{0}\rangle}{Dt}+\text{div}\{\bar{\boldsymbol{q}}_{R}+\tilde{\boldsymbol{q}}^{\text{turb}}+p_{\text{turb}}^{M}\langle\boldsymbol{u}\rangle-\boldsymbol{R}^{K}\cdot\langle\boldsymbol{u}\rangle\}=\frac{\partial\bar{p}}{\partial t}+\bar{\boldsymbol{j}}\cdot\bar{\boldsymbol{E}}. \tag{9.86}$$

In practice, for various cases, the averaged equation for the internal energy of the plasma can be used in form (9.54*), (9.85), or (9.86).

The above derived differential equations of the mean motion for a turbulized, electrically conducting medium should be complemented by the averaged equation of state for pressure. Applying statistical averaging to (9.34) yields the exact expression

$$\bar{p}=\Re\bar{\rho}\langle T\rangle, \tag{9.87}$$

which is commonly used in simple models of MHD turbulence based on the gradient hypothesis of closure.

9.3 Derivation of the Defining Relations or Turbulent Flows of an Electrically Conducting Medium in the Presence of a Magnetic Field

In the previous section, based on the classical single-fluid MHD equations with a simplified form of the Ohm law, we derived using the weighted-mean Favre averaging the MHD equations for the scale of mean motion. These equations are designed to describe the averaged turbulent flows of an electrically conducting medium in the presence of a magnetic field. In general, they have the same form as the MHD equations for the regular (predictable) regime of motion of a conducting fluid, as it is presented in Appendix. At the same time, the main system of averaged MHD equations consisting of (9.36), (9.45*), (9.52), (9.54*), (9.78), and (9.87) is non-closed, since it contains the uncertain correlation second moments $\boldsymbol{q}^{\mathrm{turb}}, \boldsymbol{R}^{K}$, and $\boldsymbol{R}^{M}$ which appeared as a result of averaging the original nonlinear MHD equations, along with the mean values of such hydrodynamic parameters of state as $\bar{\rho}$, $\langle \boldsymbol{u} \rangle, \bar{p},$, $\langle T \rangle$, and $\bar{\boldsymbol{B}}$ and their derivatives. Apart from these correlation functions, we need information about the turbulent flow of the specific volume $\boldsymbol{J}^{\mathrm{turb}}_{(1/\rho)}$, related to the density fluctuations and about the correlation functions that include the pressure fluctuations. This raises the well known problem of closure in turbulence related to the need for constructing the defining relations for the uncertain correlation quantities appearing in the averaged MHD equations, which has its own peculiar features in the presence of a magnetic field. Thus, our next objective is to derive these closure relations for the turbulent flows related to the heat, momentum, and mass transfer. For this purpose, we use the methods of extended non-equilibrium thermodynamics (see Jou et al. 2001).

9.3.1 The Balance Equation for the Averaged Entropy of a Conducting Medium

We perform our thermodynamic analysis of a turbulized electrically conducting medium by assuming that the one-point correlations $\langle \mathcal{A}'' \mathcal{B}'' \rangle$ for any (but not equal to the hydrodynamic flow velocity $\boldsymbol{u}$) pulsating thermodynamic parameters $\mathcal{A}$ and $\mathcal{B}$ are small compared to the first-order terms $\langle \mathcal{A} \rangle \langle \mathcal{B} \rangle$ and can be omitted, i.e., we assume below that the inequality $\langle \mathcal{A}'' \mathcal{B}'' \rangle / \langle \mathcal{A} \rangle \langle \mathcal{B} \rangle << 1, \quad (\mathcal{A} \neq \boldsymbol{u}, \mathcal{B} \neq \boldsymbol{u})$ holds. For this case, as it was shown earlier (Kolesnichenko 1998), averaging of the Gibbs fundamental identity written along the trajectory of the center of mass of a physically elementary volume (which is valid for micromotions of a medium) leads to the following equation for the weighted-mean specific entropy $\langle S \rangle(\boldsymbol{r}, t)$ and specific internal energy $\langle E \rangle(\boldsymbol{r}, t)$ of a conducting fluid:

$$\bar{\rho} \langle T \rangle \frac{D\langle S \rangle}{Dt} = \bar{\rho} \frac{D\langle E \rangle}{Dt} + \bar{\rho} \bar{p} \frac{D\langle 1/\bar{\rho} \rangle}{Dt}. \tag{9.88}$$

Let us now eliminate the substantial derivatives of the parameters $\langle E \rangle$ and $(1/\bar{\rho})$ from the right-hand side of this equation using averaged continuity (9.36) and energy (9.55) equations; as a result, we obtain a substantial balance equation for the averaged entropy $\langle S \rangle$ of the medium in the following explicit form:

$$\bar{\rho}\frac{D\langle S\rangle}{Dt} + \operatorname{div}\boldsymbol{J}_{\langle S\rangle} = \sigma_{\langle S\rangle} = \sigma_{\langle S\rangle}^{(i)} + \sigma_{\langle S\rangle}^{(e)}, \tag{9.89}$$

where

$$\boldsymbol{J}_{\langle S\rangle} \equiv \boldsymbol{J}_{\langle E\rangle}/\langle T\rangle \equiv \left\{\frac{\bar{\boldsymbol{q}} + \bar{\boldsymbol{q}}_R + \bar{\boldsymbol{q}}^{\text{turb}}}{\langle T\rangle}\right\}, \tag{9.90}$$

$$\begin{cases} \sigma_{\langle S\rangle}^{(i)} \equiv \frac{1}{\langle T\rangle}\left\{-\left(\boldsymbol{J}_{\langle S\rangle}\cdot\frac{\partial\langle T\rangle}{\partial\boldsymbol{r}}\right) + \left(\bar{\Pi} : \frac{\partial\langle\boldsymbol{u}\rangle}{\partial\boldsymbol{r}}\right) + \frac{\nu_M}{4\pi\mu}\left|\operatorname{rot}\bar{\boldsymbol{B}}\right|^2\right\} \geq 0, \\ \sigma_{\langle S\rangle}^{(e)} \equiv \frac{1}{\langle T\rangle}\left\{-\overline{p'\operatorname{div}\boldsymbol{u}''} + \left(\boldsymbol{J}_{(1/\rho)}^{\text{turb}}\cdot\frac{\partial\bar{p}}{\partial\boldsymbol{r}}\right) + \bar{\rho}\varepsilon_\Sigma\right\} \equiv \frac{\Im}{\langle T\rangle}. \end{cases} \tag{9.91}$$

Here, the quantity $\sigma_{\langle S\rangle}^{(i)}(\boldsymbol{r},t)$, which defines the rate of local production of the averaged entropy $\langle S\rangle(\boldsymbol{r},t)$ of the mixture (attributable to the irreversible transfer processes inside the subsystem of averaged motion) is always positive. However, the quantity $\sigma_{\langle S\rangle}^{(e)}(\boldsymbol{r},t)$ which pertains to the sink or inflow of averaged entropy can be different in sign and, as is clear from the subsequent analysis, reflects the entropy exchange between the subsystems of averaged motion and the so-called subsystem of turbulent chaos (Kolesnichenko and Marov 1999). Thus, it follows from (9.89) that the averaged entropy $\langle S\rangle$ alone is insufficient to adequately describe all features of the turbulent flow of an electrically conducting fluid, since the second law of thermodynamics does not hold for it. Furthermore, the entropy $\langle S\rangle$ is not related explicitly to any parameters of the pulsating turbulent chaos that characterize its internal structure, in particular, to such key turbulence characteristics as the kinetic turbulence energy $\langle b\rangle$ and the magnetic-field turbulence energy $\langle b_M\rangle$. In fact, for this reason, it seems necessary to consider the subsystem of turbulent chaos when constructing an adequate thermodynamic model of MHD turbulence.

9.3.2 Entropy Balance Equations and Entropy Production for the Subsystem of Turbulent Chaos in a Conducting Medium

In what follows, we describe the turbulized motion of an electrically conducting fluid within the framework of a two-fluid continuum consisting of two interrelated open subsystems: the subsystem of mean motion, which results from the probability-theoretical averaging of the MHD equations for an instantaneous (actual) fluid flow, and the subsystem of turbulent chaos (the so-called turbulent superstructure), which is associated with the pulsational motion of an electrically conducting medium (Kolesnichenko 1998). We assume that a thermodynamically elementary volume $\mathrm{d}\boldsymbol{r}$ of turbulent gas-field chaos can be characterized by extensive state variables, such as the internal-energy density $E_{\text{turb}}(\boldsymbol{r},t)$ and the generalized entropy $S_{\text{turb}}(\boldsymbol{r},t)$ of the matter and magnetic field turbulization, and intensive state variables, such as the generalized turbulization temperature $T_{\text{turb}}(\boldsymbol{r},t)$ (a quantity that characterizes the degree of turbulent pulsation intensity) and the turbulization pressure $p_{\text{turb}}(\boldsymbol{r},t)$ (cf. Blackadar 1955). It is important to bear in mind that such generalized thermodynamic parameters of the state of turbulent chaos as the turbulization entropy S_{turb} and energy E_{turb} (which are considered below as the primary concepts) are introduced here a priori to ensure the coherence of the theory and, generally, not need to have an exact physical interpretation (Jou et al. 2001).

Let us now turn to the corollaries of this formal approach. Following the Gibbs method, we select the Gibbs fundamental equation for the generalized entropy $S_{\text{turb}}(\boldsymbol{r},t) = S_{\text{turb}}(E_{\text{turb}}(\boldsymbol{r},t), 1/\bar{\rho}(\boldsymbol{r},t))$ as a local characteristic function that contains all of the thermodynamic information about the subsystem of turbulent gas-field chaos in a steady state. This functional relation is assumed to be given a priori. Let us now take, as is commonly done in the formalized construction of classical local-equilibrium thermodynamics, the following definitions of the conjugate variables $T_{\text{turb}}(\boldsymbol{r},t)$ and $p_{\text{turb}}(\boldsymbol{r},t)$ (assuming all of the derivatives noted are positive):

$$\frac{1}{T_{\text{turb}}} \equiv \left\{\frac{\partial S_{\text{turb}}}{\partial E_{\text{turb}}}\right\}_{1/\bar{\rho}}, \quad \frac{p_{\text{turb}}}{T_{\text{turb}}} \equiv \left\{\frac{\partial S_{\text{turb}}}{\partial(1/\bar{\rho})}\right\}_{E_{\text{turb}}}. \tag{9.92}$$

In this case, the meaning of the generalized (turbulization) temperature and pressure can be assigned to the intensive variables $T_{\text{turb}}(\boldsymbol{r},t)$ and $p_{\text{turb}}(\boldsymbol{r},t)$, respectively. The corresponding differential form of the Gibbs fundamental equation written along the trajectory of the center of mass of a physically elementary volume takes the form (cf. Marov and Kolesnichenko 2006)

$$T_{\text{turb}}\frac{DS_{\text{turb}}}{Dt} = \frac{DE_{\text{turb}}}{Dt} - \frac{1}{3}R^K_{ij}\delta_{ij}\frac{D}{Dt}\left(\frac{1}{\bar{\rho}}\right). \tag{9.93}$$

Below, we identify the quantity E_{turb} with the total turbulence energy of an electrically conducting fluid

$$\bar{\rho}E_{\text{turb}} \equiv \bar{\rho}b_{\Sigma} = \overline{\rho|\boldsymbol{u}''|^2/2} + \overline{|\boldsymbol{B}'|^2}/8\pi\mu; \tag{9.94}$$

then,

$$p_{\text{turb}} \equiv \frac{1}{3}(\boldsymbol{R}^K : \boldsymbol{U}) = \frac{2}{3}\bar{\rho}b_{\Sigma}. \tag{9.95}$$

We derive the corresponding balance equation for the turbulization entropy S_{turb} of a conducting medium from (9.93) by the above method using, for this purpose, the equation for the specific volume $(1/\bar{\rho})$ and (9.78) for the total turbulent energy b_{Σ}. This yields

$$\bar{\rho}\frac{DS_{\text{turb}}}{Dt} + \text{div}\left\{\frac{\boldsymbol{J}_{b_{\Sigma}}}{T_{\text{turb}}}\right\} = \sigma_{(S_{\text{turb}})} \equiv \sigma^{(i)}_{(S_{\text{turb}})} + \sigma^{(e)}_{(S_{\text{turb}})}, \tag{9.96}$$

where

$$J_{b_\Sigma} = \left\{ \overline{\rho(|u''|^2/2 + p'/\rho)u''} - \overline{\left(\Pi' + \frac{B'B'}{4\pi\mu}\right) \cdot u''} - \frac{1}{4\pi\mu}\left[\overline{(B' \cdot u'')}\bar{B} + \overline{u''(B \cdot \bar{B})}\right] \right\}, \tag{9.97}$$

$$\begin{cases} 0 \le \sigma^{(i)}_{(S_{\text{turb}})} = \dfrac{1}{T_{\text{turb}}} \left\{ -J_{b_\Sigma} \cdot \dfrac{\partial \ln T_{\text{turb}}}{\partial r} + \left(\left[R^K - \frac{1}{3}(R^K : U)U \right] : \dfrac{\partial \langle u \rangle}{\partial r} \right) \right. \\ \left. \quad + \dfrac{1}{4\pi\mu}\left(R^M : \dfrac{\partial \bar{B}}{\partial r} \right) \right\}, \\ \sigma^{(e)}_{(S_{\text{turb}})} \equiv \dfrac{1}{T_{\text{turb}}} \left\{ \overline{p' \operatorname{div} u''} - \left(J^{\text{turb}}_{(1/\rho)} \cdot \dfrac{\partial \bar{p}}{\partial r} \right) - \bar{\rho}\varepsilon_\Sigma \right\} \equiv -\dfrac{\Im}{T_{\text{turb}}}. \end{cases} \tag{9.98}$$

Here, $J_{S_{\text{turb}}} \equiv \{J_{b_\Sigma}/T_{\text{turb}}\}$ is the substantial flux of the entropy S_{turb} for the subsystem of turbulent chaos, while the quantities $\sigma^{(i)}_{(S_{\text{turb}})}$ and $\sigma^{(e)}_{(S_{\text{turb}})}$ mean the rates of local production and the sink of the pulsational entropy S_{turb}, respectively.

Note that referring the individual terms of (9.96) to the turbulent flux $J_{S_{\text{turb}}}$ or to the production $\sigma_{(S_{\text{turb}})}$ of an averaged entropy is, to a certain extent, ambiguous because a number of alternative formulations using various definitions of the turbulent entropy flux S_{turb} different from (9.97) are also possible. This kind of reasoning is discussed in more detail by de Groot and Mazur (1962) and Gyarmati (1970).

9.3.3 Balance Equation for the Total Entropy

Thus, introducing two entropies, $\langle S \rangle$ and S_{turb}, specifies our view of the original turbulized conducting continuum as a thermodynamic complex that consists of two mutually open subsystems filling up the same coordinate space continuously: the subsystem of mean motion of the conducting fluid and the subsystem of turbulent gas-field chaos. Combining (9.90) and (9.96), we find the balance equation for the total entropy of the system

$$S_\Sigma \equiv \langle S \rangle + S_{\text{turb}} \frac{\partial}{\partial t}(\bar{\rho} S_\Sigma) + \operatorname{div}\left\{ \bar{\rho} S_\Sigma \langle u \rangle + \frac{J_{\langle E \rangle}}{\langle T \rangle} + \frac{J_{b_\Sigma}}{T_{\text{turb}}} \right\} = \sigma_\Sigma \ge 0, \tag{9.99}$$

where

$$\sigma_\Sigma \equiv \sigma^{(i)}_{\langle S \rangle} + \sigma^{(i)}_{(S_{\text{turb}})} + \Im \frac{T_{\text{turb}} - \langle T \rangle}{T_{\text{turb}} \langle T \rangle}$$

is the total entropy production related to the irreversible processes inside the turbulized plasma; and

$$\Im \equiv \left\{ -\overline{p' \operatorname{div} \boldsymbol{u}''} + \left(\boldsymbol{J}^{\text{turb}}_{(1/\rho)} \cdot \frac{\partial \bar{p}}{\partial \boldsymbol{r}} \right) + \bar{\rho} \varepsilon_{\Sigma} \right\} \tag{9.100}$$

is the energy flux of the transition between the subsystems of averaged motion and turbulent chaos. The positive quantity σ_Σ written using (9.91), (9.92), and (9.98) has the following bilinear structure:

$$\begin{aligned} 0 \leq \sigma_\Sigma \equiv \Im \frac{T_{\text{turb}} - \langle T \rangle}{T_{\text{turb}} \langle T \rangle} + \frac{1}{\langle T \rangle} \left\{ - \left(\boldsymbol{J}_{\langle E \rangle} \cdot \frac{\partial \ln \langle T \rangle}{\partial \boldsymbol{r}} \right) + \left(\bar{\Pi} : \frac{\partial \langle \boldsymbol{u} \rangle}{\partial \boldsymbol{r}} \right) + \frac{\nu_M}{4\pi\mu} |\operatorname{rot} \bar{\boldsymbol{B}}|^2 \right\} \\ + \frac{1}{T_{\text{turb}}} \left\{ - \left(\boldsymbol{J}_{b_\Sigma} \cdot \frac{\partial \ln T_{\text{turb}}}{\partial \boldsymbol{r}} \right) + \left(\left[\boldsymbol{R}^K - \frac{1}{3} (\boldsymbol{R}^K : \boldsymbol{U}) \boldsymbol{U} \right] : \frac{\partial \langle \boldsymbol{u} \rangle}{\partial \boldsymbol{r}} \right) + \frac{1}{4\pi\mu} \left(\boldsymbol{R}^M : \frac{\partial \bar{\boldsymbol{B}}}{\partial \boldsymbol{r}} \right) \right\}. \end{aligned} \tag{9.101}$$

We see from this expression that the local production σ_Σ of the total entropy S_Σ related to the irreversible processes inside the turbulized conducting continuum is defined by the set of thermodynamic fluxes

$$\boldsymbol{J}_{\langle E \rangle}, \boldsymbol{J}_{b_\Sigma}, \bar{\Pi}, \left[\boldsymbol{R}^K - \tfrac{1}{3} (\boldsymbol{R}^K : \boldsymbol{U}) \boldsymbol{U} \right], \boldsymbol{R}^M, \Im \tag{9.102}$$

and conjugate thermodynamic forces

$$\frac{\partial}{\partial \boldsymbol{r}} \left(\frac{1}{\langle T \rangle} \right), \frac{\partial}{\partial \boldsymbol{r}} \left(\frac{1}{T_{\text{turb}}} \right), \frac{1}{\langle T \rangle} \frac{\partial \langle \boldsymbol{u} \rangle}{\partial \boldsymbol{r}}, \frac{1}{T_{\text{turb}}} \frac{\partial \langle \boldsymbol{u} \rangle}{\partial \boldsymbol{r}}, \frac{1}{T_{\text{turb}}} \frac{\partial \bar{\boldsymbol{B}}}{\partial \boldsymbol{r}}, \frac{(T_{\text{turb}} - \langle T \rangle)}{T_{\text{turb}} \langle T \rangle}. \tag{9.103}$$

According to the main postulate of generalized non-equilibrium thermodynamics, in the case where the thermodynamic system is near its local equilibrium or near a stable stationary non-equilibrium state, the thermodynamic fluxes can be represented as linear functions of the conjugate macroscopic forces $J_{\gamma i} = \sum_\delta L^{ij}_{\gamma\delta} X_{\delta j} (\gamma, \delta = 1, 2, \ldots, f)$. This allows us to derive the necessary defining relations for the as yet indeterminate averaged molecular and turbulent fluxes appearing in the averaged MHD equations. It is important to note that (9.101) makes it possible to obtain the defining relations for all three main flow regimes of an electrically conducting turbulized medium, namely, for the averaged regular regime; for the regime of developed turbulence, when the turbulent transfer fluxes are much more effective than the corresponding averaged molecular fluxes; and, finally, in the general case where the processes of averaged molecular and turbulent transfer are comparable in effectiveness (e.g., near a boundary wall). It can be also seen from (9.101) that the spectrum of the possible cross effects for the turbulent flow regime of a conducting fluid is broadened significantly compared to its regular flow regime. However, unfortunately, there are currently no reliable experimental data that would quantitatively describe many cross effects in a conducting turbulized medium. Furthermore, the contribution from any cross effects to the total rate of the transfer processes is known to be generally an order of magnitude smaller than that from the direct effects. Taking into account these two circumstances, below we use the positivity condition for the production of the total entropy σ_Σ for each individual term, i.e., we assume that $\sigma^{(i)}_{\langle S\rangle} \geq 0$, $\sigma^{(i)}_{S_{\text{turb}}} \geq 0$, and $\sigma_{\langle S\rangle, S_{\text{turb}}} \geq 0$. We also omit a number of cross effects in the linear constitutive relations without any special reservations.

To conclude this section, we note that the first term on the right-hand side of (9.101), which describes the entropy production inside the complete system through the irreversible entropy exchange between the subsystems of turbulent chaos and averaged motion, is always positive in view of the second law of thermodynamics,

$$\sigma_{\langle S\rangle, S_{\text{turb}}} \equiv \Im \cdot \frac{T_{\text{turb}} - \langle T\rangle}{T_{\text{turb}}\langle T\rangle} \geq 0 \tag{9.104}$$

and, hence, the "direction" of the thermodynamic flux $\Im$ is determined by the sign of the state function $X_\Im \equiv (1/\langle T\rangle - 1/T_{\text{turb}})$. This function should be considered as the conjugate thermodynamic force (a macroscopic factor) that produces precisely this entropy flux. This kind of entropy exchange between two mutually open subsystems is known to be an indispensable condition for a structured collective behavior, i.e., it can be a source of self-organization in one of them (Prigogine and Stengers 1994).

9.3.4 Stationary-Nonequilibrium Regime of the Subsystem of Turbulent Chaos. Derivation of the Defining Relations

Let us now show that, in the case of a stationary non-equilibrium state, the dissipative activity of the subsystem of turbulent chaos is just determined by the inflow of negative entropy ($\sigma^e_{(S_{\text{turb}})} \equiv -\Im/T_{\text{turb}} < 0$) from the subsystem of averaged motion. For this purpose, we shall analyze the important regime of the developed turbulent motion of a conducting fluid—the regime of stationary non-equilibrium turbulence where a quasi-steady state in which $\mathrm{d}S_{\text{turb}}/\mathrm{d}t \cong 0$ and the turbulization entropy flux $\boldsymbol{J}_{S_{\text{turb}}} \equiv \boldsymbol{J}_{b_\Sigma}/T_{\text{turb}}$ is constant, i.e., $\boldsymbol{J}_{S_{\text{turb}}} \cong \text{const}$ (Kondepudi and Prigogine 1998), is established in the structure of the subsystem of turbulent chaos. This condition means that the production $\sigma^i_{(S_{\text{turb}})}$ of the turbulization energy S_{turb} is offset by its outflow $\sigma^e_{(S_{\text{turb}})}$ to such an extent that the total production of the entropy S_{turb} is almost absent, $\sigma_{(S_{\text{turb}})} \equiv \sigma^e_{(S_{\text{turb}})} + \sigma^i_{(S_{\text{turb}})} \cong 0$. Since $\sigma^i_{(S_{\text{turb}})} \geq 0$, the relation $0 > \sigma^e_{(S_{\text{turb}})} \cong -\sigma^i_{(S_{\text{turb}})}$ is valid. Thus, the subsystem of turbulent chaos in this regime of turbulence should export the entropy into the "external medium" (i.e., give up the amount $\sigma^{e}_{(S_{\text{turb}})}$ to the system of averaged motion) in order to compensate for its production (the amount $\sigma^{i}_{(S_{\text{turb}})}$ at the expense of irreversible internal processes). In other words, an inflow of negative entropy (negentropy) from the "external medium" (i.e., the subsystem of averaged motion), $\sigma^{e}_{(S_{\text{turb}})} \equiv -\Im/T_{\text{turb}} = -\langle T\rangle\sigma^{e}_{\langle S\rangle}/T_{\text{turb}}<0$, is needed to maintain this kind of quasi-steady state inside the subsystem of turbulent chaos. It is this negentropy entering the subsystem of turbulent chaos that is spent on maintaining and perfecting its internal structure. But in such a case the relation $0 \le \sigma^{e}_{\langle S\rangle} = -T_{\text{turb}}\sigma^{e}_{(S_{\text{turb}})}/\langle T\rangle \cong T_{\text{turb}}\sigma^{i}_{(S_{\text{turb}})}/\langle T\rangle$ is then valid and balance equation (9.89) for the averaged entropy $\langle S\rangle$ of the system takes the form

$$\bar{\rho}\frac{D\langle S\rangle}{Dt} + \operatorname{div}\left\{\frac{\boldsymbol{J}_{\langle E\rangle}}{\langle T\rangle}\right\} = \sigma_{\langle S\rangle} = \sigma^{(i)}_{\langle S\rangle} + \sigma^{(e)}_{\langle S\rangle} = \sigma^{(i)}_{\langle S\rangle} + \frac{T_{\text{turb}}}{\langle T\rangle}\sigma^{i}_{(S_{\text{turb}})}, \qquad (9.105)$$

where the following relation is valid for the local energy scattering $\langle T\rangle\sigma_{\langle S\rangle}$:

$$0 \le \langle T\rangle\sigma_{\langle S\rangle} \equiv -\left(\tilde{\boldsymbol{q}}^{\text{turb}} \cdot \frac{\partial \ln\langle T\rangle}{\partial \boldsymbol{r}}\right) + \frac{\nu_M}{4\pi\mu}|\operatorname{rot}\bar{\boldsymbol{B}}|^2 + \left(\left[\boldsymbol{R}^K - \frac{1}{3}(\boldsymbol{R}^K : \boldsymbol{U})\boldsymbol{U}\right] : \frac{\partial\langle \boldsymbol{u}\rangle}{\partial \boldsymbol{r}}\right) + \frac{1}{4\pi\mu}\left(\boldsymbol{R}^M : \frac{\partial \bar{\boldsymbol{B}}}{\partial \boldsymbol{r}}\right). \qquad (9.106)$$

Because in the stationary non-equilibrium state the entropy outflow from the subsystem of the averaged motion is positive, $0 \le \sigma^{e}_{\langle S\rangle} = \Im/\langle T\rangle$, the rate of $\Im$ entropy (heat) exchange between the averaged and turbulent motions is also positive, $\Im \ge 0$. However, it then follows from inequality (9.104) that the turbulization temperature T_{turb} is higher than the averaged temperature of the turbulized continuum ($T_{\text{turb}}>\langle T\rangle$), which is in complete agreement with the basic synergetic principle on self-organization of the dissipative system. Basically, in accordance with this principle, the formation of ordered structures (in our case, the eddy structures of various scales in the subsystem of turbulent chaos) when heat is removed from the system, i.e., when passing to lower temperatures, is a universal property of matter.

Formula (9.106) includes the fluxes and thermodynamic forces of the zeroth-, first-, and second-tensor dimensions. If the gradient of the averaged velocity $\partial\langle \boldsymbol{u}\rangle/\partial \boldsymbol{r}$ (second-rank tensor) in (9.106) is broken down into symmetric and anti-symmetric components,

$$\frac{\partial\langle \boldsymbol{u}\rangle}{\partial \boldsymbol{r}} = \left(\frac{\partial\langle \boldsymbol{u}\rangle}{\partial \boldsymbol{r}}\right)^{s} + \left(\frac{\partial\langle \boldsymbol{u}\rangle}{\partial \boldsymbol{r}}\right)^{a} = \left(\frac{\partial\overset{\circ}{\langle \boldsymbol{u}\rangle}}{\partial \boldsymbol{r}}\right)^{s} + \frac{1}{3}\boldsymbol{U}\operatorname{div}\langle \boldsymbol{u}\rangle + \frac{1}{2}\operatorname{rot}\langle \boldsymbol{u}\rangle, \qquad (9.107)$$

where

$$D_{jk} \equiv (\partial\langle \boldsymbol{u}\rangle/\partial \boldsymbol{r})^{s}_{jk} = \frac{1}{2}\left(\partial\langle u_k\rangle/\partial x_j + \partial\langle u_j\rangle/\partial x_k\right),$$

$(\partial\langle \boldsymbol{u}\rangle/\partial \boldsymbol{r})^a_{jk} = \frac{1}{2}\left(\partial\langle u_k\rangle/\partial x_j - \partial\langle u_j\rangle/\partial x_k\right)$ are the symmetric (strain-rate tensor) and anti-symmetric (skew-symmetric gradient tensor of velocity) parts of the shear-rate tensor for the averaged motion $\overset{\circ}{\boldsymbol{D}} \equiv (\partial\langle \overset{\circ}{\boldsymbol{u}}\rangle/\partial \boldsymbol{r})^s = \boldsymbol{D} - \frac{1}{3}\delta_{ik}\mathrm{div}\langle \boldsymbol{u}\rangle$, respectively, then the scalar product of the total turbulent stress tensor of an electrically conducting medium $\boldsymbol{R}^K \equiv \boldsymbol{R} + \boldsymbol{T}^{\mathrm{turb}}$ and the gradient of the averaged velocity can be written as

$$\left(\left[\boldsymbol{R}^K - \frac{1}{3}(\boldsymbol{R}^K : \boldsymbol{U})\boldsymbol{U}\right] : \frac{\partial\langle \boldsymbol{u}\rangle}{\partial \boldsymbol{r}}\right) = \left(\overset{\circ}{\boldsymbol{R}}{}^K : \overset{\circ}{\boldsymbol{D}}\right). \tag{9.108}$$

Here, $\overset{\circ}{\boldsymbol{R}}{}^K \equiv \overset{\circ}{\boldsymbol{R}} + \overset{\circ}{\boldsymbol{T}}{}^M_{\mathrm{turb}}$; the symmetric tensors of the turbulent Reynolds stresses $\boldsymbol{R}$ and the magnetic tensions $\boldsymbol{T}^M_{\mathrm{turb}}$ for the pulsational magnetic-field component are represented as

$$\overset{\circ}{\boldsymbol{R}} \equiv \boldsymbol{R} - \frac{1}{3}(\boldsymbol{R} : \boldsymbol{U})\boldsymbol{U} = \boldsymbol{R} + \frac{2}{3}\bar{\rho}\langle b\rangle \boldsymbol{U}, \tag{9.109}$$

$$\overset{\circ}{\boldsymbol{T}}{}^M_{\mathrm{turb}} \equiv \boldsymbol{T}^M_{\mathrm{turb}} - \frac{1}{3}(\boldsymbol{T}^M_{\mathrm{turb}} : \boldsymbol{U})\boldsymbol{U} = \boldsymbol{T}^M_{\mathrm{turb}} - \frac{2}{3}\bar{\rho}\langle b_M\rangle \boldsymbol{U}. \tag{9.110}$$

When writing (9.108) we took advantage of the fact that the scalar product of symmetric and anti-symmetric tensors is always equal to zero. Similarly, if (9.41*) is used, the double product $\left(\boldsymbol{R}^M : \partial\bar{\boldsymbol{B}}/\partial \boldsymbol{r}\right)$ can be represented as

$$\frac{1}{4\pi\mu}\left(\boldsymbol{R}^M : \frac{\partial\bar{\boldsymbol{B}}}{\partial \boldsymbol{r}}\right) = \frac{1}{4\pi\mu}\left(\boldsymbol{R}^M : \left\{\left(\frac{\partial\bar{\boldsymbol{B}}}{\partial \boldsymbol{r}}\right)^s + \frac{1}{2}\mathrm{rot}\,\boldsymbol{B}\right\}\right) = -\mathsf{G}\cdot\bar{\boldsymbol{j}}. \tag{9.111}$$

Therefore, in the linear approximation when using the Curie-Prigogine principle (according to which no relation is possible between tensors of different ranks in an isotropic medium (de Groot and Mazur 1962)), the following generalized rheological equations can be written for the turbulent fluxes and conjugate thermodynamic forces for a conducting medium:

$$\begin{cases} \tilde{\boldsymbol{q}}^{\mathrm{turb}} \equiv (\boldsymbol{q}^{\mathrm{turb}} - \overline{p'\boldsymbol{u}''}) = -\lambda^{\mathrm{turb}}\partial\langle T\rangle/\partial \boldsymbol{r}, \\ \boldsymbol{R}^K = -\frac{2}{3}\bar{\rho}(\langle b\rangle - \langle b_M\rangle)\boldsymbol{U} + 2\bar{\rho}\nu_K^{\mathrm{turb}}\left(\partial\langle \overset{\circ}{\boldsymbol{u}}\rangle/\partial \boldsymbol{r}\right)^s - 2\nu_{K,M}^{\mathrm{turb}}(\partial\bar{\boldsymbol{B}}/\partial \boldsymbol{r})^s, \\ \boldsymbol{R}^M = -2\nu_{M,K}^{\mathrm{turb}}(\partial\langle \boldsymbol{u}\rangle/\partial \boldsymbol{r})^a + 2\nu_M^{\mathrm{turb}}(\partial\bar{\boldsymbol{B}}/\partial \boldsymbol{r})^a, \end{cases} \tag{9.112}$$

or, in the coordinate form,

$$\begin{cases} J_{\langle E\rangle k} = -\lambda^{\mathrm{turb}}\partial\langle T\rangle/\partial x_k, \\ R^K_{jk} = -\frac{2}{3}\bar{\rho}\Gamma\delta_{jk} + \bar{\rho}\nu_K^{\mathrm{turb}}\left\{\left(\frac{\partial\langle u_j\rangle}{\partial x_k} + \frac{\partial\langle u_k\rangle}{\partial x_j}\right) - \frac{2}{3}\delta_{jk}\mathrm{div}\langle \boldsymbol{u}\rangle\right\} - \nu_{K,M}^{\mathrm{turb}}\left(\frac{\partial\bar{B}_j}{\partial x_k} + \frac{\partial\bar{B}_k}{\partial x_j}\right), \\ R^M_{jk} = -\nu_{M,K}^{\mathrm{turb}}\left(\frac{\partial\langle u_j\rangle}{\partial x_k} - \frac{\partial\langle u_k\rangle}{\partial x_j}\right) + \nu_M^{\mathrm{turb}}\left(\frac{\partial\bar{B}_j}{\partial x_k} + \frac{\partial\bar{B}_k}{\partial x_j}\right), \end{cases} \tag{9.113}$$

where $\Gamma \equiv \langle b \rangle - \langle b_M \rangle$; and λ^{turb}, ν_K^{turb}, and ν_M^{turb} are the coefficients of turbulent thermal conductivity, turbulent kinematic viscosity, and turbulent magnetic-field diffusion, respectively. Note that, in the general case of inhomogeneous turbulence, these coefficients must be anisotropic (to be second- or fourth-rank tensors (Moffat 1978; Yoshizawa 1990; Biskamp 2003)); indeed, a strong magnetic field makes the turbulence anisotropic, since the motion along the magnetic field is preferred (see, e.g., Branover and Tsinover 1970). In the case of developed turbulence, these empirical turbulent transport coefficients depend on the following parameters of an electrically conducting medium: $\bar{\rho}, \partial\langle \boldsymbol{u} \rangle / \partial \boldsymbol{r}, \quad \boldsymbol{B}/4\pi\mu, L$. Note that the parameter L is not the actual turbulence in a flow with a magnetic field, but just some geometric characteristic of the position of a point equal to the ordinary "mixing length" (see Ievlev 1975). It is important to keep in mind that defining relations (9.112)–(9.114) correspond to the regime of a stationary non-equilibrium state of the turbulent field. At the same time, in the general case of strongly developed turbulence, the first relation in (9.112), which describes the turbulent heat exchange processes, takes a slightly different form:

$$\boldsymbol{q}^{\text{turb}} = \overline{p'\boldsymbol{u}''} - \lambda^{\text{turb}} \left(\frac{\partial \langle T \rangle}{\partial \boldsymbol{r}} - \frac{1}{\bar{\rho}\langle c_p \rangle} \frac{\partial \bar{p}}{\partial \boldsymbol{r}} \right),$$

which can be derived for this case from bilinear function (3.143)—see derivation of the formula (3.148). In what follows, we need the components of kinetic Reynolds tensor (9.105) written in cylindrical coordinates:

$$R_{rr}^K \equiv -\bar{\rho}\langle |u_r''|^2 \rangle + \frac{\overline{|B_r'|^2}}{4\pi\mu} = -\frac{2}{3}\bar{\rho}\Gamma + 2\bar{\rho}\nu_K^{\text{turb}} \left(\frac{\partial \langle u_r \rangle}{\partial r} - \frac{1}{3}\text{div}\langle \boldsymbol{u} \rangle \right),$$

$$R_{\varphi\varphi}^K \equiv -\bar{\rho}\left\langle \left|u_\varphi''\right|^2 \right\rangle + \frac{\overline{\left|B_\varphi'\right|^2}}{4\pi\mu} = -\frac{2}{3}\bar{\rho}\Gamma + 2\bar{\rho}\nu_K^{\text{turb}} \left(\frac{1}{r}\frac{\partial \langle u_\varphi \rangle}{\partial \varphi} + \frac{\langle u_r \rangle}{r} - \frac{1}{3}\text{div}\langle \boldsymbol{u} \rangle \right),$$

$$R_{zz}^K \equiv -\bar{\rho}\langle |u_z''|^2 \rangle + \frac{\overline{|B_z'|^2}}{4\pi\mu} = -\frac{2}{3}\bar{\rho}\Gamma + 2\bar{\rho}\nu_K^{\text{turb}} \left(\frac{\partial \langle u_z \rangle}{\partial z} - \frac{1}{3}\text{div}\langle \boldsymbol{u} \rangle \right),$$

$$R_{r\varphi}^K \equiv -\bar{\rho}\langle u_r'' u_\varphi'' \rangle + \frac{\overline{B_r' B_\varphi'}}{4\pi\mu} = R_{\varphi r}^K = \bar{\rho}\nu_K^{\text{turb}} \left(\frac{1}{r}\frac{\partial \langle u_r \rangle}{\partial \varphi} + r\frac{\partial}{\partial r}\left(\frac{\langle u_\varphi \rangle}{r} \right) \right),$$

$$R_{\varphi z}^K \equiv -\bar{\rho}\langle u_\varphi'' u_z'' \rangle + \frac{\overline{B_\varphi' B_z'}}{4\pi\mu} = R_{z\varphi}^K = \bar{\rho}\nu_K^{\text{turb}} \left(\frac{\partial \langle u_\varphi \rangle}{\partial z} + \frac{1}{r}\frac{\partial \langle u_z \rangle}{\partial \varphi} \right),$$

$$R_{zr}^K \equiv -\bar{\rho}\langle u_z'' u_r'' \rangle + \frac{\overline{B_z' B_r'}}{4\pi\mu} = R_{rz}^K = \bar{\rho}\nu_K^{\text{turb}} \left(\frac{\partial \langle u_z \rangle}{\partial r} + \frac{\partial \langle u_r \rangle}{\partial z} \right). \qquad (9.114)$$

The following expression for the turbulent electromotive force for an isotropic medium follows from (9.93) and (9.112):

$$cG = -\nu_M^{\text{turb}} \operatorname{rot} \bar{\boldsymbol{B}} + \nu_{M,K}^{\text{turb}} \operatorname{rot}\langle \boldsymbol{u} \rangle. \tag{9.115}$$

Here, the coefficients ν_M^{turb} and $\nu_{M,K}^{\text{turb}}$ determined by the velocity field $\boldsymbol{u}''$ are assumed to be constant. Since $\bar{\boldsymbol{B}}$ is a pseudovector (i.e., it changes sign when the spatial coordinates are inverted) and G is a true (polar) vector and must be formed from different true vectors, the coefficient ν_M^{turb} must be a scalar, while the coefficients $\nu_{M,K}^{\text{turb}}$ and $\nu_{K,M}^{\text{turb}}$ are pseudoscalars. We keep in mind that the terms with the coefficients $\nu_{M,K}^{\text{turb}}$ and $\nu_{K,M}^{\text{turb}} \cong \frac{7}{5}\nu_{M,K}^{\text{turb}}$ (Yoshizawa 1990), which describe the cross effects, can often be omitted.

Equation (9.115) takes the form $cG = -\nu_M^{\text{turb}} \operatorname{rot} \bar{\boldsymbol{B}} = -\nu_M^{\text{turb}}(4\pi\mu/c)\bar{\boldsymbol{j}}$, where $\bar{\boldsymbol{j}}$ is the mean current. Substitution of this expression into (9.45*) yields the following induction equation for the mean fields:

$$\bar{\rho}\frac{D}{Dt}\left(\frac{\bar{\boldsymbol{B}}}{\bar{\rho}}\right) = \left(\bar{\boldsymbol{B}} \cdot \frac{\partial}{\partial \boldsymbol{r}}\right)\langle \boldsymbol{u} \rangle + (\nu_M + \nu_M^{\text{turb}})\nabla^2 \bar{\boldsymbol{B}}.$$

Obviously, it thus follows that the combined effect of the electromotive force G in (9.45*) is just reduced to a change in the effective magnetic-field diffusion coefficient, i.e., ν_M transforms into $(\nu_M + \nu_M^{\text{turb}})$. In the case where the coefficient ν_M^{turb} is positive, it might be expected that random mixing (produced by the $\boldsymbol{u}''$ field) should strengthen rather than weaken the molecular diffusion process.

Let us now show that within the framework of the approach developed here, the turbulent diffusion coefficient ν_M^{turb} must always be positive. We should recall, however, that this assertion is in conflict with the well-known viewpoint of Moffatt who surmised that the coefficient ν_M^{turb} under certain extreme conditions could be negative (Moffat 1978). To clarify the point let us substitute (9.113) and (9.114) into (9.106). This gives rise to an expression for the entropy production in the form

$$0 \le \langle T \rangle \sigma'_{\langle S \rangle} \equiv \left(\overset{\circ}{\boldsymbol{R}}{}^K : \overset{\circ}{\boldsymbol{D}}\right) - G \cdot \bar{\boldsymbol{j}} = 2\bar{\rho}\nu_K^{\text{turb}} \overset{\circ}{\boldsymbol{D}} : \overset{\circ}{\boldsymbol{D}} + \nu_M^{\text{turb}}(4\pi\mu/c^2)|\bar{\boldsymbol{j}}|^2. \tag{9.106*}$$

Here, the first term represents the increase in entropy due to the kinematic turbulent viscosity and the second term represents the increase in entropy $\langle S \rangle$ due to the magnetic viscosity. The second law of thermodynamics requires that entire sum (9.106*) be positive. On the other hand, the second term of this sum can be nonzero even when the first term is equal to zero (the symmetric tensor $\overset{\circ}{\boldsymbol{D}}$ which serves as a measure of the relative motion between various fluid elements vanishes, for example, when the matter is locally at rest ($\langle \boldsymbol{u} \rangle = 0$) or when it rotates as a rigid body at a constant angular velocity Ω ($\langle \boldsymbol{u} \rangle = \Omega \times \boldsymbol{r}$)). Therefore, each of these terms must always be positive. In particular, it thus follows that the magnetic coefficient of turbulent viscosity (diffusion) ν_M^{turb} must inevitably and under all conditions be positive, implying that random mixing (produced by the $\boldsymbol{u}''$-field) always strengthens the magnetic molecular-diffusion process. The above reasoning is the thermodynamic substantiation of the impossibility of the appearance of a negative magnetic viscosity in isotropic, mirror-symmetric turbulent flows of an electrically conducting fluid.

We derived (9.115) for isotropic (in the hydrodynamic sense) turbulence where the field of pulsating velocities $\boldsymbol{u}''$ possesses a mirror symmetry, under which any mean quantities obtained from it are invariant when the field is reflected in an arbitrary plane. However, the left-handed rotations in turbulent motion may happen to be more probable than the right-handed ones, or vice versa. In this case, there is no mirror symmetry of the $\boldsymbol{u}''$ field and the turbulence is helical (see Chap. 8). Equation (9.115) generalized to this case takes the form (see Steenbeck et al. 1966)

$$c\mathrm{G} = \alpha\bar{\boldsymbol{B}} - \nu_M^{\text{turb}}\,\text{rot}\,\bar{\boldsymbol{B}} + \nu_{M,K}^{\text{turb}}\,\text{rot}\langle\boldsymbol{u}\rangle, \tag{9.116}$$

where the dimensionless coefficient α is a pseudoscalar. Let us show that the coefficient α must be zero for an isotropic and mirror-symmetric velocity field $\boldsymbol{u}''$. Indeed, for an isotropic medium, both some given realization of the ensemble of this field and the realization obtained from it by a mirror reflection are equiprobable. Then, on the one hand, the quantity α should not be changed if this reflection is performed, since the ensemble did not change, but, on the other hand, the coefficient α should change its sign, since it is a pseudoscalar; therefore, $\alpha = 0$. Thus, in contrast to gyrotropic (reflection non-symmetric) turbulence, reflection-symmetric isotropic turbulence can only produce turbulent magnetic field diffusion.

Returning now to the Maxwell equations for the mean fields (9.40) and Ohm law (9.41) and substituting (9.51) into them, we obtain the final expression for the averaged current

$$\bar{\boldsymbol{j}} = \sigma_e^{\text{turb}}\left(\boldsymbol{E}^* + \alpha c^{-1}\,\bar{\boldsymbol{B}}\right). \tag{9.117}$$

Here, the turbulent conductivity σ_e^{turb} is defined by the formula

$$\sigma_e^{\text{turb}} = \frac{\sigma_e}{1 + 4\pi\mu\nu_M^{\text{turb}}\sigma_e/c^2} = \frac{\sigma_e\nu_M}{\nu_M + \nu_M^{\text{turb}}}, \tag{9.118}$$

from which follows that the turbulent conductivity σ_e^{turb} is lower than the molecular conductivity σ_e.

To conclude this section let us list the basic differential equations and finite relations which characterize a relatively simple model of the turbulized multicomponent continuum. These are, first of all, the hydrodynamic equations of mean motion (3.21), (3.23), (3.28), (3.58), the averaged equation of state for pressure in the form (3.71), the relation (3.139) defining turbulent flux of specific volume of mixture $\boldsymbol{J}_{(1/\rho)}^{\text{turb}}$, as well as the rheological relations (3.127), (3.128), and (3.177) for the turbulent fluxes of diffusion $\boldsymbol{J}_\alpha^\Sigma$, heat $\boldsymbol{q}^\Sigma$, and Reynolds stress tensor $\boldsymbol{R}$.

9.3.5 Derivation of the Correction Function to the Turbulent Viscosity Coefficient for a Conducting Medium with a Variable Density

Let us now consider a semiempirical method for determining the turbulent transport coefficients in an electrically conducting medium in the presence of magnetic field. In the case of isotropic turbulence, both coefficients ν_K^{turb} and ν_M^{turb} are known to be close to the product $w_{\text{turb}} l_{\text{cor}}$ of the velocity of turbulent eddies $w_{\text{turb}} \cong \sqrt{|\boldsymbol{u}''|^2}$ and their correlation length l_{cor}, while the coefficient α is by an order of magnitude $\alpha \cong -1/3\overline{\boldsymbol{u}'' \cdot \text{rot}\, \boldsymbol{u}''}\tau_{\text{cor}} = -1/3h\tau_{\text{cor}}$, where h is the hydrodynamic helicity (pseudoscalar) and τ_{cor} is a scale that characterizes the variations in turbulent velocity $\boldsymbol{u}''$ field with time (Krause and Radler 1980). In particular, if we assume, according to the standard hypothesis by Shakura and Sunyaev (1973), that l_{cor} is the accretion-disk half thickness and w_{turb} can be expressed in terms of the speed of sound, then turbulent diffusion leads to a characteristic decay time of the magnetic field (or, rather, its components that change noticeably on the scale of the disk thickness) on the order of the Keplerian rotation period. In this case, the "effective" magnetic Reynolds number is $\boldsymbol{Re}_M \propto 1$ and the turbulent transport is important.

Generalizing the well-known Kolmogorov formula for a non-conducting fluid to the case of MHD turbulence, we can assume, with an accuracy sufficient for practical applications, that the kinetic coefficient of turbulent viscosity ν_K^{turb} can be calculated using the formula $\nu_K^{\text{turb}} = L\sqrt{b_\Sigma}$, where L is the Prandtl mixing length (here, the numerical factor is included in L). However, this formula does not explicitly include the magnetic field effect on the turbulent flow of a conducting fluid via the mixing length, which is unacceptable for many practically important cases of the developed MHD turbulence (e.g., for large-scale perturbations). Therefore, a correction that takes into account both the magnetic-field effect on the mixing length and the reverse effects of the heat transfer on the development of turbulence should generally be applied to the above relation for ν_K^{turb}.

Below, we assume that the hydrodynamic plasma velocity shear is the main source of turbulence. Assuming the stationary non-equilibrium regime of turbulence in an electrically conducting medium, we use balance equation (9.96) for the turbulization entropy S_{turb} to determine the above correction factor to the turbulent viscosity coefficient. In the developed regime of turbulence, we assume (based on considerations of the similarity theory at the Reynolds number $\boldsymbol{Re} \to \infty$) that the total dissipation of the turbulent kinetic and turbulent magnetic energies into heat ε_Σ is proportional to the cube of the pulsational velocity w_{turb} and is inversely proportional to the scale length L—the mixing length scale (Landau and Lifshitz 1982). In the stationary-nonequilibrium regime of the subsystem of turbulent chaos, (9.96) takes the form

$$\bar{\rho}\frac{DS_{\text{turb}}}{Dt} + \operatorname{div}\boldsymbol{J}_{S_{\text{turb}}} = \frac{1}{T_{\text{turb}}}\left\{\left(\overset{\circ}{\boldsymbol{R}}^{K} : \overset{\circ}{\boldsymbol{D}}\right) - \mathsf{G}\cdot\bar{\boldsymbol{j}} - \left(\boldsymbol{J}^{\text{turb}}_{(1/\rho)}\cdot\frac{\partial\bar{p}}{\partial\boldsymbol{r}}\right) - \bar{\rho}\varepsilon_{\Sigma}\right\} \cong 0. \tag{9.119}$$

In writing this equation we, as it is often done, neglected the term $\overline{p'\operatorname{div}\boldsymbol{u}'}$, which describes the effect of the generation of pulsation entropy S_{turb} under pressure forces (otherwise, this term may be assumed to be included in the left-hand side of (9.119)); in this case, we have only to change the definition of the substantial flux $\boldsymbol{J}_{S_{\text{turb}}}$ of the entropy S_{turb}. The right side of this equation can be represented as

$$2\bar{\rho}\nu_K^{\text{turb}}\left(\overset{\circ}{\boldsymbol{D}} : \overset{\circ}{\boldsymbol{D}}\right) - \frac{\nu_M^{\text{turb}}}{4\pi\mu}\left(\frac{c\alpha}{\nu_M^{\text{turb}}}\bar{\boldsymbol{B}}\cdot\operatorname{rot}\bar{\boldsymbol{B}} - \operatorname{rot}^2\bar{\boldsymbol{B}}\right)$$
$$+\frac{\lambda^{\text{turb}}}{\langle c_p\rangle\langle T\rangle}\boldsymbol{g}\cdot\left(\frac{\partial\langle T\rangle}{\partial\boldsymbol{r}} - \frac{\boldsymbol{g}}{\langle c_p\rangle}\right) - \bar{\rho}\varepsilon_{\Sigma} \cong 0, \tag{9.120}$$

if the corresponding expressions for its individual terms are used:

$$\left(\overset{\circ}{\boldsymbol{R}}^{K} : \overset{\circ}{\boldsymbol{D}}\right) = 2\bar{\rho}\nu_K^{\text{turb}}\left(\overset{\circ}{\boldsymbol{D}} : \overset{\circ}{\boldsymbol{D}}\right), \tag{9.121}$$

$$-\,\mathbf{G}\cdot\bar{\boldsymbol{j}} = \frac{\nu_M^{\text{turb}}}{4\pi\mu}\left(l_{\text{cor}}^{-1}\bar{\boldsymbol{B}}\cdot\operatorname{rot}\bar{\boldsymbol{B}} + \operatorname{rot}^2\bar{\boldsymbol{B}}\right), \tag{9.122}$$

$$\left(\boldsymbol{J}^{\text{turb}}_{(1/\rho)}\cdot\frac{\partial\bar{p}}{\partial\boldsymbol{r}}\right) \cong -\frac{\lambda^{\text{turb}}}{\langle c_p\rangle\langle T\rangle}\boldsymbol{g}\cdot\left(\frac{\partial\langle T\rangle}{\partial\boldsymbol{r}} - \frac{\boldsymbol{g}}{\langle c_p\rangle}\right), \tag{9.123}$$

where $l_{\text{cor}} \equiv -(\nu_M^{\text{turb}}/\alpha)$ is the correlation scale that characterizes the variation of the turbulent velocity $\boldsymbol{u}''$ field in space. Here, we used (9.113) and (9.116) for the fluxes $\boldsymbol{R}^K$ and G, as well as by the expression

$$\boldsymbol{J}^{\text{turb}}_{(1/\rho)} = -\overline{p'\boldsymbol{u}''}/\bar{p} + \tilde{\boldsymbol{q}}^{\text{turb}}/\bar{\rho}\langle c_p\rangle\langle T\rangle \cong \tilde{\boldsymbol{q}}^{\text{turb}}/\bar{\rho}\langle c_p\rangle\langle T\rangle \tag{9.124}$$

for the turbulent specific-volume flux $\boldsymbol{J}^{\text{turb}}_{(1/\rho)}$ (see (3.139) and (3.83)) and the approximate relation $\partial\bar{p}/\partial\boldsymbol{r} \cong \bar{\rho}\boldsymbol{g} - \partial\,(p^M_{av} + p^M_{\text{turb}})/\partial\boldsymbol{r} \cong \bar{\rho}\boldsymbol{g}$ for the pressure gradient of the matter (the mechanical quasi-equilibrium condition for plasma).

Using the designation w_{turb} for the characteristic pulsation velocity of a conducting medium and L for the Prandtl mixing length, in the absence of a field, we may write

$$\nu_K^{\text{turb}} = Lw_{\text{turb}}, \quad \nu_M^{\text{turb}} = \frac{Lw_{\text{turb}}}{\boldsymbol{Pr}_M^{\text{turb}}}, \quad \frac{\lambda^{\text{turb}}}{\bar{\rho}\langle c_p\rangle} = \frac{Lw_{\text{turb}}}{\boldsymbol{Pr}_K^{\text{turb}}}, \quad \varepsilon_{\Sigma} = \frac{1}{\alpha_{ss}^2}\frac{w_{\text{turb}}^3}{L}. \tag{9.125}$$

To a first approximation, we assume that the empirical constant α_{ss} and the turbulent Prandtl–Schmidt numbers (kinetic $\boldsymbol{Pr}_K^{\text{turb}} = \bar{\rho}\langle c_p\rangle \nu_K^{\text{turb}}/\lambda^{\text{turb}}$ and magnetic $\boldsymbol{Pr}_M^{\text{turb}} = \nu_K^{\text{turb}}/\nu_M^{\text{turb}}$) are constant and that $\nu_M^{\text{turb}}>0$ and $\boldsymbol{Pr}_M^{\text{turb}} \cong \boldsymbol{Pr}_K^{\text{turb}}$. Substituting these expressions into (9.120) we obtain for the stationary regime:

$$w_{\text{turb}}\left\{2L\left(\overset{\circ}{\boldsymbol{D}}:\overset{\circ}{\boldsymbol{D}}\right)+\frac{L}{\boldsymbol{Pr}_M^{\text{turb}}}\frac{\left(l_{\text{cor}}^{-1}\bar{\boldsymbol{B}}\cdot\operatorname{rot}\bar{\boldsymbol{B}}+\operatorname{rot}^2\bar{\boldsymbol{B}}\right)}{4\pi\mu\bar{\rho}}+\frac{L}{\boldsymbol{Pr}_K^{\text{turb}}}\frac{\boldsymbol{g}}{\langle T\rangle}\cdot\left(\frac{\partial\langle T\rangle}{\partial\boldsymbol{r}}-\frac{\boldsymbol{g}}{\langle c_p\rangle}\right)\right.$$
$$\left.-\frac{1}{\alpha_{ss}^2}\frac{w_{\text{turb}}^2}{L}\right\}\cong 0. \tag{9.126}$$

Equation (9.126) breaks up into two equations: the equation $w_{\text{turb}}=0$ corresponding to the laminar flow regime and the equation

$$w_{\text{turb}}^2=\alpha_{ss}^2L^2\left\{2\left(\overset{\circ}{\boldsymbol{D}}:\overset{\circ}{\boldsymbol{D}}\right)+\frac{\left(l_{\text{cor}}^{-1}\bar{\boldsymbol{B}}\cdot\operatorname{rot}\bar{\boldsymbol{B}}+\operatorname{rot}^2\bar{\boldsymbol{B}}\right)}{4\pi\mu\bar{\rho}\,\boldsymbol{Pr}_M^{\text{turb}}}+\frac{1}{\boldsymbol{Pr}_K^{\text{turb}}}\frac{\boldsymbol{g}}{\langle T\rangle}\cdot\left(\frac{\partial\langle T\rangle}{\partial\boldsymbol{r}}-\frac{\boldsymbol{g}}{\langle c_p\rangle}\right)\right\}, \tag{9.127}$$

describing the steady turbulent regime. Equation (9.127) has a real solution when

$$2\left(\overset{\circ}{\boldsymbol{D}}:\overset{\circ}{\boldsymbol{D}}\right)+\frac{1}{\boldsymbol{Pr}_K^{\text{turb}}}\left\{\frac{\left(l_{\text{cor}}^{-1}\bar{\boldsymbol{B}}\cdot\operatorname{rot}\bar{\boldsymbol{B}}+\operatorname{rot}^2\bar{\boldsymbol{B}}\right)}{4\pi\mu\bar{\rho}}+\frac{\boldsymbol{g}}{\langle T\rangle}\cdot\left(\frac{\partial\langle T\rangle}{\partial\boldsymbol{r}}-\frac{\boldsymbol{g}}{\langle c_p\rangle}\right)\right\}\geq 0;$$

hence, we obtain

$$\boldsymbol{Ri}_\Sigma\equiv\boldsymbol{Ri}_K-\boldsymbol{Ri}_M\leq\left(\boldsymbol{Ri}_\Sigma\right)_{cr}=\boldsymbol{Pr}_K^{\text{turb}}, \tag{9.128}$$

where the following notations are used:

$$\boldsymbol{Ri}_K\equiv\frac{\frac{1}{\langle T\rangle}\left(\tilde{\boldsymbol{q}}^{\text{turb}}\cdot\frac{\partial\bar{p}}{\partial\boldsymbol{r}}\right)}{\lambda^{\text{turb}}2\left(\overset{\circ}{\boldsymbol{D}}:\overset{\circ}{\boldsymbol{D}}\right)}=-\frac{\frac{\boldsymbol{g}}{\langle T\rangle}\cdot\left(\frac{\partial\langle T\rangle}{\partial\boldsymbol{r}}-\frac{\boldsymbol{g}}{\langle c_p\rangle}\right)}{2\left(\overset{\circ}{\boldsymbol{D}}:\overset{\circ}{\boldsymbol{D}}\right)}, \tag{9.129}$$

$$\boldsymbol{Ri}_M\equiv-\frac{1}{\bar{\rho}\nu_M^{\text{turb}}}\frac{\mathsf{G}\cdot\bar{\boldsymbol{j}}}{2(\overset{\circ}{\boldsymbol{D}}:\overset{\circ}{\boldsymbol{D}})}=\frac{1}{4\pi\mu\bar{\rho}}\frac{\operatorname{rot}^2\bar{\boldsymbol{B}}+l_{\text{cor}}^{-1}\bar{\boldsymbol{B}}\cdot\operatorname{rot}\bar{\boldsymbol{B}}}{2(\overset{\circ}{\boldsymbol{D}}:\overset{\circ}{\boldsymbol{D}})}. \tag{9.130}$$

Here, $\boldsymbol{Ri}_K$ and $\boldsymbol{Ri}_M$ are, respectively, the gradient MHD Richardson number (the dimensionless quantity that defines the relative contribution from the thermal convection of the matter to the turbulent energy generation in a conducting medium compared to the energy transfer from the averaged motion) and the magnetohydrodynamic Richardson number (which takes into account the effect of the magnetic field and, in particular, the turbulent electromotive force on the generation of turbulence in the flow). It is easy to see that $\boldsymbol{Ri}_M$ is proportional to the Alfven number $\boldsymbol{Al} \equiv \boldsymbol{N}/\boldsymbol{Re}_M \equiv B_0^2/4\pi\mu\rho V_0^2$ and characterizes the magnetic-to-kinetic energy ratio for the plasma.

If $\boldsymbol{Ri}_\Sigma = \boldsymbol{Pr}_K^{\text{turb}}$, then there exists the unique real solution $w_{\text{turb}} = 0$ corresponding to the laminar regime. Let the turbulent regime take place and, hence, $\boldsymbol{Ri}_\Sigma < (\boldsymbol{Ri}_\Sigma)_{cr}$; for the turbulent viscosity coefficient of an electrically conducting fluid, we then obtain

$$\nu_K^{\text{turb}} = \alpha_{ss} L^2 \sqrt{2\left(\overset{\circ}{\boldsymbol{D}} : \overset{\circ}{\boldsymbol{D}}\right)} \sqrt{1 - \boldsymbol{Ri}_\Sigma/\boldsymbol{Pr}_K^{\text{turb}}} = \alpha_{ss} L^{*2} \sqrt{2\left(\overset{\circ}{\boldsymbol{D}} : \overset{\circ}{\boldsymbol{D}}\right)}. \tag{9.131}$$

Thus, having considered the balance equation for the chaos entropy, we established the form of the dimensionless function

$$\varphi\left(\frac{\boldsymbol{Ri}_\Sigma}{(\boldsymbol{Ri}_\Sigma)_{cr}}\right) = \sqrt{1 - \frac{\boldsymbol{Ri}_K - \boldsymbol{Ri}_M}{\boldsymbol{Pr}_K^{\text{turb}}}}, \quad L^* \equiv L\left(1 - \frac{\boldsymbol{Ri}_\Sigma}{\boldsymbol{Pr}_K^{\text{turb}}}\right)^{1/4}, \tag{9.132}$$

which includes the magnetic-field effect and the reverse effect of the heat transfer on the development of turbulence via the mixing length. Concurrently, we obtained an approximate estimate for the critical Richardson number $(\boldsymbol{Ri}_\Sigma)_{cr} = \boldsymbol{Pr}_K^{\text{turb}} = \bar{\rho}\langle c_p\rangle \nu_K^{\text{turb}}/\lambda^{\text{turb}}$. Here, L^* is the mixing length for the flow in a magnetic field.

To calculate the mixing length L in the absence of a magnetic field, we can use the Prandtl–Nikuradze formula with the correction suggested by Van Driest (1956) that takes into account the faster decay of turbulence near a rigid wall $L = L_{00}\{1 - \exp(-zV^*/A\nu)\}$; here, A is a numerical constant, V^* is the dynamic velocity on the wall, and L_{00} is the mixing length in the absence of a magnetic field calculated according to Prandtl and Nikuradze (see, e.g., Lapin 1982) which, for example, for a disk yields

$$\frac{L_{00}}{h_{\text{disk}}} = 0.14 - 0.08\left(1 - \frac{z}{h_{\text{disk}}}\right)^2 - 0.06\left(1 - \frac{z}{h_{\text{disk}}}\right)^4. \tag{9.133}$$

Generally, a special differential equation should be invoked to determine L.

In conclusion, note that the magnetic-field effect on the flow of electrically conducting media depends significantly on the relative orientation of the magnetic field and the mean-flow velocity. The magnetic field can be generated not only by currents in the plasma, but also by external (with respect to the moving medium under consideration) sources. Therefore, in many cases, it is convenient to represent the field strength as the sum of two quantities: the strength of the external magnetic field and the strength of the field generated by currents in the plasma.

9.4 Modeling the Turbulent Transport Coefficients in a Thin Accretion Disk

The key problem in astrophysics is related to the mechanism of angular momentum exchange between accretion disc and protostar. The developed thermodynamic model of MHD-turbulence in magnetic field is applicable to numerical modeling of the structure and evolution of accretion discs, in particular those around T-Tauri stars giving birth to planets or binary stars with mass transfer (of dwarfs novae systems). Such discs formed by a matter rotating around compact stellar objects and are retained on nearly circular orbit by combined action of gravity and centrifugal forces. Independent of what caused protoplanetary disc to set up, at the early stage of evolution its matter possesses angular momentum sufficient to reside at a circular orbit. At the same time, every disc mass element spirals slowly (accretes) towards mother star, which should result in the angular momentum transfer from the inner regions outward.

Indeed, the existing mass and angular momentum distribution in the Solar system, as well as in other numerous systems of young stars with discs, argue for the necessity to invoke an efficient mechanism of mass and angular momentum redistribution in due course of the planetary system formation. Classical ideas are based on friction mechanism of the angular momentum transfer in the accretion disc from protostar outward and its subsequent distribution in space or, in a case of the binary stars, on mechanism of tidal interaction responsible for angular momentum return back to its original source—satellite star. Note that friction mechanism can be operational only if the disc kinematic viscosity exceeds by many orders of magnitude molecular viscosity to provide the necessary mass and accompanying angular momentum transfer for T-Tauri stars.

Such a great viscosity can be accomplished in a turbulent protoplanetary disc, the turbulent state of disc matter caused by its large-scale shift of velocity in the differential rotation relative gravity center being the key concept in astrophysics (see, e.g., Zel'dovich 1981; Fridman 1989; Dubrulle 1993; Gor'kaviy and Fridman 1994; Balbus and Hawley 1991; Richard and Zahn 1999; Bisnovaty-Kogan and Lovelace 2001). Besides, MHD turbulence can contribute significantly to the angular momentum redistribution (see Eardley and Lightman 1975; Galeev et al. 1979; Coroniti 1981; Tout and Pringle 1992; Brandenburg et al. 1996; Lesch 1996), its efficiency as dissipation mechanism depending on magnetic lines reconnection (Kadomtsev 1987).

It is reasonable to assume that chaotic magnetic fields caused by the turbulent dynamo mechanism or brought by accreting interstellar plasma, occurred in the disc at early stage of its evolution. These fields having energy compared to that of hydrodynamic turbulence, mixed owing to disc matter differential rotation and experienced reconnections at the disc boundary, could contribute substantially to the turbulent viscosity not only in the inner disc regions but also in the outer parts of its corona where the matter is ionized. Let us recall that such a fundamental physical process in space plasma as magnetic lines reconnection responsible for exhibiting many features of plasma activity, is only possible when in an electro-conductive matter magnetic lines of opposite direction come close together. This process results in the high electric current density set up. Note also that before reconnection, in plasma having a definite storage of the magnetic energy the so called break (tearing) instability develops which ultimately is responsible for the reconnection and transfer of the magnetic energy into the plasma kinetic or thermal energy.

Alongside with chaotic magnetic fields, large scale well ordered magnetic field in some parts of accretion disc can exist. Due to turbulent transfer such a field extends to at least the disc inner edge. This field penetrates to both sides of the disc vicinity. These outer regions experience magnetic strengths influence for which both small scale disturbances connected with turbulence and large scale shear flows are responsible. This results in not only turbulent viscosity and turbulent diffusion parameters variations but also affects connection with electrodynamics of mean fields. In particular, since in the rotating electro-conductive medium effective magnetic diffusion is accompanied by the turbulent electromotive $\alpha \boldsymbol{B}$ force set up (the so-called α-effect related ultimately to spirality influence on the magnetic field induction generation, see Chap. 8), one may expect significant action of turbulent dynamo on the accretion disc structure and evolution. It is known that small scale girotropic (reflection non-invariant turbulence) in the fast rotating disc creates "loops" (α-effect) when any magnetic field force tube acquires the form of a distorted Ω letter (see Parker 1955). This magnetic loop is accompanied by the current having either parallel or anti-parallel component relative to the applied mean magnetic field for left or right screw accidental spiral motions, respectively. Energy of Joule heat release due to such currents is a powerful source of heating responsible for the disc corona formation that is comparable by its thickness to the disk itself (Galeev et al. 1979). This is just the size of "primary loops" produced by the buoyancy force although one should keep in mind that in reality the corona can be even thicker. The reason is that in the process of small loops reconnection, loops of progressively larger size can be formed. Basically, this is the well known inverse cascade in the three-dimensional MHD motions when turbulence possesses both kinematic and magnetic spirality. At the same time, large scale force lines running through the corona and closed in the disc, maintain magnetic connection between distant disc regions. Such a connection may be also responsible for an additional strengths source and hence the disc heating.

We therefore see that because of viscous stresses due to differential disc rotation and turbulent dynamo in plasma the disk corona is heated similar to what occurs in the solar corona. In turn, hot corona gives rise to outflow (streams) of matter and field. Actually, such a stream represents magnetized rotating wind flowing from the accretion disc. Together with matter and magnetic field the wind transfers to infinity significant part of the disc angular momentum allowing the disk to contract slowly. In other words, rotating wind outflow, jointly with viscous friction mechanism, provides another opportunity to transport an excess angular momentum from the inner part of protostar-disc system outward. Note that an effective focusing of the disc matter flow (jets) can be formed by magnetic stresses in the wind.

In order to reconstruct structure and evolution of the protoplanetary accretion disk that surrounded the proto-Sun the representative model involving numerous processes in a conducting disk and its corona is required (Kolesnichenko and Marov 2006, 2007). As in the case of a non-conducting disk, modeling the turbulent viscosity coefficient ν_K^{turb} in a plasma disk medium under the action of a magnetic field is one of the problems of paramount importance. This section is devoted to studying this problem.

9.4.1 Viscosity Law in Thin Keplerian Disks

Below, we use an inertial cylindrical coordinate system (r, φ, z) with the origin at the center of gravity; the $z = 0$ plane is assumed to coincide with the disk midplane. We analyze the disk system under the following assumptions:

1. A model of slowly evolving protoplanetary gas cloud that rotates around the z axis fixed in space with an angular velocity $\Omega(\boldsymbol{r})$ is under study.
2. Disk configuration is assumed to be in a state of quasi-stationary rotation in an inertial reference frame with the origin at the center of mass of the proto-Sun; hence, it necessarily possesses an axial symmetry $(\partial/\partial\varphi = 0)$: $\bar{s} = \bar{s}(r, z), \bar{\rho} = \bar{\rho}(r, z), \bar{p} = \bar{p}(r, z), \langle T\rangle = \langle T\rangle(r, z), \Omega = \Omega(r, z)$, etc. (see, e.g., Tassoul 1979).
3. Ratio of the disk half thickness $h_{\text{disk}}(r)$ to its radius r is assumed to be much smaller than unity, $h_{\text{disk}}(r)/r << 1$ (the disk thinness condition).
4. Self-gravity of the disk material is neglected compared to the gravitation field of the proto-Sun.
5. Radiation pressure in the disk is assumed to be much smaller than the gas pressure, $p_R << p_g$.
6. Rotation is assumed to be so slow that the meridional circulation of the protoplanetary cloud material may be neglected, i.e., basically, for accretion disks, the averaged motion of the cosmic fluid is realized only in the azimuthal direction

$$\langle u_r\rangle = 0, \quad \langle u_\varphi\rangle = r\Omega(r, z), \quad \langle u_z\rangle = 0, \tag{9.134}$$

while the true velocity of the flow of the gas–dust mixture pulsates randomly about this mean value, varying highly irregularly in the meridional and azimuthal directions.

Under these assumptions, averaged mass conservation law (9.36) always holds, since the motions are axisymmetric and there are no meridional flows, $\text{div}\langle \boldsymbol{u}\rangle = 0$. The components of kinetic Reynolds tensor (9.113) and the strain-rate tensor $\overset{\circ}{\boldsymbol{D}}$ in cylindrical coordinates then take the form

$$\begin{cases} R_{rr}^K = R_{\varphi\varphi}^K = R_{zz}^K = -\frac{2}{3}\bar{\rho}\Gamma, \quad R_{zr}^K = R_{rz}^K = 0, \\ R_{\varphi z}^K = R_{z\varphi}^K = \bar{\rho}\nu_K^{\text{turb}} r \frac{\partial\Omega(r,z)}{\partial z}, \quad R_{r\varphi}^K = R_{\varphi r}^K = \bar{\rho}\nu_K^{\text{turb}} r \frac{\partial\Omega(r,z)}{\partial r}, \end{cases} \tag{9.135}$$

$$\overset{\circ}{\boldsymbol{D}} = \boldsymbol{i}_r\boldsymbol{i}_\varphi \frac{r}{2}\frac{\partial\Omega(r,z)}{\partial r} + \boldsymbol{i}_\varphi\boldsymbol{i}_r \frac{r}{2}\frac{\partial\Omega(r,z)}{\partial r} + \boldsymbol{i}_\varphi\boldsymbol{i}_z \frac{r}{2}\frac{\partial\Omega(r,z)}{\partial z} + \boldsymbol{i}_z\boldsymbol{i}_\varphi \frac{r}{2}\frac{\partial\Omega(r,z)}{\partial z}. \quad (9.136)$$

Hence, for the dissipative function, we have

$$2\overset{\circ}{\boldsymbol{D}} : \overset{\circ}{\boldsymbol{D}} = 4D_{r\varphi}^2 + 4D_{z\varphi}^2 = r^2\left\{\left(\frac{\partial\Omega(r,z)}{\partial r}\right)^2 + \left(\frac{\partial\Omega(r,z)}{\partial z}\right)^2\right\}. \quad (9.137)$$

Let us first consider the case of a Keplerian differentially rotating thin disk lying in the $r-\varphi$ plane (located at $z=0$ in cylindrical coordinates). Each disk layer with radius r moves almost exactly according to Kepler's third law, i.e., it rotates increasingly fast as the central body (with mass $M_\odot$) is approached: the Keplerian orbital velocity is $\langle u_\varphi\rangle = r\Omega_K(r) = r\sqrt{GM_\odot/r^3}$, where G is the gravitational constant, while the angular velocity of orbital rotation $\Omega_K(r)$ increases as $r^{-3/2}$. Such motion is typical for a large-scale shear flow that can be considered the main source of turbulence for the protoplanetary disk system.

For such an astrophysical disk, the turbulent viscosity coefficient ν_K^{turb} was first estimated by Shakura and Sunyaev (1973) who constructed a simple model of a thin disk. This approach based on the Kolmogorov's concept of $\nu_K^{\text{turb}} = u_{\text{turb}} l_{\text{turb}}$ for the viscosity generated by MHD turbulence in the disk where u_{turb} is the root-mean-square velocity of the turbulent pulsations limited by the thermal speed of sound $c_s|_{z=0}$ calculated in the disk midplane, $u_{\text{turb}} \le c_s|_{z=0} \cong \sqrt{\gamma\bar{p}/\bar{\rho}}|_{z=0}$ and the so-called Prandtl mixing length l_{turb} limited, according to the assumption, by the disk half thickness h_{disk}

$$l_{\text{turb}} \le h_{\text{disk}} = \left(\frac{p}{\rho\Omega_K^2}\right)^{1/2} \approx \frac{c_s|_{z=0}}{\Omega_K}, \quad (9.138)$$

which can be estimated using the balance of forces in the z direction: $p/z \cong \rho g z/r \cong \rho\Omega_K^2 z$ (Shapiro and Teukolsky 1983). Using also the mass density $\bar{\rho}$ and the gas pressure $\bar{p}$ in the disk they derived the following relation

$$R_{r\varphi}^K = \bar{\rho}\nu_K^{\text{turb}} r\,\partial_r\left(\frac{\langle u_\varphi\rangle}{r}\right) = -\frac{3}{2}\bar{\rho}u_{\text{turb}}l_{\text{turb}}\Omega_K(r) \le -\frac{3}{2}\bar{\rho}c_s^2|_{z=0} \cong -\bar{p}|_{z=0}, \quad (9.139)$$

between the r,φ component of the Reynolds turbulent stress tensor $R_{r\varphi}$ and the gas pressure $\bar{p}$. Introducing a dimensionless free-viscosity parameter α_{ss}, one can then write

$$R_{r\varphi}^K = -\frac{3}{2}\bar{\rho}\nu_K^{\text{turb}}\Omega_K(r) = -\alpha_{ss}\bar{p}. \quad (9.140)$$

The relation between the Shakura–Sunyaev parameter α_{ss} and the turbulent kinematic viscosity coefficient in (9.140) is

$$\alpha_{ss} = -\frac{R^K_{r\varphi}}{\bar{p}} \cong \frac{-\bar{\rho}\langle u''_r u''_\phi\rangle + \overline{B'_r B'_\varphi}/4\pi\mu}{\bar{\rho}c_s^2} \cong \frac{3\nu_K^{\text{turb}}\Omega_K(r)}{2c_s^2}. \tag{9.141}$$

This parameter, which, in general, defies any accurate calculation, satisfies the constraint

$$\alpha_{ss} \leq 1. \tag{9.142}$$

The astrophysical models constructed using (9.139) pertain to the so-called viscous α disks. The main advantage of such a heuristic description of the disk turbulence consists in its relative simplicity. In particular, in the opinion of many researchers, it suffices to substitute the turbulent viscosity coefficient for the molecular viscosity coefficient in the equations of stellar magnetohydrodynamics to somehow take into account the processes of the turbulization of the conducting medium in astrophysical disks. In models of this kind, α_{ss} generally remains a free parameter in the disk-structure equations. Its determination based on various assumptions about the nature of the physical processes that take place in the disk was the subject of numerous studies (see, e.g., the vast bibliography to the review by Makalkin 2004). In particular, the value of the Shakura–Sunyaev disk parameter α_{ss}, which characterizes the degree of excitation of the turbulent motions, can be calibrated empirically, in particular, using the time-dependent spectra obtained during the observations of outbursts in binary systems with a mass transfer containing dwarf novae. For this case, the values of α_{ss} were found in the interval $0.1 \leq \alpha_{ss} \leq 1$ (Lynden-Bell and Pringle 1974; Bath and Pringle 1981). These values coincide with the estimates from Eardley et al. (1978), who considered the viscosity produced by the velocity shear and the reconnection of chaotic magnetic-field lines. However, the values of $0.01 \leq \alpha_{ss} \leq 1$ were obtained for this case. A completely different approach to the role of magnetic fields in the mechanics of momentum removal was presented by Blandford (1976), who argued that, if an ordered magnetic field with a large perpendicular component existed in a disk, then both the energy and angular momentum could be carried away by magnetized plasma jets moving perpendicularly to the disk plane what is actually observed.

Coroniti (1981) and Tout and Pringle (1992) obtained a relationship between the viscosity in a disk and the process of magnetic reconnection inside the disk. It is well known that the reconnection rate can be characterized by the quantity $M_A = u/c_A$, where u is the matter velocity ahead of the discontinuity and $c_A = |B|/\sqrt{4\pi\mu\rho}$ is the Alfvenic velocity ahead of the discontinuity (Priest and Forbes 2000). Both models use a shear flow inside the disk to enhance the magnetic field and use MHD turbulence as the mechanism of radial-mass transfer. In the model by Coroniti, the Keplerian motion in the disk produces a magnetic field in the course of time that lies in the disk plane, forming elliptical cells. These magnetic cells are continuously produced and destroyed in a turbulent process that leads to the radial plasma diffusion in the disk. The Shakura–Sunyaev viscosity parameter obtained in this paper can be expressed in terms of the reconnection parameter as $\alpha_{ss} \approx M_A^{2/3}$. In the model by Tout and Pringle, no special magnetic-field geometry is assumed in advance, but the sources and sinks for its various components are just estimated. In this model, it was shown that there exists a vertical-field reconnection triggered by strong radial-shear flows, which is a more important process than the azimuthal field reconnection, as assumed in the model by Coroniti. The model by Tout and Pringle gives the following expression for the Shakura–Sunyaev parameter: $\alpha_{ss} \approx 0.6 M_A$. For the accretion consistent with the observations of various astrophysical phenomena to be produced, the Mach number M_A must be ~0.1, which suggests very high reconnection rates. However, at present, there is no reason to believe that such a rapid reconnection is possible in the regime of turbulent MHD motion arising in the disk. Numerical simulations that would consider the turbulent dynamo and the reconnection process self-consistently are probably needed to obtain the actual picture of the relationship between accretion and reconnection. Other analytical models that allow α_{ss} to be calculated are also known, but none of them can be recognized to be final, since in all these studies, modeling the Shakura–Sunyaev parameter is basically reduced to expressing the unknown quantity α_{ss} in terms of another unknown quantity (e.g., M_A).

At the same time, the Shakura–Sunyaev approach developed by its authors specially for thin accretion disks disregards the dynamical processes of the gas–magnetic field interaction and, in particular, disregards the magnetic field effect on the turbulent flow via the mixing length and the inverse effect of the heat transfer on the development of disk turbulence. In addition, the original disk accretion model by Shakura and Sunyaev is known to be unstable to both thermal and slow perturbations. In connection with the development of a numerical model for the structure and evolution of the solar protoplanetary disk and its corona, it seems appropriate to depart from the α_{ss} formalism and to generalize (9.140) to the case of stratified (in density) matter of a finite-thickness subdisk.

9.4.2 Modeling the Turbulent Viscosity Coefficient in a Protoplanetary Disk of Finite Thickness

Let us first of all address the question of what magnetic-field configuration can be in a disk. If there were no disk, then the dipole magnetic-field lines would have no φ component (i.e., they would be poloidal), while the $\bar{B}_z$ component would predominate near the equatorial plane. However, if there is a disk around a star, then the magnetic-field lines that were initially "frozen" into the rotating plasma are displaced in the φ direction, which eventually gives rise to a nonzero $\bar{B}_\varphi$ component. If the field lines, being displaced, are assumed to remain continuous when crossing the disk surface, then the $\bar{B}_\varphi$ components must be equal in magnitude, but opposite in direction above and below the disk plane. Thus, the change in magnetic field inside the disk will be much larger in height z than in its radius r (e.g., $\bar{B}_\varphi$ changes from $+\bar{B}_\varphi$ to $-\bar{B}_\varphi$ as z changes from $-h_{\text{disk}}$ to $+h_{\text{disk}}$, where $h_{\text{disk}} << r$ is the disk half thickness). Furthermore, it is obvious that $\bar{B}_r << \bar{B}_\varphi, \bar{B}_z$, since the disk is located near the equatorial plane, while the gas moves in nearly circular orbits.

Thus, in the axisymmetric case under consideration, (9.131) for the turbulent viscosity coefficient can be written as

$$\nu_K^{\text{turb}} = \alpha_{ss} L^{*2} r \sqrt{\left(\frac{\partial \Omega(r,z)}{\partial r}\right)^2 + \left(\frac{\partial \Omega(r,z)}{\partial z}\right)^2}, \tag{9.143}$$

$$L^*(z) \equiv L(z) \left\{ 1 - \frac{\boldsymbol{Ri}_K - \boldsymbol{Ri}_M}{\boldsymbol{Pr}_K^{\text{turb}}} \right\}^{0.25}, \tag{9.143*}$$

where, in view of the above assumptions, the Richardson numbers $\boldsymbol{Ri}_K$ and $\boldsymbol{Ri}_M$ are

$$\boldsymbol{Ri}_K = \frac{\Omega_{K,\text{mid}}^2(r) z}{r^2} \frac{1}{\langle T \rangle} \frac{\partial \langle T \rangle / \partial z + G_a}{((\partial \Omega(r,z))/\partial r)^2 + ((\partial \Omega(r,z))/\partial z)^2}, \tag{9.144}$$

$$\boldsymbol{Ri}_M \cong \frac{1}{4\pi \mu \bar{\rho}} \frac{(\partial \bar{B}_\varphi / \partial z)^2}{r^2 \left\{ ((\partial \Omega(r,z))/\partial r)^2 + ((\partial \Omega(r,z))/\partial z)^2 \right\}}. \tag{9.145}$$

In writing these relations, we used the following representation for the effective gravity $\boldsymbol{g} = \{0, 0, -g_z\}$ (per unit mass with a correction for the centrifugal acceleration), where

$$g_z = \frac{G M_\odot z}{r^3} \left(1 + \frac{z^2}{r^2}\right)^{-3/2} \cong \Omega_{K,\text{mid}}^2(r) z; \tag{9.146}$$

$$\Omega_K(r,z) \equiv \sqrt{\frac{G\mathcal{M}_\odot}{(r^2 + z^2)^{3/2}}} \tag{9.147}$$

is the Keplerian angular velocity; $\Omega_{K,\text{mid}}(r) \equiv \Omega_K(r,0) = \sqrt{G\mathcal{M}_\odot / r^3}$ is the Keplerian angular velocity in the disk midplane;

$$G_a \equiv \frac{g_z}{\langle c_P \rangle} = -\frac{1}{\langle c_P \rangle} \frac{G M_\odot z}{r^3} \left(1 + \frac{z^2}{r^2}\right)^{-3/2} \cong \frac{1 - \gamma}{\gamma} \frac{1}{\Re} \Omega_{K,\text{mid}}^2 z \tag{9.148}$$

is the adiabatic temperature gradient in the protoplanetary accretion disk; and $\gamma = \langle c_P \rangle / (\langle c_P \rangle - \Re)$ is the adiabatic index for the averaged continuum. We see from

(9.144) that for an adiabatic temperature distribution in height (indifferent stratification)

$$-\frac{\partial\langle T\rangle}{\partial z}=\left(-\frac{\partial\langle T\rangle}{\partial z}\right)_{ad}=\frac{\Omega^2_{K,\mathrm{mid}}(r)}{\langle c_P\rangle}z,$$

the Richardson number $\boldsymbol{Ri}_K=0$, i.e., the temperature gradient in the disk has no effect on the turbulent transport coefficients. However, for an unstable thermal stratification of the accretion disk, where the temperature gradients are superadiabatic,

$$\frac{\partial\langle T\rangle}{\partial z}-\frac{\Omega^2_{K,\mathrm{mid}}(r)}{\langle c_P\rangle}z=f\frac{\partial\langle T\rangle}{\partial z},\tag{9.149}$$

$\boldsymbol{Ri}_K<0$ (here, the factor f, which characterizes the excess of the vertical temperature gradient in the disk over the adiabatic gradient, can reach $f=0.2$ for $r=10$ AU (Makalkin and Dorofeeva 1995)) and the turbulence energy increases at the expense of the instability energy in the direction perpendicular to the equatorial plane of the disk (convective source of turbulence); concurrently, the turbulent viscosity coefficient increases. At the same time, the spatial non-uniformity (in height) of the averaged magnetic field causes the turbulent energy to increase, since the magnetic Richardson number is always greater than zero, $\boldsymbol{Ri}_M>0$. The inverse turbulent Prandtl–Schmidt number $1/\boldsymbol{Pr}_K^{\mathrm{turb}}$ in (9.143) may be taken to be equal to unity in the case where the shear stresses in a differentially rotating disk are the main turbulence mechanism; however, it can be by a factor of 2 or 3 larger when the thermal convection in the vertical direction is responsible for the turbulence (Shakura et al. 1978).

It can be shown (Marov and Kolesnichenko 2006) that the vertical gradient in angular velocity $\partial\Omega(r,z)/\partial z$ can occasionally be neglected in the dissipative function $\Phi_{\langle\boldsymbol{u}\rangle}=2\bar{\rho}\nu_K^{\mathrm{turb}}\overset{\circ}{\boldsymbol{D}}:\overset{\circ}{\boldsymbol{D}}$ for an axisymmetric disk compared to its radial gradient $\partial\Omega(r,z)/\partial r$. For the bulk of the disk (except for the regions close to the proto-Sun), the following approximate expression is then valid for the turbulent viscosity coefficient:

$$\nu_K^{\mathrm{turb}}(r,z)=\alpha_{ss}L^{*2}r\left|\frac{\partial}{\partial r}\Omega(r,z)\right|,\quad L^*(z)\equiv L(z)\left(1-\frac{\boldsymbol{Ri}-\boldsymbol{Ri}_M}{\boldsymbol{Pr}^{\mathrm{turb}}}\right)^{0.25}.\tag{9.150}$$

Fin order for this expression to formally coincide with Shakura–Sunyaev formula (9.140) which is valid for gas-phase accretion disks, we must set the parameters $\boldsymbol{Ri}=0$ and $\boldsymbol{Ri}_M=0$ and substitute the angular velocity of Keplerian rotation $\Omega_{K,\mathrm{mid}}(r)=(GM_\odot)^{1/2}r^{-3/2}$ in (9.150). If we now use $L=h_{\mathrm{disk}}=\sqrt{(\bar{p}/\bar{\rho})|_{z=0}}/\Omega_{K,\mathrm{mid}}$ as the turbulence scale length, then the following expressions follow

$$v_K^{\text{turb}} = \frac{3}{2}\alpha_{ss}\left(\frac{\bar{p}}{\bar{\rho}}\right)|_{z=0}/\Omega_{K,\text{mid}},$$

$$R_{r\varphi} = \bar{\rho}v_K^{\text{turb}}r\left(\frac{\partial\Omega_{K,\text{mid}}(r)}{\partial r}\right) = -\frac{9}{4}\alpha_{ss}\bar{p}|_{z=0},$$

which coincides with (9.140) (since the factor 9/4 has no fundamental importance for the free parameter α_{ss}).

To conclude this chapter let us emphasize again that the development of adequate cosmogony models is related to the comprehensive study of dynamical and thermal evolution of the heterogeneous gas–dust matter of the differentially rotating protoplanetary disk with allowance made for MHD, turbulent, and radiation effects and with the involvement of phase transitions, chemical reactions, and coagulation processes. The aggregate state of the main components of the protoplanetary matter, the location of their condensation–sublimation fronts, and ultimately, the chemical composition of the planets, their satellites, and minor bodies depend on the space–time distribution of thermohydrodynamic parameters of the disk medium. The cosmochemical data obtained by directly studying the extraterrestrial matter should serve as an important constraint and as a clue in determining the extent to which these models are close to reality.

Here we extended our earlier study of heterogeneous medium (Kolesnichenko and Marov 1999; Marov and Kolesnichenko 2002) over electric conductive disks taking into account MHD effects. Basically, theoretical treatment of the numerous problems involved highlights the most comprehensive approach for further refinement of the models development.

The main results obtained in this chapter can be summarized as follows:

1. Formulation of the basic system of MHD equations for the instantaneous (actual) parameters of a conducting cosmic fluid flow designed to numerically simulate the circumsolar protoplanetary disk at various stages of its evolution and in spatial zones located at various distances from a protostar.
2. Probability-theoretical Favre averaging of the stochastic equations of motion and the MHD equations with the goal of phenomenologically describing the averaged plasma flow and the turbulent heat-and-mass transfer processes in a conducting disk.
3. Derivation of the defining relations for the correlation characteristics of a turbulent, electrically conducting fluid flow in the presence of a magnetic field that are needed to close the MHD equations for the scale of mean motion.
4. Development of a semiempirical method for modeling the turbulent transport coefficients in an electrically conducting disk with allowance made for the magnetic-field effect and the inverse effect of the heat transfer on the development of turbulence in a rotating, electrically conducting disk.

Appendix
Elements of Tensor Calculus

A. Vector and Tensor Designations

The following tensor designations are used in the book:

a tensor of zero rank (scalar),
$\boldsymbol{a}$ (a_{k}) tensor of first rank (vector),
$\boldsymbol{A}$ (A_{kj}) tensor of second rank,
$\boldsymbol{U}$ (δ_{kj}) unit tensor (δ_{kj} — Kronecker symbol),
$\boldsymbol{J}$ (J_{ijk}) tensor of third rank.

Simmetric and Antisymmetric Tensors

Simmetric and antisymmetric tensors are defined as follows:
simmetric:

$$\boldsymbol{A} = \boldsymbol{A}^{\mathrm{transp}} \; (A_{kj} = A_{jk}) \tag{A.1}$$

antisymmetric:

$$\boldsymbol{A} = -\boldsymbol{A}^{\mathrm{transp}} \; (A_{kj} = -A_{jk}) \tag{A.2}$$

Trace of tensor is defined as the sum of its diagonal elements:

$$\mathrm{Sp}\boldsymbol{A} = \sum_k A_{kk} \tag{A.3}$$

M.Y. Marov and A.V. Kolesnichenko, *Turbulence and Self-Organization: Modeling Astrophysical Objects*, Astrophysics and Space Science Library 389,
DOI 10.1007/978-1-4614-5155-6, © Springer Science+Business Media New York 2013

Scalar and Tensor (Internal) Product

Scalar product:

$$\begin{array}{lll} \text{of two vectors} & \boldsymbol{a}\cdot\boldsymbol{b}=\sum_k a_k b_k & \text{(scalar)} \\ \text{of vector and tensor} & \mathbf{A}\cdot\boldsymbol{b}=\sum_k A_{jk} b_k & \text{(vector)} \\ \text{of tensor and vector} & \boldsymbol{b}\cdot\boldsymbol{A}=\sum_k b_k A_{\mathrm{kj}} & \text{(vector)} \\ \text{of two tensors} & \boldsymbol{A}\cdot\boldsymbol{B}=\sum_k A_{jk} B_{ki} & \text{(tensor)} \end{array} \tag{A.4}$$

Dual scalar product oftensors:

$$\boldsymbol{A}:\boldsymbol{B}=\sum_{k,j} A_{jk}B_{kj} \quad \text{(scalar)} \tag{A.5}$$

Internal (dyad) tensor product:

$$\begin{array}{lll} \text{of two vectors} & (\boldsymbol{ab})_{jk}=a_j b_k & \text{(tensor of second rank)} \\ \text{of vector and tensor} & (\boldsymbol{aB})_{ijk}=a_i B_{jk} & \text{(tensor of third rank)} \\ \text{of tensor and vector} & (\boldsymbol{Ba})_{ijk}=B_{ij} a_k & \text{(tensor of third rank)} \\ \text{of two tensors} & (\boldsymbol{AB})_{ijkl}=A_{ij} B_{kl} & \text{(tensor of fourth rank).} \end{array} \tag{A.6}$$

Vector product of twovectors and tensor andvector:

$$(\boldsymbol{a}\times\boldsymbol{b})_k=\sum_{i,j}\varepsilon_{ijk}a_i b_j \text{ (vector)}, \quad (\boldsymbol{B}\times\boldsymbol{a})_{ik}=\sum_{j,l}\varepsilon_{jkl}B_{ij}a_l \text{ (tensor)} \tag{A.7}$$

where symbol of permutation $\varepsilon_{\mathrm{ijk}}$ takes the values:

$$\varepsilon_{ijk}=\begin{cases} +1 \text{ under even permutation of indexes (i.e. } 123,\ 231,\ 312) \\ -1 \text{ under odd permutation of indexes (i.e. } 321,\ 132,\ 213) \\ 0 \text{ under recurring indexes.} \end{cases} \tag{A.8}$$

B. Cylindrical Coordinates

Expressions for the different operators used in the equations of heterogeneous mechanics are represented here, for the convenience, in the cylindrical coordinate system r, φ, z (in axisymmetric case, $\partial/\partial\varphi = 0$). Here are the expressions of operators acting on:

(1) scalars:

$$\frac{\mathrm{d}\mathcal{B}}{\mathrm{d}t} \equiv \frac{\partial \mathcal{B}}{\partial t} + \mathbf{u}\cdot\nabla\mathcal{B} = \frac{\partial \mathcal{B}}{\partial t} + u_r\frac{\partial \mathcal{B}}{\partial r} + u_z\frac{\partial \mathcal{B}}{\partial z}, \quad \nabla\mathcal{B} = \mathbf{i}_r\frac{\partial \mathcal{B}}{\partial r} + \mathbf{i}_z\frac{\partial \mathcal{B}}{\partial z} \tag{B.1}$$

$$\nabla^2\mathcal{B} = \frac{1}{r}\frac{\partial}{\partial r}\left(r\frac{\partial \mathcal{B}}{\partial r}\right) + \frac{\partial^2\mathcal{B}}{\partial z^2}, \tag{B.2}$$

(2) vectors:

$$\nabla\cdot\mathbf{A} = \frac{1}{r}\frac{\partial(r\mathcal{A}_r)}{\partial r} + \frac{\partial \mathcal{A}_z}{\partial z}, \tag{B.3}$$

$$\nabla\mathbf{A} = \mathbf{i}_r\mathbf{i}_r\frac{\partial A_r}{\partial r} + \mathbf{i}_r\mathbf{i}_\varphi\frac{\partial A_\varphi}{\partial r} + \mathbf{i}_r\mathbf{i}_z\frac{\partial A_z}{\partial r} + \mathbf{i}_\varphi\mathbf{i}_r\frac{A_\varphi}{r} + \mathbf{i}_\varphi\mathbf{i}_\varphi\frac{A_r}{r} + \mathbf{i}_z\mathbf{i}_r\frac{\partial A_r}{\partial z} + \mathbf{i}_z\mathbf{i}_\varphi\frac{\partial A_\varphi}{\partial z} + \mathbf{i}_z\mathbf{i}_z\frac{\partial A_z}{\partial z}, \tag{B.4}$$

(3) dyads:

$$\mathbf{P} = \mathbf{i}_r\mathbf{i}_r P_{rr} + \mathbf{i}_r\mathbf{i}_\varphi P_{r\varphi} + \mathbf{i}_r\mathbf{i}_z P_{rz} + \mathbf{i}_\varphi\mathbf{i}_r P_{\varphi r} + \mathbf{i}_\varphi\mathbf{i}_\varphi P_{\varphi\varphi} + \mathbf{i}_\varphi\mathbf{i}_z P_{\varphi z} + \mathbf{i}_z\mathbf{i}_r P_{zr} + \mathbf{i}_z\mathbf{i}_\varphi P_{z\varphi} + \mathbf{i}_z\mathbf{i}_z P_{zz}, \tag{B.5}$$

$$\begin{aligned}\nabla\cdot\mathbf{P} &= \mathbf{i}_r\left[\frac{1}{r}\frac{\partial(r\,\mathrm{P}_{rr})}{\partial r} - \frac{\partial \mathrm{P}_{zr}}{\partial z} - \frac{\mathrm{P}_{\varphi\varphi}}{r}\right]\\ &\quad + \mathbf{i}_\varphi\left[\frac{1}{r}\frac{\partial(r\,\mathrm{P}_{r\varphi})}{\partial r} + \frac{\partial \mathrm{P}_{z\varphi}}{\partial z} + \frac{\mathrm{P}_{\varphi r}}{r}\right] + \mathbf{i}_z\left[\frac{1}{r}\frac{\partial(r\,\mathrm{P}_{rz})}{\partial r} + \frac{\partial \mathrm{P}_{zz}}{\partial z}\right].\end{aligned} \tag{B.6}$$

Then for strain tensor and strain velocity tensor we obtain, respectively:

$$\begin{aligned}\mathbf{D} &\equiv \frac{1}{2}\left(\nabla\mathbf{u} + (\nabla\mathbf{u})^{\mathrm{transp}}\right)\\ &= \mathbf{i}_r\mathbf{i}_r\frac{\partial \mathrm{u}_r}{\partial r} + \mathbf{i}_r\mathbf{i}_\varphi\frac{1}{2}\left(\frac{\partial u_\varphi}{\partial r} - \frac{u_\varphi}{r}\right) + \mathbf{i}_r\mathbf{i}_z\frac{1}{2}\left(\frac{\partial u_z}{\partial r} + \frac{\partial u_r}{\partial z}\right)\\ &\quad + \mathbf{i}_\varphi\mathbf{i}_r\frac{1}{2}\left(\frac{\partial u_\varphi}{\partial r} - \frac{u_\varphi}{r}\right) + \mathbf{i}_\varphi\mathbf{i}_\varphi\frac{u_r}{r} + \mathbf{i}_\varphi\mathbf{i}_z\frac{1}{2}\frac{\partial u_\varphi}{\partial z}\\ &\quad + \mathbf{i}_z\mathbf{i}_r\frac{1}{2}\left(\frac{\partial u_z}{\partial r} + \frac{\partial u_r}{\partial z}\right) + \mathbf{i}_z\mathbf{i}_\varphi\frac{1}{2}\frac{\partial u_\varphi}{\partial z} + \mathbf{i}_z\mathbf{i}_z\frac{\partial u_z}{\partial z},\end{aligned} \tag{B.7}$$

$$\begin{aligned}\mathring{\mathbf{D}} &\equiv \frac{1}{2}\left(\nabla\mathbf{u} + (\nabla\mathbf{u})^{\mathrm{transp}}\right) - \frac{1}{3}\mathbf{I}\nabla\cdot\mathbf{u}\\ &= \mathbf{i}_r\mathbf{i}_r\left(\frac{\partial u_r}{\partial r} - \frac{1}{3}\mathbf{I}\nabla\cdot\mathbf{u}\right) + \mathbf{i}_r\mathbf{i}_\varphi\frac{1}{2}\left(\frac{\partial u_\varphi}{\partial r} - \frac{u_\varphi}{r}\right) + \mathbf{i}_r\mathbf{i}_z\frac{1}{2}\left(\frac{\partial u_z}{\partial r} + \frac{\partial u_r}{\partial z}\right)\\ &\quad + \mathbf{i}_\varphi\mathbf{i}_r\frac{1}{2}\left(\frac{\partial u_\varphi}{\partial r} - \frac{u_\varphi}{r}\right) + \mathbf{i}_\varphi\mathbf{i}_\varphi\left(\frac{u_r}{r} - \frac{1}{3}\mathbf{I}\nabla\cdot\mathbf{u}\right) + \mathbf{i}_\varphi\mathbf{i}_z\frac{1}{2}\frac{\partial u_\varphi}{\partial z}\\ &\quad + \mathbf{i}_z\mathbf{i}_r\frac{1}{2}\left(\frac{\partial u_z}{\partial r} + \frac{\partial u_r}{\partial z}\right) + \mathbf{i}_z\mathbf{i}_\varphi\frac{1}{2}\frac{\partial u_\varphi}{\partial z} + \mathbf{i}_z\mathbf{i}_z\left(\frac{\partial u_z}{\partial z} - \frac{1}{3}\mathbf{I}\nabla\cdot\mathbf{u}\right).\end{aligned} \tag{B.8}$$

Double-point Gibbs multiplication serves as operator widely used in hydrodynamics. If, following the Gibbs notations,**a**, **b**, **c**, **d**are arbitrary vectors, then**ab** : **cd** $= (\mathbf{a} \cdot \mathbf{c})(\mathbf{b} \cdot \mathbf{d})$. In particular, for unit vectors we have:

$$\mathbf{i}_j\mathbf{i}_k : \mathbf{i}_l\mathbf{i}_m = \mathbf{i}_j \cdot \mathbf{i}_l)\,(\mathbf{i}_k \cdot \mathbf{i}_m) = \delta_{jl}\delta_{km}, \tag{B.9}$$

and for two dyads we have:

$$\begin{aligned}\mathbf{D}^{(1)}\!:\mathbf{D}^{(2)} &= \left(\sum_j\sum_k \mathbf{i}_j\mathbf{i}_k D^{(1)}_{jk}\right):\left(\sum_l\sum_m \mathbf{i}_l\mathbf{i}_m D^{(2)}_{lm}\right)\\ &= \sum_j\sum_k\sum_l\sum_m \delta_{jl}\delta_{km} D^{(1)}_{jk} D^{(2)}_{lm} = \sum_j\sum_k D^{(1)}_{jk} D^{(2)}_{jk},\end{aligned} \tag{B.10}$$

or

$$2\overset{\circ}{\mathbf{D}}:\overset{\circ}{\mathbf{D}} = 2D^2_{rr} + 2D^2_{\varphi\varphi} + D^2_{zz} + 4D^2_{r\varphi} + 4D^2_{rz} + 4D^2_{z\varphi} - \frac{2}{3}(\nabla\cdot\mathbf{u})^2. \tag{B.11}$$

Conclusion

The natural dynamical systems and their physical properties, specifically those observed in outer space, draw much attention, and constant progress is made in studying them. The fundamental patterns of the formation of macromolecular structures from a homogeneous non-differentiable (chaotic) medium toward ordering structures becomes a paradigm of how modern science broadens the horizons of knowledge. The goal of our mathematical models, in particular those that model cosmic objects that are inaccessible to direct investigation, is understanding the genesis and evolution of the key mechanisms that underlie natural processes in our environment and in the Universe at large.

We focused on the theoretical approach of developed turbulence in multicomponent mixtures of reacting gases and heterogeneous gas-dust media as well as on the construction of continuum models for turbulized hydrodynamic systems, with application to space objects. Because these systems, as a rule, have complicated physical and chemical characteristics, their mathematical modeling presents serious problems. Generally it requires taking into account the compressibility of flow, the variability of thermophysical properties, heat and mass exchange, chemical reactions, phase transitions and radiation transfer, and the processes occurring under the influence of gravitational and electromagnetic forces.

We considered a number of complex astrophysical problems based on the methods of continuum mechanics using the proposed stochastic-thermodynamic approach to semi-empirical modeling of the developed turbulence in reacting multicomponent gases and gas-dust media, incorporating structured turbulence properties in a homogeneous fluid. This theoretical advancement serves the purpose of reconstructing space object evolution based on the refined numerical models.

The topical theory of self-organization in space environment was thoroughly discussed in support of the general concept of stochastic dynamics related to the emerging ordered structures in the case of significant deviation from equilibrium. Great attention was paid to the theory of self-organization in the developed hydrodynamic turbulence that involves cooperative vortex structure formation. We considered the origin and evolution of coherent vortex structures in turbulent flows by analyzing the relationship between order and chaos in open dissipative

M.Y. Marov and A.V. Kolesnichenko, *Turbulence and Self-Organization: Modeling Astrophysical Objects*, Astrophysics and Space Science Library 389,
DOI 10.1007/978-1-4614-5155-6, © Springer Science+Business Media New York 2013

systems from the standpoint of stochastic nonlinear thermodynamics of irreversible processes. This study focuses on applications to specific phenomena, first of all, the origin and evolution of protoplanetary accretion gas-dust disks and the emerging primary dust clusters on the background of turbulent transport and self-organization.

Basically, examples of self-organizing dynamical systems can be consciously expanded from accretion disks to the observed peculiar features of numerous objects in the Universe that underlie cosmology and cosmogony, but we are aware that this approach to the structure formation and evolution of processes in nature may not be shared by other researchers. This problem was specifically addressed in the first chapter, where different dynamical systems (nature of stars and planets, stellar-galactic and planetary evolution, protoplanetary accretion disk formation and evolution, structure and evolution of the Universe) were discussed, though this problem extends far beyond the narrower problems of modeling structured turbulence. Essentially, the subsequent chapters supported the general basic concept of natural ordering and, in particular, the formation of highly organized nonequilibrium structures in cosmic media.

References

Chapter 1

Acuna, M.H., Connerney, J.E.P., Ness, N.F., et al.: Global distribution of crystal magnetism discovered by the Mars Global Surveyor MAG/ER experiment. Science **284**, 790 (1999)

Afanasenko, T.S., Rodin, A.V.: Influence of collisional line broadening on the spectra and fluxes of thermal radiation in the lower atmosphere of venus. Astron. Vestn. **39**, 214 (2005)

Alexander, C.M.O.D., Boss, A.P., Keller, L.P., et al.: Astronomical and meteoritic evidence for the nature of interstellar dust and its processing in Protoplanetary Discs. In: Protostars and Planets V. Arizona Press, AZ (2007)

Alpher, R.A., Bethe, H., Gamov, G.: The origin of chemical elements. Phys. Rev. **73**, 803 (1948)

Amelin, Y., Krot, A.N., Hutcheon, I.D., Ad Ulyanov, A.A.: Pb isotopic ages of chondrules and Ca, Al-rich inclusions. Science **297**, 1678–1683 (2002)

Amelin, Y., Kaltenbach, A., Iizuka, T., Stirling, C.H., Ireland, T.R., Petaev, M., Jacobsen, S.B.: Importance of uranium isotope variations for chronology of the Solar System's first solids. 41th Lunar Planetary Science Conference, Abstract #1648 (2010)

Anders, E., Owen, T.: Origin and abundances of volatiles. Science **198**, 453 (1977)

Andersen, J.: Accurate masses and radii of normal stars. Astron. Astrophys. Rev. **3**, 91 (1991). Dirac House, Temple Back, Bristol

Anderson, J.D., Lau, E.L., Sjogren, W.L., et al.: Gravitational constraints on the internal structure of ganymede. Nature **384**, 541 (1996)

Anderson, J.D., Schubert, G., Jacobsen, R.A., et al.: Distribution of rock, metals, and ices in Callisto. Science **280**, 1573 (1998)

Anderson, R.C., Dohm, J.M., Golombek, M.P., et al.: Primary centers and secondary concentrations of tectonic activity through time in the western hemisphere of mars. J. Geophys. Res. **106**, 20563 (2001)

Anishchenko, A.A.: Complex Oscillations in Simple Systems. Nauka, Moscow (1990) (in Russian)

Anishchenko, A.A., Vadivasova, T.E., Shimanskii-Gaier, L.: Dynamic and static description of oscillatory systems. In: Regular and Chaotic Dynamics. Research Center, Institute of Computer Research, Moscow-Izhevsk (in Russian) (2005)

Armitage, P.J., Livio, M., Pringle, J.E.: Episodic accretion in magnetically layered protoplanetary disks. Mon. Not. Roy. Astron. Soc **324**, 705 (2001)

Arnold, V.I.: Small denominators and the problem of stability of motion in classical and celestial mechanics. Usp. Mat. Nauk **18**, 85 (1963a)

Arnold, V.I.: Small denominators and the problems of stability of motion in classical and celestial mechanics. Usp. Mat. Nauk. **18**(6), 91–192 (1963b)

M.Y. Marov and A.V. Kolesnichenko, *Turbulence and Self-Organization: Modeling Astrophysical Objects*, Astrophysics and Space Science Library 389,
DOI 10.1007/978-1-4614-5155-6, © Springer Science+Business Media New York 2013

Artymowicz, P.: Beta pictoris: An early solar system. Annu. Rev. Earth Planet. Sci. **25**, 175 (1997)
Atreya, S.K., Pollack, J.B., Matthews, M.S.: Origin and Evolution of Planetary and Satellite Atmospheres. University of Arizona, Tucson (1989)
Atreya, S.K., Mahaffy, P.R., Niemann, H.B., et al.: Composition and origin of the atmosphere of Jupiter—An update, and implications for the extrasolar giant planets. Planet. Space Sci. **51**, 105 (2003)
Avduevsky, V.S., Borodin, N.F., Kuznetsov, V.V., et al.: Temperature, pressure, and density of the venusian atmosphere from venura-4 measurements. Dokl. Akad. Nauk SSSR **179**, 310 (1968)
Avduevsky, V.S., Marov, M.Y., Rozhdestvensky, M.K.: A tentative model of the atmosphere of the planet venus based on the results of measurements of venera 5 and 6. J. Atmos. Sci. **27**, 561 (1970)
Avduevsky, V.S., Zavelevich, F.S., Marov, M.Y., et al.: Numerical simulation of radiative-convective heat exchange in the venusian atmosphere. Kosmich. Issled. **9**, 280 (1971)
Avduevsky, V.S., Golovin, Y.M., Zavelevich, F.S., et al.: Preliminary results of the study of the light regime in the atmosphere and on the surface of venus. Kosmich. Issled. **14**, 735 (1976)
Ayres, T.R.: Thermal bifurcation in the solar outer atmosphere. Astrophys. J. **244**, 1064 (1981)
Balbus, S.A., Hawley, J.F.: Instability, turbulence, and enhanced transport in accretion disks. Rev. Mod. Phys. **70**, 1 (1998)
Ballesteros-Paredes, J., Klessen, R.S., Mac Low, M.-M., Vazquez-Semadeni, E.: Molecular cloud turbulence and star formation. In: Protostars and Planets V. Arizona Press, AZ (2007)
Barenblatt, G.I.: On the motion of suspended particles in a turbulent flow occupying a half-space or a plane open channel of finite depth. Prikl. Mat. Mekh. **19**, 61 (1955)
Barrow-Green, J.: Poincare and the Three Body Problem. Princeton University, Princeton (1996)
Barsukov, V.L., Marov, M.Y. (eds.): The Satellites of Jupiter. Mir, Moscow (1985) (in Russian)
Basri, G.: Observations of brown dwarfs. Annu. Rev. Astron. Astrophys. **38**, 485 (2000)
Batchelor, G.K.: The Theory of Homogeneous Turbulence. Cambridge University, Cambridge (1953)
Belotserkovskii, O.M.: Numerical Simulation in Continuum Mechanics. Nauka, Moscow (1994) (in Russian)
Belotserkovskii, O.M.: Numerical Experiment in Turbulence: From Order to Chaos. Nauka, Moscow (1997) (in Russian)
Benner, L.A.M.: Triton. In: Shirley, J.H., Fairbridge, R.W. (eds.) Encyclopedia of Planetary Sciences. Chapman and Hall, London (1997)
Bergin, E.A., Aikawa, Yu., Blake, G.A., van Dishoeck, E.F.: The chemical evolution of protoplanetary disks. In: Protostars and Planets V. Arizona Press, AZ (2007)
Bershadskii, A.G.: Virtual statistical equilibrium and quasi-stationary spectra of homogeneous turbulence. Zh. Eksp. Teor. Fiz. **94**, 117 (1988)
Bershadskii, A.G.: Fractal structure of turbulent vortices. Zh. Eksp. Teor. Fiz. **96**, 625 (1989)
Bethe, H.A.: The theory of supernovae. In: Masevich, A.G. (ed.) Essays in Nuclear Astrophysics, p. 418. Mir, Moscow (1966)
Bezard, B., de Bergh, C., Fegley, B., et al.: The abundance of sulfur dioxide below the clouds of venus. Geophys. Res. Lett. **20**, 1587 (1993)
Bibring, J.-P., Langevin, Y., Gendrin, A., The Omega Team: Mars surface diversity as revealed by the OMEGA. Mars Express observations. Science **307**, 1576 (2005)
Binney, J., Merrifield, M.: Galactic Astronomy. Princeton University, Princeton (1998)
Bisikalo, D.V., Marov, M.Y., Shematovich, V.I., Strel'nitskij, V.S.: The flow of the subliming gas in the near-nuclear (knudsen) layer of the cometary coma. Adv. Space Res. **9**, 53 (1989)
Bisnovaty-Kogan, G.S., Lovelace, R.V.E.: Advective accretion disks and related problems including magnetic fields. New Astron. Rev. **45**, 663 (2001)
Black, D.C.: Extra-solar planets: Searching for other planetary systems. In: Weissman, P.R., McFadden, L.-A., Johnson, T.V. (eds.) Encyclopedia of the Solar System, p. 941. Academic, San Diego (1999)

Blum, J., Wurm, G.: Experiments of sticking, restructuring and fragmentation of preplanetary dust aggregates. Icarus **143**, 138

Boice, D.C., Huebner, W.: Physics and chemistry of comets. In: Weissman, P.R., McFadden, L.-A., Johnson, T.V. (eds.) Encyclopedia of the Solar System, p. 519. Academic, San Diego (1999)

Bonnell, I.A., Larson, R.B., Zinnecker, H.: The origin of the initial mass function. In: Protostars and Planets V. Arizona Press, AZ (2007)

Borner, G.: The Early Universe. Springer, Heidelberg (1993)

Bouvier, A., Wadhva, M., Janey, P.: Pb–Pb isotope systematics in the Allende chondrule. Goldshmidt Conference Abstracts, July 13–18, Vancouver, Canada, A11 (2008)

Bouvier, A., Wadhwa, M.: In: Synchronizing the absolute age and relative clocks: Pb–Pb and Al–Mg systematics in CAIs from the Allende and NWA 2364 CV3 chondrites. 40th Lunar and Planet. Sci. Conf. Abstract #2184 (2009)

Bouvier, A., Wadhwa, M.: Pb–Pb isotope dating of the unique basaltic achondrite NWA 2976. 41th Lunar and Planet. Sci. Conf. Abstract #1489 (2010)

Bouvier et al.: Geochim. Cosmochim. Acta. **71**, 1583–1604 (2007)

Bouwman, J., Meeus, G., de Koter, A., et al.: Processing of silicate dust grains in herbig Ae/Be systems. Astron. Astrophys. **375**, 950 (2001)

Bouwman, J., van Boekel, R., Min, M., et al.: 10 μm Spectroscopic survey of herbig Ae star disks: Grain growth and crystallization. Astron. Astrophys. **437**, 189 (2005)

Brain, D.A., Jakosky, B.M.: Atmospheric loss since the onset of the martian geologic record: Combined role of impact erosion and sputtering. J. Geophys. Res. **103**, 22689 (1998)

Britt, D.T., Lebofsky, L.A.: Asteroids. In: Weissman, P.R., McFadden, L.-A., Johnson, T.V. (eds.) Encyclopedia of the Solar System, p. 585. Academic, San Diego (1999)

Brown, G.L., Roshko, A.: On density effects and large structures in turbulent mixing layers. J. Fluid Mech. **64**, 775 (1974)

Bruno, A.D.: Restricted Three-Body Problem. Nauka, Moscow (1990) (in Russian)

Bryden, G., Chen, X., Lin, D., et al.: Tidally induced gap formation in protostellar disks: Gap clearing and suppression of protoplanetary growth. Astrophys. J. **514**, 344 (1999)

Buratti, B.J.: Outer planet icy satellites. In: Weissman, P.R., McFadden, L.-A., Johnson, T.V. (eds.) Encyclopedia of the Solar System, p. 435. Academic, San Diego (1999)

Burns, J.A., Matthews, M.S.: Satellites. University of Arizona, Tucson (1986)

Busse, F.H.: A simple model of convection in the jovian atmosphere. Icarus **29**, 255 (1976)

Bylov, B.F., Vinograd, R.I., Grobman, D.M., Nemytskii, V.V.: The Theory of Lyapunov Exponents and Its Applications to the Questions of Stability. Nauka, Moscow (1966) (in Russian)

Cameron, A.G.W., Ward, W.: The origin of the moon. Paper presented at the Sci. Proc. Lunar. Conf., 7th. Houston, p. 120 (1976)

Cantwell, B.J.: Organized motions in turbulent flow. Annu. Rev. Fluid Mech. **13**, 457 (1981)

Canup, R.M., Asphaug, E.: Origin of the moon in a giant impact near the end of the earth's formation. Nature **412**, 708 (2001)

Carr, M.H.: The Surface of Mars. Yale University, New Heaven and London (1981)

Carr, M.H.: D/H on Mars: Effects of floods volcanism impacts and polar processes. Icarus **87**, 210 (1990)

Carr, M.H.: Water on Mars. Oxford University, Oxford (1996)

Carr, M.H.: Mars surface and interior. In: Weissman, P.R., McFadden, L.-A., Johnson, T.V. (eds.) Encyclopedia of the Solar System, p. 291. Academic, San Diego (1999a)

Carr, M.H.: Retention of an atmosphere on early mars. J. Geophys. Res. **104**, 21897 (1999b)

Carr, M.H.: Martian oceans, valleys and climate. Astron. Geophys. **41**, 3.20 (2000)

Carr, M.H., Belton, M.J.S., Chapman, C.R., et al.: Evidence for a subsurface ocean on Europa. Nature **39**, 363 (1998)

Cassen, P.M., Summers, A.: Models of the formation of the solar nebula. Icarus **53**(1), 26–40 (1983)

Cassen, P.M., Woolum, D.S.: The origin of the solar system. In: Weissman, P.R., McFadden, L.-A., Johnson, T.V. (eds.) Encyclopedia of the Solar System, p. 35. Academic, San Diego (1999)

Chamberlain, J.W., Hunten, D.M.: Theory of Planetary Atmospheres. An introduction to Their Physics and Chemistry, 2nd edn. Academic, San Diego (1987)

Chapman, D.R.: Computational aerodynamics and prospects for its development: Dryden lecture. Raketn. Tekhn. Kosmonavt. **18**, 3 (1980)

Charney, J.G., Stern, M.E.: On the stability of internal baroclinic jets in a rotating atmosphere. J. Atmos. Sci. **19**, 159 (1962)

Chiang, E.I.: In: Witt, A.N., Clayton, G.C., Draine, B.T. (eds.) Proceedings of the Dust in Protoplanetary Disks, 26–30 May 2003. Estes Park Colorado. Astrophysics of Dust, ASP Conference Series V. 309, p. 213 (2004)

Chiang, E.I., Lithwick, Y.: Neptune trojans as a test bed for planet formation. Astrophys. J. **628**, 520 (2005)

Chiosi, C., Bertelli, G., Bressan, A.: New development in understanding the HR diagram. Annu. Rev. Astron. Astrophys. **30**, 235 (1992)

Christensen, P.R., Bandfield, J.L., Clark, R.N., et al.: Detection of crystalline hematite mineralization on Mars by the thermal emission spectrometer: Evidence for near-surface water. J. Geophys. Res. **105**, 9623 (2000)

Chyba, C.F., Owen, T.C., Ip, W.-H.: Impact delivery of volatiles and organic molecules to earth. In: Gehrels, T., Matthews, M.S., Schumann, A. (eds.) Hazards Due to Comets and Asteroids, p. 99. University of Arizona, Tucson (1994)

Clifford, S.M., Parker, T.J.: The evolution of the martian hydrosphere: Implications for the fate of a primordial ocean and the current state of the northern plains. Icarus **154**, 40 (2001)

Colaprete, A., 16 co-authors: Detection of water in the LCROSS ejecta plum. Science. **330**, 463–468 (2010)

Contopoulos, G.: Order and Chaos in Dynamical Astronomy. Springer, Berlin (2002)

Coustenis, A., Lorenz, R.: Titan. In: Weissman, P.R., McFadden, L.-A., Johnson, T.V. (eds.) Encyclopedia of the Solar System, p. 377. Academic, San Diego (1999)

Crisp, D., Titov, D.: In: Bougher, S.W., Hunten, D.M., Phillips, R.J. (eds.) The Thermal Balance of the Venus Atmosphere in Venus II, p. 353. University of Arizona, Tucson (1997)

Crow, S.C., Champagne, F.H.: Orderly structures in jet turbulence. J. Fluid Mech. **48**, 547 (1971)

Cruikshank, D.P., 78 collaborating authors: Neptune and Triton. University of Arizona, Tucson (1995)

Cruikshank, D.P., Barucci, M.A., Emery, J.P., et al.: Physical Properties of Trans-Neptunian Objects. In: Protostars and Planets V. Arizona Press, AZ (2007)

Cuzzi, J.N.: Blowing in the wind: III. Accretion of dust rims by chondrule-sized particles in a turbulent protoplanetary nebula. Icarus **168**, 484 (2004)

Cuzzi, J.N., Weidenschilling, S.J.: In: Lauretta, D., Leshin, L.A., McSween, H. (eds.) Particle-Gas Dynamics and Primary Accretion in Meteorites and the Early Solar System II, p. 353. University of Arizona, Tuscon (2006)

Cuzzi, J.N., Davis, S.S., Dobrovolskis, A.R.: Blowing in the wind. II. Creation and redistribution of refractory inclusions in a turbulent protoplanetary nebula. Icarus **166**, 385 (2003)

Cuzzi, J.N., Ciesla, F.J., Petaev, M.I., et al.: Nebula evolution of thermally processed solids: Reconciling models and meteorites chondrites and the protoplanetary disk, ASP Conference Series, vol. 341. In: Krot, A.N., Scott, E.R.D, Reipurth, B. (eds.) Proceedings of a workshop held 8–11 November 2004 in Kauai, Hawaii, Astronomical Society of the Pacific, San Francisco, p. 732 (2005)

Dalal, N., Kochanek, C.S.: Direct detection of cold dark matter substructure. Astrophys. J. **572**, 25 (2002)

de Groot, S.R., Mazur , P.: Non–Equilibrium Thermodynamics. North-Holland Publishing Company. Amsterdam. (1962)

de Kampe Feriet, J.: Statistical mechanics and theoretical models of diffusion processes. In: Frenkiel, F.N., Sheppard, P.A. (eds.) Atmospheric Diffusion and Air Pollution, p. 134. Academic, New York and London (1959)

Del Genio, A.D.: Atmospheres. In: Shirley, J.H., Fairbridge, R.H. (eds.) Encyclopedia of Planetary Sciences. Chapman and Hall, London (1997)

Delsemme, A.H.: The deuterium enrichment observed in recent comet is consistent with the cometary origin of sea water. Planet. Space Sci. **47**, 125–131 (1999)

Diagu, F., Holmes, P.: Celestial Encounters: The origins of Chaos and Stability. Princeton University, Princeton (1997)

Dominik, C., Blum, J., Cuzzi, J., Wurm, G.: Growth of dust as the initial step toward planet formation. In: Protostars and Planets V. Arizona Press, AZ (2007)

Donahue, T.M., Hoffman, J.H., Hodges, R.R., Watson, A.J.: Venus was wet: A measurement of the ratio of deuterium to hydrogen. Science **216**, 630 (1982)

Donahue, T.M., Grinspoon, D.H., Hartle, R.E., Hodges, R.R.: In: Bougher, S.W., Hunten, D.M., Phillips, R.J. (eds.) Ion/Neutral Escape of Hydrogen and Deuterium: Evolution of Water in Venus II, p. 385. University of Arizona, Tucson (1997)

Dorofeeva, V.A., Makalkin, A.B..: Evolution of the Early Solar System. Cosmochemical and Physical Aspects. Editorial URSS, Moscow (2004) (in Russian)

Doroshkevich, A.G., Zel'dovich, Y.B., Sunyaev, R.A.: Adiabatic theory of galaxy formation. In: Pikelner, S.B. (ed.) Origin and Evolution of Galaxies and Stars, p. 65. Nauka, Moscow (1976) (in Russian)

Dubrulle, B.: Differential rotation as a source of angular momentum transfer in the solar nebula. Icarus **106**, 59 (1993)

Dullemond, C.P., Dominik, C.: Dust coagulation in protoplanetary disks: A rapid depletion of small grains. Astron. Astrophys. **434**, 971 (2005)

Dullemond, C.P., Hollenbach, D., Kamp, I., D'Alessio, P.: Models of the structure and evolution of protoplanetary discs. In: Protostars and Planets V. Arizona Press, AZ (2007)

Duncan, M., Levison, H.: A scattered comet disk and the origin of Jupiter family comet. Science **276**, 1670 (1997)

Duncan, M.J., Lissauer, J.J.: Solar system dynamics. In: Weissman, P.R., McFadden, L.-A., Johnson, T.V. (eds.) Encyclopedia of the Solar System, p. 809. Academic, San Diego (1999)

Durisen, R.H., Boss, A.P., Mayer, L., et al.: Gravitational instabilities in gaseous protoplanetary disks and implications for giant planet formation. In: Protostars and Planets V. Arizona Press, AZ (2007)

Dutrey, A., Guilloteau, S., Ho, P.: Interferometric spectro-imaging of molecular gas in protoplanetary discs. In: Protostars and Planets V. Arizona Press, AZ (2007)

Dyson, Freeman, A.M.: Terraforming Venus. Corres. J. Brit. Interplan. Soc. **42**, 593 (1989)

EAA.: Encyclopedia of Astronomy and Astrophysics. Nature Publishing Group and Institute of Physics Publishing, Dirac House, Temple Back, Bristol

Eardley, D.M., Lightman, A.P., Payne, D.G., Shapiro, S.L.: Accretion discs around massive black holes: Persistent emission spectra. Astrophys. J. **234**, 53 (1978)

Ebeling, W., Feistel, R.: Chaos and cosmos. Evolution synergetics. In: Regular and Chaotic Dynamics. Research Center, Institute of Computer Research, Moscow-Izhevsk (in Russian) (2005)

Eidelman, S., et al.: Particle data group. Phys. Lett. B **592**, 1 (2004) (URL: http://pdg.lbl.gov)

Elgaroy, O., Lahav, O., Percival, W.J., et al.: New upper limit on the total neutrino mass from the 2 degree field galaxy redshift survey. Phys. Rev. Lett. **89**, 61 (2002)

Elliot, J.L., Nicholson, P.D.: The rings of uranus. In: Greenberg, R., Brahic, A. (eds.) Planetary Rings, p. 25. University of Arizona, Tucson (1984)

Eneev, T.M.: On the possible structure of the outer (Trans-Neptunian) regions of the solar system. Pis'ma Astron. Zh. **6**, 295 (1981)

Eneev, T.M., Kozlov, N.N.: A model for the accumulation formation of planetary systems. I. Numerical experiments. Astron. Vestn. **15**, 80 (1981)

Esposito, L.W., Knollenberg, R.G., Marov, M.Y., et al.: The clouds and hazes of venus. In: Hunten, D.M., Colin, L., Donahue, T.M., Moroz, V.I. (eds.) Venus, pp. 484–564. University of Arizona, Tucson (1983)

Esposito, L., Cuzzi, J.N., Holdberg, J.B., et al.: Saturn's rings: Structure, dynamics and particle properties. In: Gehrels, T., Matthews, M.S. (eds.) Saturn, p. 463. University of Arizona, Tucson (1984)
Esposito, L., Brahic, A., Burns, J.A., Marouf, E.A.: Particle properties and processes in uranus' rings. In: Bergstrahl, J.T., Miner, E.D., Matthews, M.S. (eds.) Uranus, p. 410. University of Arizona, Tucson (1991)
Fanale, F.P.: Mars: Atmosphere and volatile history. In: Weissman, P.R., McFadden, L.-A., Johnson, T.V. (eds.) Encyclopedia of the Solar System, p. 277. Academic, San Diego (1997)
Feigenbaum, M.J.: Universal behavior in nonlinear system. Physica **7D**, 16 (1978)
Fernandez, J.A.: Dynamics of comets: Recent developments and new challenges. In: Milani, A., Martino, M.D., Cellino, A. (eds.) Asteroids, Comets, Meteors 1993, p. 223. Kluwer, Dordrecht (1994)
Fernandez, J.A.: Cometary dynamics. In: Weissman, P.R., McFadden, L.-A., Johnson, T.V. (eds.) Encyclopedia of the Solar System, p. 537. Academic, San Diego (1999)
Ferraz-Mello, S.: Dynamics of the Galilean Satellites of Jupiter. IAG-USP, Sao Paulo, Brazil (1979)
Fixsen, D.J., Mather, J.C.: The spectral results of the far-infrared absolute spectrophotometer instrument on COBE. Astrophys. J. **581**, 817 (2002)
Fowler, R.H.: On dense matter. Mon. Not. Roy. Astron. Soc. **87**, 114 (1926)
French, R., Nicholson, P.D., Porco, C.C., Marouf, E.A.: Dynamics and structure of the uranian rings. In: Bergstrahl, J.T., Miner, E.D., Matthews, M.S. (eds.) Uranus, p. 327. University of Arizona, Tucson (1991)
Fridman, F.M.: To the dynamics of a viscous differentially rotating gravitating medium. Pis'ma Astron. Zh. **15**, 1122 (1989)
Fridman, A.M., Polyachenko, V.L.: Physics of Gravitating System. Springer, New York (1984)
Fridman, A.M., Sagdeev, R.Z., Horuzhyi, O.V., Polyachenko, E.V.: Observed manifestations of chaos in spiral galaxies. In: Modern Problems of Mechanics and Space Physics, p. 12. Fizmat., Moscow, (in Russian) (2003)
Gaidos, E., Selsis, F.: From protoplanets to protolife: The emergence and maintenance of life. In: Protostars and Planets V. Arizona Press, AZ (2007)
Galileo Orbiter Team: J. Geophys. Res. **103**(Special Issue), E10 (1998)
Galileo Orbiter Team.: Icarus **135**(Special Issue) (1998)
Galileo Probe Team: J. Geophys. Res. Planets **103**(Special Issue), E9 (1998)
Galimov, E.M.: The Phenomenon of Life. Between Equilibrium and Nonlinearity. The Origin and Principles of Evolution. URSS, Moscow (2001) (in Russian)
Galimov, E.M.: On the origin of the Moon's material. Geokhimiya **4**, 691 (2004)
Galimov, E.M.: Prerequisites and conditions for the emergence of life. Research problems. Geokhimiya **5**, 1 (2005)
Galimov, E.M.: The Moon–Helium-3 Project. Nauka, Moscow (2006) (in Russian)
Galimov, E.M.: Concept of stable ordering and ATP-dependent mechanism of life origin. In: Galimov, E.M. (ed.) Problems of the Biosphere Origin and Evolution, pp. 23–32. URSS, Moscow (2008)
Galimov, E.M., Krivtsov, A.M., Zabrodin, A.V., et al.: Dynamical Model for the formation of the Earth-Moon System. Geokhimiya **11**, 1139 (2005)
Galimov, E.M.: The Moon and Earth formation from the common supra-planetary gas-dust cluster. Paper presented at the XIX Russian Symposium on isotopes geochemistry, Moscow (in Russian), 16 November (2010)
Gallagher, J.S., Hunter, D.A., Tutukov, A.V.: Star formation histories of irregular galaxies. Astrophys. J. **284**, 544 (1984)
Gawiser, E., Silk, J.: Extracting primordial density fluctuations. Science **280**, 1405 (1998)
Gillett, S.L.: Establishment and stabilization of earthlike conditions on venus. J. British Interplanet. Soc. **44**, 151 (1991)

Girardi, M., Borgani, S., Giuricin, G., et al.: Optical luminosities and mass-to-light ratios of nearby galaxy clusters. Astrophys. J. **530**, 62 (2000)
Gladstone, G.R., 19 co-authors: LRO-LAMP observations of the LCROSS impact plume. Science. **330**, 472–476 (2010)
Glansdorff, P., Prigogine, I.: Thermodynamic Theory of Structure, Stability and Fluctuation. Wiley-Interscience, New York (1973)
Glatzmaier, G.A., Evonuk, M., Rogers, T.M.: Differential rotation in giant planets maintained by density-stratified turbulent convection. Geophys. Astrophys. Fluid Dyn. **103**(1), 31–51 (2009)
Gledzer, E.B., Dolzhanskii, F.V., Obukhov, A.M.: Hydrodynamic Systems and Their Applications. Nauka, Moscow (1981) (in Russian)
Goldrich, P., Ward, W.R.: The formation of planetesimals. Astrophys. J. **183**, 1051 (1973)
Golitsyn, G.S.: Introduction to the Dynamics of Planetary Atmospheres. Gidrometeoizdat, Leningrad (1973) (in Russian)
Golombek, M.P., Bridges, N.T.: Erosion rates on mars and implications for climate change: constraints from the pathfinder landing site. J. Geophys. Res. **105**, 1841 (2000)
Gor'kavyi, N.N., Fridman, A.M.: Physics of Planetary Rings. Nauka, Moscow (1994) (in Russian)
Gounelle, M., Shu, F.H., Shang, H., et al.: The irradiation origin of beryllium radioisotopes and other short-lived radionuclides. Astrophys. J. **640**, 1163 (2006)
Green, B.: The Elegant Universe. Superstrings, Hidden Dimensions, and the Quest for the Ultimate Theory. Vintage Books, New York (1999)
Green, B.: Tha Fabric of the Cosmos. Space, Time, and the Texture of Reality. Alfred A. Knopf, New York (2004)
Greenberg, R., Brahic, A.: Planetary Rings. University of Arizona, Tucson (1984)
Grinspoon, D.H., Pollack, J.B., Sitton, B.R., et al.: Probing Venus's cloud structure with Galileo NIMS. Planet. Space Sci. **41**, 515 (1993)
Gulick, V.C.: Origin of the valley networks on mars: A hydrological perspective. Geomorphology **37**, 241 (2001)
Haken, H.: Advanced Synergetics. Instability Hierarchies of Self-Organizing Systems and Devices. Springer, Berlin (1983)
Haken, H.: Information and Self-Organization. A Macroscopic Approach to Complex Systems. Springer, Berlin (1988)
Hansen, C.J., Kawaler, S.D.: Stellar Interiors. Physical Principles, Structure, and Evolution. Springer, Berlin (1994)
Hartmann, W.K.: A Traveller's Guide to Mars. The Mysterious Landscapes of the Red Planet. Workman Publ, New York (2003)
Hartmann, W.K., Berman, D.C.: Elysium planitia lava flows: Crater count chronology and geological implications. J. Geophys. Res. **105**, 15011 (2000)
Hartmann, W.K., Davies, D.R.: Satellite-sized planetesimals and lunar origin. Icarus **24**, 504–515 (1975)
Hartmann, W.K., Neukum, G.: Cratering chronology and the evolution of mars. Space Sci. Rev. **96**, 165 (2001)
Hawley, J.F., Balbus, S.A.: A powerful local shear instability in weakly magnetized disk. II nonlinear evolution. Astrophys. J. **376**, 223 (1991)
Hayne, P.O., 5 co-authors: Diviner lunar radiometer observations of the LCROSS impact. Science. **330**, 477–479 (2010)
Head, J.W., Basilevsky, A.T.: In: Weissman, P.R., McFadden, L.-A., Johnson, T.V. (eds.) Venus: Surface and Interior in Encyclopedia of the Solar System, p. 161. Academic, San Diego (1999)
Head III, J.W., Hiesinger, H., Ivanov, M.A., et al.: Possible ancient oceans on mars: evidence from mars orbiter laser altimeter data. Science **286**, 2134 (1999)
Hildebrand, R.: The determination of cloud masses and dust characteristics from submillimeter thermal emission. Astron. Soc. **24**, 267 (1983)

Holtzman, J.: Microwave background anisotropies and large-scale structure in universes with cold dark matter, baryons, radiation, and massive and massless neutrinos. Astrophys. J. Suppl. Ser. **71**, 1 (1989)

Holtzman, J., Primack, J.R.: Cluster correlations for cold and hot dark matter and other models. Astrophys. J. **405**, 428 (1993)

Horsthemke, W., Lefever, R.: Noise Induced Transitions. Springer, Berlin-Heidelberg-New York-Tokyo (1984)

Hunten, D.M.: Venus: Atmosphere. In: Weissman, P.R., McFadden, L.-A., Johnson, T.V. (eds.) Encyclopedia of the Solar System, p. 147. Academic, San Diego (1999)

Hunten, D.M., Tomasko, M.G., Flasar, F.M., et al.: Titan. In: Gehrels, T., Matthews, M.S. (eds.) Saturn, p. 671. University of Arizona, Tucson (1984)

Ingersoll, A.P., Pollard, D.: Motion in the interiors and atmospheres of Jupiter and Saturn: scale analysis, inelastic equations, barotropic stability criterion. Icarus **52**, 62 (1982)

Ingersoll, A.P., Vasalada, A.R., Galileo Imaging Team: In: Dynamics of jupiter's atmosphere in SPSI: The Galileo Mission to the Jupiter System. In: Andersen, J (ed.) Highlights of Astronomy, v. 11B, p. 1046. Kluwer Academic Publishers (1998)

Ingersoll, A., Beebe, R., Conrath, B., Hunt, G.: Structure and dynamics of saturn's atmosphere. In: Gehrels, T., Matthews, M.S. (eds.) Saturn, p. 195. University of Arizona, Rucson (1984)

Ingersoll, A.P., Barnet, C.D., Beebe, R.F., et al.: Dynamic meteorology of neptune. In: Cruikshank, D. (ed.) Neptune and Triton, p. 613. University of Arizona, Tucson (1995)

Ipatov, S.I.: Migration of Celestial Bodies in the Solar System. Editorial URSS, Moscow (2000) (in Russian)

Ipatov, S.I., Mather, J.C., Marov, M.Ya.: Migration of Icy Bodies to the Terrestrial Planets, Joint Assembly AGU, GS, MAS, MSA, SEG, and UGM, 23–26 May, Baltimore, Maryland, USA, Eos Trans.AGU, 87(36) (2006a)

Ipatov, S.I., Marov, M.Y., Mather, J.C.: Delivery of volatiles to the terrestrial planets. Astrobiology **6**, 241 (2006b)

Ivanov, B.A.: Comparison of the Sizes of Asteroids and Impact Craters in Modern Problems of Mechanics and Space Physics. Nauka, Moscow (2003) (in Russian)

Ivanov, M.A., Head, J.W.: Chryse planitia mars: Topographic configuration outflow channel continuity and sequence and tests for hypothesized ancient bodies of water using Mars Orbiter Laser Altimeter (MOLA) Data. J. Geophys. Res. **106**, 3275 (2001)

Izakov, M.N., Marov, M.Y.: The upper atmosphere. In: Barsukov, V.L., Volkov, V.P. (eds.) The Planet Venus, pp. 114–132. Nauka, Moscow (1989) (in Russian)

Jakosky, B.M., Jones, J.H.: The history of Martian volatiles. Rev. Geophys. **35**, 1 (1997)

Jenkins, J.M., Kolodne, B.J., Butler, B.J., et al.: Microwave Remote Sensing of the Temperature and Distribution of Sulfur Compounds in the Lower Atmosphere of Venus. Icarus **158**, 312 (2002)

Jewitt, D.: Kuiper Belt Objects. Annu. Rev. Earth Planet. Sci. **27**, 287 (1999)

Jewitt, D., Luu, J., Trujillo, C.: Large Kuiper Belt Objects: The Mauna Kea 8K CCD Survey. Astron. J. **115**, 2125 (1998)

Jewitt, D., Chizmadia, L., Grimm, R., Prialnik, D.: Water in the small bodies of the solar system. In: Protostars and Planets V. Arizona Press, AZ (2007)

Johnson, T.V., Soderblom, L.A.: Io. Sci. Am. **249**, 56 (1983)

Kadomtsev, B.B., Ryazanov, A.I.: What is synergetics? Priroda **8**, 2 (1983)

Kahn, R.: The spatial and seasonal distribution of Martian clouds and some meteorological implications. J. Geophys. Res. **89**, 6671 (1984)

Kaplan, J., Yorke, J.A.: Springer lecture notes in mathematics. **730**, 228 (1978)

Kasting, J.F., Toon, O.B., Pollack, J.B.: How climate evolved on the terrestrial planets. Sci. Am. **256**, 90 (1988)

Keizer, J.: Statistical Thermodynamics of Nonequilibrium Processes. Springer, New York (1987)

Kemper, F., Vriend, W.J., Tielens, A.G.G.M.: The absence of crystalline silicates in the diffuse interstellar medium. Astrophys. J. **609**, 826 (2004)

Kerzhanovich, V.V., Marov, M.Y.: The atmospheric dynamics of venus according to Doppler measurements by the venera entry probes. In: Hunten, D.M., Colin, L., Donahue, T.M., Moroz, V.I. (eds.) Venus, p. 766. University of Arizona, Tucson (1983)

Keszthelyi, L., McEwen, A.: Magmatic differentiation of Io. Icarus **130**, 437 (1997)

Khlopkov, Y.I., Zharov, V.A., Gorelov, S.L.: Coherent Structures in a Turbulent Boundary Layer. Moscow Physicotechnical Institute, Moscow (2002) (in Russian)

Khurana, K.K., Kivelson, M.G., Stevenson, D.J., et al.: Induced magnetic fields as evidence for subsurface oceans in Europa and Callisto. Nature **395**, 777 (1998)

Kieffer, H.H., Jakosky, B.M., Snyder, C.W., Matthews, M.S.: Mars. University of Arizona, Tucson (1992)

Kippenhahn, R., Weigert, A.: Stellar Structure and Evolution. Springer, Berlin (1990)

Kita, N.T., Huss, G.R., Tachibana, S., et al.: Constraints on the origin of chondrules and cais from short-lived and long-lived radionuclides. In: Krot, A.N., Scott, E.R.D., Reipurth, B. (eds.) Chondrites and the Protoplanetary Disk, p. 558. ASP Conference Series, San Francisco (2005)

Kivelson, M.G., Khurana, K.K., Russell, C.T., et al.: Discovery of Ganymede's magnetic field by the Galileo spacecraft. Nature **384**, 537 (1996)

Kivelson, M.G., Khurana, K.K., Stevenson, D.J., et al.: Europa and Callisto: Induced or intrinsic fields in a periodically varying plasma environment. J. Geophys. Res. **104**, 4609 (1999)

Klein, T., Touboul, M., Burkhardt, C., Bourdon, B.: Dating the first ~100 Ma of the solar system: From the formation of CAIs to the origin of the Moon. Goldschmidt Conference Abstracts, Vancouver, Canada, A480, 13–18 July (2008a)

Klein, T., Touboul, M., Van Orman, J.A., Bourdon, B., Maden, C., Mezger, K., Halliday, A.N.: Hf-W thermocronometry: Closure temperature and constraints on the accretion and cooling history of H chondrites parent body. Earth Planet. Sci. Lett **270**, 106–118 (2008b)

Klein, T., Touboul, M., Bourdon, B., Nimmo, F., Mezger, K., Palme, N., Jacobsen, S., Yin, Q.-Z., Halliday, A.N.: Hf-W chronology of the accretion and early evolution of asteroids and terrestrial planets. Geochim. Cosmochim. Acta **73**, 5150–5188 (2009)

Klimontovich, Y.L.: Turbulent Motion and the Structure of Chaos: A New Approach to the Statistical Theory of Open Systems. Nauka, Moscow (1990) (in Russian)

Klimontovich, Y.L.: Introduction to the Physics of Open Systems. Janus-K, Moscow (2002) (in Russian)

Kliore, A.J., Patel, R.: The vertical structure of the atmosphere of Venus from pioneer Venus orbiter radio occultation. J. Geophys. Res. **85**, 7957 (1980)

Klypin, A., Holtzman, J., Primack, J.R., Regos, E.: Structure formation with cold plus Hot dark matter. Astrophys. J. **416**, 1 (1993)

Kolb, E.W., Turner, M.S.: The Early Universe. Addison-Wesley, New York (1990)

Kolesnichenko, A.V., Marov, M.Y.: Turbulence of Multicomponent Media. MAIK-Nauka, Moscow (1999) (in Russian)

Kolmogorov, A.N.: Local structure of turbulence in an incompressible fluid at very large reynolds numbers. Dokl. Akad. Nauk SSSR **30**, 299 (1941)

Kolmogorov, A.N.: On the conservation of conditionally periodic motions at a small change of the Hamiltonian. Dokl. Akad. Nauk SSSR **98**, 572 (1954)

Kolmogorov, A.N.: On the entropy per unit time as a metric invariant of automorphisms. Dokl. Akad. Nauk SSSR **124**, 754 (1959)

Kolmogorov, A.N.: Refining the views about the local structure of turbulence in an incompressible viscous fluid at large reynolds numbers, Mecanique de la turbulence: Colloq. Intern. CNRS, Marseille, aout, Sept. 1961, 447 (in Russian) (1962)

Krestyanikova, M.A., Shematovich, V.I.: Stochastic models of hot planetary and satellite coronas: The hot oxygen corona of mars. Astron. Vestn. **38**, 1 (2006)

Krimigis, S.M., Mitchell, D.G., Hamilton, D.C., et al.: A nebula of gases from Io surrounding Jupiter. Nature **415**, 994 (2002)

Krylov, N.S.: Works on the Substantiation of Statistical Physics. Akad. Nauk USSR, Moscow-Leningrad (1950) (in Russian)

Kuskov, O.L., Dorofeeva, V.A., Kronrod, V.A., Makalkin, A.B..: Systems of Jupiter and Saturn. URSS, Moscow (2009) (in Russian)
Kuzmin, A.D., Marov, M.Y.: Physics of the Planet Venus. Nauka, Moscow (1974) (in Russian)
Kuzmin, P.O., Zabalueva, E.V., Mitrofanov, I.G., Litvak, M.L., et al.: Seasonal redistribution of water in the surface layer of Martian regolith based on the neutron detector HEND on board Mars Odyssey. Sol. Sys. Res. **41**(2), 99 (2007)
Kwok, S.: The Origin and Evolution of Planetary Nebulae. Cambridge University Press, Cambridge (2000)
Lagrange, A.-M., Backman, D., Artymowicz, P.: Planetary material around main sequence stars. In: Mannings, V., Boss, A.P., Russell, S.S. (eds.) Protostars and Planets IV, p. 639. University of Arizona, Tucson (2000)
Landau, L.D., Lifshitz, E.M.: Continuum Mechanics. Gostekhteorizdat, Moscow-Leningrad (1944) (in Russian)
Landau, L.D., Lifshitz, V.M.: Hydrodynamics. Nauka, Moscow (1988) (in Russian)
Laskar, J.A.: Numerical experiment on the chaotic behavior of the solar system. Nature **338**, 237 (1989)
Laskar, J.A.: The chaotic motion of the solar system: A numerical estimate of the size of the chaotic zones. Icarus **88**, 266 (1990)
Levison, H., Duncan, M.: From the kuiper belt to jupiter family comets: The spatial distribution of ecliptic comets. Icarus **127**, 13 (1997)
Levison, H.F., Weissman, P.R.: The Kuiper belt. In: Weissman, P.R., McFadden, L.-A., Johnson, T.V. (eds.) Encyclopedia of the Solar System, p. 557. Academic, San Diego (1999)
Levison, H.F., Morbidelli, A., Gomes, R., Backman, D.: Planet migration in planetesimal discs. In: Protostars and Planets V. Arizona Press, AZ (2007)
Levy, E.H., Lunine, J.L.: Books-received—protostars and planets III. British Astron. Assoc. J. **103**, 266 (1993)
Lewis, J.S.: Physics and Chemistry of the Solar System. Academic, New York (1997). Revisd Edn
Liddle, A., Rand, L.: The cold dark matter density perturbations. Phys. Rep. **231**, 1 (1993)
Lieske, J.H.: Theory of motion of Jupiter's Galilean satellites. Astron. Astrophys. **56**, 333 (1977)
Limaye, S.S., Grund, C.J., Burre, S.P.: Zonal mean circulation at the cloud level of Venus: Spring and Fall 1979 OCPP observations. Icarus **51**, 416 (1982)
Linde, A.D.: Particle Physics and Inflationary Cosmology. Harwood Academic, Chur (1990)
Linde, A.D.: The self-reproducing inflationary universe. Sci. Am. **271**, 48 (1994)
Lissauer, J.J.: Planet formation. Annu. Rev. Astron. Astrophys. **31**, 129 (1993)
Lissauer, J.J., Stevenson, D.J.: Formation of giant planets. In: Protostars and Planets V. Arizona Press, AZ (2007)
Litvak, M.L., Mitrofanov, I.G., Kozyrev, A.S., et al.: Comparison between polar regions of Mars from HEND/Odyssey data. Icarus **180**(1), 23 (2006)
Liu, J., Schneider, T.: Mechanisms of jet formation on the giant planets. J. Atmos. Sci. **67**, 3652–3672 (2010)
Lorenz, E.N.: Deterministic nonperiodic flow. J. Atmos. Sci. **20**, 130 (1963)
Loskutov, A.Y.: Dynamical chaos. Systems of classical mechanics. Usp. Fiz. Nauk**177**, 989 (2007)
Lucchitta, B.K.: Antarctic ice streams and outflow channels on mars. Geophys. Res. Lett. **28**, 403 (2001)
Lucey, P.: The poles of the Moon. Elements **5**(1), 41–46 (2009)
Luhman, K.L., Joergen, V., Lad, C., et al.: The formation of Brown Dwarfs: Observations. In: Protostars and Planets V. Arizona Press, AZ (2007)
Lunine, J.: The atmospheres of Uranus and Neptune. Annu. Rev. Astron. Astrophys. **31**, 217 (1993)
Lyne, A.G., Graham-Smith, F.: Pulsar Astronomy. Cambridge University Press, Cambridge (1998)
Lyra, W., Johansen, A., Zsom, A., Klahr, H., Piskunov, N.: Planet formation bursts at the borders of the dead one in 2D numerical simulations of circumstellar disks. Astron. Astrophys. **1265**, 1–20 (2009)

Makalkin, A.B..: Radial compaction of the dust subdisk in a protoplanetary disk as possible way to gravitational instability. Lunar Planet. Sci. **25**, 827 (1994)
Makalkin, A.B..: Problems of the evolution of protoplanetary disks. In: Modern Problems of Mechanics and Space Physics. Nauka, Moscow (in Russian) (2003)
Malhotra, R.: The origin of Pluto's orbit: Implications for the solar system beyond Neptune. Astron. J. **110**, 420 (1995)
Malin, M.C., Edgett, K.S.: Evidence for recent ground water seepage and surface runoff on mars. Science **288**, 2330 (2000)
Malin, M.C., Edgett, K.S.: Sedimentary rocks of early mars. Science **290**, 1927 (2001)
Mandelbrot, B.B.: On the geometry of homogeneous turbulence, with stress on the fractal dimension of the iso-surfaces of scalars. J. Fluid Mech. **72**, 401 (1975)
Mangold, N., Allemand, P.: Topographic analysis of features related to ice on mars. Geophys. Res. Lett. **28**, 407 (2001)
Marchal, C.: The Three-Body Problem. Elsevier, Amsterdam (1990)
Marcy, G., Cochran, W., Mayor, M.: Extrasolar planets around main sequence stars. In: Mannings, V., Boss, A., Russell, S. (eds.) Protostars and Planets IV, p. 1285. University of Arizona, Arizona (2000)
Marov, M.Y.: A model of the venusian atmosphere. Dokl. Akad. Nauk SSSR **196**, 67 (1971)
Marov, M.Y.: Venus: A perspective at the beginning of planetary exploration. Icarus **16**, 415 (1972)
Marov, M.Y.: Results of venus missions. Annu. Rev. Astron. Astrophys. **16**, 141 (1978)
Marov, M.Y.: Planets of the Solar System. Nauka, Moscow (1986) (in Russian)
Marov, M.Y.: Physical properties and models of comets. Sol. Sys. Res. **28**, 5 (1994)
Marov, M.Y.: Small bodies and problems of cosmogony. Usp. Fiz. Nauk **175**, 668 (2005)
Marov, M.Y., Grinspoon, D.H.: The Planet Venus. Yale University Press, New Heaven and London (1998)
Marov, M.Y., Ipatov, S.I.: Volatile inventory and early evolution of the planetary atmospheres. In: Marov, M.Y., Rickman, H. (eds.) Collisional Processes in the Solar System, p. 223. Kluwer Acad. Publ, Dordrecht-Boston-London (2001)
Marov, M.Y., Ipatov, S.I.: Migration of dust particles and volatiles delivery to the terrestrial planets. Astron. Vestn. **40**, 5 (2006)
Marov, M.Y., Kolesnichenko, A.V.: Introduction to Planetary Aeronomy. Nauka, Moscow (1987) (in Russian)
Marov, M.Y., Kolesnichenko, A.V.: Mechanics of Turbulence of Multicomponent Gases. Kluwer Academic Publishers, Dordrecht-Boston-London (2001)
Marov, M. Ya.: Mars, atmosphere. In: Maran, S.P. (ed.) The Astronomy and Astrophysics Encyclopedia, pp. 407–411. Van Norstrand Reinhold and Cambridge Univ. Press (1994)
Marov, M.Ya., Galtsev, A.P., Shari V.P.: Estimation of the Influence of Sulfur Dioxide on Heat Transfer in the Venus Atmosphere. In: Proceedings of the 19th Lunar and Planetary Sci. Conf., p. 349 (1989)
Marov, M.Ya., Shematovich, V.I., Bisikalo, D.V., Gerard, J.-C.: Non-Equilibrium Kinetics in the Aeronomy Problems: Theory and Applications. Kluwer Academic Publishers, Dordrecht, 274, 377 (1996)
Marov, M.Ya, Kolesnichenko, A.V., Makalkin, A.B.., et al.: Modeling the evolution of a protoplanetary gas-dust disk and the mechanism of the development of gravitational instability. Report submitted to the Conference "Origin of the Biosphere and Photonics of Nucleic Acids", Terskol, 6–10 Aug 2007 (in Russian) (2007b)
Marov, M.Ya., Ipatov, S.I.: Volatiles and biogenic matter delivery by small bodies and dust particles. II International Conference "Biosphere Origin and Evolution", 28 Oct–2 Nov 2007, p. 13, Loutraki, Greece, Abstracts Book (2007a)
Marov, M.Ya., Ipatov, S.I.: Migration of dust particles and volatiles delivery to the inner planets. In: Abstracts of 35th Lunar and Planetary Science Conference, League City, TX, USA, 14–18 March, p. 1268 (2005)

Marov, M.Ya., Ipatov, S.I.: Volatiles Delivery to the Terrestrial Planets, 26th IAU General Assembly, Aug. 14–25, 2006, Prague, Czech Republic, Abstract no. JD 11–24, 359 (2006)
Marov, M.Ya., Rahe, J.: The Jupiter system. In: Andersen, J (ed.) Highlights of Astronomy, v. 11B, p. 1041. Kluwer Academic Publishers (1998)
Marov, M.Y., Shari, V.P.: Long-Wavelength Radiation Transfer in the Lower Atmosphere of Venus, Preprint no. 23. Keldysh Institute of Applied Mathematics, Academy of Sciences of USSR, Moscow (1973) (in Russian)
Marov, M.Y., Avduevsky, V.S., Borodin, N.F., et al.: Preliminary results on the venus atmosphere from the venera 8 descent module. Icarus **20**, 407 (1973)
Marov, M.Y., Byvshev, B.V., Baranov, Y.P., et al.: Nephelometric measurements on Venera-9 and Venera-10. Kosmich. Issled. **14**, 729 (1976)
Marov, M.Y., Lystsev, V.E., Lebedev, V.N., et al.: The structure and microphysical properties of the venus clouds: Venera 9, 10, and 11 data. Icarus **44**, 608 (1980)
Marov, M.Y., Galtsev, A.P., Shari, V.P.: H_2O profile in the lower atmosphere of Venus from effective flux measurements. Kosmich. Issled. **22**, 267 (1984)
Marov, M.Y., Galtsev, A.P., Shari, V.P.: Thermal radiation transfer and water content in the venusian atmosphere. Sol. Sys. Res. **19**, 15 (1985)
Marov, M.Y., Volkov, V.P., Volkov, V.P., Surkov, Y.A., Rivkin, M.L.: The lower atmosphere. In: Barsukov, V.L., Volkov, V.P. (eds.) The Planet Venus, pp. 25–68. Nauka, Moscow (1989a) (in Russian)
Marov, M.Y., Galtsev, A.P., Shari, V.P.: Thermal regime of the venusian atmosphere. In: Barsukov, V.L., Volkov, V.P. (eds.) The Planet Venus, pp. 94–132. Nauka, Moscow (1989b) (in Russian)
Marov, M.Y., Kolesnichenko, A.V., Makalkin, A.B.., et al.: From protosolar cloud to the planetary system: Model of gas-dust disk evolution. In: Galimov, E.M. (ed.) Problems of the Biosphere Origin and Evolution, p. 233. URSS, Moscow (2008)
Marsden, J.E., McCracken, M.: The Hopf Bifurcation and Its Applications. Springer, New York (1976)
Maser, J.K.: From the big bang to a Nobel Prize and further. Usp. Fiz. Nauk. **177**, 1278 (2007)
Masevich, A.G., Tutukov, A.V.: Evolution of Stars: Theory and Observations. Nauka, Moscow (1988) (in Russian)
Masser, D.: Four clues to cosmology. In: Kapitsa, S.P. (ed.) Cosmos: The Almanac, p. 13. V mire nauki, Moscow (2006) (in Russian)
Mather, J.C., Fixsen, D.J., Shafer, R.A., et al.: Calibrator design for the COBE far-infrared absolute spectrophotometer (FIRAS). Astrophys. J. **512**, 511 (1999)
Mathew, K.J., Marti, K.: Early evolution of Martian volatiles: Nitrogen and Noble Gas components in ALH84001 and chassigny. J. Geophys. Res. **106**, 1401 (2001)
Mathis, J.S.: Interstellar dust and extinction. Annu. Rev. Astron. Astrophys. **28**, 37 (1990)
Matson, D., Blaney, D.: Io. In: Weissman, P.R., McFadden, L.-A., Johnson, T.V. (eds.) Encyclopedia of the Solar System, p. 357. Academic, San Diego (1999)
Mayer, L., Governato, F., Kaufmann, T.: The Formation of Disk Galaxies in Computer Simulations, eprint arXiv: 0801.3845 (2008)
Mayor, M., Queloz, D.A.: Jupiter-mass companion to a solar-type star. Nature **378**, 355 (1995)
McCinnon, W.B., Kirk, R.L.: Triton. In: Weissman, P.R., McFadden, L.-A., Johnson, T.V. (eds.) Encyclopedia of the Solar System, p. 405. Academic, San Diego (1999)
McEwen, A.S., Keszthelyi, L., Geissler, P., et al.: Active volcanism on Io as seen by Galileo SSI. Icarus **135**, 181 (1998a)
McEwen, A.S., Keszthelyi, L., Spencer, J.R., et al.: High-temperature silicate volcanism on Jupiter's Moon Io. Science **281**, 87 (1998b)
McFadden, L.-A.: Near-Earth asteroids. In: Weissman, P.R., McFadden, L.-A., Johnson, T.V. (eds.) Encyclopedia of the Solar System, p. 607. Academic, San Diego (1999)
McKay, C.P., Davis, W.L.: Planets and the origin of life. In: Weissman, P.R., McFadden, L.-A., Johnson, T.V. (eds.) Encyclopedia of the Solar System, p. 899. Academic, San Diego (1999)

McKeegan, K.D., Davis, A.M.: In: Davis, A.M. (ed.) Meteorites, Planets, and Comets, Vol. 1, Treatise on Geochemistry, p. 431. Holland, Turekian, K.K. (eds.) Elsevier-Pergamon, Oxford (2003)

McSween Jr., H.Y.: What we have learned about mars from SNC meteorites. Meteoritics **29**, 757 (1994)

McSween Jr., H.Y.: Meteorites and Their Parent Planets. Cambridge University Press, Cambridge (1999)

Meier, R., Owen, T.C.: Cometary deuterium. Space Sci. Rev. **90**, 33 (1999)

Meier, R., Smith, B., Owen, T., Terrile, R.J.: The surface of titan from NICMOS observations with the hubble space telescope. Icarus **145**, 462 (2000)

Mellon, M.T., Feldman, W.C., Prettyman, T.H.: The presence and stability of ice in the southern hemisphere of Mars. Icarus **169**(2), 324 (2004)

Melosh, H.J., Sonett, C.R.: In: Hartman, W.K., Phillips, R.J., Taylor, G.J. (eds.) When Worlds Collide: Jetted Vapor Plumes and the Moon's Origin//Origin of the Moon, p. 621. Lunar Planet. Inst, Houston (1986)

Meyer, M.R., Backman, D.E., Weinberger, A., Wyatt, M.C.: Evolution of circumstellar discs around normal stars: Placing our solar system in context. In: Protostars and Planets V. Arizona Press, AZ (2007)

Millan-Gabet, R., Malbet, F., Akeson, R., et al.: The circumstellar environments of young stars at AU scales. In: Protostars and Planets V. Arizona Press, AZ (2007)

Mitchell, J.F.B.: The 'Greenhouse' effect and climate change. Rev. Geophys. **27**, 115 (1989)

Mitrofanov, I.G., Litvak, M.L., Kozyrev, A.S., et al.: Estimate of water abundance in the Martian soil based on the data of HEND neutron measurements by spacecraft 2001 Mars Odyssey. Sol. Sys. Res. **38**(4), 291 (2004)

Mitrofanov, I.G., 28 co-authors: Hydrogen mapping of the lunar south pole usingthe LRO neutron detector experiment LEND. Science. **330**, 483–485 (2010)

Monin, A.S.: On the turbulence spectrum in a temperature-inhomogeneous medium. Izv. Akad. Nauk SSSR Ser. Geogr. Geophys. **3**, 397 (1962)

Monin, A.S.: Weather Forecast as a Problem of Physics. Nauka, Moscow (1969) (in Russian)

Monin, A.S.: Theoretical Foundations of Geophysical Hydrodynamics. Gidrometeoizdat, Leningrad (1988) (in Russian)

Monin, A.S., Yaglom, A.M.: Statistical Hydrodynamics, vol. 1. Gidrometeoizdat, St. Petersburg (1992) (in Russian)

Monin, A.S., Yaglom, A.M.: Statistical Hydrodynamics, vol. 2. Gidrometeoizdat, St. Petersburg (1996) (in Russian)

Morbidelli, A.: New insights on the kuiper belt. Science **280**, 2071 (1998)

Morbidelli, A., Chambers, J., Lunine, J.I.: Source regions and timescales for the delivery of water to the earth. Meteor. Planet. Sci. **35**, 1309 (2000)

Morbidelli, A., Levison, H., Tsiganis, K., Gomes, R.: Chaotic capture of Jupiter's Trojan asteroids in the early solar system. Nature **435**, 462 (2005)

Moroz, V.I., Moshkin, B.E., Ekonomov, A.P., et al.: Spectrophotometric experiment onboard the venera-13 and venera-14 landers. Preliminary analysis of the absorption in H_2O bands. Kosmich. Issled **21**, 246 (1983)

Moroz, V.I., Ekonomov, A.P., Moshkin, B.Ye., et al.: In: Kliore, A.J., et al. (eds.) Venus International Reference Atmosphere, Adv. Space Res. **5**, 197 (1985)

Morrison, D. (ed.): Satellites of Jupiter. University of Arizona, Tucson (1982)

Morrison, D., Mattheus, M.S.: Satellites of Jupiter. University of Arizona, Tucson, AZ (1982)

Moser, J.: On invariant curves of area-preserving mappings on an annulus. Nachr. Akad. Wiss. Göettingen Math. Phys. **K1**, 1 (1962)

Moser, J.: Lectures on Hamiltonian Systems. American Mathematical Society, Providence (1968)

Murray, C.D.: Chaotic motion in the solar system. In: Weissman, P.R., McFadden, L.-A., Johnson, T.V. (eds.) Encyclopedia of the Solar System, p. 825. Academic, San Diego (1999)

Murray, C.D., Dermott, S.F.: Solar System Dynamics. Cambridge University Press, Cambridge (1999)

Nagasawa, M., Thommes, E.W., Kenyon, S.J., et al.: The diverse origins of terrestrial-planet systems. In: Protostars and Planets V. Arizona Press, AZ (2007)

Najita, J.R., Carr, J.S, Glassgold, A.E., Valenti, J.A.: Gaseous inner discs. In: Protostars and Planets V. Arizona Press, AZ (2007)

Nakagawa, Y., Nakagawa, K., Hayashi, C.: Growth and sedimentation of dust grains in the primordial solar nebula. Icarus **45**, 517 (1981)

Narlikar, J.V.: Introduction to Cosmology. Cambridge University Press, Cambridge (1993)

Nash, D.B., Carr, M.H., Gradie, J., et al.: In Satellites, 629 (1986)

Natta, A., Testi, L., Calvet, N., et al.: Dust in proto-planetary discs: properties and evolution. In: Protostars and Planets V. Arizona Press, AZ (2007)

Nicolis, G., Prigogine, I.: Self-Organization in Nonequilibrium Systems. Wiley, New York (1977)

O'Dell, C.R.: Observational properties of the orion proplyds. Astron. J. **115**, 263 (1998)

Obukhov, A.M.: On the energy distribution in the spectrum of turbulent flow. Izv. Akad. Nauk SSSR Ser. Geogr. Geophys **5**, 453 (1941)

Obukhov, A.M.: On the geostrophic wind. Izv. Akad. Nauk SSSR Ser. Geogr. Geophys **13**, 281 (1949)

Obukhov, A.M.: Some specific features of atmospheric turbulence. J. Fluid Mech. **13**, 77 (1962)

Ockert-Bell, M.E., Burns, J.A., Danbar, I.J., et al.: The structure of Jupiter's ring system as revealed by the Galileo imaging experiment. Icarus **138**, 188 (1999)

Offen, J.R., Kline, S.J.: A proposed model of the bursting process in turbulent boundary layer. J. Fluid Mech. **70**, 209 (1975)

Owen, T., Bar-Nun, A.: Comets, impacts and atmospheres. Icarus **116**, 215 (1995)

Owen, T.C., Bar-Nun, A.: From the interstellar medium to planetary atmospheres via comets. In: Marov, M.Y., Rickman, H. (eds.) Collisional Processes in the Solar System, p. 249. Kluwer Acad. Publ, Dordrecht-Boston-London (2001)

Owen, T., Mahaffy, P., Niemann, H.B., et al.: A Low-temperature origin for the planetesimals that formed Jupiter. Nature **402**, 269 (1999)

Oyama, V.I., Carle, G.C., Woeller, F., et al.: Pioneer venus gas chromotography of the lower atmosphere of venus. J. Geophys. Res. **85**, 7891 (1980)

Ozernoy, L.M.: Vortex theory of the origin of galaxies and their systems. In: Pikelner, S.B. (ed.) Origin and Evolution of Galaxies and Stars, p. 105. Nauka, Moscow (1976) (in Russian)

Ozernoy, L.M., Chibisov, G.V.: Dynamical parameters of galaxies as a consequence of cosmological turbulence. Sov. Astron. J. **47**, 469 (1970)

Padgett, D., Brander, W., Stapelfeldt, S.S., et al.: HST/NICMOS imaging of disks and envelopes around very young stars. Astron. J. **117**, 1490 (1999)

Paige, D.A., 26 co-authors: Diviner lunar radiometer observations of cold traps in the Moon's south polar region. Science. **330**, 479 (2010)

Papaloizou, J.C.B., Nelson, R.P., Kley, W., et al.: Disk-planet interactions during planet formation. In: Protostars and Planets V. Arizona Press, AZ (2007)

Peale, S.J., Cassen, P., Reynolds, R.T.: Melting of Io by tidal dissipation. Science **203**, 892 (1979)

Peebles, P.J.E.: Principles of Physical Cosmology. Princeton University Press, Princeton (1993)

Penzias, A.A., Wilson, R.W.: A measurement of excess antenna temperature at 4080 mc/s. Astrophys. J. **142**, 419 (1965)

Phillips, R.J., Zuber, M.T., Solomon, S.C., et al.: Ancient geodynamics and global scale hydrology on mars. Science **291**, 2587 (2001)

Pieters, C.M., 28 co-authors: Character and spatial distribution of OH/H_2O on the surface of the Moon seen by M3on Chandrayaan-1. Science. **326**(5952), 568–572 (2009)

Pikelner, S.B., Kaplan, S.A.: Foundations of the theory of star formation. The origin of first-generation stars. In: Pikelner, S.B. (ed.) Origin and Evolution of Galaxies and Stars, p. 190. Nauka, Moscow (1976) (in Russian)

Poincare, H.: Calcul des Probabilites. Gauthier-Villars, Paris (1912)

Poincare, A.: New Methods of Celestial Mechanics in Poincare A. Selected Works, vol. 1 and 2. Nauka, Moscow (1972) (in Russian)
Pollack, J.B., Sagan, C.: Planetary engineering. In: Lewis, J., Matthews, M. (eds.) Resources of Near-Earth Space. University of Arizona, Tucson (1991)
Pollack, J.B., Toon, O.B., Boese, R.: Greenhouse models of venus' high surface temperature, as constrained by pioneer venus measurements. J. Geophys. Res. **85**, 8223 (1980)
Pollack, J.B., Dalton, J.B., Grinspoon, D.H., et al.: Near infrared light from Venus' nightside: A spectroscopic analysis. Icarus **103**, 1 (1993)
Pollack, J.B., Hollenbach, D., Beckwith, S., et al.: Composition and radiative properities of grains in molecular clouds and accretion disks. Astrophys. J. **421**, 615 (1994)
Pomeau, Y., Manneville, P.: Intermittent transition to turbulence in dissipative dynamical systems. Commun. Math. Phys. **74**, 189 (1980)
Porco, C.C.: Planetary rings. In: Weissman, P.R., McFadden, L.-A., Johnson, T.V. (eds.) Encyclopedia of the Solar System, p. 457. Academic, San Diego (1999)
Porco, C.C., Nicholson, P.D., Lissauer, J.J., et al.: Neptune and triton. In: Cruikshank, D. (ed.) Neptune's Rings System, p. 703. University of Arizona, Tucson (1995)
Prigogine, I.: The End of Certainty. Time, Chaos and the New Laws of Nature. Free Press, New York (1997)
Prigogine, I., Defay, R.: Chemical Thermodynamics. Longmans Green and Co., London-New York-Toronto (1954)
Prigogine, I., Stengers, I.: Order out of Chaos. Bantam, New York (1984)
Primack, J.R., Gross, M.A.K.: In: Caldwell, D.O. (ed.) Hot Dark Matter in Cosmology. Current Aspects of Neutrino Physics. Springer, Berlin (2001)
Primack, J.R., Holtzman, J., Klypin, A., Caldwell, D.O.: Cold and hot dark matter cosmology. Phys. Rev. Lett. **74**, 2160 (1995)
Rees, D., Roper, R.G., Lloye, K.H., Low, C.H.: Determination of the structure of the atmosphere between 90 and 250 km by means of contaminant releases at Woomera, May, 1968. Philos. Trans. R. Soc. Lond. A **A271**, 631 (1972)
Reipurth, B., Jewitt, D., Keil, K. (eds.): Protostars and Planets V. University of Arizona, Tucson (2007)
Revercomb, H.E., Sromovsky, L.A., Suomi, V.E.: Net thermal radiation measurements in the atmosphere of Venus. Icarus **61**, 521 (1985)
Richard, D., Zahn, J.-P.: Turbulence in differentially rotating flow. What can be learned from the Couette-Taylor experiment. Astron. Astrophys. **347**, 734 (1999)
Richardson, L.F.: Atmospheric diffusion shown on a distance–neighbour graph. Proc. R. Soc. Lond. A **110**(756), 709 (1926)
Rieder, R.R., Gellert, R., Anderson, R.C., Bruckner, J.: Chemistry of rocks and soils at Meridian planum from the alpha particle X-ray spectrometer. Science **306**, 1746 (2004)
Rodin, A.V., Wilson, R.: Seasonal cycle of the Mars climate: Experimental data and numerical modeling. Space Res. **44**(4), 1 (2006)
Rossow, W.B., Del Genio, A.D., Limaye, S.S., et al.: Cloud morphology and motions from pioneer venus images. J. Geophys. Res. **85**, 8107 (1980)
Ruelle, D., Takens, F.: On the nature of turbulence. Commun. Math. Phys. **20**(167), 343 (1971). 23
Russell, S.S., Hartmann, L.A., Cuzzi, J.N., et al.: Timescales of the solar protoplanetary disk. In: Lauretta, D., Leshin, L.A., McSween, H. (eds.) Meteorites and the Early Solar System II, p. 233. University of Arizona, Tucson (2006)
Safronov, V.S.: Evolution of the Protoplanetary Cloud and Formation of the Earth and Planets. Nauka, Moscow (1969) (in Russian)
Safronov, V.S.: Current status of the theory of the Earth's origin. Dokl. Akad. Nauk SSSR. **6**, 5 (1982). Physics of the Earth
Safronov, V.S.: Evolution of the dust component of the circumsolar protoplanetary disk. Astron. Vestn. **21**, 216 (1987)

Sagan, C.: Reducing greenhouses and the temperature history of Earth and Mars. Nature **269**, 224 (1977)

Sagan, C., Khare, B.N., Lewis, J.S.: Organic matter in the Saturn system. In: Gehrels, T., Matthews, M.S. (eds.) Saturn, p. 788. University of Arizona, Tucson (1984)

Salpeter, E.E.: The luminosity function and stellar evolution. Astrophys. J. **121**, 161 (1955)

Samuelson, R.E.: Atmospheric thermal structure. In: Shirley, J.H., Fairbridge, R.H. (eds.) Encyclopedia of Planetary Sciences. Chapman and Hall, London (1997)

Schmidt, O.Y.: Four Lectures on the Earth's Origin, 3rd edn. USSR Academy of Sciences, Moscow (1957) (in Russian)

Schneider, N.M., Trauger, J.T.: The structure of the Io torus. Astrophys. J. **450**, 450 (1995)

Schneider, G., Smith, B.A., Becklin, E.E., et al.: NICMOS imaging of the HR 4796A circumstellar disk. Astrophys. J. **513**, L127 (1999)

Schröder, M.: Fractals, Chaos, Power Laws (Minutes from an Infinite Paradise). Freeman Company, New York (1990)

Schlichting, H., Grenzschicht-Theorie. Verlag G.: Braun. Karlsruhe. (1964)

Schubert, G.: Inside the solid planets and moons. Phys. World **10**, 45 (1997)

Schubert, G., Covey, C., Del Genio, A.D., et al.: Structure and circulation of the venus atmosphere. J. Geophys. Res. **85**, 8007 (1983)

Schubert, G., Zhang, K., Kivelson, M.G., Anderson, J.D.: The magnetic field and internal structure of Ganymede. Nature **384**, 544 (1996)

Sedov, L.I.: Similarity and Dimensional Methods in Mechanics. Nauka, Moscow (1965) (in Russian)

Seiff, A.: Thermal structure of the atmosphere of venus. In: Hunten, D.M., Colin, L., Donahue, T.M., Moroz, V.I. (eds.) Venus, p. 215. University of Arizona, Tucson (1983)

Seiff, A., Shofield, J.T., Kliore, A.J., et al.: Models of the structure of the atmosphere of venus from the surface to 100 kilometers altitude. In: Kliore, A.J., et al. (eds.) Venus International Reference Atmosphere, Adv. Space Res. **5**, 3 (1985)

Sekiya, M., Takeda, H.: Were planetesimals formed by dust accretion in the solar nebula? Earth Planets Space. **55**, 263 (2003)

Sekiya, M., Takeda, H.: Does the gas flow through a porous dust aggregate help is growth in a protoplanetary disk? Icarus. **176**, 220 (2005)

Shakura, N.I., Sunyaev, R.A.: Black holes in binary systems. Observational appearance. Astron. Astrophys. **24**, 337 (1973)

Shematovich, V.I., Tsvetkov, G.A., Krestyanikova, M.A., Marov, M.Y.: Stochastic models of hot planetary and satellite coronas: General losses of water in the Matrian atmosphere. Astron. Vestn. **41**, 113 (2007)

Shklovsky, I.S.: Stars: Their Birth, Life, and Death. Nauka, Moscow (1975) (in Russian)

Shklovsky, I.S.: The Universe, Life, and Intelligence. Nauka, Moscow (1987) (in Russian)

Shultz, P.H., 5 co-authors: The LCROSS cratering experiment. Science. **330**, 468–472 (2010)

Silk, J., Szalay, A.S., Zel'dovich, Y.B.: The large-scale structure of the universe. Sci. Am. **249**, 56 (1983)

Sinai, Y.G.: Dynamical systems with elastic reflections. Ergodic properties of scattering billiards. Usp. Mat. Nauk **25**, 141 (1970)

Sinai, Ya.G.: To the substantiation of the ergodic hypothesis for one dynamical system of statistical mechanics. Dokl. Akad. Nauk. SSSR. **6**(1963)

Sirotkin, F., Karetnikov, V.F.: Formation of protoplanetary systems through the coalescence of binary components and contraction to the main sequence. Astron. Rep. **50**, 655 (2006)

Skorov, Y.V., Marov, M.Y., Korolev, A.E.: Mass transfer in the near-surface layer of a cometary nucleus: A gas-kinetic approach. Sol. Sys. Res. **36**, 87 (2002)

Smoot, J.F.: Anisotropy of the cosmic microwave background radiation: Discovery and scientific significance. Usp. Fiz. Nauk **177**, 1294 (2007)

Smoot, G.F., Bennett, C.L., Kogut, A., et al.: Structure in the COBE differential microwave radiometer first-year maps. Astrophys. J. **396**, L1 (1992)

Spencer, J.: Europa. In: Encyclopedia of Astronomy and Astrophysics. Nature Publishing Group and Institute of Physics Publishing, Dirac House, Temple Back, Bristol (2001)
Spirin, A.S.: The RNA world and its evolution. Mol. Biol. **39**(4), 466–472 (2005)
Spirin, A.S.: When, where, and in what environment could the RNA world appear and evolve? Paleontol. J. **41**(5), 481–488 (2007)
Spirin, A.S.: Ancient RNA world. Paleontol. J. **44**(7), 1–10 (2010)
Spudis, P.D.: The Once and Future Moon. Smithsonian Institution Press, Washington (1996)
Squyres, S.W., Kasting, J.F.: Early Mars: How warm and how wet? Science **265**, 744 (1994)
Squyres, S.W., Grotzinger, J.P., Aridson, R.E., et al.: In situ evidence for an ancient aqueous environment at Meridiani Planum, Mars. Science **306**, 1709–1714 (2004)
Stratonovich, R.L.: Nonlinear Nonequilibrium Thermodynamics. Nauka, Moscow (1985) (in Russian)
Strom, S.E., Edwards, S., Skrutskie, M.F.: Evolutionary time scales for circumstellar disks associated with intermediate- and solar-type stars. In: Protostars and Planets III (A93-42937 17-90), 837 (1993)
Sunyaev, R.A., Zeidovich, Y.B.: Formation of clusters of galaxies: Protocluster fragmentation and intergalactic gas heating. Astron. Astrophys. **20**, 189 (1972)
Sunyaev, R.A., Zeldovich, Y.B.: Formation of clusters of galaxies: Protocluster fragmentation and intergalactic gas heating. Astron. Astrophys. **20**, 189 (1972)
Suomi, V.E., Sromovsky, L.A., Revercomb, H.E.: Net radiation in the atmosphere of venus: Initial results from Magellan. J. Geophys. Res. **85**, 8200 (1980)
Sussman, G.J., Wisdom, J.: Numerical evidence that the motion of pluto is chaotic. Science **241**, 433 (1988)
Sussman, G.J., Wisdom, J.: Chaotic evolution of the solar system. Science **257**, 56 (1992)
Taylor, S.R.: The Moon. In: Weissman, P.R., McFadden, L.-A., Johnson, T.V. (eds.) Encyclopedia of the Solar System, p. 247. Academic, San Diego (1999)
Team, G.P.: Galileo at Jupiter: Results from the probe. Science **272**, 837 (1996)
Tegmark, M.: Parallel universes. In: Kapitsa, S.P. (ed.) Cosmos: The Almanac, p. 21. V mire nauki, Moscow (2006) (in Russian)
Tikhonov, A.N., Arsenin, V.Y.: Methods of Solving Ill-Posed Problems. Nauka, Moscow (1986) (in Russian)
Titov, D.V., Svedhem, H., Trillard, D., Rodriguez-Canabal, J.: The Venus Express Team. 35th COSPAR Scientific Assembly, Paris, France (2004). 3963
Tokano, T.: Water cycle in atmosphere and shallow subsurface. In: Tokano, T. (ed.) Water on Mars and Life, pp. 191–216. Springer, Berlin-Heidelberg (2005)
Tomasko, M.G.: The thermal balance of the lower atmosphere of venus. In: Hunten, D.M., Colin, L., Donahue, T.M., Moroz, V.I. (eds.) Venus. University of Arizona, Tucson (1983)
Toomre, A.: On the gravitational stability of a disk of stars. Astrophys. J. **139**, 1217 (1964)
Townsend, A.A.: The Structure of Turbulent Shear Flow. Cambridge University Press, Cambridge (1956)
Trimble, V.: Existence and nature of dark matter in the universe. Annu. Rev. Astron. Astrophys. **25**, 425 (1987)
Tutukov, A.V.: Early stages of dynamical evolution of star cluster models. Astron. Astrophys. **70**, 57 (1978)
Tutukov, A.V., Pavlyuchenkov, Y.: A model of astrophysical accretion diffusion disks. Astron. Zh. **82**, 881 (2004)
Tutukov, A.V., Dremov, V.V., Dremova, G.N.: Dynamical evolution of galaxy clusters in the context of the problem of supermassive CD-galaxy formation. Astron. Zh. **84**, 427 (2007)
Van Boekel, R., Min, M., Leinert, C., et al.: The building blocks of planets within the 'Terrestrial' region of protoplanetary disks. Nature **432**, 479 (2004)
Van Mieghem, J.: Atmospheric Energetics. Clarendon Press-Oxford. (1973)
Vashkov'yak, M.A., Teslenko, N.G.: Evolution of the orbital characteristics of saturnian, uranian, and neptunian outer satellites. Astron. Vestn. **42**(2008)

Vinogradov, A.P.: Introduction to Ocean Geochemistry. Nauka, Moscow (1967) (in Russian)
Vinogradov, A.P. (ed.): Cosmochemistry of the Moon and Planets. Nauka, Moscow (1975) (in Russian)
Vinogradov, A.P. (ed.): Cosmochemistry of the Moon and Planets, Nauka, Moscow (in Russian)
Vinogradov, A.P., Volkov, V.P.: On the Wollastonite equilibrium as a mechanism defining the composition of the venusian atmosphere. Geokhimiya **7**, 755 (1971)
Volkov, V.P., Marov, M.Y., Lebedev, V.N., Sidorov, Y.I., Shari, V.P.: The clouds. In: Barsukov, V.L., Volkov, V.P. (eds.) The Planet Venus, pp. 68–94. Nauka, Moscow (1989) (in Russian)
Wadhwa, M., Russell, S.S.: Timescales of accretion and differentiation in the early solar system: The meteoritic evidence. In: Mannings, V., Boss, A.P., Russell, S.S. (eds.) Protostars and Planets IV, p. 995. University of Arizona, Tucson (2000)
Wadhwa, M., Amelin, Y., Davis, A.M., et al.: From Dust to Planetesimals: Implications for the Solar Protoplanetary Disk from Short-Lived Radionuclides in Protostars and Planets V. Arizona, AZ (2007)
Ward, W.R., Hahn, M.: Neptune's eccentricity and the nature of the kuiper belt. Science **280**, 2104 (1998)
Ward, W.R., Rudy, D.: Resonant obliquity of mars? Icarus **94**, 160 (1991)
Ward, W.R., Murray, B.C., Malin, M.C.: Climatic variations on mars. 2. Evolution of carbon dioxide atmosphere and polar caps. J. Geophys. Res. **79**, 3387 (1974)
Wasserburg, G.J.: Short-lived nuclei in the early solar system. In: Black, D.C., Matthews, M.S. (eds.) Protostars and Planets II, p. 703. University of Arizona, Tucson (1985)
Wasserburg, G.J., Papanastassiou, D.A.: Some short-lived nuclides in the early solar system—connection with the original interstellar medium. In: Barnes, C.A., et al. (eds.) Essays in Nuclear Astrophysics, p. 85. Cambridge University Press, Cambridge (1982)
Watson, L.L., Hutcheon, I.D., Epstein, S., Stolper, E.M.: Water on mars: Clues from deuterium/hydrogen and water contents of hydrous phases in SNC meteorites. Science **265**, 86 (1994)
Watson, A.M., Stapelfeldt, K.R., Wood, K., Menard, F.: Multi-wavelength imaging of young stellar object discs: Toward an understanding of disc structure and dust grain evolution. In: Protostars and Planets V. Arizona Press, AZ (2007)
Weidenschilling, S.J.: The origin of comets in the solar nebula: A unified model. Icarus **127**, 290 (1997)
Weidenschilling, S.J.: Formation of planetesimals in the solar nebula. In: Levy, E.H., Lunine, J.I. (eds.) Protostars and Planets III. University of Arizona, Tucson (2000)
Weidenschilling, S.J.: Formation of planetesimals and accretion of the terrestrial planets. Space Sci. Rev. **92**, 295 (2000)
Weissman, P.R.: In: T.W. Rettig, T.W., Hahn, J.M. (eds.) The Oort Cloud in Completing the Inventory of the Solar System. ASP Conference Series 107, 265 (1996)
West, R.A.: Atmospheres of the giant planets. In: Weissman, P.R., McFadden, L.-A., Johnson, T.V. (eds.) Encyclopedia of the Solar System, p. 315. Academic, San Diego (1999)
West, R.A., Strobel, D.F., Tomasko, M.G.: Clouds, aerosols and photochemistry in the Jovian atmosphere. Icarus **65**, 161 (1986)
Wetherill, G.W.: Formation of the earth. Annu. Rev. Earth Planet. Sci. **18**, 205 (1990)
Wetherill, G.W., Stewart, G.R.: Accumulation of a swarm of small planetesimals. Icarus **77**, 330 (1989)
Whipple, F.L.: A comet model. II. Physical relations for comets and meteors. Astrophys. J. **113**, 464 (1951)
White, S.D.M., Frenk, C.S., Davis, M.: Clustering in a neutrino-dominated universe. Astrophys. J. **274**, L1 (1983)
Whitworth, A., Bate, M.R., Nordlund, A., et al.: The formation of brown dwarfs: Theory. In: Protostars and Planets V. Arizona Press, AZ (2007)
Wilhelms, D.E.: To a Rocky Moon: a Geologist's History of Lunar Exploration. University of Arizona Press, Tucson (1993)
Wilson, L.: Planetary volcanism. In: Weissman, P.R., McFadden, L.-A., Johnson, T.V. (eds.) Encyclopedia of the Solar System, p. 877. Academic, San Diego (1999)
Wisdom, J.: Urey prize lecture: Chaotic dynamics in the solar system. Icarus **72**, 241 (1987)

Woo, R., Armstrong, J.W., Kliore, A.J.: Small scale turbulence in the atmosphere of venus. Icarus **52**, 12 (1982)
Wooden, D., Desch, S., Harker, D., et al.: Comet grains and implications for heating and radial mixing in the protoplanetary disk. In: Protostars and Planets V. Arizona Press, AZ (2007)
Youdin, A.N., Shu, F.: Planetesimal formation by gravitational instability. Astrophys. J. **580**, 494 (2002)
Zaslavsky, G.M.: Physics of chaos in hamiltonian systems. In: Regular and Chaotic Dynamics. Research Center, Institute of Computer Research, Moscow-Izhevsk (in Russian) (2004)
Zel'dovich, Y.B.: On the friction of fluids between rotating cylinders. Proc. R. Soc. Lond. A **A374**, 299 (1981)
Zel'dovich Ya.B.: Structure of the universe. In: Sunaev, R.A. (ed.) Results of Sciences and Engineering. Astronomy, 22, 4 (in Russian) (1983)
Zel'dovich, Y.B., Raizer, Y.P.: Physics of Shock Waves and High-Temperature Hydrodynamic Phenomena. Nauka, Moscow (1966) (in Russian)
Zharkov, V.N., Gudkova, T.V.: Construction of Martian interior model. Sol. Sys. Res. **39**, 343 (2005)
Zhithik, I.A., Kuzin, S.V., Sobelman, I.I., et al.: Main results of the SPIRIT experiment onboard the CORONAS-F satellite. Sol. Sys. Res. **39**, 442 (2005)
Zuber, M.T., Solomon, S.C., Phillips, R.J., et al.: Internal structure and early thermal evolution of mars from mars global surveyor topography and gravity. Science **287**, 1788 (2000)
Zuckerman, B., Becklin, E.E.: Submillimeter studies of main-sequence stars. Astrophys. J. **414**, 793 (1993)
Zurek, R.W., Barnes, J.R., Haberle R.M., et al.: Dynamics of the atmosphere of Mars. In: Kieffer, H.H., et al. (eds.) Mars, 835 pp. University of Arizona Press, Tucson (1992)

Chapter 2

Anishchenko, V.S., Astakhov, V.V., Vadivasova, T.E., et al.: Nonlinear Effects in Chaotic and Stochastic Systems. Institute of Computer Research, Moscow-Izhevsk (in Russian) (2003)
Chapman, S., Cowling, T.G.: The Mathematical Theory of Non-uniform Gases. Cambridge University, Cambridge (1952)
Coleman, B.D.: Arch. Rat. Mech. Anal **17**, 1 (1964)
Curtiss, C.F.: Symmetric Gaseous Diffusion Coefficients. J. Chem. Phys. **49**, 2917 (1968)
de Groot, S., Mazur, P.: Non-equilibrium Thermodynamics. North-Holland, Amsterdam (1962)
Ferziger, J., Kaper, H.: Mathematical Theory of Transport Processes in Gases. North-Holland, Amsterdam (1972)
Frank-Kamenetskii, D.A.: Diffusion and Heat Transfer in Chemical Kinetics. Nauka, Moscow (in Russian) (1987)
Gyarmati, I.: Non-equilibrium Thermodynamics. Field Theory and Variational Principles. Springer, New York (1970)
Hirschfelder, D., Curtiss, C., Bird, R.: Molecular Theory of Gases and Liquids. John Wiley and Sons, New York and London (1954)
Jou, D., Casas-Vazquez, J., Lebon, G.: Extended Irreversible Thermodynamics. Springer, Berlin (2001)
Keizer, J.: Statistical Thermodynamics of Nonequilibrium Processes. Springer, New York (1987)
Kolesnichenko, A.V., Marov, M.Y.: Turbulence of Multicomponent Media. MAIK-Nauka, Moscow (in Russian) (1999)
Kolesnichenko, A.V., Tirsky, G.A.: Stefan-Maxwell relations and heat flux for imperfect multicomponent continuous media. Chisl. Metod. Mekh. Splosh. Sredy **7**, 4 (1976)

Kondepudi, D., Prigogine, I.: Modern Thermodynamics, From Heat Engines to Dissipative Structures. John Wiley and Sons, New York (1998)
Lapin, Yu.V., Strelets, N. Kh., Shchur, N.L.: Calculation of Heat and Mass Transfer in Flows of Viscous Nonequilibrium Gas Mixtures with Multicomponent Diffusion Effects in Heat and Mass Transfer VII, vol. 3, p. 121. Institute of Heat and Mass Transfer of the Academy of Sciences of BSSR, Minsk (in Russian) (1984)
Marov, M.Y., Kolesnichenko, A.V.: Introduction to Planetary Aeronomy. Nauka, Moscow (in Russian) (1987)
Mason, E.A.: The Onsager Reciprocal Relations. Experimental Evidence in Foundations of Continuum Thermodynamics. Mac-Millan, London and Basingstoke (1974)
Mason, E.A., Saxena, S.C.: Approximate formula for the conductivity of gas mixtures. Phys. Fluids **1**, 361 (1958)
Maxwell, J.C.:In: Niven, W.D. (ed.) The Scientific Papers of James Clerk Maxwell, vol. 1, p. 377. Cambridge University, Cambridge, (1890)
Miller, D.G.: The Onsager Relations. Experimental Evidence in Foundations of Continuum Thermodynamics. Mac-Millan, London and Basingstoke (1974)
Muckenfuss, C., Curtiss, C.F.: Thermal conductivity of multicomponent gas mixtures. J. Chem. Phys. **29**, 1273 (1958)
Munster, A.: Chemical Thermodynamics. Editorial URSS, Moscow (in Russian) (2002)
Nigmatullin, R.I.: Dynamics of Multiphase Media. Nauka, Moscow (in Russian) (1987)
Noll, W.: The Foundations of Mechanics and Thermodynamics. Springer, Berlin (1974)
Oran, E., Boris, J.: Numerical Simulation of Reactive Flow. Cambridge University, Cambridge (2001)
Prigogine, I., Defay, R.: Chemical Thermodynamics. Longmans Green and Co., London–New York–Toronto (1954)
Sedov, L.I.: Introduction to Continuum Mechanics. Fizmatlit, Moscow (in Russian) (1962)
Sedov, L.I.: Continuum Mechanics. Nauka, Moscow (in Russian) (1984)
Stefan, J.: Wien Sitzungsber **63**, 63 (1871)
Stiller, W.: Arrhenius Equation and Non-equilibrium Kinetics. Teubner, Leipzig (1989)
Truesdell, C.: Rational Thermodynamics. McGraw-Hill, New York (1984)

Chapter 3

Barenblatt, G.I.: Similarity, Self-Similarity, Intermediate Asymptotics (The Theory and Applications to Geophysical Hydrodynamics). Gidrometeoizdat, Moscow (in Russian) (1978)
Blackadar, A.K.: Extension of the Laws of Thermodynamics to Turbulent System, J. Meteorology **12**, 165 (1955)
Boussinesq, J. Essai sur la theoriedes eaux courantes. Memoires presentees par diverses Savants a l'Acad. d. Sci. Paris, 23, 46 (1977)
Bradshaw, P.: The analogy between streamline curvature and buoyancy in turbulent shear flow. J. Fluid Mech. **36**, 117 (1969)
Brasseur, G., Solomon, S.: Aeronomy of the Middle Atmosphere. Reidel, Dordrecht (1984)
Bruyatsky, E.V.: Turbulent Stratified Jet Flows. Naukova Dumka, Kiev (in Russian) (1986)
Davydov, B.I.: On statistical turbulence. Dokl. Nauk USSR **127**, 980 (1959)
Davydov, B.I.: On statistical dynamics of incompressible turbulent fluid. Dokl. Akad. Nauk USSR **136**, 47 (1961)
de Groot, S., Mazur, P.: Non-equilibrium thermodynamics. North-Holland, Amsterdam (1962)
Favre, A.: Statistical equations of turbulent gases in problems of hydrodynamics and continuum mechanics. SIAM, Philadelphia (1969). 231
Frost, W., Moulden, T. (eds.): Handbook of Turbulence, Plenum, New York and London (1977)

Gyarmati, I.: Non-equilibrium Thermodynamics. Field Theory and Variational Principles. Springer, Berlin (1970)
Haken, H.: Advanced Synergetics. Instability Hierarchies of Self-Organizing Systems and Devices. Springer, Berlin (1983)
Haken, H.: Information and Self-Organization. A Macroscopic Approach to Complex Systems. Springer, Berlin (1988)
Hinze, J.O.: Turbulence. Fizmatlit, Moscow (in Russian) (1963)
Ievlev, V.M.: Turbulent Motion of High-Temperature Continuous Media. Nauka, Moscow (in Russian) (1975)
Ievlev, V.M.: Numerical Simulation of Turbulent Flows. Nauka, Moscow (in Russian) (1990)
Jou, D., Casas-Vazquez, J., Lebon, G.: Extended Irreversible Thermodynamics. Springer, Berlin (2001)
Kolesnichenko, A.V.: Stefan-Maxwell Relations and Heat Flow for Turbulent Multicomponent Continuous Media in Problems of Modern Mechanics, p. 52. Moscow State University, Moscow (in Russian) (1998)
Kolesnichenko, A.V., Marov, M.Y.: Turbulence of Multicomponent Media. MAIK-Nauka, Moscow (in Russian) (1999)
Kolesnichenko, A.V. and Marov, M.Ya. (1979) Influence of Turbulence on the Structure and Energetics of the Lower Thermosphere of a Planet, Preprint no. 175, Institute of Applied Mathematics, Academy of Sciences of USSR, Moscow (in Russian).
Kolesnichenko, A.V.: Methods of Nonequilibrium Thermodynamics for Describing Turbulent Multicomponent Hydrothermodynamic Systems with Chemical Reactions, Preprint no. 66, Institute of Applied Mathematics, Academy of Sciences of USSR, Moscow (in Russian) (1980)
Kolesnichenko, A.V., Marov, M.Ya.: Modeling Turbulent Transport Coefficients in Aeronomy Problems, Preprint no. 20, Institute of Applied Mathematics, Academy of Sciences of USSR, Moscow (in Russian) (1980)
Kolesnichenko, A.V., Marov, M.Ya.: Second-Order Modeling of Turbulent Exchange Coefficients for Shear Flows of Multicomponent Compressible Media, Preprint no. 31, Institute of Applied Mathematics, Academy of Sciences of USSR, Moscow (in Russian) (1984)
Kolmogorov, A.N.: Local structure of turbulence in an incompressible fluid at very large Reynolds numbers. Dokl. Akad. Nauk USSR **30**, 299 (1941)
Kolmogorov, A.N.: Equations of turbulent motion for an incompressible fluid. Izv. Akad. Nauk. USSR Ser. fiz. **6**, 56 (1942)
Kompaniets, V.Z., Ovsyannikov, A.A., Polack, L.S.: Chemical Reactions in Turbulent Gas and Plasma Flows. Nauka, Moscow (in Russian) (1979)
Laikhtman, G.L.: Physics of the Atmospheric Boundary Layer. Gidrometeoizdat, Leningrad (in Russian) (1970)
Lapin, Y.V., Strelets, N.K.: Internal Flows of Gas Mixtures. Nauka, Moscow (in Russian) (1989)
Launder, B.E.: On the effects of gravitational field on the turbulent transport of heat and momentum. J. Fluid Mech. **67**, 569 (1975)
Launder, B.E., Spalding, D.B.: Mathematical Model of Turbulence. Acadamic, London (1972)
Launder, B.E., Morse, A.:In: Durst F. et al. (eds.) Numerical Calculation of Axisymmetrical Free Shear Flows with the Use of Short Closings for Stresses in Turbulent Shear Flows I, p. 291. Springer, Berlin (1979)
Lewellen, W.: In: Frost, W., Moulden, T (eds.) Method of Invariant Simulation in Handbook of Turbulence, p. 262. Plenum, New York and London (1977)
Libby, P.A., Williams, F.A.: Turbulent Reacting Flows. Academic Press, New York (1994)
Loitsyanskii, L.G.: Mechanics of Liquids and Gases. Nauka, Moscow (in Russian) (1978)
Lumley, J.L., Panofsky, H.A.: The Structure of Atmospheric Turbulence. John Wiley and Sons, New York (1964)
Marov, M.Y., Kolesnichenko, A.V.: Introduction to Planetary Aeronomy. Nauka, Moscow (in Russian) (1987)

Methods for Calculating Turbulent Flows (1984), Ed. by V. Kollman, Mir, Moscow (in Russian).
Monin, A.S., Yaglom, A.M.: Statistical Hydrodynamics. Gidrometeoizdat, St. Petersburg (in Russian) (1992)
Morkovin, M.V.: Effects of Compressibility on Turbulent flow in Mechanics of Turbulence. Gordon and Breach, New York (1961). 367
Munk, W.H., Anderson, E.R.: Notes on the Theory of the Thermocline, J. Marine Res. **1**, 276 (1948)
Munster, A.: Chemical Thermodynamics. Editorial URSS, Moscow (in Russian) (2002)
Nevzglyadov, V.G.: On the phenomenological theory of turbulence. Dokl. Akad. Nauk USSR **47**, 169 (1945)
Nevzglyadov, V.G.: On the statistical theory of turbulence. Dokl. Akad. Nauk USSR **47**, 482 (1945)
Nikuradze (1936)
Prandtl, L.: Bericht uber untersuchungen zur ausgebildeten turbulenz. Z. Angew. Math. Mech. **5**, 136 (1925)
Prandtl, L.: Bemerkungen zur theorie der freien turbulenz. Z. Angew. Math. Mech. **22**, 241 (1942)
Prandtl, L.: Uber rein neyes formelsystem fur die ausgebildete turbulenz. Nachr. Akad. Wiss. Gottingen, Math.-Phys. **K1**, 6 (1945)
Problems of Turbulence. Collection of Works, Institute of Computer Research; "Regular and Chaotic Dynamics" Research Center, Moscow-Izhevsk (in Russian) (2006)
Rosten, H., Spalding, B.: PHOENICS: Beginners Guide; User Manual; Photo User Guide. Concentration Heat and Momentum Ltd, London (1987)
Sedov, L.I.: On Promising Directions and Problems in Continuum Mechanics in Thoughts about Science and Scientists, p. 173. Nauka, Moscow (in Russian) (1980)
Sedov, L.I.: Continuum Mechanics. Nauka, Moscow (in Russian) (1984)
Taylor, G.I.: The transport of vorticity and head through fluids in turbulent motion. Proc. Roy. Soc. London **135**, 685 (1932)
Townsend, A.A.: The Structure of Turbulent Shear Flow. Cambridge University, Cambridge (1956)
Van Mieghem, J.: Atmospheric Energetics. Clarendon, Oxford (1973)

Chapter 4

Andre, J.C., De Moor, G., Vachat, R.: Turbulence approximation for inhomogeneous flow: Part II. The numerical simulation of a penetrative convection experiment. J. Atmos. Sci. **33**, 482 (1976)
Borgi, R.: Models for Numerical Calculations of Turbulent Combustion in Prediction Methods for Turbulent Flows. Hemisphere Publ, Corp. (1980). 329
Deardorff, J.W.: The use of subgrid transport equations in a three-dimensional model of atmospheric turbulence. J. Fluids Eng. **95**, 429 (1973)
de Groot, S., Mazur, P.: Non-equilibrium Thermodynamics. North-Holland, Amsterdam (1962)
Donaldson, S.R.: Calculation of turbulent flows in the atmosphere and an isolated vortex. Raketn. Tekhn. Kosmonavt. **10**, 4 (1972)
Frost, W., Moulden, T. (eds.): Handbook of Turbulence, Plenum, New York and London (1977)
Ievlev, V.M.: Turbulent Motion of High-Temperature Continuous Media. Nauka, Moscow (in Russian) (1975)
Ievlev, V.M.: Numerical Simulation of Turbulent Flows. Nauka, Moscow (in Russian) (1990)
Keller, L.V.: Theory of Convection and Turbulence, Trudy GGO, 4 (1930)

Keller, L.V., Friedman, A.A.: Differentialgleichungen fur die turbulente Bewegung einer kompressiblen Flussigkeit. In: Proceedings of the Ist International Congress Applied Mechanics, Delft, 395. 1924

Kolesnichenko, A.V.: Methods of continuum mechanics for describing turbulent multicomponent mixtures with chemical reactions and heat and mass transfer. In: Proceedings of V USSR Meeting on Theoretical and Applied Mechanics, Abstracts of Reports, Alma-Ata, 123. 1981

Kolesnichenko, A.V., Marov, M.Ya.: To the Closure Problem in the Theory of Turbulent Shear Flows of Multicomponent Mixtures of Chemically Active Gases, Preprint no. 31, Institute of Applied Mathematics, Academy of Sciences of USSR, Moscow (in Russian) (1984)

Kolesnichenko, A.V., Marov, M.Y.: Turbulence of Multicomponent Media. MAIK-Nauka, Moscow (in Russian) (1999)

Kolmogorov, A.N.: Local structure of turbulence in an incompressible fluid at very large reynolds numbers. Dokl. Akad. Nauk USSR **30**, 299 (1941)

Kolmogorov, A.N.: Equations of turbulent motion for an incompressible fluid. Izv. Akad. Nauk. USSR Ser. fiz. **6**, 56 (1942)

Kuznetsov, V.P.: Influence of Temperature and Concentration Fluctuations on the Mean Chemical Reaction Rate in a Turbulent Flow in the Second USSR Symposium on Combustion and Detonation, Institute of Chemical Physics, Academy of Sciences of USSR, Moscow, 99 (in Russian) (1969)

Launder, B.E.: On the effects of gravitational field on the turbulent transport of heat and momentum. J. Fluid Mech. **67**, 569 (1975)

Launder, B.E.: Heat and mass transport in turbulence. In: Bradshaw (ed.) Topics in Applied Physics. Springer, New York (1976). 12, 232

Launder, B.E., Morse, A.: In: Durst et al. (eds.) Numerical Calculation of Axisymmetrical Free Shear Flows with the Use of Short Closings for Stresses in Turbulent Shear Flows I. Springer, Berlin, 291 (1979)

Launder, B.E., Spalding, D.B.: Mathematical Model of Turbulence. Acadamic, London (1972)

Lewellen, W.: Method of Invariant Simulation. In: Frost, W., Moulden, T. (eds.) Handbook of Turbulence. Plenum Press, New York and London (1977). 262

Lykosov, V.N.: On the Antigradient Moment Transfer in a Low-Level Jet Flows. Izv. Akad. Nauk USSR, Phys. Atmosphere Ocean **27**, 325 (1991)

Marov, M.Y., Kolesnichenko, A.V.: Introduction to Planetary Aeronomy. Nauka, Moscow (in Russian) (1987)

Mellor, G.L., Yamada, T.: A hierarchy of turbulence closure models for planetary boundary layers. J. Atmos. Sci. **31**, 1791 (1974)

Mellor, G.L., Yamada, T.: Development of a turbulence closure model for geophysical fluid problems. Rev. Geophys. Space Phys. **20**, 851 (1982)

Monin, A.S., Yaglom, A.M.: Statistical Hydrodynamics. Gidrometeoizdat, St. Petersburg (in Russian) (1971)

O'Brien, E.E.: In: Libby, P.A., Williams, F.A. (eds.) Method of Probability Density Functions in the Theory of Turbulent Flows with Chemical Reactions in Turbulent Reacting Flows. Springer, Berlin (1980)

Prigogine, I., Defay, R.: Chemical Thermodynamics. Longmans Green and Co., London–New York–Toronto (1954)

Rotta, J.: Statistische theorie nichthomogener turbulenz teil 1. Physik **129**, 547 (1951)

Rotta, J.: Statistische theorie nichthomogener turbulenz teil 2. Physik **131**, 51 (1951)

Libby, P.A., Williams, F.A. (eds.): Turbulent Reacting Flows. Springer, Berlin (1980)

Durst, F.et al. (eds.): Turbulent Shear Flows I. Springer, Berlin (1979)

Turbulence: Principles and Applications, Moscow, MIR PH (1980) (original in: Frost, W., Moulden, T. (eds.) Handbook of Turbulence. Plenum Press, New York and London (1977))

Williams, F.A.: Combustion Theory. Addison-Wesley, Palo Alto–London (1965)

Chapter 5

Batchelor, G.K.: The Theory of Homogeneous Turbulence. Cambridge University Press, Cambridge (1953)
Batchelor, G.K., Townsend, A.A.: The nature of turbulent motion at high wave number. Proc. Roy. Soc. London **A199**, 238 (1949)
Belotserkovskii, O.M., Oparin, A.M.: Numerical Experiment in Turbulence: From Order to Chaos. Nauka, Moscow (in Russian) (2000)
Berge, P., Pomeau, Y., Vidal, C.: Order within Chaos. Hermann, Paris (1984)
Blackadar, A.K.: Extension of the laws of thermodynamics to turbulent system. J. Meteorol. **12**, 165 (1955)
Boiko, A.V., Grek, G.R., Dovgal, A.V., Kozlov, V.V.: Physical Mechanisms of the Transition to Turbulence in Open Flows. "Regular and Chaotic Dynamics" Research Center, Institute of Computer Research, Moscow-Izhevsk (in Russian) (2006)
Brown, G.L., Roshko, A.: On density effects and large structures in turbulent mixing layers. J. Fluid Mech. **64**, 775 (1974)
Crow, S.C., Champagne, F.H.: Orderly structures in jet turbulence. J. Fluid Mech. **48**, 547 (1971)
Crownover, R.M.: Introduction to Fractals and Chaos. Jones and Bartlett, Boston (1995)
de Groot, S., Mazur, P.: Non-equilibrium Thermodynamics. North-Holland Publishing Company, Amsterdam (1962)
Demidovich, B.P.: Lectures on the Mathematical Theory of Stability. Moscow State University, Moscow (in Russian) (1998)
Druden, H.I.: Recent advances in the mechanics of boundary layer flow. Adv. Appl. Mech. **1**, 1 (1948)
Ebeling, W.: Formation of Structures in Irreversible Processes: An Introduction to the Theory of Dissipative Structures. Institute of Computer Research, "Regular and Chaotic Dynamics" Research Center, Moscow-Izhevsk (in Russian) (2004)
Ebeling, W., Engel, A., Feistel, R.: Physics of Evolution Processes. A Synergetic Approach, Editorial USSR, Moscow (in Russian) (2001)
Ferziger, J., Kaper, H.: Mathematical Theory of Transport Processes in Gases. North-Holland Publishing Company, Amsterdam-London (1972)
Frenkiel, F.N., Klebanoff, P.S.: On the log-normality of the small-scale structure of turbulence. Bound. Layer Meteorol. **8**, 173 (1975)
Frisch, U.: The Legacy of A.N. Kolmogorov. Cambridge University Press, Cambridge (1995)
Frisch, U., Sulem, P.L., Nelkin, M.: A simple dynamical model of intermittent fully developed turbulence. J. Fluid Mech. **87**, 719 (1978)
Graham, R.: Stochastic Methods in Nonequilibrium Thermodynamics in Stochastic Nonlinear Systems in Physics, Chemistry, and Biology. In: Arnold, L., Lefever, R. (eds.) Proceedings of the Workshop, Bielefeld, Fed. Rep. of Germany, Springer, Heidelberg, 202, 5–11 October 1980 (1981)
Haken, H.: Synergetics. Springer, Heidelberg-Berlin-New York (1981)
Haken, H.: Advanced Synergetics. Instability Hierarchies of Self-Organizing Systems and Devices. Springer, Berlin (1983)
Haken, H.: Information and Self-Organization. Springer, Berlin-Heidelberg-New York (1988)
Jou, D., Casas-Vazquez, J., Lebon, G.: Extended Irreversible Thermodynamics. Springer, Verlag Berlin Heidelberg (2001)
Keizer, J.: Statistical Thermodynamics of Nonequilibrium Processes. Springer, Verlag New York (1987)
Khlopkov, Y.I., Zharov, V.A., Gorelov, S.L.: Coherent Structures in a Turbulent Boundary Layer. Moscow Physicotechnical Institute, Moscow (in Russian) (2002)
Klimontovich, Y.L.: Statistical Theory of Open Systems. Janus, Moscow (in Russian) (1995)
Klimontovich, Y.L.: Introduction to the Physics of Open Systems. Janus-K, Moscow (in Russian) (2002)

Klyatskin, V.I., Tatarskii, V.I.: The approximation of a diffusion random process in some nonstationary stochastic problems of physics. Usp. Fiz. Nauk **110**, 499 (1973)
Kolesnichenko, A.V.: Synergetic approach to describing developed turbulence. Astron. Vestn. **36**, 121 (2002)
Kolesnichenko, A.V.: Thermodynamic modeling of developed structured turbulence with energy dissipation fluctuations. Astron. Vestn. **38**, 144 (2004)
Kolesnichenko, A.V., Marov, M.Y.: Methods of nonequilibrium thermodynamics for description of multicomponent turbulent gas mixtures. Arch. Mech. **37**, 3 (1985)
Kolesnichenko, A.V., Marov, M.: Ya. Turbulence of Multicomponent Media, MAIK-Nauka, Moscow (in Russian) (1999)
Kolesnichenko, A.V.: Methods of Nonequilibrium Thermodynamics for Describing Turbulent Multicomponent Hydrothermodynamic Systems wth Chemical Reactions, Preprint No. 66, Institute of Applied Mathematics, Academy of Sciences of USSR, Moscow (in Russian) (1980)
Kolesnichenko, A.V.: Stefan-Maxwell Relations and Heat Flow for Turbulent Multicomponent Continuous Media in Problems of Modern Mechanics, Moscow State University, Moscow, 52 (in Russian) (1998)
Kolesnichenko, A.V.: Synergetic Approach to Describing Stationary-Nonequilibrium Turbulence of Astro- and Geophysical Systems, Preprint No. 6, Institute of Applied Mathematics, Academy of Sciences OF USSR, Moscow (in Russian) (2003)
Kolmogorov, A.N.: Precisions sur la structure locale de la turbulence dans un fluide visqueux nombres de Reynolds eleves. In: La turbulence en mecanique des fluids. A. Favre, L.S.G. Kovasznay, R. Dumas, J. Gaviglio and M. Coantic. (Eds.) Gauthier-Villars, Paris (1961) pp. 447–451
Kolmogorov, A.N.: Local structure of turbulence in an incompressible fluid at very large Reynolds numbers. Dokl. Akad. Nauk USSR **30**, 299 (1941)
Kolmogorov, A.N.: Refining the Views about the Local Structure of Turbulence in an Incompressible Viscous Fluid at Large Reynolds Numbers, Mecanique de la turbulence: Colloq. Intern. CNRS, Marseille, aout - sept. 1961, 447 (in Russian) (1962)
Kondepudi, D., Prigogine, I.: Modern Thermodynamics, from Heat Engines to Dissipative Structures. John Wiley and Sons, New York, Weinheim, Brisbane, Toronto, Singapore (1999)
Landau, L.D., Lifshitz, V.M.: Hydrodynamics. Nauka, Moscow (in Russian) (1988)
Mandelbrot, B.B.: Intermittent turbulence in self-similar cascades: Divergence of high moments and dimension of the carrier. J. Fluid Mech. **62**, 401 (1974)
Mandelbrot, B.: The Fractal Geometry of Nature. Freeman and Company, New York (1983)
Mandelbrot, D.D., Van Ness, J.W.: Fractional Brownian Motions. Fractional Noises and Applications, SIAM Review **10**, 422 (1987)
Marov, M.Y., Kolesnichenko, A.V.: Mechanics of Turbulence of Multicomponent Gases. Kluwer Academic, Dordrecht–Boston–London (2001)
Marsden, J.E., McCracren, M.: The Hopf Bifurcation and Its Applications, Applied Math. Sci., Vol. 19. Springer, Berlin–Heidelberg–New York (1976)
Millionshchikov, M.D.: To the theory of homogeneous and isotropic turbulence. Izv. Akad. Nauk USSR, Ser. Geogr. Geoph. **5**, 433 (1941)
Monin, A.S., Yaglom, A.M.: Statistical Fluid Mechanics, vol. 2. Cambridge, Mass., USA: MIT Press (1975)
Monin, A.S., Yaglom, A.M.: Statistical Hydrodynamics, vol. 1. Gidrometeoizdat, St. Petersburg (in Russian) (1992)
Monin, A.S., Yaglom, A.M.: Statistical Hydrodynamics, vol. 2. Gidrometeoizdat, St. Petersburg (in Russian) (1996)
Monin, A.S., Polubarinova-Kochina, P.Y., Khlebnikov, V.I.: Cosmology, Hydrodynamics, Turbulence: A.A. Fridman and the Development of his Legacy. Nauka, Moscow (in Russian) (1989)
Munster, A.: Chemical Thermodynamics. Editorial URSS, Moscow (in Russian) (2002)
Nakhushev, A.N.: Fractional Calculus and Its Applications. Fizmatlit, Moscow (in Russian) (2003)

Nigmatullin, R.R.: The realization of generalized transfer equation in a medium with fractal geometry. Phys. Status Solidi B **133**, 425 (1986)
Nigmatullin, R.R.: Fractional integral and its physical interpretation. Teor. Mat. Fiz. **90**, 354 (1992)
Novikov, E.A., Stewart, R.U.: Intermittency of turbulence and the spectrum of energy dissipation fluctuations. Izv. Akad. Nauk USSR, Ser. Geoph. **3**, 408 (1964)
Obukhov, A.M.: On the energy distribution in the spectrum of turbulent flow. Izv. Akad. Nauk USSR, Serg. Geogr. Geophys. **5**, 453 (1941)
Obukhov, A.M.: Some specific features of atmospheric turbulence. J. Fluid Mech. **13**, 77 (1962)
Oldham, K.B., Spanier, J.: The Fractional Calculus (Theory and Applications of Differentiation and Integration to Arbitrary Order). Acadamic, New York (1974)
Olemskii, A.I.: Theory of stochastic systems with singular multiplicative noise. Usp. Fiz. Nauk **168**, 287 (1998)
Olemskii, A.I., Flat, A.Y.: Using the concept of a fractal in condensed matter physics. Usp. Fiz. Nauk **163**, 1 (1993)
Onsager, L.: Statistical hydrodynamics. Nuovo Cimento **6**((Suppl.)), 279 (1949)
Prigogine, I.: Introduction to Thermodynamics of Irreversible Processes. John Wiley & Sons, New York (1967)
Prigogine, I., Stengers, I.: Order out of chaos. Heinemann, London (1984)
Prigogine, I., Stengers, I.: Time. Chaos. Quantum. Toward Solving the Paradox of Time. Progress Publishing House, Moscow (in Russian) (1994)
Rabinovich, M.I., Sushchik, M.M.: Regular and chaotic dynamics of structures in fluid flows. Usp. Fiz. Nauk **160**, 1 (1990)
Samko, S.G., Kilbas, A.A., Marichev, O.I.: Fractional-Order Integrals and Derivatives and Their Applications, Nauka I Tekhnika (in Russian) (1987)
Sedov, L.I.: On Promising Directions and Problems in Continuum Mechanics in Thoughts about Science and Scientists. Nauka, Moscow (in Russian) (1980). 173
Shkhanukov, M.K.: On the convergence of difference schemes for differential equations with a fractional derivative. Dokl. Akad. Nauk USSR **348**, 746 (1996)
Shkhanukov-Lafishev, M.Kh., Nakhusheva, F.M.: Boundary-Value Problems for a Fractional-Order Diffusion Equation and Grid Methods of Their Solution in Nonclassical Equations of Mathematical Physics, Institute of Mathematics, Siberian Branch of the Russian Academy of Sciences, Novosibirsk, 37 (in Russian) (1998)
Sinai, Ya.G., Shilnikov, L.P. (eds.). Strange Attractors, a Collection of Papers, Moscow (in Russian) (1981)
Stratonovich, R.L.: Nonlinear Nonequilibrium Thermodynamics. Nauka, Moscow (in Russian) (1985)
Tikhonov, V.I., Mironov, M.A.: Markovian Processes. Sov. Radio, Moscow (in Russian) (1977)
Uchaikin, V.V.: Self-similar anomalous diffusion and stave laws. Usp. Fiz. Nauk **173**, 847 (2003)
Yorke, J.A., Yorke, E.D.: Chaotic behavior and hydrodynamics. In: Swinney, H., Gollub, J. (eds.) Hydrodynamic Instabilities and the Transition to Turbulence. Springer, Berlin (1981)
Zaslavsky, G.M.: Physics of Chaos in Hamiltonian Systems. Institute of Computer Research, Moscow-Izhevsk (in Russian) (2004)

Chapter 6

Akopyan, I.G.: On the establishment of a synchronous regime in a vacuum-tube oscillator. Radiotekhnika i Elektronika **11**, 34 (1966)
Anishchenko, V.S., Astakhov, V.V., Vadivasova, T.E., et al.: Nonlinear Effects in Chaotic and Stochastic Systems. Institute of Computer Research, Moscow-Izhevsk (in Russian) (2003)

Anselmet, F., Gagne, Y., Hopfinger, E.J., Antonia, R.A.: High-order velocity structure functions in turbulent shear flow. J. Fluid Mech. **140**, 63 (1984)
Bogolyubov, N.N., Mitropolskii, Y.A.: Asymptotic Methods in the Theory of Nonlinear Oscillations. Fizmatgiz, Moscow (in Russian) (1963)
Bostondzhian, V.A.: A Manual on Statistical Distributions, Chernogolovka (2000)
Brown, G.L., Roshko, A.: On density effects and large structures in turbulent mixing layers. J. Fluid Mech. **64**, 775 (1974)
Buckingham, A.: Mechanics of turbulence of multicomponent gases. Appl. Mech. Rev. **56**, 811 (2003)
Crow, S.C., Champagne, F.H.: Orderly structures in jet turbulence. J. Fluid Mech. **48**, 547 (1971)
de Groot, S., Mazur, P.: Non-equilibrium Thermodynamics. North-Holland, Amsterdam (1962)
Dubrulle, B.: Intermittency in fully developed turbulence: Log-Poisson statistics and generalized scale covariance. Phys. Rev. Lett. **73**, 959 (1994)
Frenkiel, F.N., Klebanoff, P.S.: On the log-normality of the small-scale structure of turbulence. Bound. Layer Meteorol. **8**, 173 (1975)
Frisch, U.: Turbulence. The Legacy of A.N. Kolmogorov. Cambridge University, Cambridge (1995)
Gardiner, C.W.: Handbook of Stochastic Methods for Physics, Chemistry and the Natural Sciences. Springer-Verlag, Berlin Heidelberg, New York, Tokyo (1983)
Gikhman, I.I., Skorokhod, A.V.: Stochastic Differential Equations. Naukova Dumka, Kiev (in Russian) (1968)
Hahn, G., Shapiro, S.: Statistical Models in Engineering. John Wiley & Sons, Inc. New York London Sydney (1967)
Horsthemke, W., Lefever, R.: Noise Induced Transitions. Springer, Berlin–Heidelberg–New York–Tokyo (1984)
Horsthemke, W., Malek Mansour, M.: The influence of external noise on nonequilibrium phase transitions. Zs. Phys. **B24**, 307 (1976)
Janke, E., Emde, F., Loesch, F.: Special Functions. Formulas, Graphs, Tables. Nauka, Moscow (in Russian) (1977)
Kadomtsev, B.B., Ryazanov, A.I.: What is synergetics? Priroda **8**, 2 (1983)
Khlopkov, Y.I., Zharov, V.A., Gorelov, S.L.: Coherent Structures in a Turbulent Boundary Layer. Moscow Physicotechnical Institute, Moscow (in Russian) (2002)
Klimontovich, Y.L.: Turbulent Motion and the Structure of Chaos: A New Approach to the Statistical Theory of Open Systems. Nauka, Moscow (in Russian) (1990)
Klimontovich, Y.L.: Statistical Theory of Open Systems. Janus, Moscow (in Russian) (1995)
Kolesnichenko, A.V.: Synergetic approach to describing developed turbulence. Astron. Vestn. **36**, 121 (2002)
Kolesnichenko, A.V.: Thermodynamic modeling of developed structured turbulence with energy dissipation fluctuations. Astron. Vestn. **38**, 144 (2004)
Kolesnichenko, A.V., Marov, M.Y.: Turbulence of Multicomponent Media. MAIK-Nauka, Moscow (in Russian) (1999)
Kolesnichenko, A.V., Marov, M.Y.: Fundamentals of mechanics of heterogenic media with application to circumsolar protoplanetary cloud: Influence of solid paticles on the disc turbulence. Solar System Res. **40**(1), 2–62 (2006)
Kolmogorov, A.N.: Local structure of turbulence in an incompressible fluid at very large Reynolds numbers. Dokl. Akad. Nauk USSR **30**, 299 (1941)
Kolmogorov, A.N.: Refining the Views about the Local Structure of Turbulence in an Incompressible Viscous Fluid at Large Reynolds Numbers, Mecanique de la turbulence: Colloq. Intern. CNRS, Marseille, aout - sept. 1961, 447 (in Russian) (1962)
Kullback, S.: Information Theory and Statistics. John Wiley & Sons, Inc. London-Chapman and Hall, Limited New York (1959)
Kuramoto, Y.: Chemical Oscillations. Waves and Turbulence. Springer, Berlin (1984)
La Frieda, J.R., Lindsey, W.C.: Transient analysis of phase-locked tracking systems in the presence of noise. Trans. IEEE **IT-19**, 36 (1973)
Landa, P.S.: Self-Oscillations in Distributed Systems. Nauka, Moscow (in Russian) (1983)

Landau, L.D., Lifshitz, V.M.: Hydrodynamics. Nauka, Moscow (in Russian) (1988)
Leontovich, M.A.: Introduction to Thermodynamics. Gostekhizdat, Moscow (in Russian) (1954)
Marov, M.Y., Kolesnichenko, A.V.: Mechanics of Turbulence of Multicomponent Gases. Kluwer Academic, Dordrecht–Boston–London (2002)
Migdal, A.A.: Vortex Fokker-Planck Equation in Nonlinear Waves. Structures and Bifurcations. Nauka, Moscow (in Russian) (1987). 159
Monin, A.S., Yaglom, A.M.: Statistical Hydrodynamics, vol. 2. Gidrometeoizdat, St. Petersburg (in Russian) (1996)
Obukhov, A.M.: Some specific features of atmospheric turbulence. J. Fluid Mech. **13**, 77 (1962)
Osipov, G., Pikovsky, A., Rosenblum, M., Kurths, J.: Phase synchronization effects in a lattice of nonidentical rossler oscillators. Phys. Rev. E **55**, 2353 (1997)
Pikovsky, A., Rosenblum, M., Kurths, J.: Synchronization in a population of globally coupled chaotic oscillators. Europhys. Lett. **34**, 165 (1996)
Pikovsky, A.S., Rosenblum, M.G., Kurths, J.: Synchronization. A Universal Concept in Nonlinear Sciences. Cambridge University Press, Cambridge (2001)
Prigogine, I., Stengers, I.: Time. Chaos. Quantum. Toward Solving the Paradox of Time. Progress Publishing House, Moscow (in Russian) (1994)
Rabinovich, M.I.: Stochastic self-oscillations and turbulence. Usp. Fiz. Nauk **125**, 123 (1978)
Rabinovich, M.I., Sushchik, M.M.: Regular and chaotic dynamics of structures in fluid flows. Usp. Fiz. Nauk **160**, 1 (1990)
Rabinovich, M.I., Trubetskov, D.I.: Introduction to the Theory of Oscillations and Waves. Nauka, Moscow (in Russian) (1984)
Ruelle, D., Takkens, F.: On the nature of turbulence. Commun. Math. Phys. **20**, 167 (1971)
She, Z.S., Leveque, E.: Universal scaling laws in fully developed turbulence. Phys. Rev. Lett. **72**, 336 (1994)
Shilnikov, L.P., Shilnikov, A.L., Turaev, D.V., Chua, L.: Methods of the Qualitative Theory in Nonlinear Dynamics, Pt. 1. Institute of Computer Research, Moscow-Izhevsk (in Russian) (2004)
Sinai, Ya.G., Shilnikov, L.P. (eds.): Strange Attractors. A Collection of Papers, Moscow (in Russian) (1981).
Starr, V.: Physics of Negative Viscosity Phenomena. McGraw-Hill, New York (1968)
Stratonovich, R.L.: Selected Questions of the Fluctuation Theory in Radio Engineering. Sov. Radio, Moscow(in Russian) (1961)
Stratonovich, R.L.: Nonlinear Nonequilibrium Thermodynamics. Nauka, Moscow (in Russian) (1985)
Tikhonov, V.I.: Nonlinear Transformations of Random Processes. Radiol Svyaz', Moscow (in Russian) (1986)
Tikhonov, V.I., Mironov, M.A.: Markovian Processes. Sov. Radio, Moscow (in Russian) (1977)

Chapter 7

Abramovich, G.N., Girshovich, T.A.: On the diffusion of heavy particles in turbulent gas flows. Dokl. Akad. Nauk USSR **212**, 573 (1973)
Alfven, H., Arrhenius, G.: Evolution of the Solar System. Sci. & Techn. Inf. Office, NASA, Washington, D.C. (1976)
Armitage, P.J., Livio, M., Pringle, J.E.: Episodic accretion in magnetically layered protoplanetary disks. Mon. Not. R. Astron. Soc. **324**, 705 (2001)
Balbus, S.A., Hawley, J.F.: Instability, turbulence, and enhanced transport in accretion disks. Rev. Mod. Phys. **70**, 1 (1998)

Barenblatt, G.I., Golitsyn, G.S.: Local Structure of Developed Dust Storms. Moscow State University, Moscow (in Russian) (1973)
Barenblatt, G.I., Golitsyn, G.S.: Local structure of mature dust storms. J. Atmos. Sci. **31**, 1917 (1974)
Barge, P., Sommeria, J.: Did planet formation begin inside persistent gaseous vortices? Astron. Astrophys **295**, L1 (1995)
Beckwith, S.V.W., Henning, T., Nakagawa Y. In: Mannings, V., Boss, A.P., Rassell, S.S. (eds.) Dust Properties and Assembly of Large Particles in Protoplanetary Disks in Protostars and Planets IV, Tucson Univ. Ariz. Press, 533 (2000)
Bisnovaty-Kogan, G.S., Lovelace, R.V.E.: Advective accretion disks and related problems including magnetic fields. New Astron. Rev **45**, 663 (2001)
Bryden, G., Chen, X., Lin, D., et al.: Tidally induced gap formation in protostellar disks: Gap clearing and suppression of protoplanetary growth. Astrophys. J. **514**, 344 (1999)
Busroyd, R.: Flow of Gas with Suspended Particles, Mir, Moscow (in Russian) (1975)
Cabot, W., Canuto, V.M., Hubickyj, O., Pollack, J.B.: The role of turbulent convection in the primitive solar nebula. Icarus **69**, 423 (1987)
Cameron, A.G.W.: Accumulation processes in the primitive solar nebula. Icarus **18**, 407–450 (1973)
Cameron, A.G.W.: Formation and evolution of the primitive solar nebula. In: Black, D.C., Matthews, M.S. (eds.) Protostars and Planets II, pp. 1073–1099. Univ. of Arizona Press, Tucson, AZ (1985)
Cameron, A.G.W.: The primitive solar accretion disc and formation of planets. In: Dermott, S.F. (ed.) The Origin of the Solar System. pp. 49–75. NY (1978)
Chapman, S., Cowling, T.G.: The Mathematical Theory of Non-uniform Gases. Cambridge University Press, Cambridge (1952)
Chavanis, P.-H.: Trapping of Dust by Coherent Vortices in the Solar Nebula,//arXiv:astro-ph/9912087, 16, 1. (1999)
Cuzzi, J.N.: Blowing in the Wind: III. Accretion of dust rims by chondrule-sized particles in a turbulent protoplanetary nebula. Icarus **168**, 484 (2004)
Cuzzi, J.N., Dobrovolskis, A.R., Champney, J.M.: Particle-gas dynamics in the midplane of a protoplanetary nebula. Icarus **106**, 102 (1993)
Cuzzi, J.N., Davis, S.S., Dobrovolskis, A.R.: Blowing in the Wind. II. Creation and Redistribution of Refractory Inclusions in a Turbulent Protoplanetary Nebula. Icarus **166**, 385 (2003)
Cuzzi, J.N., Ciesla, F.J., Petaev, M.I., et al.: Nebula Evolution of Thermally Processed Solids: Reconciling Models and Meteorites Chondrites and the Protoplanetary Disk, ASP Conference Series, Vol. 341. In: Krot, A.N., Scott, E.R.D., Bo Reipurth (eds.) Proceedings of a Workshop held 8-11 November 2004 in Kaua'i, Hawaii, Astronomical Society of the Pacific, San Francisco, 732 (2005)
Cuzzi, J.N., Weidenschilling, S.J. In: Lauretta, D., Leshin, L.A., McSween, H. (eds.) Particle-Gas Dynamics and Primary Accretion in Meteorites and the Early Solar System II, Univ. of Arizona Press, Tucson, 353. (2006)
Danon, H., Wolfshtein, M., Hetsroni, G.: Numerical calculation of two—phase turbulent round jet. Int. Multiphase Flow **3**, 223 (1977)
de Groot, S., Mazur, P.: Non-equilibrium Thermodynamics. North-Holland, Publishing Company, Amsterdam (1962)
Derevich, I.V.: Influence of an admixture of large particles on turbulent characteristics of a gas suspension in channels. PMTF **2**, 70 (1994)
Dominik, C., Blum, J., Cuzzi, J., Wurm, G.: Growth of dust as the initial step toward planet formation in protostars and planets V. Arizona, AZ (2007)
Dorofeeva, V.A., Makalkin, A.B..: Evolution of the Early Solar System. Cosmochemical and Physical Aspects. Editorial URSS, Moscow (in Russian) (2004)
Dubrulle, B.: A turbulent closure model for thin accretion disks. Astron. Astrophys. **266**, 592 (1992)

Dubrulle, B.: Differential rotation as a source of angular momentum transfer in the solar nebula. Icarus **106**, 59 (1993)
Dubrulle, B., Morfill, G., Sterzic, M.: The dust subdisk in the protoplanetary nebula. Icarus **114**, 237 (1995)
Dullemond, C.P., Dominik, C.: Dust coagulation in protoplanetary disks: A rapid depletion of small grains. Astron. Astrophys. **434**, 971 (2005)
Eardley, D.M., Lightman, A.P.: Magnetic viscosity in relativistic accretion discs. Astrophys. J. **200**, 187 (1975)
Eardley, D.M., Lightman, A.P., Payne, D.G., Shapiro, S.L.: Accretion disks around massive black holes: Persistent emission spectra. Astrophys. J. **234**, 53 (1978)
Elghobashi, S.E., Abou-Arab, T.W.: A second-order turbulence model for two-phase flows. Heat Transfer **5**, 219 (1982)
Elghobashi, S.E., Abou-Arab, T.W.: A two-equation turbulence model for two-phase flows. Phys. Fluids **26**, 931 (1983)
Epstein, P.S.: On the resistance experienced by spheres in their motion through gases. Phys. Rev. **23**, 710 (1924)
Favre, A.: Statistical equations of turbulent gases in problems of hydrodynamics and continuum mechanics. SIAM, Philadelphia (1969). 231
Fortov, V.E., Khrapak, A.G., Yakubov, I.T.: Physics of Strongly Coupled Plasma. Clarendon Press, Oxford (2006)
Fridman, F.M.: To the dynamics of a viscous differentially rotating gravitating medium. Pis'ma Astron. Zh. **15**, 1122 (1989)
Fridman, A.M., Boyarchuck, F.F., Bisikalo, D.V., et al.: The collective mode and turbulent viscosity in accretion disks. Phys. Lett. A **317**, 181 (2003)
Fuks, N.A.: Mechanics of Aerosols. Akad. Nauk USSR, Moscow (in Russian) (1955)
Garaud, P., Barriere-Fouchet, L., Lin, D.N.C.: Individual and collective behavior of dust particles in a protoplanetary nebula. Astroph. J. **603**, 292 (2005)
Gavin, L.B., Naumov, V.A., Shor, V.V.: Numerical analysis of a gas jet with heavy particles based on a two-parameter turbulence model. PMTF **1**, 62 (1984)
Goldrich, P., Ward, W.R.: The formation of planetesimals. Astrophys. J. **183**, 1051 (1973)
Goodmann, J., Pindor, B.: Secular instability and planetesimal formation in the dust layer. Icarus **148**, 537 (2000)
Gor'kavyi, N.N., Fridman, A.M.: Physics of Planetary Rings. Nauka, Moscow (in Russian) (1994)
Gorbis, Z.R.: Heat Exchange and Hydromechanics of Disperse Through Flows. Energiya, Moscow (in Russian) (1970)
Gore, R.A., Crowe, C.T.: Effect of particle size on modulating turbulent intensity. Int. J. Multiphase Flow **15**, 279 (1989)
Grad, H.: On the kinetic theory of rarefied gases. Commun. Pure Appl. Math. **2**, 331 (1949)
Hahn, G., Shapiro, S.: Statistical Models in Engineering. John Wiley & Sons, New York, London Sydney (1967)
Hayashi, C., Nakazawa, K., Nakagawa, Y.: In: Black, D.C., Matthews, M.S. (eds.) Formation of the Solar System In Protostars and Planets II, University of Arizona, Tucson, 1100. (1985)
Hazlehurst, J., Sargent, W.L.W.: Astrophys J. **130**, 276 (1959)
Hersant, F., Dudrulle, B., Hure, J.-M.: Turbulence in Circumstellar Disks, Astron. Astrophys. (2004) Manuscript No. 3549, 1
Hirschfelder, D., Curtiss, C., Bird, R.: Molecular Theory of Gases and Liquids. John Wiley and Sons, New York and London (1954)
Hunter, S.C., Cherry, S.S., Kliegel, J.R., Waldman, C.H.: Gas-particle nozzle flow with reaction and particle size change. AIAA Paper **37**, 14 (1981)
Ievlev, V.M.: Approximate equations of turbulent motion for an incompressible fluid. Izv. Akad. Nauk USSR Ser MZhG **1**, 91 (1970)
Ievlev, V.M.: Turbulent Motion of High-Temperature Continuous Media. Nauka, Moscow (in Russian) (1975)

Kartushinskii, A.I.: Transport of an inertial admixture in a two-phase turbulent jet. Izv. Akad. Nauk USSR Ser MZhG **1**, 36 (1984)

Kolesnichenko, A.V.: To the theory of turbulence in planetary atmospheres: numerical simulation of structural parameters. Astron. Vestn. **29**, 133 (1995)

Kolesnichenko, A.V.: Stefan-Maxwell Relations and Heat Flow for Turbulent Multicomponent Continuous Media in Problems of Modern Mechanics. Moscow State University, Moscow (in Russian) (1998). 52

Kolesnichenko, A.V.: Modeling turbulent transport coefficients in a gas-dust accretion disk. Astron. Vestn. **34**, 516 (2000)

Kolesnichenko, A.V.: Hydrodynamics aspects of modeling mass transfer and coagulation in a turbulent accretion disk. Astron. Vestn. **35**, 139 (2001)

Kolesnichenko, A.V.: Synergetic Approach to Describing Stationary-Nonequilibrium Turbulence of Astrophysical Systems in Modern Problems of Mechanics and Space Physics. Fizmatlit, Moscow (in Russian) (2003). 123

Kolesnichenko, A.V.: On the synergetic formation mechanism of coherent structures in the continuum theory of developed turbulence. Astron. Vestn. **38**, 405 (2004)

Kolesnichenko, A.V., Maksimov, V.M.: Generalized darcy filtering law as a corollary of the Stefan-Maxwell relations for a heterogeneous medium. Mat. Mod. **13**, 3 (2001)

Kolesnichenko, A.V., Marov, M.Y.: Turbulence of Multicomponent Media. MAIK-Nauka, Moscow (in Russian) (1999)

Kolesnichenko, A.V., Marov, M.Y.: Fundamentals of mechanics of heterogenic media with application to circumsolar protoplanetary cloud: Influence of solid paticles on the disc turbulence. Solar System Res. **40**(1), 2–62 (2006)

Kolmogorov, A.N.: Equations of turbulent motion for an incompressible fluid. Izv. Akad. Nauk. USSR Ser. fiz. **6**, 56 (1942)

Kolmogorov, A.N.: On a new version of the gravitational theory of motion of suspended sediments, vestn. MGU **3**, 41 (1954)

Kusaka, T., Nakano, N., Hayashi, C.: Growth of Solid Particles in the Primordial Solar Nebula. Progr. Theor. Phys **44**, 1580 (1970)

Laikhtman, G.L.: Physics of the Atmospheric Boundary Layer. Gidrometeoizdat, Leningrad (in Russian) (1970)

Landau, L.D., Lifshitz, V.M.: Hydrodynamics. Nauka, Moscow (in Russian) (1988)

Lee, K.W.: Change of particle size distribution during brownian coagulation. J. Colloid Interface Sci. **92**, 38 (1983)

Leonard, A.: Energy cascade in large eddy simulations of turbulent fluid flows. Adv. Geophys. **A18**, 237 (1974)

Lin, D.N.C., Papaloizou, J.: On the structure and evolution of the primordial solar nebula. Mon. Not. Roy. Astron. Soc. **191**, 37–48 (1980)

Lissauer, J.J., Stewart, G.R.: In: Levy, E.H., Lunine, I.J. (eds.) Growth of Planets from Planetesimals in Protostars and Planets III, Univ. Arizona Press, Tucson, 1061 (1993)

Loginov, V.I.: Dehydration and Desalinization of Oil. Khimiya, Moscow (in Russian) (1979)

Lynden-Bell, D., Pringle, J.E.: The evolution of viscous discsand the origin of the nebular variables. Mon. Not. Roy. Astron. Soc. **168**, 603–637 (1974)

Makalkin, A.B..: Radial compaction of the dust subdisk in a protoplanetary disk as possible way to gravitational instability. Lunar Planet. Sci. **25**, 827 (1994)

Makalkin, A.B..: Problems of the Evolution of Protoplanetary Disks in Modern Problems of Mechanics and Space Physics. Fizmatlit, Moscow (in Russian) (2003). 402

Makalkin, A.B..: Peculiarities of the evolution of a viscous circumsolar protoplanetary disk. Astron. Vestn. **38**, 559 (2004)

Makalkin, A.B.., Dorofeeva, V.A.: Structure of the protoplanetary accretion disk around the sun at the T tauri phase. I Initial data, equations, and modeling methods. Astron. Vestn. **29**, 99 (1995)

Makalkin, A.B.., Dorofeeva, V.A.: Structure of the protoplanetary accretion disk around the sun at the T tauri phase. II. Results of model calculations. Astron. Vestn. **30**, 496 (1996)

Marov, M.Y., Kolesnichenko, A.V.: Introduction to Planetary Aeronomy. Nauka, Moscow (in Russian) (1987)

Marov, M.Y., Kolesnichenko, A.V.: Mechanics of Turbulence of Multicomponent Gases. Kluwer Academic Publishers, Dordrecht–Boston–London (2002)

Mazin, I.P.: Theoretical estimation of the coagulation coefficient of droplets in clouds. Trudy TsAO **95**, 12 (1971)

Mednikov, E.P.: Turbulent Transport and Sedimentation of Aerosols. Nauka, Moscow (in Russian) (1981)

Melville, W.K., Bray, K.N.C.: A Model of the Two-Phase Turbulent Jet. Int. J. Heat Mass Transfer **22**, 647 (1979)

Monin, A.S., Yaglom, A.M.: Statistical Fluid Mechanics. Vol. 1. Cambridge (Mass., USA): MIT Press (1971)

Monin, A.S., Yaglom, A.M.: Statistical Hydrodynamics, vol. I. Gidrometeoizdat, St. Petersburg (in Russian) (1992)

Monin, A.S., Yaglom, A.M.: Statistical Hydrodynamics. Gidrometeoizdat, St. Petersburg (in Russian) (1996)

Morkovin, M.V.: Effects of Compressibility on Turbulent Flow in Mechanics of Turbulence. Gordon and Breach, New York (1961). 367

Nakagawa, Y., Nakagawa, K., Hayashi, C.: Growth and sedimentation of dust grains in the primordial solar nebula. Icarus **45**, 517 (1981)

Nakagawa, Y., Sekiya, M., Hayashi, C.: Settling and growth of dust particles in a laminar phase of a low-mass solar nebula. Icarus **67**, 375 (1986)

Natta, A., Testi, L., Calvet, N., et al.: Dust in Protoplanetary Disks: Properties and Evolution in Protostars and Planets V. Arizona, AZ (2007)

Nigmatulin, R.I.: Foundations of Mechanics of Heterogeneous Media. Nauka, Moscow (in Russian) (1978)

Nomura, H.: Structure and instabilities of an irradiated viscous protoplanetary disks. Astrophys. J. **567**, 587 (2002)

Pollack, J.B., McKay, C.P., Cristofferson, B.M.: A calculation of a rosseland mean opacity of dust grains in primordial solar system nebulae. Icarus **64**, 473 (1985)

Prigogine, I., Defay, R.: Chemical Thermodynamics. Longmans Green and Co., London, New York, Toronto (1954)

Prigogine, I., Stengers, I.: Time. Chaos. Quantum. Toward Solving the Paradox of Time. Progress Publishing House, Moscow (in Russian) (1994)

Richard, D., Zahn, J.-P.: Turbulence in differentially rotating flow. What can be learned from the Couette-Taylor experiment. Astron. Astrophys. **347**, 734 (1999)

Ruden, S.P., Pollack, J.B.: The dynamical evolution of the protosolar nebula. Asrophys. J. **375**, 740 (1991)

Russell, S.S., Hartmann, L.A., Cuzzi, J.N., et al.: In: Lauretta, D., Leshin, L.A., McSween, H. (eds.) Timescales of the Solar Protoplanetary Disk in Meteorites and the Early Solar System II. University of Arizona, Tucson, 233 (2006).

Safronov, V.S.: Evolution of the Protoplanetary Cloud and Formation of the Earth and Planets. Nauka, Moscow (in Russian) (1969)

Safronov, V.S.: Current status of the theory of the Earth's origin. Dokl. Akad. Nauk USSR, Phys. Earth **6**, 5 (1982)

Safronov, V.S.: Evolution of the dust component of the circumsolar protoplanetary disk. Astron. Vestn. **21**, 216 (1987)

Safronov, V.S., Guseinov, K.M.: On the possibility of in situ comet formation. Astron. Vestn. **1990** (24), 248 (1990)

Schlichting, H.: Grenzschicht-Theorie. Verlag G. Braun. Karlsruhe (1964)

Schlichting, H.: Boundary-Layer Theory. McGraw-Hill, New York (1968)

Schmidt, O.Y.: Four Lectures on the Earth's Origin, 3rd edn. USSR Academy of Sciences, Moscow (in Russian) (1957)

Schmitt, W., Henning, T., Mucha, R.: Dust evolution in protoplanetary accretion disks. Astron. Astrophys. **325**, 569 (1997)
Sekiya, M., Nakagawa, Y.: Settling of dust particles and formation of planetesimals. Prog. Theor. Phys. Suppl. **96**, 141 (1988)
Shakura, N.I., Sunyaev, R.A.: Black holes in binary systems. Observational appearance. Astron. Astrophys. **24**, 337 (1973)
Shakura, N.I., Sunyaev, R.A., Zilitinkevich, S.S.: On the turbulent energy transport in accretion disk. Astron. Astrophys. **62**, 179 (1978)
Shapiro, S., Teukolsky, S.: Black Holes, White Dwarfs, and Neutron Stars. John Wiley & Sons, New York (1983)
Shraiber, A.A., Gavin, L.B., Naumov, V.A., Yatsenko, V.P.: Hydromechanics of Two-Component Flows with Polydisperse Material. Naukova Dumka, Kiev (in Russian) (1980)
Shraiber, A.A., Milyutin, V.N., Yatsenko, V.P.: Turbulent Flows of a Gas Suspension. Naukova Dumka, Kiev (in Russian) (1987)
Smoluchowski, M.: Three Lectures on Diffusion, Brownian Molecular Motion, and Coagulation of Colloidal Particles. Brownian Motion in Colloid Coagulation. ONTI, Moscow (in Russian) (1936)
Soo, S.L., Ihrig, H.K., Kouh, A.F.: Experimental determination of statistical properties of two-phase turbulent motion. Trans. ASME J. Basic Engng. **82**, 609 (1960)
Stepinski, T.F., Valageas, P.: Global evolution of solid matter in turbulent protoplanetary disks. I. Aerodynamics of solid particles. Astron. Astrophys. **309**, 301 (1996)
Stepinski, T.F., Valageas, P.: Global evolution of solid matter in turbulent protoplanetary disks. II. Development of icy planetesimals. Astron. Astrophys. **319**, 1007 (1997)
Sternin, L.E., Shraiber, A.A.: Multiphase Flows of Gas with Particles. Mashinostroenie, Moscow (in Russian) (1994)
Sternin, L.E., Maslov, B.N., Shraiber, A.A., Podvysotskii, A.M.: Two-Phase Mono- and Polydisperse Flows of Gas with Particles. Mashinostroenie, Moscow (in Russian) (1980)
Stokes, G.G.: On the effect of the internal friction of fluids on the motion of pendulums. Trans. Camb. Phil. Soc. **9**, 8 (1851)
Takeuchi, T., Lin, D.N.C.: Radial flow of dust particles in accretion disks. Astrophys. J. **581**, 1344 (2002)
Takeuchi, T., Lin, D.N.C.: Surface out in optically thick dust disks by the radiation pressure. Astrophys. J. **593**, 524 (2003)
Tanga, P., Babiano, A., Dubrulle, B.: Forming planetesimals in vortices. Icarus **121**, 158 (1996)
Tassoul, J.-L.: Theory of Rotating Stars. Princeton University Press, Princeton, New Jersey (1978)
Toomre, A.: On the gravitational stability of a disk of stars. Astrophys. J. **139**, 1217 (1964)
Townsend, A.A.: The Structure of Turbulent Shear Flow. Cambridge University, Cambridge (1956)
Van Mieghem, J.: Atmospheric Energetics. Clarendon Press, Oxford (1973)
Varaksin, A.Y.: Turbulent Flows of Gas with Solid Particles. Fizmatlit, Moscow (in Russian) (2003)
Vereshchagin, I.P., Levitov, V.I., Mirzobekyan, G.Z., Pashin, M.M.: Foundations of Electrogasdynamics of Disperse Systems. Energiya, Moscow (in Russian) (1974)
Voloshchuk, V.M.: Kinetic Theory of Coagulation. Gidrometeoizdat, Moscow (in Russian) (1984)
Voloshchuk, V.M., Sedunov, Y.S.: Coagulation Processes in Disperse Systems. Gidrometeoizdat, Leningrad (in Russian) (1975)
Wadhwa, M., Amelin, Y., Davis, A.M., et al.: From Dust to Planetesimals: Implications for the Solar Protoplanetary Disk from Short-Lived Radionuclides in Protostars and Planets V. Arizona, AZ (2007)
Wasson, J.T.: Meteorites: Their Record of Early Solar-System History. Freeman, New York (1985)
Watanabe, S., Nakagawa, Y., Nakazawa, K.: Cooling and quasi-static contraction of the primitive solar nebula after gas accretion. Astrophys. J. **358**, 282 (1990)

Weidenschilling, S.J.: Aerodynamics of solid bodies in the solar nebula. Mon. Not. R. Astron. Soc. **180**, 57 (1977)
Weidenschilling, S.J.: Dust to planetesimals: Settling and coagulation in the solar nebula. Icarus **44**, 172 (1980)
Weidenschilling, S.J.: Evolution of grains in a turbulent solar nebula. Icarus **60**, 553 (1984)
Whipple, F.L.: From Plasma to Planet, London (1972)
Willacy, K., Klahr, H.H., Millar, T.J., Henning, T.: Gas and Grain Chemistry in a Protoplanetary Disk. Astron. Astrophys. **338**, 995 (1998)
Woods, J.D., Drake, J.C., Goldsmith, P.: Coalescence in a turbulent cloud. Quart. J. Roy. Met. Soc. **98**, 135 (1972)
Yarin, L.P., Hetsroni, G.: Turbulence intensity in dilute two-phase flow -3. The particles-turbulence interaction in dilute two-phase flow. Int. J. Multiphase Flow **20**, 27 (1994)
Youdin, A.N., Goodman, J.: Streaming Instabilities in Protoplanetary Disks, arXiv: Astro-ph/0409263, 1, 1 (2004)
Youdin, A.N., Shu, F.: Planetesimal formation by gravitational instability. Astrophys. J. **580**, 494 (2002)
Zaichik, L.I., Varaksin, A.Y.: Influence of the wake behind large particles on the intensity of carrier-flow turbulence. TVT **37**, 1004 (1999)
Zel'dovich, Y.B.: On the friction of fluids between rotating cylinders. Proc. Roy. Soc. London **A374**, 299 (1981)
Zuev, Y.V., Lepeshinskii, I.F.: Mathematical model of a two-phase turbulent jet. Izv. Akad. Nauk USSR Ser. MZhG **6**, 69 (1981)

Chapter 8

Alekseenko, S.V., Kuibin, P.A., Okulov, V.L.: Introduction to the Theory of Concentrated Vortices. Institute of Computer Research, Moscow-Izhevsk (in Russian) (2005)
Andre, J.D., Lesieur, M.: Evolution of high reynolds number isotropic three-dimensional turbulence: Influence of helicity. J. Fluid Mech. **81**, 187 (1977)
Armitage, P.J., Livio, M., Pringle, J.E.: Episodic accretion in magnetically layered protoplanetary disks. Mon. Not. R. Astron. Soc. **324**, 705 (2001)
Balbus, S.A., Hawley, J.F.: Instability, turbulence, and enhanced transport in accretion disks. Rev. Mod. Phys. **70**, 1 (1998)
Barge, P., Sommeria, J.: Did planet formation begin inside persistent gaseous vortices? Astron. Astrophys. **295**, L1 (1995)
Batchelor, G.K.: The Theory of Homogeneous Turbulence. Cambridge University, Cambridge (1953)
Bodenheimer, P.: Angular momentum evolution of young stars and disks. Ann. Rev. Astron. Astrophys **33**, 199 (1995)
Brissaund, A., Frisch, U., Leorat, J., et al.: Helicity cascade in fully developed turbulence. Phys. Fluids **16**, 1366 (1973)
Chavanis, P.-H.: Trapping of Dust by Coherent Vortices in the Solar Nebula,/arXiv:astro-ph/9912087, 16, 1 (1999).
de Groot, S., Mazur, P.: Non-equilibrium Thermodynamics. North-Holland, Amsterdam (1962)
Dubrulle, B.: Differential rotation as a source of angular momentum transfer in the solar nebula. Icarus **106**, 59 (1993)
Dubrulle, B., Frisch, U.: The eddy viscosity of parity-invariant flow. Phys. Rev. **43**, 5355 (1991)
Eardley, D.M., Lightman, A.P.: Magnetic viscosity in relativistic accretion discs. Astrophys. J. **200**, 187 (1975)

Edwards, S.F.: Statistical mechanics with topological constraints: I. Proc. Phys. Soc. **91**, 513 (1967)
Edwards, S.F.: Statistical Mechanics with Topological Constraints: II, J. Phys. A (Proc. Phys. Soc.) ser. 2, **1**, 15 (1968).
Eltayeb, I.A.: Hydromagnetic convection in a rapidly rotating fluid layer. Proc. Roy. Soc. **A326**, 229 (1972)
Fannjiang, A., Papanicolaou, G.: Convection—enhanced diffusion for periodic flows. SIAM J. Appl. Math. **54**, 333 (1994)
Ferrari, C.: On the Differential Equations of Turbulent Flow in Continuum Mechanics and Related Problems of Analysis. Nauka, Moscow (1972)
Fridman, F.M.: To the dynamics of a viscous differentially rotating gravitating medium. Pis'ma Astron. Zh. **15**, 1122 (1989)
Fridman, A.M., Boyarchuck, F.F., Bisikalo, D.V., et al.: The collective mode and turbulent viscosity in accretion disks. Phys. Lett. A **317**, 181 (2003)
Frisch, U.: Turbulence.The Legacy of A.N. Kolmogorov. Cambridge University, Cambridge (1995)
Galimov, E.M.: The Phenomenon of Life. Between Equilibrium and Nonlinearity. The Origin and Principles of Evolution. URSS, Moscow (in Russian) (2001)
Gama, S., Vergassola, M., Frisch, U.: Negative eddy viscosity in isotropically forced two-dimensional flow: Linear and nonlinear dynamics. J. Fluid Mech. **260**, 95 (1994)
Gilbert, A., Frisch, U., Pouquet, A.: Helicity is unnecessary for alpha-effect dynamos, but it helps. Geophys. Astrophys. Fluid Dyn. **42**, 151 (1988)
Goldrich, P., Ward, W.R.: The formation of planetesimals. Astrophys. J. **183**, 1051 (1973)
Gor'kavyi, N.N., Fridman, A.M.: Physics of Planetary Rings. Nauka, Moscow (in Russian) (1994)
Haken, H.: Advanced synergetics. Instability hierarchies of self-organizing systems and devices. Springer, Berlin (1983)
Hide, R.: A note on helicity. Geophys. Fluid Dyn. **7**, 157 (1976)
Karman, T.: Some Questions of the Turbulence Theory in Turbulence Problems. Onti NKTP USSR, Moscow (in Russian) (1936). 35
Khlopkov, Y.I., Zharov, V.A., Gorelov, S.L.: Coherent Structures in a Turbulent Boundary Layer. Moscow Physicotechnical Institute, Moscow (in Russian) (2002)
Klahr, H.H., Bodenheimer, P.: Turbulence in accretion disks: Vorticity generation and angular momentum transport via the global baroclinic instability. Astrophys. J. **582**, 869 (2003)
Kolesnichenko, A.V.: Synergetic Approach to Describing Developed Turbulence. Astron. Vestn. **36**, 121 (2002)
Kolesnichenko, A.V.: Synergetic Approach to Describing Stationary-Nonequilibrium Turbulence of Astrophysical Systems in Modern Problems of Mechanics and Space Physics. Fizmatlit, Moscow (in Russian) (2003). 123
Kolesnichenko, A.V.: Thermodynamic modeling of developed structured turbulence with energy dissipation fluctuations. Astron. Vestn. **38**, 144 (2004)
Kolesnichenko, A.V.: On the possibility of synergetic birth of mesoscale coherent structures in the macroscopic theory of developed turbulence. Mat. Mod. **17**, 47 (2005)
Kolesnichenko, A.V.: On the role of noise-induced nonequilibrium phase transitions in structuring hydrodynamic turbulence. Astron. Vestn. **39**, 243 (2005)
Kolesnichenko, A.V., Marov, M.Y.: Fundamentals of mechanics of heterogeneous media in the circumsolar protoplanetary disk: The influence of solid particles on turbulence in the disk. Astron. Vestn. **40**, 1 (2006)
Kolmogorov, A.N.: Local structure of turbulence in an incompressible fluid at very large Reynolds numbers. Dokl. Akad. Nauk USSR **30**, 299 (1941)
Kolmogorov, A.N.: Refining the Views about the Local Structure of Turbulence in an Incompressible Viscous Fluid at Large Reynolds Numbers, Mecanique de la turbulence: Colloq. Intern. CNRS, Marseille, aout - sept. 1961, 447 (in Russian) (1962)
Korenev, G.V.: Tensor Calculus. Moscow Physicotechnical Institute, Moscow (in Russian) (1996)

Kraichnan, R.H.: Inertial ranges in two-dimensional turbulence. Phys. Fluids **10**, 1417 (1967)
Kraichnan, R.H.: Helical turbulence and absolute equilibrium. J. Fluid Mech. **59**, 745 (1973)
Kraichnan, R.H.: Diffusion of passive scalar and magnetic fields by helical turbulence. J. Fluid Mech. **77**, 753 (1976a)
Kraichnan, R.H.: Eddy viscosity in two and three dimensions. J. Atmos. Sci. **33**, 1521 (1976b)
Krause, F., Raedler, K.-H.: Mean-Field Magnetohydrodynamics and Dynamo Theory. Pergamon, New York (1980)
Makalkin, A.B..: Problems of the Evolution of Protoplanetary Disks in Modern Problems of Mechanics and Space Physics. Fizmatlit, Moscow (in Russian) (2003). 402
Marov, M.Y., Kolesnichenko, A.V.: Mechanics of Turbulence of Multicomponent Gases. Kluwer Academic Publishers, Dordrecht–Boston–London (2002)
Marov, M.Ya., Kolesnichenko, A.V.: In: Fridman, A.M., Marov, M.Ya. (eds.) Chaotic and Ordered Structures in the Developed Turbulence in Astrophysical Disks: Collective and Stochastic Phenomena, Springer, the Netherlands. 23. (2006)
Moffat, H.K.: The degree of knottedness of tangled vortex lines. J. Fluid Mech. **35**, 117 (1969)
Moffat, H.K.: Magnetic Field Generation in Electrically Conducting Fluids. Cambridge University, Cambridge (1978)
Moffat, H.K.: Some development in the theory of turbulence. J. Fluid Mech. **106**, 35 (1981)
Moffat, H.K. Simple Topological Aspects of Turbulent Vorticity Dynamics. In: Tatsume T. (ed.) Proceedings of the IUTAM Symposium on Turbulence and Chaotic Phenomena in Fluids, Elseviere, 223 (1984)
Monin, A.S., Yaglom, A.M.: Statistical Hydrodynamics. Gidrometeoizdat, St. Petersburg (in Russian) (1996)
Monin, A.S., Polubarinova-Kochina, P.Y., Khlebnikov, V.I.: Cosmology, Hydrodynamics, Turbulence: A.A. Fridman and the Development of his Legacy. Nauka, Moscow (in Russian) (1989)
Moreau, J.J.: Constantes d'un ilot tourbillonnaire en fluide parfait barotrope. C.R. Acad. Sci. Paris **252**, 2810 (1961)
Nakagawa, Y., Sekiya, M., Hayashi, C.: Settling and growth of dust particles in a laminar phase of a low-mass solar nebula. Icarus **67**, 375 (1986)
Nikolaevskiy, V.N.: Angular Momentum in Geophysical Turbulence. Kluwer Academic Publishers, Dordrecht (2003)
Parker, E.N.: Cosmical Magnetic Fields, Their Origin and Activity. Clarendon, Oxford (1979)
Prigogine, I., Stengers, I.: Order out of Chaos. Bantam, New York (1984)
Richard, D., Zahn, J.-P.: Turbulence in Differentially Rotating Flow. What Can Be Learned from the Couette-Taylor Experiment. Astron. Astrophys. **347**, 734 (1999)
Saffman, P.G.: Vortex Dynamics. Cambridge University, Cambridge (1992)
Safronov, V.S.: On the Formation and Evolution of Protoplanetary Dust Clumps, Vopr. Kosmogonii **7**, 121 (1960)
Safronov, V.S.: Evolution of the Protoplanetary Cloud and Formation of the Earth and Planets. Nauka, Moscow (in Russian) (1969)
Safronov, V.S.: Current Status of the Theory of the Earth's Origin. Dokl. Akad. Nauk USSR, Phys. Earth **6**, 5 (1982)
Safronov, V.S.: Evolution of the dust component of the circumsolar protoplanetary disk. Astron. Vestn. **21**, 216 (1987)
Schmidt, O.Y.: Four Lectures on the Earth's Origin, 3rd edn. USSR Academy of Sciences, Moscow (in Russian) (1957)
Sivashinsky, G.I., Frenkel, A.L.: On negative eddy viscosity under conditions of isotropy. Phys. Fluids **A4**, 1608 (1992)
Starr, V.: Physics of Negative Viscosity Phenomena. McGraw-Hill, New York (1968)
Steenbeck, M., Krause, F., Radler, K.-H.: A calculation of the mean electromotive force in an electrically conducting fluid in turbulent motion under the influence of coriolis forces. Z. Naturforsch. **21a**, 369 (1966)

Tanga, P., Babiano, A., Dubrulle, B.: Forming planetesimals in vortices. Icarus **121**, 158 (1996)
Tassoul, J.-L.: Theory of Rotating Stars. Princeton University, Princeton (1979)
Taunsend, A.A.: The Structure of Turbulent Shear Flow. Cambridge University, Cambridge (1969)
Taylor, G.I.: Eddy motion in atmosphere. Philos. Trans. Roy. Soc. London **215**, 1 (1915)
Toomre, A.: On the gravitational stability of a disk of stars. Astrophys. J. **139**, 1217 (1964)
Vainshtein, S.I., Zeldovich, Y.B., Ruzmaikin, A.A.: Turbulent Dynamo in Astrophysics. Nauka, Moscow (in Russian) (1980)
Van Mieghem, J.: Atmospheric Energetics. Clarendon, Oxford (1973)
Vergassola, M., Gama, S., Frisch, U.: In: Proctor, M.R.E., Mathews, P.C., Rucklidge, A.M. (eds.) Proving the Existence of Negative Isotropic Eddy Viscosity in NATO-ASI: Solar and Planetary Dynamos, Cambridge University Press, Cambridge, 321 (1993)
Von Weizsacker, C.F.: Rotation Kosmischer Gasmassen. Z. Naturforch. **3a**, 524 (1948)
Von Karman, T.: Some Questions of the Theory of Turbulence. In: Problems of Turbulence. pp. 35–74. ONTI, Moscow (in Russian) (1936)
Wasiutynski, J.: Studies in hydrodynamics and structure of stars and planets. Astrophys. Norv. **4**, 86 (1946)
Youdin, A.N., Shu, F.: Planetesimal formation by gravitational instability. Astrophys. J. **580**, 494 (2002)
Zel'dovich, Y.B.: On the friction of fluids between rotating cylinders. Proc. Roy. Soc. London **A374**, 299 (1981)

Chapter 9

Akhiezer, A.I., Akhiezer, I.A., Polovin, R.V., et al.: Plasma Electrodynamics. Nauka, Moscow (1974) (in Russian)
Alfven, H.: Cosmical Electrodynamics. Clarendon Press, Oxford (1950)
Balbus, S.A., Hawley, J.F.: A powerful local shear instability in weakly magnetized disks: 1. Linear analysis. Astrophys. J. **376**, 214 (1991)
Bath, G.T., Pringle, J.E.: The evolution of viscous discs: I. Mass transfer variations. Mon. Not. Roy. Astron. Soc. **194**, 967 (1981)
Biskamp, D.: Magnetohydrodynamic Turbulence. Cambridge University Press, Cambridge (2003)
Bisnovaty-Kogan, G.S., Lovelace, R.V.E.: Advective accretion disks and related problems including magnetic fields. New Astron. Rev. **45**, 663 (2001)
Blackadar, A.K.: Extension of the laws of thermodynamics to turbulent system. J. Meteorol. **12**, 165–175 (1955)
Blandford, R.D.: Accretion disk electrodynamics—A model for double radio sources. Mon. Not. Roy. Astron. Soc. **176**, 465 (1976)
Brandenburg, A., Nordlund, A., Stein, R.F., Torkelsson, U.: The disk accretion rate for dynamo-generated turbulence. Astrophys. J. **458**, 145 (1996)
Branover, G.G., Tsinover, A.B..: Magnetohydrodynamics of Incompressible Media. Nauka, Moscow (1970) (in Russian)
de Groot, S., Mazur, P.: Non-Equilibrium Thermodynamics. North-Holland Publ. Co., Amsterdam (1962)
Dubrulle, B.: Differential rotation as a source of angular momentum transfer in the solar nebula. Icarus **106**, 59 (1993)
Eardley, D.M., Lightman, A.P.: Magnetic viscosity in relativistic accretion disks. Astrophys. J. **200**, 187 (1975)
Eardley, D.M., Lightman, A.P., Payne, D.G., Shapiro, S.L.: Accretion disks around massive black holes: Persistent emission spectra. Astrophys. J. **234**, 53 (1978)

Fridman, F.M.: To the dynamics of a viscous differentially rotating gravitating medium. Pis'ma Astron. Zh. **15**, 1122 (1989)

Galeev, A.A., Rosner, R., Viana, G.S.: Structured coronae of accretion disks. Astrophys. J. **229**, 318 (1979)

Gor'kavyi, N.N., Fridman, A.M.: Physics of Planetary Rings. Nauka, Moscow (1994) (in Russian)

Gyarmati, I.: Non-Equilibrium Thermodynamics. Field Theory and Variational Principles. Springer, Berlin (1970)

Hersant, F., Dudrulle, B., Hure, J.M.: Turbulence in circumstellar disks. Astron. Astrophys. **3549**, 1 (2004)

Ievlev, V.M.: Turbulent Motion of High-Temperature Continuous Media. Nauka, Moscow (1975) (in Russian)

Jou, D., Casas-Vazquez, J., Lebon, G.: Extended Irreversible Thermodynamics. Springer, Berlin (2001)

Kadomtsev, B.B.: Collective Phenomena in Plasma. Nauka, Moscow (1976) (in Russian)

Kadomtsev, B.B.: Reconnection of magnetic field lines. Usp. Fiz. Nauk **151**, 3–29 (1987)

Kolesnichenko, A.V.: To the theory of turbulence in planetary atmospheres: Numerical simulation of structural parameters. Astron. Vestn. **29**, 133 (1995)

Kolesnichenko, A.V.: Stefan-Maxwell Relations and Heat Flow for Turbulent Multicomponent Continuous Media in Problems of Modern Mechanics, p. 52. Moscow State University, Moscow (1998) (in Russian)

Kolesnichenko, A.V.: Modeling turbulent transport coefficients in a gas-dust accretion disk. Astron. Vestn. **34**, 516 (2000)

Kolesnichenko, A.V.: Hydrodynamics aspects of modeling mass transfer and coagulation in a turbulent accretion disk. Astron. Vestn. **35**, 139 (2001)

Kolesnichenko, A.V., Marov, M.Y.: Turbulence of Multicomponent Media. MAIK-Nauka, Moscow (1999) (in Russian)

Kolesnichenko, A.V., Marov, M.Y.: Foundations of mechanics of heterogeneous media in the circumsolar protoplanetary disk: The influence of solid particles on turbulence in the disk. Astron. Vestn. **40**, 1 (2006)

Kolesnichenko, A.V., Marov, M.Y.: On the influence of helicity on the evolution of turbulence in the solar protoplanetary cloud. Astron. Vestn. **41**, 3 (2007)

Kondepudi, D., Prigogine, I.: Modern Thermodynamics. From Heat Engines to Dissipative Structures. Wiley, New York (1998)

Krause, F., Raedler, K.-H.: Mean-Field Magnetohydrodynamics and Dynamo Theory. Pergamon, New York (1980)

Kulikovskii, A.G., Lyubimov, G.A.: Magnetohydrodynamics. Fiz.-Mat. Lit, Moscow (1962) (in Russian)

Landau, L.D., Lifshitz, E.M.: Electrodynamics of Continuous Media. Nauka, Moscow (1982) (in Russian)

Lapin, Y.V.: Turbulent Boundary Layer in Supersonic Gas Flows. Nauka, Moscow (1982) (in Russian)

Lesch, H.: Magnetic reconnection in accretion disk coronae. In: Tsinganos, C. (ed.) Solar and Astrophysical Magnetohydrodynamic Flows, p. 673. Kluwer, Dordrecht (1996)

Lynden-Bell, D., Pringle, J.E.: The evolution of viscous discs and the origin of the nebular variables. Mon. Not. Roy. Astron. Soc. **168**, 603 (1974)

Makalkin, A.B..: Peculiarities of the evolution of a viscous circumsolar protoplanetary disk. Astron. Vestn. **38**, 559 (2004)

Makalkin, A.B.., Dorofeeva, V.A.: Structure of the protoplanetary accretion disk around the Sun at the T Tauri phase. I. Initial data, equations, and modeling methods. Astron. Vestn. **29**, 99 (1995)

Marov, M.Y., Kolesnichenko, A.V.: Chaotic and ordered structures in the developed turbulence. In: Fridman, A.M., Marov, M.Y. (eds.) Astrophysical Disks: Collective and Stochastic Phenomena, p. 23. Springer, The Netherlands (2006)

Marov, M.Y., Kolesnichenko, A.V.: Mechanics of Turbulence of Multicomponent Gases. Kluwer Academic Publishers, Dordrecht-Boston-London (2002)

Moffat, H.K.: Magnetic Field Generation in Electrically Conducting Fluids. Cambridge University Press, Cambridge (1978)

Parker, E.N.: Hydromagnetic dynamo model. Astrophys. J. **122**, 293 (1955)

Priest, E., Forbes, T.: Magnetic Reconnection. MHD Theory and Applications. Cambridge University Press, New York (2000)

Prigogine, I., Stengers, I.: Time. Chaos. Quantum. Toward Solving the Paradox of Time. Progress Publishing House, Moscow (1994) (in Russian)

Richard, D., Zahn, J.P.: Turbulence in differentially rotating flow. What can be learned from the Couette-Taylor Experiment. Astron. Astrophys. **347**, 734 (1999)

Shakura, N.I., Sunyaev, R.A.: Black holes in binary systems. Observational appearance. Astron. Astrophys. **24**, 337 (1973)

Shakura, N.I., Sunyaev, R.A., Zilitinkevich, S.S.: On the turbulent energy transport in accretion disk. Astron. Astrophys. **62**, 179 (1978)

Shapiro, S., Teukolsky, S.: Black Holes, White Dwarfs, and Neutron Stars. Wiley, New York (1983)

Steenbeck, M., Krause, F., Rädler, K.-H.: Berechnung der mittleren Lorentz-Feldstärke $\overline{V \times B}$ für ein elektrisch leitendes Medium in turbulenter, durch Coriolis-Kräfte beeinflußter Bewegung. Z. Naturforsch. **21a**, 369–376 (1966)

Syrovatskii, S.I.: Magnetohydrodynamics. Usp. Fiz. Nauk **62**, 247 (1957)

Tamm, I.E.: Foundations of the Electricity Theory. Nauka, Moscow (1989) (in Russian)

Tassoul, J.L.: Theory of Rotating Stars. Princeton University, Princeton (1979)

Tout, C., Pringle, J.E.: Accretion disk viscosity—A simple model for a magnetic dynamo. Mon. Not. Roy. Astron. Soc. **259**, 604 (1992)

Van Driest, E.R.: On turbulent flow near a wall. J. Aeronaut. Sci. **23**, 107 (1956)

Yoshizawa, A.: Self-consistent turbulent dynamo modeling of reversed field pinches and planetary magnetic fields. Phys. Fluids B **2**, 1589 (1990)

Zel'dovich Ya, B.: On the friction of fluids between rotating cylinders. Proc. R. Soc. London A **374**, 299 (1981)

Coroniti, F.V.: On the magnetic viscosity in keplerian accretion disks. Astrophys. J. **244**, 587 (1981)

Index

M.Y. Marov and A.V. Kolesnichenko, *Turbulence and Self-Organization: Modeling Astrophysical Objects*, Astrophysics and Space Science Library 389, DOI 10.1007/978-1-4614-5155-6, © Springer Science+Business Media New York 2013

E

F

G

H

I

K

L

MIX
Papier aus verantwortungsvollen Quellen
Paper from responsible sources
FSC® C105338
FSC
www.fsc.org

If you have any concerns about our products,
you can contact us on
ProductSafety@springernature.com

In case Publisher is established outside the EU,
the EU authorized representative is:
Springer Nature Customer Service Center GmbH
Europaplatz 3, 69115 Heidelberg, Germany

Printed by Libri Plureos GmbH
in Hamburg, Germany